ENZYMOLOGY AND MOLECULAR BIOLOGY OF CARBONYL METABOLISM 6

ADVANCES IN EXPERIMENTAL MEDICINE AND BIOLOGY

ENZYMOLOGY AND MOLECULAR BIOLOGY OF CARBONYL METABOLISM 6

Edited by

Henry Weiner
Purdue University
West Lafayette, Indiana

Ronald Lindahl
University of South Dakota School of Medicine
Vermillion, South Dakota

David W. Crabb
Indiana University School of Medicine
Indianapolis, Indiana

and

T. Geoffrey Flynn
Queen's University
Kingston, Ontario, Canada

Springer Science+Business Media, LLC

Library of Congress Cataloging-in-Publication Data

Enzymology and molecular biology of carbonyl metabolism 6 / edited by
  Henry Weiner ... [et al.].
      p.   cm. -- (Advances in experimental medicine and biology ; v.
  414)
    "Proceedings of the 8th International Workshop on Enzymology and
  Molecular Biology of Carbonyl Metabolism, held June 29-July 3, 1996,
  in Deadwood, South Dakota"--T.p. verso.
    Includes bibliographical references and index.
    ISBN 978-0-306-45509-4   ISBN 978-1-4615-5871-2 (eBook)
    DOI 10.1007/978-1-4615-5871-2

    1. Aldehyde dehydrogenase--Congresses. 2. Aldose reductase-
  -Congresses. 3. Carbonyl reductase--Congresses. 4. Alcohol
  dehydrogenase--Congresses. 5. Carbonyl compounds--Metabolism-
  -Congresses.   I. Weiner, Henry. II. International Workshop on the
  Enzymology and Molecular Biology of the Carbonyl Metabolism (8th :
  1996 : Deadwood, S.D.) III. Series.
    [DNLM: 1. Aldehyde Dehydrogenase--physiology--congresses.
  2. Alcohol Dehydrogenase--physiology--congresses. 3. Aldehyde
  Reductase--physiology--congresses. 4. Alcohol Oxidoreductases-
  -physiology--congresses.   W1AD559 v.414 1997 / QU140E597 1997]
  QP603.A35E575   1997
  572'.791--dc21
  DNLM/DLC
  for Library of Congress   96-54477
                                                                  CIP

Proceedings of the 8th International Workshop on Enzymology and Molecular Biology of Carbonyl
Metabolism, held June 29 – July 3, 1996, in Deadwood, South Dakota

ISBN 978-0-306-45509-4

© 1997 Springer Science+Business Media New York
Originally published by Plenum Press, New York in 1997

http://www.plenum.com

10 9 8 7 6 5 4 3 2 1

The aldehyde dehydrogenase section of this symposium is
dedicated to the memory of

Professor H. Werner Goedde
born July 9, 1927, died January 4, 1996

Professor Goedde was a pioneer in investigating the relationships between
isozyme patterns of aldehyde dehydrogenase and people's alcohol
consumption. He will be missed by his many friends and colleagues.

PREFACE

Since 1982, our ever-expanding group of investigators has been meeting in exotic parts of the world to discuss aspects of three enzyme systems. The 1996 meeting was no exception. Nearly 90 scientists from 15 countries met in the small city of Deadwood, South Dakota, for four days of stimulating talks and posters and incredible scenery. Once more this meeting reflected the changing trends in biochemical research. At the 1982 meeting most of the speakers discussed isolating new enzymes and trying to characterize them. At this meeting many speakers discussed interpretations of three-dimensional structure or regulatory elements of the genes controlling for the tissue-specific expression of the enzyme. Hopefully, readers will find the proceedings of the meeting to be of interest. Though they reflects the scientific information that was presented at the meeting, they do not indicate the level of personal interactions that went on during the meeting. Once again, the willingness of the participants to discuss unpublished data and to share thoughts about the future directions of their research helped make this, like our previous seven meetings, a special scientific experience for those who attended.

One of my most important jobs as the scientific organizer of these meetings is to ask friends to help me organize the meeting. I am fortunate to have worked with two more. On behalf of all the participants I wish to express our sincere thanks to Ronald Lindahl (University of South Dakota) and Dennis Petersen (University of Colorado) for serving as the local co-organizers. In addition to doing a magnificent job, they were also able to secure some funding for our non-society. Our meeting is different from most in that everyone pays his or her own way, so the extra funds help keep this an affordable experience for all. I wish to acknowledge the support from the Nutrasweet Company and the University of South Dakota School of Medicine and University Physicians. The State of South Dakota provided a generous grant from the State of South Dakota Experimental Program to Stimulate Competitive Research (EPSCoR) as part of their Rushmore Conference series. I wish to thank Dianna Olson who handled many of the important local details that make a meeting successful. I also wish to thank Beth Zurzulo and Mary Walker for helping me at Purdue. My three co-editors reviewed with me all the manuscripts, and my sincere thanks to them for their help. Lastly, I wish to thank all the participants for making this another scientifically exciting and socially pleasant experience.

Our ninth meeting will be held in Italy from approximately July 4 to July 8, 1998. I invite scientists from around the world to contact me if they are interesting in attending and presenting at this meeting.

Henry Weiner
Purdue University
West Lafayette, Indiana

CONTENTS

CRYSTAL STRUCTURE OF A CLASS 3 ALDEHYDE DEHYDROGENASE AT 2.6Å RESOLUTION

Zhi-Jie Liu,[1] John Hempel,[2] Julie Sun,[3] John Rose,[1] David Hsiao,[3] Wen-Rui Chang,[1] Yong-Je Chung,[3] Ingrid Kuo,[2] Ronald Lindahl,[4] and Bi-Cheng Wang[1]

[1]Department of Biochemistry and Molecular Biology
University of Georgia
Athens, Georgia 30602
[2]Department of Biological Sciences and
[3]Department of Crystallography
University of Pittsburgh
Pittsburgh, Pennsylvania 15260
[4]Department of Biochemistry and Molecular Biology
University of South Dakota
Vermillion, South Dakota 57069

1. INTRODUCTION

Class 3 ALDHs prefer NAD as coenzyme, but can use NADP effectively *in vitro*. They function as dimers of identical ~50kDa monomers, and share about 30% sequence identity with either of the tetrameric class 1 or 2 enzymes. Interest in the Class 3 ALDH (ALDH-3) has focused on its diversity of expression (Lindahl, 1992). No ALDH-3 activity is detectable in normal mammalian liver, but high levels are obtained after exposure to certain xenobiotics. Many neoplasms possess elevated ALDH-3 activities, while many normal tissues, such as cornea and stomach constitutively express ALDH-3. ALDH-3 enzymes prefer aromatic aldehydes and medium chain-length (C-6 to C-10) aliphatic aldehydes as substrate. Known substrates include benzaldehyde and hexenal as well as 4-hydroxynonenal derived from membrane lipid peroxidation (Lindahl and Peterson, 1991). The microsomal ALDH-3 has also received clinical attention from the recent demonstration that Sjögren-Larsson Syndrome is the result of mutations inactivating this "fatty aldehyde dehydrogenase" (DeLaurenzi, et al, 1996).

We report here the tertiary structure of the rat liver class 3 ALDH-NAD complex which was determined at 2.6 Å resolution by X-ray crystallography. Each molecule (single chain with 452 residues) forms two open α/β domains which self-associate into a homodimer. The NAD(P)-binding domain shows novel dinucleotide-binding motif and protein-NAD interactions: the Rossmann Fold (Rossmann, et al., 1974) has 5 β-strands instead of 6, and the pyrophosphate group of NAD(P) and the corresponding 'Gly-X-Gly-X-X-Gly' motif are near the N-terminal end of helix αD instead of helix αA as found in other known NAD-dependent dehydrogenases. The adenine ribose, however forms two hydrogen bonds with Glu-140 from strand β2, which is similar to other NAD-dependent dehydrogenases.

2. RESULTS AND DISCUSSION

2.1. Structure Determination

Crystals of ALDH were grown as described previously (Rose et al, 1991). The crystals belong to the monoclinic space group $P2_1$ with cell dimensions a=64.9Å, b=170.9Å, c=47.2Å and β=110.3^0. There are two molecules per asymmetric unit related by non-crystallographic 2-fold symmetry. Two useful ρ-chloromercuribenzoic acid (PCMB) heavy atom derivatives, a 1-day soak (PCMB-1) and a 3-day soak (PCMB-3), were obtained after an extensive search. The PCMB-1 derivative has two heavy atom sites, while the PCMB-3 derivative has 7 sites with the two major sites identical to the PCMB-1 derivative (Sun, 1995). Both native and derivative data sets used for structure determination were collected on a Siemens X100 multi-wire area detector using mirror (Supper) focused 4.05 kW Cu Kα X-rays generated on a Rigaku RU200 rotating anode. The data were indexed, integrated and scaled using XENGEN 2.1 (Howard, et al., 1987). Phases were calculated by the single isomorphous replacement with anomalous scattering (SIRAS) method in combination with noise filtering (Wang, 1985) and 2-fold non-crystallographic symmetry averaging as incorporated in PHASES (Furey and Swaminathan, 1996). Sequence assignment was based on one mercury binding site (Cys-251) and one Tryptophan (Trp-80). The model was completed through several cycles of model building with O (Jones, et al, 1991) using both noise-filtered SIRAS-phased and model-phased maps.

The model was refined against data between 15.0 and 2.6Å by simulated annealing and positional refinement using Engh and Huber force fields (Engh and Huber, 1991) as incorporated in X-PLOR 3.1 (Brünger et al., 1991) Individual atoms were assigned isotropic B-factors which were refined during the latter stages of the refinement. The final model includes 894 amino acids, two NAD molecules and 279 water molecules for the dimer. No clear electron density was observed for residue 1 and residues 448 - 452 of each monomer. The R factor is 17.8% for all 2σ data between 15.0 and 2.6Å resolution (22,059 reflections). Using an 8.5% reflection test set (1875 reflections) the R_{free} (Brünger, 1992) value is 27.9%.

2.2. Description of the Structure

The dimeric enzyme with pseudo 2-fold symmetry has average dimensions approximately 90Å x 60Å x 40Å (Figure 1). The secondary structure is 42% α-helix and 17% β-strand (Table 1). Each monomer contains three domains, an NAD-binding domain (residues 1–81, 104–212 and 397–423), a catalytic domain (residues 213–396) and a bridging domain (82–103 and 424–452).

Starting from the N-terminus the first 81 residues form four helices (α1–α4) and constitute parts of the NAD-binding domain. The polypeptide chain then loops out, and on

Figure 1. Structure of an ALDH-3 dimer with α-helices as lettered tubes and β-strands as numbered sheets as viewed along the pseudo 2-fold axis.

its return to the NAD-binding domain it forms a single β-strand (β0). This loop and β0 are part of the "arm-like" bridging-domain which forms important contacts with the other monomer of the dimer. From Pro-103, the chain begins the characteristic α/β structure associated with dinucleotide-binding "Rossmann fold" (Rossmann et al., 1974). In ALDH-3, the dinucleotide-binding fold is made up of 5 β-strands (β1-β5) connected by four α-helices (αA-αD, here the secondary structure is named following the convention of Brändén and Tooze, 1991). This is in contrast to the 6 β-strands usually found in the dinucleotide-binding fold of other NAD-dependent dehydrogenases (Lesk, 1995). The chain then leaves the NAD-binding domain and forms the catalytic domain. This transition occurs with the Gly-Gly segment at positions 211–212.

The catalytic domain also exhibits an α/β structure which begins at Lys-213 and ends at Val-396. It consists of seven parallel β-strands (β6-β7 and β9-β12), one antiparallel β-strand (β8) and five helices (α9-α13). The catalytic thiol, Cys-243, is located on the turn between the first helix (α9) and the second β-strand (β7), and is at the interface between the NAD-binding and catalytic domains. The last strand in this region (β12) begins and ends with the strictly conserved residues Gly-383 and Asn-388. The final 56-residue segment passes back over the NAD-binding domain forming a seven-residue helix (α14) from Gly-414 to Ser-420, and covers an otherwise exposed portion of the central sheet.

Table 1. Secondary structure elements of ALDH-3

Helix	Residues	Length	Domain[a]	Strand	Residues	Length	Domain[a]
α1	1-15	15	N	β0	96-103	8	B
α2	22-51	30	N	β1	106-110	5	N
α3	55-61	7	N	β2	133-137	5	N
α4	63-81	19	N	β3	162-166	5	N
αA	118-129	12	N	β4	182-186	5	N
αB	143-156	14	N	β5	205-209	5	N
αC	169-177	9	N	β6	215-218	4	C
αD	189-202	14	N	β7	248-251	4	C
α9	224-236	13	C	β8	301-304	4	C
α10	253-271	19	C	β9	318-321	4	C
α11	287-297	11	C	β10	338-343	6	C
α12	347-355	9	C	β11	362-366	5	C
α13	370-379	10	C	β12	383-388	6	C
α14	414-420	7	N	β13	422-430	9	B

a: N - NAD-binding domain; C - catalytic domain; B - bridging domain

The segment continues into the bridging-domain, forming β-strand (β13) which completes an antiparallel β-ribbon with β0. This ribbon hydrogen bonds with strand β12' of the catalytic domain of the other monomer (denoted by ') becoming part of the central sheet of the open α/β structure. Thus this region is composed of nine β-strands - seven central parallel strands with single antiparallel strands at each end. Strands β6-β12 come from one molecule and strands β0' and β13' come from the other molecule in the dimer. The final 22 C-terminal residues form a random coil over the catalytic domain of the other molecule.

2.3. The Nucleotide-Binding Site

Almost all dehydrogenase structures (NAD-binding oxidoreductases) now known contain a nucleotide-binding motif which consists of a core of two sets of β-α-β units, together forming six parallel β–strands flanked by α–helices. Although the ALDH and alcohol dehydrogenase (ADH) families are believed to have arisen independently (Eklund, et al, 1976), a search using program DALI (Holm and Sander, 1993) for structural similarity to the ALDH nucleotide-binding motif showed a good agreement with the nucleotide-binding motif of 3α,20β-hydroxysteroid dehydrogenase (E.C.1.1.1.53, PDB entry 1hdc), a short-chain ADH (Ghosh, 1991), with a root mean square deviation (Cα's) of 2.3 Å and a "medium-chain" Zn dependent alcohol dehydrogenase (E.C.1.1.1.1, PDB entry 2ohx) with a root mean square deviation (Cα's) of 2.4 Å (Fig. 2 a). However, the expected last β-strand (β6) of the nucleotide-binding motif is absent in ALDH.

In most NAD-dependent dehydrogenases, a highly conserved 'G-X-G-X-X-G' sequence motif exists, which identifies the dinucleotide binding site (Wierenga et al., 1986). In ADHs, lactate dehydrogenases (LDHs), glyceraldehyde-3-phosphate dehydrogenases (GAPDHs) and most other known NAD-dependent dehydrogenase structures, the 'GXGXXG' motif occurs at the N-terminus of αA. A multiple sequence alignment (Hempel et al., 1993) suggested that the strictly conserved Gly-187 (Gly-250 in the class 1 and 2 ALDHs) could begin at the equivalent 'GXGXXG' motif in ALDHs. However, one could argue that this motif may not be revelant to ALDHs since in ALDH the "second Gly" is usually replaced by Thr, and one substitution was noted at the third Gly position in the sequence alignment. In ALDH-3 this motif occurs at the beginning of αD not α-A as

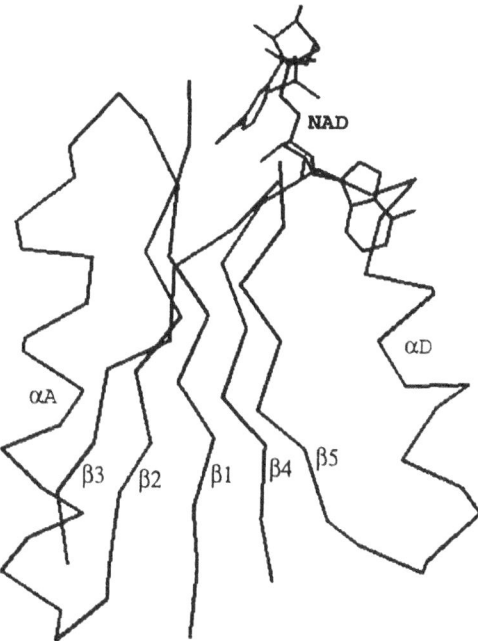

Figure 2. Location of NAD in relation to the Rossmann fold in ALDH-3. Helices αB and αC have been omitted for clarity.

in the other NAD-dependent dehydrogenase structures. In addition, the pyrophosphate bridge of NAD in ALDH-3 is no longer located at the N-terminus of αA, but is now located in proximity to, but not directly at the N-terminus of αD (Figure 2) consistant with the location of the 'GXTXXG' motif in ALDH and results in a new mode of NAD binding.

Another element of the expanded sequence motif of Rossmann folds is an acidic residue some 18 amino acids downstream of the last Gly in the 'GXGXXG' motif. The purpose of the acidic side-chain, located in β2, is attributed to a specificity for NAD, since it points toward the 2'OH of the adenine ribose (phosphorylated in NADP) in other dehydrogenase structures. Despite the significant rearrangement of NAD within the "Rossmann fold" in ALDH-3, the adenine ribose forms two hydrogen bonds with Glu-140 (residue 197 in Classes 1 and 2) from β2, which is similar to other NAD-dependent dehydrogenases.

2.4. The Catalytic Channel

A catalytic cysteine residue, forming the thiolester intermediate, was first postulated by Racker in 1955 (Racker, 1955) and Cys-243 (Cys-302 in class 1and 2 ALDHs), the only strictly conserved cysteine residue in the ALDH family (Hempel et al., 1993) has generally been accepted as the catalytic thiol. In the ALDH-3 structure, this residue is located inside a funnel-shaped channel formed at the interface of the NAD-binding and catalytic domains. The channel extends completely through the protein terminating at the NAD-binding site. The opening of the channel, about 6Å X 12 Å, is flanked by residues from the bridging domain. Cys-243 is located within the channel some 15Å from the opening. The side chain of Cys-243 extends into the center of the channel with the catalytic thiol positioned 6.8Å from NC4 of the nicotinamide ring of NAD. The surface of the channel between Cys-243 and the NAD-binding site contains a number of highly conserved residues including Asn-114, Leu-119, Thr-186, Gly-187, Glu-209, Glu-333 and Phe-335. This concentration of highly conserved residues suggests similar catalytic environments for the class 1 and 2 enzymes.

3. CONCLUSIONS

In the ALDH-3 structure, the coenzyme-binding motif has high similarity with both the "short chain" and "medium chain" ADH families, with two major exceptions: strand β6 is absent in ALDH and the 'GXGXXG' motif is not at its usual location (if it can be considered to be present at all). The ALDH-3 structure shows a novel mode of NAD-binding within the Rossmann fold. Glu-140 from β2 is an important residue, attributed to a specificity for NAD, but opens an interesting question as to how the enzyme can bind NADP. A deep channel, found at the interface between the NAD-binding and the catalytic domains, is proposed to be the catalytic channel. Cys-243 is located inside this channel. The narrow portion of the catalytic channel, between Cys-243 and C4 of nicotinamide ring, contains a large proportion of highly conserved residues and is interpreted as the catalytic site. This structure should prove to be a valid catalog of the essential features of all aldehyde/NAD(P) oxidoreductases.

4. ACKNOWLEDGMENTS

This work was supported by AA06985 from the National Institute on Alcohol Abuse and Alcoholism and resources from the Georgia Research Alliance.

5. REFERENCES

Bränden, C. & Tooze, J. Introduction to Protein Structure. Garland, New York (1991).

Brünger, A.T., Karplus, M. & Petsko, G.A. Crystallographic refinement by simulated annealing: application to crambin. *Acta Crystallogr.* **A45**, 50–61 (1989).

Brünger, A.T. Free R value: A novel statistical quantity for assessing the accuracy of crystal structures. *Nature (London)* **355**, 472–475 (1992).

De Laurenzi, V., Rogers, GR, Hanrock, D.J., Marekov, L.N., Steinert, PM, Compton, J.G., Markova, N, Rizzo, W.B. Sjogren-Larsson syndrome is caused by mutations in the fatty aldehyde dehydrogenase gene. *Nature Genetics* **12**, 52–57 (1996).

Eklund, H. Nordström, B., Zeppezauer, E., Söderlund, G., Ohlsson, I, Boiwe, T., Söderberg, B-O., Tapia, O., Bränden, C-I., Åkeson, Å. Three dimensional structure of horse alcohol dehydrogenase at 2.4Å resolution. *J. Mol. Biol.* **102**, 27–59 (1976).

Engh, R.A. & Huber, R. Accurate bond and angle parameters for X-ray protein structure refinement. *Acta Cryst.* **A47**, 392–400 (1991).

Furey, W.F.J. & Swaminathan, S. Phases95: a program package for the processing and analysis of diffraction data from macromolecules. *Methods Enzymol* in press.

Ghosh, D., Weeks, CM., Grochulski, P., Duax, WL., Erman, M., Rimsay, R.L., Orr, J.C. Three-dimensional structure of holo 3 alpha, 20 beta hydroxysteroid dehydrogenase: a member of the short chain dehydrogenase family. *Proc. Natl. Acad. Sci USA* **88**, 10064–10068 (1991).

Hempel, J., Nicholas, H. & Lindahl, R. Aldehyde dehydrogenases: widespread structural and functional diversity within a shared framework. *Protein Science* **2**, 1890–1900 (1993).

Holm, L. & Sander, C. Protein structure comparison by alignment of distance matrices. *J. Mol. Biol.* **233**, 123–138 (1993).

Howard, A.J., Gilliland, G. L., Finzel, B. C., Poulos, T. L., Ohlendorf, D. H. & Salemme, F. R. The use of an imaging proportional counter in macromolecular crystallography. *J. Appl. Crystallogr.* **20**, 383–387 (1987).

Jones, T.A., Zou, J.Y., Cowan, S.W. & Kjeldgaard, M. Improved methods for building protein models in electron density maps and the location of errors in these models. *Acta Cryst.* `A47, 110–119 (1991).

Lesk, A.M. NAD-binding in dehydrogenases. *Curr. Opin. Struct. Bio.* **5**, 775–783 (1995).

Lindahl, R. Aldehyde dehydrogenases and their role in carcinogenesis.*Crit. Rev. Biochem. Mol. Biol.* **27**, 283–335 (1992).

Lindahl, R. & Petersen, D. Lipid aldehyde oxidation as a physiological role for class 3 aldehyde dehydrogenases. *Biochem. Pharm.* **41**, 1583–1587 (1991).

Plapp, B.V., Eklund, H., Jones, T.A. & Brändén, C.-I. Three-dimensional structure of isonicotinimidylated liver alcohol dehydrogenase. *J. Biol. Chem.* **258**, 5537–5547 (1983).

Racker, E. Actions and properties of pyridine-nucleotide linked enzymes. *Phys. Rev.* **35**, 1–56 (1955).

Rose, J.P., Hempel, J., Kuo, I., Lindahl, R. & Wang, B.C. Preliminary crystallographic analysis of class 3 rat liver aldehyde dehydrogenase. *Proteins* **8**, 305–308 (1991).

Rossmann, M.G., Moras, D. & Olsen, K.W. Chemical and biological evolution of a nucleotide binding protein. *Nature* **250**, 194–199 (1974).

Wang, B.-C. Resolution of phase ambiguity in macromolecular crystallography. *Methods Enzymol.* **115**, 90–112 (1985).

Wierenga, R.K., Terpstra, P. & Hol, W.G.J. Prediction of the occurance of the ADP-binding bab-fold in proteins, using an amino acid fingerprint. *J. Mol. Biol.* **187**, 101–107 (1986).

CONSERVED RESIDUES IN THE ALDEHYDE DEHYDROGENASE FAMILY

Locations in the Class 3 Tertiary Structure

John Hempel,[1] Zhi-Jie Liu,[2] John Perozich,[3] John Rose,[2] Ronald Lindahl,[4] and Bi-Cheng Wang[2]

[1]Department of Biological Sciences
[3]Department of Molecular Genetics and Biochemistry
University of Pittsburgh
Pittsburgh, Pennsylvania 15260
[2]Department of Biochemistry and Molecular Biology
University of Georgia
Athens, Georgia 30602
[4]Department of Biochemistry and Molecular Biology
University of South Dakota
Vermillion, South Dakota 57069

1. INTRODUCTION

As described in the preceding chapter (Liu et al., 1996), features of the class 3 ALDH Rossmann fold were totally unexpected. While Gly-187 is involved in binding NAD, its role is completely different from that observed for the first glycine residue of the GXGXXG motif in other dehydrogenases with traditional Rossmann folds. It was thus of interest, once the sequence was fit into the electron density, to examine the locations of strictly conserved residues from our earlier multiple sequence alignment (Hempel et al., 1993). A total of 23 residues were strictly conserved in that alignment. This chapter will confine itself to an attempt to correlate the role of the strictly conserved residues in the ALDH family with the class 3 ALDH (binary complex with NAD) tertiary structure.

The (rat) class 3 structure lacks the first 56 residues relative to class 1 and 2 sequences. In addition, several short indels (term: Sankoff and Kruskal, 1983) are required to make the sequence alignment, thus no simple correction factor can be used to convert the class 3 position numbering to that of other classes. For this reason, a cross-index to the position numbering of mammalian class 1 and 2 sequences is provided (Table 1).

Table 1. Cross-reference, class 3 ALDH to class 1 or 2 positional numbering

Class 3	Class 1,2 equivalent	Class 3	Class 1,2 equivalent
Arg-25	84	Thr-318	384
Pro-103	158	Glu-333	399
Gly-105	160	Phe-335	401
Gly-131	186	Gly-336	402
Lys-137	192	Pro-337	403
Gly-167	223	Asn-355	421
Gly-187	245	Gly-383	449
Gly-211	270	Asn-388	454
Gly-240	299	Gly-403	467
Cys-243	302	Ser-407	471
Gly-305	370	Gly-410	474
Pro-317	383		

2. RESULTS AND DISCUSSION

2.1. Glycine and Proline Residues

Glycine residues constitute nearly half (eleven) of the 23 strict conservations, with proline (three) the next most frequent type. It has been recognized that tertiary structural similarities are generally retained at the expense of all else during evolution (Chothia and Lesk, 1986). Glycine residues, which lack a side-chain and thus permit otherwise disallowed ϕ and ψangles in the mainchain, and proline residues, by their relatively fixed ϕ angles, stand out in their ability to respectively allow and fix particular conformational features. Thus, when the structure was examined, glycine and proline residues were expected to reflect locations important to the overall folding. Figure 1 shows that the Ramachandran coordinates of almost all strictly conserved glycine residues are found outside of regions allowed for residues with side-chains (Bränden & Tooze, (1991)).

Some correlations between individual consensus glycines and structural features can be easily noted. Gly-105 marks the beginning of the Rossmann fold, while Gly-211 punctuates its end. In between, Gly-131 and Gly-167 are found at turns after αA and β3, respectively (see Table 1, preceding chapter). The interaction of Gly-131 with Arg-25, discussed in more detail below, may be significant in positioning the N-terminal helical cluster with the coenzyme-binding domain, while Gly-167 is near the NAD molecule. Although Gly-167 is beyond hydrogen-bonding distance in the structure (6Å from the closest atom of NAD), the lack of a side-chain at this position may allow needed flexibility or steric access in formation of other complexes (e.g., E-NAD-aldehyde). One other strictly conserved glycine occurs in the Rossmann fold: Gly-187; the interaction of this residue with coenzyme was described in the preceding chapter.

In the catalytic domain, Gly-240 follows α9, apparently providing a needed bend in the main chain to position the catalytic thiol (below). Gly-305 terminates β8, while Gly- 336, following β-9, separates Phe-335 in the stem of the active site (see below) from Pro-337, all of which are strictly conserved. It is tempting to speculate that the latter two residues of this strictly conserved tripeptide are needed to place the ring of Phe-335 in a special orientation to allow stacking with the nicotinamide ring, facilitating hydride transfer (Fig 2). Further downstream in the sequence, Gly-383 immediately precedes β-12, a strand with residues that are generally well conserved in the alignment. The final two, Gly-403 and Gly-410 begin and end

Figure 1. Ramachandran plot of all 11 strictly conserved glycine residues from the class 3 structure. Allowed regions for residues with sidechains are shaded.

a prominent and unusual loop, which we have called the "U-turn". Gly-410 is characterized by a particularly unusual pair of f and y angles (Figure 1).

2.2. Catalytic Residues

Cys-243, the only strictly conserved cysteine residue in the alignment, and the expected catalytic thiol based both on its conservation and also on modification studies (summarized in Hempel et al., 1993), clearly lies at a strategic position in a deep cavity which passes through each subunit, a length of ~20Å. The cavity is roughly funnel-shaped, with the wide end (mouth) tapering over a ~15Å distance from the exterior to this cysteine residue. No strict conservations are noted lining this portion of the cavity. Continuing, from Cys-243, the narrow part of the funnel (stem) ends at NC4 of the nicotinamide ring. Strict conservations lining the stem of the funnel include Gly-187, Gly-211, Glu-333 and Phe-335.

2.3. Charged Residues

The only strictly conserved acidic residue in the alignment is Glu-333, noted above. Otherwise, there are only two other strictly-conserved charged residues in the alignment,

Table 2. Summary of residues lining the class 3 ALDH active site

Location		Upper Pocket		Lower Pocket	
α	W56 S	Y59 L			
α3-α4	E61 S	E62 D			
α4	E68 K	E69 T			
α4-β0	Q91 I	T92 D	D95 F		
β1-αA	Y115 F			A112 P	N114 N
αA	N118 V	L119 M	Q122 W		
β4				T186 T	**G187 G**
β5				E209 E	
β5-β6				L210 L	**G211 G**
α9-β7				**C243 C**	
β9-β10				**E333 E**	**F335 F**
α12-β11	L361 L				
β12-α14	I 391 I	V392 C	I 394 G		
" "	T395 V	V396 V	P398 S		
" "	F401 F	Y412 E	H413 L		
β13	K430 K	H438 *	A440 *		
"	P441 *				

The display follows the secondary structural elements, with equivalent residues from the human class 1 sequence indicated after each class 3 residue number. Asterisks denote a corresponding gap. Residues strictly conserved in an ALDH family alignment (Hempel et al., 1993) are in bold.

Arg-25 and Lys-137. Arg-25 lies in the first turn of α2 and its NE is within hydrogen-bonding distance (2.9Å) of the main chain carbonyl oxygen of Thr-19, a residue which is poorly conserved in the ALDH family, and which lies at the turn between the first two helices. This atom also lies 3.4Å from the amide nitrogen of Ala-129, at the end of αA in the coenzyme-binding domain, suggesting a compound hydrogen bond between Arg-25 and these two residues. Furthermore, one of the guanidino nitrogens (NH1) of Arg-25 is within hydrogen bonding distance of the amide nitrogen of Gly-131, in the turn between αA and β2. Gly-131 is strictly conserved, suggesting that this interaction is maintained in all ALDHs, and that it provides an important link, positioning the N-terminal helical cluster to the coenzyme-binding domain.

Lys-137 is close in sequence to the highly conserved Glu-140, which hydrogen bonds with the 2'OH of the adenine ribose of NAD, as described in the preceding chapter. Lys-137 is itself within 5.2Å of this hydroxyl. However, inspection of the structure suggests the possibility that Lys-137 and Glu-140 could form a salt bridge in the absence of cofactor since the NE and closest carboxyl oxygen are 4Å apart. Rotation about the Ca-Cb bond of Glu-140 would bring the side-chain amino and carboxylate close enough to form such an interaction, which could enable better access for NAD docking to the apoenzyme.

2.4. Residues Stabilizing Adjacent Secondary Structural Elements

The side-chain of Thr-318, which follows the strictly conserved Pro-317, is 3.18Å from (within H-bonding distance to) the carbonyl oxygen of His-304, a residue which is not well conserved but adjacent to the strictly conserved Gly-305. His-304 lies at the end of β9 while Thr-318 lies at the beginning of the subsequent (antiparallel) β10. Thus, it appears that Thr-318 provides an additional and perhaps strategic hydrogen bond between these two strands.

Similarly, the amide nitrogen of Asn-355, at the end of α12, is 2.84Å from the carbonyl oxygen of Arg-379 (also not well conserved), at the end of α13, presumably providing additional stabilization of the relative positions of these two helices.

The side chain amide nitrogen and carbonyl oxygen of Asn-388, at the end of β12, contribute contacts to the main chain carbonyl oxygen and amide nitrogen of Ser-367, (3.0 and 2.8Å, respectively). Ser-367 immediately follows β11, thus these interactions clearly provide additional coordination between these two strands. Perhaps more significantly, the side chain amide nitrogen of Asn-388 is also 3.2Å from each of the main chain carbonyl oxygens of Lys-430 and Ser-431 *from the opposite subunit*, suggesting that this may be an important inter-subunit contact.

Finally, the hydroxyl oxygen of Ser-407, in the "U-turn" noted above, is 2.7Å from both the amide nitrogen and carbonyl oxygen of Val-404, suggesting a compound hydrogen bonding scheme which could explain why serine is strictly conserved at this position. While Val-404 is not a well conserved residue, this apparent hydrogen bonding to the main chain at this position probably provides necessary extra stabilization to this structure. Interestingly, the main-chain carbonyl oxygen of residue 404 is within 3.1Å of the amide nitrogen of 407, while the carbonyl oxygens of residues 405 and 406 are each within 3.1Å of the amide nitrogen atom of residue 408. Clearly, this structure is stabilized in a highly coordinated manner.

3. CONCLUSIONS

The strict conservations generally fall into two general categories: a) residues required to maintain special structrural features (main-chain f and y angles of glycine residues in non-permissive Ramachandran regions, fixed f angles of proline residues particularly in relation to other conserved residues, and interactions apparently tethering adjacent segments of secondary structure through hydrogen bonding), and b) residues required for catalysis (the conservations in the stem of the catalytic funnel, between the catalytic Cys-243 and the nicotinamide ring). While these categories would have been expected at the outset; we are now able to provide specific details.

4. ACKNOWLEDGMENTS

Supported by AA06985 and the Georgia Research Alliance

5. REFERENCES

Bränden, C. & Tooze, J. Introduction to Protein Structure. New York: Garland, (1991).

Chothia, C., & Lesk, A. M. The relation between the divergence of sequence and structure in proteins. EMBO J 5, 823–826 (1986).

Hempel, J., Nicholas, H., & Lindahl, R. Aldehyde dehydrogenases: widespread structural and functional diversity within a shared framework. Protein Science 2, 1890–1900 (1993).

Liu, Z.-J., Hempel, J., Sun, J., Rose, J., Hsiao, D., Chang, W.-R., Chung, Y.-J., Kuo, I., Lindahl, R. & Wang, B. C. Crystal structure of class 3 aldehyde dehydrogenase at 2.6Å resolution. Adv Exp Med Biol (This volume, 1-7) (1996).

Sankoff, D. D., & Kruskal, J. B. *Time warps, string edits and macromolecules.* Addison Wesley (1983).

3

CLASS 3 ALDEHYDE DEHYDROGENASE

A View from the Hills

Ronald Lindahl, Yiqiang Xie, Josette S. Boesch, Keith Miskimins, and Robin Miskimins

Department of Biochemistry and Molecular Biology
University of South Dakota School of Medicine
Vermillion, South Dakota 57069

1. INTRODUCTION

Mammalian Class 3 aldehyde dehydrogenase includes both constitutively-expressed and xenobiotic-inducible forms encoded by at least 2 separate genes (Lindahl, 1992; Petersen and Lindahl, 1997). The rat microsomal form of Class 3 ALDH, encoded by the ALDH4 gene, is constitutively expressed. Depending on the tissue, the product of the rat ALDH3 gene is either constitutively expressed or inducible. In liver, no constitutive expression of *ALDH3* is detectable (Lindahl, 1992). However, following exposure to either polycyclic aromatic hydrocarbons such as 3-methylcholanthrene (3-MC) or dioxins such as 2,3,7,8-tetrachlorodibenzo-p-dioxin (TCDD), increases in both hepatic *ALDH3* transcript and Class 3 ALDH enzyme activity occur (Lindahl, 1992). Constitutive expression of *ALDH3* occurs in a variety of normal tissues, including the urinary bladder, lung, stomach and particularly the cornea of the eye (Lindahl, 1992). Class 3 ALDH activity is also elevated in both chemically-induced and spontaneous neoplasms from a variety of tissues, such as liver, colon, breast and leukemias as well as a number of tumor-derived cell lines (Lindahl, 1992).

ALDH3 has been cloned and characterized from both rat and human. The rat ALDH3 gene is approximately 9 Kb in length and contains 11 exons, the first of which is non-protein coding (Asman et al., 1993). The major transcription start site is 45 base pairs upstream from the translation start site in xenobiotic-treated liver, the cornea and in hepatoma cell lines differing in ALDH3 activity (Takimoto et al., 1994; Boesch et al., 1996). Deletion analysis of 5.5 Kb of the 5' flanking region of rat *ALDH3* reveals that these sequences can be divided into a strong promoter region, a more distal negative regulatory region and a region responsible for xenobiotic inducibility centered around -3.2 Kb (Takimoto et al., 1994). More distal to the xenobiotic responsive region is a second negative regulatory region. Polycyclic aromatic hydrocarbon and dioxin inducibility require the presence of functional aromatic hydrocarbon (Ah) receptor and Ah receptor nuclear translocator (Arnt) protein (Takimoto et al., 1994).

2. THE *ALDH3* PROMOTER REGION

Previous reporter gene assays identified a region within the proximal 1 Kb of the rat *ALDH3* 5' flanking region as a strong constitutive promoter in a number of cell lines regardless of the Class 3 ALDH enzyme activity (Takimoto et al., 1994). Computer analysis of this proximal 1 Kb region identified numerous putative consensus transcription factor binding sites (Asman et al., 1993). Located by this analysis were 2 xenobiotic response elements (XRE), 1 drug response element (DRE), a liver activating protein (LAP) site, several Ap1 sites, a GC box and a putative CAAT box. Several other putative elements could be identified by visual inspection, but did not approach consensus levels.

In order to determine which cis-acting sequences within the proximal region of the rat ALDH3 5' flanking region were responsible for the strong promoter activity, a series of studies were performed to examine regulation of basal *ALDH3* expression. DNase I footprint analysis identified 4 regions that interact with DNA-binding proteins within the first 1 Kb above the transcription start site (Fig. 1). The same footprinted regions were identified in nuclear extracts from normal rat liver, 3-MC-treated rat liver or from HTC and McARH7777 rat hepatoma cell lines. HTC cells and 3-MC treated rat liver express high levels of Class 3 ALDH enzyme activity (Lindahl, 1992). Normal rat liver and 7777 cells are examples of *ALDH3* non-expressing tissues (Lindahl, 1992).

Chloramphenicol acetyltransferase reporter gene assays using various *ALDH3* promoter region deletion constructs transfected into HTC or 7777 cells indicated that the 2 more proximal footprinted sites conferred strong promoter activity to the CAT gene. A construct containing both proximal footprint sites possessed approximately twice the CAT activity in both cell lines, compared to a construct containing only the most proximal site. The level of CAT activity was three times greater than an SV40-CAT reporter construct in each cell line. Addition of the distal footprinted regions reduced CAT expression in both cell lines. Inclusion of the third site reduced activity to levels of the construct containing only the most proximal site. Adding the most distal site reduced CAT activity to less that 50% of that seen in the most proximal site-only construct.

Gel mobility shift assays under a variety of conditions (competition and supershift assays) indicate that members of the Sp1 family of transcription factors interact with the 2 proximal footprint sites. For the -35 to -65 region, this is consistent with the computer analysis. That the -220- to -243 region also interacts with Sp1 was unexpected, as this region contains a known Ap1 site, but no consensus GC box. Since no evidence exists that the cryptic CAAT and TATA motifs identified earlier are functional and since nuclear extracts from cell lines and tissues differing greatly in ALDH3 enzyme activity show identi-

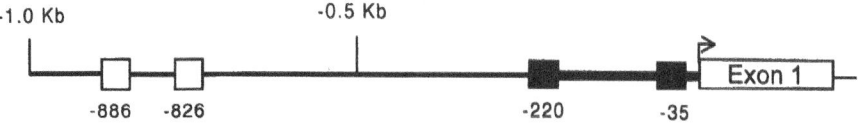

Figure 1. Organization of the rat *ALDH3* promoter. Solid boxes denote the Sp1 Sites. Open boxes are the NF1-like/novel protein binding sites. Thick bar defines the strong promoter region. Numbers are base pairs upstream of the transcription start site (arrow).

cal binding patterns to these 2 regions, we propose that the two Sp1 sites serve as the major sites of general transcription machinery alignment for *ALDH3*.

Similar gel shift assays using the 2 distal footprint regions indicate that they both interact with members of NF1 family of DNA-binding proteins. However, the complexity of the gel mobility shift assay binding patterns indicate that another protein(s), heretofore unidentified, may also be interacting with a novel cis-acting element found in these 2 regions. Further, yet a second additional protein may be involved in DNA-binding to the most distal site. While NF-1 proteins are commonly positive transcriptional regulators of genes in which they have been identified, it has been reported that NF1 sites in the retinol-binding protein and HMG CoA reductase genes as act transcriptional repressors (Gil et al., 1988; Colantuoni et al., 1987).

Consistent with earlier CAT reporter gene assays (Takimoto et al., 1994), no evidence for a significant role of any of the putative XRE or DRE elements identified by computer analysis was provided by the footprint, reporter gene or gel shift assays of the proximal 1 Kb of the rat *ALDH3* 5' flanking region from cells and tissues differing greatly in Class 3 ALDH activity. Although an approximate 2-fold increase in CAT activity in the presence of TCDD was observed in both HTC and H4IIEC3 hepatoma cells, this increase was minimal compared to the 15-fold increase observed with the -3.2 Kb construct (Takimoto et al., 1994). Further, the present extensive functional promoter analysis is consistent with our earlier observations that xenobiotics acting via electrophile response elements (AhRE), such as tert-butylhydroquinone, are without effect in reporter gene assays (Takimoto et al., 1994). Thus, we conclude that the proximal 1 Kb of the rat ALDH3 gene functions to regulate the constitutive expression of *ALDH3* in a variety of tissues. This region is not significantly involved in either the tissue-specific or xenobiotic-mediated expression of this gene.

Finally, our results confirm the importance of a thorough functional analysis of any putative cis-acting elements or trans-acting factors identified by other means. Of the approximately 15 cis-acting elements first identified by computer analysis of the rat *ALDH3* 5' flanking region, only the Sp1 site at -35 to -65 was confirmed. The more proximal Sp1 site was identified by sequence analysis as a putative Ap1 site. The -826 to -851 NF1 site has been variously characterized as putative Ap1 and/or DRE binding sites. The more distal NF1 site has been undescribed or an Ap1 site previously. None of various putative XRE, AhRE or DRE elements previously identified have been shown be functionally active.

3. CONSTITUTIVE EXPRESSION OF *ALDH3*: THE CORNEA

A second area of interest has been to understand the mechanisms underlying constitutive *ALDH3* expression using our recently developed rat corneal epithelium culture model (Boesch et al., 1996). Transfection of the set of rat *ALDH3* 5.5Kb 5' flanking region deletion-CAT reporter gene constructs into corneal epithelium reveals that the same general pattern of functional cis-acting elements regulate constitutive ALDH3 expression in the cornea as regulate xenobiotic-inducible expression in the liver. The same strong promoter, negative regulatory region, strong positive regulatory region followed by an even more distal negative region functioning in TCDD-treated hepatoma cells also are involved in corneal *ALDH3* expression. Interestingly, the region near -3.0 Kb responsible for xenobiotic-induced hepatic ALDH3 gene expression is also critical for constitutive cornea *ALDH3* expression. This 700 base pair region contains a single XRE buried among multiple repeats of a unique CA-rich sequence.

An extensive series of gel mobility shift assays, including competition and super-shift assays, indicate that corneal *ALDH3* expression uses the same cis-acting elements in the -3.0 Kb region as does xenobiotic induction. However, different trans-acting factors interact with these sequences. For example, although the XRE sequence is important, we find no evidence for Ah receptor and/or Arnt involvement in constitutive corneal *ALDH3* expression. Multiple other DNA-binding proteins interact with these DNA sequences. The exact nature of the trans-acting factors responsible for constitutive ALDH3 gene expression remain to be defined. Whether they act alone or together and whether they displace Ah and/or Arnt also remains to be determined. However, it is clear that regulation of rat Class 3 ALDH gene expression involves a limited number of cis-acting elements, but a diversity of trans-acting factors to modulate expression under a variety of conditions.

4. ACKNOWLEDGMENT

This research is supported by NIH grant CA-21103 to R. L.

5. REFERENCES

Asman, D. C., Takimoto, K., Pitot, H. C., Dunn, T. J. and Lindahl, R.: Organization and characterization of the rat class 3 aldehyde dehydrogenase gene. J. Biol. Chem. 268 (1993) 12530–12536.

Boesch, J. S., Lee, C. and Lindahl, R.: Constitutive expression of class 3 aldehyde dehydrogenase in cultured rat corneal epithelium. J. Biol. Chem. 271 (1996) 5150–5157.

Colantuoni, V., Pirozzi, A., Blance, C. and Cortese, R.: Negative control of liver- specific gene expression: cloned human retinol-binding protein gene is repressed in HeLa cells. EMBO J. 6 (1987) 631–636.

Gil, G., Smith, J. R., Goldstein, J. L., Slaughter, C. A., Orth, K., Brown, M. S. and Osborne, T. F.: Multiple genes encode nuclear factor 1-like proteins that bind to the promoter for 3-hydroxy-3-methylglutaryl-coenzyme A reductase. Proc. Natl. Acad. Sci. 85 (1988) 8963–8967.

Lindahl, R.: Aldehyde dehydrogenases and their role in carcinogenesis. Crit. Rev. Biochem. Molec. Biol. 27 (1992) 283–355.

Petersen, D. and Lindahl, R.: Aldehyde Dehydrogenases. In **Comprehensive Toxicology** Vol. 3. Biotransformation. Guengerich, F., Ed., Elsevier Science, Oxford, 1997, In press.

Takimoto, K., Lindahl, R., Dunn, T. J. and Pitot, H. C.: Structure of the 5' flanking region of class 3 aldehyde dehydrogenase in the rat. Arch. Biochem. Biophys. 312 (1994) 538–546.

4

HUMAN CORNEAL AND LENS ALDEHYDE DEHYDROGENASES

Purification and Properties of Human Lens ALDH1 and Differential Expression as Major Soluble Proteins in Human Lens (ALDH1) and Cornea (ALDH3)

Gordon King[1] and Roger Holmes[2]

[1]Griffith University
Brisbane Qld 4111 Australia
[2]The University of Newcastle
Callaghan NSW 2308 Australia

1. INTRODUCTION

Aldehyde dehydrogenases (ALDHs) are now recognised as a complex gene family., which includes a group of NAD-dependent ALDH (EC 1.2.1.3) isozymes. Seven human ALDHs have been reported, of which ALDH1, 2, 3, 5, 6, and 7 (Hsu et al,1994), and a related enzyme, γ-amino butyraldehyde dehydrogenase (GABADH) (Kurys et al,1993), have been thus far cloned and sequenced (Hsu et al., 1994). Sequence analysis verifies that these are closely related enzymes, and optimised alignments show that 62 amino acids are conserved, including the catalytically significant Gly-245, Gly-250, Glu-268 and Cys-302 (Hsu et al., 1994). Human ALDHs 1, 2 and 3 have been further classified according to their genetic identity as Class 1(ALDH1, liver cytosolic), Class 2 (ALDH2, liver mitochondrial) and Class 3(ALDH3, stomach cytosolic) isozymes. In addition, a gene locus designated ALDHx, which shares 85% homology with Class 2 ALDH, has also been reported (Hsu and Chang, 1991).

These ALDHs catalyse the NAD-dependent oxidation of a wide range of aldehydes to their corresponding carboxylic acids. Functional roles have been assigned to individual ALDH isozymes which reflect their relative efficiencies with particular aldehyde substrates (reviewed in Kurys et al., 1993). For example, liver mitochondrial ALDH2 apparently plays a key role in ethanol metabolism as a result of its very low Km with acataldehyde as substrate (Greenfield and Pietruszko,1977). A wide range of biological aldehydes are recognized as 'natural' substrates for members of this gene family, including aldehydes derived from the metabolism of biogenic amines, steroids, retinoids, carbohydrates, alcohols, monoamines, diamines and polyamines (Pietruszko,1983; Algar and Holmes,1988; Ambroziak and Pietruszko, 1991; Dockham et al,1992), as well as peroxidic aldehydes (Algar and Holmes,1989; Evces and Lindahl,1989; Abedinia et al,1990). In addition, Cooper at al (1993) have observed that members of this gene family have a number of other functional roles , including serving as a major soluble protein in cornea (Abedinia et al,1990; Verhagen et al,1991) and lens (Wistow et al,1991); functioning as an androgen receptor in human genital skin fibroblasts (Pereira et al., 1991); and as thiol esterases (Sidhu and Blair, 1975).

The cornea of the cow, pig, sheep, baboon, opossum and human contain very high levels of ALDH activity (Holmes and VandeBerg, 1986; Holmes, 1988; Holmes et al., 1989), and corneal ALDH has now been purified from the rat (Evces and Lindahl, 1989), cow (Abedinia et al., 1990), baboon (Algar and Holmes, 1989) and human (Gondhowiardjo at al, 1991; King and Holmes, 1993). Although corneal ALDH from different sources varied in their catalytic efficencies, each of these purified enzymes exhibited the following similarities: both NADP and NAD can act as coenzyme; 'high-Km' for acetaldehyde as substrate; 'low-Km' for benzaldehyde and peroxidic aldehydes; and with a dimeric 54 kD subunit structure. Thus, these enzymes resemble ALDH3 (or stomach Class 3 ALDH) in these and other properties. Molecular genetic studies (Alexander et al., 1981; Cooper et al., 1991) have shown that BCP54 is 78 percent homologous at the nucleotide level with the tumour-associated rat ALDH Class 3 (Jones et al., 1988). Based on its kinetic properties, a role for corneal ALDH3 in the oxidation of aldehyde products of lipid metabolism, including lipid peroxidation products, has been proposed (Holmes and VandeBerg, 1986; Holmes, 1988; Holmes et al., 1989; Evces and Lindahl, 1989; Boesch et al., 1996). The high protein levels of this enzyme in bovine cornea, suggest a further function in the direct absorption of UVB radiation (Abedinia et al., 1990). More recently, corneal ALDH3 has been described as a corneal crystallin (Cooper et al., 1993).

The mammalian cornea absorbs much of the UVB radiation from both solar and terrestrial sources, and a significant portion of UVA, that strikes it, whereas the lens absorbs essentially all the remaining UVB and UVA radiation that enters the eye (Boettner and Wolters, 1962; Zigman,1983). In the human lens, chromophores responsible for UVA absorption have been identified as kyurenine and its derivatives (van Heyningen, 1970), however no chromophore has been identified as responsible for the UVB absorption in the cornea (Boettner and Wolters, 1962; Zigman 1983). The absorption of light quanta may led to the production of major oxidizing chemical species, which may react with lipids, generating peroxidic aldehydes, such as malondialdehyde, trans-2-hexenal, and 4-hydroxynonenal, which in turn may interact with proteins and form cross-linked species (Kagan et al., 1973). Thus, the rapid removal of aldehydes arising from peroxidic processes may be especially critical for these tissues whose function is to transmit light without interference. Here we report further studies of the distribution and properties of ALDHs in the human eye, including evidence for the differential distribution of ALDH1 and ALDH3 as major soluble proteins in human lens and cornea,respectively.

2. MATERIALS AND METHODS

2.1. Tissue Sources and Extraction

Human corneas and lenses were obtained from the Queensland Eye Bank and stored frozen at -70 oC until used. Homogenates were prepared in 0.05 M Tricine buffer (pH 7.5,

0.1 mM in EDTA, 1.0 mM in DTT, and 0.1% (w/v) deoxycholate) at 0 oC, and then cen-trifuged at 45,000 for 20 mins at 4° C. Antibodies. Rabbit antihuman corneal ALDH3 polyclonal antibodies were produced from purified human corneal ALDH3 (King and Holmes,1993). Rabbit antihuman liver ALDH1 and ALDH2 polycolonal antibodies and purified human liver ALDH1 were gifts from Drs A.Yoshida, L. Hsu and A. Shibuya of the Beckman Research Institute. Sodium Dodecyl Sulfate (SDS) Gel Electrophoresis. Polyacrylamide gels (12% separating and 4% stacking) containing 1% SDS were prepared in a mini-Protean slab gel system, according to the procedure of Laemmli (Laemmli, 1970), as modified by Ames (Ames, 1974). After electrophoresis, the gels were stained us-ing a standard Coomassie technique.

2.2. Immunoblotting

Western blots of IEF and SDS-polyacrylamide gels were prepared using standard procedures. Reaction with the primary antibodies was performed at room temperature us-ing antibody diluted into 5% w/v Dutch Jug milk powder in Tris buffered saline at pH 7.5 for one hour. Reaction with the secondary antibody was performed at room temperature with the peroxidase-linked anti IgG in 5% w/v Dutch Jug milk powder in Tris buffered sa-line at pH 7.5 for one hour. Development of the antibody reactions was performed using a standard diaminobenzidine/nickel chloride procedure.

2.3. Enzyme Assays

ALDH activity was determined using a spectrophotometric assay procedure (Holmes and VandeBerg, 1986), performed in 0.1M pyrophosphate buffer at pH 8.5. ALDH assays at low subsrate concentrations were conducted as for the spectrophotometric assays, ex-cept that the rate of production of NADH was followed by fluorescence at 458 nm. All re-agents used were of analytical grade. Propionaldehyde was synthesised from n-propanol and redistilled prior to assay.

2.4. Protein Estimations

Protein concentrations were estimated using a bicinchoninic acid-based procedure (Smith, et al., 1985). The standard curve was produced using bovine serum albumin.

2.5. Purification Procedure

ALDH1 and ALDH3 were purified from single human lens and corneas by a combi-nation of ion-exchange and affinity chromatography in filtered N2-saturated buffers at 4°C. The ocular tissue was homogenised in five volumes of 0.05 M Tricine buffer (pH 7.5, 0.1 mM in EDTA, 1.0 mM in DTT, and 0.1% (w/v) deoxycholate) at 0°C. The centrifuged homogenate was diluted (4-fold) into loading buffer and adjusted to pH 5.8 prior to load-ing onto CM-cellulose (10 mL). ALDH1 eluted with the wash, 10 mM phosphate buffer (pH 5.8, 1.0 mM EDTA, 1.0 mM DTT),whereas ALDH3 eluted with 100 mM phosphate buffer (pH 7.4, 1.0 mM EDTA, 1.0 mM DTT). The ALDH in both fractions was further purified using AMP-Sepharose (5 mL) equilibrated with 50 mM phosphate buffer (pH 7.0, 0.25M NaCl 1.0 mM EDTA, 1.0 mM DTT). After unbound protein was eluted, highly pu-rified enzyme was then eluted with loading buffer containing 0.1 mM NAD. Two ALDH-containing fractions were pooled and concentrated.

Figure 1. An analysis of activity and protein content distribution for ALDH1 and ALDH3 of anterior tissues of the human eye. The data was determined from purification studies, except in the case of lens ALDH3. ALDH3 activities were determined with benzaldehyde as substrate and ALDH1 activities were determined with acetaldehdye as substrate.

3. RESULTS

Human cornea and the lens extracts exhibited two major forms of ALDH activity, which showed similar properties to ALDH1 and ALDH3 in terms of their kinetic properties, pI values, subunit molecular weight and antibody reactivity (Figure 1). Human corneal extracts exhibited very high levels of ALDH3 activity, as well as minor ALDH1 activity (~ 1% of total). The total lens ALDH activity observed was ~ 20% of the total corneal ALDH activity, and was shown to be mainly due to ALDH1, with a minor contribution from ALDH3 activity (< 3% of total). Table 1 provides a summary of the results of a typical purification of ALDH1 from the human lens. Under our assay conditions, lens ALDH1 exhibited a range of specific activities from 0.2 to 0.59 I.U./mg (mean value of 0.37). Table 2 compares the kinetic properties observed for human lens ALDH1 with those previously reported for human liver ALDH1 (Greenfield and Pietruszko, 1977). The results show good agreement between these enzymes in these and other properties.In addi-

Table 1. Purification of ALDH1 from human lens

Purification Step	Activity (I.U.)	Protein (mg)	Specific Activity (I.U. mg-1)	Yield %
Homogenate	0.26	46	0.006	100
CM-Cellulose	0.18	14	0.013	70
AMP-Sepharose	0.06	0.1	0.6	23

Table 2. Kinetic and biochemical properties of human lens ALDH and
human liver ALDH1

Property	Human lens ALDH	Human liver ALDH1
Km acetaldehyde	111µM[1]	120µM[2]
Km propionaldhyde	8µM[1]	11µM[2]
Disulfiram inhibition	fully inhibited @ 4.2µM[1]	Ki value 0.02µ[2]
Apparent subunit weight (kD)	53.8[1]	54.8[2], 56.2[3]
pI	4.9-5.1[1]	5.1[3,4]
Cross-reactivity with anti- human liver ALDH1	+++[1]	+++[5]

(1) This work; (2) Greenfield & Pietruszko (1977); (3) Dockham et al. (1993); (4) Yoshida et al. (1993); (5) Ikawa et al.

tion, human lens ALDH1 exhibited activity with both malondialdehyde and retinaldehyde as substrates.

Figure 2 illustrates an SDS-PAGE gel, stained with Coomassie Blue, resolving protein subunits for human corneal and lens extracts, as well as for purified human cornea ALDH3 and lens ALDH1 isozymes. As a result of the differential specific activities for these enzymes, ALDH1 is present in higher concentrations of protein in human lens, as compared with human corneal ALDH3. In this study, corneal ALDH3 was observed at a level of ~ 0.1 mg/g wet tissue, whereas lens ALDH1 was observed at 1.5 mg/g wet tissue. Further, owing to the very high protein concentration in the lens compared to the cornea,when these values are expressed as a percentage of total soluble protein, the situation reverses itself. Corneal ALDH3 represents 5% of the soluble protein, whereas lens ALDH1 represents only 1–2% of the soluble protein in this tissue.

Figure 3 illustrates a Western blot of an SDS-polyacrylamide gel on which both human lens and corneal soluble extracts were reacted with rabbit antihuman liver ALDH1 and/or antihuman corneal ALDH3. Major cross-reactive species for ALDH3 were observed only in the cornea, whereas ALDH1 cross-reactive species were most prominent in the lens, with some activity also being observed in corneal extracts.

Figure 2. SDS-polyacryamide gel stained with Coomassie Blue. Lane 1: standards; Lane 2: Purified human lens ALDH1; Lane 3: human lens homogenate; Lane 4: purified human lens ALDH1 and corneal ALDH3; Lane 5: human corneal homogenate.

Figure 3. Western blot of a SDS-polyacryamide gel. Lane 1: human lens homogenate reacted with rabbit antihuman liver ALDH1.; Lane 2: partially purified human liver ALDH1 reacted with rabbit antihuman liver ALDH1; Lane 4: human corneal homogenate reacted with both rabbit antihuman liver ALDH1 and rabbit antihuman corneal ALDH3.

4. DISCUSSION

This study appears to be the first report that ALDH1 is present in sufficient quantities in the human lens to be described as a major soluble protein, and represents 1–2 % of human lens soluble protein, or ~ 0.002% of wet tissue. The latter represents a comparable level for human liver ALDHs on a wet weight basis, of which 40% exists as ALDH1 (based on studies by Hempel et al, 1982). Another study has also reported high levels of ALDH activity in the human lens (Crabbe and Hoe,1991).

Purified human lens ALDH1 exhibited similar properties to human liver ALDH1 in respect to its kinetic characteristics with acetaldehyde and propionaldehyde, its ability to oxidise malondialdehyde and retinaldehyde, its sensitivity to disulfiram, and its apparent pI and subunit molecular weight (Table 2). In addition, human lens ALDH1 cross reacted with rabbit antihuman liver ALDH1 polyclonal antibodies, confirming that this enzyme is identical with,or closely homologous to, human liver ALDH1. ALDH1 activity was also observed in the human cornea but at ~ 10% the level of the lens. Moreover, human corneal ALDHs exhibited different activity profiles, with ALDH3 representing ~ 99% of total activity.

Soluble proteins present at high levels (10% or more of total soluble protein) in the ocular lens of a range of phyla have been termed lens crystallins . One group of lens crystallins is the taxon-specific crystallins (see Tomarev and Piatigorsky, 1996) which are a group of proteins, identical to or closely related to metabolic enzymes, expressed at high levels in ocular lenses. The crystallins appear to be proteins recruited for specific purposes to the lens. This selection is reported to have occurred by two different mechanisms. For the a-crystallins and b/g-crystallins, gene duplication (one for α-crystallins and multiple for β/γ-crystallins) has occurred. However, for the taxon-specific crystallins, τ-crystallin (α-enolase) and ε-crystallin (lactate dehydrogenase B4), only one gene has been found. This mechanism of crystallin recruitment has been termed gene sharing (Piatigorsky et al., 1988; Wistow et al., 1991). If α crystallin evolves from a metabolic enzyme without gene duplication, it retains its original enzymatic activity (as it is essential for its metabolic role), while perhaps changing in other properties to meet more closely the requirements of its new role. Human lens ALDH1, while present at less than one fifth the level expected of a protein that would be termed a crystallin, is never the less present at high levels, and this high level of expression must have occured by a tissue-specific mechanism.

The presence of ALDH3 in bovine cornea at ~30% of the soluble protein has led to it's description as a corneal crystallin (Cooper et al., 1993). Thus, the crystallin concept originally proposed to describe the very high level of expression of specific proteins by differentiated lens cells (lens fibres), is now being applied to corneal proteins. Although the nature of the lens and cornea are quite distinct, this conceptual leap provides a means to investigate the presence of ALDH in the cornea in terms other than its catalytic functioning. It has been suggested that coenzyme binding capacity, thermal stability and UV absorption as an E-NAD(H) binary complex, may represent important properties consistent with a separate role in mammalian cornea. Only one gene copy has been reported for ALDH3, which means if ALDH3 is functioning as a crystallin, it is evolving in accordance with the process called gene sharing. It may be expected to possess metabolic activity but have acquired a capacity for enhanced tissue-specific expression. A recent study of the constitutive expression of rat ALDH3 in cultured corneal epithelial cells has provided an insight into this enhanced expression, which may be controlled by a light inducible-light maintenance mechanism of gene regulation (Boesch et al,1996).

In summary, these studies have demonstrated a substantial differential expression of ALDH isozymes in anterior tissues of the human eye. In the human cornea, ALDH3 is present in very high levels of both activity and protein, and constitutes ~ 5% of total soluble protein. In addition, ALDH1 is also present in significant levels. In the human lens, the situation is reversed, with ALDH1 constituting 1–2% of soluble protein and exhibiting high levels of activity, whereas ALDH3 constitutes <1% of the total ALDH activity. These high ALDH activity levels in the anterior tissues of the human eye may serve as an efficient means of oxidising aldehydes arising from lipid peroxidation following UVR absorption, as well as serving as photoreceptors for UVR, possibly by way of E-NAD(H) complexes.

5. REFERENCES

Abedinia, M., Pain, T., Algar, E.M. and Holmes, R.S.: Bovine corneal aldehyde dehydrogenase: the major soluble corneal protein with a possible dual protective role for the eye. Exp. Eye Res. 51 (1990) 419–426.

Alexander, R.J., Silverman, B. and Henley, W.L.: Isolation and characterization of BCP54 the major soluble protein of bovine cornea. Exp. Eye Res. 32 (1981) 205–216.

Algar, E.M. and Holmes, R.S.: Kinetic properties of murine liver aldehyde dehydrogenases. In Weiner, H. and Flynn, T.G. (Eds.), Enzymology and Molecular Biology of Carbonyl Metabolism 2. Alan R. Liss, N.Y., 1989 pp >93–103.

Ambroziak,W. and Pietruszko,R.: Human aldehyde dehydrogenase activity with aldehyde metabolites of monoamine, diamines and polyamines. J. Biol. Chem. 266 (1991) 13011–13018.

Ames, G.F.-l.: Resolution of bacterial proteins by polyacrylamide gel electrophoresis on slabs. J. Biol. Chem. 249 (1974) 634–644.

Boesch,J.S., Lee, C. and Lindahl,R.G.: Constitutive expression of class 3 aldehyde dehydrogenase in cultured rat epithelium. J. Biol. Chem.271 (1996) 5150–5157.

Boettner E.A. and Wolters J.R.: Transmittance of the ocular media. Invest. Ophthalmol. 1 (1962) 776–783.

Cooper, D.L., Bapist, E.W., Enghild, J., Lee, H., Isola, N. and Klintworth, G. K.: Bovine corneal protein 54K (BCP54) is a homologue of the tumour-associated (Class 3) rat aldehyde dehydrogenase (RATALD). Gene 98 (1991) 201–207.

Cooper D.L., Isola N.R., Stevenson K. and Baptist E.W.: Members of the ALDH gene family are lens and corneal crystallins. In Weiner, H., Crabb,D.W., and Flynn,G.F. (Eds.), Enzymology and Molecular Biology of Carbonyl Metabolism 4. Plenum Press, New York, 1993, pp 169–179.

Crabbe, M.J.C. and Hoe, S.T.: Aldehdye dehydrogenase, aldose reductase and free radical scavengers in cataract. Enzyme 45 (1991)188–193

Dockham,P.A., Lee, M.-O. and Sladek,N.E.: Identification of human liver aldehyde dehydrogenases that catalyse the oxidation of aldophosphamide and retinaldehyde. Biochem. Pharmacol. 43 (1992) 2453–2469.

Evces, S. and Lindahl, R.: Characterisation of rat cornea aldehyde dehydrogenase. Arch. Biochem. Biophys. 274 (1989) 518–529.

Feldman, R.I. and Weiner, H.: Horse liver aldehyde dehydrogenase 1. Purification and characterization. J. Biol. Chem. 247 (1972) 260–266.

Gondhowiardjo,T.D., van Haeringen,N.J., Hoekzema,R., Pels,L., and Kijlstra,A.: Detection of aldehyde dehydrogenase activity in human corneal extracts. Current Eye Research 10 (1991) 1001–1007.

Greenfield, N.J. and Pietruszko, R.: Two aldehyde dehydrogenases from human liver isolation via affinity chromatography and characterization of the isozymes. Biochim. Biophys. Acta 483 (1977) 35–45.

Hempel J.D., Reed, D.N. and Pietruszko, R.: Human aldehyde dehydrogenase: improved purification procedure and comparison of homogeneous isoenzymes E1 and E2. Alcoholism: Clinical and Experimental Research 6 (1982) 417–425.

Holmes, R.S. and VandeBerg, J.L.: Ocular NAD-dependent alcohol dehydrogenase and aldehyde dehydrogenase in the baboon. Exp. Eye Res. 43 (1986) 383–396.

Holmes, R.S.: Alcohol dehydrogenases and aldehyde dehydrogenases of anterior eye tissues from humans and other mammals. In Kuriyama, K., Takada, A. and Ishii, H. (Eds.), Biomedical and Social Aspects of Alcohol and Alcoholism. Elsevier Science Publishers, Amsterdam (1988), pp 51–57.

Holmes, R.S., Cheung, B., and Vandeberg, J.L.: Isoelectric focusing studies of aldehyde dehydrogenases, alcohol dehydrogenases and oxidases from mammalian anterior eye tissues. Comp. Biochem. Physiol. 93B (1989). 271–277.

Hsu, L.C. and Chang, W-C.: Cloning and characterization of a new functional aldehyde dehydrogenase gene. J. Biol. Chem. 266, (1991)12257–12265.

Hsu L.C., Chang W.-C. and Yoshida A.: Cloning of a cDNA encoding human ALDH7, a new member of the aldehyde dehydrogenase family. Gene 151 (1994) 285–289.

Jones, D.E., Brennan, M.D., Hempel, J. and Lindahl, R.: Cloning and complete nucleotide sequence of a full-length cDNA encoding a catalytically functional tumour-associated aldehyde dehydrogenase. Proc. Natl. Acad. Sci. USA 85 (1988) 1782–1786.

Kagan,V.,Schvedova,A.,Novikov,K and Kozlov,Y.: Light induced free radical oxidation of membrane lipid photoreceptors of frog retina.Biochem. Biophys.Acta 330 (1973) 76–79.

King,G. and Holmes.R.: Human corneal aldehyde dehydrogenase. Purification,kinetic characterization and phenotypic variation. Biochem. and Mol. Biol. Int. 31 (1993) 49–63.

Kurys, G., Shah, P.C., Kikonyogo, A., Reed, D., Ambroziak, W., and Pietruszko, R.: Human aldehyde dehydrogenase cDNA cloning and primary structure of the enzyme that catalyzes dehydrogenation of 4-aminobutyraldehyde. Eur. J. Biochem. 218 (1993) 311–320.

Laemmli, U.K.: Cleavage of structural protein during the assembly of the head of bacteriophage T4. Nature 227 (1970) 680–685.

Pereira, F., Rosenmann, E., Nylen, E., Kaufman, M., Pinsky, L. and Wrogemann, K.: The 56 kDa androgen binding protein is an aldehyde dehydrogenase. Biochem. Biophys. Res. Comm. 175 (1991) 831–838.

Piatigorsky, J., O'Brien, W.E., Norman, B.L., Kalumuck, K., Wistow, G.J., Borras, T., Nickerson, J.M. and Wawrousek. E.F.: Gene sharing by δ-crystallin and argininosuccinate lyase. Proc. Nat. Acad. Sci. USA 85 (1988) 3479–3483.

Pietruszko,R.: Aldehyde dehydrogenase isozymes. In Isozymes: Current Topics in Biological and Medical Research 8. Alan R. Liss,Inc. New York,1983,pp 195–217.

Scopsi, L., Larsson, L.-I.: Increased sensitivity in peroxidase immunochemistry: A comparative study of a number of peroxidase visualisation methods employing a model system. J. Histochem. 84 (1986) 221–230.

Sidhu, R.S. and Blair, A.H.: Human liver aldehyde dehydrogenase esterase activity. J. Biol. Chem. 250 (1975) 7894–7898.

Smith, P.-K., Krohn, R.I., Hermanson, G.T., Mallia, A.K., Gartner, F.H., Provenzana, M.D., Fujimoto, E.K., Goeke, N.M., Olsen, B.J., and Klenk,D.C.: Measurement of protein using bicinchoninic acid. Anal. Biochem. 150, (1985) 76–85.

Tomarev, S.I. and Piatigorsky, J.: Lens crystallins of invertebrates: diversity and recruitment from detoxification enzymes and novel proteins. Eur. J. Biochem. 235 (1996) 449–465.

van Heyningen, R.: Fluorescent glucosides in the human lens. Nature 230 (1970) 393–394.

Verhagen, C., Hoekzema, R., Verjans, G.M.G.M. and Kijlstra, A.:Identification of bovine corneal protein 54 (BCP 54) as an aldehyde dehydrogenase. Exp. Eye Res. 53 (1991) 283–284.

Wistow, G. and Kim, H.: Lens proteins expression in mammals: taxon-specificity and the recruitment of crystallins. J. Mol. Evol. 32 (1991) 262–269.

Zigman S.: The role of sunlight in human cataract formation. Surv. Ophthalmol. 27 (1983) 317- 326.

Zigman, S.: Photobiology of the lens. In Maisel, H. (ed.), The Ocular Lens. Marcel Dekker Inc., 1985, pp301–347.

5

REGULATION OF RAT ALDH-3 BY HEPATIC PROTEIN KINASES AND GLUCOCORTICOIDS

Russell A. Prough,[1] K. Cameron Falkner,[1] Gong-Hua Xiao,[1] and
Ronald G. Lindahl[2]

[1]Departments of Biochemistry and Molecular Biology
University of Louisville School of Medicine
Louisville, Kentucky 40292
[2]University of South Dakota School of Medicine
Vermillion, South Dakota 57069

1. INTRODUCTION

The aldehyde dehydrogenases (ALDH, aldehyde $NAD(P)^+$ oxidoreductase E.C. 1.2.1.3) are a superfamily of $NAD(P)^+$-dependent enzymes which are comprised of at least three classes or families based on sequence similarity. The Class 3 aldehyde dehydrogenase gene is expressed in a tissue-specific manner in microsomal and cytosolic fractions of rodents (Dunn *et al.*, 1988; Boesch *et al.*, 1996), with the highest level of constitutive expression occurring in corneal epithelium, stomach, and heart. This gene is also expressed at high levels in neoplastic tissue and some cell lines. The gene product also displays induced expression in liver, lung, bladder, colon, spleen and thymus after exposure of rodents to polycyclic aromatic hydrocarbons (PAH[1]). Takimoto *et al.* (1994) have characterized the 5'-flanking region of the *ALDH*-3 gene and demonstrated that it contains at least three major functional regulatory domains: a strong promoter proximal to the transcription start site, an inhibitory region just upstream of the promoter, and a PAH-responsive enhancer region approx. 3.5 kb upstream of the promoter. Expression of this gene appears to be controlled by interaction of at least these three functional domains.

ALDH-3 is a member of a gene battery which is regulated by action of the arylhydrocarbon receptor (*Ah*R), a cytosolic protein capable of binding PAH as ligands (Whitlock *et al.*, 1996). The *Ah* receptor-ligand complex is translocated into the nucleus after ligand activation and forms a heterodimeric complex with the arylhydrocarbon nuclear transporter protein (ARNT); subsequently they interact with specific DNA sequences denoted xenobiotic responsive elements (XREs) to increase the transcription of these genes. *Ah*R mediates induction of a number of xenobiotic metabolizing enzymes (termed the *Ah* gene battery), including cytochromes P4501A/B, glutathione *S*-transferase Ya_1, NAD(P)H:quinone oxidoreductase, UDP-glucuronosyl transferase 1*6 and Class 3 aldehyde dehydrogenase (Nebert *et al.*, 1990). The best understood response regulated by the *Ah* receptor (Whitlock *et al.*, 1996) is the induction of P4501A1. Multiple functional XREs have been identified in the 5'-upstream regulatory region of the *CYP*1A1 gene. The 5'-flanking region of the other members of the *Ah* receptor gene battery also contain one or more copies of a functional XRE-like sequence (Takimoto *et al.*, 1994; Rushmore *et al.*, 1990; Favreau and Pickett, 1991). Although there are sequences of the *ALDH*-3 PAH-responsive enhancer nearly identical to the XREs in the *CYP*1A1 flanking sequences, one functional *ALDH*-3 XRE is located much further upstream than those seen in *CYP*1A1, *glutathione S-transferase* Ya1, and *NADPH-quinone oxidoreductase* genes (Takimoto *et al.*, 1994).

A second characteristic of some genes in the Ah battery is their low level of constitutive expression possibly due to negative regulatory mechanisms (RayChaudhuri *et al.*, 1990). Several different mechanisms have been described for the negative regulation of P4501A1: putative regulatory genes by products of P4501A1 itself (RayChaudhuri *et al.*, 1990; Whitlock *et al.*, 1996), regulation by nucleosome structure (Morgan and Whitlock, 1992; Whitlock *et al.*, 1996), and specific negative regulatory transcription factors (Boucher *et al.*, 1995). While evidence exists for negative regulation of other genes in the Ah gene battery, no studies have provided mechanistic details for the *ALDH*-3 gene.

PKC has been shown to be involved in the regulation of *CYP*1A1 gene expression (Carrier *et al.*, 1992; Reiners *et al.*, 1993), since phosphorylation of proteins acting at some step in transactivation by the *Ah*R was shown to apparently be dependent upon protein kinase C (PKC) activity. Our previous experiments with cultured rat hepatocytes demonstrated that the PKC inhibitors H7 and staurosporine concomitantly inhibited PAH-induction of all five genes of the *Ah* gene battery we tested (Xiao *et al.*, 1995), including *ALDH*-3, *CYP*1A1, *glutathione S-transferase* Ya$_1$, *NAD(P)H:-quinone oxidoreductase*, *UDP glucuronosyltransferase* 1*06. The protein kinase A inhibitors H8 and HA1004 had no effect on all of these genes at equivalent concentrations, except PAH-induction of ALDH-3 mRNA and protein which was stimulated 3–4 fold (Xiao *et al.*, 1995).

Our past studies (Mathis *et al.*, 1986, 1989; Prough *et al.*, 1996; Sherratt *et al.*, 1990) on regulation of these genes by glucocorticoids (GC) demonstrated a potentiation of PAH-inducible enzyme activity associated with *CYP*1A1 and UDP-glucuronosyltransferase (UGT) in cultured adult rat hepatocytes using defined media lacking GC. Although GC alone had little or no effect on the basal expression of CYP1A1 or UGT activity in cultured rat hepatocytes, the presence of GC significantly enhanced PAH-dependent induction of these enzymes. PAH resulted in over a 40–60-fold induction of 7-ethoxyresorufin *O*-deethylase (EROD) activity in the absence of added glucocorticoids and treatment with BA plus dexamethasone resulted in 2–3-fold higher EROD activity. A similar response was seen for UDP-glucuronosyltransferase activity. However, expression of the subunits of glutathione *S*-transferase Ya$_1$ or NAD(P)H:quinone oxidoreductase after exposure to PAH was repressed by GC in cultured rat hepatocytes (Xiao *et al.*, 1995). This suggests that while GC may regulate all members of the *Ah* gene battery, the mechanisms of regulation are not identical.

The concentration-dependence and kinetics of induction of ALDH-3 and P4501A1 by PAH are distinctly different both *in vivo* (Dunn *et al.*, 1988) and *in vitro* (Asman *et al.*, 1993). Induction of ALDH-3 by PAH requires a slightly longer period of time after administration to obtain maximal induction and about 10-fold higher concentrations of PAH to reach a maximal steady-state levels. In this study, we sought to characterize the effects of protein kinase A and glucocorticoids on either basal or PAH induction of *ALDH*-3 gene expression. We sought to characterize the mechanism whereby PKA and GC regulate levels of the ALDH-3 enzyme using either cultured adult rat hepatocytes and/or HepG2 tumor cells for transient transfection experiments utilizing reporter genes containing the 5'-flanking region of the *ALDH*-3 gene. Our results suggest the existence of two (cAMP- and glucocorticoid receptor-dependent) distinct negative *cis*-acting elements located in the 5'-flanking region of the *ALDH*-3 gene.

2. NEGATIVE REGULATION BY GLUCOCORTICOIDS

As reported in past studies (Prough *et al.*, 1996) on regulation of these genes by GC, these adrenal steroids caused a potentiation of PAH-inducible enzyme activity associated with *CYP*1A1 in cultured adult rat hepatocytes using media lacking GC (Figure 1). Although GC alone had little or no effect on the basal expression of *CYP*1A1 in cultured rat hepatocytes, the presence of GC significantly enhanced PAH-dependent induction of the P4501A1 enzyme activity. 1,2-Benzanthracene (BA) caused an 60-fold induction of 7-ethoxyresorufin *O*-deethylase (EROD) activity in the absence of added glucocorticoids and treatment with BA plus dexamethasone resulted in 2–3-fold higher EROD activity (Figure 1). ALDH-3 activity measured using $NADP^+$ and benzaldehyde was not effected by inclusion of DEX in the culture medium, but was increased 35-fold in the presence of BA. In contrast to P450/A1 activity, cells treated with both DEX and BA displayed decreased induction of ALDH-3 enzyme activity by 80% (Figure 1). This result demonstrates that *ALDH*-3 gene activation is regulated quite differently by GC than is *CYP*1A1.

2.1. Presence of Canonical Glucocorticoid Responsive Elements in the *CYP*1A1 and *ALDH*-3 Genes

In order to address the role of glucocorticoids in regulation of the activation of these two genes, the presence of glucocorticoid responsive elements (GREs) in the 5'-flanking

Figure 1. Effect of Dexamethasone on PAH-induction of P4501A1 and ALDH-3 enzyme activity in cultured rat hepatocytes. Adult rat hepatocytes were prepared as described by Xiao *et al.* (1995) and treated with either DMSO (CON), 1 μM dexamethasone (DEX), 50 μM 1,2-benzanthracene (BA), or a combination of these agents (BA+DEX). The cell protein was harvested after 2–3 days and the enzyme activity measured for three pooled sets of cells. The enzyme activities were normalized to the activity seen in the presence of 50 μM BA; 64 pmol resorufin formed from 7-ethoxyresorufin per min per mg cell protein ± S.D. for P4501A1 and 20.7 nmol benzoic acid formed from benzaldehyde per min per mg cell protein ± S.D. for ALDH-3.

regions of the *CYP*1A1 and *ALDH*-3 genes were determined using information in the Gen-Bank Nucleic acid databank. Initially, we looked for canonical glucocorticoid responsive elements (GRE halfsite: TGT-C/T-CT) in these genes which might suggest direct control of the ligand-activated glucocorticoid receptor. Hines *et al.* (1985) sequenced the first intron of that gene which is 2451 bp in length and noted three GRE half-sites within a 360 basepair fragment (XbaI/Sau3a) of that intron. The first exon of the *CYP*1A1 gene is non-coding with the translation initiation site located in exon II. Interestingly, the intronic GRE sequences are conserved among the human, murine, and rat *CYP*1A1. We also evaluated the rat aldehyde dehydrogenase 3 gene characterized by Takimoto *et al.* (1994) and observed several GRE halfsites and an imperfect palindromic GRE at position -1166 bp in the 5'-flanking region of the gene.

2.2. Analysis of Functional GRES in the *CYP*1A1 and *ALDH*-3 Genes

Mathes *et al.* (1989) demonstrated that a CAT construct containing 1 kb of 5'-flanking region, exon I, intron I, and a small portion of exon II of the *CYP*1A1 gene was PAH-inducible in cultured HepG2 cells. Interestingly, this construct was responsive to inclusion of DEX in the culture media with PAH. This demonstrates that when CAT constructs containing portions of the 5'-flanking region, exon I, and intron I of *CYP*1A1 gene are used, we can recapitulate the GC potentiation of PAH-induction of CYP1A1. When Mike Mathis prepared a construct with the XbaI fragment deleted (which contains the three intronic GREs), GC responsiveness was lost. Another CAT construct we prepared contain-

Figure 2. Effect of mutation of the palindromic GRE in the 5'-flanking region of the ALDH3CAT reporter plasmid. Reporter gene constructs containing approx. 3.5 kb of the 5'-flanking region of the *ALDH*-3 gene were prepared with normal or mutated palindromic GRE near position -1166. After transfection of HepG2 cells with an expression plasmid for glucocorticoid receptor, an expression plasmid for β-galactosidase, and the respective reporter gene (ALDH3CAT or mutated ALDH3CAT), the cells were treated with 50 μM 1,2-benzanthracene (BA) or 50 μM BA plus 1 μM dexamethasone (BA+DEX). Cell protein was harvested and the expression of β-galactosidase and chloramphenicol acetyltransferase was measured by specific enzyme assay.

ing 360 bp of the first intron which contain the 3 GREs of the *CYP*1A1 gene was shown to be unresponsive to glucocorticoids (Linder and Prough, unpublished results). These results demonstrate that GC potentiation of PAH-induction of expression of *CYP*1A1 is apparently dependent upon the intronic GRE half-site sequences, but the construct containing only this enhancer, like the native gene, is apparently unaffected by glucocorticoids alone.

In a similar manner, we have prepared several CAT constructs of the 5'-flanking region of the ALDH3 gene. We used PCR methods to generate a CAT construct containing a mutated GRE instead of the native palindromic GRE (data not shown). As seen in Figure 2, DEX strikingly represses the expression of the wild type CAT construct in the same manner as was observed with the native gene in Figure 1 (Xiao *et al.*, 1995). Mutation of the perfect halfsite of the palindromic GRE (AGGACA changed to ACGGCA) in the 5'-flanking region of the ALDH3CAT construct obviated the negative regulation of this construct. This result suggests that the GRE in the regulatory region of the *ALDH*-3 is critical for modulation of gene expression by glucocorticoid receptor (GR). Inclusion of the GR antagonist, RU-38486, in the media of cells transfected with the ALDH3CAT construct prevented the repression of PAH-induction of this reporter genes in HepG2 cells by glucocorticoids (data not shown).

The ALDH3 gene is not expressed constitutively in cultured rat hepatocytes or *in vivo* in the absence of PAH; however, the ALDH3 reporter gene was expressed at measurable levels in cultured HepG2 cells. We observed that in the presence of DEX, basal expression of ALDH3CAT was also regressed (data not shown), which suggests that glucocorticoid action is independent of AhR function. Together, these data indicate that the regulation of the *ALDH*-3 gene by glucocorticoids is independent of its gene activation by *Ah*R. Most genes with functional GREs are regulated directly by the glucocorticoid receptor and its cognate responsive elements have been shown to be positively activated. However, the negative regulation we observed also requires the presence of the palindromic GRE in the 5'-flanking region of the *ALDH*-3 gene and probably requires interaction with some other *cis*-acting element in the gene.

3. REGULATION OF *ALDH*-3 GENE EXPRESSION BY PROTEIN KINASE A ACTION

3.1. Effect of Protein Kinase A Inhibitors on PAH-Induction of the *ALDH*-3 Gene in Cultured Rat Hepatocytes

Previously, we demonstrated that protein kinase C inhibitors concomitantly inhibit the function of the Ah receptor in regulation of genes of the Ah gene battery (Xiao *et al.*, 1995). However, only one gene of the battery, *ALDH*-3, was affected by inhibitors of protein kinase A. The PKA inhibitors are structural analogs of PKC inhibitors and are often used by workers in the field as negative controls for regulation by protein kinase C (Reiners *et al.*, 1993; Xiao *et al.*, 1995). As seen in Figure 3, addition of the PKA inhibitor H8 stimulated expression of the *ALDH*-3 gene in cultured hepatocytes treated with PAH, but not expression of *CYP*1A1. Other experiments were performed in cultured rat hepatocytes using compounds, such as dibutryl-cAMP or forskolin, which enhance production of cAMP which in turn stimulates PKA activity (data not shown). These two compounds potently suppressed PAH-induced expression of this gene and suppressed the potentiation of PAH-induction by protein kinase A inhibitors (data not shown).

Figure 3. Effect of the Protein Kinase A inhibitor, H8, on PAH-induction of P4501A1 and ALDH-3 enzyme activity in cultured rat hepatocytes. Adult rat hepatocytes were prepared as described by Xiao *et al.* (1995) and treated with either DMSO (CON), 50 μM H8 (H8), 50 μM 1,2-benzanthracene (BA), or a combination of these agents (BA+H8). The cell protein was harvested after 2–3 days and the enzyme activity measured for three pooled sets of cells. The enzyme activities were normalized to the activity seen in the presence of 50 μM BA; 40 pmol resorufin formed from 7-ethoxyresorufin per min per mg cell protein ± S.D. for P4501A1 and 8 nmol benzoic acid formed from benzaldehyde per min per mg cell protein ± S.D. for ALDH-3.

To test the effect of PKA inhibitors on the reporter construct ALDH3CAT, we performed transient transfection experiments in the absence or presence of H8. As was seen with the native gene (Figure 3), H8 stimulated the PAH-dependent induction of this reporter construct (Figure 4), as well as the basal expression of this reporter gene in HepG3 cells. These results suggest that the regulation of the *ALDH*-3 gene by protein kinase A is due to the action of this protein kinase in phosphorylating a transcription factor which renders the factor inactive. Alternatively, the transcription factor could act as a negative regulatory factor when phosphorylated. We have also identified a 66 bp region which appears to contain the *cis*-acting element responsible for this negative regulatory phenomenon by

Figure 4. Effect of the Protein Kinase A Inhibitor, H8, on Basal and PAH-induced Expression of the ALDH3CAT Reporter Plasmid. The CAT construct containing 3.5 kb of the 5'-flanking region of the wild type rat *ALDH*-3 gene was transfected into HepG2 cells with an expression plasmid for β-galactosidase. The cells were treated with dimethyl sulfoxide (Untreated) or 50 μM 1,2-benzanthracene (BA Treated) in the absence (-H8) or presence (+H8) of 10 μM H8. Cell protein was harvested and the expression of β-galactosidase and chloramphenicol acetyltransferase was measured by specific enzyme assay.

protein kinase A. Interestingly, deletion of this 66 bp region does not affect either PAH-dependent induction of expression or the negative regulation of the reporter gene by glucocorticoids, but does increase the level of basal expression. As anticipated from these results, the 66 bp region does not contain the palindromic GRE described above.

4. CONCLUSION

This report documents the presence of two *cis*-acting elements in the 5'-flanking region of the *ALDH*-3 gene which are involved in its negative regulation in the rat. Since deletion of the cAMP-regulated element increases basal expression, but does not affect either PAH-induced expression, or regulation by glucocorticoids, it appears that the three regulatory *cis*-elements (XRE for binding the Ah receptor, GRE for binding the glucocorticoid receptor, and the responsive elements for the putative cAMP-regulated transcription factor) must function independently. Characterization of the protein(s) which serves as the cAMP-regulated transcription factor and the other *cis*-acting elements which cause the GRE to function as a negative element is the subject of further research in our laboratories.

5. ACKNOWLEDGMENTS

This research was supported in part by U.S.P.H.S. Grants R01 ES04244 (RAP) and CA 21103 (RGL).

6. REFERENCES

Asman, D.C., Takimoto, K., Pitot, H.C., Dunn, T.J., and Lindahl, R.: Organization and characterization of the rat class 3 aldehyde dehydrogenase gene. J. Biol. Chem. 268: (1993) 12530–12536.

Boesch, J.S., Lee, C., and Lindahl, R.G.: Constitutive expression of class 3 aldehyde dehydrogenase in cultured rat corneal epithelium. J. Biol. Chem. 271: (1996) 5150–5157.

Boucher, P.D., Piechocki, M.P., and Hines, R.N.: Partial characterization of the human CYP1A1 negatively acting transcription factor and mutational analysis of its cognate DNA recognition sequence. Mol. Cell. Biol. 15 (1995) 5144–5151.

Carrier, F., Owens, R.A., Nebert. D.W., and Puga, A.: Dioxin-dependent activation of murine Cyp1a-1 gene transcription requires protein kinase C-dependent phosphorylation. Mol. Cell. Biol. 12 (1992) 1856–1963.

Dunn, T.J., Lindahl, R., and Pitot, H.C.: Differential gene expression in response to 2,3,7,8-tetrachlorodibenzo-p-dioxin (TCDD). Noncoordinate regulation of a TCDD-induced aldehyde dehydrogenase and cytochrome P-450c in the rat. J. Biol.

Favreau, L.V. and Pickett, C.B.: Transcriptional regulation of the rat NAD(P)H:quinone reductase gene. Identification of regulatory elements controlling basal level expression and inducible expression by planar aromatic compounds and phenolic antioxidants. J. Biol. Chem. 266 (1991) 4556–4561.

Hines, R. N., Levy, J. B., Conrad, R. D., Iversen, P. L., Shen, M., Renli, A. M. J., and Bresnick, E.: Gene structure and nucleotide sequence for rat cytochrome P-450c. Arch. Biochem. Biophys. 237 (1985) 465–476.

Mathis, J. M., Prough, R. A., Hines, R. N., Bresnick, E., and Simpson, E. R.: Regulation of cytochrome P-450c by glucocorticoids and polycyclic aromatic hydrocarbons in cultured fetal rat hepatocytes. Arch. Biochem Biophys. 246 (1986) 439–448.

Mathis, J. M., Houser, W. H., Bresnick, E., Cidlowski, J. A., Hines, R. N., Prough, R.A., and Simpson, E. R.: Glucocorticoid regulation of the rat cytochrome P450c (P450IA1) gene: Receptor binding within intron I. Arch. Biochem Biophys. 269 (1989) 93–105.

Morgan, J.E. and Whitlock, J.P,. Jr.: Transcription-dependent and transcription-independent nucleosome disruption induced by dioxin. Proc. Natl. Acad. Sci., U.S.A. 89 (1992) 11622–11626.

Nebert, D.W., Petersen, D.D., and Fornance, A.J., Jr.: Cellular responses to oxidative stress: the [Ah] gene battery as a paradigm. Environ. Health Perspect. 88 (1990) 13–25.

RayChaudhuri, B., Nebert, D.W. and Puga, A.: The murine Cyp1a-1 gene negatively regulates its own transcription and that of other members of the aromatic hydrocarbon-responsive [Ah] gene battery. Mol. Endocrinol. 4 (1990) 1773–1781.

Reiners, J.J. Jr., Scholler, A. Bischer, P., Cantu, A.R., and Pavone, A.: Dioxin-dependent activation of murine Cyp1a-1 gene transcription requires protein kinase C-dependent phosphorylation. Arch. Biochem. Biophys. 301 (1993) 449–454.

Rushmore, T.H., King, R.G., Paulson, K.E., and Pickett, C.B.: Regulation of glutathione S-transferase Ya subunit gene expression: Identification of a unique xenobiotic-responsive element controlling inducible expression by planar aromatic compounds. Proc. Natl. Acad. Sci. USA 87 (1990) 3826–3830.

Sherratt, A.J., Banet, D.E., and Prough, R.A.: Glucocorticoid regulation of polycyclic aromatic hydrocarbon induction of cytochrome P450IA1, glutathione S-transferases, and NAD(P)H:Quinone oxidoreductase in cultured fetal rat hepatocytes. Mol. Pharmacol. 37 (1990) 198–205.

Takimoto, K., Lindahl, R., Dunn, T.J. and Pitot, H.C.: Structure of the 5' flanking region of class 3 aldehyde dehydrogenase in the rat. Arch. Biochem. Biophys. 312 (1994) 539–546.

Whitlock, J.P., Jr., Okino, S.T., Dong, L. Ko, H.P., Clarke-Katzenberg, R., Ma, Q., and Li, H.: Induction of cytochrome P4501A1: A model for analyzing mammalian gene transcription. FASEB J. 10 (1996) 809–818.

Xiao, G-H., Pinaire, J.A., Rodrigues, A.D., and Prough, R.A.: Regulation of the *Ah* receptor gene battery in cultured adult rat hepatocytes via *Ah* receptor-dependent and independent processes. Drug Metab. Dispos. 19 (1995) 793–804.

6

MOUSE DIOXIN-INDUCIBLE *Ahd4* GENE

Structure of the 5' Flanking Region and Transcriptional Regulation

Vasilis Vasiliou,[1,2] Steven F. Reuter,[2] Te-Yen Shiao,[1] Alvaro Puga,[2] and Daniel W. Nebert[2]

[1]Department of Pharmaceutical Sciences
University of Colorado Health Sciences Center
4200 East Ninth Avenue, Denver, Colorado 80262
[2]Department of Environmental Health
University of Cincinnati Medical Center
P.O. Box 670056, Cincinnati, Ohio 45267–0056

1. INTRODUCTION

Aldehyde dehydrogenases (ALDHs; EC 1.2.1.3) oxidize various aliphatic and aromatic aldehyde substrates to the corresponding carboxylic acids (Lindahl, 1992; Yoshida, 1992). Substrates for ALDHs are diverse and include acetaldehyde (Harrington et al., 1987), biogenic amines (Mackerell et al., 1986) and other neurotransmitters (Tipton et al., 1987), retinoic acid (Lee et al., 1991), corticosteroids (Monder et al., 1982), and aldehyde products of lipid peroxidation (Lindahl and Petersen 1992). A superfamily for the mammalian ALDHs has been proposed on the basis of divergent evolution (Vasiliou et al., 1995c). Within this superfamily, at least three genes are inducible by foreign chemicals. The cytosolic ALDH1 and ALDH3 genes are inducible by phenobarbital and 2,3,7,8-tetrachlorodibenzo-*p*-dioxin (dioxin; TCDD), respectively (*reviewed in* Vasiliou et al., 1995c). The microsomal ALDH3 is inducible by TCDD and peroxisome proliferators like clofibrate (Vasiliou et al., 1996). Among these three genes, the dioxin-inducible ALDH3 is the most thoroughly studied. The gene encoding ALDH3 has been cloned in rat (Asman et al., 1993), human (Hsu et al., 1992), and mouse (termed as *Ahd4*; Vasiliou et al., 1994). The mouse *Ahd4* gene has been mapped to Chr 11 (Vasiliou et al., 1993b), and the human *ALDH3* gene to chromosome 17 (Santisteban et al., 1985).

The *Ahd4* gene has been identified as a member of the mouse [*Ah*] battery, which represents one of the best-characterized examples of coordinately regulated genes encoding drug-metabolizing enzymes (Figure 1; Vasiliou et al., 1992; Vasiliou et al., 1993a; 1993b; *reviewed in detail in* Nebert et al., 1993). The definition of a gene battery in eu-

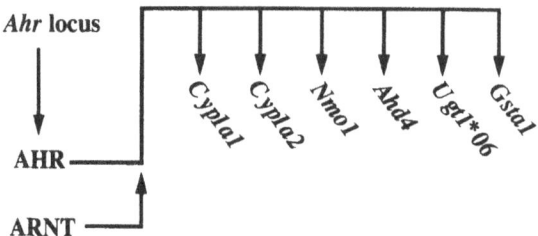

Figure 1. The mouse [*Ah*] gene battery.

karyotes is "a group of (nonlinked) genes cross-talking with one another, regarding their up- and down-regulation to endogenous or exogenous signals, via specific proteins whose binding may be combinatorial in nature (Britten and Davidson, 1969; Mcknight and Tjian 1986). The [*Ah*] battery contains at least six genes that are coordinately induced by TCDD and polycyclic aromatic hydrocarbons such as benzo[a]pyrene. In addition to two P450 genes--*Cyp1a1* and *Cyp1a2*--we have demonstrated that four Phase II genes are also members of the [*Ah*] battery: NAD(P)H:menadione oxidoreductase (*Nmo1*); a cytosolic "Class 3" aldehyde dehydrogenase (*Ahd4*); a UDP glucuronosyltransferase having 4-methylumbelliferone as substrate (*Ugt1a6*); and a glutathione transferase having 2,4-dinitro-1-chlorobenzene as substrate (*Gsta1*) (Nebert et al., 1993).

   To date we have identified at least three mechanisms by which the mouse *Ahd4* gene is regulated (*reviewed in* Vasiliou et al., 1995a). **(i)** The gene is inducible by TCDD, a process requiring functional Ah receptor (AHR) and Ah receptor nuclear translocator (ARNT), the heterodimer of which interacts with one or more aromatic hydrocarbon response elements (AhREs; also called DREs and XREs) in the 5' flanking region of the *Ahd4* gene (Vasiliou et al., 1992). The AHR is encoded by the mouse *Ahr* locus on proximal Chr 12, and ARNT is encoded by the *Arnt* locus on Chr 3. **(ii)** *Ahd4* gene expression is marked elevated in untreated *14CoS/14CoS* mouse cell lines, which have the homozygous deletion of a 3,800-kb region on Chr 7. It now appears most likely that deletion of the fumarylacetoacetate hydrolase (*Fah*) gene, located in this deleted region of Chr 7, leads to the accumulation of particular endogenous electrophiles which, in turn, activate the *Ahd4* gene as part of the oxidative stress response (*reviewed in* Vasiliou et al., 1995a). This Chr 7-mediated process has recently been shown to involve an *e*lectro*p*hile *r*esponse *e*lement (EpRE; also called ARE) and the AHR (Vasiliou et al., 1995b). **(iii)** The *Ahd4* gene is repressed by functional CYP1A1/1A2 enzymes. We have shown that AHD4 mRNA levels in the untreated *c37* mutant line are strikingly increased in the absence of endogenous CYP1A1 enzyme (Vasiliou et al., 1992; 1993a). Introduction of CYP1A1 or CYP1A2 functional enzymes--expressed by appropriate transfected cDNA-containing plasmids in *c37* stable transformants--restores the wild-type phenotype without affecting TCDD inducibility. This process appears to be dependent on a *n*egative *r*esponse *e*lement (NRE) in the regulatory region of the *Ahd4* gene (*reviewed in* Nebert et al., 1993).

   Several recent reviews about the [*Ah*] gene battery from other laboratories are incorrect when they include all dioxin- and electrophile-inducible genes as members of this gene battery. *It is this latter criterion described above (repression by CYP1A1/1A2 metabolism) that qualifies a gene as a member of the [Ah] battery.* There are many mouse dioxin-inducible genes (*e.g. Ahd3*, *Gstp1*, *Cyp1b1*, *Pai2*, *Fos* and *Jun*) and electrophile-inducible genes (*e.g. Hpxn* encoding hemopexin; *Mt1*, *Mt2*, *Mt3* and *Mt4* coding for the metallothioneins; *Sod1*, *Sod2* and *Sod3* encoding superoxide dismutases) which might be regarded as members of a "dioxin-inducible panel" and a "electrophile-inducible panel," respectively, but which have not been rigorously shown to qualify as members of the [*Ah*] gene battery.

In this chapter, we describe the gene structure and the putative transcription factor-binding sites found in a 3.2-kb fragment upstream of exon 1 of the *Ahd4* gene that might be important for positive and negative regulation of this gene. Finally, we present recent data about the transcriptional regulation of this gene with regard to TCDD- and electrophile-inducibility, as well as negative regulation.

2. MATERIALS AND METHODS

2.1. Isolation of the *Ahd4* Genomic Clones

A lambda FIX 129/SV mouse genomic library (Stratagene) was screened with a full-length AHD4 cDNA previously cloned by us (Vasiliou et al., 1993b). We prepared double-stranded DNA probes by random priming (Sambrook et al., 1989), using [α-^{32}P]dCTP (3000 Ci/mmole, New England Nuclear/DuPont) as the labeled precursor, and added these probes to hybridization solutions at 5–10 x 10^6 cpm/ml. Prehybridizations and hybridizations were carried out at 65 °C in a solution containing 3X SSC, 2.5X Denhardt's solution, 0.5% sodium dodecylsulfate (SDS), and denatured salmon sperm DNA (0.1 mg/ml). After hybridization for 16–20 h, the filters were washed twice in 2X SSC and 0.1% SDS for 10 min at room, and then in 0.3X SSC and 0.1% SDS at 60 °C for 1 h. When needed, oligonucleotide probes were end-labeled with [γ-^{32}P]dATP and T4 kinase (Sambrook et al., 1989) and prehybridizations and hybridizations were carried out at 42 °C in the solution described above. The filters were washed twice in 2X SSC and 0.1% SDS for 10 min at room temperature and then twice at 42 °C. Hybridizations were followed by autoradiography at -70 °C for 24–48 h in Kodak XAR-5 film with intensifying screens. After four rounds of screening, overlapping clones believed to carry the *Ahd4* gene were selected and aligned by restriction endonuclease digestion and Southern blot hybridization. Two positive clones, subsequently determined to contain the entire gene with flanking regions, were subloned into the *Not I* site of Bluescript SK II⁻ vector (designed as 3cG1 and 3cG6) for refined mapping and nucleotide sequencing.

2.2. Determination of Exon-Intron Junctions

Several oligonucleotides (18 bp in length) selected from the AHD4 cDNA sequence were synthesized and used as primers to sequence the *Ahd4* gene fragments that had been subcloned into Bluescript SK II⁻ (Stratagene), in order to determine the exon- intron boundaries. Intron distances were determined by PCR, utilizing synthetic oligonucleo-tide primers synthesized from the 3' and 5' ends of adjacent exons.

2.3. Determination of Nucleotide Sequence of the 5' flanking Region of the *Ahd4* Gene

An 8.0-kb *Hind III* fragment from the 3cG6 clone--containing 3.2 kb of 5' flanking region, the first and second exons, and a portion of intron 2--was subcloned into Bluescript SK II⁻ plasmid and designed as 3cG6H11. Each strand of the 3.2-kb fragment upstream of exon 1 was sequenced at least twice by the dideoxy chain termination method

(Sanger et al., 1977), using modified T7 DNA Sequenase. Nucleotide alignments were performed with the Genetics Computer Group (GCG) Sequence Analysis software package (Madison, WI).

2.4. Construction of *Ahd4*-CAT Recombinant Plasmids

A 3.2-kb DNA fragment was excised by digesting with *Xba I* the 3cGH11 plasmid, which contained the 6-kb *Hind III* segment of the *Ahd4* gene. This fragment, which held 3.2 kb of the 5' flanking region and 45 bp of the first exon, was cloned into *Xba I* site of the pCAT basic plasmid (Promega). The orientation of the pAhd4cat1 was confirmed by restriction digests and sequencing. The deletion mutants pAhd4p1 and pAhd4p2 were constructed by digestion of pAhd4cat1 with *Pst I* in separate experiments. In the pAhd4p1, two *Pst I* fragments were removed--creating a deletion of 2,124 bp from the 5' prime end of the *Ahd4* promoter. In the pAhd4p2 plasmid, the *Pst I* fragments were removed--creating a deletion of 1,364 bp from the 5' prime end of the *Ahd4* promoter. The deletion mutant pAhd4p4 was made by removing a 575-bp *Pst I-Acc I* fragment from the *Ahd4* 5' flanking region from the pAhd4p1 plasmid. The pAhd4p5 recombinant plasmid was generated by removing 420 bp from the pAhd4p1 plasmid after digestion with *Psp 5II* and repairing the ends by Klenow fragment and blunt-end ligation.

2.5. Transfection and Expression of Ahd4-CAT Plasmids in Hepa-1 Cells

The Ahd4cat chimeric DNAs were transiently transfected into the mouse hepatoma Hepa-1c1c7 wild-type cell line by the calcium phosphate procedure (Parker and Stark, 1979). *CAT* gene expression in the presence or absence of TCDD (10 nM for 18 h) was monitored by measuring CAT activity (Davies et al., 1980). The pCAT.basic is a promoterless plasmid and was used as negative control. The Rous sarcoma virus β-galactosidase plasmid was contransfected in each case, in order to normalize transfection efficiency by measuring β-galactosidase activity (Guarente, 1983).

3. RESULTS AND DISCUSSION

3.1. Structure of the Mouse Ahd4 Gene

We have determined the exon-intron organization of the mouse *Ahd4* gene (Table 1). The length of the entire gene is estimated to be approximately 10 kb. The mouse *Ahd4* gene has 11 exons interrupted by 10 introns. The first exon comprises 169 bp, is noncoding, and is followed by a 3.2-kb first intron. A similar organization has been reported for the rat orthologous gene (Asman et al., 1993). It seems that the human orthologous gene differs in structure from both the mouse *Ahd4* and rat *ALDH3* genes. Specifically, the human stomach ALDH3 was reported to lack the noncoding first exon, and a putative CAATT box (-117 through -114) and TATAA box (-79 through -76) regulatory elements were described (Hsu et al., 1992).

By primer extension analysis using extracts from several mouse cell lines following treatment with TCDD, we have determined the transcriptional start site of the *Ahd4* gene (Vasiliou et al., 1995c). The *Ahd4* gene promoter contains no TATA box; these data are consistent with studies on the rat *ALDH3* gene. It is possible that the rodent *ALDH3* gene

Table 1. Intron-exon organization of the mouse (*Ahd4*), rat and human *ALDH3* genes*

Exon number	Mouse		Rat		Human	
	Exon (bp)	Intron (kb)	Exon (bp)	Intron (kb)	Exon (bp)	Intron (kb)
1	169	3.2	161	3.0	-----	----
	noncoding		noncoding		not present?	not present?
2	167	1.2	167	1.5	211	1.55
3	232	0.6	232	0.6	232	0.81
4	86	0.3	86	0.3	86	0.35
5	209	0.7	209	0.7	209	0.8
6	121	0.6	121	0.8	118	0.68
7	139	0.8	139	0.7	142	0.63
8	167	0.5	100	0.8	167	0.4
9	100	0.3	167	0.3	100	0.56
10	131	0.2	131	0.2	131	0.09
11	201		267		247	

Numbers are based on Asman et al., 1993 (rat), Hsu et al., 1992, Schuuring et al., 1991 (human)

has multiple transcription start sites, which are tissue-specific and that the large first intron contains promoter activity. However, the likelihood that there exists an as-yet unidentified noncoding first exon in the human *ALDH3c* gene, due to a relatively large first
intron (~3 kb) in the mouse and rat orthologous genes, cannot be ruled out.

3.2. Nucleotide Sequence Analysis of the 5' Flanking Region

To study regulation of *Ahd4* gene expression, we sequenced 3.18 kb of the 5' flanking region. Nucleotide sequence analysis revealed that the *Ahd4* gene promoter lacks the
generic TATAA box. The generic TATAA box, usually found 20–30 bp upstream of eukaryotic genes but not always found, is known to play a key role in correct initiation of
transcription and engagement of RNA polymerase II (Maniatis et al., 1989).

Within the region of -59 and -300, we have identified two Sp1-binding sites. Sp1 is
known to play a role in the transcription initiation machinery--especially in genes lacking
generic TATAA boxes (*reviewed in* Goodrich et al., 1996). In addition, seven AhREs,
which contain the core sequence of GCGTG, are located at -93, -377, -392, -883, -1195 -
1667 and -2601. One EpRE is found at -751. Furthermore, four AP-1 sites were identified at
-290, -709, -3005, and -3071, and two NF-κB sites were found at -638 and -2555. Finally,
a putative C/EBPβ (NF-IL6) site was located at -361, and a putative C/EBPα site was
found at -2700. These are summarized in Figure 2.

3.3. Basal and Inducible Ahd4 Gene Expression

When the 3.2-kb *Ahd4* gene promoter attached to the CAT gene was transfected into
Hepa-1 cells, CAT activity was expressed several-fold higher, as compared with that of
the promotorless pCAT.Basic plasmid (Figure 3). This backround level, determined with
the latter plasmid, was subtracted from all CAT values. The CAT activity was increased
by 8.5-fold after TCDD treatment of the transfected cells. TCDD is known to induce the
transcription of all [*Ah*] battery genes (Nebert et al., 1993). It should be noted that the 8.5-
fold increase found in our CAT assays after TCDD does not correspond to the approximately 1,000-fold increase observed by Northern blot analysis (Vasiliou et al., 1992). One
explanation for this difference in the induction levels might be that additional AhREs or

Figure 2. Putative transcription factor-binding sites in 3.2 kb of the 5' flanking region of the mouse *Ahd4* gene. Sequence-specific transcription factors were identified by the GCG program MAP, utilizing the file which is derived from the SITES table of TFD, a database of transcription factors maintained by David Ghosh (National Center for Biotechnology Information, National Library of Medicine, NIH, Bethesda, Maryland). Identification of a particular DNA motif need not have any relationship to functionality. AhRE, aromatic hydrocarbon response element (also called DRE and XRE). EpRE, electrophile response element (also called ARE).

Figure 3. Transient expression in untreated control and TCDD-treated and t-BHQ-treated Hepa-1 cells containing Ahd4.cat recombinant plasmids. The Ahd4.cat chimeric plasmid DNAs from 3.2 kb to 0.55 bp upstream of the transcription start site contain different lengths of the *Ahd* gene promoter, as denoted by the number of base pairs in each construct. The various CAT plasmids were transiently transfected, and their expression in the absence or following treatment with 10 nM TCDD or 50 mM *t*-BHQ (18 h) was monitored by measuring the CAT activity. The RSV-bGal plasmid was cotransfected in each case to normalize the transfection efficiency. The pCat.Basic plasmid was used as a control. The mean activities of the experimental groups are compared to the CAT activity in untreated Hepa-1 cells transfected with Ahd4cat1 plasmid, which represents the 100 arbitrary units. The standard deviations of three experiments in each case were always less than 22% of the means.

other DNA elements located either upstream or downstream of the 3.2-kb fragment contribute to the dioxin-induced transcription of the intact *Ahd4* gene in mouse cell culture. Further support of this hypothesis comes from the CAT activity in cells transfected with Ahd4p2 and Ahd4p1 plasmids. These data show that AhRE5, AhRE6 and AhRE7 appear to have a synergistic effect on CAT induction following TCDD treatment (Figure 3). However, countless experimental systems have shown enormous differences between the fold-induction in CAT-containing constructs in transfected cell lines and the fold-induction of the intact gene in cell culture or in the whole animal.

As shown in Figure 3, deletion of the segment between -3.2 kb and -501 bp of the *Ahd4* promoter resulted in a 9-fold increase of CAT activity in untreated cells, when compared with basal activities found in cells transfected with Ahdcat1, Ahd4p2 or Ahd4p1 activities. Moreover, deletion of the section between -1,076 bp and -501 bp caused an 18-fold increase of CAT activity in untreated cells. These results clearly indicate the presence of an NRE in the region between -1,071 and -501. Among the putative transcription factor-binding sites found in this DNA fragment are a nuclear factor-1 (NF-1) site (TGGANNNNNGCCA located at -897 to -885, just two bases from the AhRE4 core sequence (TG**GCGTG**TG). This NF-1 site has been found to footprint in the rat gene (Lindahl et al., 1997), and it has been shown that proteins that bind to NF-1 response elements can act either as a transactivator or negative regulator of gene transcription (Jones et al., 1987; Eskild et al., 1994). The high CAT basal activity in Ahd4p4 was further induced by TCDD--indicating the functionality of AhRE1 in the dioxin induction process.

We have previously shown that treatment of Hepa-1 cells with 50 μM tBHQ or 30 μM menadione produces large increases in AHD4 mRNA levels and enzyme activities (Vasiliou et al., 1995b). Based on these data, we had predicted that the EpRE found in the 5' flanking region of the *Ahd4* gene should be functional. To confirm this, we used the Ahdcat1, Ahd4p1 (carrying the EpRE), and Ahd4p5 (lacking the EpRE) plasmids in transiently-transfected Hepa-1 cells. The CAT activity increased by 2-fold after treatment with 50 μM tBHQ--only in cells transfected with the EpRE-carrying plasmids (Figure 3). These data confirm our earlier prediction that the EpRE found in the 5' flanking region of the *Ahd4* gene is functional in the electrophile induction process.

Table 2. Transient expression of AHD4-3.2LUC recombinant plasmids in Hepa-1 and *c37* cells

Cell line	Relative luciferase activity	
	Control	TCDD
Hepa-1	100*	919
c37	4,917	54,056

The AHD4-3.2LUC chimeric plasmid DNA contains 3.2 kb of the mouse *Ahd4* promoter upstream of the luciferase reporter gene. The AHD4-3.2LUC plasmid was transiently transfected, and its expression in the absence or presence of 10 nM TCDD (18 h) was monitored by measuring the luciferase activity. RSV-bGal plasmid was cotransfected in each case to normalize the transfection efficiency. The pGL3.Basic (promotorless) and pGL3.Promoter (containing SV40 promoter) luciferase plasmids (Promega) were used as negative and positive controls, respectively. *. The mean activity of the experimental groups were compared to the luciferase activity in untreated Hepa-1 cells transfected with AHD4-3.2LUC plasmid, which represents the 100 arbitrary units.

3.4. Elevated Expression in c37 Cells

As mentioned earlier, we have found that AHD4 mRNA levels and enzyme activities are strikingly increased in the untreated *c37* mutant line, which lacks a functional CYP1A1 enzyme (Vasiliou et al., 1992; 1993a). The capacity of the 3.2-kb fragment to drive the expression of luciferase activity in Hepa-1 wild-type and *c37* cells was determined by transient transfection (Table 2). As has been found in the CAT assays, we observed about a 10-fold increase in luciferase activity in TCDD-treated Hepa-1 cells, as compared with that in untreated cells. The luciferase activity in both untreated and TCDD-treated *c37* cells was found to be 4- to 5-fold higher than the activity in TCDD-treated

Figure 4. Illustration as to how the *Ahd4* gene participates in the mouse [*Ah*] battery. *Cyp1a1*, murine gene encoding CYP1A1. CYP1A1/1A2 metabolism of an unknown endogenous substrate (ES) to an endogenous product (EP) down-regulates the *Cyp1a1* and *Ahd4* genes via a negative response element (NRE). TCDD binds to the Ah receptor (AHR) and up-regulates the *Cyp1a1* and *Ahd4* genes via aromatic hydrocarbon response elements (AhREs). CYP1A1/1A2 metabolism of endogenous as well as foreign chemicals (R) to reactive oxygenated metabolites (ROMs) up-regulates the *Ahd4* gene via the electrophile response element (EpRE). *Fah*, gene which encodes fumaryl-acetoacetate hydrolase (FAH), the final step in the tyrosine degradative pathway that ends in fumarate (F) and acetoacetone (AA); absence of FAH leads to a build-up of fumarylacetoacetate (FAA), succinylacetoacetate (SAA) and succinylacetone, which are ROMs capable of up-regulating the *Ahd4* gene via the EpRE. The *large bold numbers* refer to the mouse chromosomes on which these genes are located. Although the *arrows* are drawn as single steps, in many cases the actual number of steps denoted by one *arrow* is not known (discussed in Vasiliou et al., 1995a).

Hepa-1 wild-type cells. These data suggest that putative derepression of the *Ahd4* gene in *c37* cells might be regulated at the transcriptional level.

The fold differences in luciferase expression between TCDD-treated Hepa-1 cells and untreated *c37* cells are not in agreement, however, with our Northern blot analyses. Northern blots show that AHD4 mRNA levels in TCDD-treated Hepa-1 cells are almost equal to those found in untreated *c37* cells. The fold increase in CAT or luciferase activity in TCDD-treated Hepa-1 cells transfected with a reporter plasmid driven by the 3.2-kb 5' flanking region of the *Ahd4* gene (shown in this report) is less than the fold increase observed by Northern blot analysis (Vasiliou et al., 1992). It is likely that other *cis*-acting elements present upstream or/and downstream of the 3.2-kb fragment contribute to the dioxin induction process of the *Ahd4* gene.

4. CONCLUSIONS

The mouse *Ahd4* gene spans about 10 kb and consists of 11 exons interrupted by 10 introns. Numerous putative transcription factor-binding sites have been found in the 3.2-kb 5' flanking region immediately upstream of exon 1. Functional analysis of this 3.2-kb fragment suggests that at least three distinct functional domains exist in this region: a *negative regulatory element* (NRE), dioxin inducibility via the AhREs, and electrophile inducibility via the EpRE. These data are consistent with the *Ahd4* gene qualifying as a member of the [*Ah*] battery, as discussed earlier and as illustrated in Figure 4.

5. ACKNOWLEDGMENTS

We thank our colleagues for valuable discussions and a critical reading of this manuscript. This work was supported in part by NIH Grants R01 AG09235 and P30 ES06096.

6. REFERENCES

Asman, D.C., Takimoto, K., Pitot, H.C. and Lindahl, R.: Organization of the rat class 3 aldehyde dehydrogenase gene. J. Biol. Chem. 268 (1993) 12530–12536.

Britten, R. J. and Davidson, E.H.: Gene regulation for higher cells: a theory. Science 165 (1969) 349–357.

Davies, L.G., Dibner, M.D. and Battey, J.F.: *Basic Methods in Molecular Biology*, Elsevier Science Publishers B.V, Amsterdam (1980), pp. 244–248, 298–300.

Eskild, W., Simard, J., Hansson, V. and Guerim, S.L.: Binding of a member of the NF-1 family of transcription factors two distict *cis*-acting elements in the promoter and 5' flanking region of the human cellular retinal binding protein 1 gene. Mol. Endocrinol. 8 (1994) 732–745.

Goodrich, J.A., Cutler, G. and Tjian, R.: Contacts in context: promoter specificity and macro- molecular interactions in transcription. Cell 84 (1996) 825–830.

Guarente, L.: Yeast promoters and *lac* Z fusions designed to study expression of cloned genes in yeast. Meth. Enzymol. 101 (1983) 181–191.

Harrington, M.C., Heneham, G.T.M. and Tipton, K.F.: The role of human aldehyde dehydrogenase isoenzymes in ethanol metabolism. Progr. Clin. Biol. Res. 232 (1987) 111- 125.

Hsu, L.C., Chang, W-C., Shibuya, A. and Yoshida, A.: Human stomach aldehyde dehydrogenase cDNA and genomic cloning, primary structure, and expression in *Escherichia coli*. J. Biol. Chem. 267 (1992) 3030–3037.

Jones, K.A., Kadonaga, G.T., Rosenfeld, P.J., Kelly, T.J. and Tjian, R.: A cellular DNA- binding protein that activates eukaryotic transcription and DNA replication. Cell 48 (1987) 79–89.

Lee, M-O., Manthey, C.L. and Sladek, N.E.: Identification of mouse liver aldehyde dehydrogenases that catalyze the oxidation of retinaldehyde to retinoic acid. Biochem. Pharmacol. 42 (1991) 1279–1285.

Lindahl, R.: Aldehyde dehydrogenases and their role in carcinogenesis. CRC Crit. Rev. Biochem. Mol. Biol. 27 (1992) 283–335.

Lindahl, R. and Petersen, D.R.: Lipid aldehyde oxidation as a physiological role for class 3 aldehyde dehydrogenases. Biochem. Pharmacol. 41 (1991) 1583–1587.

Lindahl, R., Boesch, J., Xie, Y., Miskimins, K. And Miskimins, R.: Class 3 aldehyde dehydrogenase: a view from the hills. Advanc. Exp. Med. Biol. (1997) in press.

Mackerell, A.D. Jr., Blatter, E.E. and Pietruszko, R.: Human aldehyde dehydrogenase: kinetic identification of the isoenzyme for which biogenic aldehydes and acetaldehyde compete. Alcohol Clin. Exp. Res. 10 (1986) 266–270.

Maniatis, T., Goodbourn, S. and Fischer, J.A.: Regulation of inducible and tissue-specific gene expression. Science 236 (1987) 1237–1245.

Marks-Hull, H., Shiao, T.Y., Araki-Sasaki, K., Traver, R. and Vasiliou, V.: Expression of ALDH3 and NMO1 in human corneal epithelial and breast adenocarcinoma cells. Advanc. Exp. Med. Biol. (1997) in press.

McKnight, S. and Tjian, R.: Transcriptional selectivity of viral genes in mammalian cells. Cell 46 (1986) 795–805.

Monder, C., Purkaystha, A.R. and Pietruszko, R.: Oxidation of the 17-aldol (20-β-hydroxy-21- aldehyde) intermediate of corticosteroid metabolism to hydroxy acids by homogeneous human liver aldehyde dehydrogenase. J. Steroid Biochem. 17 (1982) 41–49.

Nebert, D.W., Puga, A. and Vasiliou, V.: Role of Ah receptor and the dioxin-inducible [*Ah*] gene battery in toxicity, cancer and signal transduction. Ann. N. Y. Acad. Sci. 658 (1993) 624- 640.

Parker, B. And Stark, G.: Regulation of simian virus 40 transcription: Sensitive analysis of the RNA species present early in infections by virous or viral DNA. J. Virol. 31 (1979) 360–369.

Santisteban, I., Povey, S., West, L.F., Parrington, J.M. and Hopkinson, D.A.: Chromosome assignment, biochemical and immunological studies on a human aldehyde dehydrogenase ALDH3. Ann. Hum. Genet. 49 (1985) 87–100.

Sambrook, J., Fritsch, E.F. and Maniatis, T.: Molecular cloning: A laboratory manual (1989), Cold Spring Harbor Laboratory Press, Cold Spring Harbor, NY.

Sanger, F., Nicklen, S. and Coulsen, A.R.: DNA sequencing with chain-terminating inhibitors. Proc. Natl. Acad. Sci. U.S.A. 74 (1977) 5463–5467.

Schuuring, E.M.D., Verhoeven, E., Eckey, R., Vos, H.L and Michalides, R.J. A.M.: Cloning and complete nucleotide sequence of a cDNA encoding full-length open reading frame of the human aldehyde dehydrogenase type III gene. Genbank Accession number M 74542, 1991.

Tipton, K.F., Houslay, M.D. and Turner, A.J: The catalytic behaviour of monoamine oxidase. Essays Neurochem. Pharmacol. 1 (1987) 103–138.

Vasiliou, V., Puga, A. and Nebert, D.W.: Negative regulation of the murine cytosolic aldehyde dehydrogenase-3 (Aldh-3c) gene by functional CYP1A1 and CYP1A2 proteins. Biochem. Biophys. Res. Commun. 187 (1992) 413–419.

Vasiliou, V., Puga, A. and Nebert, D.W.: Mouse class 3 aldehyde dehydrogenases: positive and negative regulation of gene expression. Advanc. Exp. Med. Biol. 328 (1993a) 131–139.

Vasiliou V., Reuter S.F., Kozak C.A. and Nebert, D.W.: Mouse dioxin-inducible cytosolic aldehyde dehydrogenase: AHD4 cDNA sequence, genetic mapping, and differences in gene expression. Pharmacogenetics 3 (1993b) 281–290.

Vasiliou, V., Reuter, S.F. and Nebert, D.W.: Organization and characterization of the murine cytosolic TCDD-inducible aldehyde dehydrogenase (*Ahd4*) gene. The Toxicologist 14 (1994) 410.

Vasiliou, V., Puga, A., Chang, C.Y., Tabor, M.W. and Nebert, D.W.: Interaction between the Ah receptor and proteins binding to the AP-1-like electrophile responsive element (EpRE) during the murine Phase II [*Ah*] battery gene expression. Biochem. Pharmacol. 50 (1995a) 2057–2068.

Vasiliou, V., Shertzer, H.G., Liu, R.M., Sainsbury, M. and Nebert, D.W.: Response of the [*Ah*] battery genes to compounds that protect against menadione toxicity. Biochem. Pharmacol. 50 (1995b) 1885–1891.

Vasiliou, V., Weiner, H., Marselos, M. and Nebert, D.W.: Mammalian aldehyde dehydrogenase genes: classification based on evolution, structure and regulation. Eur. J. Drug Metab. Pharmacokinet. 20 (1995c) 53–64.

Vasiliou, V., Kozak, C.A., Lindahl, R. and Nebert, D.W.: Mouse microsomal Class 3 aldehyde dehydrogenase: AHD3 cDNA sequence, chromosomal mapping, and dioxin inducibility. DNA Cell Biol 15 (1996) 235–245.

Yoshida, A.: Molecular genetics of human aldehyde dehydrogenase. Pharmacogenetics 2 (1992) 139–147.

7

CHANGES IN ALDEHYDE DEHYDROGENASE ISOZYMES EXPRESSION IN LONG-TERM CULTURES OF HUMAN HEMATOPOIETIC PROGENITOR CELLS[*]

Doris Meier-Tackmann,[1,3] Dharam P. Agarwal,[1] William Krueger,[2] Caroline Dereskewitz,[2] Hassan Tawhid Hassan,[2] and Alex Rolf Zander[2]

[1]Institute of Human Genetics and
[2]Bone Marrow Transplantation Centre
University Hospital Eppendorf
20246 Hamburg, Germany
[3]ASTA Medica AG
60314 Frankfurt/M, Germany

1. INTRODUCTION

Oxazaphosphorines (e.g., 4-hydroperoxycyclophosphamide (4-HC), mafosfamide, ifosfamide) are commonly used to purge malignant blood cells from autologous bone marrow samples before reinfusion (Beran et al., 1987; Yeager et al., 1990; Carlo-Stella et al., 1994). Recently published studies from many laboratories including our own have unequivocally established that there is an associative as well as causative inverse relationship between cellular content of ALDH - whether of normal or malignant cell origin - and sensitivity to oxazaphosphorines (Kohn et al., 1987; Russo and Hilton, 1988; von Eitzen et al., 1994; Agarwal et al., 1995; Sreerama and Sladek, 1995; Sladek et al., 1995; Moreb et al., 1995). Until now, only class 1 (ALDH1 and ALDH2) and class 3 (ALDH3) isozymes (constitutive and inducible forms) have been implicated in the detoxification of cyclophosphamide derivatives such as 4-HC and mafosfamide (Dockham et al., 1992; Bunting and Townsend, 1996).

Our preliminary studies with human blood subfractions have shown that whereas erythrocytes contain substantial ALDH1 activity using mafosfamide-derived aldophosphamide as substrate, leukocytes are much less active in this respect (Metzenthin et al., 1996). Kastan et al. (1990) had earlier reported that the cytosolic ALDH isozyme form

---

[*] A part of this study will be included in the M.D. thesis of C. Dereskewitz, Medical Faculty, University of Hamburg.

(ALDH1) is the only enzyme expressed in human hematopoietic progenitors cells. This isozyme was shown to be highly expressed in CD34+ progenitors and erythroids, while the lymphoid cells expressed the lowest level. Moreover, a direct role of ALDH1 in the protection against 4-HC cytotoxicity was recently demonstrated by vector-mediated transfection of ALDH1 in K562 cells which resulted in an increased resistance to 4-HC compared with wild type (Moreb et al., 1996). In contrast, we have observed that progenitor mononuclear cells (MNCs) prepared from bone marrow samples of healthy donors contain mostly the class 2 ALDH isozyme (ALDH2) (unpublished results). However, the role of ALDH2 isozyme in oxidation of aldophosphamide under *in vivo* conditions is not known. Thus, the specific ALDH isozyme form which mainly protects the primitive progenitor cells against the cytotoxic effects of purging with cyclophosphamide derivatives remains to be identified.

In the present study, we have examined the ALDH isozyme profile of hematopoietic progenitor cells from chronic myeloid leukemia (CML) patients in remission and healthy controls before and after incubating the cells with varying doses of mafosfamide followed by long-term cultures (LTC). Moreover, we have investigated the ALDH isozyme profile of immortalised cells of a chronic myeloid leukemia cell line (K-562) treated with mafosfamide followed by LTC.

2. MATERIALS AND METHODS

2.1. Chemicals and Mediums

Mafosfamide (MF) was kindly provided by ASTA Medica AG, Frankfurt/M. A stock solution of MF was prepared by dissolving 10 mg MF/ml phosphate-buffered saline (PBS, pH 7.2). Myeloid long term culture medium (Myelocult H5100), Iscove's methylcellulosecomplete ready-mix with 10% agar leucocyte conditioned medium and erythropoetin (Methocult H4431) were obtained from Stem Cell Technologies Inc., Vancouver, Canada. The composition of both the media were as decribed in detail by Hassan et al. (1996).

2.2. Human Bone Marrow Cells

Human bone marrow samples from healthy donors as well as from chronic myeloid leukemia (CML) patients in remission for autologous transplantation were obtained from the posterior iliac crest at the Bone Marrow Transplantation Centre, University Hospital Eppendorf, Hamburg. Mononuclear cells (MNCs) were separated from granulocytes and erythrocytes on Ficoll-Paque density gradient (1077g/cm^3). MNCs at the interface were harvested, washed twice with Hank's balanced salt solution (HBSS, pH 7.2) and were suspended in H5100 medium for 12 hrs at 37°C in a fully humified atmosphere containing 5% CO_2 for depletion of the adherent cells. The non-adherent monocyte-depleted MNCs were used throughout the studies. The cell count refers to the number of viable cells present in a given aliquot.

2.3. Cell Line

The human chronic myeloid leukemia cell line K-562 was obtained from the German Collection of Microorganisms and Cell Cultures (DMS, Braunschweig, Germany). Before use, the cells were subcultured in 90% RPMI 1640 medium (Gibco BRL, Germany) supplemented with 10% fetal calf serum (FCS) as recommended by the supplier. The cell count refers to the number of viable cells present in a given aliquot.

2.4. *In Vitro* Treatment with Mafosfamide

For *ex vivo* dose-response analysis, 10^7 MNCs or K-562 cells suspended in 1 ml PBS were exposed to 20 to 60 µg MF for 30 min at 37°C under constant shaking. The cells were washed free of the drug with HBSS at 4°C. Cells treated in the same way but without MF were used as controls. Cells were divided into aliquots and the number of viable cells per aliquot was determined. Cell aliquots were used for colony forming cell assays, as inoculum in long-term cultures, or frozen at -80°C immediately until used for ALDH isozyme assays and IEF.

2.5. Long-Term Cell Culture (LTC)

MF-treated and untreated MNCs and K-562 cells were grown in Myelocult H5100 supplemented with 10^{-6} mol/l hydrocortisone (Sigma, Germany). At an initial concentration of 10^6 MNCs or 10^5 K-562 cells/ml medium, the cells were inoculated in plastic tissue culture flasks (Falcon, 25 cm^2 or 75 cm^2 surface area) and cultured at 37°C under 5% CO_2 in the air and 90% humidity for 10 days. Adherent and non-adherent cells, harvested and washed twice with PBS at 20°C, were used for colony forming cell assays and ALDH isozyme analysis. The survival of MF-untreated cells in LTC was set as 100%.

2.6. Assay of Clonogenic Progenitor Cells and Colony Forming K562 Cells

Aliquots of MF-treated and untreated cells containing 4×10^4 MNCs or 4×10^3 K-562 cells were plated on 35 mm petri dishes in 1 ml Methocult H4431 medium to assay the myeloid and erythroid progenitor cells, granulocyte-macrophage progenitor cells (CFUGM) and burst forming units erythroid (BFU-E), respectively, or the colony forming unit of K-562 (CFU-L). All cultures were done in quadruplicate. Culture dishes were incubated at 37°C in a humidified atmosphere containing 5% CO_2 for 7 or 14 days. On day 7, colonies of K-562, and on day 14, colonies from MNCs containing over 50 cells were counted using an inverted microscope. The colonies were categorised *in situ* as CFU-GM myeloid colonies, and as CFU-L colonies if they were composed of tightly clustered aggregates in translucent appearance, and as BFU-E erythroid colonies if haemoglobinized cells (as red elements) were present. The CFU recovery rate of the MF-untreated cells (controls) was set as 100%.

2.7. Preparation of Cell Lysate

MNCs and K-562 cells were thawed, resuspended in 100 µl aqua dest. and lysed by sonication (30 sec at 100 watts). The lysates were centrifuged for 20 min at 27,000g. The supernatants concentrated by ultrafiltration through an anisotropic membrane filter (Centricon 30, Amicon, USA) were used for ALDH isozyme analysis.

2.8. Analytical Isoelectric Focusing (IEF) and ALDH Activity Staining

IEF was carried out on ultrathin layers (0.1 mm) of polyacrylamide gel by mixing 4.7 ml acrylamide/bisacrylamide (20%/0.9%), 13.8 ml 15% sucrose, 1 ml ampholine, pH

3.5–10, 0.5 ml ampholine, pH 4.5 - 5.4 and 10 mg ammonium persulfate in a final volume of 20 ml. One M NaOH and 1 M H_3PO_4 were used as electrode solutions. After pre-run on a cooling plate (4°C) for 20 min at 2500 V with a maximum of 15 mA and 15 W, samples of 20 μl each were placed at the cathodic end of the gel using an applicator strip (Pharmacia). Focusing was carried out for 90 min at 2500 V with a maximum of 50 mA and 25 W and for additional 10 min with a maximum of 30 W.

The gels were stained for ALDH activity with an agarose overlay gel. A 1% agarose gel was prepared in 25 ml 0.1 M pyrophosphate/0.01 M diethanolamine, pH 8.0, which contained 10 mg MTT, 60 mg NAD^+, 50 mg pyruvate, 0.5 mg meldolablue and 125 mM propionaldehyde for ALDH1 and ALDH2 or 13.2 mM 3-nitrobenzaldehyde for ALDH3. After incubation at 37°C for 2 h in the dark the stained isozyme bands were evaluated.

2.9. Densitometric Measurement

A semiquantitative method was applied to measure ALDH activity on the gels by using a laser densitometer (LKB, Ultroscan 2202). The values (area under the curve) are expressed as relative activities of ALDH1 and ALDH2 isozymes as percent of the total ALDH. However, it was not possible to correlate individual activity levels of ALDH1 and ALDH2 with the number of MNCs used in the lysates due to the small number of cells available for analysis.

3. RESULTS

3.1. ALDH Isozyme Profile of Normal and Leukemic Cells

IEF profile of ALDH isozymes from various cells and human tissue extracts is shown in Fig. 1. Individual ALDH isozymes were identified on the basis of substrate specificity and pI values. As assessed by laser densitometry, a semiquantitative difference was noted between the activity bands of various isozymes. MNCs (6×10^6), prepared from either bone marrow cells of CML patients or from healthy donors, showed the presence of both ALDH1 and ALDH2 isozymes; the activity band intensity of the latter being considerably stronger than that of the former (Fig. 1, lane 4). It was however not possible to determine the exact ratio of the two isozymes due to a very faint ALDH1 activity band using an aliquot representing 10^6 MNCs (Fig. 1, lane 6). Though there was no qualitative difference between the ALDH isozyme pattern of MNCs from healthy donors and CML patients, the band intensity of both the ALDH isozymes was weaker in CML patients.

Immortalised cells from human chronic myeloid leukemia (K-562) predominantly contained the ALDH1 isozyme. No ALDH2 activity band was detectable when a lysate aliquot representing about 10^6 cells was subjected to IEF (Fig. 2, lanes 4–8).

3.2. The Effect of Long-Term Culture on ALDH Activity

To investigate whether long-term culturing of MNCs and K-562 cells alters the ALDH isozyme expression, the cells were subjected to LTC as described in the Methods section. After LTC of MNCs from both patients and controls, there was an increase in the content of ALDH1 and ALDH2 (Fig. 1, lane 7). However, the ALDH2 content was still relatively higher as compared with the ALDH1 isozyme level. When lysates representing about 10^6 MNCs from healthy controls and CML patients were subjected to IEF, the activ-

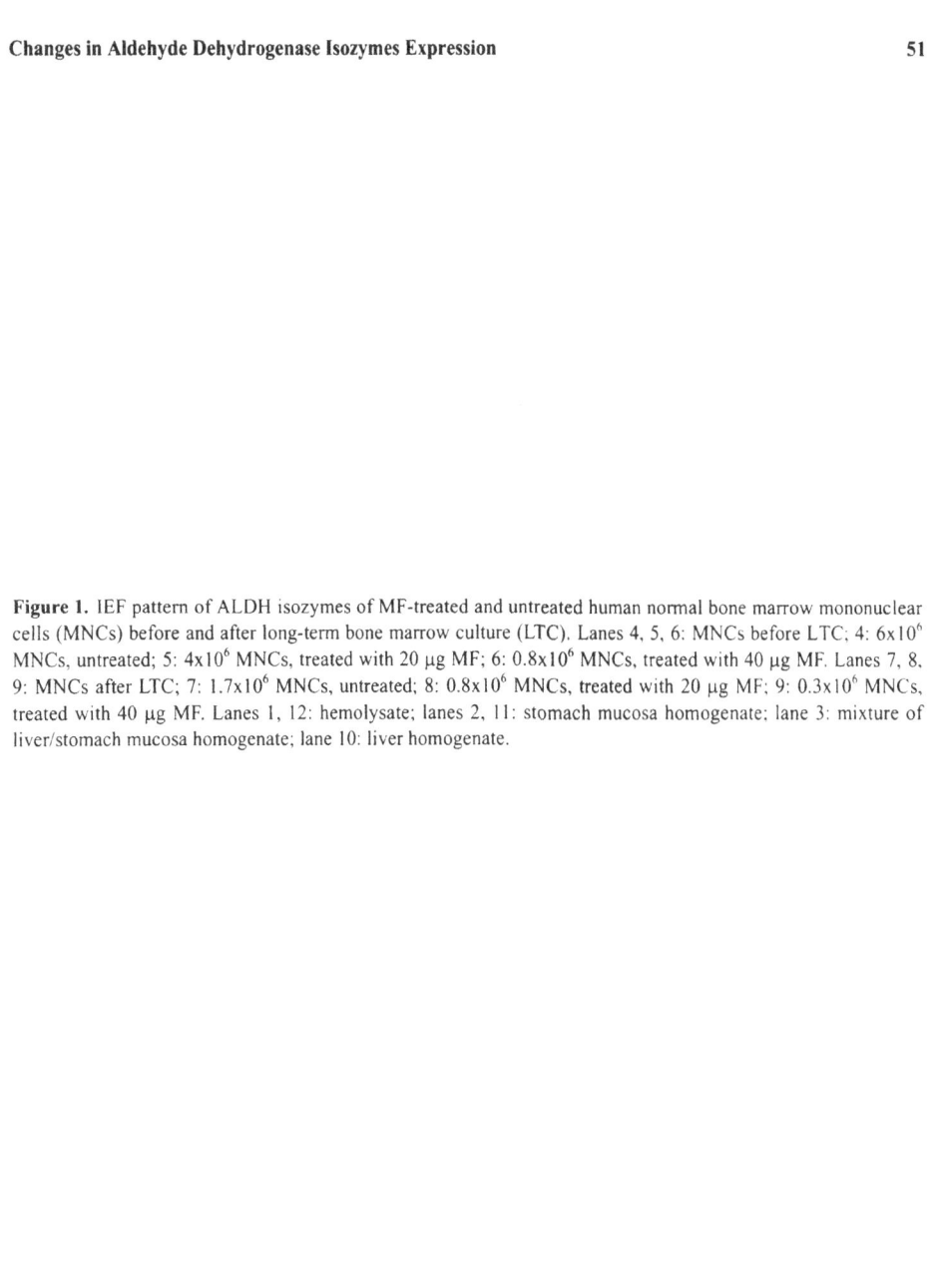

Figure 1. IEF pattern of ALDH isozymes of MF-treated and untreated human normal bone marrow mononuclear cells (MNCs) before and after long-term bone marrow culture (LTC). Lanes 4, 5, 6: MNCs before LTC; 4: 6×10^6 MNCs, untreated; 5: 4×10^6 MNCs, treated with 20 μg MF; 6: 0.8×10^6 MNCs, treated with 40 μg MF. Lanes 7, 8, 9: MNCs after LTC; 7: 1.7×10^6 MNCs, untreated; 8: 0.8×10^6 MNCs, treated with 20 μg MF; 9: 0.3×10^6 MNCs, treated with 40 μg MF. Lanes 1, 12: hemolysate; lanes 2, 11: stomach mucosa homogenate; lane 3: mixture of liver/stomach mucosa homogenate; lane 10: liver homogenate.

Figure 2. IEF pattern of ALDH isozymes of K-562 cells grown in RPMI 1640 medium before and after MF-treatment. Each lysate sample represents 10^6 cells. Lane 4: cells before treatment with MF; lane 5: MF-untreated cells (control); lanes 6, 7 and 8: cells treated with 20, 40, or 60 μg MF, respectively. Lane 1: hemolysate; lane 2: stomach mucosa homogenate; lane 3: liver homogenate.

Figure 3. IEF pattern of ALDH isozymes of MF-treated and untreated K-562 cells after LTC in Myelocule medium supplimented with hydrocortisone. Each lysate sample represents 10^6 cells. Lane 5: MF-untreated cells (control); lanes 6, 7 and 8: cells treated with 20, 40, or 60 µg MF, respectively. Lane 1: phycocyanin (IEF standards, Bio-Rad); lane 2: hemolysate; lane 3: stomach mucosa homogenate; lane 4: liver homogenate.

ity ratio for ALDH2 and ALDH1 was found to be about 70:30. As demonstrated in Fig. 3, lanes 5 & 6, there was an increase of ALDH activity in K-562 cells after LTC (ALDH1 to ALDH2 ratio = 84:16).

3.3. The Effect of Treatment of Cells with Mafosfamide

MNCs from healthy donors and CML patients, as well as K-562 cells, when treated with various amounts of MF before LTC, showed no significant change in the ALDH isozyme profile (see e.g., Fig. 1, lane 5 and Fig. 2, lanes 6–8). However, an increase in ALDH activity was observed in MNCs treated with various doses of MF followed by LTC (Fig. 1, lanes 8 & 9).

There was a dose-dependent reduction in the number of colonies in MF-treated CFC-GM from healthy controls as well as in CML patients and K-562 cells before LTC (Fig. 4A). When these cells were subjected to LTC, a marked increase in CFU-GM recovery was observed only in healthy controls. In MNCs from a healthy donor, the maximum recovery (100%) was noted with a MF concentration of 60 µg/ml cell suspension (Fig. 4B).

We have further determined the effect of treatment of various cells with MF followed by LTC on the cell survival rate. Significant differences in cell survival rate were observed in MNCs of healthy donors and CML patients as well as in K-562 cells (Fig. 5). Normal MNCs, treated with 60 µg MF/ml cell suspension, showed a 94% loss in cell survival as compared with MF-untreated MNCs from healthy controls, whereas the corresponding loss in MNCs from CML patients and K-562 cells was only between 40 to 75%. A coincidence was apparent

Figure 4. CFU recovery rate (%) and relative distribution of ALDH1 and ALDH2 (%) in normal and immortalised leukemic cells treated with MF before and after LTC. A: Recovery rate of K-562 cells treated with increasing doses of MF. B: ALDH isozyme activity and CFU recovery of normal MNCs and K-562 after treatment with 60 µg MF.

between the expression of ALDH activity and survival of MF-treated normal MNCs and K-562 cells (Fig. 6). Moreover, a similar relationship was observed between CFU recovery rate of these cells and ALDH activity after LTC (Fig. 4B). While both the isozymes showed an elevated cellular content, in MNCs there was a relative increase in ALDH1, whereas in K-562 cells there was a relative increase of ALDH2.

4. DISCUSSION

One of the major clinical goals during *ex vivo* purging of leukemic cells from marrow samples with active derivatives of CP is to achieve a maximum antileukemic effect of

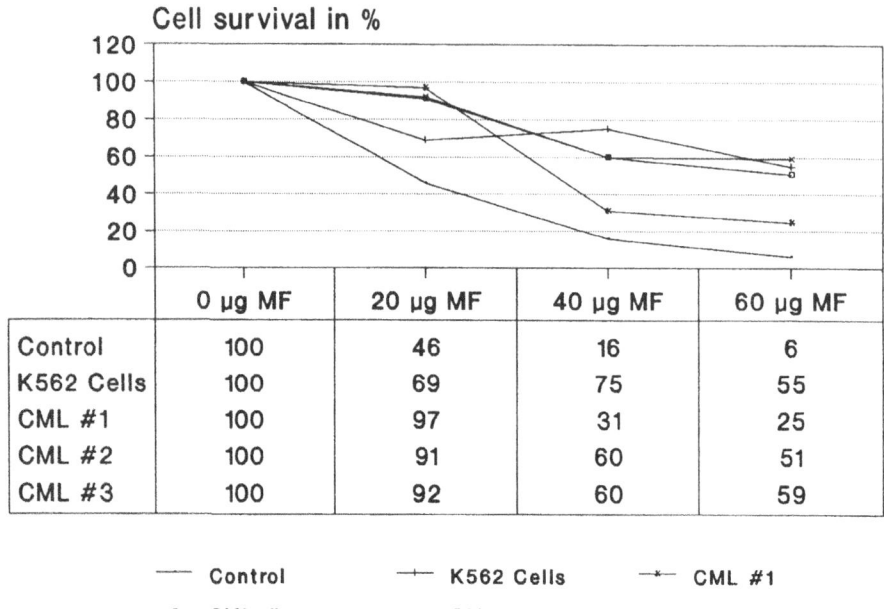

	0 µg MF	20 µg MF	40 µg MF	60 µg MF
Control	100	46	16	6
K562 Cells	100	69	75	55
CML #1	100	97	31	25
CML #2	100	91	60	51
CML #3	100	92	60	59

--- Control   +- K562 Cells   *- CML #1

-o- CML #2   *- CML #3

Figure 5. Survival rate of MF-treated MNCs (healthy control and CML patients) and K-562 cells grown in long-term culture medium for 10 days.

the drug while preserving the normal hematopoietic stem cells (GM-CFC) from the toxic damage. In addition, *ex vivo* expansion of human bone marrow cells prior to transplantation has been suggested in order to achieve a long-lasting restoration of normal hematopoiesis (Hassan et al., 1996). This cellular expansion is done by longterm bone marrow cultures (LTC) of the MNCs. Moreover, LTC of bone marrow can result in a rapid decline of leukemic stem cells (Carlo-Stella et al., 1994).

Earlier studies have documented that 4-HC exposed bone marrow depleted of GM-CFC is capable of restoring hematopoiesis (see Beran et al., 1987). A potential protective role of intracellular ALDH activity against chemotherapeutic toxicity has been suggested both in human tumor cell lines and leukemic cells of patients. Intrinsic and inducible cellular ALDH activity was found to correlate with insensitivity to activated CP (Rekha et al., 1994; Agarwal et al., 1995; Sreerama et al, 1995). Thus, agents such as interleukin-1 and tumor necrosis factor alpha which induce the expression of cellular ALDH activity in normal human hematopoietic progenitors, but not in several leukemic cell lines, have a protective effect against 4-HC toxicity (Moreb et al., 1995).

In this context, it is of interest to know whether human leukemic cells differ from normal marrow cells in their ALDH content. Moreover, it is not fully clear whether primitive pluripotent progenitor cells possess higher ALDH activity than the committed progenitor cells. Furthermore, it remains to be shown as to which of the two major ALDH isozymes, ALDH1 or ALDH2, is primarily responsible for the oxidation of aldehyde metabolite (aldophosphamide) of the active drug. It is also not known whether LTC of bone marrow MNCs leads to an alteration in the ALDH profile of the progenitor cells. In the present study we have attempted to address these aspects.

Figure 6. Comparision of ALDH activity and cell survival rate of MF-treated (60 µg/ml) cells. A: percent distribution of ALDH activity (area under the curve/million cells); B: cell survival rate (%) in long-term cultures.

One of the unique findings of the present study is the presence of very high ALDH2 activity levels in MNCs prepared from normal bone marrow and from CML patients. The presence of the mitochondrial ALDH2 in bone marrow cells has not been reported before. Earlier, Kastan et al. (1990) have reported that normal hematopoietic progenitors cells express only ALDH1, its level being higher in primitive cells than in lymphocytes and erythroids. While a direct role of ALDH1 in the resistance to 4HC has been demonstrated (Kohn and Sladek, 1985; Russo and Hilton, 1988; Kastan et al., 1990; Moreb et al., 1996), a possible role of ALDH2 isozyme in protecting the stem cells against toxicity caused by cyclophosphamide derivatives is not known.

A higher content of ALDH1 than that of ALDH2, as noted by us in K-562 cells, has been observed previously (Agarwal et al., 1995) in other tumor cell lines such as HepG2 (hepatocellular carcinoma), A549 (lung carcinoma) and UMSCC2 (pharynx carcinoma). After long-term culture, healthy bone marrow cells showed a significant increase in both ALDH1 and ALDH2 content as well as in cell survival rate whether or not the cells were treated with mafosfamide prior to LTC. Though CML cells showed a very little increase in ALDH activity after LTC, the cell survival rate was considerably high.

An increased ALDH activity of MNCs after LTC may offer an increased protection to normal hematopoietic progenitor cells against CP toxicity. However, the mechanism of induction of ALDH activity in MNCs after LTC is not known. An increased overall hematopoiesis during LTC or a selected induction of cellular $ALDH_2$ activity due to some specific chemicals or hormones present in LTC medium may be responsible for this change. The LTC medium contains, among other ingredients, hydrocortisone. Indeed, when K-562 cells were cultivated in another medium containing no hydrocortisone, no induction of ALDH activity was observed (Fig. 2, lanes 4 and 5). It has been recently shown that certain chemicals and hormones do influence the expression of ALDH in hepatoma cells (Crabb et al., 1995).

Taken together, our preliminary findings reported here indicate that both ALDH1 and ALDH2 isozymes of bone marrow cell populations are not only constitutive but also are inducible as their level may be influenced under different physiological conditions such as long-term cultures.

5. ACKNOWLEDGMENTS

This study was supported by Wilhelm Sander-Stiftung, Munich. Mafosfamide was a generous gift of ASTA Medica, Frankfurt/M. We thank Mrs. Elly Losenhausen for skillful technical assistance.

6. REFERENCES

Agarwal, D.P., von Eitzen, U., Meier-Tackmann, D. and Goedde, H.W.: Metabolism of cyclophosphamide by aldehyde dehydrogenases. Adv. Exp. Med. Biol. 372 (1995) 115–122.

Beran, M., Zander, A.R., Andersson, B.S. and McCredie K.B.: Regrowth of granulocyte-macrophage progenitor cells (GM-CFC) in suspension cultures of bone marrow depleted of GM-CFC with 4-hydroperoxycyclophosphamide (4-HC). Eur. J. Haematol. 39 (1987) 118–124.

Bunting, K.D. and Townsend, A.J.: Protection by transfected rat or human class 3 aldehyde dehydrogenase against the cytotoxic effects of oxazaphosphorine alkylating agents in hamster V79 cell lines. J. Biol. Chem. 271 (1996) 11891–11896.

Carlo-Stella, C., Mangoni, L., Almici, C., Caramatti, C., Cottafavi, L., Dotti, G.P. and Rizzoli, V.: Autologous transplant for chronic myelogenous leukemia using marrow treated *ex vivo* with mafosfamide. Bone Marrow Transplant. 14 (1994) 425–432.

Crabb, D.W., Stewart, M.J. and Xiao, Q.: Hormonal and chemical influences on the expression of class 2 aldehyde dehydrogenases in rat H4IIEC3 and human HuH7 hepatoma cells. Alcoholism Clin. Exp. Res. 19 (1995) 1414–1419.

Dockham, P.A., Lee, M.-O. and Sladek, N.E.: Identification of human liver aldehyde dehydrogenases that catalyze the oxidation of aldophosphamide and retinaldehyde. Biochem. Pharmacol. 43 (1992) 2453–2469.

Hassan, H.T., Biermann, B. and Zander, A.R.: Maintenance and expansion of erythropoiesis in human long-term bone marrow cultures in presence of erythropoietin plus stem cell factor and interleukin-3 or interleukin-11. Eur. Cytokine Netw. 7 (1996) 129–136.

Kastan, M.B., Schlaffer, E., Russo, J.E., Colvin, O.M., Civin, C.I. and Hilton, J.: Direct demonstration of elevated aldehyde dehydrogenase in human hematopoietic progenitor cells. Blood 75 (1990) 1947–1950.

Kohn, F.R., Landkamer, G.J., Manthey C.L., Ramsay, N.K.C. and Sladek N.E.: Effect of aldehyde dehydrogenase inhibitors on the *ex vivo* sensitivity of human multipotent and committed hematopoietic progenitor cells and malignant blood cells to oxazaphosphorines. Cancer Res. 47 (1987) 3180–3185).

Metzenthin, M., Meier-Tackmann, D., Agarwal, D.P., Zschaber, R. and Weh, H.-J.: Aldehyde dehydrogenase-mediated metabolism of acetaldehyde and mafosfamide in blood of healthy subjects and patients with malignant lymphoma. In: Enzymology and Molecular Biology of Carbonyl Metabolism 6. Ed. H. Weiner (1996, in press).

Moreb, J.S., Turner, C., Sreerama, L., Zucali, J.R., Sladek, N.E. and Schweder, M.: Interleukin-1 and tumor necrosis factor alpha induce class 1 aldehyde dehydrogenase mRNA and protein in bone marrow cells. Leukemia & Lymphoma 20 (1995) 77–84.

Moreb, J., Schweder, M., Suresh, A. and Zucali, J.R.: Overexpression of the human aldehyde dehydrogenase class-1 results in increased resistance to 4-hydroperoxycyclophosphamide. Cancer Gene Therapy (1996, in press).

Rekha, G.K., Sreerama, L. and Sladek, N.E.: Intrinsic cellular resistance to oxazaphosphorines exhibited by a human colon carcinoma cell line expressing relatively large amounts of a class-3 aldehyde dehydrogenase. Biochem. Pharmacol. 48 (1994) 1943–1952.

Russo, J.E. and Hilton, J.: Characterization of cytosolic aldehyde dehydrogenase from cyclophosphamide resistant L1210 cells. Cancer Res. 48 (1988) 2963–2968.

Sladek, N.E., Sreerama, L. and Rekha, G.K.: Constitutive and overexpressed human cytosolic class-3 aldehyde dehydrogenases in normal and neoplastic cells/secretions. Adv. Exp. Med. Biol. 372 (1995) 103–114.

Sreerama, L. and Sladek, N.E.: Human breast adenocarcinoma MCF-7/0 cells electroporated with cytosolic class 3 aldehyde dehydrogenases obtained from tumor cells and a normal tissue exhibit differential sensitivity to mafosfamide. Drug Metabolism and Disposition 23 (1995) 1080–1084.

Sreerama, L., Rekha, G.K. and Sladek N.E.: Phenolic antioxidantinduced overexpression of class-3 aldehyde dehydrogenase and oxazaphosphorine-specific resistance. Biochem. Pharmacol. 49 (1995) 669–675.

von Eitzen, U., Meier-Tackmann, D., Agarwal, D.P., and Goedde, H.W.: Detoxification of cyclophosphamide by human aldehyde dehydrogenase isozymes. Cancer Lett. 76 (1994) 45–49.

Yeager, A.M., Rowley, S.D., Kaizer, H. and Santos, G.W.: Ex vivo chemopurging of autologous bone marrow with 4-hydroperoxycyclophosphamide to eliminate occult leukemic cells. Am. J. Pediatr. Hematol. Oncol. 12 (1990) 245–256.

8

EXPRESSION OF ALDH3 AND NMO1 IN HUMAN CORNEAL EPITHELIAL AND BREAST ADENOCARCINOMA CELLS

Heather Marks-Hull,[1] Te-Yen Shiao,[1] Kaoru Araki-Sasaki,[2] Robert Traver,[1] and Vasilis Vasiliou[1]

[1]Department of Pharmaceutical Sciences
University of Colorado Health Sciences Center
Denver, Colorado 80262
[2]Division of Ophthalmology
Kinki Central Hospital, 3–1 Kurumazuka
Itami City, Hyogo, 664 Japan

1. INTRODUCTION

Global atmospheric changes like ozone depletion in the stratosphere contribute to enhanced chronic exposure of human eye tissue to UV light (Hendee, 1989). The hazards of UV light exposure to the eye have been recognized since the early 1900's (van der Hoeve, 1920). Exposure to UV radiation has been linked to acute keratitis and other disorders of the ocular surface (Schein, 1992). Protein modification, extensive DNA damage and generation of reactive oxygen metabolites, which initiate lipid peroxidation, are associated with UV irradiation. Aldehydes are the major products of lipid peroxidation. Interestingly, under physiologic conditions corneal epithelium expresses high constitutive levels of a cytosolic Class 3 aldehyde dehydrogenase (ALDH3) which represents 10–40% of the total soluble corneal protein in various mammalian and non-mammalian species (Holmes and VandeBerg, 1986; Downes and Holmes 1992). It has been suggested that ALDH3 plays a dual role to protect the cornea from the UV-light by catalyzing the oxidation of lipid peroxidic aldehydes generated by UV-light and by absorbing UV-light itself (*ref. in* Boesch et al., 1996).

The ALDH3 enzyme is encoded by a single gene which is a member of the aromatic hydrocarbon [*Ah*] battery. This battery, well established in mouse, consists of at least six genes that are coordinately induced by 2,3,7,8-tetrachlorodibenzo-p-dioxin (TCDD) and polycyclic aromatic hydrocarbons such as benzo[a]pyrene (reviewied in Nebert et al., 1993). In addition to two P450 genes, *Cyp1a1* and *Cyp1a2*, four Phase II genes are also members of the [Ah] battery: NA(P)H:menadione oxidoreductase (*Nmo1*); the cytosolic "Class 3" aldehyde dehydrogenase (*Ahd4*); a UDP glucuronosyltransferase (*Ugt1*06*);

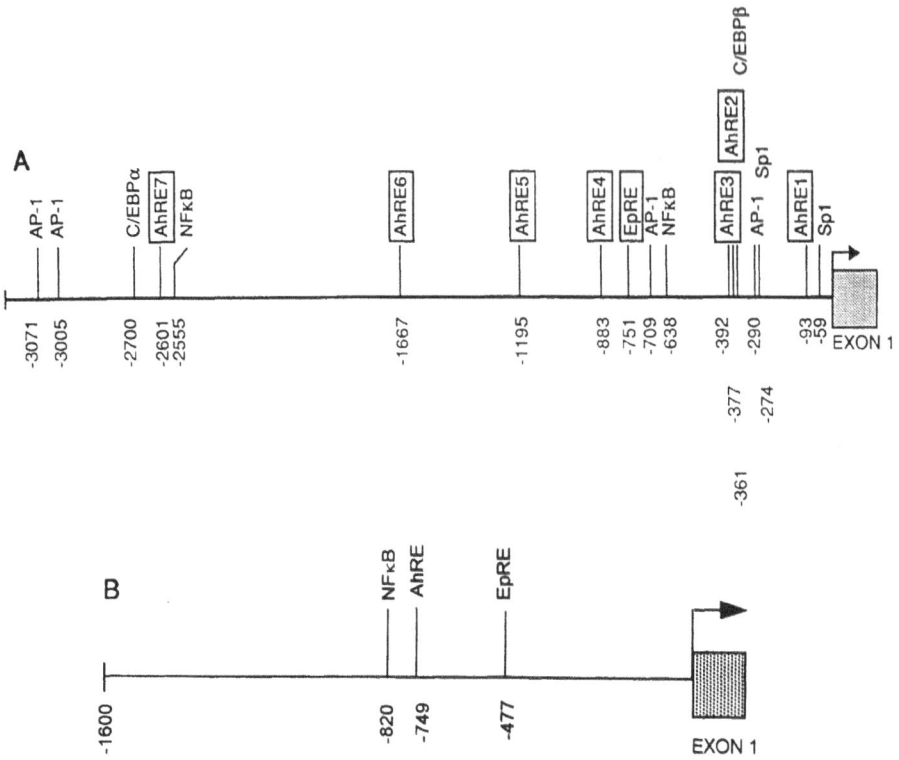

Figure 1. Putative transcription factor binding sites in (**A**) 3.2 kb of the 5' flanking region of the mouse *Ahd4* gene and (**B**) 1.6 kb of the 5' flanking region of the human *NMO1* gene.

and a glutathione transferase (*Gsta1*). Induction of all six [Ah] genes by TCDD requires the Ah receptor (AHR) and Ah receptor nuclear translocator protein (ARNT), and operates via nuclear translocation followed by binding to *a*romatic *h*ydrocarbon *r*esponse *e*lements (AhREs). Elevated expression of Phase II genes serves as a defense mechanism against endogenous or exogenous oxidative stress caused by reactive oxygen metabolites (ROM's) generated by physical agents (UV irradiation) or chemical agents (electrophiles) (Shertzer et al., 1995; Vasiliou et al., 1995a). Increased expression of NMO1 and ALDH3 genes in conditions of oxidative stress involves the *e*lectro*p*hile *r*esponsive *e*lement (EpRE; Jaiswal 1994; Vasiliou et al., 1995b). Both AhREs and EpRE are present in the 5' flanking region of both ALDH3 and NMO1 (Figure 1A & 1B; Jaiswal, 1994; Vasiliou et al., 1995c).

Expression of ALDH3 is tissue-specific and can be either constitutive or inducible by xenobiotics (*ref. in* Lindahl 1992; Vasiliou et al., 1995d). Besides cornea, ALDH3 activity or mRNA have been found in rat stomach, urinary bladder, lung and skin (Dunn et al., 1988; Vasiliou and Marselos, 1989; Pappas et al., 1997). A similar pattern has recently been found for the NMO1 expression in rat tissues (P. Pappas, et al., manuscript in preparation). Further evidence for tissue-specific regulation of *ALDH3* expression was shown by differential ALDH3 inducibility in various rat tissues after treatment with methylcholanthrene or TCDD (Dunn et al., 1988; Vasiliou and Marselos, 1989). Due to high constitutive expression in the cornea, it is obvious to conclude that ALDH3 plays a major role in the protection of the eye from UV light. What is the mechanism of the high constitutive ALDH3 expression in the corneal epithelium? Induction of the ALDH3 via a UV-light stimulated signal transduction pathway is one possibility. It has become clear that UV-light induces transcription of several genes such as transcription factors (AP-1, c-Fos, c-Jun, and NF-kB), stress associated proteins, proteases, and several growth factors (Sachsenmaier et al., 1994). Since the AP-1, EpRE and NF-κB and UV response element putative binding sites are present in the 5' flanking region of the mouse *Ahd4* (orthologous of the rat and human *ALDH3*; Vasiliou et al., 1995c), it is possible that this gene is activated by UVR in the cornea producing a state of constitutive up-regulation.

In this chapter we present the regulation of both ALDH3 and NMO1 in human corneal epithelial (HCE; high ALDH3 expression) and breast adenocarcinoma (MCF-7; low ALDH3 expression) cell lines. The molecular mechanisms of ALDH3 and NMO1 expression and their induction by TCDD were investigated using 3.2 and 1.6 kb 5' flanking regions of the mouse *Ahd4* (orthologous to human *ALDH3*) and human *NMO1* gene respectively, linked to luciferase reporter gene. Our results show that both *ALDH3* and *NMO1* genes are inducible in human MCF-7 cells, but only *NMO1* is inducible in human corneal epithelial cells. The lack of ALDH3 inducibility in HCE cells suggests that ALDH3 gene might be under control of a cornea-specific promoter or *cis*-acting factor located upstream or downstream of the 3.2 kb 5' flanking region of this gene.

2. MATERIALS AND METHODS

2.1. Treatment of Animals

C57BL/6J male mice, were purchased from The Jackson Laboratory, Bar Harbor, ME, and acclimated in our mouse colony for 1–2 weeks before experimentation. As needed, TCDD (100 μg/kg for 24 h) in *p*-dioxane (0.5 ml/kg) was given intraperitoneally.

2.2. Cell Cultures

The immortalized human corneal epithelial cell (HCE) line (high ALDH3 activity) was cultured in supplemented hormone epithelial medium (SHEM) (Araki-Sasaki et al., 1995). The human breast cancer MCF-7 cell line was cultured in Dulbecco's modified Eagle's medium (DMEM), supplemented with 10% fetal calf serum. The mouse hepatoma (Hepa-1) wild-type (*wt*) cell culture (Bernard et al., 1974; Benedict et al., 1973) was routinely grown in DMEM supplemented with 5% fetal calf serum. All cells were cultured in a 37°C incubator injected with 5% CO_2.

2.3. RNA Extraction and Northern Blot Analysis

Total RNA from mouse tissue samples (either control or treated with TCDD) was extracted by the acid guanidinium thiocyanate method (Chomczynski and Sacchi, 1987). RNA was separated by formaldehyde-agarose gel electrophoresis, transferred to Nytran membranes and UV cross-linked. Radioactively labeled cDNA probes were prepared by random priming (Sambrook et al., 1989), using [α-^{32}P]dCTP as the labeled precursor. Pre-hybridization and hybridization were carried out at 52°C in a solution containing 50% deionized formamide, 6X SSC, 2.5X Denhardt's solution, 0.5% sodium dodecylsulfate (SDS), and denatured salmon sperm DNA (0.1 mg/mL). After hybridization for 16–20 hours, the filters were washed twice under progressively more stringent conditions and exposed to X-ray film at -70°C with intensifying screens for 24–48 hours. Probes included mouse AHD4 (Vasiliou et al., 1993) and CYP1A1 (Kimura et al., 1984) full-length cDNAs.

2.4. Functional Analysis

We have developed constructs of 3.2 and 1.6 kb 5' flanking regions of the mouse *Ahd4* and human *NMO1* gene respectively, linked to the luciferase reporter gene. AHD4.LUC and NMO1.LUC chimeras were transiently transfected into HCE, MCF-7, and HEPA-1 cells by the calcium phosphate procedure (Parker and Stark, 1979). The LUC gene expression activity, in the presence and absence of 1 or 10 nm TCDD, was monitored by measuring luciferase activity. The pGL3.basic (promotorless) and pGL3.promoter (SV40 promoter) luciferase plasmids (PROMEGA) were used as negative and positive controls, respectively. The Rous sarcoma virus β-Gal plasmid was co-transfected in each case to normalize the transfection efficiency by measuring β-galactosidase activity (Guarente, 1983). HCE cells were also treated with varying doses of UVA (0.22, 0.43, 0.65 mJ/cm^2), UVB (0.45, 0.90, 1.35 mJ/cm^2), and UVC (0.23, 0.46, 0.69 mJ/cm^2) 14–18 hours before measuring luciferase activity.

2.5. Lipid Peroxidation

Cells were washed with phosphate buffered saline (PBS) and exposed to UV in PBS solution. After treatment, PBS was replaced with cell media. Three hours later, cells were harvested in 0.5 mL PBS containing 10% trichloroacetic acid (TCA) and 0.02% butylated hydroxytoluene (BHT). After a 15 minute incubation at 70°C, the mixture was cooled to 35°C and centrifuged at 14,000 rpm for 5 minutes. To the supernatant, 0.5 mL of saturated thiobarbituric acid (TBA) was added and heated for 30 minutes at 70°C, cooled to 35°C and centrifuged at 14,000 rpm for 5 minutes. 0.25 mL of butanol was added to the supernatant, mixed by vortexing, and the phases were separated by centrifugation at 3,000 x g for 5 minutes. The butanol phase was combined with one volume of 50% mobile phase (15% acetonitrile/0.6% tetrahydrofuran/1 mM potassium phosphate pH~7) and 50% methanol, and then subjected to HPLC analysis using a propylamine column (5μM, 200 x 4.5 mm) and the mobile phase at 1.0 mL/min. Under these conditions, malondialdehyde (MDA) elutes at 4.68 minutes and was monitored by flourometric detection at 550 emission wavelength and 515 excitation wavelength. Standard curves were prepared using synthetic MDA. Treatment with iron sulfate plus ascorbate, known to induce lipid peroxidation, was the positive control for lipid peroxidation experiments.

3. RESULTS AND DISCUSSION

3.1. Tissue-Specific Expression of AHD4

In mammalian species such as human (King and Holmes, 1993), baboon, pig, sheep, cow (Holmes et al., 1989), rat (Messiha and Price, 1983) and mouse (Holmes et al., 1989), AHD4 represents one of the major corneal soluble proteins (10–40% of the total soluble protein) with high enzymatic activity. Our northern blot analysis using liver, eye, and stomach mRNA from untreated and TCDD-treated C57BL/6J mice, showed that AHD4 is

Figure 2. Northern blot analysis of liver, eye and stomach AHD4 and CYP1A1 mRNAs in control and TCDD treated C57BL/6J mice. "-" and "+" denote control and TCDD (12 hours) treatment, respectively. Comparing the AHD4 mRNA levels versus the amount of RNA loaded in each lane as indicated by the ethidium bromide stained 18S rRNA and 28S rRNA, we conclude that the eye--cornea-- has the highest expression of AHD4 compared to the liver and stomach.

highly expressed in the eye (Figure 2). In addition to the AHD4 cDNA probe, a mouse CYP1A1 probe was used as a positive marker for TCDD induction of gene expression.

In liver tissue, both CYP1A1 and AHD4 mRNA basal levels were negligible and CYP1A1, but not AHD4 mRNA, was highly induced by TCDD. We have previously demonstrated that AHD4 mRNA is inducible in mouse cell cultures by TCDD or benzo[a]pyrene (Vasiliou et al., 1992; 1993). The lack of TCDD-inducible AHD4 mRNA in mouse liver has been previously reported --even at very high doses of TCDD (200 μg/kg for 24 h) or benzo[a]pyrene (500 mg/kg for 24 h) (Vasiliou et al., 1993). This finding differs from that in the rat, where the orthologous *ALDH3c* gene is expressed and induced by TCDD or benzo[a]pyrene in both cell cultures and the intact animal (Marselos and Vasiliou, 1991; *reviewed in* Lindahl, 1992). Based on these results one might conclude that, in mouse hepatic or hepatoma cell cultures, either an inhibitor of *Ahd4* gene expression is absent, or an activator of *Ahd4* gene expression is present, as compared with hepatic *Ahd4* gene expression in the intact mouse.

In the stomach, a tissue that has been reported to have elevated AHD4 enzyme activity, we found elevated basal AHD4 mRNA levels which were not induced by TCDD; the CYP1A1 basal mRNA levels were undetectable but highly increased after TCDD treatment.

In the eye, CYP1A1 mRNA basal levels were nil but induced by TCDD; however, AHD4 mRNA basal levels were elevated and not further induced by TCDD. These data suggest that *Ahd4* gene expression in the eye is not effected by TCDD. We have found similar results for the rat eye ALDH3 after treatment with β-naphthoflavone (P. Pappas, et al., manuscript in preparation). The fact that CYP1A1 is inducible indicates that both AHR and ARNT are present and functional in the eye. Similarly, this laboratory has found that NMO1 is expressed in mouse and rat eye and is inducible by TCDD or β-naphthofla-vone (data not shown). Zhao and Shichi (1995) have recently shown CYP1A1 inducibility in the eye of C57BL/6J mouse after treatment with β-naphthoflavone. This induction has been shown to be under the same genetic regulation as CYP1A1 in liver tissue which supports our data described above. In conclusion, in addition to enzyme activities, the AHD4 mRNA levels are highly expressed in the eye but not induced by TCDD indicating a different mechanism of *Ahd4* regulation in the eye than this found in liver.

3.2. Functional Analysis of AHD4 and NMO1

To examine the molecular mechanism of *Ahd4* expression in cornea, we have used an immortalized human corneal epithelial cells (HCE), which has high ALDH3 enzyme activity (Araki-Sasaki et al., 1995). HCE cells transiently transfected with AHD4.LUC chimera showed nil luciferase activity (Figures 3 and 4). In addition, we have found no induction after treatment with 1 and 10 nM of TCDD. Since the AHD4 induction is AHR-dependent and the AHR is polymorphic in humans (high and low affinity for TCDD; *reviewed* in Nebert et al., 1993) we tested a 100 nM dose of TCDD; the luciferase activity was found to be nil (data not shown). When NMO1.LUC was transiently transfected, basal and TCDD-inducible luciferase activity was detected in HCE cells (Figure 3 and 4). These data indicate the presence and the functionality of both AHR and ARNT in HCE cells. Further evidence comes from the induction of HCE cells transiently transfected with the pAhRD plasmid which contains the AhR domain of the mouse *Cyp1a1* gene upstream of the SV40 promoter of the pGL3. promoter plasmid (Figure 4).

Figure 3. Transient expression and TCDD induction of AHD4.LUC and NMO1.LUC recombinant plasmids in HCE, MCF-7, and Hepa-1 cells. RSV- βGal plasmid was cotransfected in each to normalize the transfection efficiency. The pGL3.promoter and pGL3.basic luciferase plasmids were used as negative and positive controls (data not shown).

Figure 4. Expression of pAHD4–3F.LUC, pNMO1–1.6F.LUC, and pAhRD.LUC recombinant plasmids in HCE cells. The chimeric plasmid DNAs were transiently transfected and their expression was monitored by measuring the luciferase activity. The pGL3.promoter, a positive control and RSV-βGal plasmid was cotransfected in each to normalize the transfection efficiency. The pGL3.Basic luciferase plasmids was used as negative control.

To test the functionality of the mouse AHD4 promoter region in human cells, MCF-7 cell line was used. In the MCF-7 cell line transiently transfected with either AHD4.LUC or NMO1. LUC, a 3–5 fold increase in luciferase activity was observed after treatment with TCDD (Figure 3). The Hepa-1 cell line was used as a positive control for the mouse AHD4 promoter and to test the functionality of the human NMO1 promoter region in mouse cells. A 2- and 9-fold increase in expression was observed in AHD4.LUC and NMO1.LUC activity respectively, after treatment with TCDD. According to our data the human *NMO1* and mouse *Ahd4* promoter regions are functional in both mouse and human cells.

These results show that both *ALDH3* and *NMO1* genes are inducible in human MCF-7 and mouse Hepa-1 cells, but only *NMO1* is inducible in human corneal epithelial cells. The inducibility of *NMO1* by TCDD provides further evidence of the existence and functionality of AHR and ARNT proteins in HCE cells. The lack of *ALDH3* inducibility by TCDD in HCE cells suggests that the *ALDH3* gene might be under control of a cornea-specific promoter or transcription factor present either downstream or upstream of the 3.2 kb 5' flanking region of the *Ahd4* gene.

3.3. Is the Corneal *Ahd4* Regulated by UV-Light?

It has become evident that AHD4 may serve to protect the eye against UV-light by oxidizing the aldehydes generated by lipid peroxidation and by direct absorption of UV-B light (Downes et al., 1994). Schmitz and colleagues (1995) have indicated that UVA induces lipid peroxidation in human skin fibroblasts, which is associated with depletion of glutathione (GSH) as a part of the antioxidant defense mechanism. It is tempting to speculate that UV-light induces lipid peroxidation and the expression of *Ahd4* gene in cornea plays a role in this defense mechanism. The AP-1, EpRE and NF-κB and UVR putative transcription factor binding sites are present in the 5' flanking region of the mouse *Ahd4* (Vasiliou et al., 1995b), therefore, and as suggested by Boesch et al., (1995) *Ahd4* gene

Figure 5. Effects of various doses of UVA, UVB and UVC on lipid peroxidation in HCE cells as measured by MDA production. See text for details.

might be a light activated gene. To study this possibility we have exposed HCE cells to various doses of UVA, UVB and UVC irradiation and lipid peroxidation was measured by monitoring the levels of MDA production. Our data show for the first time that UVA, UVB and UVC induced lipid peroxidation in corneal cells with the order of effectiveness to be UVA<UVB<UVC (Figure 5). These same doses of UVR were used to expose HCE cells transiently transfected with AHD4.LUC chimera. Our data show that UVR failed to induce luciferase activity, indicating that UVR does not transactivate *Ahd4* gene expression via this 3.2 kb 5' flanking region (data not shown). Again, a cornea-specific promoter or transcription factor, UV-dependent or independent, present at either downstream or upstream of the 3.2 kb 5' flanking region might control corneal *Ahd4* gene expression.

In conclusion, our results show that although *ALDH3* and *NMO1* genes are inducible in human MCF-7 cells, only *NMO1* is inducible in human corneal epithelial cells. Lack of TCDD inducibility was also observed in mouse eye. The lack of eye ALDH3 inducibility and basal expression of luciferase driven by the 3.2 kb *Ahd4* 5' flanking region suggest that the *ALDH3* gene might be under control of a cornea-specific promoter or a cornea-specific DNA element outside of this region. The latter possibility is currently under investigation in our laboratory by using larger fragments of the 5' flanking region and the first intron, in an effort to determine the regulation of the high constitutive AHD4 expression in cornea.

4. ACKNOWLEDGMENTS

We thank our colleagues, for valuable discussions and a critical reading of this manuscript. We thank Dr. Dennis R. Petersen for his help on the lipid peroxidation assays.

This work was supported by a Department of Pharmaceutical Sciences Seed grant (University of Colorado, School of Pharmacy).

5. REFERENCES

Araki-Sasaki, K., Ohashi, Y., Sasabe, T., Hayashi, K., Watanabe, H., Tano, Y. and Handa, H.: An SV40-immortalized human corneal epithelial cell line and its characterization. Inv. Opthalmol. Vis. Sci. 36 (1995) 614–621.

Benedict, W.F., Gielen, J.E., Owens, I.S., Niwa, A. and Nebert, D.W.: Aryl hydrocarbon hydroxylase induction in mammalian liver cell culture. IV. Stimulation of the enzyme activity in established cell lines derived from rat or mouse hepatoma and from normal rat liver. Biochem. Pharmacol. 22 (1973) 2766–2769.

Bernard, H.P., Darlington, G.J. and Ruddle, F.H.: Expression of liver phenotypes in cultured mouse hepatoma cells: Synthesis and secretion of serum albumin. Dev. Biol. 35 (1974) 83–96.

Boesch, J.S., Lee, C. and Lindahl, R.G.: Constitutive expression of class 3 aldehyde dehydrogenase in cultured rat corneal epithelium. J. Biol. Chem. 271 (1995) 5150–5157.

Chomczynski, P. and Sacchi, N.: Single-step method of RNA isolation by acid guanidinium thiocyanate-phenol-chloroform extraction. Anal. Biochem. 12 (1987) 156–159.

Downes, J. and Holmes, R.: Development of aldehyde dehydrogenase and alcohol dehydrogenase in mouse eye: Evidence for light-induced changes. Biol.Neonate 61 (1992) 118–123.

Downes, J.E., Swann, P.G. and Holmes, R.S.: Differential corneal sensitivity to ultraviolet light among inbred strains of mice. Correlation of ultraviolet B sensitivity with aldehyde dehydrogenase deficiency. Cornea 13 (1994) 67–72.

Dunn, T.J., Lindahl, R. and Pitot, H.C.: Differential gene expression to 2,3,7,8- tetrachlorodibenzo-*p*-dioxin (TCDD). Noncoordinate regulation of a TCDD-induced aldehyde dehydrogenase and cytochrome P-450c in the rat. J.Biol.Chem. 263 (1988) 10878–10886.

Guarente, L.: Yeast promoters and lac Z fusions designed to study expression of cloned genes in yeast. Methods in Enzymology 101 (1983) 181–191.

Hendee, W.R.: Harmful effects of ultraviolet radiation. J. Am. Med. Assoc. 262 (1989) 380–384.

Holmes, R.S. and Vanderberg, J.L.: Ocular NAD-dependent alcohol dehydrogenase and aldehyde dehydrogenase in the baboon. Exp. Eye Res. 43 (1986) 383–396.

Holmes, R.S., Cheung, B. and VandeBerg, J.L.: Isoelectric focusing studies of aldehyde dehydrogenases, alcohol dehydrogenases and oxidases from mammalian anterior eye tissues. Comp. Biochem. Physiol. 93 B (1989) 271–277.

Jaiswal, A.K.: Jun and fos regulation of NAD(P)H:quinone oxidoreductase gene expression. Pharmacogenetics 4 (1994) 1–10.

Kimura, S., Gonzalez, F.J. and Nebert, D.W. : The murine Ah locus. comparison of the complete cytochrome P1–450 and P2–450 cDNA nucleotide and amino acid sequences. J. Biol. Chem. 259 (1984) 10705–10713.

King, G. and Holmes, R.S.: Human corneal aldehyde dehydrogenase: purification, kinetic characterization and phenotypic variation. Biochem. Mol. Biol. Int. 31 (1993) 49–63.

Lindahl, R.: Aldehyde dehydrogenases and their role in carcinogenesis. CRC Crit.Rev.Biochem.Mol.Biol. 27 (1992) 283–335.

Marselos, M. and Vasiliou, V.: Effect of various chemicals on the aldehyde dehydrogenase activity of the rat liver cytosol. Chem-Biol. Interact. 79 (1991) 79–89.

Messiha, F.S. and Price, J.: Properties and regional distribution of ocular aldehyde dehydrogenase in the rat. Neurobehav. Toxicol. Teratol. 5 (1983) 251–254.

Nebert, D.W., Puga, A. and Vasiliou, V.: Role of the Ah receptor and the dioxin-inducible [Ah] gene battery in toxicity, cancer and in signal transduction. Ann. N. Y. Acad. Sci. 685 (1993) 624–640.

Pappas, P., Stephanou, P., Vasiliou, V., Karamanakos, P. and Marselos, M.: Ontogenesis and expression of ALDH activity in the skin and eye of the rat. Adv. Exp. Med. Biol. *In press* (1997).

Parker, B. and Stark, G.: Regulation of simian virus 40 transcription: Sensitive analysis of the RNA species present early in infections by virous or viral DNA. J. Virol. 31 (1979) 360- 369.

Sachsenmaier, C., Radler-Pohl, A, Muller, A., Herrlich, P. and Rahmsdorf, H.J.: Damage to DNA by UV light and activation of transcription factors. Biochem. Pharmacol. 47 (1994) 129–136.

Sambrook, J., Fritsch, E.F. and Maniatis, T.: Molecular Cloning: A Laboratory Manual, 2nd Ed. Cold Spring Harbor Laboratory Press, Cold Spring Harbor, New York (1989).

Schein, O.D.: Phototoxicity and the cornea. J. Natl. Med. Assoc. 84 (1992) 579–583.

Schmitz, S., Garbe, C., Jimbow, K., Wulff, A., Daniels, H., Eberle, J. and Orfanos, C.E.: Photodynamic action of ultraviolet A: Induction of cellular hydroperoxides. Recent Results in Canc. Res. 139 (1995) 43–55.

Shertzer, H.G., Vasiliou, V., Liu, R.M., Tabor, M.W. and Nebert, D.W.: Enzyme induction by L-buthionine(S,R)-sulfoximine in cultured hepatoma cells. Chem. Res. Toxicol. 8 (1995) 431–436.

Van der Hoeve, J.: Eye lesions produced by light rich in ultraviolet rays, senile cataract, senile degeneration of macula. Am. J. Opthalmol. 3 (1920) 178–194.

Vasiliou, V. and Marselos, M.: Tissue distribution of inducible aldehyde dehydrogenase activity in the rat after treatment with phenobarbital or methylcholanthrene. Pharmacol. Toxicol. 64 (1989) 39–42.

Vasiliou, V., Puga, A. and Nebert, D.W.: Negative regulation of the murine cytosolic aldehyde dehdrogenase-3 (ALDH-3c) gene by functional CYP1A1 and CYP1A2 proteins. Biochem. Biophys. Res. Commun. 187 (1992) 413–419.

Vasiliou, V., Reuter, S.F., Kozak, C.A. and Nebert, D.W.: Mouse dioxin-inducible cytosolic aldehyde dehydrogenase-3: AHD4 cDNA sequence, genetic mapping, and differences in mRNA levels. Pharmacogenetics 3 (1993) 281–290.

Vasiliou, V., Shertzer, H.G., Liu, R-M., Sainsbury, M. and Nebert, D.W.: Response of the [*Ah*] battery genes to compounds that protect against menadione toxicity. Biochem. Pharmacol. 50 (1995a) 1885–1891.

Vasiliou, V., Puga, A., Chang, C.Y., Tabor, W.M. and Nebert, D.W.: Interaction between the Ah receptor and proteins binding to the AP-1 like electrophile response element (EpRE) during murine phase II [Ah] battery gene expression. Biochem. Pharmacol. 50 (1995b) 2057–2068.

Vasiliou, V., Reuter, S.F., Kozak, C.A. and Nebert, D.W.: Mouse Class 3 Aldehyde Dehydrogenases. Adv. Exp. Med. Biol. 372 (1995c) 151–158.

Vasiliou, V., Weiner, H., Marselos, M. and Nebert, D.W.: Mammalian aldehyde dehydrogenase genes: classification based on evolution, structure and regulation. Eur. J. Drug Metab. Pharmacokin. 20 (1995d) 53–64.

Zhao, C. and Shichi, H.: Immunocytochemical study of cytochrome P450 (1A1/1A2) induction in murine ocular tissue. Exp. Eye Res. 60(2) (1995) 143–152.

9

A PRELIMINARY REPORT ON THE CLONING OF A CONSTITUTIVELY EXPRESSED RAT LIVER CYTOSOLIC ALDH cDNA BY PCR

Eva C. Kathmann and James J. Lipsky

Clinical Pharmacology Unit
Department of Pharmacology
Mayo Clinic and Foundation
Rochester, Minnesota 55905

1. INTRODUCTION

Aldehyde dehydrogenase (ALDH, EC 1.2.1.3) is a ubiquitous enzyme that cata-lyzes the oxidation of acetaldehyde to acetic acid during ethanol metabolism. ALDH is located in virtually all mammalian tissues and found in all major subcellular organelles. The hepatic cDNAs for phenobarbital(PB)-inducible cytosolic ALDH, mitochondrial ALDH, microsomal ALDH, and tumor specific cytosolic ALDH have been cloned in the rat (Dunn, *et al.*, 1989; Farres, *et al.*, 1989; Miyauchi, *et al.*, 1991; Jones, *et al.*, 1988). However, the cDNA for a constitutively expressed rat liver cytosolic ALDH has not been cloned. While constitutively expressed cytosolic ALDHs have been characterized in the livers of numerous species, there is controversy in the literature as to whether this he-patic isoform of ALDH exists in the rat. To address this question, a PCR-based approach to cloning a constitutively expressed cytosolic ALDH from a rat liver cDNA library was employed.

2. METHODS

2.1. Materials

The rat liver cDNA library (Mature Sprague Dawley, 6 months old, male) in the Lambda Zap II vector was from Stratagene (La Jolla, CA); rat mitochondrial and human cytosolic ALDH cDNA in pT7–7 were gifts from Dr. Henry Weiner (Purdue University, West Lafayette, IN); PCR SuperMix was obtained from GibcoBRL (Grand Island, NY).

Enzymology and Molecular Biology of Carbonyl Metabolism 6
edited by Weiner *et al.* Plenum Press, New York, 1996

2.2. Primer Design

Primers were designed to amplify rat liver cytosolic ALDH cDNA sequences. Five mammalian liver cytosolic ALDH cDNA sequences (human, mouse, sheep, long-ear elephant shrew, and short-ear elephant shrew) were aligned. The aligned sequences were examined for regions of homology consisting of at least 20 nucleotides that were not conserved in the rat mitochondrial, microsomal, and tumor specific ALDH cDNA sequences. The 5'-ends of the two primers that were used correspond to nucleotides 89 and 921 of the human cytosolic ALDH cDNA sequence. Primers were made by the Mayo Molecular Core Facility using an ABI model 394 DNA synthesizer (Perkin Elmer-Applied Biosystems Inc., Foster City, CA).

2.3. PCR

PCR SuperMix, primers, and template DNA were combined to give PCR reaction conditions consisting of 2 U of Taq DNA polymerase, 180 μM of each dNTP, 230 nM of each primer, and 10 ng or 2×10^8 pfu of template DNA. Magnesium chloride was added to a final concentration of 4.5 mM, 3.5 mM, or 1.5 mM to amplify the middle, 5' end, or 3' end of the cDNA, respectively. After a one-minute initial denaturation at 94°C, 30 cycles of a one-minute denaturation at 94°C, a one-minute annealing at 62°C, and a two-minute extension at 72°C were performed.

2.4. Sequence Analysis

Fluorescent DNA sequencing was done by the Mayo Molecular Core Facility using the Dye Terminator Cycle Sequencing Ready Reaction and the Perkin Elmer ABI Prism model 377 DNA Sequencer (Perkin Elmer-Applied Biosystems Inc., Foster City, CA) . Nucleotide sequences were analyzed with FASTA, PILEUP, BESTFIT, and FRAGMENT ASSEMBLY programs of the University of Wisconsin Genetic Computer Group (GCG) Program, version 8.1.

3. RESULTS

PCR was performed using primers designed to be highly specific for cytosolic ALDH sequences. A rat liver cDNA library (6-month-old male Sprague-Dawley) was used as template. Rat liver mitochondrial ALDH cDNA and human liver cytosolic ALDH cDNA, both in plasmid pT7–7, were used as negative and positive control templates, respectively. While there was no PCR amplification of the negative control template, an 800-bp PCR product was amplified from both the positive control template and the rat liver cDNA library template. The sequence of the 800-bp PCR product amplified from the rat liver cDNA library was compared to the sequences in the GenBank database. The five highest homologies are shown in Table 1. Since the 800-bp PCR product is most highly homologous to cytosolic ALDH cDNA sequences, it appears that the PCR product is a portion of a rat liver cytosolic ALDH cDNA. The sequence of the 800-bp PCR product corresponds to nucleotides 121 to 888 of the human cytosolic ALDH cDNA sequence.

Using the sequence of the 800-bp PCR product and the sequence of the Lambda Zap II vector in which the library was prepared, primers were designed to amplify the 5' and 3' ends of the cDNA. The amplified regions of the cDNA were aligned based on sequence

Table 1. Identities between the nucleotide sequence of the
800-bp PCR product and sequences in the
GenBank database

ALDH cDNA Sequence	% Identity
Rat kidney cytosolic ALDH	98.2
C56BL/6 mouse liver cytosolic ALDH	92.8
BALB/c mouse liver cytosolic ALDH	92.7
Rat liver PB-inducible cytosolic ALDH	89.0
Human liver cytosolic ALDH	84.2
Rat liver mitochondrial ALDH	66.3

The first five sequences listed were the highest homologies in the Gen-
Bank Database.

overlap to obtain a 1750-bp preliminary consensus sequence. The consensus sequence was compared to the sequences in the GenBank database. The six highest homologies are shown in Table 2. Since cytosolic ALDH cDNA sequences show the highest homology to the preliminary consensus sequence, it appears that this sequence is that of a rat liver cytosolic ALDH cDNA.

4. DISCUSSION

We have designed PCR primers that show high specificity for liver cytosolic ALDH cDNA sequences. While these primers are theoretically capable of amplifying rat hepatic PB-inducible cytosolic ALDH, this was not a concern since the rat liver cDNA library used was not constructed from a PB-induced rat. The PCR primers were designed to amplify nucleotides 89 to 921 based on the human cytosolic ALDH cDNA sequence. We amplified an 800-bp product from both human cytosolic ALDH cDNA and the rat liver cDNA library, but did not amplify rat liver mitochondrial ALDH cDNA. The nucleotide sequence of the 800-bp PCR product amplified from the rat liver cDNA library corresponds to nucleotides 121 to 888 of the human cytosolic ALDH cDNA sequence. The sequence of the 800-bp PCR product was most homologous to rat kidney cytosolic ALDH, mouse liver cytosolic ALDH, rat liver PB-inducible cytosolic ALDH, and human liver cytosolic ALDH (Table 1). By amplifying the 5' and 3' ends of the cDNA, we have obtained

Table 2. Identities between the nucleotide sequence of
the preliminary 1750-bp consensus sequence and
sequences in the GenBank database

ALDH cDNA Sequence	% Identity
Rat kidney cytosolic ALDH	93.7
BALB/c mouse liver cytosolic ALDH	89.7
C56BL/6 mouse liver cytosolic ALDH	85.9
Human liver cytosolic ALDH 3' end	84.8
Rat liver PB-inducible cytosolic ALDH	83.4
Human liver cytosolic ALDH	78.2
Rat liver mitochondrial ALDH	64.5

The first six sequences listed were the highest homologies in the GenBank
Database.

a preliminary 1750-bp consensus sequence which is also most similar to rat kidney cytosolic ALDH, mouse liver cytosolic ALDH, rat liver PB-inducible cytosolic ALDH, and human liver cytosolic ALDH (Table 2).

The existence of constitutively expressed ALDH in the rat liver cytosol is controversial. Lindahl and Evces (1984) found little or no ALDH activity in the liver cytosol of Buffalo, Fischer 344, Long-Evans, Sprague-Dawley, Wistar and Purdue/Wistar rats. However, Weiner and his colleagues have identified up to five isozymes of ALDH in Purdue/Wistar rat liver cytosolic fractions based on isoelectric focusing (Berger and Weiner, 1977; Tank *et al.*, 1981; Truesdale-Mahoney, *et al.*, 1981; Cao, *et al.*, 1989). Our preliminary results lend support to the existence of constitutively expressed rat liver cytosolic ALDH.

In conclusion, it appears that we have amplified portions of a constitutively expressed rat liver cytosolic ALDH cDNA based on nucleotide sequence analysis. These initial results provide a strategy employing PCR for cloning a full length cDNA for a constitutively expressed cytosolic ALDH from a rat liver cDNA library.

5. ACKNOWLEDGMENTS

This work was supported by grants NIH AA-09543 and FDA-000–886. We wish to thank Dr. Henry Weiner for his generous gifts of rat mitochondrial and human cytosolic ALDH cDNAs.

6. REFERENCES

Berger, D. and Weiner, H. Relationship between alcohol preference and biogenic aldehyde metabolizing enzymes in rats. Biochem. Pharmacol. 26 (1977) 841–846.

Cao, Q.-N., Tu, G.-C. and Weiner, H. Presence of cytosolic aldehyde dehydrogenase isozymes in adult and fetal rat liver. Biochem. Pharmacol. 38 (1989) 77–83.

Dunn, T. J., Koleske, A. J., Lindahl, R. and Pitot, H. C.: Phenobarbital-inducible aldehyde dehydrogenase in the rat: cDNA sequence and regulation of the mRNA by phenobarbital in responsive rats. J. Biol. Chem. 264 (1989) 13057–13065.

Farres, J., Guan, K. L. and Weiner, H.: Primary structures of rat and bovine liver mitochondrial aldehyde dehydrogenases deduced from cDNA sequences. Eur. J. Biochem. 180 (1989) 67–74.

Jones, D. E., Brennan, M. D., Hempel, J. and Lindahl, R.: Cloning and complete nucleotide sequence of a full-length cDNA encoding a catalytically functional tumor-associated aldehyde dehydrogenase. Proc. Natl. Acad. Sci. U.S.A. 85 (1988) 1782–1786.

Lindahl, R. and Evces, S.: Comparative subcellular distribution of aldehyde dehydrogenase in rat, mouse and rabbit liver. Biochem. Pharmacol. 33 (1984) 3383–3389.

Miyauchi, K., Masaki, R., Taketani, S., Yamamoto, A., Akayama, M. and Tashiro, Y.: Molecular cloning, sequencing and expression of cDNA for rat liver microsomal aldehyde dehydrogenase. J. Biol. Chem. 266 (1991) 19536–19542.

Tank, A. W., Weiner, H. and Thurman, J. A.: Enzymology and subcellular localization of aldehyde oxidation in rat liver. Oxidation of 3,4-dihydroxyphenylacetaldehyde derived from dopamine to 3,4-dihydroxyphenylacetic acid. Biochem. Pharmacol. 30 (1981) 3265–3275.

Truesdale-Mahoney, N., Doolittle, D. P. and Weiner, H. Genetic basis for the polymorphism of rat liver cytosolic aldehyde dehydrogenase. Biochem. Genet. 19 (1981) 1275–1282.

ONTOGENESIS AND EXPRESSION OF ALDH ACTIVITY IN THE SKIN AND THE EYE OF THE RAT

Perikles Pappas,[1] Panayiotis Stephanou,[1] Vasilis Vasiliou,[2]
Petros Karamanakos,[1] and Marios Marselos[1]

[1]Department of Pharmacology
Medical School, University of Ioannina
451 10 Ioannina, Greece
[2]Department of Pharmaceutical Sciences, School of Pharmacy
University of Colorado Health Sciences Center
4200 East Ninth Avenue, Denver, Colorado 80262

1. INTRODUCTION

As the body's first line of defense against external insult, both the eye and the skin are exposed routinely to chemical agents, serving as a portal of entry for topical contactants. In addition, global atmospheric changes like ozone depletion in the stratosphere contribute to enhanced chronic exposure of the skin and the eye of human tissue to UV light (Hendee, 1989). The hazardous effects of UV light on the eye were already recognized in 1920 by van der Hoeve. By the 1920s, basal cell carcinomas of the skin were also noted.

One mechanism that is thought to play an important role in the UV toxicity involves generation of oxygen radicals, leading to lipid peroxidation and protein modification. This process is also thought to be related to aging and many human diseases, like cancer, atherosclerosis and arthritis (Marx, 1987; Cross et al., 1987). Major products of lipid peroxidation are aldehydes which are known to be cytotoxic, carcinogenic and mutagenic (Esterbauer et al., 1991). Various metabolic pathways exist for the metabolism of aldehydes to less reactive forms. One of the most important pathways for aldehyde metabolism is their oxidation to carboxylic acids by aldehyde dehydrogenases (ALDHs). Several ALDH enzymes are present in mammals and other species, and the ALDH activity is detectable practically in all tissues (Deitrich, 1966; Vasiliou & Marselos, 1989).

Two rat liver cytosolic aldehyde dehydrogenases, ALDH1 and ALDH3, are of particular interest because are inducible by different classes of foreign chemicals (Marselos, 1976; Marselos et al., 1979; Lindahl, 1992). Phenobarbital-type inducers increase ALDH1 enzyme activity about 20-fold in rats having the "responsive" genotype *R/R*, but only 2-

fold in rats having the "nonresponsive" genotype *r/r* (Deitrich, 1971; Marselos, 1976). Dioxin (2,3,7,8-tetrachlorodibenzo-*p*-dioxin; TCDD) and polycyclic aromatic hydrocarbons (PAHs) such as benzo-a-pyrene (BaP) increase the cytosolic ALDH3 enzyme activity in all rat species tested today. In addition, ALDH3 is constitutively expressed in cornea, stomach urinary bladder and lung (Vasiliou, 1988; Dunn et al., 1988).

The *ALDH3* gene is a member of the *a*romatic *h*ydrocarbon-responsive [*Ah*] gene battery. TCDD and polycyclic hydrocarbons coordinately up-regulate the transcription of genes in the [*Ah*] battery, which, in addition to *ALDH3*, also includes three other "Phase II" genes--*GSTA1*, *UGT1*06* and *NMO1*--and at least two cytochrome P450 genes, *CYP1A1* and *CYP1A2* (Nebert and Gonzalez, 1987; Vasiliou et al., 1992; 1993; Nebert et al., 1993). This induction process requires functional cytosolic Ah receptor (AHR) and Ah receptor nuclear translocator (ARNT) protein (Swanson and Bradfield, 1993). After binding to the ligand, the AHR-inducer complex translocates into the nucleus, forms a heterodimer with the ARNT, and binds to one or more *a*romatic *h*ydrocarbon-*r*esponse *e*lements (AhREs) that have been identified upstream of all six of the above-mentioned mammalian [*Ah*] battery genes. Although the interrelationship and "cross-talk" amongst [*Ah*] battery genes have been established in the mouse (Nebert et al., 1993), it is presumed that similar [*Ah*] gene interactions exist in the rat and human.

ALDH1 and *ALDH3* are two distinct genes encoding enzymes with different substrate specificities, coenzyme requirements, and sensitivity to inhibitors (Lindahl, 1992). The ALDH1 enzyme activity is measured with propionaldehyde as substrate and NAD^+ as coenzyme, whereas measurement of ALDH3 enzyme activity is best carried out with benzaldehyde and $NADP^+$ (Torronen et al., 1981). Specificity of ALDH3 expression can be determined from the ratio of B/NADP- to P/NAD-dependent ALDH activities. Under physiologic conditions, or following phenobarbital treatment, the B/NADP to P/NAD ratio is <**1.0**, whereas treatment of the rat with AHR ligands such as BaP or TCDD produces a ratio of >**1.0** (Vasiliou and Marselos, 1989; Marselos and Vasiliou, 1991).

In the resent study, the ALDH1 and ALDH3 isozymes were studied during tissue development of rat eye and skin as well as their inducibility in these tissues after local application of BaP.

2. MATERIALS AND METHODS

2.1. Animals

Albino rats (weighing 200–250 g) of the Wistar/Mol/Io/RR substrain (Marselos, 1976) were used. The rats were kept in plastic cages (Makrolon) with wood chip bedding (*Populus sp.*), and the animals were given free access to tap water and pelleted chow. For the ontogenic experiments fetuses of gestational day 18, new-born (on day 1) weanlings (1-, 2- and 3-week-old), and adult male animals (16-, 24- and 36-week old) were used.

2.2. Treatment

BaP (200 mg/kg, in olive oil) was instilled to the eye (two drops in the left eye) and applied to the skin (0.5 ml at saved small area of the skin) of the rats. Twelve male rats (weighing 250–280g) were selected and separated in two groups of six. The first group was treated with BaP every other day for one week (4 times, total); the second group received BaP every three days for one month (10 times, total). No irritations or toxicity signs were observed at the treated tissues after either weekly or monthly BaP-administration. Controls received the vehicle only, given on the same time schedule to match the experimental groups.

2.3. Preparation of the Tissues

Soluble fractions --combined from three litters (approximately 24 fetuses of gestational day 18), and from six 6-week-old and six 12-week-old females--were prepared for ALDH enzyme assays according to procedures previously described by us (Vasiliou & Marselos, 1989). Briefly, after the rats had been killed by decapitation, tissues were homogenized with a mechanical homogenizer in 3 vol (w/v) of ice-cold 0.25 M sucrose solution. The homogenate was centrifuged at 15,000 x g for 30 min; the resultant soluble fraction was used for the assays. In the case of liver, this supernatant were further centrifuged (105,000 x g for 60 min) at 4°C to separate cytosol and microsomes which were used for the biochemical assays. All preparations were performed at 4°C.

2.4. Enzyme Assays

The cytosolic ALDH isozymes were measured according to the methods described in Vasiliou and Marselos (1989), using benzaldehyde and NADP for ALDH3 activity, and propionaldehyde and NAD for ALDH1 (or P/NAD ALDH) activity. The microsomal ethoxyresorufin-O-dealkylation (EROD) activity, which represents the cytocrome P-4501A1 isoform, and cytosolic glutathione-S-transferases (GSTs) activity were measured using ethoxyresorufin and 1-chloro-2,4-dinitrobenzene as substrates, respectively, according to the methods described by Burke and Mayer (1975), and Habig et al. (1974). Protein determinations were carried out by the biuret method (Gornall et al., 1949) using bovine serum albumin as a standard. Statistical analyses of the results were performed by Student's two-tailed t-test.

3. RESULTS AND DISCUSSION

3.1. Ontogenesis of ALDH1 and ALDH3 in the Rat Eye and Skin

The ontogenic experiments showed that ALDH activities in both eye and skin are gradually increased right after birth (Figures 1 & 2). The eye ALDH3 activity was found in traces before and during the day of birth. High increases of ALDH activity were observed during the 1st and 2nd week of age which are associated with the eye opening. The activity at first week was more than 50-fold increased, and at second week about 160-fold, compared with the enzyme activity on 1st day of birth. After the first 3 weeks of age, ALDH3 activity reached a plateau which remained until the 36th week of age. Is the ALDH3 light activated? This might be one possibility. Downes and Holmes (1992) have found a similar developmental profile for the mouse ocular ALDH3, as well as a significant difference in the ocular ALDH between mice exposed to light and animals maintained in darkness at all ages examined. Their observation of significant high levels of ocular ALDH activity in the light exposed animals further supports the significance of this enzyme in the eye function and indicates a relation between ocular ALDH and light exposure. Previous studies on mammalian ocular ALDH3 have demonstrated that this enzyme is localized in the cornea for which high constitutive levels have been reported. Although the physiological substrate for this enzyme has not been identified, it has been found that

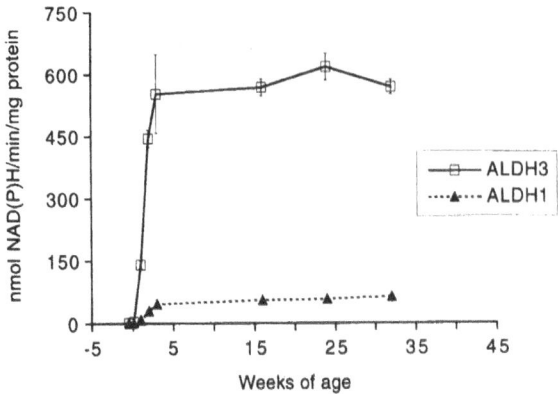

Figure 1. Ontogenesis of ALDH3 and ALDH1 activities in the eye of Wistar rat.

ALDH3 catalyzes efficiently medium chain aldehydic products of lipid peroxidation (Lindahl and Petersen, 1991; Marselos and Lindahl, 1988). Boech et al. (1996) have shown that ALDH3 is the only ALDH enzyme expressed in rat cornea. A dual role has been proposed for the corneal ALDH3 which includes direct absorption by the UV light and the detoxification of UV induce peroxidic aldehydes. Accordingly, neonatal rats (this study) and mice (Downes and Holmes, 1992) dramatically increase their levels of corneal ALDH during the period following eye opening and maturation of the eye, and are thereby adapted for light exposure and are potentially protected against UV light tissue damage.

Similar developmental patterns have been found for ALDH1; however, ALDH1 specific activities were 10 times lower to those of ALDH3 activities. ALDH1 has been detected in lens and retina (Saari et al., 1995; Godbout et al., 1993). ALDH1 has also been identified in embryonic mouse retina through protein microsequencing (McCaffery et al., 1991). A cDNA encoding for the 56-kDa androgen protein expressed in human genital fibroblasts has recently been isolated, sequenced and identified as an ALDH1 (McCammon et al., 1993). More interestingly, the 11-thromboxane B2 dehydrogenase has been identified as an ALDH1 (Westlund et al., 1994). Retinaldehyde appears to be a substrate for the ALDH1 enzyme (Lee et al., 1991; Dockman et al., 1992; Chen et al., 1994). Oxidation of retinol and metabolic cleavage of beta-carotene give rise to retinaldehyde which can be further metabolized to retinoic acid by ALDH1 (Sladek & Lee, 1993). Our data on the developmental pattern of the ocular ALDH1 further support the significance of this enzyme in the eye development.

The skin ALDH3 and ALDH1 activities were enhanced right after birth and their ontogenic profile was found to be similar with that observed in the eye, reaching a plateau after the 3rd week of age (Figure 2). However, the eye ALDH1 and ALDH3 specific activities were about 5- and 12-fold higher compared to those of the skin in all developmental stages. To our knowledge this is the first time that constitutive expression of ALDH3 in skin is reported. We have recently found that high ALDH3 mRNA levels are present in the rat skin (Pappas and Vasiliou, manuscript in preparation).

3.2. Effect of BaP in the ALDH3 and ALDH1 in the Rat Eye and Skin

The dioxin or/and PAH-treatment increases the ALDH3 activity and the mRNA levels in rat liver as well as in a number of extrahepatic tissues including heart, lungs and uri-

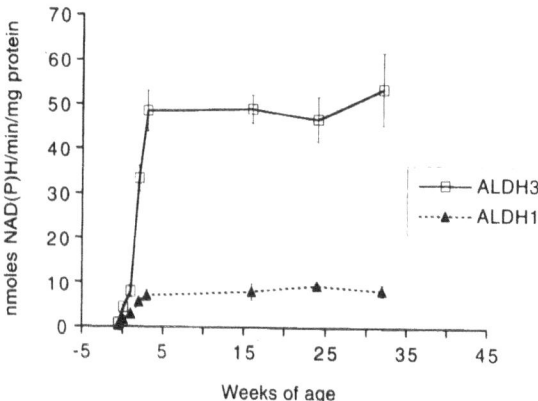

Figure 2. Ontogenesis of ALDH3 and ALDH1 activities in the skin of Wistar rat.

nary bladder (*reviewed in* Vasiliou et al., 1995). In the present series of experiments we aimed to study the inducibility of ALDH3 activity in eye and skin. Our data in Table 1 show that the high constitutive eye ALDH3 activity was not affected by either one week or one month treatment with BaP. Similarly, the P/NAD ALDH activity was unchanged after BaP-treatment (Table 1). The ALDH3 activity is constitutively expressed in rat skin; and further induced (1.5-fold, p<0.025) after local application of BaP (Table 1) on skin one week or one month after treatment. The P/NAD ALDH activity were not affected by both BaP-treatment protocols tested (Table 1).

To address the question of whether BaP enters the bloodstream and reaches other tissues than eye and skin, we studied a number of hepatic enzymes including ALDH3, EROD, GST and P/NAD ALDH. Local application of BaP to the eye and skin enhanced both hepatic cytosolic ALDH3 (2-fold) and P/NAD ALDH isozyme (1.5-fold), as well as the hepatic microsomal CYP1A1 activity (5-fold), in a statistically significant way (Figure 3). No significant changes were observed for total hepatic cytosolic GST activity between treated and untreated animals (Figure 3). The mechanism of the enhancement of hepatic enzyme activities by BaP local application, remains unknown. The transport of the ALDH3 inducer (BaP or its metabolites) to other tissues through the bloodstream is most likely the case. However, since skin inflammation has been reported as a result of topical

Table 1. Effects of subchronic BaP-treatment on ALDH3 and ALDH1 activities of the eye and skin in Wistar rat

Treatment/Tissue			ALDH3	P/NAD ALDH
One week	eye	C	618.4 ± 33.3	57.7 ± 4.1
		E	583.7 ± 11.4	55.3 ± 1.0
	skin	C	46.9 ± 5.1	9.4 ± 0.7
		E	68.1 ± 4.7 *	9.6 ± 0.4
One month	eye	C	594.0 ± 38.3	64.9 ± 4.2
		E	580.5 ± 14.8	62.7 ± 3.8
	skin	C	38.3 ± 3.5	5.1 ± 0.5
		E	59.2 ± 7.1 *	5.3 ± 0.7

Results are expressed in mean (±SE); ALDH3 and P/NAD ALDH activities are expressed in: nmol NAD(P)H/min/mg protein.
* Means statistically different from the control group, at P<0.025.

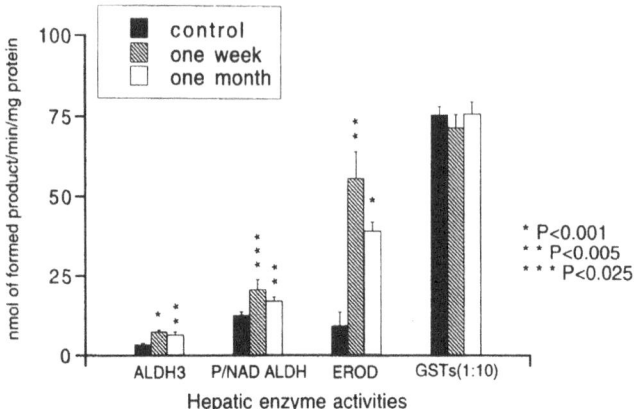

Figure 3. Effects of subchronical BaP-treatment in eye and skin on the rat hepatic ALDH3, P/NAD ALDH, EROD and GST enzyme activities.

application of PAHs (Taylor, 1971), the posibility of hepatic ALDH3 induction as a response to inflammatory signals transferred to the liver can not be rouled out at this time. Moreover, inhibition of BaP induction of the [*Ah*] battery enzymes by nonsteroidal anti-inflammatory drugs has been recently reported by us (Pappas et al.,1995).

In conclusion, our results indicate that ALDH3 is constitutive in both eye and skin and that the skin ALDH3 is inducible by BaP. The lack of inducibility of ALDH3 in the eye has also been observed in the mouse and human corneal epithelial cells (Marks-Hull et al., 1997). The authors have also shown that CYP1A1 and NMO1 genes are transcriptionally activated in cornea concluding that AHR and its partner ARNT are present and functional in the eye. All these data indicate a different mechanism of ALDH3 gene regulation in the eye compared to those of liver and skin. The eye, despite its small mass, contains two relatively large avascular areas (the cornea and the lens), which are bounded by unique active transport systems responsible for steady state of hydration and hence transparency. On the other hand, the skin protects the body against external insults to maintain internal homeostasis. These mechanisms of tissue-specific regulation of ALDH3 gene are currently under investigation in our laboratories.

4. ACKNOWLEDGMENTS

We thank our colleagues, for valuable discussions and a critical reading of this manuscript. The authors are grateful to Mrs. Olga Tsoumani for her excellent technical assistance. This work was supported in part by a Department of Pharmaceutical Sciences (University of Colorado, School of Pharmacy) Seed grant (VV) and a Hellenic State Agency of Research and Technology grant (MM-VV).

5. REFERENCES

Boesch, J.S., Lee, C. and Lindahl, R.: Constitutive expression of class 3 aldehyde dehydrogenase in cultured rat corneal epithelium. J.Biol.Chem. 271 (1996) 5150–5157.

Burke, M.D. and Mayer, R.T.: Inherent specificities of purified cytochromes P-450 and P-448 toward biphenyl hydroxylation and ethoxyresorufin deethylation. Drug Metab.Dispos. 3 (1975) 245–253.

Chen, M., Achkar, C. and Gudas, L.J.: Enzymatic conversion of retinaldehyde to retinoic acid by cloned murine cytosolic and mitochondrial aldehyde dehydrogenases. Mol.Pharmacol. 46 (1994) 88–96.

Cross, C.E., Halliwell, B., Borish, E.T., Pryor, W., Ames, B.N., Saul, R.L., McCord, J.M. and Harman, D.: Oxygen radicals in human disease. Ann.Intern.Med. 107 (1987) 526–545.

Deitrich, R.A.: Tissue and subcellular distribution of mammalian aldehyde-oxidazing capacity. Biochem.Pharmacol. 15 (1966) 1911–1922.

Deitrich, R.A.: Genetic aspects of increase in rat liver aldehyde dehydrogenase induced by phenobarbital. Science 173 (1971) 334–336.

Dockham, P.A., Lee, M.O. and Sladek, N.E.: Identification of human liver aldehyde dehydrogenases that catalyze the oxidation of aldophosphamide and retinaldehyde. Biochem.Pharmacol. 43 (1992) 2453–2469.

Downes, J. and Holmes, R.: Development of aldehyde dehydrogenase and alcohol dehydrogenase in mouse eye: Evidence for light-induced changes. Biol.Neonate 61 (1992) 118–123.

Dunn, T.J., Lindahl, R. and Pitot, H.C.: Differential gene expression to 2,3,7,8- tetrachlorodibenzo-*p*-dioxin (TCDD). Noncoordinate regulation of a TCDD-induced aldehyde dehydrogenase and cytochrome P-450c in the rat. J.Biol.Chem. 263 (1988) 10878–10886.

Esterbauer, H., Schaur, R.J. and Zollner, H.: Chemistry and biochemistry of 4-hydroxynonenal malondialdehyde and related aldehydes. FreeRad.Biol.Med. 11 (1991) 81–128.

Godbout, R.: Identification and characterization of transcripts present at elevated levels in the undifferentiated chick retina. Exp.EyeRes. 56 (1993) 95–106.

Gornall, A.G., Bardawill, C.J. and David, M.M.: Determination of serum proteins by means of the biuret reaction. J.Biol.Chem. 177 (1949) 751–766.

Habig, W.H., Pabst, M.J. and Jakoby, W.B.: Glutathione-S-transferases. The first enzymatic step in mercapturic acid formation. J.Biol.Chem. 249 (1974) 7130–7139.

Hendee, W.R.: Harmful effects of ultraviolet radiation. J.Am.Med.Assoc. 262 (1989) 380–384.

Lindahl, R. and Petersen, D.R.: Lipid aldehyde oxidation as physiological role for class 3 aldehyde dehydrogenases. Biochem Pharmacol 41 (1991) 1583–1587.

Lindahl, R.: Aldehyde dehydrogenases and their role in carcinogenesis. CRC Crit.Rev.Biochem.Mol.Biol. 27 (1992) 283–335.

Lee, M.O., Manthey, C.L. and Sladek, N.E.: Identification of mouse liver aldehyde dehydrogenases that catalyze the oxidation of retinaldehyde to retinoic acid. Biochem.Pharmacol. 42 (1991) 1279–1285.

Marks-Hull, H., Shiao, T-Y., Araki-Sasaki, K., Traver, R. and Vasiliou, V.: Expression of ALDH3 and NMO1 in human corneal epithelial and breast adenocarcinoma cells. Adv.Exp.Med.Biol. (1997) *in press.*

Marselos, M.: Genetic variation of drug metabolizing enzymes in the Wistar rat. Acta Pharmacol.Toxicol. 39 (1976) 186–197.

Marselos, M., Torronen, R., Koivoula, T. and Koivusalo, M.: Comparison of phenobarbital and carcinogen-induced aldehyde dehydrogenases in the rat. Biochem.Pharmacol. 27 (1979) 110–118.

Marselos, M. and Lindahl, R.: Substrate preference of a cytosolic aldehyde dehydrogenase inducible in rat liver by treatment with 3-methylcholanthrene. Toxicol Appl Pharmacol 95 (1988) 339–345.

Marselos, M. and Vasiliou, V.: Effect of various chemicals on the aldehyde dehydrogenase activity of the rat liver cytosol. Chem-Biol.Interact. 79 (1991) 79–89.

Marx, J.L.: Oxygen free radicals linked to many diseases. Science 235 (1987) 529–531.

McCaffery, P., Tempst, P., Lara, G. and Drager, U.: Aldehyde dehydrogenase is a posiional marker in the retina. Development 112 (1991) 693–702.

McCammon, D.K., Zhou, P., Turney, M.K., McPhaul, M.J. and Kovacs, W.J.: An androgenic affinity ligand covalently binds to cytosolic aldehyde dehydrogenase from human genital skin fibroblasts. Mol.Cellul.Endocr. 91 (1993) 177–183.

Nebert, D.W. and Gonzalez, F.J.: P450 genes: Structure, evolution and regulation. Annu.Rev.Biochem. 56 (1987) 945–993.

Nebert, D.W., Puga, A. and Vasiliou, V.: Role of Ah receptor and the dioxin-inducible [*Ah*] gene battery in toxicity, cancer and signal transduction. Ann.N.Y.Acad.Sci. 685 (1993) 624- 640.

Pappas, P., Vasiliou, V., Karageorgou, M., Stefanou, P. and Marselos, M.: Studies on the induction of rat class 3 aldehyde dehydrogenase. Adv.Exp.Med.Biol. 372 (1995) 143- 149.

Saari, J.C., Champer, R.J., Asson-Batres, M.A., Garwin, G.G., Huang, J., Crabb, J.W. and Milam, A.H.: Characterization and localization of an aldehyde dehydrogenase to amacrine cells of bovine retina. Visual Neuroscience 12 (1995) 263–272.

Sladek, N.E. and Lee, M.O.: The use of immortalized mouse L1210/OAP cells established in culture to study the major class 1 aldehyde dehydrogenase-catalyzed oxidation of aldehydes in intact cells. Adv.Exp.Med.Biol. 328 (1993) 51–62.

Swanson, H.I., and Bradfield, C.A.: The AH receptor: genetics, structure and function. Pharmacogenetics 3 (1993) 213–230.

Taylor, B.A.: Strain distribution and linkage tests of 7,12-dimethylbenzanthracene (DMBA) inflammatory response in mice. Life Sci. 10 (1971) 1127–1134.

Torronen, R., Nousiainen, U. and Hänninen, O.: Induction of aldehyde dehydrogenase activity by polycyclic aromatic hydrocarbons. Chem-Biol.Interact. 36 (1981) 33–34.

Vasiliou, V.: Effect of various chemicals on ALDH activities in the Wistar rat. Ph.D. Thesis, University of Ioannina, Greece, 1988.

Vasiliou, V. and Marselos, M.: Tissue distribution of inducible aldehyde dehydrogenase activity in the rat after treatment with phenobarbital or methylcholanthrene. Pharmacol.Toxicol. 64 (1989) 39–42.

Vasiliou, V., Puga, A. and Nebert, D.W.: Negative regulation of the murine cytosolic aldehyde dehydrogenase-3 (Aldh-3c) gene by functional CYP1A1 and CYP1A2 proteins. Biochem.Biophys.Res.Commun. 187 (1992) 413–419.

Vasiliou, V., Puga, A. and Nebert, D.W.: Mouse class 3 aldehyde dehydrogenases: Positive and negative regulation in gene expression. Advanc. Exp. Med. Biol. 4 (1993) 131–139.

Vasiliou, V., Weiner, H., Marselos, M. and Nebert, D.W.: Mammalian aldehyde dehydrogenase genes: clasification based on evolution, structure and regulation. Eur. J. Drug Metab. Pharmacokinet. 20 (1995) 53–64.

Westlund, P., Fylling, A.C., Cederlund, E. and Jornvall, H.: 11-Hydroxythromboxane B2 dehydrogenase is identical to cytosolic aldehyde dehydrogenase. FEBS Lett. 345 (1994) 99–103.

11

CLASS 1 AND CLASS 3 ALDEHYDE DEHYDROGENASE LEVELS IN THE HUMAN TUMOR CELL LINES CURRENTLY USED BY THE NATIONAL CANCER INSTITUTE TO SCREEN FOR POTENTIALLY USEFUL ANTITUMOR AGENTS

Lakshmaiah Sreerama and Norman E. Sladek

Department of Pharmacology
University of Minnesota Medical School
Minneapolis, Minnesota 55455

1. INTRODUCTION

With the ultimate goal of discovering new anticancer agents, the National Cancer Institute (NCI), through its Developmental Therapeutics Program, uses a semiautomatic procedure (Monks et al., 1991) to annually evaluate thousands of compounds for their ability to inhibit the growth of each of a panel of 60 human tumor cell lines (reviewed in Boyd and Paull, 1995). The panel includes nine subpanels, each representing a specific tumor type, viz., leukemia, melanoma, and cancers of the brain, breast, colon, kidney, lung (non-small cell), ovary, and prostate, and each consisting of at least six tumor cell lines, except for the prostate cancer subpanel which consists of only two cell lines. The results of these tests, viz., GI_{50}, TGI and LC_{50} values (the concentrations of an agent required to effect 50% growth inhibition, total growth inhibition and 50% cell-kill, respectively) have been, and are being, stored in an electronic database.

Positive and negative numbers termed deltas ($\log_{10}$ of individual values minus the mean of all $\log_{10}$ values) and reflecting the differential sensitivities of the 60 tumor cell lines to a given compound can be generated from the GI_{50}, TGI or LC_{50} values that quantify the cytotoxic action of that compound against each of the 60 tumor cell lines. Plotting these values along a vertical reference line that corresponds to the mean value results in a unique profile or fingerprint. Visual inspection of fingerprints, aka mean graphs, obtained for an antitumor agent of known clinical utility and for various test compounds, allows the identification of agents in the latter group that exhibit fingerprints similar to those exhibited by the former and that may thus be worthy of further testing. Such evaluations can also be made electronically and, indeed, a computer program that computes correlation co-efficients quantifying degrees of similarity, viz., COMPARE, has been written for this purpose (Paull et al., 1989).

A recently utilized variation of the foregoing is based on the molecular charac-terization of the 60 tumor cell lines used in drug evaluation. Thus, cellular levels of a known molecular determinant of cellular sensitivity to a therapeutically useful anticancer agent, e.g., a transport protein or an enzyme, is quantified, this information is used to con-struct a fingerprint or seed profile, and COMPARE analysis is conducted as before (Lee et al., 1994; Fitzsimmons et al., 1996). This type of analysis enabled NCI investigators to identify >20 agents previously not suspected of being potential passengers for a transport protein known to be an important determinant of cellular sensitivity to a number of thera-peutically useful anticancer agents, viz., the P-glycoprotein (P170) coded for by the *mdr-1* gene (Lee et al., 1994).

Class 1 and class 3 aldehyde dehydrogenases (ALDH-1 and ALDH-3, respectively) are established molecular determinants of cellular sensitivity to a subgroup of nitrogen mustards, viz., cyclophosphamide, 4-hydroperoxycyclophosphamide, mafosfamide, ifosfa-mide and 4-hydroperoxyifosfamide, that are of substantial value in the treatment of sev-eral cancers and that, collectively, are known as oxazaphosphorines (Sladek, 1993, 1994). This is because they catalyze the detoxification of these agents (Fig. 1). As compared to nitrogen mustards that are not detoxified by these enzymes, e.g., melphalan, chlorambucil and mechlorethamine, and, in the case of certain cancers, other antitumor agents that are also not detoxified by these enzymes, oxazaphosphorines usually exhibit a more favorable therapeutic index. This is largely because a) ALDH-1 and/or ALDH-3 are present in the cytosol of certain, otherwise vulnerable, critical normal cells, e.g., bone marrow stem cells and those of the stomach mucosa, and b) these enzymes are ordinarily absent, or present at very low levels, in certain types of tumors (Sladek 1993, 1994; Uckun et al., 1994; Sladek and Sreerama, 1995; Sreerama and Sladek, 1996).

Quantification of ALDH-1 and ALDH-3 expression in each of the NCI's panel of 60 human tumor cell lines and submission of this information to COMPARE analysis may lead to the identification of cytotoxic agents bioinactivated (as well as bioactivated) by ALDH-1 and/or ALDH-3. Such agents could be equally, or even more, therapeutically useful than are the oxazaphosphorines. Thus, with the ultimate objective of using COM-PARE analysis to identify compounds of this type that may be present amongst the more than 45,000 agents that the NCI has already tested for antitumor activity, we quantified ALDH-1 and ALDH-3 levels in each of the 60 human tumor cell lines that the NCI uses to screen for such activity. The results of this effort are presented herein.

2. MATERIALS AND METHODS

The 60 human tumor cell lines used by the NCI's Developmental Therapeutics Pro-gram to screen for anticancer agents were supplied by Dr. Anne Monks, Program Re-sources, Inc., NCI-Frederick Cancer Research and Development Center, Frederick, MD. Freshly harvested cells were washed (centrifugation) once with phosphate-buffered saline (PBS; an aqueous solution of 138 mM sodium chloride, 2.7 mM potassium chloride, 5.4 mM sodium phosphate and 3.2 mM potassium phosphate, pH 7.4) and the cell pellets thus obtained were then snap-frozen in liquid nitrogen and shipped to us in dry ice. Soluble (105,000 g supernatant) fractions were obtained by placing the cell pellets into ice-cold aqueous solutions of 1.15% KCl (w/v) and 1 mM EDTA, pH 7.4 (1×10^7 cells/ml), break-

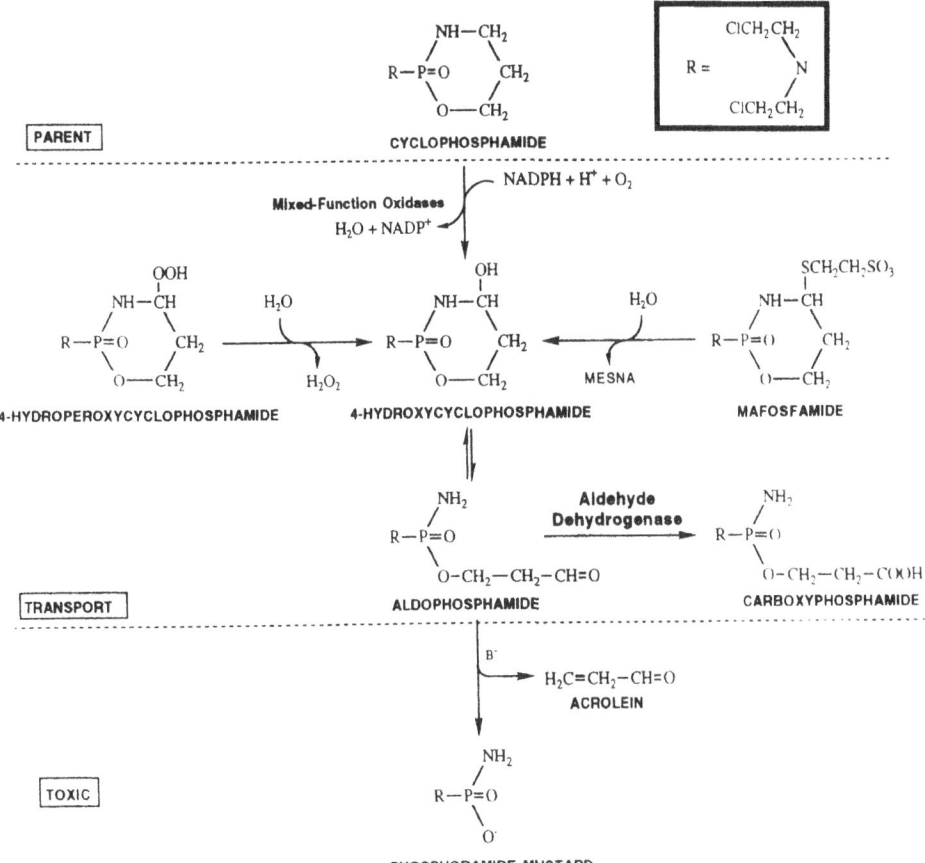

Figure 1. Salient features of oxazaphosphorine metabolism. Cyclophosphamide, mafosfamide and 4-hydroperoxycyclophosphamide are collectively known as oxazaphosphorines. Each is a prodrug and each gives rise to 4-hydroxycyclophosphamide which exists in equilibrium with its ring-opened tautomer, aldophosphamide. 4-Hydroxycyclophosphamide and aldophosphamide are, themselves, also without cytotoxic activity. However, aldophosphamide gives rise to acrolein and phosphoramide mustard, each of which is cytotoxic; the latter effects the bulk of the therapeutic action effected by the oxazaphosphorines. Alternatively, aldophosphamide can be further oxidized to carboxyphosphamide by certain aldehyde dehydrogenases, e.g.. ALDH-1 and ALDH-3. Carboxyphosphamide is without cytotoxic activity nor does it give rise to a cytotoxic metabolite. Aldehyde dehydrogenase-catalyzed oxidation of aldophosphamide to carboxyphosphamide is, therefore, properly viewed as an enzyme-catalyzed bioinactivation (detoxification) of the oxazaphosphorines. Ifosfamide and 4-hydroperoxyifosfamide (not shown) are also oxazaphosphorines. They are close structural analogs of cyclophosphamide and 4-hydroperoxycyclophosphamide, respectively, and undergo analogous reactions (Sladek, 1994).

ing up the suspended cells with the aid of a Dounce homogenizer, and then submitting the resultant homogenates to centrifugation at 105,000 g and 4°C for 1 hour (Sreerama and Sladek 1993).

4-Hydroperoxycyclophosphamide was supplied by Dr. J. Pohl, Asta Medica AG, Frankfurt, Germany. Aldophosphamide was generated in aqueous solution by chemical reduction of 4-hydroperoxycyclophosphamide; methyl sulfide was used as the reducing agent (Dockham et al., 1992).

Preparation of purified human stomach mucosa ALDH-1 and ALDH-3, and chicken antibodies specific for human liver ALDH-1 (anti-ALDH-1 IgY) and human stomach mucosa ALDH-3 (anti-ALDH-3 IgY), was as described previously (Dockham et al., 1992; Sreerama and Sladek, 1993).

Direct quantification of ALDH-1 and ALDH-3 activities in soluble (105,000 g supernatant) fractions prepared from 1 to 30×10^6 cells was by spectrophotometric assay as described by Dockham et al. (1992) and Sreerama and Sladek (1993). Acetaldehyde and NAD, 4 mM each, were used as the substrate and cofactor, respectively, to quantify ALDH-1 activity. Assuming that ALDH-1 and ALDH-3 were the only aldehyde dehydrogenases present in the 105,000 g supernatant fractions, most of the NAD-linked enzyme-catalyzed oxidation of acetaldehyde to acetic acid effected by the 105,000 g supernatant fractions obtained from the 60 human tumor cell lines was almost certainly due to ALDH-1 because, relative to ALDH-1, ALDH-3 catalyzes this reaction only very poorly (specific activities of the purified ALDH-1 and ALDH-3 preparations were, respectively, 2,850, and 102, mIU/mg protein) under the assay conditions that were used. Supporting the assumption were our observations that these were the only aldehyde dehydrogenases detected when 105,000 g supernatant fractions obtained from each of 14 tumor cell lines belonging to the lung and colon cancer subgroups were submitted to polyacrylamide gel isoelectric focusing and the developed gels were stained for aldehyde dehydrogenase activity (acetaldehyde and benzaldehyde, 4 mM in each case, were the substrates and 4 mM NAD was the cofactor). Benzaldehyde and NADP, 4 mM each, were used as the substrate and cofactor, respectively, to quantify ALDH-3 activity. Making the same assumption as above, essentially all of the NADP-linked enzyme-catalyzed oxidation of benzaldehyde to benzoic acid effected by the 105,000 g supernatant fractions was almost certainly due to ALDH-3 because, relative to ALDH-3, ALDH-1 catalyzes this reaction extremely poorly (specific activities of the purified ALDH-1 and ALDH-3 preparations were, respectively, 10, and 60,500, mIU/mg protein) under the assay conditions that were used. Aldophosphamide (160 μM) and NAD (4 mM) were used as substrate and cofactor, respectively, when the ability of enzymes present in 105,000 g supernatant fractions (largely, if not exclusively, ALDH-1 and ALDH-3) to catalyze the oxidative detoxification of oxazaphosphorines was quantified directly.

Indirect quantification of ALDH-1 and ALDH-3 activities in soluble (105,000 g supernatant) fractions was by an enzyme-linked immunosorbent assay (ELISA) essentially as described by Hornbeck et al. (1991). Briefly, 105,000 g supernatant fractions prepared from 0.025 to 10×10^6 cells were placed into a coating buffer (100 mM sodium carbonate, pH 9.6) and the proteins present in 100 μl of these preparations were coated onto the well surfaces of 96-well microtitration plates by incubating for two hours at 37°C. In order to avoid non-specific binding of antibodies to the well surfaces in the subsequent step, the plates were then thoroughly washed with washing buffer (PBS containing 0.05% Tween-20), 200 μl of blocking solution (PBS containing 0.25% BSA and 0.05% Tween-20) was added to each of the washed protein-coated wells, and the incubation was continued at 37°C for two more hours. The blocking solution was then emptied from the wells, 100 μl aliquots of antibodies (diluted 1:1000 with blocking solution) raised against human ALDH-1 (anti-ALDH-1 IgY) or human ALDH-3 (anti-ALDH-3 IgY) were added to the blocked wells, and incubation at 37°C was continued for two more hours. At the end of this incubation, the wells were washed with washing buffer, 100 μl aliquots of secondary antibody (anti-chicken IgG) conjugated to alkaline phosphatase (diluted 1:1000 with blocking solution) were added to the wells, and incubation at 37°C was continued for yet another two hours. The wells were then washed thoroughly with washing buffer, 100 μl aliquots of the substrate solution (0.1 mM p-nitrophenyl phosphate dissolved in Tris-buff-ered saline [138 mM sodium chloride, 2.7 mM potassium chloride and 25 mM Tris-HCl, pH 9.0]) were added to the wells, and incubation at 37°C was continued for 10 min. Absorbance due to the formation of p-nitrophenol was recorded at 405 nm with the aid of a microtitre plate reader connected to a Macintosh computer equipped with the SOFTmax program (UV-Max Kinetic Microplate Reader, Molecular Devices Corporation, CA). Cellular levels of ALDH-1 and ALDH-3 in terms of catalytic activities were estimated from standard curves generated with purified ALDH-1 and ALDH-3; the specific activities used for this purpose were 2,850, and 60,500, mIU/mg protein for ALDH-1 and ALDH-3, respectively.

LC$_{50}$ values quantifying the sensitivities of each of the NCI's 60 human tumor cell lines to cyclophosphamide, ifosfamide, 4-hydroperoxyifosfamide, melphalan, chlorambucil and mechlorethamine were downloaded from the NCI's database and mean graphs intended to highlight potential differential sensitivities on the part of the 60 human tumor cell lines to a given agent were constructed as described by Paull et al. (1989) with the aid of the Macintosh-based Cricket graph-III computer program (Computer Associates International, Inc., Islandia, NY)

Construction of mean graphs for each of four catalytic activities, viz., ALDH-1, ALDH-3, ALDH-1 + ALDH-3, and NAD-linked enzyme-catalyzed oxidation of aldophosphamide, was as follows. First, the catalytic activities of a given enzyme or combination of enzymes in each of the cell lines were each expressed as a $\log_{10}$ value and these values were averaged. Second, the mean $\log_{10}$ value for a given enzyme or combination of enzymes was subtracted from each of the individual $\log_{10}$ values for that enzyme or combination of enzymes to create positive and negative values termed deltas. Deltas were then plotted along a vertical reference line representing the mean $\log_{10}$ value of enzyme activity in the 60 cell lines. Positive values, projecting to the left of the vertical reference line, represent cellular levels of the enzyme or combination of enzymes that exceeded the corresponding mean value. Negative values, projecting to the right of the vertical reference line, represent cellular levels of the enzyme or combination enzymes that were less than the corresponding mean value. Thus, a bar projecting 1 unit to the left of the reference line denotes that the enzyme level in that cell line is 10 times greater than the corresponding mean value for the 60 cell lines.

Pearson correlation coefficients quantifying the relationships between cellular sensitivities to various agents and cellular levels of aldehyde dehydrogenase were calculated with the aid of the Macintosh-based Cricket graph-III computer program.

3. RESULTS AND DISCUSSION

Detectable levels of ALDH-1 and ALDH-3 were present in each of the 105,000 g supernatant fractions obtained from the panel of 60 human tumor cell lines. Values quantifying the cellular levels of either ALDH-1 or ALDH-3 activities were essentially independent of the assay, viz., direct (spectrophotometric) or indirect (ELISA), used to obtain them (r > 0.98) thus substantiating the contention made in Materials and Methods that, in the 60 human tumor cell lines used in this investigation, ALDH-1 and ALDH-3, respectively, account for virtually all of the NAD-linked enzyme-catalyzed acetaldehyde oxidation and NADP-linked enzyme-catalyzed benzaldehyde oxidation effected by the 105,000 g supernatant fractions obtained from these cell lines. ALDH-1 and ALDH-3 levels in the 60 human tumor cell lines ranged from 0.03 to 183, and 0.2 to 1407, mIU/10^7 cells, re-

Figure 2. ALDH-1 and ALDH-3 activity levels in each of the 60 human tumor cell lines currently used by the NCI to screen for potentially useful anticancer agents. Quantification of ALDH-1 and ALDH-3 activities present in soluble (105,000 g supernatant) fractions obtained from each of the 60 human tumor cell lines was by ELISA as described in Materials and Methods. Mean graphs of this data are presented in Fig. 3.

spectively (Fig. 2). Highest levels of each were found in the non-small cell lung cancer cell line subpanel. Lowest levels of each were found in the leukemia, and brain and breast cancer, cell line subpanels. Enzyme levels varied considerably within any given subpanel, e.g., ALDH-3 levels ranged from 0.3 to 1407 mIU/10^7 cells in the 9 cell lines comprising the non-small cell lung cancer subpanel. Interestingly, the relative amounts of ALDH-1 and ALDH-3 in each of the cell lines appeared to be somewhat related (r = 0.66). Mean

Figure 3. Mean graphs of ALDH-1, ALDH-3 and ALDH-1 + ALDH-3 activity levels in the 60 human tumor cell lines currently used by the NCI to screen for potentially useful anticancer agents. Shown are mean graphs, constructed as described in Materials and Methods, of the data presented in Fig. 2.

graphs of the data presented in Fig. 2 are shown in Fig. 3. Levels of NAD-linked enzyme-catalyzed oxidation of aldophosphamide, the pivotal oxazaphosphorine metabolite (Fig. 1), to carboxyphosphamide in 105,000 g supernatant fractions obtained from each of the 60 human tumor cell lines, and the mean graph thereof, are shown in Fig. 4. Each of these profiles was essentially identical to the corresponding profiles for ALDH-1-cata-lyzed oxidation of acetaldehyde (Fig. 2). This was as expected because it is highly un-

Figure 4. Levels of NAD-linked enzyme-catalyzed oxidation of aldophosphamide to carboxyphosphamide in each of the 60 human tumor cell lines currently used by the NCI to screen for potentially useful anticancer agents. Direct (spectrophotometric) quantification of NAD-linked enzyme-catalyzed oxidation of aldophosphamide to carboxyphosphamide by soluble (105,000 g supernatant) fractions obtained from each of the 60 human tumor cell lines (**A**), and construction of the mean graph of the results thereof (**B**), were as described in Materials and Methods.

likely that there are any relevant enzymes other than ALDH-1 and ALDH-3 present in the 105,000 g supernatant, or any other, fractions of these cells, and because ALDH-1 catalyzes the oxidation of aldophosphamide much more readily (Km = 52 μM, Vmax = 4,800 mIU/mg and Vmax/Km = 92 mIU/mg/μM) than does ALDH-3 (Km = 526 μM, Vmax = 405 mIU/mg and Vmax/Km = 0.77 mIU/mg/μM) (Dockham et al., 1992; Sreerama and Sladek, 1994).

The mean graphs shown in Figs. 3 and 4 can be used as seed profiles in COMPARE analysis to identify compounds in the NCI database that exhibit mean graphs of cellular sensitivities to such compounds (GI_{50}, TGI or LC_{50}) that resemble those of the seed profiles and, therefore, are likely to be detoxified by ALDH-1 and/or ALDH-3 and, consequently, to be selectively toxic to tumor cells. Thus, use of the ALDH-1 or the NAD-linked enzyme-catalyzed oxidation of aldophosphamide profile is expected to identify cytotoxic compounds that may be inactivated by ALDH-1, use of the ALDH-3 profile is expected to identify cytotoxic compounds that may be inactivated by ALDH-3, and use of the ALDH-1 + ALDH-3 profile is expected to identify cytotoxic compounds that may be inactivated by each of these enzymes.

A shortcoming of the NCI's drug evaluation protocol is that it is unable to identify potentially useful prodrugs that are activated in cells other than the target tumor cells. Cyclophosphamide and ifosfamide are prodrugs. They must first be oxidized (toxified) before they can effect any cytotoxic action or before irreversible inactivation (detoxification) catalyzed by ALDH-1 and/or ALDH-3 becomes a determinant of cellular sensitivity to them (Fig. 1). Oxidation of these agents is catalyzed by any of several mixed-function oxidases present largely in the liver (Chang et al., 1993). These enzymes are not found in most tumor cells. This is the reason why analogues of cyclophosphamide, e.g., 4-hydroperoxycyclophosphamide and mafosfamide, and of ifosfamide, e.g., 4-hydroperoxyifosfamide, that give rise to 4-hydroxycyclophosphamide and 4-hydroxyifosfamide, respectively, without the need of enzyme involvement (Fig. 1) must ordinarily be used in *in vitro* studies dealing with sensitivity of normal and tumor cells to the oxazaphosphorines (Sladek, 1994). As expected then, visual comparisons of the mean graphs of cellular levels of ALDH-1, ALDH-3, ALDH-1 + ALDH-3 (Fig. 3) or NAD-linked enzyme-catalyzed oxidation of aldophosphamide (Fig. 4) with those of the sensitivities (LC_{50} values) of the 60 human tumor cell lines to cyclophosphamide or ifosfamide (Fig. 5) did not suggest any correlation and statistical analysis of the regression line generated by these comparisons (Table 1) substantiated the visual impressions. Visual inspection of the mean graphs presented in Fig. 5 strongly suggests that a mixed-function oxidase that catalyzes the oxidation of both cyclophosphamide and ifosfamide, e.g., an isoform of the cytochrome P450 2C family (Chang et al., 1993), is, or two mixed-function oxidases, one of which catalyzes the oxidation of cyclophosphamide but not that of ifosfamide, e.g., cytochrome P450 2B6 (Chang et al., 1993), and the other of which catalyzes the oxidation of ifosfamide but not that of cyclophosphamide, e.g., cytochrome P450 3A4 (Chang et al., 1993), are, present in the renal carcinoma RXF-393 cell line, and that a mixed-function oxidase that catalyzes the oxidation of ifosfamide, but not that of cyclophosphamide, e.g., cytochrome P450 3A4, is present in the non-small cell lung carcinoma cell lines NCI-H522 and NCI-H226.

Although chlorambucil, melphalan and mechlorethamine, like the oxazaphosphorines, are chemically classified as nitrogen mustards and kill cells by cross-linking DNA, they and their metabolites, unlike the oxazaphosphorines, are not substrates for ALDH-1 and/or ALDH-3 and, thus, are not subject to detoxification catalyzed by these enzymes. Again as expected then, visual comparisons of the mean graphs of cellular levels of

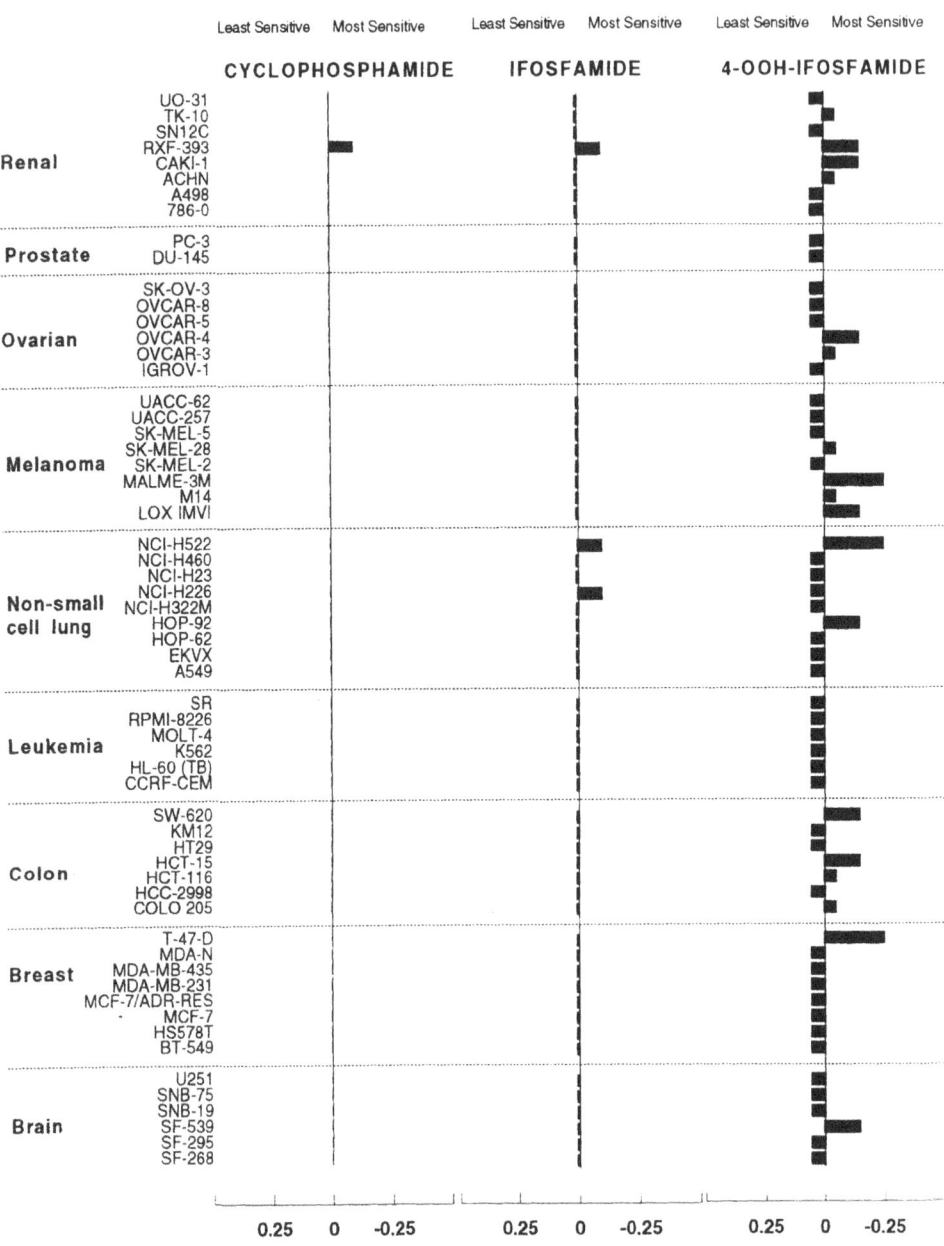

Figure 5. Mean graphs of LC_{50} values quantifying cellular sensitivities of the 60 human tumor cell lines currently used by the NCI to screen for potentially useful anticancer agents to cyclophosphamide, ifosfamide and 4-hydroperoxyifosfamide (4-OOH-ifosfamide). LC_{50} values were obtained from the NCI's database, and mean graphs thereof were constructed, as described in Materials and Methods.

Table 1. Lack of correlation between LC_{50} values quantifying the cellular sensitivities of the 60 human tumor cell lines currently used by the NCI to screen for potentially useful anticancer agents to selected anticancer agents of known clinical value, and cellular levels of aldehyde dehydrogenase

	Pearson Correlation Coefficient, r		
Anticancer Agent	ALDH-1	ALDH-3	ALDH-1 + ALDH-3
Cyclophosphamide	0.24	0.15	0.19
Ifosfamide	0.01	0.01	-0.01
4-Hydroperoxyifosfamide	0.00	0.06	0.01
Melphalan	0.14	0.22	0.18
Chlorambucil	0.09	0.13	0.10
Mechlorethamine	0.18	0.17	-0.21

*Publically available LC_{50} values ($\log_{10}$ M) quantifying the sensitivities of each of the 60 tumor cell lines to the anticancer agents listed in the table were downloaded from the electronically stored NCI database into a text format acceptable for Macintosh-based Cricket graph-III computer program analysis. Deltas ($\log_{10}$ of individual values minus the mean of all $\log_{10}$ values) reflecting the differential sensitivities of the 60 tumor cell lines were generated from the sets of LC_{50} values thus obtained (Figs. 5 and 6). Cellular levels of ALDH-1 and ALDH-3 activities in the panel of 60 tumor cell lines were determined by indirect assay, i.e., ELISA; deltas reflecting differential cellular levels of these enzymes were generated from these values (Figs. 3 and 4). Pearson correlation coefficients were calculated to quantify the relationships between cellular sensitivities to the listed agents and cellular levels of aldehyde dehydrogenase (essentially "COMPARE" analysis).

ALDH-1, ALDH-3, ALDH-1 + ALDH-3 (Fig. 3) or NAD-linked enzyme-catalyzed oxidation of aldophosphamide (Fig. 4) with those of the sensitivities (LC_{50} values) of the 60 tumor cell lines to chlorambucil, melphalan or mechlorethamine (Fig. 6) did not suggest any correlation and statistical analysis of the regression lines generated by these comparisons (Table 1) substantiated the visual impression. For the same reason, and particularly relevant to the notion presented herein, no correlation between cellular sensitivities to phosphoramide mustard or ifosfamide mustard and cellular levels of ALDH-1, ALDH-3, ALDH-1 + ALDH-3 and/or NAD-linked enzyme-catalyzed oxidation of aldophosphamide is expected. Testing of this expectation awaits the determination and/or release by the NCI of the relevant values quantifying cellular sensitivities to these agents.

ALDH-1 and ALDH-3 are established determinants of cellular sensitivity to the oxazaphosphorines. Assuming that no other major determinants of cellular sensitivity to these drugs are operative, the expectation was that visual comparisons of the mean graphs of cellular levels of ALDH-1, ALDH-3, ALDH-1 + ALDH-3 (Fig. 3) or NAD-linked enzyme-catalyzed oxidation of aldophosphamide (Fig. 4) with those of the sensitivities (LC_{50} values) of the 60 human tumor cell lines to 4-hydroperoxyifosfamide would suggest strong correlations and that statistical analysis of the regression lines generated from these comparisons would substantiate the visual impressions. These expectations were not realized (Figs. 3, 4 and 5, and Table 1). There are at least two possible reasons why they were not. The most obvious is that our assumption was incorrect, i.e., that, in addition to ALDH-1 and/or ALDH-3, there is at least one other major determinant of cellular sensitivity to the oxazaphosphorines operative in these cells. Alternatively, and much more likely, the *in vitro* LC_{50} values generated by the NCI may reflect cellular sensitivity to acrolein rather than to phosphoramide mustard, the oxazaphosphorine metabolite that largely effects cell-kill *in vivo,* as well as *in vitro* when drug-exposure is for 30 min or less, even though acrolein is a potent cytotoxic agent (Sladek, 1994). The drug evaluation protocol used by the NCI is such that the 60 human tumor cell lines are always continuously ex-

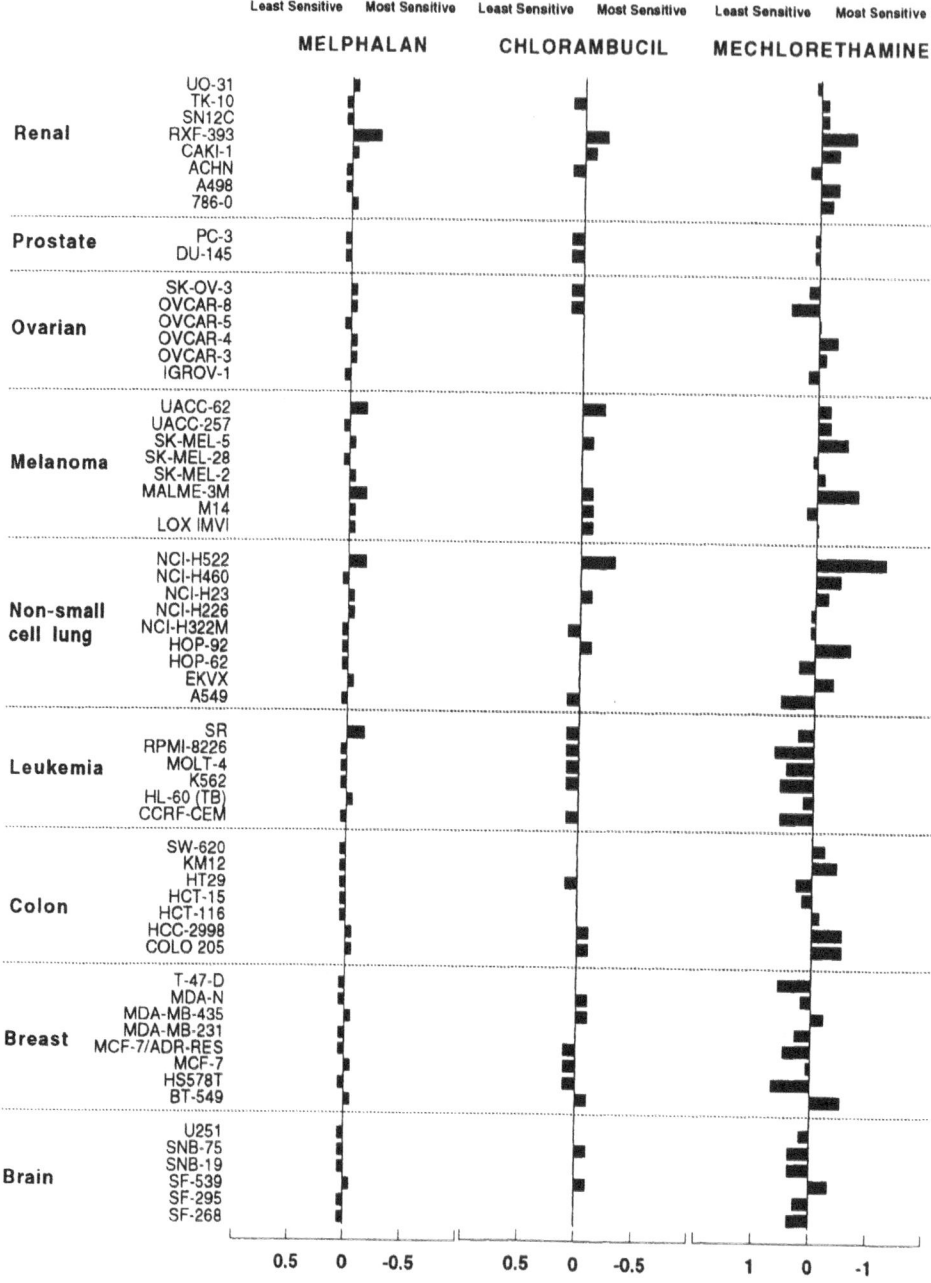

Figure 6. Mean graphs of LC_{50} values quantifying cellular sensitivities of the 60 human tumor cell lines currently used by the NCI to screen for potentially useful anticancer agents to melphalan, chlorambucil and mechlorethamine. LC_{50} values were obtained from the NCI's database, and mean graphs thereof were constructed, as described in Materials and Methods.

posed to the test agent, e.g., 4-hydroperoxyifosfamide, for 48 hours and the cells are then tested for viability (Monks et al., 1991). Whereas the fluid bathing tumor cells is in a continuous state of flux during drug exposure *in vivo*, it remains static during the 48-hour drug-exposure period that is standard in the NCI's *in vitro* drug evaluation protocol. As a consequence, breakdown products generated extracellularly are neutralized/removed indefinitely *in vivo*, whereas *in vitro* they are not. The small amount of acrolein that is released extracellularly in 30 min binds nonspecifically to various macromolecules and is, consequently, effectively all neutralized; however, as drug-exposure time increases beyond about 30 min, increasingly more of the oxazaphosphorine breaks down to acrolein and increasingly greater amounts of it are not neutralized. Thus, *in vitro*, cell-kill effected by extracellularly-generated acrolein becomes increasingly contributory to total cell-kill as drug-exposure time beyond 30 min increases. The foregoing deliberations illustrate yet another potential shortcoming in the NCI's drug evaluation protocol. Testing of the "NCI LC_{50} values reflect cellular sensitivity to acrolein" possibility awaits the determination and/or release by the NCI of the relevant values quantifying cellular sensitivities to acrolein. It follows, however, that, if cell-kill effected by acrolein is the explanation for the lack of correlation between cellular sensitivities (LC_{50} values) to 4-hydroperoxyifosfamide and aldehyde dehydrogenase levels in the 60 human tumor cell lines, mean graphs of the cellular sensitivities to 4-hydroperoxycyclophosphamide, mafosfamide or 4-hydroperoxyifosfamide (GI_{50}, TGI or LC_{50} values) as historically and presently determined by the NCI cannot be used as seed profiles in COMPARE analysis when the desire is to identify cytotoxic agents that are influenced by the same molecular determinants of cellular sensitivity that are operative *in vivo* in the case of the oxazaphosphorines. However, the use of mean graphs of cellular levels of ALDH-1, ALDH-3, ALDH-1 + ALDH-3 and/or NAD-linked enzyme-catalyzed oxidation of aldophosphamide as seed profiles in COMPARE analysis is likely to achieve this objective, and even if it doesn't (because additional determinants of cellular sensitivity to the oxazaphosphorines are operative), it should identify cytotoxic agents that are detoxified by these enzymes and thus have the potential of being selectively toxic to certain tumor cells.

COMPARE analyses using ALDH-1, ALDH-3, ALDH-1 plus ALDH-3 and/or NAD-linked enzyme-catalyzed oxidation of aldophosphamide seed profiles is also expected to identify agents in the NCI database that are activated (toxified) as a consequence of reactions catalyzed by ALDH-1 and/or ALDH-3, e.g., retinaldehyde which is oxidized, in a reaction catalyzed by ALDH-1, to retinoic acid, an agent that promotes cellular differentiation, therefore, cessation of indefinite proliferation (Dockham et al., 1992; Sladek and Lee, 1993). For such agents, a strong negative correlation between GI_{50}, TGI or LC_{50} values and cellular levels of ALDH-1, ALDH-3, ALDH-1 + ALDH-3 and/or NAD-linked enzyme-catalyzed oxidation of aldophosphamide would be obtained. The therapeutic potential of such agents appears to be substantially more problematic, however, because they are not likely to be selectively toxic to tumor cells; indeed, they are likely to be selectively toxic to critical normal cells, e.g., bone marrow stem cells, *vide supra*.

Further testing of the type described below is anticipated in those cases where a cytotoxic agent exhibiting a mean graph profile that is similar to one or more of the aldehyde dehydrogenase seed profiles is identified. Early on, whether such compounds are indeed substrates for ALDH-1 and/or ALDH-3 should be ascertained. *In vitro* drug sensitivity testing in an additional culture model, viz., a cell line that constitutively expresses none or only small amounts of the relevant enzyme but that can be induced to express large amounts of it, and, subsequently, *in vivo* drug sensitivity testing in relevant animal models, should follow.

4. ACKNOWLEDGMENTS

The authors gratefully acknowledge the assistance of Dr. George S. Johnson, Program Director, Biochemistry and Pharmacology, Grants and Contracts Operations Branch, National Cancer Institute, Bethesda, MD, in the procurement of the 60 human tumor cell lines. These investigations were supported by USPHS Grant CA 21737 and Department of Defense Grant DAMD 17–94-J-4057.

5. REFERENCES

Boyd, M.R. and Paull, K.D.: Some practical considerations and applications of the National Cancer Institute in vitro anticancer drug discovery screen. Drug Devel. Res., 34 (1995) 91–109.

Chang, T.K.H., Weber, G.F., Crespi, C.L. and Waxman, D.J.: Differential activation of cyclophosphamide and ifosphamide by cytochromes P-450 2B and 3A in human liver microsomes. Cancer Res., 53 (1993) 5629–5637.

Dockham, P.A., Lee, M.-O. and Sladek, N.E.: Identification of human liver aldehyde dehydrogenases that catalyze the oxidation of aldophosphamide and retinaldehyde. Biochem. Pharmacol., 43 (1992) 2453–2469.

Fitzsimmons, S. A., Workman, P., Grever, M., Paull, K., Camalier, R. and Lewis, A. D.: Reductase enzyme expression across the National Cancer Institute tumor cell line panel: correlation with sensitivity to mitomycin C and EO9. J. Natl. Cancer Inst., 88 (1996) 259–269.

Hornbeck, P., Winston, S.E. and Fuller, S.A.: Enzyme-linked immunosorbent assays (ELISA). In Ausubel, F.M., Brent, R., Kingston, R.E., Moore, D.D., Seidman, J.G., Smith, J.A. and Struhl, K. (Eds.), Current Protocols in Molecular Biology, Greene Publishing Associates and Wiley-Interscience. N. Y., 1991, pp 11.2.1–11.2.22.

Lee, J.-S., Paull, K., Alvarez, M., Hose, C., Monks, A., Grever, M., Tito Fojo, A. and Bates, S.E.: Rhodamine efflux patterns predict P-glycoprotein substrates in the National Cancer Institute drug screen. Mol. Pharmacol., 46 (1994) 627–638.

Monks, A., Scudiero, D., Skehan, P., Shoemaker, R., Paull, K., Vistica, D., Hose, C., Langley, J., Cronise, P., Vaigro-Wolff, A., Gray-Goodrich, M., Campbell, H., Mayo, J. and Boyd, M.: Feasibility of a high-flux anticancer drug screen using a diverse panel of cultured human tumor cell lines. J. Natl. Cancer Inst., 83 (1991) 757–766.

Paull, K.D., Shoemaker, R.H., Hodes, L., Monks, A., Scudiero, D.A., Rubinstein, L., Plowman, J. and Boyd, M.R.: Display and analysis of patterns of differential activity of drugs against human tumor cell lines: development of mean graph and COMPARE algorithm. J. Natl. Cancer Inst., 81 (1989) 1088–1092.

Sladek, N.E.: Oxazaphosphorine-specific acquired cellular resistance. In Teicher, B.A. (Ed.), Drug Resistance in Oncology, Marcel Dekker, N.Y., 1993, pp. 375–411.

Sladek, N.E.: Metabolism and pharmacokinetic behavior of cyclophosphamide and related oxazaphosphorines. In Powis, G. (Ed.), Anticancer Drugs: Reactive Metabolism and Drug Interactions, Pergamon Press, United Kingdom, 1994, pp. 79–156.

Sladek, N.E. and Lee, M.-O.: The use of immortalized mouse L1210/OAP cells established in culture to study the major class 1 aldehyde dehydrogenase-catalyzed oxidation of aldehydes in intact cells. Adv. Exp. Med. Biol., 328 (1993) 51–62.

Sladek, N.E. and Sreerama, L.: Cytosolic class-3 and class-1 aldehyde dehydrogenase activities (ALDH-3 and ALDH-1, respectively) in human primary breast tumors. Proc. Am. Assoc. Cancer Res., 36 (1995) 325.

Sreerama, L. and Sladek, N.E.: Identification and characterization of a novel class 3 aldehyde dehydrogenase overexpressed in a human breast adenocarcinoma cell line exhibiting oxazaphosphorine-specific acquired resistance. Biochem. Pharmacol., 45 (1993) 2487–2505.

Sreerama, L. and Sladek, N.E.: Identification of a methylcholanthrene-induced aldehyde dehydrogenase in a human breast adenocarcinoma cell line exhibiting oxazaphosphorine-specific acquired resistance. Cancer Res., 54 (1994) 2176–2185.

Sreerama, L. and Sladek, N.E.: Overexpression of glutathione S-transferases, DT-diaphorase and an apparently tumor-specific cytosolic class-3 aldehyde dehydrogenase by Warthin tumors and mucoepidermoid carcinomas of the parotid gland. Arch. Oral Biol., 41 (1996) In press.

Uckun, F.M., Chandan-Langlie, M., Dockham, P.A., Aeppli, D. and Sladek N.E.: Sensitivity of primary clonogenic blasts from acute lymphoblastic leukemia patients to an activated cyclophosphamide, viz., mafosfamide. Leukemia and Lymphoma, 13 (1994) 417–428.

12

SEX DIFFERENCES IN ENDOGENOUS RETINOID RELEASE IN THE POST-EMBRYONIC SPINAL CORD OF THE WESTERN MOSQUITOFISH, *Gambusia affinis affinis*

E. Rosa-Molinar,[1] P. J. McCaffery,[2] and B. Fritzsch[3]

[1]Department of Cell Biology and Anatomy
University of Nebraska Medical Center
Omaha, Nebraska 68198–6395
[2]Division of Developmental Neuroscience
E. K. Shriver Center
Waltham, Massachusetts, 02254
[3]Department of Biomedical Sciences
Creighton University
Omaha, Nebraska 68102

1. INTRODUCTION

Retinoic acid, a potent transcriptional activator, is believed to be an important factor in the regulation of vertebrate development (for review see Eichele, 1989; Gudas, 1994; Tabin, 1991; Boncinelli, et al., 1991; Kessel and Gruss, 1991a; 1991b; Kessel, 1992; Gudas, 1994). Retinaldehyde dehydrogenases catalyze the last step of retinoic acid synthesis and these enzymes are known to belong to a larger aldehyde dehydrogenases family (for review see Petersen and Lindahl, 1997). The distribution of endogenous retinaldehyde dehydrogenases in the developing animal sets up patterns of endogenous gradients of retinoic acid which modulate gene transcription and has been suggested to aid in establishing the developmental framework which organizes both the segmented body plan and the structure of individual organs (for review see Kessel and Gruss, 1991a; 1991b; Marsh-Armstrong, et al., 1995; 1994; Hyatt, et al., 1992; Watterson, et al., 1954; McCaffery, et al., 1991; 1992; McCaffery and Dräger, 1993; 1994; 1995; Dräger and McCaffery, 1995). This is clearly seen during the anteroposterior patterning of the trunk (Marsh-Armstrong, et al., 1994) in the Zebrafish, *Danio rerio*, in which an endogenous gradient of retinaldehyde dehydrogenase creates a matching gradient of retinoic acid along the anterior-to-posterior axis of the trunk, such as the pectoral, pelvic and anal fin. This occurs during the critical periods in embryogenesis when axial and appendicular structures are forming. Transcriptional regulation of the family of homeobox (Hox) genes, which are normally expressed in sequence along the length of the trunk (for review see Krumlauf, 1994; Krumlauf, et al., 1993; Kenyon, 1994; McGinnis and Krumlauf, 1992; Kessel and Gruss, 1991; Kessel, 1992), has been shown to be activated during development by retinoic acid in a sequential order, with 3' end genes being activated more rapidly after exposure to retinoic acid, and 5' end genes responding at progressively later times following retinoic acid exposure (for reviews see Krumlauf, 1994; Krumlauf, et al., 1993; Kenyon, 1994; McGinnis and Krumlauf, 1992; Kessel and Gruss, 1991; Kessel, 1992). Normal patterns of developmental gene expression, such as in the case of Hox genes, are altered in conditions of exogenous retinoic acid excess and deficiency (for review see Eichele, 1989; Gudas, 1994; Tabin, 1991; Boncinelli, et al., 1991; Kessel and Gruss, 1991a; 1991b; Kessel and Gruss, 1991; Kessel, 1992; Langston and Gudas, 1994), resulting in "homeotic" transformations, which change the normal order of vertebrae (changing the phenotype of anterior vertebra to that of posterior vertebrae), or the neurons in the hindbrain (Manns and Fritzsch, 1992).

However, retinoic acid has not been thought to play a role in the regulation of sexual dimorphism, a role considered to be the domain of steroids, specifically, testosterone and estrogen. The Western Mosquitofish, *Gambusia affinis affinis* (Baird and Girard, 1854), a sexually dimorphic internal fertilizing species of the family Poeciliidae (order Atherinomorpha; Parenti, 1981; 1993), shows distinct sexual dimorphism in the hemal spines of vertebrae 14–16 (Fig. 1).

Figure 1. Alcian blue and alizarin red S cleared and stained whole-mount preparation of the anal fin and its appendicular support of a 30.0 mm standard length (SL) mature female, scale bar=40μ (a) and a 30.0 mm SL mature male (b) *G. a. affinis*, scale bar=40μ; HS, hemal spine; PR, pleural ribs; PT, proximal pterygiophores; DT, distal pterygiophores; LIG, ligastyle. The orientation of the specimens are defined as follows: D, dorsal; V, ventral; C, caudal, R, rostral.

The development of the vertebral differences (Fig. 2) of first posteriorizing more anterior segments (formation of hemal spines more anteriorly) then anteriorizing more posterior segments (loss of hemal spines and formation of parapophysis with pleural ribs) is reminiscent of effects obtained by disturbing developmental programs by treatment with retinoic acid (Marshall, et al., 1992; Krumlauf, et al., 1993; Krumlauf. 1992; Slack and Tannahill, 1992).

For example, mice can transform cervical vertebrae (no ribs) into thoracic vertebrae (with ribs) if treated with retinoic acid, and this effect has been correlated with changes in the expression of homeobox genes (Krumlauf, et al., 1993; Krumlauf, 1992; Krumlauf, et al., 1993). Moreover, recent evidence suggests that lumbar vertebrae (no ribs) can be transformed into thoracic vertebrae (with ribs) by either over or underexpression of a single homeobox gene, either Hoxc1 {Hox 3.1} or Hoxc3 {Hox 3.3} (Krumlauf, et al., 1993; Krumlauf, 1992; Krumlauf, et al., 1993; Jegalian and DeRobertis, 1992; Pollock, et al., 1992).

We speculated that the homeotic transformation of caudal into precaudal vertebrae in *G. a. affinis* may represent a natural example of differential activation of one or more homeobox genes (Rosa-Molinar et al., 1994). A natural regulator of these may be retinoic acid. Could male/female differences in endogenous retinoic acid synthesis during post-embryonic development modulate the hormone-dependent sexual dimorphism of the genital area observed in *G. a. affinis*?

Figure 2. This scheme represents the changes in the post-anal region (trunk) from the embryonic to adult stages. An anteriorizing process creates in 6.5 mm SL fish an additional hemal spine in the 11th vertebral segment whereas a posteriorizing process produces an additional pleural rib at the 10th vertebral segment. In 12 mm SL fish, the hemal spines at the 11th and 12th vertebral segments are reabsorbed and pleural ribs are formed. During sexual maturation, the hemal spine on the 13th vertebrae is reabsorbed and a pleural rib is formed at this spine. This resorption occurs, if at all, much later in females (19 mm SL). In the adult male, the posteriorization has added a pleural rib at the 14th vertebral segment; however this occurs without the resorption of a hemal spine at this segment. Females may bear two segments with caudal (hemal spines) and precaudal (pleural ribs) characteristics (13 and 14) depending on whether or not hemal spine 13 was ever reabsorbed. The arrows underneath each drawing indicate the direction of the anteriorizing (from the right) and posteriorizing (from the left) process (from *Acta Anat* 151:20–35;1994, with permission).

The first step in answering this question is to determine if there are differences in endogenous levels of retinoic acid and the enzyme responsible for the synthesis of retinoic acid in female and male *G. a. affinis*. The second step in answering this question is to determine if gonadal hormones, specifically androgens, modulate endogenous levels of retinoic acid and the enzyme(s) responsible for the synthesis of retinoic acid. The present study demonstrates sex differences in the endogenous retinoic acid levels in the post-embryonic spinal cord during the development of the sexually dimorphic genital area of male and female *G. a. affinis*. Further, the androgen 17 α-methyltestosterone is able to regulate the activity of the spinal cord retinaldehyde dehydrogenase and alter the synthesis of retinoic acid.

2. SPATIAL AND TEMPORAL PATTERNS OF ENDOGENOUS RETINOIC ACID SYNTHESIS IN THE POST-EMBRYONIC SPINAL CORD OF IMMATURE *G. a. affinis*

To asses the spatial patterns of endogenous retinoic acid synthesis, the retinoic acid levels and the enzyme responsible for the synthesis of retinoic acid were measured in euthanized immature female (n=10) and male (n=10) *G. a. affinis* as described by McCaffery, et al (1992). We focused our attention on spinal cord segments 8–16 since the sexually dimorphic vertebrae lie in this area (Rosa-Molinar, et al., 1994). Work with *D. rerio* (Marsh-Armstrong, et al., 1995) indicated that the spinal cord expressed high levels of a retinaldehyde dehydrogenase and that this was the most likely source of retinoic acid acting on vertebrae development. Briefly, spinal cord segments 8–16 were dissected and placed in 200 µl of L15 media containing 20% fetal calf serum. The spinal cord segments were cultured overnight in L15 medium and later pelleted by centrifugation. The tissues were then homogenized and assayed for protein levels. Retinoic acid released into the culture medium (supernatant) was measured in volumes that were adjusted for equal tissue protein. The supernatants were titrated onto F9 teratocarcinoma cells transfected with a construct carrying the retinoic acid inducible response element (RARE) to drive *lacZ* (Wagner, et al., 1992). Equivalent dilutions along the linear part of the dose response curve were plotted. The intensity of the blue reaction was measured with a microtiter plate reader (ELISA reader, Fisher Biotech BT100; accuracy of the colorimetric readings ±1%). The enzyme(s) that may be responsible for the synthesis of retinoic acid was also examined using a zymography bioassay which involves examining protein fractions, which were separated by isoelectric focusing (McCaffery, et al., 1992). To correlate the developmental changes with the retinoic acid levels, female and male *G. a. affinis* were cleared following the Dingerkus and Uhler (1977) whole-mount procedure for the Alcian Blue staining of cartilage and Alizarin Red S staining of bone. All observations and photomicrography were performed using a SHZ Olympus dissecting microscope or an Olympus BHZ microscope with or without differential interference contrast optics.

Both the F9 reporter cell assay and the zymograpghic bioassay revealed that the endogenous retinoic acid levels and the levels of the endogenous enzyme(s) responsible for retinoic acid synthesis were higher in spinal cord segment 8–16 of immature males (Fig. 3a-3b) than in immature females (Fig. 4a-4b).

These levels of retinoic acid synthesis are significantly lower than those detected in embryonic spinal cord, either in *D. rerio* (Marsh-Armstrong, et al., 1995) or mouse (McCaffery and Dräger, 1994). Low retinaldehyde dehydrogenase activity could be detected and this was also higher in immature males (Fig. 3b) than in immature females (Fig. 4b).

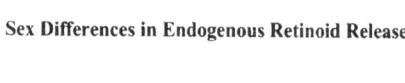

Figure 3. Endogenous retinoic acid levels (a), enzyme activity (b), and alcian blue and alizarin red S cleared and stained whole-mount preparation of the anal fin and its appendicular support of a 13.0 mm SL immature male, scale bar=40μ (c) *G. a. affinis*; HS, hemal spine; PR, pleural ribs; PT, proximal pterygiophores; DT, distal pterygiophores; LIG, ligastyle.

Figure 4. Endogenous retinoic acid levels (a), enzyme activity (b), and alcian blue and alizarin red S cleared and stained whole-mount preparation of the anal fin and its appendicular support of a 13.0 mm SL immature female, scale bar=40µ (c) *G. a. affinis*; HS, hemal spine; PR, pleural ribs; PT, proximal pterygiophores; DT, distal ptery-giophores. Note that the hemal spine of the 13th vertebrae is still attached (arrow).

The higher levels of enzyme(s) activity and endogenous retinoic acid in the immature male (Fig. 3a-3b) may, in part, regulate the transformation and transposition of the anal fin and its appendicular support. This developmental event is revealed in the alcian blue and alizarin red S cleared and stained whole-mount preparations of the genital area (Fig. 3c-4c). The developmental changes observed in the anal fin and its appendicular support of the immature males and females were similar to those previously described (Rosa-Molinar, et al., 1994); alcian blue and alizarin red S cleared and stained whole-mount *G. a. affinis* revealed that the tip of hemal arch of vertebra 13 was disconnected in males, and this free element (ligastyle) will be later transposed first to vertebra 12 and later 11 (Fig. 3c).

3. SPATIAL AND TEMPORAL PATTERNS OF ENDOGENOUS RETINOIC ACID SYNTHESIS IN THE POST-EMBRYONIC SPINAL CORD OF IMMATURE *G. a. affinis* FOLLOWING ANDROGEN EXPOSURE

We previously described in both female and male *G. a. affinis* a developmental program that creates 6 vertebrae (11–16) which are markedly different (at least at one stage in development) from each other and from pre-caudal and caudal vertebrae (Rosa-Molinar, et al., 1994). Most of these unique vertebrae in this area have both pleural ribs and hemal spines (at least at one point in their development) and most are responsive to androgens. We termed this third area, including these vertebrae and the associated anal fin, the genital area. The development of the third region, the genital area, begins, as observed by Turner (Turner, 1941; 1942; 1960), with the resorption of the hemal spine of vertebra 13 (Rosa-Molinar, et al., 1994). This is only one of three hemal spines to be resorbed in the process of anteriorizing caudal vertebrae (Rosa-Molinar, et al., 1994). Resorption of the 13th hemal arch in the male coincides with the onset of a sex-specific developmental program that causes the addition of material at the anterior tips of the hemal arches at vertebrae 14–16. We suspected that the growth and elongation of the 14th-16th hemal spines as well as the transposition of the anal fin and its appendicular support are influenced by male gonadal hormone (Rosa-Molinar, et al., 1994).

Therefore, we investigated the effects of 17α-methyltestosterone on the spatial patterns of the enzyme(s) responsible for retinoic acid synthesis as well as levels of endogenous retinoic acid. Immature female and male *G. a. affinis* were given *per os* fish diets coated with either 17α-methyltestosterone (40μg/gm diet) or cod-liver oil. The treatment was given by feeding the fish at 5% of their body weight, and the ration was divided into three feedings over a five day period. The fish were then euthanized, and the zymography bioassay, F9 reporter cell assay, and alcian blue and alizarin red S clear and stain procedures were performed as described previously (McCaffery et al., 1992; Dingerkus and Uhler, 1977). The F9 reporter cells revealed that endogenous retinoic acid levels were relatively unchanged in spinal cord segments 8–16 of immature males treated with 17α-methyltestosterone (Fig. 5a).

However, the F9 reporter cells revealed that the endogenous retinoic acid levels in spinal cord segments 8–16 of immature females treated with 17α-methyltestosterone were markedly higher than levels observed in the immature males (Fig. 6a).

Retinaldehyde dehydrogenase activity was also determined and in both immature males (Fig. 5b) and immature females (Fig. 6b) the endogenous enzyme activity correlated with levels of endogenous retinoic acid (Fig. 5b-6b). The activity of the enzyme(s) respon-

.

Figure 5. Endogenous retinoic acid levels (a), enzyme activity (b), and alcian blue and alizarin red S cleared and stained whole-mount preparation of the anal fin and its appendicular support of a 13.0 mm SL immature male, scale bar=40μ (c) *G. a. affinis* treated with 17α-methyltestosterone; HS, hemal spine; PR, pleural ribs; PT, proximal pterygiophores; DT, distal pterygiophores. Note that the hemal spine of the 13th vertebrae is still attached (arrow).

Figure 6. Endogenous retinoic acid levels (a), enzyme activity (b), and alcian blue and alizarin red S cleared and stained whole-mount preparation of the anal fin and its appendicular support of a 13.0 mm SL immature female, scale bar=40µ (c) *G. a. affinis* treated with 17α-methyltestosterone; HS, hemal spine; PR, pleural ribs; PT, proximal pterygiophores; DT, distal pterygiophores. Note that the hemal spine of the 13th vertebrae is still attached (arrow).

sible for the synthesis of endogenous retinoic acid was higher in spinal cord segments 8–16 of immature females treated with 17α-methyltestosterone (Fig. 6b) than in immature males treated with 17α-methyltestosterone (Fig. 5b).

Alcian blue and alizarin red S cleared and stained whole-mount preparations of immature male and female *G. a. affinis* treated with 17α-methyltestosterone for five days revealed the masculinization of immature females (Fig. 6c) as demonstrated by the growth, elongation, and bending of the hemal spines of vertebra 14–16 (Fig. 6c). 17α-methyltestosterone treatment induced typical but precocious growth, elongation and bending (Fig. 5c) of the hemal spines of vertebrae 14–16 in immature males.

Following extended treatment with 17α-methyltestosterone (26 days), immature female *G. a. affinis* were completely masculinized (Fig. 7). Alcian blue and alizarin red S cleared and stained whole-mount preparations of immature female *G. a. affinis* showed extensive transformation (growth and elongation) of anal fin lepidotrichia, fusion of the proximal pterygiophores 3–5, and the elongation and bending of the 14th–16th vertebral hemal spines (Fig. 7). Alcian blue and alizarin red S cleared and stained whole-mount preparations of immature male *G. a. affinis* showed that extended treatment with 17α-methyltestosterone had no effect on the elements of the anal fin and its appendicular support. However, the transposition of the anal fin appendicular support was prevented in both male and female *G. a. affinis* treated with 17α-methyltestosterone due to the partial or complete mineralization of the 13th hemal spine which blocked the anterior shifting of the anal fin (Fig. 7).

Figure 7. Alcian blue and alizarin red S cleared and stained whole-mount preparation of the anal fin and its appendicular support of a 19.0 mm SL immature female, scale bar = 25μ *G. a. affinis* following 26 days treatment with 17α-methyltestosterone. Note the calcified hemal spine on the 13th vertebra (arrow); HS, hemal spine; PR, pleural ribs; PT, proximal pterygiophores; DT, distal pterygiophores.

4. POSSIBLE ROLES OF ENDOGENOUS RETINOIC ACID IN THE POST-EMBRYONIC SPINAL CORD

In this report the spatial and temporal patterns of endogenous levels of retinoic acid and the endogenous enzyme(s) responsible for the synthesis of retinoic acid within the post-embryonic spinal cord, specifically spinal cord segment 8–16 of immature male and female *G. a. affinis* were described. This spinal cord segment corresponds to a sexually dimorphic region of the *G. a. affinis* axial body plan, the genital area (Rosa-Molinar, et al., 1994). Within spinal cord segment 8–16, endogenous levels of retinoic acid and the enzyme(s) activity responsible for the synthesis of retinoic acid are higher in immature males than in immature females. The differences in the endogenous retinoic acid levels and the enzyme(s) correspond with a sex-specific transformation and transposition of the males' anal fin and its appendicular support. This event is essential in the development of the male sex-specific genital system and is an androgen dependent event. This finding led to the examination of the endogenous retinoic acid levels of immature females and males treated with 17α-methyltestosterone. The data show that endogenous retinoic acid levels were much higher in immature females treated with 17α-methyltestosterone than in immature males. Levels of retinaldehyde dehydrogenase activity increase following 17 α-methyltestosterone treatment, although not to the same extent as the rise in retinoic acid synthesis. This may suggest an additional contributing factor accounting for the increase in retinoic synthesis such as a higher availability of retinaldehyde substrate. This has been previously shown to account for high retinoic acid synthesis in the trunk of *D. rerio* (Marsh-Armstrong, et al., 1995).

The retinaldehyde dehydrogenase described in the *G. a. affinis* spinal cord appears to be very similar to that of *D. rerio* based on its isoelectric point (Fig. 8). However, enzyme activity in post-embryonic *G. a. affinis* is substantially lower than that present in embryonic *D. rerio*. This likely reflects the general decrease in retinaldehyde dehydrogenase activity during development of the vertebrate spinal cord (McCaffery and Dräger, 1994).

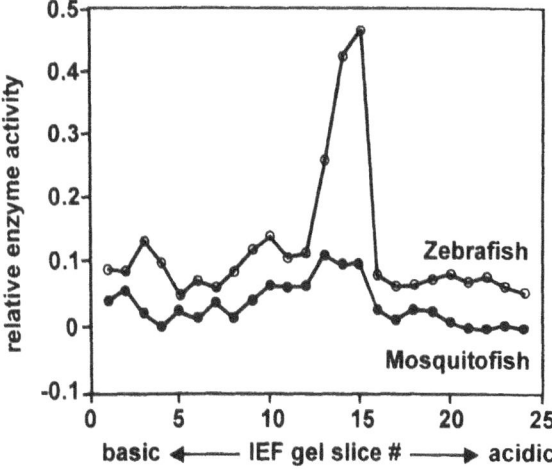

Figure 8. Zymography bioassay comparing the isoelectric points of endogenous enzyme(s) responsible for the synthesis of retinoic acid in trunk (n=40) of *D. rerio* and spinal cord (n=10) of *G. a. affinis.*

The influence of exogenous 17 α-methyltestosterone on the development of the sexually dimorphic area was shown using alcian blue and alizarin red S cleared and stained immature female and male *G. a. affinis*. Our data show that exogenous treatment of immature females with 17 α-methyltestosterone not only results in the 1.) masculinization of the axial and anal fin and its appendicular support elements and 2.) premature calcification of the 13th hemal spine which results in blocking the transposition of the anal fin and its appendicular support, but also that there is masculinization of the retinoic acid synthesis and enzyme activity. In addition, the blockage of the transposition of the anal fin and its appendicular support in both male and female *G. a. affinis* treated with 17α-methyltestosterone support our previous hypothesis (Rosa-Molinar, et al., 1994) that the timing of androgen exposure is critical to the development of a functional male genital area in *G. a. affinis*.

5. SUMMARY

In this study we have shown sex differences in endogenous retinoic acid synthesis and retinaldehyde dehydrogenase activity in the post-embryonic spinal cords of immature female and male *G. a. affinis*. The F9 reporter cell assay and the zymography bioassay showed that the endogenous retinoic acid levels correlated with the levels of the endogenous enzyme(s) responsible for retinoic acid synthesis. These data also showed that both the endogenous retinoic acid levels and the enzyme(s) were higher in spinal cord segments 8–16 of immature males than in immature females. We have also shown that exogenous treatment of 17 α-methyltestosterone results in the masculinization of the immature female's anal fin and its appendicular support elements as well as the endogenous synthesis of retinoic acid and retinaldehyde dehydrogenase activity. The F9 reporter cell assay showed that the endogenous retinoic acid levels were relatively unchanged in spinal cord segments 8–16 of immature males treated with 17α-methyltestosterone. However, the F9 reporter cell assay showed that the endogenous retinoic acid levels in spinal cord segments 8–16 of immature females treated with 17α-methyltestosterone were markedly higher than levels observed in the immature males. The data also showed that the activity of the enzyme(s) responsible for the synthesis of endogenous retinoic acid was higher in spinal cord segments 8–16 of immature females treated with 17α-methyltestosterone than in immature males treated with 17α-methyltestosterone. Currently under investigation is the question of what role the endogenous enzyme(s) responsible for the synthesis of retinoic acid plays either alone or in concert with androgen in organizing hormone-dependent sexually dimorphic areas in the teleost body plan.

6. ACKNOWLEDGMENTS

The animal use, care, and procedures described in this study were conducted under guidelines established by the National Institutes of Health and the University of Nebraska Medical Center/University of Nebraska at Omaha Institutional Animal Care and Use Committee (IACUC # 91–020–09 and 92–068–01). All fish were collected under a state permit (Scientific permit No's. 96–114) issued by the Nebraska Game and Parks Commission. We thank Drs. J. Taylor, R. McCue, and D. Nichols for their editorial comments on this manuscript; Dr. J. Johar and the Departments of Chemistry and Biology at Wayne State College for providing valuable laboratory space; Dr. Michael Wagner and Dr.

Thomas Jessell for the retinoic acid reporter cell line; B. Proskocil, J. Cooper, D. Jacques, D. Fitzpatrick, J. Schneider, and M. Guzman de Christensen for technical assistance. E. R-M is supported by an APA/MFP Pre-doctoral Fellowship in Neuroscience (NIM-HIT32MH18882), Blanche Widaman Fellowship of the University of Nebraska Medical Center, and a grant by NSF (EPSCOR/OSR9–255225). Additional funds were provided by the University of Nebraska Medical Center Student Research Forum and the Creighton-Nebraska Combined Department of Psychiatry, University of Nebraska Medical Center. P. J. M. is supported by a grant from NIH (HD05515).

7. REFERENCES

Baird, S. F., Girard G.: Descriptions of new species of fishes collected by Mr. J. H. Clark on the U. S. and Mexican boundary survey under Lt. Col. J. D. Graham. *Proc Acad Nat Sci Phila* 6 (1854) 390.

Boncinelli, E., Simeone, A., Acampora, D., and Mavillo, F.: Hox gene activation by retinoic acid. *Trends in Genetics* 7 (1991) 329–3341.

Dräger, U. C. and McCaffery, P.: Retinoic acid synthesis in the developing spinal cord. In Weiner, H., Holmes, R. and Wermuth, B. (eds), *Enzymology and Molecular Biology of Carbonyl Metabolism*. Plenum Press, N. Y., 1995, pp. 185–192.

Dinkergus, G. and Uhler, L. D.: Enzyme clearing of alcian blue stained whole small vertebrates for the demonstration of cartilage. *Stain Techol* 52 (1977) 229–232.

Eichele, G.: Retinoids and vertebrate limb pattern formation. *Trends in Genetics* 5 (1989) 246–251.

Goodrich E. S.: *Studies on the structure and development of vertebrates*. Chicago, University of Chicago Press, 1930, pp. 113–158.

Gudas, L. J.: Retinoids and vertebrate development. *J Biol Chem* 269 (1994) 15399–14402.

Hyatt, G., Schmitt, E. A., Marsh-Armstrong, N. R., and Dowling, J. E.: Retinoic acid-induced duplication and the zebrafish retina. *Proc Natl Acad Sci* 89 (1992) 8293–8297.

Jegalian B. G., De Robertis E. M.: Homeotic transformation in the mouse induced by overexpression of a human Hox 3.3 transgene. *Cell* 71 (1992) 901–910.

Kenyon C.: If birds can fly, why can't we? Homeotic genes and evolution. *Cell* 78 (1994) 175–180.

Kessel M., Gruss P.: Homeotic transformations of murine vertebrae and concomitant alteration of Hox-codes induced by retinoic acid. *Cell* 67 (1991a) 89–104.

Kessel, M.: Respecification of vertebral identities by retinoic acid. *Development* 115 (1992) 487–501.

Kessel, M. and Gruss, P: Homeotic transformation of murine vertebrae and concomitant alteration of Hox codes induced by retinoic acid. *Cell* 67 (1991b) 89–104.

Krumlauf R.: Hox genes in vertebrate development. *Cell* 78 (1994) 191–201.

Krumlauf R., Marshall H., Studer M., Nonchev S, Sham MH, Lumsden, A: Hox homeobox genes and regionalization of the nervous system. *J Neurobiol* 24 (1993) 1328–1340.

Krumlauf R.: Evolution of the vertebrate Hox homeobox genes. *Bioessays* 14 (1992) 245–251.

Langston, A. W. and L. J. Gudas: Retinoic acid and homeobox gene regulation. *Curr Opin Genetics and Dev* 4 (1994) 550–555.

Manns, M. and B. Fritzsch: Retinoic acid affects the organization of reticulospinal neurons in developing *Xenopus*. *Neurosci Letters* 139 (1992) 253–256.

Marshall H, Nonchev S, Sham MH, Muchamore I, Lumsden A, Krumlauf R: Retinoic acid alters the hindbrain Hox code and induces the transformation of rhombomeres 2/3 into a rhombomere 4/5 identity. *Nature* 360 (1992) 737–741.

Marsh-Armstrong, N., McCaffery, P., Hyatt, G., Alonso, L. Dowling, J. E., Gilbert, W., and Dräger, U. C.: Retinoic acid in the anteroposterior patterning of the zebrafish trunk. *Roux's Arch Dev Biol* 205 (1995) 103–113.

Marsh-Armstrong, N., McCaffery, P., Gilbert, W., Dowling, J. E. and Dräger, U. C. : Retinoic acid is necessary for the development of the ventral retina in zebrafish. *Proc Natl Acad Sci* 91 (1994) 7286–7290.

McCaffery, P. and Dräger, U. C.: Retinoic acid synthesizing enzymes in the embryonic and adult vertebrate. In Weiner, H., Crabb, D.W., and Flynn, T.G. (eds), *Enzymology and Molecular Biology of Carbonyl Metabolism*. Plenum Press, N. Y., 1995, pp. 173–183.

McCaffery, P. and Dräger, U. C.: Retinoic acid synthesis in the developing retina. In Weiner, H., Crabb, D.W., and Flynn, T.G. (eds), *Enzymology and Molecular Biology of Carbonyl Metabolism*. Plenum Press, N.Y., 1993, pp. 181–190.

McCaffery, P. and Dräger, U. C.: Hot spots of retinoic acid synthesis in the developing spinal cord. *Proc Natl Acad Sci* 91 (1994) 7194–7197.

McCaffery, P., Lee, M., Wagner, M. A., Sladek, N. E., and Dräger, U. C.: Asymmetrical retinoic acid synthesis in the dorsoventral axis of the retina. *Development* 115 (1992) 371–382.

McCaffery, P., Tempst, P., Lara, G., and Dräger, U. C.: Aldehyde dehydrogenase is a positional marker in the retina. *Development* 112 (1991) 693–702.

McGinnis, W. and Krumlauf, R.: Homeobox genes and axial patterning. *Cell* 68 (1992) 283–302.

Parenti, L. R.: Relationship of atherinomorph fishes (Teleostei). *Bull Mar Sci* 52 (1993) 170–196.

Parenti, L. R.: A phylogenetic and biogeographic analysis of cyprinodontiform fishes (Teleostei, Atherinomorpha). *Bull Am Mus Nat Hist* 168 (1981) 335–557.

Pereira, F. A., Rosenmann, E., Nylen. E. G., and Wrogemann, K.: Human cytosolic aldehyde dehydrogenase in androgen insensitivity syndrome. In Weiner, H., Crabb, D.W., and Flynn, T.G. (eds), *Enzymology and Molecular Biology of Carbonyl Metabolism*. Plenum Press, N.Y., 1993, pp. 45–50.

Petersen, D. and Lindahl, R.: Aldehyde dehydrogenase. In F. P. Guengerich (ed), *Comprehensive Toxicology. Vol. 3 Biotransformation*, Elisever Science, Oxford, 1997.

Pollock, R. A., Jay, G., and Biebrich C. J.: Altering the boundaries of Hox 3.1 expression: Evidence for antipodal gene regulation. *Cell* 71 (1992) 911–923.

Rosa-Molinar E., Hendricks, S. E., Rodriguez-Sierra, J. F., and Fritzsch, B: Development of the Anal Fin Appendicular Support in the Western Mosquitofish, *Gambusia affinis affinis* (Baird and Girard, 1854): A Reinvestigation and Reinterpretation. *Acta Anat* 151 (1994) 20–35.

Slack J.M.W., Tannahill D.: Mechanism of anteroposterior axis specification in vertebrates. *Development* 114 (1992) 285–302.

Tabin, C. J.: Retinoids, homeoboxes, and growth factors, Towards molecular models for limb development. *Cell* 66 (1991) 199–217.

Turner C. L.: Morphogenesis of the gonopodium in *Gambusia affinis affinis*. *J Morphol* 69 (1941)161–185.

Turner C. L.: Morphogenesis of the gonopodial suspensorium in *Gambusia affinis affinis* and the induction of male suspensorial characters in the female by androgenic hormones. *J Exp Zool* 91 (1942) 167–193.

Turner C. L.: The effects of steroid hormones on the development of some secondary sexual characters in cyprinodont fishes. *Trans Am Microsc Soc* 79 (1960) 320–333.

Wagner, M., Han, B. and Jessell, T. M.: Regional differences in retinoid release from embryonic neural tissue detected by an in vitro reporter assay. *Development* 116 (1992) 55–66.

Watterson, R. L., Fowler, I. and Fowler, B. J.: The role of the neural tube and the notochord in development of the axial skeleton of the chick. *Am J Anat* 95 (1954) 337–399.

13

ORGAN DISTRIBUTION OF ALDEHYDE DEHYDROGENASE ACTIVITY IN THE NEWT, *Notophthalmus viridescens*

Harry E. Settles,[1] Josette S. Boesch,[2] and Ronald G. Lindahl[2]

[1]Department of Anatomy and Structural Biology and
[2]Department of Anatomy Biochemistry and Molecular Biology
The University of South Dakota School of Medicine
Vermillion, South Dakota 57069

1. INTRODUCTION

Aldehyde dehydrogenases are NAD-dependent enzymes that catalyze the oxidation of aldehydes to carboxylic acids. Three major classes of mammalian aldehyde dehydrogenase (ALDH) are recognized (Lindahl, 1992). Class 1 is cytosolic and NAD-specific. Class 2 is mitochondrial, also uses NAD, and preferentially functions at micromolar concentrations of small aliphatic aldehydes. Class 1 and 2 are tetramers of identical subunits. The monomers have a molecular weight of 55 kD. Class 3 ALDH preferentially uses NAD, but can use NADP. Class 3 oxidizes aromatic aldehydes such as benzaldehyde. The enzyme is a dimer of identical monomers with molecular weights of 50kD.

Aldehyde dehydrogenases have been widely studied in mammals and in several lower animals. While many phylogenetic gaps are being filled with regard to studies of aldehyde dehydrogenase enzymology and molecular biology (Hempel et al., 1993), we know of only one study in an amphibian (Nilsson, 1989). We have begun studies in the newt, *Notophthalmus viridescens*, in order to characterize its ALDH phenotypes and to isolate their corresponding genes. We report here the specific activity of ALDH in a variety of newt tissues and the results of western immunoblots for classes 1 and 2 aldehyde dehydrogenase. The specific activities with propionaldehyde and benzaldehyde indicate the presence of class 1 ALDH. The western immunoblots are also consistent with class 1 ALDH. We also include here data on histochemistry in newt liver, stomach and intestine which show the locations of ALDH activity in these tissues.

2. MATERIALS AND METHODS

Adult newts, 8–10cm in length, were maintained in a 2 gallon aquarium. The newts were fed freeze dried tubifex daily and the water was changed daily. Newts were anesthe-

Table 1. Aldehyde dehydrogenase activity in several newt tissues

	Intestine		Stomach		Kidney		Lung		Liver	
	P/NAD[2]	B/NADP[3]	P/NAD	B/NADP	P/NAD	B/NADP	P/NAD	B/NADP	P/NAD	B/NADP
Mean specific Activity[1]	29.5	11.0	12.6	4.0	12.0	2.5	10.8	1.8	10.5	2.4
Assays	17	3	37	7	8	3	8	2	33	9
Standard deviation	9.9	4.8	6.2	2.0	3.0	0.7	8.8	0.8	5.6	1.5

[1]mIU/mg protein.
[2]Propionaldehyde as substrate with NAD as cofactor.
[3]Benzaldehyde as substrate with NADP as cofactor.

tized in 1% 3-aminobenzoic acid ethyl ester prior to obtaining tissues. Specific activities were assayed spectrophotometrically using propionaldehyde or benzaldehyde as substrates and NAD or NADP as cofactors (Lindahl and Evces, 1984). Assays were also done with propionaldehyde diluted 1 to 100. Western immunoblots used anti-class 1 and anti-class 2 antibodies which were raised against rat ALDH. Histochemical localization was done using propionaldehyde as the substrate and NAD as the cofactor according to the protocol described by Lindahl et al. (1983).

3. RESULTS

3.1. Specific Activities

Of the tissues examined, intestine possessed the highest ALDH specific activity (Table 1). The activity of the intestine was more than double the activities of the next group of tissues, including stomach, kidney, lung and liver. ALDH activity was from 2 to 5 times higher with propionaldehyde as substrate than with benzaldehyde as substrate in all tissues assayed. Newt ALDH activity was largely NAD dependent. In assays with micromolar concentrations of propionaldehyde, the activities were reduced to approximately one-third compared to assays done at millimolar substrate concentrations.

3.2. Western Immunoblots

Western immunoblot analysis of various newt tissues detected a polypeptide of approximately 61 kD that cross reacted with anti-rat class 1 ALDH antibodies. No cross reaction was detectable between any newt tissue and anti-rat class 2 ALDH antibodies.

3.3. Histochemistry

Histochemical analysis of newt tissues indicated that ALDH activity was located in the hepatocytes of the liver and was particulate in the hepatocytes (Fig. 1A). The ALDH activity was located in the epithelial lining of the stomach and intestine (Fig. 1B and 1C). Interestingly, the activity was located basally and was diffuse in the stomach epithelium while the activity was located apically and was particulate in the intestine epithelium.

Figure 1. Cryostat sections of newt liver(A), stomach(B) and intestine(C) incubated in the presence of propionaldehyde and NAD. ALDH activity is indicated by the dark stain. A. Newt liver is highly pigmented and the larger dark blotches are melanocytes. The activity is particulate and evenly distributed throughout the hepatocytes. B. The stain is diffuse and located in the basal region of the epithelial cells. C. The stain is particulate and located in the apical region of the epithelial cells.

4. DISCUSSION

In the frog (Nilsson, 1989), the greatest ALDH activity was in the liver, followed by the kidney with considerably less activity in the intestine. Our results indicated that in the newt the intestine has the greatest activity, followed by the stomach, kidney, lung and liver. The 2 to 5 times higher activity with propionaldehyde and NAD over benzaldehyde and NADP is consistent with the major newt ALDH being class 1 or 2. That there was significant ALDH activity at micromolar substrate concentrations is also consistent with class 1 and 2 ALDH.

Consistent with the assay data, anti-rat class 1 antibodies cross reacted against a newt polypeptide of approximately 61kD in liver, stomach and intestine. This is slightly larger than the 55kD molecular weight of mammalian ALDH monomers. The lack of cross-reaction between the newt polypeptide and mammalian class 2 ALDH antibody may indicate that the newt class 2 ALDH polypeptide is sufficiently different from its rat counterpart that the antibodies do not recognize the newt ALDH.

The histochemical localization of ALDH activity has been reported for the rat stomach mucosa (Maly, 1992) and for the rat liver (Lindahl et al., 1983). The epithelial cells in the rat stomach stained basally for ALDH activity just as the newt stomach epithelial cells did in our results. The particulate nature of the stain in newt liver and intestinal tissue suggests that the enzyme is associated with an organelle, possibly mitochondria, consistent with the assay data, although the Western immunoblot data suggest the newt ALDH is not closely related to mammalian class 2 ALDH.

5. CONCLUSIONS

The newt has ALDH activity in a variety of tissues. The highest activity we found was in intestine. The results of our assays are consistent with ALDH class 1 and 2. Western blot analyses are consistent with a class 1 ALDH. The activity is histochemically localized in liver hepatocytes and in the epithelial cells of the stomach and intestine.

6. REFERENCES

Hempel, J., Nicholas, H. and Lindahl, R.: Aldehyde dehydrogenases: widespread structural and functional diversity within a shared framework. Protein Science 2 (1993) 1890–1900.

Lindahl, R.: Aldehyde dehydrogenases and their role in carcinogenesis. Crit. Rev. Biochem. Mol. Biol. 27 (1992) 283–335.

Lindahl, R., Clark, R. and Evces, S.: Histochemical localization of aldehyde dehydrogenase during rat hepatocarcinogenesis. Cancer Res. 43 (1983) 5972–5977.

Lindahl, R. and Evces, S.: Comparative subcellular distribution of aldehyde dehydrogenase in rat, mouse and rabbit liver. Biochem. Pharm. 33 (1984) 3383–3389.

Maly, I.P., Arnold, M., Kreiger, K., Zalewska, M. and Sasse, D.: The intramucosal distribution of gastric alcohol dehydrogenase and aldehyde dehydrogenase activity in rats. Histochemistry 98 (1992) 311–315

Nilsson, G.E.: Organ distribution of aldehyde dehydrogenase activity in the frog, *Rana esculenta* L. Comp. Biochem. Physiol. 92C (1989) 263–265.

14

IN HEPATOMA CELL LINES RESTORED LIPID PEROXIDATION AFFECTS CELL VIABILITY INVERSELY TO ALDEHYDE METABOLIZING ENZYME ACTIVITY

Rosa A. Canuto,[1] Margherita Ferro,[2] Marina Maggiora,[1] Rosanna Federa,[2] Olga Brossa,[1] Anna M. Bassi,[2] Ronald Lindahl,[3] and Giuliana Muzio[1]

[1]Dipartimento di Scienze Cliniche e Biologiche
Università di Torino, Ospedale S. Luigi
10043 Orbassano, Torino, Italy
[2]Istituto di Patologia Generale
Università di Genova, Italy
[3]Department of Biochemistry and Molecular Biology
University of South Dakota
Vermillon, South Dakota 57069

1. INTRODUCTION

Hepatoma cells are less susceptible to oxidative stress than normal hepatocytes: the decreased content of polyunsaturated fatty acids in such cells decreases their capability to undergo lipid peroxidation (Gravela et al., 1975; Canuto et al., 1991; Cheeseman et al., 1988; Feo et al., 1975; Canuto et al., 1994; Masotti et al., 1988). A consequence of this is the reduced production of aldehydes inside the cells (Poli et al., 1986). Depending on the quantity present, aldehydes have several effects on the cells. In particular, the effects of 4-hydroxynonenal (4-HNE), an important aldehyde produced by lipid peroxidation, have been studied (Esterbauer et al., 1991). At μM concentrations, the effects are positive: adenylate cyclase and phospholipase C are stimulated (Paradisi et al., 1985; Garramone et al., 1988) and differentiation is induced in HL-60 cells (Barrera et al., 1991). At higher concentrations, 4-HNE is cytotoxic (Esterbauer et al., 1991; Canuto et al., 1995).

The capability of hepatoma cells to metabolize lipid peroxidation aldehydes also differs from that of normal cells. Differences in the aldehyde metabolizing systems of hepatoma cells have been correlated with their different degrees of deviation. In almost all the hepatoma cells examined, aldehyde dehydrogenase (ALDH) and aldehyde reductase (AL-RED) increase with the degree of deviation, whereas glutathione transferase decreases in hepatoma cell lines, but increases in chemically-induced hepatoma; NADH-dependent al-

Enzymology and Molecular Biology of Carbonyl Metabolism 6
edited by Weiner *et al.* Plenum Press, New York, 1996

cohol dehydrogenase decreases in hepatoma cell lines and remains unchanged in chemically-induced hepatoma.

ALDH and ALRED are probably more important than other enzymes in metabolizing lipid peroxidation products in tumor cell lines, and also in hepatoma defence mechanisms. When 4-HNE is externally added to cultured hepatoma cells with different degrees of deviation and different ALDH and ALRED contents, it induces cytotoxic effects that are inversely proportional to the content of these two enzymes (Canuto et al., 1994).

In the light of the above, and since hepatoma cells have both a reduced capability to undergo lipid peroxidation (Canuto et al., 1995) and increased ALDH and ALRED activities, we decided to increase arachidonic acid in hepatoma cell lines so as to restore lipid peroxidation to an approximately normal level, and to stimulate the cells to produce cytotoxic or cytostatic aldehydes, in order to examine the effects of these aldehydes on cell growth, viability and on the activities of aldehyde metabolizing enzymes.

7777 and JM2 hepatoma cell lines were chosen for these experiments; these cells have a lower content of arachidonic acid (Canuto et al., 1995) and are less susceptible to lipid peroxidation than hepatocytes. Their content of enzymes metabolizing lipid peroxidation products differs: the more deviated JM2 has more ALDH and ALRED than the less deviated 7777.

2. MATERIALS AND METHODS

2.1. Culture Conditions

Hepatoma cell lines (7777 and JM2) were seeded (day 0) and maintained for 24h in a medium A [DMEM/F12 plus 2 mM glutammine, 1% antibiotic/antimycotic] plus 10 % of newborn calf serum; 24h later (day 1) they were put into medium B [medium A plus 0.4 % of albumin, 1% ITS (insulin, transferrin, sodium selenite), 1% nonessential amino acid, 1% vitamin solution] supplemented with arachidonic acid (300 nmoles per million cells); 24h later (day 2), the supplemented medium was removed, and the cells were put into medium B, and given 4 doses of prooxidant, (500 µM ascorbate/100 µM iron sulphate), at 12h intervals. The various parameters (lipid peroxidation, cell number and enzyme activities) were determined on cells harvested at the following times: 24h after seeding (day 1), after 24h of enrichment with arachidonic acid (day 2), 4h after the 1st dose of ascorbate/iron sulphate (day 2.15), 12h after the 2nd dose of ascorbate/iron sulphate (day 3), 4h after the 3rd dose of ascorbate/iron sulphate (day 3.15), 12h after the 4th dose of ascorbate/iron sulphate (day 4).

2.2. Lipid Peroxidation

Lipid peroxidation was measured as production of malondialdehyde (MDA) as described by Canuto et al. (1995).

2.3. Viability Test

Cell viability was determined as lactate dehydrogenase (LDH) release. LDH was evaluated in the medium separated from monolayers, as described by Kornberg (1955); results were expressed as nmoles of NADH consumed per ml of the medium.

2.4. Enzyme Assays

The cells were resuspended in a medium containing 250 mM sucrose and 20 mM Tris-HCl, so as to have 1 g of cells in 10 ml, broken by three freeze-thaw cycles in liquid nitrogen, and sonicated 3 times for 2 s. Diluted homogenates were centrifuged at 134000xg for 45 min (rot. 65.13 Centrikon T-2060) to obtain the cytosolic fraction. Aldehyde dehydrogenase (ALDH) and aldehyde reductase (ALRED) were assayed in the cytosol fractions (Canuto et al., 1994). Proteins were determined by a biuret procedure (Gornal et al., 1949).

2.5. Western Blot Analysis

The cells were resuspended in 50 mM sodium phosphate (pH 8.5) containing 1 mM EDTA and 1 mM dithiothreitol so as to have 2 g of cells in 10 ml, broken by three freeze-thaw cycles in liquid nitrogen, sonicated 3 times for 2 s and centrifuged at 134000xg for 45 min (rot. 65.13 Centrikon T-2060) The cytosolic fraction obtained was used for Western blot analysis. SDS-polyacrylamide gel electroforesis was performed as described by Laemmli (1970); the separating and stacking gels were, respectively, 10% and 5% acrylamide. After electrophoresis, the proteins were electrotransferred to a PVDF membrane, which was then blocked for 1h with PBS containing 10% non fat dry milk. The membranes were then rinsed and treated sequentially with class 3 ALDH specific antibody (Lindahl and Feinstein, 1976) and Sigma alkaline phosphatase conjugated anti-rabbit IgG. Color was developed using BCIP (bromo-chloro-indolyl phosphate) and NBT(nitro blue tetrazolium) as substrate for alkaline phosphatase.

2.6. Northern Blot Analysis

Total RNA was prepared by the method of Chomczynski and Sacchi (1987) and quantified by measuring the absorbance at 260 nm. Fifteen μg of RNA was heat-denatured at 65 °C for 10 min, applied to formaldehyde agarose gel, electrophoresed and transferred by capillary action to Hybond N+ (Amersham Corp.). Prehybridization was done in 50% formamide, 5x SSC, 5x Denhardt's, 200 μg/ml salmon sperm DNA at 42 °C for 18h. Hybridization conditions were identical, except for 1x Denhardt's. DNA probe for class 3 ALDH was labeled with [^{32}P]dCTP (Amersham Corp.). Hybridization was carried out at 42 °C for 24h. The blot was washed in 2x SSC plus 0.1% SDS (twice at room temperature for 5 min and once at 57 °C for 10 min) and in 0.2x SSC plus 0.1% SDS (once at 57 °C for 10 min). Kodak Hyperfilm-MP was exposed for 3 days with intensifying screens at -80 °C.

2.7. Statistical Analysis

All data are expressed as means ± S.D. The significance of differences between group means was assessed by variance analysis, followed by the Newman-Keuls test.

3. RESULTS AND DISCUSSION

At the same concentration of arachidonic acid (60 μM) and at the same doses of prooxidant, 7777 cells were more sensitive to lipid peroxidation products than JM2.

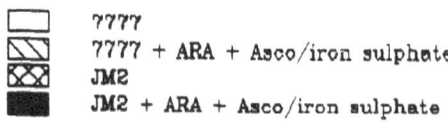

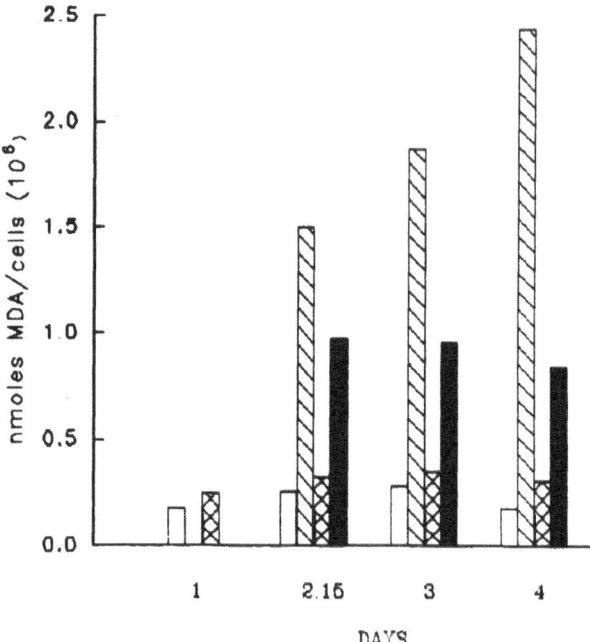

Figure 1. Lipid peroxidation in 7777 and JM2 hepatoma cells enriched with arachidonic acid and exposed to as-corbate/iron sulphate. Lipid peroxidation was determined as malondialdehyde (MDA) production per 10^6 cells. Cells were enriched with arachidonic acid incubated with ascorbate/iron sulphate and harvested as in materials and methods.

Both cell lines achieved lipid peroxidation after enrichment with arachidonic acid and exposure to ascorbate/iron sulphate, and the susceptibility of both increased with the number of doses of prooxidant, but 7777 more quickly than JM2, as shown in figure 1. In fact, 7777 cells produced 2.5 nmoles of MDA per million cells after 4 doses of prooxidant, whereas JM2 cells produced only 0.8 nmoles. Production of MDA in control cells re-mained unchanged over time.

Cell growth in the two cell lines also differed, as figure 2 shows: the number of treated 7777 cells increased, but less than controls, until the 3rd dose, after which it drasti-cally decreased. JM2 cells behaved like the controls until the 3rd dose, as shown in fig-ure 3, but after the 4th dose they decreased to 30% below normal.

After the 3rd dose, 7777 cells did not only stop growing, but also began dying: figure 4 shows lactate dehydrogenase activity released from cells against time. After the 4th dose, practically all cells were dead, whereas JM2 cells were still all alive. JM2 are not shown in this figure, because they are similar to controls in terms of lactate dehydrogenase activity.

If the concentration of the arachidonic acid administred to enrich JM2 cells was dou-bled, they behaved as 7777 cells: they produced more aldehydes and the number of cells begins to drop with the first two doses of prooxidant (data not shown).

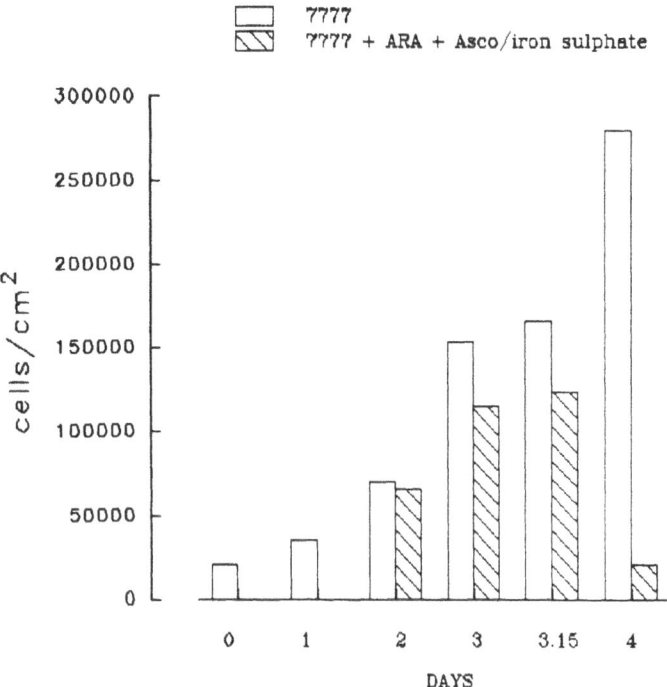

Figure 2. Number of 7777 hepatoma cells enriched with arachidonic acid and exposed to ascorbate/iron sulphate. Cells were enriched with arachidonic acid incubated with ascorbate/iron sulphate and harvested as in materials and methods.

Thus, aldehydes produced by lipid peroxidation reduce the growth of hepatoma cells in proportion to the quantity produced, and at high quantities kill the cells. The quantity of aldehydes present in the cells depends on their production, i.e. on the quantity of polyunsaturated fatty acid broken down by the prooxidant, and also on the capability of the cells to metabolize the aldehydes.

7777 and JM2 have a different content of ALDH and ALRED, higher in the latter. This might be the factor causing the differing rates of net aldehyde production and the differing effects that lipid peroxidation products have on cell growth.

During treatment with arachidonic acid and ascorbate/iron sulphate, activities of ALDH and ALRED were also determined; 0.1 mM 4-HNE was used as substrate for both enzymes, and 2.5 mM benzaldehyde for ALDH alone. 4-HNE was used because it is one of the most important aldehydes produced during lipid peroxidation; benzaldehyde was used because class 3 ALDH preferentially oxidizes this substrate (Lindahl and Evces, 1987).

Table 1 shows that the treated 7777 cells behaved as controls with regard to NAD-dependent ALDH, whereas NADP-dependent ALDH reached zero by day 3, i.e. after the 3rd dose of ascorbate/iron sulphate. This was the point in time when cell growth dropped. It should be noted that the enzyme activity increased over time, and that the values of NADP-dependent ALDH were lower than those of the NAD-dependent ALDH. The same behaviour was seen for NADP-dependent ALDH using benzaldehyde as substrate (table 2). Enzymatic activities were not determined on day 4, since the cells had almost all died.

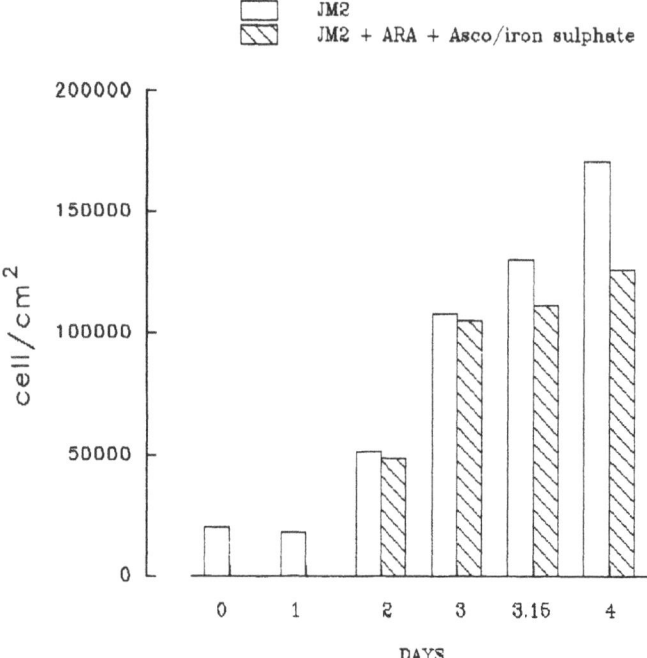

Figure 3. Number of JM2 hepatoma cells enriched with arachidonic acid and exposed to ascorbate/iron sulphate. Cells were enriched with arachidonic acid incubated with ascorbate/iron sulphate and harvested as in materials and methods.

As table 1 shows, NAD-dependent ALDH levels in treated JM2 cells behaved like those of controls with 4-HNE as substrate, and enzyme activity increased over time. For either 4-HNE or benzaldehyde, NADP-dependent ALDH increased over time, but at the end of the experiment, i.e. after the 4th dose of ascorbate/iron sulphate, it decreased; this was the point in the time when the cell number also decreased (table 2). With both substrates, inhibition of enzymatic activities was 42%.

Western-blot analysis on JM2 cells did not show any differences between controls and treated cells after the 4th dose of ascorbate/iron sulphate (figure 5); in 7777 cells, the protein quantity was very low, but after the 3rd dose of prooxidant system a decrease was evident.

Northern-blot analysis, which was only carried out on JM2 cells, showed that there were differences between controls and treated cells (figure 6). Class 3 ALDH mRNA decreased after the 2nd dose of ascorbate/iron sulphate and disappeared entirely after the 4th dose. mRNA had also decreased in the control cells by the end of the experiment, but less so than treated cells. From these data it appears that 4-HNE may interfact with protein in the cytosol, and also directly with DNA.

With regard to ALRED, in 7777 cells there were no differences between controls and treated cells. Also in this case, enzymatic activity increased over time (table 3). In JM2 cells, ALRED activity decreased after the 4th dose of ascorbate/iron sulphate, but the difference against controls was not significant (table 3).

In conclusion, lack of substrate is an important cause of decreased lipid peroxidation in hepatoma cells; restoring their susceptibility to lipid peroxidation by artificially increas-

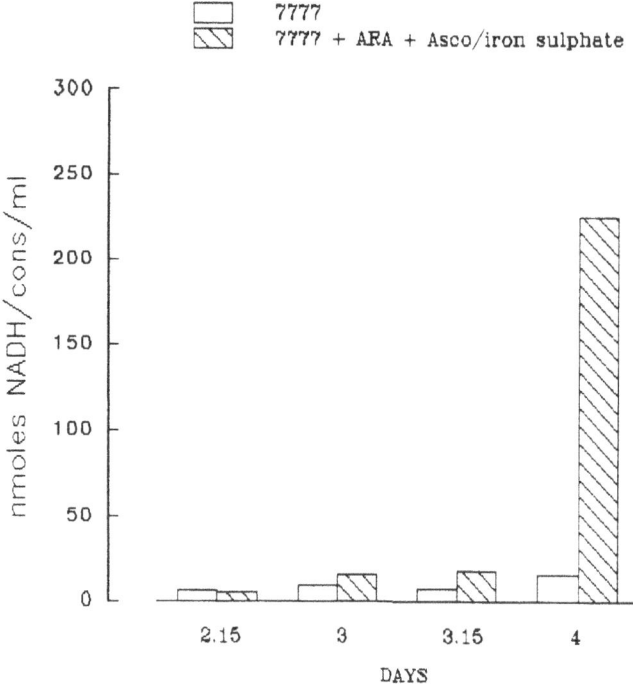

Figure 4. Lactate dehydrogenase release in 7777 hepatoma cells enriched with arachidonic acid and exposed to ascorbate/iron sulphate. LDH was determined in the medium and evaluated as NADH consumed per ml of the medium. Cells were enriched with arachidonic acid incubated with ascorbate/iron sulphate and harvested as in materials and methods.

Table 1. 4-Hydroxynonenal aldehyde dehydrogenase activity in citosol isolated from 7777 and JM2 hepatoma cells enriched with arachidonic acid and exposed to ascorbate/iron sulphate

	7777		JM2	
Day	Control	Treated	Control	Treated
NAD				
2	1.65 ± 0.68a	2.17 ± 0.62a	5.42 ± 1.08a	7.08 ± 2.31a
3	4.82 ± 1.03b	4.68 ± 0.70b	11.57 ± 2.25b	12.24 ± 1.78b
3.15	7.57 ± 1.47c	6.64 ± 1.14c		
4	12.32 ± 2.01d		16.42 ± 3.88c	16.24 ± 1.78c
NADP				
2	0.38 ± 0.10a	0.42 ± 0.09a	5.93 ± 1.67a	4.06 ± 1.23a
3	1.11 ± 0.21b	1.21 ± 0.18b	8.56 ± 1.98ab	9.95 ± 2.15b
3.15	1.40 ± 0.19b	0.04 ± 0.01c		
4	2.21 ± 0.46d		14.62 ± 3.12c	8.45 ± 1.43b

Data are espressed as nmoles of NAD(P)H produced/min/mg of protein and represent mean ± S.D. of 5 experiments. 4-hydroxynonenal concentration was 0.1 mM.

Treated: cells enriched with arachidonic acid, incubated with ascorbate/iron sulphate and harvested as in materials and methods.

For each coenzyme and for 7777 or JM2 cells, means with different letters are statistically different ($p < 0.05$) from one another as determined by variance analysis followed by the Newman-Keuls test.

Table 2. Benzaldehyde aldehyde dehydrogenase activity in citosol isolated
from 7777 and JM2 hepatoma cells enriched with arachidonic
acid and exposed to ascorbate/iron sulphate

	7777		JM2	
Day	Control	Treated	Control	Treated
3.15	1.15 ± 0.23	0*		
4	1.23 ± 0.19		704.96 ± 100.98	407.32 ± 50.53*

Data are espressed as nmoles of NADPH produced/min/mg of protein and represent mean ± S.D. of 5
experiments.
Benzaldehyde concentration was 2.5 mM.
Treated: cells enriched with arachidonic acid, incubated with ascorbate/iron sulphate and harvested as in
materials and methods.
t-test: *, p < 0.001

Figure 5. Western-blot analysis of class 3 aldehyde dehydrogenase in 7777 and JM2 hepatoma cells enriched with
arachidonic acid and exposed to ascorbate/iron sulphate. 1: Control cells for lane 2. 2: Cells harvested 24 h after
enrichment with arachidonic acid. 3: Control cells for lane 4. 4: Cells enriched with arachidonic acid and harvested 12 h after 2nd dose of ascorbate/iron sulphate. 5: 7777: Control cells for lane 6. 6: 7777: Cells enriched
with arachidonic acid and harvested 6 h after 3rd dose of ascorbate/iron sulphate. 5: JM2: Control cells for lane 6.
6: JM2: Cells enriched with arachidonic acid and harvested 12 h after 4th dose of ascorbate/iron sulphate.

Figure 6. Northern-blot anlysis of mRNA of class 3 aldehyde dehydrogenase in JM2 hepatoma cells enriched
with arachidonic acid and exposed to ascorbate/iron sulphate. 1: Control cells for lane 2. 2: Cells harvested 24 h
after enrichment with arachidonic acid. 3: Control cells for lane 4. 4: Cells enriched with arachidonic acid and
harvested 12 h after 2nd dose of ascorbate/iron sulphate. 5: Control cells for lane 6. 6: Cells enriched with arachidonic acid and harvested 12 h after 4th dose of ascorbate/iron sulphate.

Table 3. 4-Hydroxynonenal aldehyde reductase activity in cytosol isolated
from 7777 and JM2 hepatoma cells enriched with arachidonic acid and
exposed to ascorbate/iron sulphate

	7777		JM2	
Day	Control	Treated	Control	Treated
2	5.76 ± 0.24a	5.51 ± 1.05a	18.14 ± 4.27a	12.52 ± 3.96b
3	10.09 ± 2.11b	9.57 ± 0.99b	29.45 ± 4.25c	31.52 ± 5.71c
3.15	7.54 ± 1.24a	9.11 ± 1.65ab		
4	6.96 ± 0.96a		26.68 ± 3.44cd	22.15 ± 3.61d

Data are espressed as nmoles of NADPH consumed/min/mg of protein and represent mean ± S.D. of 5 experiments. 4-hydroxynonenal concentration was 0.1 mM.
Treated: cells were with arachidonic acid, incubated with ascorbate/iron sulphate and harvested as in materials and methods.
Means with different letters are statistically different (p < 0.05) from one another as determined by variance analysis followed by the Newman-Keuls test.

ing the arachidonic acid content reduces cell growth and viability. Fatty acid modification of tumour cells may thus be a useful tool to increase the efficacy of treatments affecting their viability (Spector and Burns, 1987; Sugioka et al., 1981).

However, the capability of the hepatoma cells to metabolize lipid peroxidation products, and in particular the levels of ALDH and ALRED, which increase with the degree of deviation, must be precisely known.

It is also of note that lipid peroxidation products reduce the mRNA and the activity of class 3 ALDH, and that the decrease of this enzyme corresponds to the decrease of cell growth.

4. ACKNOWLEDGMENTS

This study was supported by grants from the Italian Ministry for University, Scientific and Technological Research, the Italian National Research Council, the Italian Association for Cancer Research and from NIH grant CA21103 to RL.

5. REFERENCES

Barrera, G., Di Mauro C., Muraca, R., Ferrero, D., Cavalli, G., Fazio, V.M., Paradisi, L. and Dianzani, M.U.: Induction of differentiation in human HL 60 cells by 4-hydroxynonenal product of lipid peroxidation. Exp. Cell Res. 197 (1981) 148–152.

Canuto, R.A., Muzio, G., Biocca, M.E. and Dianzani, M.U.: Lipid peroxidation in rat AH-130 hepatoma cells enriched in vitro with arachidonic acid. Cancer Res. 51 (1991) 4603–4608.

Canuto, R.A., Ferro, M., Muzio, G., Bassi, A.M., Leonarduzzi, G., Maggiora, M., Adamo, D., Poli, G. and Lindhal, R.: Role of aldehyde metabolizing enzymes in mediating effects of aldehyde products of lipid peroxidation in liver cells. Carcinogenesis. 15 (1994) 1359–1364.

Canuto, R.A., Muzio, G., Bassi, A.M., Maggiora, M., Leonarduzzi, G., Lindahl, R., Dianzani M.U. and Ferro, M.: Enrichment of arachidonic acid increases the sensitivity of hepatoma cells to the cytotoxic effects of oxidative stress. Free Rad. Biol. Med. 18 (1995) 287–293.

Cheeseman, K.H., Emery, S., Maddix, S.P., Slater, T.F., Burton, G-W. and Ingold, K.U.: Studies on lipid peroxidation in normal and tumour tissues. The Yoshida rat liver tumour. Biochem. J. 250 (1988) 247–252.

Chomczynski, P. and Sacchi, N.: Single-step method of RNA isolation by acid guanidinium thiocynate-phenol-chloroform extraction. Anal. Biochem. 162 (1987) 156–159.

Esterbauer, H., Schauer, R.J. and Zollner, H.: Chemistry and biochemistry of 4-hydroxynonenal, malonaldehyde and related aldehydes. Free Rad. Biol. Med. 11 (1991) 81–128.

Feo, F., Canuto, R.A., Garcea, R. and Gabriel, L.: Effect of cholesterol content on some physical and functional properties of mitochondria isolated from adult rat liver, fetal liver, cholesterol-enriched liver and hepatoma AH-130, 3924A and 5123. Biochim. Biophys. Acta 413 (1975) 116–134.

Garramone, A., Cenni, M. and Rossi, M.A.: Attivazione della fosfolipasi C epatica da parte del 4-idrossinonenale. Boll. Soc. It. Biol. Sper. LXIV (1988) 809–816.

Gravela, E., Feo, F., Canuto, R.A., Garcea, R. and Gabriel, L.: Functional and structural alteration of liver ergastoplasmic membranes during D,L-ethionine hepatocarcinogenesis. Cancer Res. 35 (1975) 3041–3047.

Gornal, A.G., Bardawill, C.J. and David, M.: Determination of serum proteins by means of the biuret reaction. J. Biol. Chem. 177 (1949) 751–766.

Kornberg, A.: Lactic dehydrogenase of muscle. In S.P. Colowick and N.D. Kaplan (Eds), Methods of Enzymology, Vol. 1, Academic Press, New York, 1955, pp 441–443.

Laemmli, U.K.: Cleavage of structural proteins during the assembly of the head of bacteriophage T4. Nature 227 (1970) 680–685.

Lindahl, R. and Fenstein, R.N.: Purification and immunochemical characterization of aldehyde dehydrogenase from 2-acetylaminofluorene-induced rat hepatomas. Biochim. Biophys. Acta 452 (1976) 345–355.

Lindahl, R. and Evces, S.: Changes in aldehyde dehydrogenase activity during diethylnitrosamine-initiated rat hepatocarcinogenesis. Carcinogenesis 8 (1987) 785–790.

Masotti, L., Casali, E. and Galeotti, T.: Lipid peroxidation in tumour cells. Free Rad. Biol. Med. 4 (1988) 377–386.

Paradisi, L., Panagini, C., Parola, M., Barrera, G. and Dianzani, M.U.: Effects of 4-hydroxynonenal on adenylate cyclase and 5'-nucleotidase activities in rat liver plasma membranes. Chem. Biol. Interact. 53 (1985) 209–217.

Poli G., Cecchini, G., Biasi, F., Chiarpotto, E., Canuto, R.A., Biocca, M.E., Muzio, G., Esterbauer, H. and Dianzani, M.U.: Resistance to oxidative stress by hyperplastic rat liver tissue monitored in term of unpolar and medium polar carbonyls. Biochim. Biophys. Acta 883 (1986) 207–214

Spector, A.A. and Burns, C.P.: Biological and therapeutic potential of membrane lipid modification in tumors. Cancer Res. 47 (1987) 4529–4537.

Sugioka, K., Nakano, H., Noguchi, T., Tsuchiya, J. and Nakano, M.: Decomposition of unsaturated phospholipid by iron-ADP-Adriamycin coordination complex. Biochem. Biophys. Res. Commun. 100 (1981) 1251–1258.

15

PROFILES OF HEPATIC CELLULAR PROTEIN ADDUCTION BY MALONDIALDEHYDE AND 4-HYDROXYNONENAL

Studies with Isolated Hepatocytes

Dylan P. Hartley and Dennis R. Petersen

School of Pharmacy and Hepatobiliary Research Center
University of Colorado Health Sciences Center
Denver, Colorado 80262

1. INTRODUCTION

The peroxidative degradation of membrane lipids is initiated by free radicals originating from oxygen or in certain cases, the metabolism of xenobiotics. It is generally agreed that the peroxidation of polyunsaturated membrane lipids is a well-controlled, ongoing event associated with normal cell turnover and the overall process of aging. Conversely, the process of uncontrolled lipid peroxidation is a biochemical event associated with hepatic injury as a result of exposure to halogenated hydrocarbons, excessive alcohol ingestion, and acute or chronic iron overload (Comporti, 1985). Consistent with the autocatalytic nature of lipid peroxoidation, each of these chemical agents can result in the production of a radical species capable of initiating lipid peroxidation.

Whereas the peroxidative breakdown of membrane lipids has predictable effects on the integrity of cellular membranes, this degradative process also results in production of *alpha, beta*-unsaturated aldehydes that, because of their electrophilic nature, can elicit a diversity of adverse cellular effects ranging from disruption of calcium homeostasis to enzyme inhibition (Esterbauer *et al.*, 1991). A variety of *alpha, beta*-unsaturated carbonyl compounds derived from the peroxidation of polyunsaturated fatty acids have been identified with the most abundant products being malondialdehyde (MDA) and 4-hydroxynonenal (4-HNE). Therefore, numerous investigations have described a variety of cellular dysfunctions attributable to these more abundant products of lipid peroxidation. (Esterbauer *et al.*, 1991; Schaur *et al.*, 1990).

The cytotoxicity of MDA and 4-HNE reflects a balance between the rate at which the aldehydes are produced and their respective rates of detoxification. It is now well documented that multiple enzymatic pathways (i.e. aldehyde dehydrogenases, alcohol de-

Enzymology and Molecular Biology of Carbonyl Metabolism 6
edited by Weiner *et al.* Plenum Press, New York, 1996

hydrogenases and glutathione *S*-transferases) in the hepatocyte are very effective in metabolizing and detoxifying these products of lipid peroxidation (Mitchell and Petersen, 1987; Esterbauer *et al.*, 1985; Canuto *et al.*, 1993; Boleda *et al.*, 1993; Hartley and Petersen 1995). However, even in light of these effective detoxification pathways, the facile electrophilic reactivity of MDA and 4-HNE with cellular nucleophilic constituents represents a major threat to hepatocellular homeostasis. These intermolecular interactions can result in formation of aldehyde-protein adducts that are thought to play a central role in the cytotoxic events associated with lipid peroxidation. While the specific consequences of aldehyde adduction with certain proteins have not been determined, it is predictable that the alkylation of proteins critical for cellular homeostasis may represent important mechanisms in hepatotoxic responses to certain chemicals. To date, very little is known concerning the time course of cellular protein adduction during lipid peroxidative events or the characteristics of proteins that are adducted. The purpose of the present study was to quantitate the production of specific aldehydic products during a time course of iron-initiated lipid peroxidation in freshly isolated hepatocytes and to detect aldehyde-adducted proteins by use of polyclonal antibodies produced against MDA- or 4-HNE-protein adducts.

2. METHODS AND PROCEDURES

2.1. Chemicals and Solutions

All solutions were prepared in deionized and distilled water. The reagents used throughout this study were of analytical grade and purchased from Sigma Chemical Co. (St. Louis MO) or Mallinkrodt Specialty Chemical Co. (Paris KY). Collagenase type I used for isolating rat hepatocytes was purchased from Worthington Biochemicals, (Freehold, NJ).

2.2. Rat Hepatocyte Isolation and Experimentation

Male rats of either the High-Alcohol Sensitivity (HAS) or Low-Alcohol Sensitivity (LAS) selected lines of rats were used as hepatocyte donors. These genetic lines of rats have been established by selective breeding to establish populations of rats that differ genetically in behavioral responses to ethanol (Draski *et al.*, 1992). Given the systematic breeding scheme used to produce these selected lines, the HAS and LAS rats differ only at those genetic loci that contribute to genetic differences in specific behavioral responses to ethanol. Rats were anesthetized by intraperitoneal injection of sodium pentobarbital (65 mg/kg) and hepatocytes were isolated according to the collagenase-perfusion procedure described previously (Hartley and Petersen, 1995). Hepatocytes isolated by this procedure generally displayed viability greater than 85% as measured by trypan blue exclusion. Isolated hepatocytes were suspended in 10 ml of Kreb-Henseleit buffer supplemented with 11.5 mM HEPES (pH 7.4) at a concentration of 10^6 hepatocytes ml^{-1} and were placed in rotating 25-ml round-bottom flasks maintained at 37°C under a continuous atmosphere of 95% O_2, 5% CO_2. The isolated hepatocytes were allowed to incubate under these conditions for 30 minutes prior to initiating the experiments. Peroxidation of the preincubated hepatocytes was initiated by addition of final concentrations of 0.4mM $FeSO_4$/1mM ascorbate to each flask. One hundred microliters of the suspension were removed after 0, 2.5, 5, 10, 15, 30 and 60 minutes incubation and subjected to analysis for 4-HNE, hexanal, MDA and reduced glutathione. Hexanal and 4-HNE were quantitated following derivatization with cyclohexadione to form fluorescent products that are resolvable by reversed-phase HPLC analysis with fluorescent detection (Yoshino *et al.*, 1986). For the studies described here, the HPLC procedure was modified by employing an C_{18} ODS, 5 µm column (2.5 X 250 mm) and isocratic mobile phase system consisting of 40% acetonitrile and flow rate of 1.0 ml/min. Under these conditions, 4-HNE and hexanal eluted at 8.5 and 13.6 min. respectively. Four-hydroxyoctenal served as the internal standard and eluted at 7.0 minutes. Samples were quantitated by comparison to calibration curves prepared using synthetic 4-HNE and hexanal standards. Malondialdehyde was quantitated as thiobarbituric acid-reactive substances (TBARS) using the thiobarbituric acid procedure described elsewhere (Buege and Aust, 1984). Hepatocyte glutathione concentrations were determined using the HPLC procedure for separation and UV detection of the 2,4-dinitrofluorbenzene derivative of glutathione (Reed *et al.*, 1980).

2.3. Polyclonal Antibody Preparation, Immunoprecipitation, and Western Blotting Procedures

Polyclonal antibodies recognizing MDA or 4-HNE-adducted proteins were produced in rabbits using standard immunization and boosting protocols. For the studies described here, polyclonal antibodies were prepared against MDA-protein adducts by immunizing and boosting rabbits with MDA-adducted keyhole limpet hemocyanin (MDA-KLH). Polyclonal antibodies recognizing 4-HNE-adducted proteins were produced by immunizing rabbits with 4-HNE adducted to KLH. This adduct was prepared by a first forming Michael addition products of 4-HNE with the thiol group of cysteine and then linking the 4-HNE-cysteine adduct to KLH *via* glutaraldehyde. The polyclonal antibodies produced against MDA-KLH or 4-HNE-KLH were characterized by ELISA procedures to identify approximate dilution factors for the antisera. Evaluation for specificity of the polyclonal antibodies by ELISA or Western blotting procedures revealed no cross-reactivity; the anti-4-HNE-KLH antibodies did not recognize MDA-adducted proteins and antibodies directed against MDA-KLH displayed no specific interactions with 4-HNE-adducted proteins.

Samples were obtained from hepatocyte incubations at 0, 30, 60 and 120 minutes following addition of iron/ascorbate for detection of aldehyde-adducted proteins by immunoblotting procedures. The sampling and preparation procedure involved withdrawing a 1.0 ml sample containing 1 X 10^6 hepatocytes, rapidly removing the cells by centrifugation at 1000g, resuspending the cells in a 500 mM Tris-HCl buffer, pH 7.5, containing 5mM $MgCl_2$, 1mg/ml Rnase and Dnase and 45% glycerol. The cell suspension was then treated with 20 ul of 2.5% SDS (w/v), and thoroughly mixed with 200 ul of the immunoprecipitation buffer described below. In order to minimize the nonspecific antibody-protein interactions, samples obtained from the hepatocyte incubations were immunoprecipitated according to procedures described elsewhere (Harlow and Lane, 1988). Briefly, equivalent volumes of immunoprecipitation buffer (100 mM Tricine, pH 8.2, 300 mM NaCl, 0.75% Triton X-100, 10mM EDTA, 0.02% NaN_3) were mixed with protein samples and MDA- or 4-HNE antisera and mixed for 12 hours at 4°C. Fifty microliters of a 6% (w/v) solution of protein A-sepharose CL-4B beads was added to each sample and mixed for one hour. The protein A beads (immunoprecipitation complex) were collected by centrifugation, washed and diluted in electrophoresis running buffer. Standard SDS gel electrophoresis was employed to separate adducted proteins according on the basis of molecular weight. The samples were transferred from SDS gels to nitrocellulose membrane which was then blocked with 1% BSA. The membranes were exposed to primary antisera (1:500 anti-

body:phosphate-buffered saline) for 1 hour, washed in phosphate-buffered saline and incubated with biotin-conjugated secondary antibody for 30 minutes. The membrane was
rinsed and incubated for an additional 30 minutes with a Streptavidin-conjugated horse-
radish peroxidase (1:5,000). Immunopositive interactions were visualized with an enhanced chemiluminescence substrate.

3. RESULTS

Following isolation and preincubation, hepatocyte viability, as measured by trypan
blue exclusion, was approximately 85%. Hepatocytes maintained in control incubations
containing no iron/ascorbate displayed viabilities of 80 to 85% through the entire time
course of the experiments. The viability of hepatocytes incubated with iron/ascorbate
fluctuated between 80 to 85% throughout the experiment and was not significantly different from that observed in control incubations. The maintenance of viability suggests
that the oxidative stress induced by iron/ascorbate did not result in observable cytotoxicity during the time-course of these experiments. The addition of iron/ascorbate to hepatocyte incubations initiated significant lipid peroxidation as evidenced by marked
elevations in malondialdehyde, hexanal and 4-HNE (Fig. 1A-C). Malondialdehyde concentrations, expressed as TBARS, markedly increased within the first 30 minutes following the addition of iron/ascorbate (Fig. 1A) and continued to increase steadily to peak
concentrations at 60 minutes that were approximately 4-fold greater than those in hepatocyte incubations not containing iron/ascorbate. Within 2.5 minutes following the addition
of iron/ascorbate, hexanal (Fig. 1B) and 4-HNE concentrations (Fig. 1C) increased 33-
fold and 2.5 fold, respectively above those observed in control incubations. In hepatocyte
incubations supplemented with iron/ascorbate, the concentrations of these two aldehydes
declined rapidly and by 30 minutes returned to values not significantly different from
control incubations.

It is predictable that oxidative stress resulting in marked lipid peroxidation evident
in Fig. 1A-C, would challenge hepatocellular antioxidant systems. Iron/ascorbate-induced
changes in hepatocyte glutathione concentrations are presented in Fig. 2. Reduced glutathione concentrations in control incubations ranged from 35 to 40 nmoles/10^6 hepatocytes during the time course of the experiments. However, glutathione concentrations in
hepatocytes exposed to iron/ascorbate decreased steadily throughout the experiment to approximately 70% of control values suggesting that the prooxidant system used in these experiments was effective in altering cellular thiol status as well as initiating significant lipid
peroxidation. It is interesting to note that while exposure to iron/ascorbate resulted in a
significant increase in lipid peroxidation and reduction of glutathione content, cell viability was maintained throughout the time course of the experiments.

The immunoprecipitation-immunoblot analysis of MDA adducted proteins prepared
from control and iron/ascorbate peroxidized hepatocytes is presented in Fig. 3. The approximate molecular weight of the immunoreactive proteins was determined relative to
molecular weight standards. The very prominent 55 kDa band present in all samples represents residual IgG originating from the immunoprecipitation procedure. It is interesting
that immunoreactive proteins were detected in the 80 to 85 kDa range in control incubations. The observation that these immunoreactive proteins became more distinct throughout the time course of the experiments suggests they are sensitive to adduction with MDA
generated in response to low levels of spontaneous lipid peroxidation occurring in the control incubations. It is apparent from Fig. 3 that exposure of hepatocytes to iron/ascorbate

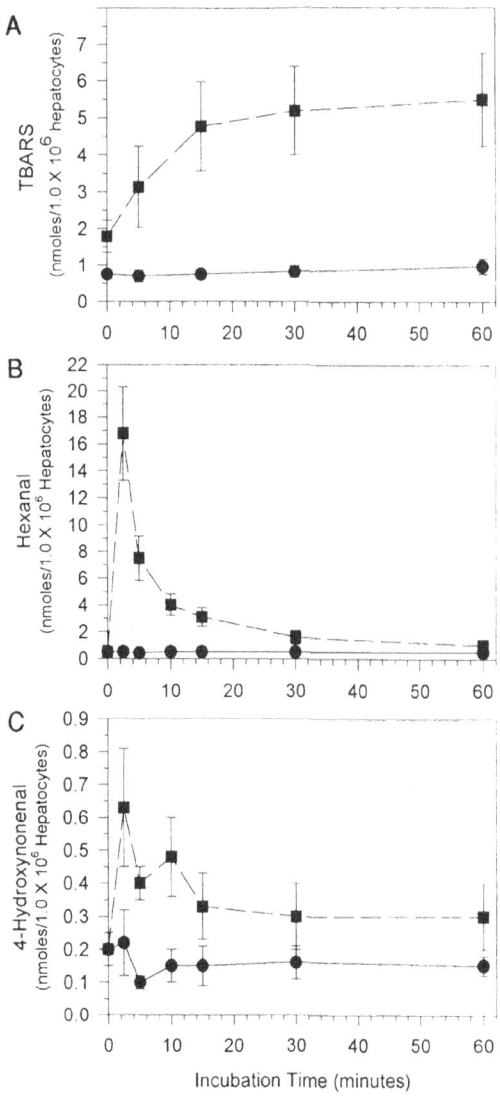

Figure 1. A-C. Profiles of thiobarbituric acid reactive substances -TBARS (1A.) hexanal (1B.) and 4-hydroxynonenal (1C.) in control (circles) and iron-ascorbate-treated (squares) hepatocytes. Cell viability was determined concomitantly at each time point. Data are presented as mean ± SEM for 5 to 6 individual experiments.

results in time-dependent increase in the MDA-adduction of several proteins. This is most apparent in the hepatocytes incubated with iron/ascorbate for 120 minutes where adducted proteins having molecular weights of 150, 130, 85, 45, 38, 33, 30, 28 and 20 kDa were detected. The immunoblots probed with 4-HNE-KLH antisera did not reveal prooxidant-dependent formation of 4-HNE-modified protein and are not shown. However, these blots did reveal 4-HNE-adducted proteins in control hepatocytes with apparent molecular weights of 80, 100 and 150 kDA that increased in intensity with time. The presence of 4-HNE adducted proteins in hepatocyte incubations not exposed to prooxidant stress sug-

Figure 2. Reduced glutathione status of isolated hepatocytes during exposure to iron/ascorbate. The circles represent the glutathione status of control hepatocytes while squares depict glutathione concentrations in hepatocytes exposed to iron/ascorbate. The values represent the mean ± SEM for 4 to 5 individual experiments.

Figure 3. Immunoprecipitation-immunoblot analysis of cellular proteins from control hepatocytes or hepatocytes exposed to iron/ascorbate (Fe/Asc). Immunoprecipitation and immunoblotting was performed using MDA-KLH antisera as outlined in the text. The molecular weight of MDA-adducted proteins was estimated relative to molecular weight standards.

gests that specific hepatocellular proteins are extremely sensitive to adduction by extremely low concentrations of 4-HNE.

4. DISCUSSION

It is well documented that iron/ascorbate is an effective prooxidant when added to isolated cellular systems. The data presented here reveals the temporal relationship between iron/ascorbate-induced hepatocellular cytotoxicity, lipid peroxidation, GSH status and aldehyde-protein adduct formation. While iron/ascorbate caused only a marginal decrease in cell viability and depletion of reduced glutathione, significant lipid peroxidation occurred as evidenced by marked increases in TBARS, hexanal and 4-HNE. Consistent with data presented here, the ability of isolated hepatocytes to withstand oxidative stress whether originating from iron (Poli *et al.*, 1985) or anoxia/reoxygenation (Grune *et al.*, 1993) has been reported. Likewise, the maintenance of cell viability in the presence of prooxidant-mediated decreases in hepatocellular glutathione has been reported (Grune *et al.*, 1993) and is confirmed in the present study.

Especially striking in the data presented here are the profiles of aldehydic products of lipid peroxidation produced following addition of iron/ascorbate to the hepatocyte incubations. Hexanal and 4-HNE appeared within 2.5 minutes following prooxidant challenge but returned to control values within 30 minutes. It is well documented (Canuto et al., 1993; Hartley and Petersen 1995; Mitchell and Petersen, 1987) that hepatocytes have multiple and effective enzymatic pathways capable of detoxifying aldehydic products of lipid peroxidation and, in this context, the profiles of appearance and disappearance of hexanal and 4-HNE presented in Fig. 1 reflect the effectiveness of these enzymatic systems. However, it is apparent from Fig. 1 that TBARS accumulated over time suggesting that MDA may not be metabolized as effectively as hexanal or 4-HNE. This notion is supported by the previous reports (Siu and Draper, 1982; Siems *et al.*, 1990) of lower $^{14}CO_2$ production from liver preparations exposed to ^{14}C -labeled MDA as compared to ^{14}C-labeled 4-HNE. Collectively, these data demonstrate the potential of iron/ascorbate to initiate peroxidative events in isolated hepatocytes resulting in significant elevations of multiple aldehydic products of lipid peroxidation. Likewise, these data confirm the effectiveness of the aldehyde dehydrogenase, alcohol dehydrogenase and glutathione S-transferase pathways in detoxifying potentially toxic aldehydic products of lipid peroxidation.

Immunoprecipitation and immunoblotting of hepatocyte lysates with MDA antisera (Fig. 3) detected proteins ranging in molecular weight from 150 to 20 kDa that are targets for adduction by endogenously-produced MDA. The intensity of the adducted proteins increased during the time course of the experiments reflecting the time-dependent accumulation of MDA in the hepatocytes incubates (Fig. 1A.). In all probability, the adduction of these proteins by MDA occurs through Schiff base formation. Presumably, there are numerous proteins in the liver cell that could participate in this amine-aldehyde interaction. The fact that there are relatively few adducted proteins detected suggest that certain proteins in hepatocytes are specific targets for MDA adduction. This is the first report to appear documenting the occurrence of specific hepatocellular proteins alkylated by aldehydic products of lipid peroxidation during prooxidant-induced injury of viable isolated hepatocytes. While one previous report has been published (Uchida *et al.*, 1993) describing the use of antibodies generated against 4-HNE-proteins haptens to identify adducted proteins in hepatocytes, these investigators treated hepatocytes that had been isolated, frozen and thawed with high concentrations of the potent oxidizing agent *tert*-butylhydroperoxide. While numerous adducted proteins were detected by these investigators, the results are difficult to compare to those presented here since their studies were performed using an extremely potent oxidizing agent and metabolically inactive hepatocytes. In the experiments presented here, the metabolic state of the hepatocytes was assessed and indicated that the cells remained viable through the experiments. While the proteins adducted by MDA have not been characterized or identified, the observation that the MDA adducted proteins appeared concurrent with the accumulation of MDA suggests that protein adduct formation could be related to inactivation of the enzymes responsible for detoxification of MDA.

In contrast to the effect of iron/ascorbate exposure on MDA accumulation and adduct formation, prooxidant exposure did not enhance significant accumulation of 4-HNE or intensify the formation of 4-HNE protein adducts in isolated hepatocytes. This observation is most likely related to the efficient enzymatic pathways that eliminate 4-HNE in hepatocytes. It was, however, noted that control and iron/ascorbate- treated hepatocytes displayed 4-HNE adducted proteins with molecular weights of 80, 100 and 150 kDa . As noted in Fig. 3, the 80 and 100 kDa proteins adducted with MDA were also detected with increasing time in control incubations. It is important to note that since there was no cross reactivity of the anti-4-HNE-KLH and anti-MDA-KLH antibodies the 80, and 100 kDa proteins are easily adducted by either MDA or 4-HNE. These data suggest that certain proteins are very susceptible to adduction by aldehydic products of lipid peroxidation and may be extremely sensitive markers for very low levels of oxidative stress.

At present, it is difficult to identify specific roles of aldehyde-adducted proteins in prooxidant hepatocellular injury. The data presented here suggest that formation of aldehyde adducted proteins does not necessarily result in immediate cell death. It is predictable that aldehyde adduction of a significant number of proteins playing a central role in cellular homeostasis would result in immediate and significant loses in cell viability. Alternatively, the cellular consequences of aldehyde-protein adduction may be cumulative and, as a result, observable only after more prolonged periods of oxidative stress. However, it is possible that there are relatively abundant lysine-or cysteine-rich cellular proteins that perform a non-catalytic, chaperon function by adducting and eliminating potentially toxic aldehydic products of lipid peroxidation. Certainly, identification and characterization of these proteins will be essential to determine their specific roles in prooxidant-induced hepatocellular injury.

5. ACKNOWLEDGMENTS

This research was supported by PHS grants AA 090300 and AA05370

6. REFERENCES

Boleda, M.D., Saubi, N., Farres, J. and Pares, X.. Physiological substrates for rat alcohol dehydrogenase classes: aldehydes of lipid peroxidation, omega-hydroxyfatty acids, and retinoids. Archiv. Biochem. Biophys. 307 (1993) 85–90.

Canuto, R.A., Muzio, G., Maggiora, M, Biocca, M.E. and Dianzani, M.U. Glutathione S-transferase, alcohol dehydrogenase and aldehyde reductase activities during diethylnitrosamine-carcinogenesis in rat liver. *Cancer Letters* 68 (1993) 177–183.

Comporti, M. Biology of Disease: Lipid peroxidation and cellular damage in toxic liver injury. *Laboratory Investigation* 53 (1985) 599–622.

Draski, L.J., Spuhler, K.P. Erwin, V.G., Baker, R.C. and Deitrich, R.A. Selective breeding of rats differing in sensitivity to the effects of acute ethanol administration. *Alcoholism: Clin. Exper. Res.* 16 (1992) 48–54.

Esterbauer, H., Zollner, H. and Lang, J. Metabolism of the lipid peroxidation product 4-hydroxynonenal by isolated hepatocytes and liver cytosolic fractions. *Biochem. J.* 228 (1985) 363–373.

Esterbauer, H., Schaur, R. J., and Zollner, H. Chemistry and biochemistry of 4-hydroxynonenal, malondialdehyde and related aldehydes. *Free Rad. Biol. Med.* 11(1991) 81–128.

Grune, T., Siems, W.G. and Schneider, W. Accumulation of aldehydic lipid peroxidation products during postanoxic reoxygenation of isolated hepatocytes. *Free Rad. Biol. Med.* 15 (1993) 125–132.

Hartley, D.P. and Petersen, D.R. The hepatocellular metabolism of 4-hydroxynonenal by alcohol dehydrogenase, aldehyde dehydrogenase and glutathione S-transferase. *Arch. Biochem. Biophys.* 316 (1995) 197–205.

Harlowe, E. and Lane, D. In: Atibodies: A Laboratory Manual (1988) pp 421–468. Cold Spring Harbor Laboratory, Cold Spring Harbor, NY.

Mitchell, D.Y. and Petersen, D.R. The oxidation of *alpha,beta*-unsaturated aldehydic products of lipid peroxidation by rat liver aldehyde dehydrogenase. *Toxicol. Appl. Pharmacol.* 87 (1987) 403–410.

Poli, G., Dianzani, M. U., Cheeseman, K. H., Slater, T. F., Lang, J., and Esterbauer, H. Separation and characterization of the aldehydic products of lipid peroxidation stimulated by carbon tetrachloride or ADP-iron in isolated rat hepatocytes and rat liver microsomal suspensions. *Biochem. J.* 227 (1985) 629–638.

Reed, D., Babson, J., Beatty, P., Ellis, A. and Potter, D. High pressure liquid chromatographic analysis of nanomole levels of glutathione, glutathione disulfides and related thiols and disulfides. *Analyt. Biochem.* 106 (1980) 55–67.

Schaur, R.J., Zollner, H. and Esterbauer, H. Biological effects of aldehydes with particular attention to 4-hydroxynonenal and malondialdehyde. 1990 pp 141–163, Vol. III ed. C. Vigo-Pelfrey, boca Raton, FL, CRC Press Inc.

Siems, W., Zollner, H., Esterbauer, H. Metabolic pathways of the lipid peroxidation product 4-hydroxynonenal in hepatocytes - quantitative assessment of an autooxidative defense system. *Free Radic. Biol. Med.* 9 (1990) 110.

Siu, G.M. and Draper, H.H. Metabolism of malondialdehyde *in vivo* and *in vitro*. *Lipids* 17 (1982) 349–355.

Uchida, K., Szweda, H-Z. C. and Stadtman, E.R. Immunochemical detection of 4-hydroxynonenal protein adducts in oxidized hepatocytes. *Proc. Natl. Acad. Sci.* 90 (1993) 8742–8746.

Yoshino, K., Matsuura, T., Sano, M, Saito, S. and Tomita, I. Fluorometric liquid chromatographic determination of aliphatic aldehydes arising from lipid peroxides. *Chem. Pharm. Bull.* 34 (1986) 1694–1700.

INHIBITION OF HUMAN CLASS 3 ALDEHYDE DEHYDROGENASE, AND SENSITIZATION OF TUMOR CELLS THAT EXPRESS SIGNIFICANT AMOUNTS OF THIS ENZYME TO OXAZAPHOSPHORINES, BY THE NATURALLY OCCURRING COMPOUND GOSSYPOL

Ganaganur K. Rekha and Norman E. Sladek

Department of Pharmacology
University of Minnesota Medical School
Minneapolis, Minnesota 55455

1. INTRODUCTION

Cytosolic class 3 aldehyde dehydrogenase (ALDH-3) is a demonstrated determinant of cellular sensitivity to the cytotoxic action of certain widely used antineoplastic pro-drugs collectively referred to as oxazaphosphorines, e.g., cyclophosphamide, ifosfamide, 4-hydroperoxycyclophosphamide, 4-hydroperoxyifosfamide and mafosfamide (cellular sensitivity to these drugs decreases as cellular levels of ALDH-3 increase) (Sladek, 1993; Sreerama and Sladek, 1993a,b, 1994; Bunting et al., 1994; Rekha et al., 1994; Sladek et al., 1995; Sreerama et al., 1995; Bunting and Townsend, 1996). Thus, tumor cells, other-wise sensitive to the oxazaphosphorines, became resistant to these drugs when electropo-rated with purified ALDH-3 protein or transfected with the cDNA coding for ALDH-3 (Bunting et al., 1994; Sreerama and Sladek, 1995; Bunting and Townsend, 1996), and, of therapeutic significance, relatively elevated levels of this enzyme can account for intrin-sic, transient acquired, and stable acquired, resistance to the oxazaphosphorines on the part of malignant cells (Sreerama and Sladek, 1993a,b, 1994; Rekha et al., 1994; Sreerama et al., 1995). Resistance to the oxazaphosphorines mediated by ALDH-3 is ostensibly due to the enzyme-catalyzed oxidative detoxification of aldophosphamide, the pivotal metabo-lite of these prodrugs (Sreerama and Sladek, 1993a, 1994; Rekha et al., 1994; Sreerama et al., 1995; Bunting and Townsend, 1996).

Inhibition of ALDH-3 would therefore be expected to sensitize otherwise relatively insensitive tumor cells to the oxazaphosphorines when relatively high cellular levels of ALDH-3 is the basis for the relative insensitivity. Thus, inhibitors of ALDH-3 could be of therapeutic value. However, to date, no inhibitor of ALDH-3 has been identified, although inhibition of ALDH-3-catalyzed oxidation of aldophosphamide to carboxyphosphamide by alternative substrates can be effected, *vide infra*. Known inhibitors of class 1 (ALDH-1) and/or class 2 (ALDH-2) aldehyde dehydrogenases, e.g., disulfiram and chloral hydrate, do not, or only minimally, inhibit ALDH-3 (Sreerama and Sladek, 1993a, 1994) and, predictably, do not sensitize tumor cells to the oxazaphosphorines when such cells are insensitive to these agents because of relatively high ALDH-3 levels (reviewed in Sladek et al., 1995). Another inhibitor of ALDH-1, viz., 4-(diethylamino)benzaldehyde, proved to be a substrate for ALDH-3 (Rekha et al., 1994; Sladek et al., 1995). Alternative substrates for ALDH-3, e.g., benzaldehyde and 4-(diethylamino)benzaldehyde, compete with oxazaphosphorines for the catalytic site and thus sensitize tumor cells that express relatively large amounts of ALDH-3 and are therefore otherwise relatively insensitive to these agents (Sreerama and Sladek, 1994; Rekha et al., 1994).

In our search for an inhibitor of ALDH-3, gossypol, a polyphenolic aldehyde (Fig. 1) found in cottonseed extracts, emerged as a possible candidate. It is known to inhibit several NAD(P)-linked dehydrogenases, e.g, lactate dehydrogenase (Lee et al., 1982; Olgiati and Toscano, 1983; Burgos et al., 1986), malate dehydrogenase (Burgos et al., 1986), glutamate dehydrogenase (Burgos et al., 1986), glyceraldehyde-3-phosphate dehydrogenase (Ikeda, 1990), 11-β-hydroxysteroid dehydrogenase (Sang et al., 1991) and alcohol dehydrogenase (Messiha, 1991a). Directly germane to our interests, gossypol has been shown to inhibit "hepatic aldehyde dehydrogenase" activity when given to mice (Messiha, 1991a,b). Therefore, we initiated studies intended to ascertain the effect of gossypol on the catalytic activity of ALDH-3.

ALDH-1, known to also catalyze the irreversible oxidation (detoxification) of aldophosphamide (Dockham et al., 1992), and ALDH-2 were included in our investigations so that the relative specificity, if any, of the inhibitory effect of gossypol towards each of the three classes of aldehyde dehydrogenases could be ascertained.

The ALDH-3 present in human tumor cells/tissues (tALDH-3), e.g., cultured breast adenocarcinoma MCF-7 cells, colon carcinoma C cells, and salivary gland Warthin tumors and mucoepidermoid carcinomas, although otherwise seemingly identical to the ALDH-3 present in human normal tissues/fluids (nALDH-3), e.g., stomach mucosa and saliva, differs from the latter in that it exhibits a much greater ability to catalyze the oxidative detoxification of the oxazaphosphorines (reviewed in Sladek et al., 1995). Hence, both tALDH-3 and nALDH-3 were included in our investigations.

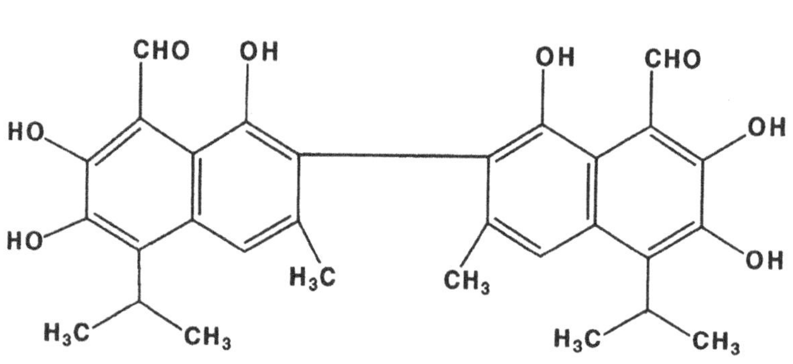

Figure 1. Structure of gossypol [1,1',6,6',7,7'-hexahydroxy-5,5'-diisopropyl-3,3'-dimethyl-(2,2'-binaphthalene)-8,8'-dicarboxyaldehyde].

Aldehyde dehydrogenases are bifunctional enzymes, i.e., they catalyze both oxidative (oxidation of aldehydes to acids) and hydrolytic (hydrolysis of esters) reactions. Thus, the effect of gossypol on each of these reactions was determined.

Gossypol has been shown to be toxic to various human tumor cells including human breast adenocarcinoma MCF-7 cells (Tuszynski and Cossu, 1984; Joseph et al., 1986; Band et al., 1989; Wu et al., 1989; Benz et al., 1990; Ford et al., 1991; Hu et al., 1993; Coyle et al., 1994; Gilbert et al., 1995) and is being tested clinically for its potential as an anticancer agent (Wu, 1989; Stein et al., 1992; Flack et al., 1993; Seidman, 1996). The sensitivity of cultured human breast adenocarcinoma MCF-7/0 (cellular levels of ALDH-3 are very low) and MCF-7/0/CAT (cellular levels of ALDH-3 are relatively high) cells to gossypol was determined in the present investigation. The ability of gossypol to negate the influence of relatively high cellular levels of ALDH-3 on the cellular sensitivity of tumor cells to oxazaphosphorines was also determined.

2. MATERIALS AND METHODS

Mafosfamide was provided by Dr. J. Pohl, Asta Medica AG, Frankfurt, Germany. Phosphoramide mustard cyclohexylamine was supplied by the Drug Development Branch, Division of Cancer Treatment, National Cancer Institute, Bethesda, MD. *E. coli* [BL21(DE3)pLysS] transfected with pET-19b vector to which human ALDH-1 cDNA [cloned from human hepatoma Hep G2 cells (Moreb et al., 1996)] was ligated, was provided by Dr. Jan Moreb, University of Florida, Gainsville, FL. A vector, viz., pT7–7, to which human ALDH-2 cDNA [cloned from human liver (Zheng et al., 1993)] was ligated, was provided by Dr. Henry Weiner, Purdue University, Lafayette, IN. Transfection of human ALDH-2 cDNA ligated to the pT7–7 vector into *E. coli* [BL21(DE3)pLysS] was by Drs. P. A. Dockham and L. Sreerama of our laboratory as described by Sambrook et al. (1989). Generation and purification of recombinant human ALDH-1 (rALDH-1) and ALDH-2 (rALDH-2) were by Dr. V. R. Devaraj of our laboratory [for details see Devaraj, V.R., Sreerama, L., Lee, M.J.C., Nagasawa, H.T. and Sladek, N.E.: Yeast aldehyde dehydrogenase sesnsitivity to inhibition by chlorpropamide analogues as an indicator of human aldehyde dehydrogenase sensitivity to these agents. In Weiner, H., Lindahl, R., Crabb, D.W. and Flynn, T.G. (Eds.), Enzymology and Molecular Biology of Carbonyl Metabolism - 6, Plenum Press, New York, 1997]. Human erythrocyte glyceraldehyde-3-phosphate dehydrogenase, human placental alkaline phosphatase type XXIV, racemic gossypol and *p*-nitrophenyl phosphate were purchased from Sigma Chemical Co., St. Louis, MO. All other chemicals and reagents were obtained from the sources listed in previous publications (Dockham et al., 1992; Sreerama and Sladek, 1993a, 1994; Sreerama et al., 1995).

Human normal stomach mucosa ALDH-3 (nALDH-3) and the ALDH-3 (tALDH-3) present in human breast adenocarcinoma MCF-7/0 cells cultured in the presence of 30 μM catechol for 5 days to induce the enzyme (MCF-7/0/CAT cells) were purified as described previously (Sreerama and Sladek, 1993b; Sreerama et al., 1995).

NAD-linked oxidation of acetaldehyde catalyzed by rALDH-1 and rALDH-2 at 37°C and pH 8.1, NAD(P)-linked oxidation of benzaldehyde catalyzed by nALDH-3 and tALDH-3 at 37°C and pH 8.1, and hydrolysis of *p*-nitrophenyl acetate catalyzed by each of these enzymes at 25°C and pH 7.5, were quantified spectrophotometrically as described previously (Dockham et al., 1992; Sreerama and Sladek, 1993a). NAD-linked oxidation of glyceraldehyde-3-phosphate catalyzed by glyceraldehyde-3-phosphate dehydrogenase at 37°C and pH 7.6, and hydrolysis of *p*-nitrophenyl phosphate catalyzed by alkaline phosphatase at 25°C and pH 9.8, were quantified spectrophotometrically as described by Lambeir et al. (1991) and Chueh et al. (1981), respectively. Preincubation of gossypol or vehicle together with the complete reaction mixture except for the substrate was for 5 min. Preliminary experiments revealed that the degree of gossypol-mediated inhibition of ALDH-catalyzed oxidation did not increase when preincubation of gossypol with the enzyme was for 20, rather than 5, min. Preincubation temperatures and pHs were the same as incubation temperatures and pHs. All reactions were started by the addition of substrate. Stock ethanol solutions of gossypol were prepared freshly each day and were stored protected from light, on ice, before addition to the reaction mixture. The final concentration of ethanol in the reaction mixture was always 0.5%; this concentration of ethanol did not inhibit any of the enzyme-catalyzed reactions under investigation.

Human breast adenocarcinoma MCF-7/0 and MCF-7/0/CAT cells were cultured (monolayer), harvested when still in exponential growth, resuspended in growth medium, and checked for viability (usually greater than 95% as judged by trypan blue exclusion) as described previously (Sreerama and Sladek, 1993a; Sreerama et al., 1995). Drug exposure and the colony-forming assay used to determine surviving fractions were also as described previously (Sreerama and Sladek, 1993a). Briefly, freshly harvested cells were diluted with drug-exposure medium to a concentration of 1×10^5 cells/ml and then exposed to drug (mafosfamide or phosphoramide mustard) or vehicle for 30 min at pH 7.4 and 37°C after which they were harvested and cultured in drug-free growth medium for 15 days. Colonies (^{3} 50 cells) were then visualized with methylene blue dye and counted. Stock solutions of mafosfamide and phosphoramide mustard were prepared by placing them into aqueous solution just before use. In some experiments, cells were preincubated with 75 μM gossypol or vehicle for 5 min at 37°C prior to the addition of mafosfamide or phosphoramide mustard. Stock ethanol solutions of gossypol were diluted with drug-exposure medium just before use. Ethanol concentrations in the drug-exposure medium never exceeded 0.5%; this concentration of ethanol did not affect the rate of cell proliferation. At the concentration used (75 μM), gossypol effected only a small amount of cell-kill (< 10%) and this was taken into account when calculating the effect of including gossypol in the drug-exposure medium on LC_{90} (concentration of drug required to effect 90% cell-kill) values for mafosfamide and phosphoramide mustard.

Double-reciprocal (Lineweaver-Burk) plots of initial catalytic rates (determined in duplicate) as a function of substrate concentrations (at least four and usually six) were used to estimate the Km and Vmax values. Ki values were determined by plotting the slopes of the lines generated by the double-reciprocal (Lineweaver-Burk) plots as a function of gossypol concentrations. In the case of double-reciprocal (Lineweaver-Burk) plots, computer-assisted Wilkinson weighted linear regression analysis (Wilkinson, 1961) effected by the MacWilkins program (Microsoft, Bellevue, WA) was used to generate best-fit lines. Computer-assisted unweighted linear regression analysis effected by the STATView statistical program (Brainpower Inc., Calabas, CA) was used to generate best-fit lines for all other straight-line functions.

3. RESULTS AND DISCUSSION

Gossypol was not a substrate for either the oxidative (oxidation of aldehydes) or the hydrolytic (hydrolysis of esters) reactions catalyzed by any of the ALDHs studied.

Oxidative reactions catalyzed by rALDH-1, rALDH-2 and the ALDH-3s were inhibited by gossypol (Fig. 2 and Table 1). As judged by the concentrations of gossypol re-

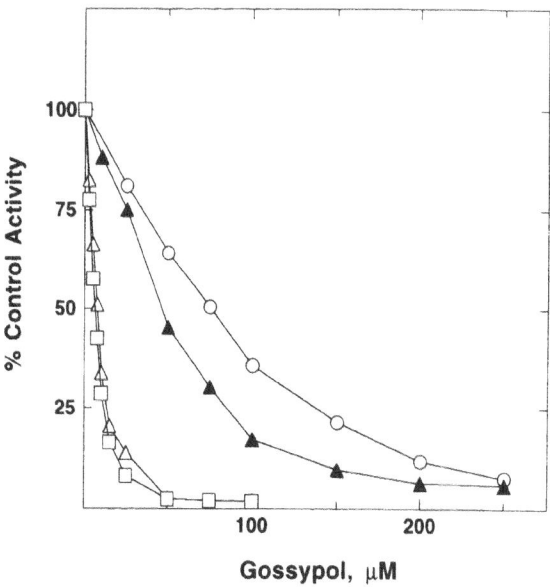

Figure 2. Inhibition by gossypol of oxidative reactions catalyzed by human aldehyde dehydrogenases. The sensitivities of rALDH-1 (O), rALDH-2 (▲), nALDH-3 (△), and tALDH-3 (□) to inhibition by gossypol were determined as described in Materials and Methods and the footnotes to Table 1. Data points are means of duplicate determinations. Control catalytic rates are listed in a footnote to Table 1. IC_{50} values were estimated directly from the plots shown in this figure and are given in Table 1.

Table 1. Inhibition by gossypol of human aldehyde dehydrogenase-catalyzed oxidation and hydrolysis: IC_{50} values[a]

Enzyme	IC_{50}, μM	
	Oxidation[b]	Hydrolysis[c]
rALDH-1	75	> 200
rALDH-2	45	> 200
nALDH-3	7.5	> 200
tALDH-3	6.6	> 200
GAPDH	70	—
Alkaline phosphatase	—	130

[a]Enzymes were incubated with vehicle or various concentrations of gossypol for 5 min, substrate was added, and initial catalytic rates were quantified as described in Materials and Methods. A minimum of seven gossypol concentrations, at least two of which effected less than 50% inhibition, and two of which effected greater than 50% inhibition, were used. Examples of plots of the primary data thus obtained are shown in Fig. 2. IC50 values were estimated directly from such plots.
[b]Substrates and cofactors were, respectively, acetaldehyde and NAD (4 mM each) for rALDH-1, acetaldehyde (2 mM) and NAD (4 mM) for rALDH-2, benzaldehyde (4 mM) and NAD (1 mM) for the ALDH-3s, and glyceraldehyde 3-phosphate and NAD (1 mM each) for glyceraldehyde-3-phosphate dehydrogenase (GAPDH). Uninhibited catalytic rates were 0.56, 3.3, 25, 31 and 56 IU/mg protein for rALDH-1, rALDH-2, nALDH-3, tALDH-3 and GAPDH, respectively.
[c]Substrates were p-nitrophenyl acetate (500 μM) for the dehydrogenases and p-nitrophenyl phosphate (10 mM) for AP. Uninhibited catalytic rates were 126, 476, 11, 9.6 and 16 IU/mg protein for rALDH-1, rALDH-2, nALDH-3, tALDH-3 and alkaline phosphatase, respectively.

Table 2. Inhibition by gossypol of human aldehyde dehydrogenase-catalyzed oxidation: Ki values[a]

Enzyme	Substrate and Cofactor		Km (μM)	Vmax (IU/mg)	Ki (μM)
	Variable (mM)	Fixed (mM)			
rALDH-1	Acetaldehyde (0.125 - 4)	NAD (4)	546	0.66	69
	NAD (0.025 - 1)	Acetaldehyde (4)	59	0.63	5.3
rALDH-2	Acetaldehyde (0.05 - 1)	NAD (4)	5.5[b]	3.4	37
	NAD (0.125 - 4)	Acetaldehyde (2)	327	3.6	7.0
nALDH-3	Benzaldehyde (0.165 - 4)	NAD (1)	413	27	10
	NAD (0.025 - 1)	Benzaldehyde (4)	33	25	0.32
	Benzaldehyde (0.165 - 4)	NADP (4)	389	64	5.8
	NADP (0.5 - 4)	Benzaldehyde (4)	765	72	0.56
tALDH-3	Benzaldehyde (0.125 - 4)	NAD (1)	349	35	4.3
	NAD (0.020 - 1)	Benzaldehyde (4)	47	32	0.10
	Benzaldehyde (0.125 - 4)	NADP (4)	319	69	4.4
	NADP (0.5 - 4)	Benzaldehyde (4)	764	72	0.19

[a]Enzymes were incubated with vehicle or various concentrations of gossypol for 5 min, substrate was added, and initial catalytic rates were quantified as described in Materials and Methods. Representative plots of the primary data from which the kinetic constants were estimated are given in Figs. 3 and 4.

[b]Unlikely to be accurate because it is difficult to ascertain Km values that are less than about 10 μM from the very flat Lineweaver-Burk plots that we generated. Thus, the Km value was determined to be < 0.1 μM when a more appropriate experimental design and method of analysis, viz., integrated Michaelis analysis of a single enzyme-progress curve, was used [Dockham et al., 1992].

quired to effect 50% inhibition (IC_{50}), ALDH-3s were, relative to rALDH-1, rALDH-2 and human glyceraldehyde-3-phosphate dehydrogenase, far more sensitive to inhibition by gossypol. Inhibition could not be reversed by passing the enzyme inhibitor complex (tALDH-3 gossypol generated by incubating tALDH-3 with an amount of gossypol that inhibited tALDH-3-catalyzed oxidation of benzaldehyde by 85%, viz., 20 μM, for 5 min) through a PD-10 (Sephadex G-25) column (data not shown). Seemingly inconsistent with our findings, others have reported that inhibition of lactate dehydrogenase-X by gossypol was reversed when the enzyme inhibitor complex was passed through a Bio Gel P4 column (Olgiati and Toscano, 1983).

Fifty-percent inhibition of a hydrolytic reaction (hydrolysis of p-nitrophenyl acetate) catalyzed by the ALDHs was not achieved at the highest concentration of gossypol, 200 μM, tested (Table 1). As expected (Feldman and Weiner, 1972; Sidhu and Blair, 1975; Sreerama and Sladek, 1993a), inclusion of 1 mM NAD in the reaction mixture increased the rates at which ALDH-1 and ALDH-2 catalyzed the hydrolysis of p-nitrophenyl acetate (3- and 6-fold, respectively), but not the rates at which the ALDH-3s did so (data not shown). As when NAD was not included in the incubation mixture, 50% inhibition of ALDH-catalyzed hydrolysis by gossypol, 200 μM, was not achieved when it was (data not shown). The hydrolytic reaction (hydrolysis of p-nitrophenyl phosphate) catalyzed by human placental alkaline phosphatase was inhibited by gossypol, albeit poorly (Table 1).

As judged by Ki values determined with respect to the substrate, as well as those determined with respect to the cofactor, ALDH-3-catalyzed oxidation was inhibited to a greater extent by gossypol than was that catalyzed by rALDH-1 and rALDH-2 (Table 2). Inhibition was always noncompetitive with respect to the aldehyde. A double-reciprocal (Lineweaver-Burk) plot illustrating this point in the case of tALDH-3 is shown in Fig. 3. Inhibition was always competitive with respect to the cofactor. A double-reciprocal

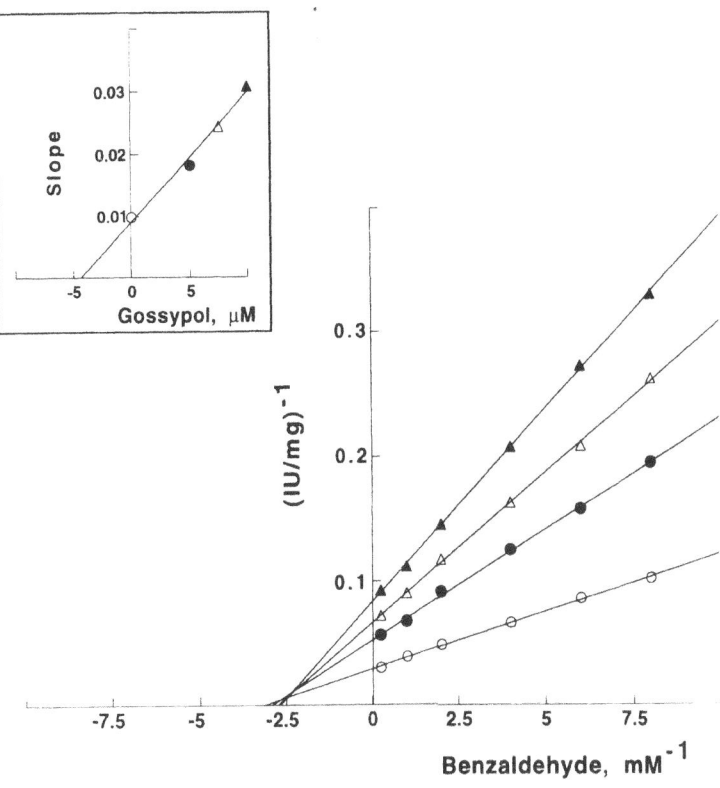

Figure 3. Inhibition of NAD-linked tALDH-3-catalyzed oxidation of benzaldehyde by gossypol: Lineweaver-Burk plot of initial catalytic rates as a function of substrate, benzaldehyde, concentrations. tALDH-3 was incubated with 0 (O), 5 (●), 7.5 (△) or 10 (▲) μM gossypol for 5 min, various concentrations of the substrate, benzaldehyde, were added, and initial catalytic rates were quantified as described in Materials and Methods. The NAD concentration was 1 mM. Data points are means of duplicate determinations. Inset: Slopes generated by double-reciprocal (Lineweaver-Burk) plots were plotted as a function of gossypol concentrations for the purpose of determining the Ki value. Km, Vmax and Ki values obtained in this experiment were 349 μM, 35 IU/mg and 4.3 μM, respectively.

(Lineweaver-Burk) plot illustrating this point in the case of tALDH-3 is shown in Fig. 4. Thus, gossypol competes with the cofactor, rather than with the aldehyde, for a binding site.

Gossypol has been shown to inhibit a variety of NAD(P)-linked enzyme-catalyzed oxidations, *vide supra*, as well as NAD(P)H-linked enzyme-catalyzed reductions, e.g., those catalyzed by aldose reductase (Deck et al., 1991) and 5α-reductase (Moh et al., 1993). Consistent with our findings, inhibition was, with two exceptions, noncompetitive with respect to the substrate and competitive with respect to the pyridine nucleotide in each of the cases where this determination was made (Burgos et al., 1986; Ikeda, 1990; Moh et al., 1993). One of the exceptions was lactate dehydrogenase-X-catalyzed reduction of α-ketobutyrate where inhibition was competitive with respect to the substrate and noncompetitive with respect to the pyridine nucleotide (Morris et al., 1986). The other was 11-β-hydroxysteroid dehydrogenase-catalyzed oxidation of corticosterone where inhibition was competitive with respect to the substrate (Sang et al., 1991).

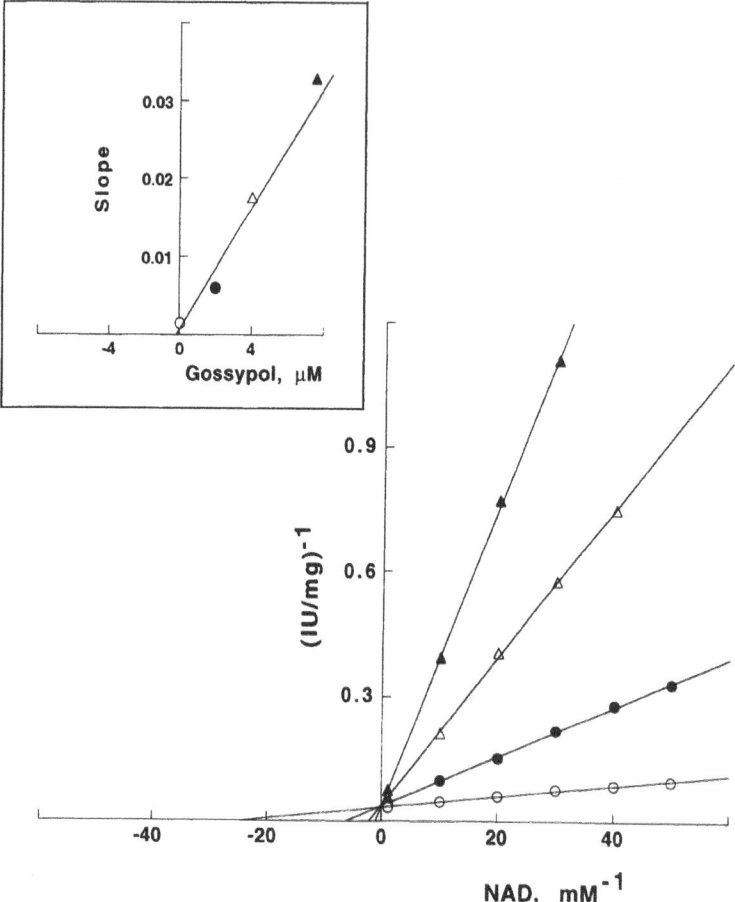

Figure 4. Inhibition of NAD-linked tALDH-3-catalyzed oxidation of benzaldehyde by gossypol: Lineweaver-Burk plot of initial catalytic rates as a function of cofactor, NAD, concentrations. tALDH-3 was incubated with 0 (O), 2 (●), 4 (△) or 7.5 (▲) μM gossypol for 5 min, the substrate, benzaldehyde (4 mM), was added, and initial catalytic rates were quantified as described in Materials and Methods. Data points are means of duplicate determinations. Inset: Slopes generated by the double-reciprocal (Lineweaver-Burk) plots were plotted as a function of gossypol concentrations for the purpose of determining the Ki value. Km, Vmax and Ki values obtained in this experiment were 47 μM, 32 IU/mg and 0.10 μM, respectively.

Gossypol has also been shown to inhibit enzymes that are not dehydrogenases or reductases, e.g., glutathione S-transferases (Lee et al., 1982; Benz et al., 1990), DNA polymerase α (Rosenberg et al., 1986), protein kinase C (Nakadate et al., 1988; Benz et al., 1990), calmodulin-stimulated cyclic AMP phosphodiesterase (Benz et al., 1990), and adenylate cyclase (Olgiati et al., 1984). Interestingly, inhibition of adenylate cyclase by gossypol was competitive and it has been inferred that gossypol may inhibit all enzymes for which pyridine nucleotides or ATP are a substrate or cofactor (Olgiati and Toscano, 1983; Olgiati et al., 1984).

Initially, gossypol was of investigative interest mainly because of its male contraceptive properties (reviewed in Wu, 1989). Subsequently, its therapeutic potential in the treatment of certain gynecological diseases, e.g., endometriosis and uterine myoma, became of interest (reviewed in Wu, 1989). Recognition, during the course of those investigations, of its inhibitory effect on a number of intracellular enzymes important for cellular growth led to considerable interest in developing other therapeutic uses for gossypol, viz., the chemotherapy of microbial, parasitic and neoplastic diseases. Thus, it was soon demonstrated that gossypol and its derivatives inhibit the *in vitro* replication of certain viruses (Wichmann, et al., 1982; Royer, et al., 1991), *in vitro* growth of *Trypanosoma cruzi*, the parasite that causes Chagas' disease (Montamat et al., 1982), and, most notably, the *in vitro* proliferation of a variety of human tumor cell lines, e.g., those derived from breast, hepatic, pancreatic, ovarian, testicular, colon, skin, adrenal, cervical and brain carcinomas (Tuszynski and Cossu, 1984; Joseph et al., 1986; Band et al., 1989; Wu et al., 1989; Benz et al., 1990; Ford et al., 1991; Hu et al., 1993; Coyle et al., 1994; Gilbert et al., 1995). Moreover, gossypol potently inhibited proliferation of doxorubicin-resistant (Benz et al., 1990; Hu et al., 1993), as well as estrogen-responsive and estrogen-nonresponsive (Gilbert et al., 1995), human breast cancer cells. Most importantly, it was demonstrated in several animal models that tumor cell-kill could be achieved *in vivo* with doses of gossypol that were not injurious to the host animal (Tso, 1984; Wu et al., 1989; Chang et al., 1993; Naik et al., 1995). These observations prompted clinical trials of gossypol for the treatment of certain cancers, viz., metastatic carcinomas of the ovary (Wu, 1989), various advanced cancers (Stein et al., 1992), metastatic adrenocortical carcinomas (Flack et al., 1993) and metastatic breast cancers (Seidman, 1996).

In our investigations, gossypol was toxic to human breast adenocarcinoma MCF-7/0 and MCF-7/0/CAT cells (Fig. 5). LC_{90} values were 175 and 205 µM, respectively. Proliferation of MCF-7/0 and MCF-7/0/CAT cells was essentially unaffected by 75 µM gossypol.

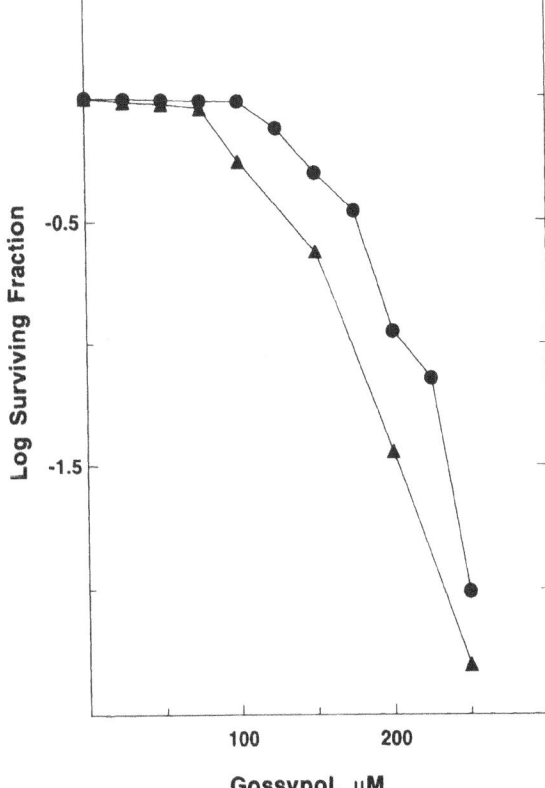

Figure 5. Sensitivity of human breast adenocarcinoma MCF-7/0 cells that had been grown in the presence and absence of catechol to gossypol. Exponentially growing MCF-7/0 cells were cultured in the presence of vehicle (▲; MCF-7/0) or 30 µM catechol (●; MCF-7/0/CAT) for 5 days after which time they were harvested and exposed to various concentrations of gossypol for 35 min at 37°C. The cells were then harvested and grown in drug-free growth medium for 15 days. The colony-forming assay described in Materials and Methods was used to determine surviving fractions. Data points are means of triplicate determinations.

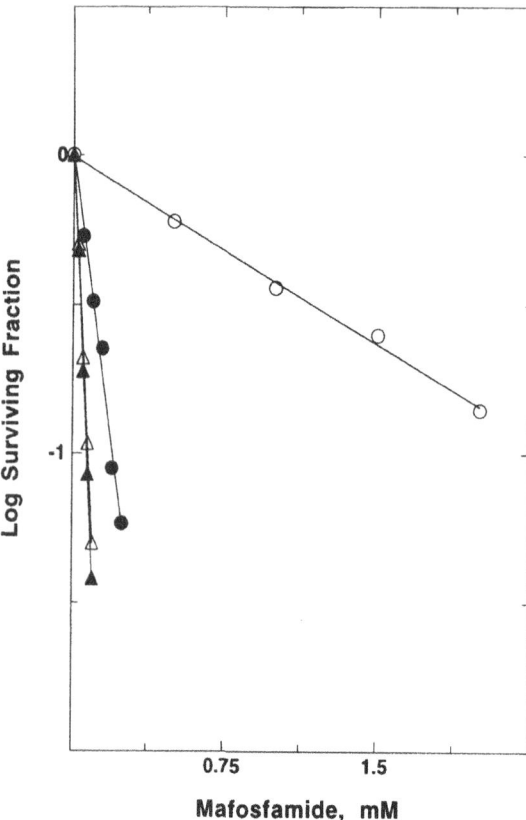

Figure 6. Sensitivities of human breast adenocarcinoma MCF-7/0 and MCF-7/0/CAT cells to mafosfamide in the presence and absence of gossypol. Exponentially growing MCF-7/0 cells were cultured in the presence of vehicle (△, ▲ MCF-7/0) or 30 µM catechol (O, ● MCF-7/0/CAT) for 5 days after which time they were each harvested and incubated with vehicle (△,O) or 75 µM gossypol (▲,●) for 5 min at 37°C. Mafosfamide was added and incubation was continued for an additional 30 min. The cells were then harvested and grown in drug-free growth medium for 15 days. The colony-forming assay described in Materials and Methods was used to determine surviving fractions. Data points are means of triplicate determinations. Cellular levels of aldehyde dehydrogenase activities (cofactor and substrate were 4 mM each of NADP and benzaldehyde, respectively) were 1.8 and 680 mIU/10^7 cells in 105,000 g supernatant fractions obtained from Lubrol-treated whole homogenates of MCF-7/0 and MCF-7/0/CAT cells, respectively.

Addition of gossypol (75 µM) to the drug-exposure medium prior to exposure to mafosfamide markedly increased the sensitivity of tumor cells that express large amounts of ALDH-3, viz., MCF-7/0/CAT, to the oxazaphosphorine (Fig. 6 and Table 3). As expected, identical treatment of tumor cells that express very small amounts of ALDH-3, viz., MCF-7/0, only very minimally increased their sensitivity to mafosfamide. Also as expected because ALDH-3 does not catalyze the detoxification of phosphoramide mustard, the ultimate cytotoxic metabolite of mafosfamide (Sladek, 1994), addition of gossypol to the drug-exposure medium prior to exposure to this agent essentially did not increase the sensitivity of MCF-7/0/CAT cells to it (Table 3).

Racemic gossypol was used in our, as well as in most other, investigations, including the clinical trials referred to above. Significant differences in the potency of the gossypol enantiomers with respect to inhibition of proliferation, fertilization and enzyme catalysis have been reported. Thus, the (-)enantiomer has been shown to be more toxic to cultured tumor cells than was the (+)enantiomer or the racemate (Joseph et al., 1986; Band et al., 1989; Benz et al., 1990; Ford et al., 1991), and some researchers have attributed the contraceptive properties of gossypol entirely to the (-)enantiomer, although others have reported that the two enantiomers are almost equipotent in that regard (reviewed in Yu, 1987). Some enzymes, viz., glutathione S-transferases α and π, have been shown to be inhibited to a greater extent by the (-)enantiomer (Benz et al., 1990), but others, viz., protein kinase C, calmodulin-stimulated cyclic AMP phosphodiesterase, DNA polymerase α and

Table 3. Sensitivity of human breast adenocarcinoma MCF-7/0 and MCF-7/0/CAT cells to mafosfamide and phosphoramide mustard in the presence and absence of gossypol[a]

Cell Line	ALDH-3 (mIU/10^7 cells)	Anticancer Agent	Gossypol (µM)	LC$_{90}$ (µM)
MCF-7/0	2	Mafosfamide	0	70
			75	65
MCF-7/0/CAT	680	Mafosfamide	0	> 2000
			75	200
		Phosphoramide mustard	0	1400
			75	1100

[a]Human breast adenocarcinoma MCF-7/0 cells were cultured in the presence of vehicle (MCF-7/0) or 30 µM catechol (MCF-7/0/CAT) for 5 days. At the end of this time, cells were harvested, washed, and resuspended in drug-exposure medium. The cells (1 x 10^5 cells/ml) were then incubated with gossypol or vehicle for 5 min at 37°C after which time various concentrations of mafosfamide, phosphoramide mustard or vehicle were added and incubation was continued as before for 30 min at 37°C. The colony-forming assay described in Materials and Methods was used to determine surviving fractions. LC$_{90}$ values (concentrations of drug required to effect 90% cell-kills) were obtained from plots of log surviving fractions versus concentrations of drug (Fig. 6). Values are means of LC$_{90}$s obtained in two experiments. Cellular levels of ALDH-3 activity (NADP-linked enzyme-catalyzed oxidation of benzaldehyde; 4 mM each of cofactor and substrate) in 105,000 g supernatant fractions obtained from Lubrol-treated whole homogenates of tumor cells were determined as described in Materials and Methods.

glutathione S-transferase µ, are reportedly each almost equally sensitive to inhibition by the two enantiomers (Rosenberg et al., 1986; Benz et al., 1990), and equisensitivity to the enantiomers on the part of rat and human lactate dehydrogenase-X (Kim et al., 1985; Morris et al., 1986; Yao et al., 1987), as well as greater sensitivity to the (−)enantiomer on the part of the hamster orthologue (Morris et al., 1986; Den Boer and Grootegoed, 1988), have been reported. The relative ability of the two enantiomers to inhibit ALDH-catalyzed oxidations remains to be determined. Seemingly relevant to any potential clinical use of gossypol, marked differences in the disposition of the two enantiomers have been documented (reviewed in Yu, 1987). For example, the plasma half-life of the (−)enantiomer is 4.55 hours whereas that of the (+)enantiomer is 133 hours in patients receiving a single oral dose of racemic gossypol (Wu et al., 1986).

Gossypol is found in the leaves, roots and seeds of the cotton plant. Commercial products originating from these plants include cottonseed meal, used extensively as a protein supplement in animal feed, and cottonseed oil, frequently used for processing human food. Federal (United States) regulation limits the concentration of free gossypol in animal feed, as well as in products intended for human consumption, to 450 ppm (FDA, 1974). Gossypol has been found to be present in milk collected from cows that had been fed the federally allowable 450 ppm gossypol for six days (Hu et al., 1994). Uncertain is whether dietary intake of gossypol by humans is sufficient to inhibit aldehyde dehydrogenases, and thus the rate at which oxazaphosphorines, ethanol (acetaldehyde), etc., are detoxified, but that would seem unlikely, at least in the United States.

In any event, the findings reported herein establish the therapeutic potential of combining gossypol with an oxazaphosphorine in the treatment of certain cancers. Moreover, given the antitumor activity that gossypol itself exhibits, it can be envisaged that in the case of some of these cancers, gossypol could be of dual therapeutic value when combined with an oxazaphosphorine in the therapeutic protocol.

4. ACKNOWLEDGMENT

These investigations were supported by USPHS Grant CA 21737 and Department of Defense Grant DAMD 17–94-J-4057.

5. REFERENCES

Band, V., Hoffer, A.P., Band, H., Rhinehardt, A.E., Knapp, R.C., Matlin, S.A. and Anderson, D.J.: Antiproliferative effect of gossypol and its optical isomers on human reproductive cancer cell lines. Gynecol. Oncol. 32 (1989) 273–277.

Benz, C.C., Keniry, M.A., Ford, J.M., Townsend, A.J., Cox, F.W., Palayoor, S., Matlin, S.A., Hait, W.N. and Cowan, K.H.: Biochemical correlates of the antitumor and antimitochondrial properties of gossypol enantiomers. Mol. Pharmacol. 37 (1990) 840–847.

Bunting, K.D. and Townsend, A.J.: Protection by transfected rat or human class 3 aldehyde dehydrogenases against the cytotoxic effects of oxazaphosphorine alkylating agents in hamster V79 cell lines. J. Biol. Chem. 271 (1996) 11891–11896.

Bunting, K.D., Lindahl, R. and Townsend, A.J.: Oxazaphosphorine-specific resistance in human MCF-7 breast carcinoma cell lines expressing transfected rat class 3 aldehyde dehydrogenase. J. Biol. Chem. 269 (1994) 23197–23203.

Burgos, C., Gerez de Burgos, N.M., Rovai, L.E. and Blanco, A.: In vitro inhibition by gossypol of oxidoreductases from human tissues. Biochem. Pharmacol. 35 (1986) 801–804.

Chang, C.J.G., Ghosh, P.K., Hu, Y.F., Brueggemeier, R.W. and Lin, Y.C.: Antiproliferative and antimetastatic effects of gossypol on Dunning prostate cell-bearing Copenhagen rats. Res. Commun. Chem. Pathol. Pharmacol. 79 (1993) 293–312.

Chueh, S.-H., Chang, G.-G., Chang, T.-C. and Pan, F.: Involvement of arginine residue in the phosphate binding site of human placental alkaline phosphatase. Int. J. Biochem. 13 (1981) 1143–1149.

Coyle, T., Levante, S., Shetler, M. and Winfield, J.: In vitro and in vivo cytotoxicity of gossypol against central nervous system tumor cell lines. J. Neurooncol. 19 (1994) 25–35.

Deck, L.M., Vander Jagt, D.L., Royer, R.E.: Gossypol and Derivatives: a new class of aldose reductase inhibitors. J. Med. Chem. 34 (1991) 3301–3305.

Den Boer, P.J. and Grootegoed, J.A.: Differential effects of (+)- and (−)-gossypol enantiomers on LDH-C$_4$ activity of hamster spermatogenic epithelium in vitro. J. Reprod. Fertil. 83 (1988) 701–709.

Dockham, P.A., Lee, M.-O. and Sladek, N.E.: Identification of human aldehyde dehydrogenases that catalyze the oxidation of aldophosphamide and retinaldehyde. Biochem. Pharmacol. 43 (1992) 2453–2469.

FDA: Food drug cosmetic law. 56 (1974) 518.94, 172.894.

Feldman, R.I. and Weiner, H.: Horse liver aldehyde dehydrogenase. II. Kinetics and mechanistic implications of the dehydrogenase and esterase activity. J. Biol. Chem. 247 (1972) 267–272.

Flack, M.R., Pyle, R.G., Mullen, N.M., Lorenzo, B., Wu, Y.W., Knazek, R.A., Nisula, B.C. and Reidenberg, M.M.: Oral gossypol in the treatment of metastatic adrenal cancer. J. Clin. Endocrinol. Metab. 76 (1993) 1019–1024.

Ford, J.M., Hait, W.N., Matlin, S.A. and Benz, C.C.: Modulation of resistance to alkylating agents in cancer cell by gossypol enantiomers. Cancer Lett. 56 (1991) 85–94.

Gilbert, N.E., O'Reilly, J.E., Chang, C.J.G., Lin, Y.C. and Brueggemeier, R.W.: Antiproliferative activity of gossypol and gossypolone on human breast cancer cells. Life Sci. 57 (1995) 61–67.

Hu, Y.-F., Chang, C.J.G., Brueggemeier, R.W. and Lin, Y.C.: Gossypol inhibits basal and estrogen-stimulated DNA synthesis in human breast carcinoma cells. Life Sci. 53 (1993) 433–438.

Hu, Y.-F., Chang, C.J.G., Brueggemeier, R.W. and Lin, Y.C.: Presence of antitumor activities in the milk collected from gossypol-treated dairy cows. Cancer Lett. 87 (1994) 17–23.

Ikeda, M.: Inhibition kinetics of NAD-linked enzymes by gossypol acetic acid. Andrologia 22 (1990) 409–416.

Joseph, A.E.A., Matlin, S.A. and Knox, P.: Cytotoxicity of enantiomers of gossypol. Br. J. Cancer 54 (1986) 511–513.

Kim, I.C., Waller, D.P. and Fong, H.H.S.: Inhibition of LDH-X by gossypol optical isomers. J. Androl. 6 (1985) 344–347.

Lambeir, A.-M., Loiseau, A.M., Kuntz, D.A., Vellieux, F.M., Michels, P.A.M. and Opperdoes, F.R.: The cytosolic and glycosomal glyceraldehyde-3-phosphate dehydrogenase from Trypanosoma brucei. Kinetic properties and comparison with homologous enzymes. Eur. J. Biochem. 198 (1991) 429–435.

Lee, C.-Y.G., Moon, Y.S., Yuan, J.H., and Chen, A.F.: Enzyme inactivation and inhibition by gossypol. Mol. Cell. Biochem. 47 (1982) 65–70.

Messiha, F.S.: Effect of gossypol on kinetics of mouse liver alcohol and aldehyde dehydrogenase. Gen. Pharmacol. 22 (1991a) 573–576.

Messiha, F.S.: Behavioral and metabolic interaction between gossypol and ethanol. Toxicol. Lett. 57 (1991b) 175–181.

Moh, P.P., Chang, G.C.J., Brueggemeier, R.W. and Lin, Y.C.: Effect of gossypol on 5α-reductase and 3α-hydroxysteroid dehydrogenase activites in adult rat testes. Res. Comm. Chem. Pathol. Pharmacol. 82 (1993) 12–26.

Montamat, E.E., Burgos, C., Gerez de Burgos, N.M., Rovai, L.E. and Blanco, A.: Inhibitory action of gossypol on enzymes and growth of *Trypanosoma cruzi*. Science 218 (1982) 288–289.

Moreb, J., Schweder, M., Suresh, A. and Zucali, J.R.: Overexpression of the human aldehyde dehydrogenase class I results in increased resistance to 4-hydroperoxycyclophosphamide. Cancer Gene Ther. 3 (1996) 24–30.

Morris, I.D., Higgins, C. and Matlin, S.A.: Inhibition of testicular LDH-X from laboratory animals and man by gossypol and its isomers. J. Reprod. Fertil. 77 (1986) 607–612.

Naik, H.R., Petrylak, D., Yagoda, A., Lehr, J.E., Akhtar, A. and Pienta, K.J.: Preclinical studies of gossypol in prostate carcinoma. Int. J. Oncol. 6 (1995) 209–213.

Nakadate, T., Jeng, A.Y. and Blumberg, P.M: Comparison of protein kinase C functional assays to clarify mechanisms of inhibitor action. Biochem. Pharmacol. 37 (1988) 1541–1545.

Olgiati, K.L. and Toscano, W.A. Jr.: Kinetics of gossypol inhibition of bovine lactate dehydrogenase X. Biochem. Biophys. Res. Commun. 115 (1983) 180–185.

Olgiati, K.L., Toscano, D.G., Atkins, W.M. and Toscano, W.A. Jr.: Gossypol inhibition of adenylate cyclase. Arch. Biochem. Biophys. 231 (1984) 411–415.

Rekha, G.K., Sreerama, L. and Sladek, N.E.: Intrinsic cellular resistance to oxazaphosphorines exhibited by a human colon carcinoma cell line expressing relatively large amounts of a class-3 aldehyde dehydrogenase. Biochem. Pharmacol. 48 (1994) 1943–1952.

Rosenberg, L.J., Adlakha, R.C., Desai, D.M. and Rao, P.N.: Inhibition of DNA polymerase α by gossypol. Biochim. Biophys. Acta 866 (1986) 258–267.

Royer, R.E., Mills, R.G., Deck, L.M., Mertz, G.J. and Vander Jagt, D.L.: Inhibition of human immunodeficiency virus type I replication by derivatives of gossypol. Pharmcol. Res. 24 (1991) 407–412.

Sambrook, J., Fritsch, E.F. and Maniatis, T.: Preparation and transformation of competent *E. coli*. In Molecular Cloning. A Laboratory Manual, 2nd edition, Cold Spring Harbor Laboratory Press, 1989, pp. 1.74–1.84.

Sang, G.W., Lorenzo, B. and Reidenberg, M.M.: Inhibitory effects of gossypol on corticosteroid 11-β-hydroxysteroid dehydrogenase from guinea pig kidney: a possible mechanism for causing hypokalemia. J. Steroid Biochem. Mol. Biol. 39 (1991) 169–176.

Seidman, A.D.: Chemotherapy for advanced breast cancer: a current perspective. Semin. Oncol. 23 (1996) 55–59.

Sidhu, R.S. and Blair, A.H.: Human liver aldehyde dehydrogenase. Esterase activity. J. Biol. Chem. 250 (1975) 7894–7898.

Sladek, N.E.: Oxazaphosphorine-specific acquired cellular resistance. In Teicher, B.A. (Ed.), Drug Resistance in Oncology, Marcel Dekker, N.Y., 1993, pp. 375–411.

Sladek, N.E.: Metabolism and pharmacokinetic behavior of cyclophosphamide and related oxazaphosphorines. In Powis, G. (Ed.), Anticancer Drugs: Reactive Metabolism and Drug Interactions, Pergamon Press, United Kingdom, 1994, pp. 79–156.

Sladek, N.E., Sreerama, L. and Rekha, G.K.: Constitutive and overexpressed human cytosolic class-3 aldehyde dehydrogenases in normal and neoplastic cells/secretions. Adv. Exp. Med. Biol. 372 (1995) 103–114.

Sreerama, L. and Sladek, N.E.: Identification and characterization of a novel class 3 aldehyde dehydrogenase overexpressed in a human breast adenocarcinoma cell line exhibiting oxazaphosphorine-specific acquired resistance. Biochem. Pharmacol. 45 (1993a) 2487–2505.

Sreerama, L. and Sladek, N.E.: Overexpression or polycyclic aromatic hydrocarbon-mediated induction of an apparently novel class 3 aldehyde dehydrogenase in human breast adenocarcinoma cells and its relationship to oxazaphosphorine-specific acquired resistance. Adv. Exp. Med. Biol. 328 (1993b) 99–113.

Sreerama, L. and Sladek, N.E.: Identification of a methylcholanthrene-induced aldehyde dehydrogenase in a human breast adenocarcinoma cell line exhibiting oxazaphosphorine-specific acquired resistance. Cancer Res. 54 (1994) 2176–2185.

Sreerama, L. and Sladek, N.E.: Human breast adenocarcinoma MCF-7/0 cells electroporated with cytosolic class 3 aldehyde dehydrogenases obtained from tumor cells and a normal tissue exhibit differential sensitivity to mafosfamide. Drug Metab. Dispos. 23 (1995) 1080–1084.

Sreerama, L., Rekha, G.K. and Sladek, N.E.: Phenolic antioxidant-induced overexpression of class-3 aldehyde dehydrogenase and oxazaphosphorine-specific resistance. Biochem. Pharmacol. 49 (1995) 669–675.

Stein, R.C., Joseph, A.E.A., Matlin, S.A., Cunningham, D.C., Ford, H.T. and Coombes, R.C.: A preliminary clinical study of gossypol in advanced human cancer. Cancer Chemother. Pharmacol. 30 (1992) 480–482.

Tso, W.-W.: Gossypol inhibits Ehrlich ascites tumor cell proliferation. Cancer Lett. 24 (1984) 257–261.

Tuszynski, G.P. and Cossu, G.: Differential cytotoxic effect of gossypol on human melanoma, colon carcinoma, and other tissue culture cell lines. Cancer Res. 44 (1984) 766–771.

Wichmann, K., Vaheri, A. and Luukkainen, T.: Inhibiting herpes simplex virus type 2 infection in human epithelial cells by gossypol, a potent spermicidal and contraceptive agent. Am. J. Obstet. Gynecol. 142 (1982) 593–594.

Wilkinson, G.N.: Statistical estimations in enzyme kinetics. Biochem. J. 80 (1961) 324–332.

Wu, D.-F.: An overview of the clinical pharmacology and therapeutic potential of gossypol as a male contraceptive agent and in gynaecological disease. Drugs 38 (1989) 333–341.

Wu, D.-F., Yu, Y.-W., Tang, Z.-M. and Wang, M.-Z.: Pharmacokinetics of (±)-, (+)-, and (–)-gossypol in humans and dogs. Clin. Pharmacol. Ther. 39 (1986) 613–618.

Wu, Y.-W., Chik, C.L. and Knazek, R.A.: An in vitro and in vivo study of antitumor effects of gossypol on human SW-13 adrenocortical carcinoma. Cancer Res. 49 (1989) 3754–3758.

Yao, K.-Q., Gu, Q.-M. and Lei, H.-P.: Effect of (±)-, (+)- and (–)-gossypol on the lactate dehydrogenase-X activity of rat testis. J. Ethnopharmacol. 20 (1987) 25–29.

Yu Y.-W.: Probing into the mechanism of action, metabolism and toxicity of gossypol by studying its (+)- and (–)-sterioisomers. J. Ethnopharmacol. 20 (1987) 65–78.

Zheng, C.-F., Wang, T.T.Y. and Weiner, H.: Cloning and expression of the full-length cDNAs encoding human liver class 1 and class 2 aldehyde dehydrogenase. Alcohol. Clin. Exp. Res. 17 (1993) 828–831.

17

ALDEHYDE DEHYDROGENASE-MEDIATED METABOLISM OF ACETALDEHYDE AND MAFOSFAMIDE IN BLOOD OF HEALTHY SUBJECTS AND PATIENTS WITH MALIGNANT LYMPHOMA[*]

Melanie Metzenthin,[1] Doris Meier-Tackmann,[1,3] Dharam P. Agarwal,[1] Reinhart Zschaber,[2] and Hans-Joseph Weh[2]

[1]Institute of Human Genetics
[2]Division of Oncology and Haematology
University Hospital Eppendorf
20246 Hamburg, Germany
[3]ASTA Medica AG
60314 Frankfurt/M, Germany

1. INTRODUCTION

Cyclophosphamide (CP) and mafosfamide (MF) are converted in various tissues to 4-hydroxycyclophosphamide, and subsequently to aldophosphamide (AP). In target cells, AP is further broken down to the toxic compounds phosphoramide mustard and acrolein. AP can be oxidized to a nontoxic metabolite (carboxyphosphamide) catalyzed by one or more types of aldehyde dehydrogenase (ALDH) isoenzymes. Thus, intracellular ALDH activity appears to be an important determinant in modulating sensitivity to CP in patients undergoing chemotherapy. The class 1 cytosolic enzyme (ALDH1) has been shown to be particularly important in the metabolism of CP and MF derivatives in bone marrow cells (Kohn and Sladek, 1985; Kastan et al., 1990; Dockham et al., 1992). The overall contribution of blood ALDH in this respect is, however, not known. In the present study, we have determined ALDH activity in human blood subfractions from healthy subjects and malignant lymphoma patients undergoing combination chemotherapy plus or minus CP.

---

[*] This work will be a part of the MD thesis of M. Metzenthin to be submitted to the Faculty of Medicine, University of Hamburg.

2. MATERIALS AND METHODS

Mafosfamide (MF) and 4-hydroperoxycyclophosphamide (4-HC) were kindly provided by Asta Medica AG, Frankfurt/M. Aldophosphamide used in various assays was prepared from MF and 4-HC as described before (von Eitzen et al., 1994).

2.1. Patients

Hodgkin and non-Hodgkin lymphoma patients (n=28) undergoing chemotherapy were included in the study. The patients received polychemotherapy plus or minus CP. Those treated with CP received in addition vincristine, procarbazine, and prednison. Patients treated with regimen without CP mainly received anthracyclin, interferone, pentostatin (deoxycoformycin), and glucocorticoid. In specific cases, when patients did not respond to other chemotherapy regimens, they were treated with Fludara® (fludarabine phosphate) as a single-agent therapy.

2.2 Preparation of Blood Fractions and Enzyme Assay

Blood was collected from healthy blood donors and patients. Leucocytes were isolated from other blood cells by using a FicollPaque density gradient centrifugation. The leucocyte layer was removed from the Ficoll-Paque, washed with 0.32 M saccharose and centrifuged. The pellet was suspended in aqua dest. and lysed by sonication. The supernatant obtained after centrifugation was passed through a mini column (Pierce, USA) filled with CM-Sephadex G-50. To elute the ALDH enzyme protein, the column was washed with 30 mM phosphate buffer, pH 6.0 containing 1 mM EDTA and 1 mM dithiothreitol. The eluate was concentrated by ultrafiltration through an anisotropic membrane filter (Centricon 30, Amicon, USA) and used for the enzyme assay.

The sedimented erythrocytes obtained by Ficoll-Paque density gradient centrifugation were washed with 0.25 M saccharose solution and suspended in 0.1 M phosphate buffer (pH 6.0). The red blood cells were hemolysed by adding an equal volume of chloroform followed by centrifugation at 27,000g for 15 min at 4°C. A hemoglobin-free lysate was obtained by passing the hemolysate through a CM-Sephadex G-50 column followed by elution with 30 mM phosphate buffer, pH 6.0 containing 1 mM EDTA and 1 mM dithiothreitol. The lysate was subsequently used for the enzyme assays.

Platelets were separated from plasma using the methods described by Faraj et al. (1987). The washed platelets were sonified and centrifuged; the supernatant was used for the activity assay. Assay of ALDH was carried out in various lysate preparations using MF or acetaldehyde as substrate. The reduction of NAD^+ was measured with a double-beam spectrophotometer. A standard reaction mixture contained pyrophosphate buffer, (32 mM, pH 8.2 for MF and 0.1 M, pH 9.5 for acetaldehyde), 1 mM EDTA, 4 mM MF or 5 mM acetaldehyde, 4 mM NAD^+ and 0.1 mM pyrazole. Glutathione was added to the reaction mixture in a final concentration of 5 mM when MF was used as substrate.

For the *ex vivo* study, intact erythrocytes from a healthy blood donor were isolated, washed and suspended in 0.25 M sucrose buffered with 60 mM sodium phosphate buffer, pH 7.4. Erythrocytes (200 µl suspension) were incubated with various MF doses (dissolved in the same buffer) for 30 min at 37°C using a water bath. After incubation, the suspension was centrifuged (1,110g; 4°C), and the supernatant containing MF was discarded. The red cells were hemolyzed and hemoglobin-free lysate was prepared as described above and used for the ALDH assay.

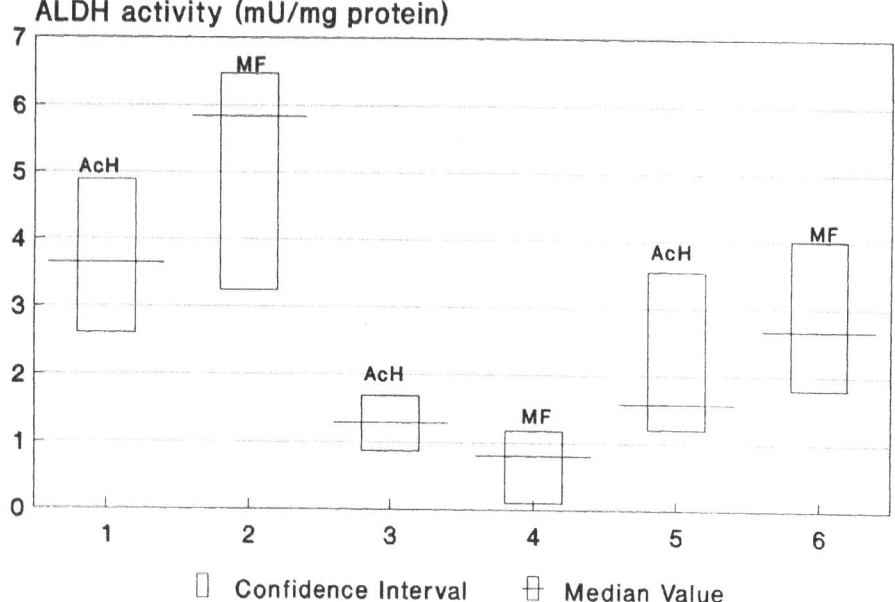

Figure 1. Specific ALDH activity in red cells of healthy controls (1, 2) and lymphoma patients undergoing poly-chemotherapy plus CP (3, 4) and minus CP (5, 6). Enzyme assay was carried out with acetaldehyde (AcH) or ma-fosfamide (MF).

As the data were not normally distributed, Wilcoxon signed rank test was applied for calculation of the group differences and statistical significance. The data are presented as medians; confidence intervals were calculated and are presented as box plots (Fig. 1).

3. RESULTS

In healthy subjects, no ALDH activity was detectable in plasma and blood platelets with both MF and acetaldehyde as substrates. On the other hand, red cell lysates exhibited a substantial ALDHmediated oxidation of MF as well as acetaldehyde. A Km value of 35.7 μmol was calculated for red cell ALDH with aldophosphamide (prepared from 4-HC) as substrate.

Median values and confidence intervals for erythrocyte ALDH activity of the control subjects and patients undergoing polychemotherapy plus or minus CP are presented in Fig. 1. The median value for red cell ALDH activity in healthy controls was significantly higher with MF than that with acetaldehyde ($p < 0.05$). The corresponding values in Hodgkin and non-Hodgkin lymphoma patients undergoing chemotherapy were significantly lower as compared to healthy controls ($p < 0.10$). However, in patients treated with CP, the level of ALDH activity with acetaldehyde was higher than that with MF, but this difference was not statistically significant ($p < 0.10$). In patients under treatment without CP, the enzyme activity values with MF were higher than those in patients treated with combination therapy plus CP ($p < 0.10$).

Leucocytes oxidized both MF and acetaldehyde, albeit at a significantly lower level than the red cells ($p < 0.10$). The oxidation of MF was significantly lower ($p < 0.05$) than

Figure 2. Specific ALDH activity in leucocytes of healthy controls and lymphoma patients undergoing poly-chemotherapy plus or minus CP. Data shown are median values (confidence intervals: * = 1.511.81; ** = 0.59–0.79). Activity assay with MF was only possible in healthy controls.

that of acetaldehyde (Fig. 2). Due to a very low ALDH activity in leucocytes, it was not possible to detect a measurable ALDH activity in leucocyte lysates from patients with MF as substrate.

Changes in red cell ALDH activity were followed in lymphoma patients during combination chemotherapy with and without CP and in remission. In a lymphoma patient (# 17) treated with chemotherapy combination plus CP, a substantial decrease in enzymatic oxidation of acetaldehyde by erythrocyte lysates was noted as compared to that observed before commencing the chemotherapy (Fig. 3). However, there was a relatively less decrease in ALDH activity when MF was used as substrate. The red cell ALDH activity of this patient increased after the chemotherapy was discontinued and showed a complete recovery in remission. In a patient (# 15) undergoing chemotherapy with Fludara® only, the ALDH activity curves remained descending until day 76 although chemotherapy was stopped at day 6 (Fig. 4).

When intact erythrocytes were pre-incubated with MF followed by hemolysis, the lysates showed a lower ALDH activity with both MF and acetaldehyde (Table 1). This effect was found to be dosedependent, especially with MF.

4. DISCUSSION

We have shown earlier (Agarwal et al., 1995) that normal and neoplastic cell lines expressing an unequal level of class 1 and/or class 3 ALDH activity also exhibit differential sensitivity to oxazaphosphorines. Cell lines lacking in ALDH isozymes were more

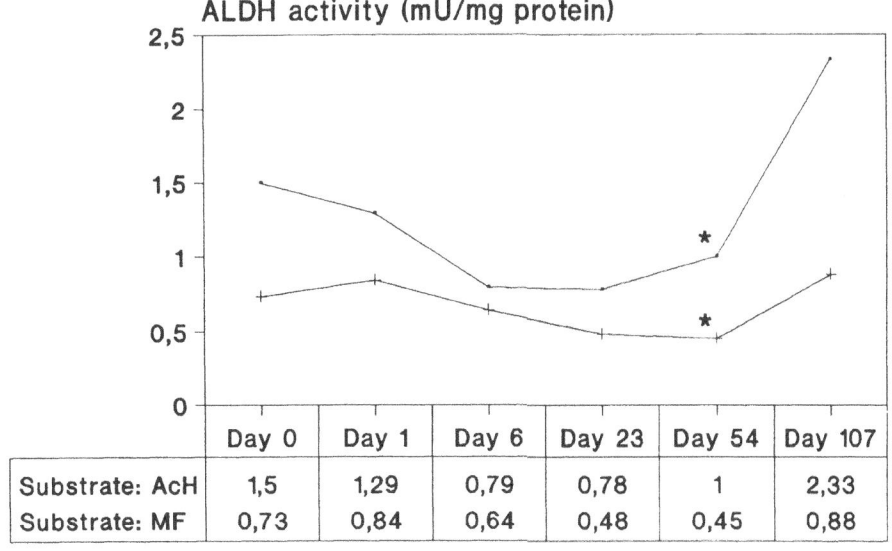

Figure 3. Red cell ALDH activity before and during polychemotherapy plus CP, as well as in remission (patient # 17). Termination of chemotherapy is indicated by an asterisk.

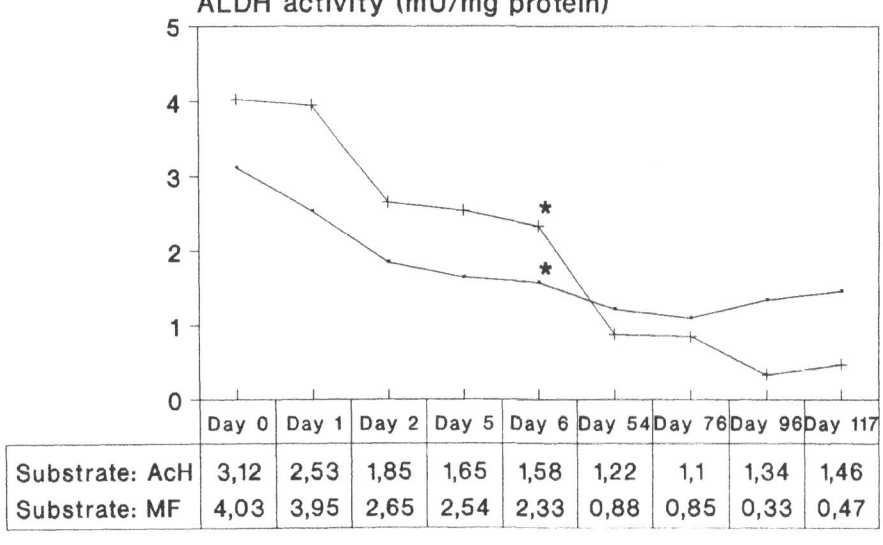

Figure 4. Red cell ALDH activity before and during chemotherapy with Fludara®, as well as in remission (patient # 15). Termination of chemotherapy is indicated by an asterisk.

Table 1. Erythrocyte ALDH activity with acetaldehyde and MF as substrates
after pre-incubation of the intact erythrocytes at 37°C
for 30 min with different concentrations of MF

Pre-incubation with mafosfamide at a concentration of	Specific ALDH activity (mU/mg protein)	
	Acetaldehyde	Mafosfamide
0	3.01	3.26
4 mM	2.32	1.15
8 mM	1.55	0.78
12 mM	0.93	0

sensitive to aldophosphamide than the cell lines which expressed ALDH activity. Apparently, blood ALDH plays also an important role in modulating sensitivity to CP and its derivatives in patients undergoing chemotherapy. It remains, however, to be proven if blood ALDH activity can indeed account for the detoxification of aldophosphamide in patients undergoing chemotherapy with oxazaphosphorines, and whether this detoxification step can lead to the development of acquired resistance to oxazaphosphorines in patients. Such information may be useful in the treatment and management of patients undergoing chemotherapy with CP derivatives.

The preliminary results presented here clearly show that MF is metabolized by human blood. As judged by ALDH activity level in various blood subfractions, the total MF oxidation capacity was, however, relatively low compared with that of liver (von Eitzen et al., 1994). Among the various blood fractions examined, erythrocytes possessed the highest ALDH activity. The results of the present study suggest that in contrast to liver ALDH1, MF is a better substrate for the erythrocyte ALDH1. Though leucocytes also exhibited a substantial ALDH activity, the overall contribution of leucocytes in MF metabolism may be of less significance as compared to erythrocytes.

Ex vivo incubation of intact erythrocytes with MF leads to a substantial inhibition of the red cell ALDH activity, particularly when MF is used as substrate for the activity assay. This substrate-specific inhibition suggests a tight binding of aldophosphamide on the substrate binding pocket of ALDH. Indeed, lymphoma patients during treatment with CP exhibited the lowest ALDH activity with MF as substrate. Moreover, the recovery of ALDH activity after cessation of the chemotherapy plus CP was less pronounced with MF than that measured with acetaldehyde. On the other hand, a patient undergoing chemotherapy without CP (only Fludara®) showed a lower ALDH activity level with acetaldehyde than with MF. In this patient, the recovery of ALDH activity was also higher with acetaldehyde than with MF. Apparently, chemotherapy with Fludara® has a more toxic effect on erythrocyte ALDH activity during the course of treatment and in remission than polychemotherapy plus CP.

5. ACKNOWLEDGMENTS

This study was financially supported by Wilhelm Sander-Stiftung, Germany. We are thankful to Dr. J. Pohl, ASTA Medica, Frankfurt/M for kindly providing MF and 4-HC used in the present study. We also thank Mrs. Elly Losenhausen for technical assistance.

6. REFERENCES

Agarwal, D.P., von Eitzen, U., Meier-Tackmann, D. and Goedde, H.W.: Metabolism of cyclophosphamide by aldehyde dehydrogenases. In: Enzymology and Molecular Biology of Carbonyl Metabolism 5. Eds. H. Weiner, R.S. Holmes and D.W. Crabb. Plenum Press, New York (1995) pp 115–122.

Dockham, P.A., Lee, M.O. and Sladek, N.E.: Identification of human liver aldehyde dehydrogenases that catalyze the oxidation of aldophosphamide and retinaldehyde. Biochem. Pharmacol. 43 (1992) 2453–2469.

Faraj, B.A., Lenton, J.D., Kutner, M., Camp, V.M., Stammers, T.W., Lee, S.R., Lolies, P.A. and Chandora, D.: Prevalence of low monoamine oxidase function in alcoholism. Alcoholism Clin. Exp. Res. 11 (1987) 464–467.

Kastan, M.B., Schlaffer, E., Russo, J.E., Colvin, O.M., Civin, C.I. and Hilton, J.: Direct demonstration of elevated aldehyde dehydrogenase in human hematopoietic progenitor cells. Blood 75 (1990) 1947–1950.

Kohn F.R. and Sladek, N.E.: Aldehyde dehydrogenase activity as the basis for the relative insensitivity of murine pluripotent hematopoietic stem cells to oxazaphosphorines. Biochem. Pharmacol. 34 (1985) 3465–3471.

von Eitzen, U., Meier-Tackmann, D., Agarwal, D.P. and Goedde, H.W.: Detoxification of cyclophosphamide by human aldehyde dehydrogenase isozymes. Cancer Lett. 76 (1994) 45–49.

18

YEAST ALDEHYDE DEHYDROGENASE SENSITIVITY TO INHIBITION BY CHLORPROPAMIDE ANALOGUES AS AN INDICATOR OF HUMAN ALDEHYDE DEHYDROGENASE SENSITIVITY TO THESE AGENTS

Varadahalli R. Devaraj,[1] Lakshmaiah Sreerama,[1] Melinda J. C. Lee,[2] Herbert T. Nagasawa,[2] and Norman E. Sladek[1]

[1]Department of Pharmacology and
[2]Department of Medicinal Chemistry
University of Minnesota
Minneapolis, Minnesota 55455

1. INTRODUCTION

Aldehyde dehydrogenase (E.C. 1.2.1.3) is a polymorphic enzyme that is relatively substrate-nonspecific. Several isoenzymes are found in human tissues. Based on primary structure or subcellular distribution and kinetic, physical and immunochemical properties, they have been placed into one of three classes, viz., class 1, e.g., ALDH-1, class 2, e.g., ALDH-2 and class 3, e.g., ALDH-3 (Anonymous, 1989; Lindahl and Hempel, 1990; Goedde and Agarwal, 1990; Lindahl, 1992). These enzymes catalyze the biotransformation (bioactivation and/or bioinactivation) of a broad spectrum of endogenous (biogenic) and exogenous (xenobiotic) aldehydes that are physiologically and/or pharmacologically important (Sladek et al., 1989, 1995; Lindahl, 1992; Sladek, 1993, 1994). For example, ALDH-1 catalyzes the oxidation of retinaldehyde to retinoic acid, the latter being a potent modulator of cell growth and differentiation; ALDH-1 and, especially, ALDH-2 catalyze the oxidation of ethanol-derived acetaldehyde to acetate, a detoxifying reaction; ALDH-3 catalyzes the oxidation of 4-hydroxynonenal and other aldehydic products of lipid peroxidation to their corresponding acids, also a detoxifying reaction; and ALDH-1 and ALDH-3 catalyze the oxidation of aldophosphamide to carboxyphosphamide, yet another detoxifying reaction, since, alternatively, aldophosphamide, a metabolite of anticancer prodrugs collectively known as oxazaphosphorines, e.g., cyclophosphamide, gives rise to phosphoramide mustard, the metabolite that effects the cytotoxic action of these prodrugs.

Enzymology and Molecular Biology of Carbonyl Metabolism 6
edited by Weiner *et al.* Plenum Press, New York, 1996

It follows that inhibitors of these enzymes would be of experimental, and, in some cases, even of clinical, value. Inhibitors of ALDH-1 and ALDH-2, e.g., chloral hydrate, disulfiram and cyanamide, have been identified, and, in the case of disulfiram and cyanamide, their clinical utility as alcohol deterrents has been established (Peachey, 1981; Sellers et al., 1981; Brien and Loomis, 1985; Petersen, 1992). However, these agents do not inhibit ALDH-3 *in vitro* or *ex vivo* at pharmacologically relevant concentrations (Sreerama and Sladek, 1993, 1994; Sladek et al., 1995) and, to date, there are, with the exception of gossypol (Rekha and Sladek, 1997), no known inhibitors of ALDH-3, although alternative substrates, e.g., benzaldehyde and diethylaminobenzaldehyde, can be used to inhibit ALDH-3-catalyzed oxidation of other aldehydes, e.g., aldophosphamide (Sreerama and Sladek, 1993, 1994, 1995; Rekha et al., 1994; Sladek et al., 1995; Bunting and Townsend, 1996).

Like disulfiram and cyanamide, the oral hypoglycemic agent chlorpropamide is thought to be a pro-inhibitor of the aldehyde dehydrogenases, most notably ALDH-2, that catalyze the oxidation of ethanol-derived acetaldehyde (Öhlin et al., 1982; Little and Petersen, 1985). Consistent with this notion, about 23% of patients who receive this agent experience, on ingesting alcohol, an adverse reaction characterized by facial flushing and general malaise. These reactions are identical to those experienced by persons who consume alcohol following administration of disulfiram as well as by functional ALDH-2-deficient individuals who consume alcohol (Logie et al., 1976; Öhlin et al., 1982; Goedde and Agarwal, 1990; Petersen, 1992). Nitroxyl (HNO) and *n*-propylisocyanate have been postulated to be the chlorpropamide metabolites that inhibit hepatic aldehyde dehydrogenase-catalyzed reactions (Nagasawa et al., 1988, 1989; Lee et al., 1992a,b).

Based on this premise, a number of N^1-substituted chlorpropamide analogues, viz., ester and alkyl derivatives, have been designed and synthesized as potential alcohol deterrents (Nagasawa et al., 1988, 1989; Lee et al., 1992a,b).

Aldehyde dehydrogenases are bifunctional enzymes in that they catalyze the hydrolysis of esters in addition to catalyzing the oxidation of aldehydes (Feldman and Weiner, 1972; Sidhu and Blair, 1975; Blatter et al., 1992; Sladek et al., 1995). Whether catalysis of hydrolytic reactions by these enzymes is of physiological or pharmacological consequence is not known.

The ester analogues were designed with the intent of exploiting the dual catalytic activities exhibited by the aldehyde dehydrogenases, viz., to release HNO, a potent inhibitor of aldehyde dehydrogenase-catalyzed oxidations, upon ester hydrolysis catalyzed by these enzymes (Lee et al., 1992b). Indeed, some of the compounds that were synthesized have been shown to undergo hydrolytic cleavage catalyzed by yeast aldehyde dehydrogenase (yALDH) and to inhibit yALDH-catalyzed acetaldehyde oxidation.

The alkyl analogues, on the other hand, were designed to release *n*-propylisocyanate, a potent inhibitor of yeast and rodent hepatic aldehyde dehydrogenases, without the necessity of any enzyme participation (Nagasawa et al., 1988, 1989; Lee et al., 1992a). Some of the compounds that were synthesized have been shown to inhibit yeast and rodent hepatic mitochondrial aldehyde dehydrogenase-catalyzed oxidation, decrease acetaldehyde clearance in rodents given ethanol, and to be devoid of a hypoglycemic effect.

Historically, we and others have used the ability of these and other compounds to inhibit yALDH-catalyzed oxidation of acetaldehyde as the initial indicator of potentially useful alcohol-deterrent activity (Watanabe et al., 1986; Nagasawa et al., 1988, 1989; Lee et al., 1992a,b). Although the catalytic site amino acids are conserved (Saigal et al., 1991), yeast and human aldehyde dehydrogenases differ substantially with regard to their primary structure and catalytic properties (Bostian and Betts, 1978; Lindahl and Hempel,

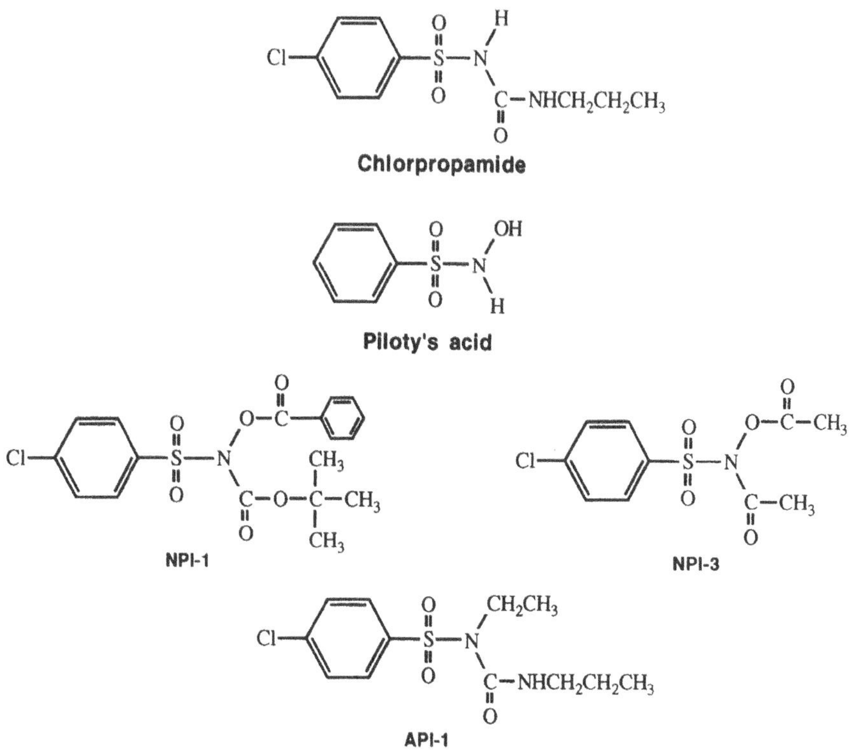

Chlorpropamide

Piloty's acid

NPI-1

NPI-3

API-1

Chlorpropamide: 1-(p-chlorobenzenesulfonyl)-3-*n*-propylurea

Piloty's acid: benzenesulfohydroxamic acid

NPI-1: (benzoyloxy)[(4-chlorophenyl)sulfonyl]carbamic acid 1,1-dimethylethyl ester

NPI-3: N-acetyl-N-(acetyloxy)-4-chlorobenzenesulfonamide

API-1: 4-chloro-N-ethyl-N-[(propylamino)carbonyl]benzenesulfonamide

Figure 1. Structures of chlorpropamide, three analogues thereof, and Piloty's acid.

1990). Thus, the sensitivity of yALDH to a candidate inhibitor may not be paralleled by the sensitivity of human aldehyde dehydrogenases to it, and the use of yALDH to screen for potential alcohol deterrents may not be predictive. Alternatives to the use of yALDH in the initial screen, albeit not commercially available and costlier to generate, are recombinant human ALDH-1 and ALDH-2 (rALDH-1 and rALDH-2, respectively).

The present investigation sought to ascertain whether sensitivity to a candidate inhibitor on the part of ALDH-1 and/or ALDH-2, and, for completeness, ALDH-3, was in fact, predicted by the sensitivity of yALDH to inhibition by that agent. Further, since the ALDH-3 present in human normal tissues/fluids (nALDH-3) is somewhat different from the ALDH-3 present in human tumor cells/tissues (tALDH-3), viz., the latter (putatively, tumor-specific) is better able to catalyze the detoxification of aldophosphamide (Sladek et al., 1995; Sreerama and Sladek, 1995), both nALDH-3 and tALDH-3 were included in our investigations. Chlorpropamide analogues of the types described above (Fig. 1) were used as test inhibitors.

2. MATERIALS AND METHODS

E. coli [BL21(DE3)pLysS] transfected with pET-19b vector to which human ALDH-1 cDNA [cloned from human hepatoma Hep G2 cells; cDNA sequence was identical to that of human liver ALDH-1 cDNA (Moreb et al., 1996)] was ligated, was provided by Dr. Jan Moreb, University of Florida, Gainsville, FL. A vector, viz., pT7–7, to which human ALDH-2 cDNA [cloned from human liver (Zheng et al., 1993)] was ligated, was provided by Dr. Henry Weiner, Purdue University, Lafayette, IN. Transfection of human ALDH-2 cDNA ligated to the pT7–7 vector into *E. coli* [BL21(DE3)pLysS] was by Drs. P. A. Dockham and L. Sreerama of our laboratory as described by Sambrook et al. (1989). (Benzoyloxy)[(4-chlorophenyl)sulfonyl]carbamic acid 1,1-dimethylethyl ester (NPI-1), N-acetyl-N-(acetyloxy)-4-chlorobenzenesulfonamide (NPI-3) and 4-chloro-N-ethyl-N-[(propylamino)carbonyl]benzenesulfonamide (API-1) were synthesized as described previously (Nagasawa et al., 1989; Lee et al., 1992b). Chromatographically purified yeast aldehyde dehydrogenase, human erythrocyte glyceraldehyde-3-phosphate dehydrogenase, human placental alkaline phosphatase type XXIV, chlorpropamide, *p*-nitrophenyl phosphate, phenylmethylsulfonylfluoride and Triton X-100 (*t*-octylphenyloxypolyethyloxyethanol) were purchased from Sigma Chemical Co., St. Louis, MO. Benzenesulfohydroxamic acid (Piloty's acid) was purchased from Aldrich Chemical Co., Milwaukee, WI. Ampicillin, chloramphenicol and isopropylthio-β-D-galactoside were purchased from USB Corp., Cleveland, OH. Centriprep-30 concentrators were purchased from Amicon Inc., Beverly, MA. Ni^{2+}-chelated Sepharose CL 6B (His-Bind resin) was purchased from Novagen Inc., Madison, WI. Luria-Bertani medium (powder) was purchased from Bio-101 Inc., Vista, CA. All other chemicals and reagents were purchased from the sources listed in previous publications (Dockham et al., 1992; Sreerama and Sladek, 1993).

E. coli transfected with ALDH-1 cDNA ligated to the pET-19b vector, or with ALDH-2 cDNA ligated to the pT7–7 vector, were cultured overnight (14–16 h) at 37°C in growth medium, viz., Luria-Bertani broth (10 g/L tryptone, 5 g/L yeast extract, 10 g/L NaCl) supplemented with ampicillin (50 and 100 mg/L, respectively, in the cases of pT7–7 and pET-19b transfected *E. coli*) and chloramphenicol (34 mg/L), subjected to vigorous shaking (220 rpm) effected by an environmental orbital-shaker (Model 4628, Lab-Line Instruments Inc., Melrose Park, IL) at 37°C. Media absorbance at 600 nm after the overnight culture was usually ~1. The overnight cultures were diluted (1:1) with growth medium, isopropylthio-β–D–galactoside (238 and 95 mg/L, respectively, in the cases of pET-19b and pT7–7 transfected *E. coli*) was added, and the resultant suspensions were cultured for an additional 4 to 5 h at 37°C to induce the expression of rALDH-1 and rALDH-2. *E.coli* were then harvested by centrifugation at 5,000 g and 4°C for 5 min. Bacterial pellets thus obtained were washed (centrifugation) once with 0.9% NaCl after which they were resuspended in a volume of lysis buffer (50 mM Tris-HCl, pH 8.0, supplemented with 2 mM EDTA, 1% Triton X-100 and 10 μM phenylmethylsulfonylfluoride) that was one tenth of the original culture volume and then lysed in an ice-bath by submitting them to sonication (Artek Dismembrator Model 300; setting of 30) for a total of 4 min (divided into 8 bursts interspersed with 1 min rest intervals). The resultant lysates were centrifuged at 5,000 g and 4°C for 10 min and aldehyde dehydrogenase activity was quantified in the supernatant fractions thus obtained. Ion exchange and affinity column chromatography was used as described below to purify the rALDH-1 and rALDH-2 present in these supernatant fractions to apparent homogeneity.

All column chromatographic procedures were performed at 4–6°C. All buffers were degassed prior to use. The protein content of the preparations loaded onto the chromatographic columns never exceeded 15 mg/ml and typically was much less. Concentration of eluates was with the aid of an Amicon Diaflo concentrator fitted with a YM-30 membrane and pressurized with nitrogen (Dockham et al., 1992; Sreerama and Sladek, 1993) except when the sample volumes were ~10 ml or smaller, in which cases, Amicon Centriprep-30 centrifugation concentrators were used. Acetaldehyde (4 and 2 mM for rALDH-1 and rALDH-2, respectively) and NAD (4 mM) were used as substrate and cofactor, respectively, to monitor rALDH-1 or rALDH-2 activity in column eluates. Protein elution from the columns was monitored at 280 nm with the aid of an ISCO UA-5 absorbance monitor. Otherwise, the method of Bradford (1976) was used to quantify protein concentrations. The Bio-Rad protein assay reagent was used for this purpose; bovine serum albumin was used as the standard.

PD-10 (Sephadex G-25) columns were used as described previously (Dockham et al., 1992; Sreerama and Sladek, 1993) to first transfer the rALDH-1 present in *E. coli* lysate supernatant fractions into a 20 mM imidazole buffer solution, pH 6.8, supplemented with 1 mM each of EDTA and dithiothreitol, and the rALDH-2 present in *E. coli* lysate supernatant fractions into a 25 mM 2-(N-morpholino)ethane sulfonic acid buffer solution, pH 6.0, supplemented with 1 mM each of EDTA and dithiothreitol. DEAE-Sephacel anion exchange chromatography (Dockham et al., 1992) followed by Ni^{2+}-chelated Sepharose CL 6B affinity chromatography (manufacturers [Novagen Inc., Madison WI] protocol) was then used to obtain apparently pure rALDH-1. CM-Sepharose CL 6B cation exchange chromatography followed by DEAE-Sephacel anion exchange chromatography and, subsequently, 5'-AMP-Sepharose CL 6B affinity chromatography (Dockham et al., 1992) was then used to obtain apparently pure rALDH-2. Overall recoveries of rALDH-1 and rALDH-2 were 50 to 60% and 25 to 30%, respectively.

pET-19b is designed to add 21 amino acids (amino acids 2 to 11 being histidines) to either the N- or C-terminal end of the protein coded for by the cDNA ligated to it, thereby enabling purification by Ni^{2+}-chelated Sepharose CL 6B affinity chromatography. Thus, rALDH-1, generated and purified as described above, is actually a fusion protein made up of a 21 amino acid peptide attached to the N-terminal end of ALDH-1. Specific activities (acetaldehyde and NAD, 4 mM each, were the substrate and cofactor, respectively) of several batches of purified rALDH-1 ranged from 600 to 780 mIU/mg protein as compared to a specific activity of 2850 mIU/mg protein obtained for the native enzyme (Sreerama and Sladek, 1997). The subunit molecular mass of rALDH-1 was determined to be 53.7 kDa (data not shown) as compared to a molecular mass of 52 kDa for the native enzyme (Sreerama and Sladek, unpublished observation). The Km value of 429 ± 61 µM (mean ± SE; Table 3) defining rALDH-1 catalyzed oxidation of acetaldehyde at 37°C was essentially identical to that, 483 µM, defining native enzyme-catalyzed oxidation of acetaldehyde under identical conditions (Dockham et al., 1992). Specific activities (acetaldehyde, 2 mM, and NAD, 4 mM, were the substrate and cofactor, respectively) of several batches of purified rALDH-2 ranged from 1700 to 2400 mIU/mg protein. The subunit molecular mass was 52 kDa (data not shown). These values are identical to those obtained for the native enzyme (Dockham et al., 1992).

Human normal stomach mucosa ALDH-3 (nALDH-3) and the ALDH-3 (tALDH-3) present in human breast adenocarcinoma MCF-7/0 cells cultured in the presence of 30 µM catechol for 5 days to induce the enzyme (MCF-7/0/CAT cells) were purified as described previously (Sreerama and Sladek, 1993; Sreerama et al., 1995). Specific activities (benzaldehyde, 4 mM, and NAD, 1 mM, were the substrate and cofactor, respectively) of nALDH-3 and tALDH-3 were 31,000 and 34,000 mIU/mg protein, respectively. The subunit molecular mass of each was 54.5 kDa (data not shown). These values are identical to those previously reported for these enzymes (Sreerama and Sladek, 1993; Sreerama et al., 1995).

Denaturing polyacrylamide gel electrophoresis (10% polyacrylamide gels containing 1% sodium dodecyl sulfate) was used as described previously (Sreerama and Sladek, 1993) to ascertain the purity of each enzyme preparation.

NAD-linked oxidation of acetaldehyde catalyzed by rALDH-1, rALDH-2 and yALDH at 37°C and pH 8.1, NAD-linked oxidation of benzaldehyde catalyzed by nALDH-3 and tALDH-3 at 37°C and pH 8.1, and hydrolysis of *p*-nitrophenyl acetate catalyzed by each of these enzymes at 25°C and pH 7.5, were quantified spectrophotometrically as described previously (Dockham et al., 1992; Sreerama and Sladek, 1993). NAD-linked oxidation of glyceraldehyde-3-phosphate catalyzed by glyceraldehyde-3-phosphate dehydrogenase at 37°C and pH 7.6, and hydrolysis of *p*-nitrophenyl phosphate catalyzed by alkaline phosphatase at 25°C and pH 9.8, were quantified as described by Lambeir et al. (1991) and Chueh et al. (1981), respectively. Preincubation of the putative inhibitor or vehicle together with the complete reaction mixture except for substrate was for 5 min except in the case of Piloty's acid where it was for 10 min. Preincubation temperatures and pHs were the same as incubation temperatures and pHs. All reactions were started by the addition of substrate. Stock solutions of chlorpropamide and its analogues were prepared in dimethyl sulfoxide. The final concentration of dimethyl sulfoxide in the reaction mixture never exceeded 5%; this concentration of dimethyl sulfoxide did not inhibit any of the enzyme-catalyzed reactions under investigation. Stock solutions of Piloty's acid were prepared in dimethylformamide. Aliquots thereof of the desired size were placed into reaction tubes, dimethylformamide was evaporated with the aid of a stream of nitrogen, and preincubation was initiated by adding the complete reaction mixture minus the substrate.

Table 1. Inhibition by chlorpropamide analogues and Piloty's acid of yeast and human aldehyde dehydrogenase-catalyzed oxidation: IC_{50} values[a]

Enzyme	IC_{50}, μM				
	Chlorpropamide	NPI-1	NPI-3	Piloty's acid	API-1
yALDH	> 2,000	13	17	10	24
rALDH-1	> 2,000	49	217	20	150
rALDH-2	> 2,000	47	900	13	23
nALDH-3	> 2,000	121	775	105	223
tALDH-3	> 1,000	45	305	117	93
GAPDH	> 2,000	> 150	303	*b*	> 500
Alkaline phosphatase	> 1,000	> 200	> 500	*b*	> 300

[a]Enzymes were incubated with vehicle or various concentrations of the putative inhibitors for 5 (chlorpropamide, NPI-1, NPI-3 and API-1) or 10 (Piloty's acid) min, substrate was added, and initial catalytic rates were quantified as described in Materials and Methods. Substrates and cofactors were, respectively, acetaldehyde (0.8 mM) and NAD (4 mM) for yALDH, acetaldehyde and NAD (4 mM each) for rALDH-1, acetaldehyde (2 mM) and NAD (4 mM) for rALDH-2, benzaldehyde (4 mM) and NAD (1 mM) for the ALDH-3s, and glyceraldehyde-3-phosphate and NAD (1 mM each) for glyceraldehyde-3-phosphate dehydrogenase (GAPDH). *p*-Nitrophenyl phosphate (10 mM) was the substrate for alkaline phosphatase. Uninhibited initial catalytic rates (mean ± SE; n = 6) were 9.1 ± 0.4, 0.72 ± 0.02, 1.7 ± 0.1, 29 ± 1, 32 ± 1, 52 ± 2 and 15 ± 1 IU/mg protein for yALDH, rALDH-1, rALDH-2, nALDH-3, tALDH-3, GAPDH and alkaline phosphatase, respectively. Examples of plots of the primary data thus obtained are given in Fig. 2. IC_{50} values were calculated from such data as described in Materials and Methods.

[b]Not determined

Computer-assisted unweighted nonlinear regression analysis effected by the STATView statistical program (Brain Power Inc., Calabas, CA) was used to generate the curves that best-fit plots of enzyme activities (% of control) as a function of inhibitor concentrations (a minimum of five, at least two of which effected less than 50% inhibition and two of which effected greater than 50% inhibition) and, subsequently, to estimate the concentration of inhibitor that effected a 50% decrease in catalytic activity (IC_{50}). Double-reciprocal (Lineweaver-Burk) plots of initial catalytic rates as a function of substrate concentrations (at least three and usually five) were used to estimate the Km and Vmax values. Except in the case of rALDH-2, Ki values were determined by plotting the slopes of the lines generated by double-reciprocal (Lineweaver-Burk) plots as a function of inhibitor concentrations. In the case of rALDH-2, Ki values were determined by plotting the reciprocals of initial catalytic rates as a function of inhibitor concentrations (Dixon plots) because Km values were relatively small and, thus, Ki values were difficult to ascertain accurately from Lineweaver-Burk plots. In the case of double-reciprocal (Lineweaver-Burk) plots, computer-assisted Wilkinson weighted linear regression analysis (Wilkinson, 1961) effected by the MacWilkins program (Microsoft, Bellevue, WA) was used to generate the best-fit lines. Computer-assisted unweighted linear regression analysis effected by the STATView statistical, or Kaleida Graph (Synergy Software, Inc., Reading, PA), programs was used to generate best-fit lines for all other straight-line functions.

3. RESULTS AND DISCUSSION

In agreement with the reports of others (Öhlin et al., 1982; Little and Petersen, 1985), and the suggestion that metabolic activation is required to effect aldehyde dehydrogenase inhibitory activity (Nagasawa et al., 1985, 1989), chlorpropamide did not inhibit any of the enzymes tested (Table 1).

Each of the three chlorpropamide analogues tested, viz., NPI-1, NPI-3 and API-1 (Fig. 1), inhibited enzyme-catalyzed oxidation of aldehydes to acids by each of the substrate-nonspecific aldehyde dehydrogenases tested (Fig. 2 and Table 1). Differential potency on the part of the chlorpropamide analogues was observed. In general, the rank order was NPI-1 > API-1 > NPI-3. Differential sensitivity to these inhibitors on the part of the aldehyde dehydrogenases was also observed. Thus, as judged by IC_{50} values, oxidative catalysis by yALDH was quite sensitive to inhibition by all three analogues whereas that by human aldehyde dehydrogenases often was not. In particular, none of the human aldehyde dehydrogenases were very sensitive to inhibition by NPI-3, nALDH-3 was relatively insensitive to inhibition by NPI-1, and rALDH-1, nALDH-3 and tALDH-3 were relatively insensitive to API-1. In general, the rank order of enzyme sensitivity to these agents was yALDH >> rALDH-2 ^{3} tALDH-3 ^{3} rALDH-1 > nALDH-3. As judged by IC_{50} values, none of these agents showed much potential as a clinically useful inhibitor of ALDH-3s since none was very potent or specific in that regard. Inhibition by chlorpropamide analogues was relatively specific for substrate-nonspecific aldehyde dehydrogenases since a substrate-specific dehydrogenase, viz., glyceraldehyde-3-phosphate dehydrogenase, and a hydrolase, viz., alkaline phosphatase, were not inhibited by reasonable concentrations of these agents.

Significant inhibition by chlorpropamide analogues of the hydrolytic reaction catalyzed by aldehyde dehydrogenases was observed only in the cases of yALDH (NPI-3 and API-1) and rALDH-2 (API-1) (Table 2).

Kinetic constants, viz., Km, Vmax and Ki values, defining the catalysis of oxidative reactions by yALDH, rALDH-1, rALDH-2, nALDH-3 and tALDH-3 and inhibition

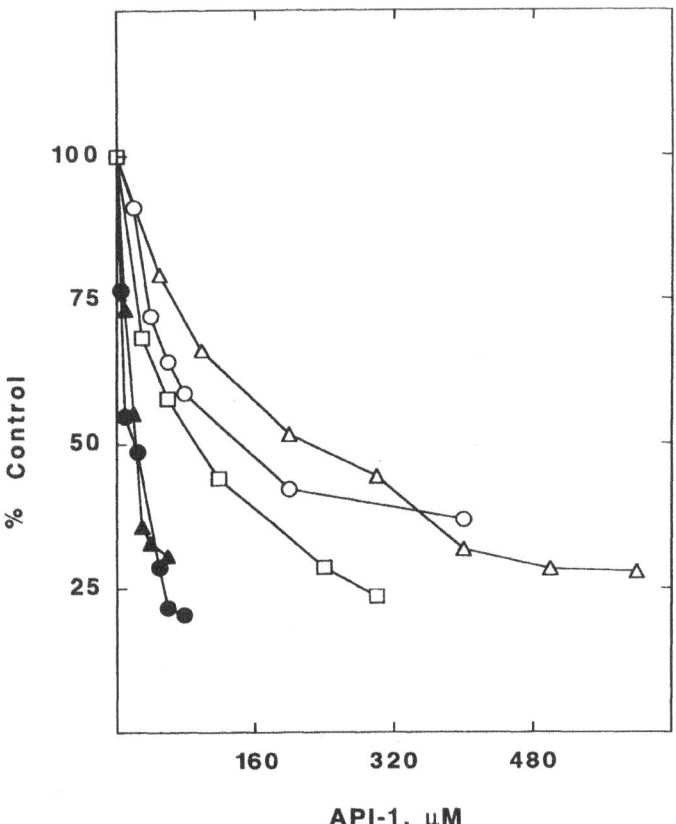

Figure 2. Inhibition by API-1 of the oxidative reaction catalyzed by yeast and human aldehyde dehydrogenases. The sensitivities of yALDH (●), rALDH-1 (○), rALDH-2 (▲), nALDH-3 (△) and tALDH-3 (□) to inhibition by API-1 were determined as described in Materials and Methods and a footnote to Table 1. Data points are means of duplicate determinations. Control catalytic rates are given in a footnote to Table 1. Best-fit curves, and IC_{50} values estimated therefrom, were generated from the data presented in this figure as described in Materials and Methods. IC_{50} values thus obtained are given in Table 1.

thereof, are given in Tables 3 and 4. Representative plots from which these values were obtained are shown in Figs. 3, 4 and 5. The Km and Vmax values obtained in the present investigation are in good agreement with those reported previously (Dickinson and Haywood, 1987; Lambeir et al., 1991; Dockham et al., 1992; Sreerama and Sladek, 1993, 1994; Sreerama et al., 1995). Irrespective of the chlorpropamide analogue tested, inhibition of yALDH- and rALDH-1-catalyzed oxidation was always noncompetitive (Fig. 3 and data not shown), inhibition of rALDH-2-catalyzed oxidation was always competitive (Fig. 5) or of the mixed type (data not shown), and inhibition of nALDH-3- and tALDH-3-catalyzed oxidation was always competitive (Fig. 4 and data not shown). The general rank order of inhibitor potency as judged by Ki values was identical to that generated on the basis of IC_{50} values, viz., NPI-1 > API-1 > NPI-3. However, the general rank order of enzyme sensitivity to these agents as judged by Ki values, viz., yALDH = rALDH-2 = tALDH-3 > nALDH-3 ³ rALDH-1, is somewhat different from that generated on the basis of IC_{50} values.

Table 2. Inhibition by chlorpropamide analogues of yeast and human aldehyde dehydrogenase-catalyzed hydrolysis: IC_{50} values[a]

Enzyme	IC_{50}, μM			
	Chlorpropamide	NPI-1	NPI-3	API-1
yALDH	> 2,000	> 100	40	7.3
rALDH-1	> 2,000	> 100	> 1,000	> 400
rALDH-2	> 2,000	> 100	> 1,000	19
nALDH-3	> 2,000	> 100	388	> 400
tALDH-3	> 2,000	> 100	> 1,000	> 400

[a]Enzymes were incubated with vehicle or various concentrations of the putative inhibitors for 5 min, substrate (500 μM p-nitrophenyl acetate) was added, and initial catalytic rates were quantified as described in Materials and Methods. Uninhibited initial catalytic rates (mean ± SE; n = 6) were 427 ± 9, 155 ± 6, 583 ± 15, 8.7 ± 0.2 and 8.7 ± 0.6 IU/mg protein for yALDH, rALDH-1, rALDH-2, nALDH-3 and tALDH-3, respectively.

Table 3. NAD-linked enzyme-catalyzed oxidation of aldehydes: Km and Vmax values[a]

Enzyme	Substrate	NAD, mM	n	Km ± SE, μM	Vmax ± SE, IU/mg
yALDH	Acetaldehyde	4	4	34 ± 3	10 ± 1
rALDH-1	Acetaldehyde	4	4	429 ± 61	0.64 ± 0.02
rALDH-2	Acetaldehyde	4	1	4[b]	1.9 ± 0.0
nALDH-3	Benzaldehyde	1	4	414 ± 14	30 ± 1
tALDH-3	Benzaldehyde	1	4	357 ± 12	32 ± 1

[a]Initial catalytic rates were quantified, and Km and Vmax values were calculated, as described in Materials and Methods. Examples of plots of primary data from which the Km and Vmax, as well as Ki (Table 4), values were estimated are given in Figs. 3 and 4.
[b]Unlikely to be accurate because it is difficult to ascertain Km values that are less than about 10 μM from the very flat Lineweaver-Burk plots that we generated. Thus, the Km value was determined to be < 0.1 μM when a more appropriate experimental design and method of analysis, viz., integrated Michaelis analysis of a single enzyme-progress curve, was used (Dockham et al., 1992).

Table 4. Inhibition by chlorpropamide analogues of yeast and human aldehyde dehydrogenase-catalyzed oxidation: Ki values[a]

Enzyme	Ki, μM		
	NPI-1	NPI-3	API-1
yALDH	7.2	8.3	23
rALDH-1	21	142	138
rALDH-2	7.0	80	3.0
nALDH-3	18	153	34
tALDH-3	3.9	26	19

[a]Enzymes were incubated with vehicle or various concentrations of the putative inhibitors for 5 min, substrate was added, and initial catalytic rates were quantified as described in Materials and Methods. Cofactor concentrations and substrates were as listed in Table 3. Km and Vmax values defining each of the enzyme-catalyzed reactions are also given in Table 3. Examples of plots of primary data from which the Ki values were estimated are given in Figs. 3, 4 and 5.

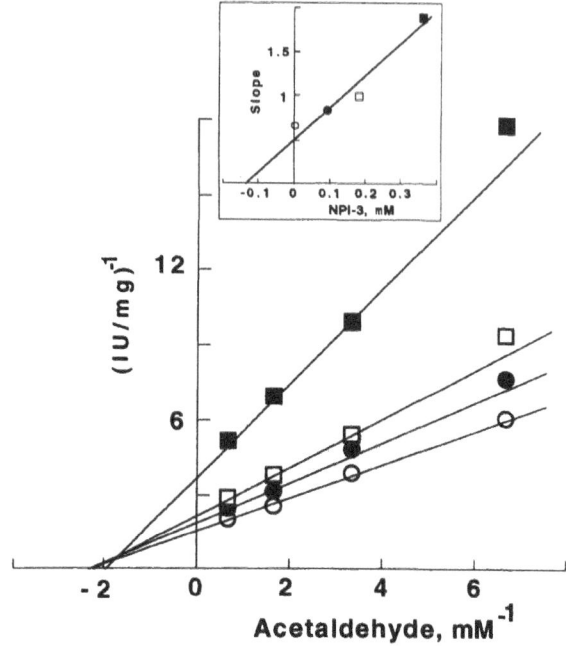

Figure 3. Inhibition of rALDH-1-catalyzed oxidation of acetaldehyde by NPI-3: Lineweaver-Burk plot. rALDH-1 was incubated with 0 (O), 90 (●), 180 (□) or 360 (■) μM NPI-3 for 5 min, various concentrations of the substrate, acetaldehyde, were added, and initial catalytic rates were quantified as described in Materials and Methods. The NAD concentration was 4 mM. Data points are means of triplicate determinations. Inset: Slopes generated by the double-reciprocal (Lineweaver-Burk) plots were plotted as a function of NPI-3 concentrations for the purpose of determining the Ki value. Km, Vmax and Ki values obtained in this experiment were 434 μM, 0.66 IU/mg and 142 μM, respectively.

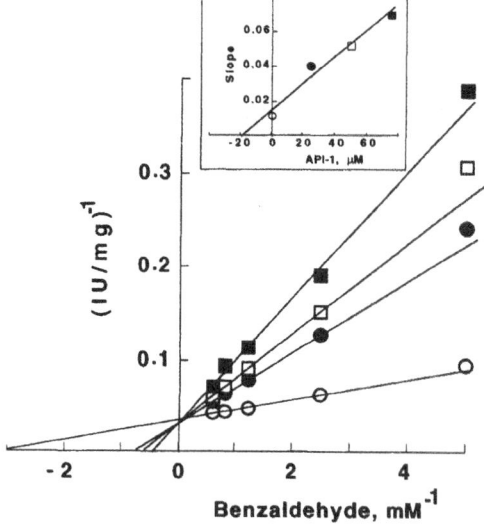

Figure 4. Inhibition of NAD-linked tALDH-3-catalyzed oxidation of benzaldehyde by API-1: Lineweaver-Burk plot. tALDH-3 was incubated with 0 (O), 25 (●), 50 (□) or 75 (■) μM API-1 for 5 min, various concentrations of the substrate, benzaldehyde, were added, and initial catalytic rates were quantified as described in Materials and Methods. The NAD concentration was 1 mM. Data points are means of triplicate determinations. Inset: Slopes generated by the double-reciprocal (Lineweaver-Burk) plots were plotted as a function of API-1 concentrations for the purpose of determining the Ki value. Km, Vmax and Ki values obtained in this experiment were 337 μM, 30 IU/mg and 19 μM, respectively.

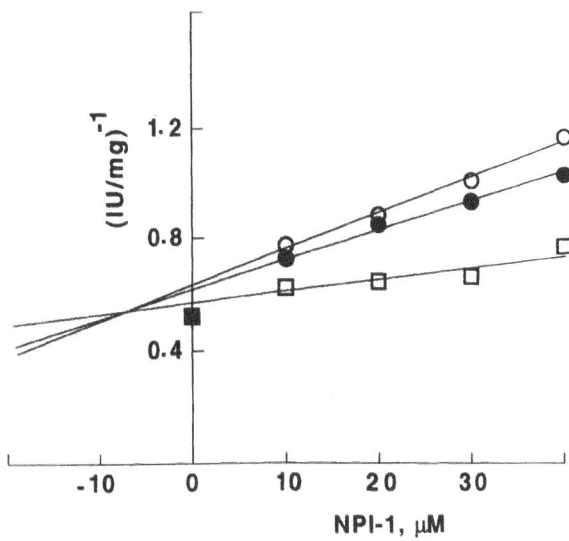

Figure 5. Inhibition of rALDH-2-catalyzed oxidation of acetaldehyde by NPI-1: Dixon plot. rALDH-2 was incubated with vehicle or various concentrations of NPI-1 for 5 min, 250 (O), 500 (●) or 1000 (□) μM acetaldehyde was added, and initial catalytic rates were determined as described in Materials and Methods. The NAD concentration was 4 mM. Data points are means of triplicate determinations. The Ki value was 7 μM. ■, Vmax determined by a Lineweaver-Burk plot (not shown) was 1.9 IU/mg. The Km as determined by a Lineweaver-Burk plot was 3.6 μM.

As judged by Ki values, too, sensitivity of yALDH to inhibition by the three chlorpropamide analogues did not consistently reflect sensitivity of the human aldehyde dehydrogenases to inhibition by them. For example, yALDH and, to a lesser extent, tALDH-3 were sensitive to inhibition by NPI-3, but rALDH-1, rALDH-2 and nALDH-3 were not. Conversely, rALDH-2 was much more sensitive to inhibition by API-1 than was yALDH.

As judged by Ki values, NPI-1 showed some potential as a clinically useful inhibitor of ALDH-3s since it was relatively potent and specific in that regard. Interestingly, but not

Table 5. Inhibition by chlorpropamide analogues of yeast and human recombinant class 2 aldehyde dehydrogenase-catalyzed hydrolysis[a]

Enzyme	Ki, μM		
	NPI-1	NPI-3	API-1
yALDH	b	73	3.5
rALDH-2	b	b	22

[a]Enzymes were incubated with vehicle or various concentrations of the putative inhibitors for 5 min, the substrate, 500 μM p-nitrophenyl acetate, was added, and initial catalytic rates were quantified as described in Materials and Methods. Km and Vmax values (mean ± SE) for the yALDH-catalyzed reaction (n = 3) were 37 ± 6 μM and 342 ± 55 IU/mg, respectively. Km and Vmax values for the rALDH-2-catalyzed reaction (n = 1) were 30 μM and 584 IU/mg, respectively.

[b]Ki values were not determined because inhibition was less than 20% at the highest concentration of inhibitor tested, Table 2.

totally unexpected, tALDH-3, as compared to nALDH-3, was significantly more sensitive to inhibition by each of the three chlorpropamide analogues (Tables 1 and 4). This observation further substantiates the notion that tALDH-3, putatively tumor-specific, is a subtle variant of nALDH-3 (Sladek et al., 1995).

Ki values defining the inhibition of yALDH- and rALDH-2-catalyzed hydrolysis by NPI-3 and API-1 are given in Table 5. Inhibition of yALDH- and rALDH-2-catalyzed hydrolysis by these agents was of the mixed type (data not shown).

A priori, the expectation was that yALDH-, rALDH-1- and rALDH-2-catalyzed oxidation would be much more sensitive to inhibition by NPI-1 and NPI-3 than would be that catalyzed by the two ALDH-3s. This was because 1) as judged by the relative rates at which they catalyze the hydrolysis of *p*-nitrophenyl acetate (see footnote to Table 2), the former were expected to catalyze the bioactivation of the pro-inhibitors at a much faster rate, and 2) the former were much more sensitive to Piloty's acid (Table 1), an agent that spontaneously gives rise to HNO (the inhibitory action of Piloty's acid is thought to be effected by HNO, rather than by the parent compound) (Nagasawa et al., 1995). The expectation was not realized (Tables 1 and 4). Why is uncertain. Hydrolytic release of HNO from NPI-3 catalyzed by yALDH has been demonstrated as has inhibition of yALDH-catalyzed oxidation of acetaldehyde to acetate by HNO, albeit, by deduction (Lee et al., 1992b). It may be that the relative rates at which human aldehyde dehydrogenases catalyze hydrolysis of *p*-nitrophenyl acetate do not reflect the relative rates at which these enzymes catalyze the hydrolysis of NPI-1 and NPI-3. Indeed, NPI-1 and NPI-3 may not even be substrates for human aldehyde dehydrogenases. Even so, HNO may still have been generated given that the pH of the enzyme assay mixture was 8.1 and that, in general, esters of hydroxamic acids readily undergo hydrolysis in aqueous alkaline solutions. NPI-1 and NPI-3 have been shown to undergo rapid ester hydrolysis when placed into a strongly alkaline (0.1 N NaOH) aqueous solution (Lee et al., 1992b). By itself, however, this eventuality, does not explain our findings given that yALDH, rALDH-1 and rALDH-2 are more sensitive to inhibition by HNO (Piloty's acid) than are the ALDH-3s (Table 1). The possibility that the parent compounds themselves effect the inhibitory action is seemingly negated by our preliminary findings that inhibition increased as the time of preincubation (enzyme and inhibitor) increased from zero to about three min, though not beyond (data not shown).

Several, collectively rather compelling, observations support the notion that, at least in the cases of ALDH-1 and ALDH-2, aldehydes and esters bind to the same amino acid residue, viz., cysteine 302 (Kitson et al., 1991; Blatter et al., 1992; Mukerjee and Pietruszko, 1992, 1994; Farrés et al., 1995). However, some observations are inconsistent with this notion (Klyosov et al., 1996 and the references cited therein). Our findings that the oxidative reactions were inhibited by chlorpropamide analogues whereas the hydrolytic reactions, in general, were not, are, seemingly, also in conflict with the single site hypothesis.

It could be argued that, with the recently made availability of recombinant human aldehyde dehydrogenases, they, rather than yALDH, should be used to screen for aldehyde dehydrogenase inhibitors ultimately expected to be used in humans, especially since the use of yALDH for this purpose can give rise to false positives, e.g., comparing IC_{50} values quantifying NPI-3 inhibition of yALDH (17 μM) and rALDH-2 (900 μM) (Table 1), and, probably less often, false negatives, e.g., comparing Ki values defining API-1 inhibition of yALDH (23 μM) and rALDH-2 (3 μM) (Table 4). In particular, structure-activity relationships generated by this model may not be useful when the goal is to design an optimal inhibitor of one of the human aldehyde dehydrogenases, e.g., comparing Ki values defining the sensitivity of yALDH to NPI-1 (7.2 μM) and NPI-3 (8.3 μM) with the sensitivity of rALDH-2 to NPI-1 (7 μM) and NPI-3 (80 μM) (Table 4). The yALDH model does, however, offer some advantages, viz., yALDH is commercially available, relatively inexpensive, and, in most cases, at least in our limited study, relatively more sensitive to candidate inhibitors.

Perhaps the most economical and rewarding approach would be to use yALDH in primary screening, thereby identifying virtually all candidates that have any possibility of being of value, and then submitting the agents thus identified to secondary screening in which inhibition of the human aldehyde dehydrogenase(s) of interest would be evaluated. Agents showing promise in the secondary screen would then be further evaluated in an appropriate culture, organ and/or whole animal model.

4. ACKNOWLEDGMENT

These investigations were supported by Department of Defense Grant DAMD 17–94-J-4057.

5. REFERENCES

Anonymous: Nomenclature of mammalian aldehyde dehydrogenases. Prog. Clin. Biol. Res. 290 (1989) xix-xxi.

Blatter, E.E., Abriola, D.P. and Pietruszko, R.: Aldehyde dehydrogenase. Covalent intermediate in aldehyde dehydrogenation and ester hydrolysis. Biochem. J. 282 (1992) 353–360.

Bostian, K.A. and Betts, G.F.: Kinetics and reaction mechanism of potassium-activated aldehyde dehydrogenase from *Saccharomyces cerevisiae*. Biochem. J. 173 (1978) 787–798.

Bradford, M.M.: A rapid and sensitive method for the quantitation of microgram quantities of protein utilizing the principle of protein-dye binding. Anal. Biochem. 72 (1976) 248–254.

Brien, J.F. and Loomis, C.W.: Aldehyde dehydrogenase inhibitors as alcohol-sensitizing drugs: a pharmacological perspective. Trends Pharmacol. Sci. (1985) 477–480.

Bunting, K.D. and Townsend, A.J.: Protection by transfected rat or human class 3 aldehyde dehydrogenases against the cytotoxic effects of oxazaphosphorine alkylating agents in hamster V79 cell lines. J. Biol. Chem. 271 (1996) 11891–11896.

Chueh, S.-H., Chang, G.-G., Chang, T.-C. and Pan, F.: Involvement of arginine residue in the phosphate binding site of human placental alkaline phosphatase. Int. J. Biochem. 13 (1981) 1143–1149.

Dickinson, F.M. and Haywood, G.W.: The role of the metal ion in the mechanism of the K⁺-activated aldehyde dehydrogenase of *Saccharomyces cerevisiae*. Biochem. J. 247 (1987) 377–384.

Dockham, P.A., Lee, M.-O. and Sladek, N.E.: Identification of human liver aldehyde dehydrogenases that catalyze the oxidation of aldophosphamide and retinaldehyde. Biochem. Pharmacol. 43 (1992) 2453–2469.

Farrés, J., Wang, T.T.Y., Cunningham, S.J. and Weiner, H.: Investigation of the active site cysteine residue of rat liver mitochondrial aldehyde dehydrogenase by site-directed mutagenesis. Biochemistry 34 (1995) 2592–2598.

Feldman, R.I. and Weiner, H.: Horse liver aldehyde dehydrogenase. II. Kinetic and mechanistic implications of the dehydrogenase and esterase activity. J. Biol. Chem. 247 (1972) 267–272.

Goedde, H.W. and Agarwal, D.P.: Pharmacogenetics of aldehyde dehydrogenase (ALDH). Pharmacol. Ther. 45 (1990) 345–371.

Kitson, T.M., Hill, J.P. and Midwinter, G.G.: Identification of a catalytically essential nucleophilic residue in sheep liver cytoplasmic aldehyde dehydrogenase. Biochem. J. 275 (1991) 207–210.

Klyosov, A.A., Rashkovetsky, L.G., Tahir, M.K. and Keung, W.-M.: Possible role of liver cytosolic and mitochondrial aldehyde dehydrogenases in acetaldehyde metabolism. Biochemistry 35 (1996) 4445–4456.

Lambeir, A.-M., Loiseau, A.M., Kuntz, D.A., Vellieux, F.M., Michels, P.A.M. and Opperdoes, F.R.: The cytosolic and glycosomal glyceraldehyde-3-phosphate dehydrogenase from *Trypanosoma brucei*. Kinetic properties and comparison with homologous enzymes. Eur. J. Biochem. 198 (1991) 429–435.

Lee, M.J.C., Elberling, J.A. and Nagasawa, H.T.: N¹-Hydroxylated derivatives of chlorpropamide and its analogs as inhibitors of aldehyde dehydrogenase *in vivo*. J. Med. Chem. 35 (1992a) 3641–3647.

Lee, M.J.C., Nagasawa, H.T., Elberling, J.A. and DeMaster, E.G.: Prodrugs of nitroxyl as inhibitors of aldehyde dehydrogenase. J. Med. Chem. 35 (1992b) 3648–3652.

Lindahl, R.: Aldehyde dehydrogenases and their role in carcinogenesis. Crit. Rev. Biochem. Mol. Biol. 27 (1992) 283–335.

Lindahl, R. and Hempel, J.: Aldehyde dehydrogenases: what can be learned from a baker's dozen sequences. Adv. Exp. Med. Biol. 284 (1990) 1–8.

Little, R.G. II and Petersen, D.R.: Effect of tolbutamide and chlorpropamide on acetaldehyde metabolism in two inbred strains of mice. Toxicol. Appl. Pharmacol. 80 (1985) 206–214.

Logie, A.W., Galloway, D.B. and Petrie, J.C.: Drug interactions and long-term antidiabetic therapy. Br. J. Clin. Pharmacol. 3 (1976) 1027–1032.

Moreb, J., Schweder, M., Suresh, A. and Zucali, J.R.: Overexpression of the human aldehyde dehydrogenase class I results in increased resistance to 4-hydroperoxycyclophosphamide. Cancer Gene Ther. 3 (1996) 24–30.

Mukerjee, N. and Pietruszko, R.: Human mitochondrial aldehyde dehydrogenase substrate specificity: comparison of esterase with dehydrogenase reaction. Arch. Biochem. Biophys. 299 (1992) 23–29.

Mukerjee, N. and Pietruszko, R.: Inactivation of human aldehyde dehydrogenase by isosorbide dinitrate. J. Biol. Chem. 269 (1994) 21664–21669.

Nagasawa, H.T., DeMaster, E.G., Kwon, C.-H., Fraser, P.S. and Shirota, F.N.: Structure vs. activity in the sulfonylurea-mediated disulfiram-ethanol reaction. Alcohol 2 (1985) 123–128.

Nagasawa, H.T., Elberling, J.A., Shirota, F.N. and DeMaster, E.G.: A nonhypoglycemic chlorpropamide analog that inhibits aldehyde dehydrogenase. Alcohol. Clin. Exp. Res. 12 (1988) 563–565.

Nagasawa, H.T., Elberling, J.A., DeMaster, E.G. and Shirota, F.N.: N^1-Alkyl-substituted derivatives of chlorpropamide as inhibitors of aldehyde dehydrogenase. J. Med. Chem. 32 (1989) 1335–1340.

Nagasawa, H.T., Kawle, S.P., Elberling, J.A., DeMaster, E.G. and Fukuto, J.M.: Prodrugs of nitroxyl as potential aldehyde dehydrogenase inhibitors vis-à-vis vascular smooth muscle relaxants. J. Med. Chem. 38 (1995) 1865–1871.

Öhlin, H., Jerntorp, P., Bergström, B. and Almér, L-O.: Chlorpropamide-alcohol flushing, aldehyde dehydrogenase activity, and diabetic complications. Brit. Med. J. 285 (1982) 838–840.

Peachey, J.E.: A review of the clinical use of disulfiram and calcium carbimide in alcoholism treatment. J. Clin. Psychopharmacol. 1 (1981) 368–375.

Petersen, E.N.: The pharmacology and toxicology of disulfiram and its metabolites. Acta Psychiatr. Scand. 86 (1992) 7–13.

Rekha, G.K. and Sladek, N.E.: Inhibition of human class 3 aldehyde dehydrogenase, and sensitization of tumor cells that express significant amounts of this enzyme to oxazaphosphorines, by the naturally occuring compound gossypol. In Weiner, H., Lindahl, R., Crabb, D.W. and Flynn, T.G. (Eds.) Enzymology and Molecular Biology of Carbonyl Metabolism - 6, Plenum Press, New York, 1997.

Rekha, G.K., Sreerama, L. and Sladek, N.E.: Intrinsic cellular resistance to oxazaphosphorines exhibited by a human colon carcinoma cell line expressing relatively large amounts of a class-3 aldehyde dehydrogenase. Biochem. Pharmacol. 48 (1994) 1943–1952.

Saigal, D., Cunningham, S.J., Farrés, J. and Weiner, H.: Molecular cloning of the mitochondrial aldehyde dehydrogenase gene of *Saccharomyces cerevisiae* by genetic complementation. J. Bacteriol. 173 (1991) 3199–3208.

Sambrook, J., Fritsch, E.F. and Maniatis, T.: Preparation and transformation of competent *E. coli*. In Molecular Cloning: A Laboratory Manual, 2nd edition, Cold Spring Harbor Laboratory Press, 1989, pp. 1.74–1.84.

Sellers, E.M., Naranjo, C.A. and Peachey, J.E.: Drugs to decrease alcohol consumption. New Eng. J. Med. 305 (1981) 1255–1262.

Sidhu, R.S. and Blair, A.H.: Human liver aldehyde dehydrogenase. Esterase activity. J. Biol. Chem. 250 (1975) 7894–7898.

Sladek, N.E.: Oxazaphosphorine-specific acquired cellular resistance. In Teicher, B.A. (Ed.), Drug Resistance in Oncology, Marcel Dekker, N.Y., 1993, pp. 375–411.

Sladek, N.E.: Metabolism and pharmacokinetic behavior of cyclophosphamide and related oxazaphosphorines. In Powis, G. (Ed.), Anticancer Drugs: Reactive Metabolism and Drug Interactions, Pergamon Press, United Kingdom, 1994, pp. 79–156.

Sladek, N.E., Manthey, C.L., Maki, P.A., Zhang, Z. and Landkamer, G.L.: Xenobiotic oxidation catalyzed by aldehyde dehydrogenases. Drug Metab. Rev. 20 (1989) 697–720.

Sladek, N.E., Sreerama, L. and Rekha, G.K.: Constitutive and overexpressed human cytosolic class-3 aldehyde dehydrogenases in normal and neoplastic cells/secretions. Adv. Exp. Med. Biol. 372 (1995) 103–114.

Sreerama, L. and Sladek, N.E.: Identification and characterization of a novel class 3 aldehyde dehydrogenase overexpressed in a human breast adenocarcinoma cell line exhibiting oxazaphosphorine-specific acquired resistance. Biochem. Pharmacol. 45 (1993) 2487–2505.

Sreerama, L. and Sladek, N.E.: Identification of a methylcholanthrene-induced aldehyde dehydrogenase in a human breast adenocarcinoma cell line exhibiting oxazaphosphorine-specific acquired resistance. Cancer Res. 54 (1994) 2176–2185.

Sreerama, L. and Sladek, N.E.: Human breast adenocarcinoma MCF-7/0 cells electroporated with cytosolic class 3 aldehyde dehydrogenases obtained from tumor cells and normal tissue exhibit differential sensitivity to mafosfamide. Drug Metab. Dispos. 23 (1995) 1080–1084.

Sreerama, L. and Sladek, N.E.: Class 1 and class 3 aldehyde dehydrogenase levels in the human tumor cell lines currently used by the National Cancer Institute to screen for potentially useful antitumor agents. In Weiner, H., Lindahl, R., Crabb, D.W. and Flynn, T.G. (Eds.) Enzymology and Molecular Biology of Carbonyl Metabolism - 6, Plenum Press, New York, 1997.

Sreerama, L., Rekha, G.K. and Sladek, N.E.: Phenolic antioxidant-induced overexpression of class-3 aldehyde dehydrogenase and oxazaphosphorine-specific resistance. Biochem. Pharmacol. 49 (1995) 669–675.

Watanabe, A., Hobara, N. and Nagashima, H.: Activation and inhibition of yeast aldehyde dehydrogenase activity by pantethine and its metabolites. Ann. Nutr. Metab. 30 (1986) 54–57.

Wilkinson, G.N.: Statistical estimations in enzyme kinetics. Biochem. J. 80 (1961) 324–332.

Zheng, C.-F., Wang, T.T.Y. and Weiner, H.: Cloning and expression of the full-length cDNAs encoding human liver class 1 and class 2 aldehyde dehydrogenases. Alcohol. Clin. Exp. Res. 17 (1993) 828–831.

19

LASER LIGHT SCATTERING AND ULTRACENTRIFUGE STUDIES ON SHEEP LIVER CYTOSOLIC ALDEHYDE DEHDYROGENASE

Jacqueline J. Harvey,[1] Paul D. Buckley,[1] James A. Lewis,[2] and Neil D. Pinder[2]

[1]Department of Chemistry
Massey University
Palmerston North, New Zealand
[2]Department of Physics
Massey University
Palmerston North, New Zealand

1. INTRODUCTION

Evidence for dissociation of the tetrameric sheep liver cytosolic aldehyde dehydrogenase at pH 7.6 was reported by Blackwell *et al.* (1987). Subsequently the existence of order of mixing effects led Buckley *et al.* (1991) to propose that the tetrameric enzyme was dissociating into an inactive dimer or monomer. Gel filtration studies at pH 7.4 and at pH 5.2 confirmed that dissociation did take place significantly at pH 5.2, with the extent of the dissociation being concentration dependent (Buckley *et al.*, 1991).

To explore the association and dissociation phenomena further ultracentrifuge studies and laser light scatter studies have been carried out on the enzyme.

2. EXPERIMENTAL METHODS

2.1. Materials

NAD$^+$ (grade III) was obtained from Sigma Chemical Co. (St.. Louis, MO, U.S.A.) and propanal solutions were prepared as described by MacGibbon *et al.* (1977).

2.2. Methods

Sheep liver cytosolic aldehyde dehydrogenase was prepared essentially by the method of MacGibbon *et al.* (1977) but with the addition of a pH gradient chromatography step (Dickinson *et al.*, 1981) to remove any mitochondrial aldehyde dehydrogenase contamination. As a final purification step the enzyme was loaded onto a 4-acetylphenoxy sepharose affinity column and eluted with 0.010 M 4-hydroxyacetophenone.

Ultracentrifuge studies were carried out on a Beckman model E analytical ultracentrifuge with an An-H rotor with Schlieren optics set at 546.1 nm and an analyzer angle set at 60° An aluminium 2.5° single sector cell was used. The rotor which was derated to a maximum of 56000 rpm was kept at 25°C throughout the experiments. Progress of each run was monitored visually and photographs were taken at appropriate intervals. The molecular weight of the enzyme was calculated via the sedimentation and diffusion method and also via the Stokes-Einstein equation.

Gel chromatography was carried out on a Sepharose 12 column The systems used were 0.025 M potassium dihydrogen phosphate buffer, pH 7.4 or 0.01 M sodium ethanoate- ethanoic acid buffer at pH 5.2 and acetonitrile/ water 2 : 1.

For the laser light scattering studies a Spectra Physics 165 argon laser was used with a wavelength of 488 nm. The laser was operated at 100 mW but this was sometimes increased up to 300 mW when the signal from the scattering was weak. All solutions were filtered into specially cleaned NMR tubes. Measurements were made at 25 ± 0.1 °C with the beam adjusted to pass just below the surface of the liquid, in order to minimize problems due to the settling of larger particles. The intensity autocorrelation function of the scattered light was sampled with a 48 channel linear digital autocorrelator. The time interval between adjacent channels (sample time) was adjustable. Data were accumulated in short runs of 0.4 to 10 seconds duration depending on the scattered intensity. These runs were then averaged. A rejection scheme discarded those runs whose mean photon count significantly exceeded the average. In this way contributions from dust particles could be excluded. Diffusion coefficients were obtained using a computer program (Daivis, 1989) which applied the method of cumulants to yield the z-average diffusion coefficient. For an ideal case of a non-disperse non-interacting species the normalized autocorrelation function is a single exponential function, so a plot of the natural logarithm of the autocorrelation function against time is linear. For a polydisperse sample the plots are curved with on the whole short time data collected from smaller scatterers and long time data from larger scatterers. The degree of non-linearity was an indicator of the non-ideality of the sample. First to fourth order polynomials were fitted to each curve and the diffusion coefficient was obtained from the initial slope of the log graph from each of these fits. Usually the second order fit was appropriate.

Laser light scattering at elevated temperatures were carried out on a sample of enzyme at pH 7.4. The measurements were made at 5°C intervals between 25°C and 40°C. A control sample was taken through the same temperature changes for the same time but without exposure to the laser light. The activity of the control sample was monitored after every significant temperature change.

3. RESULTS AND DISCUSSION

3.1. Ultracentrifuge Studies

Even when spun at the maximum available speed the enzyme samples at both pH 7.4 and pH 5.2 at all concentrations gave a single peak on ultracentrification. As expected for a species as large as aldehyde dehydrogenase the peak moved fairly quickly through the cell, but it did not move so fast as to suggest that the sample had denatured and was aggre-

Table 1. Diffusion coefficients of aldehyde dehydrogenase
at pH 7.4 from ultracentrifuge studies

Concentration (μM)	D (10^{-11} m^2 s^{-1})	D_G (10^{-11} m^2 s^{-1})
8.5	6.0 ± 0.6	5.5 ± 0.6
17.0	4.5 ± 0.4	5.1 ± 0.3
34.0	3.5 ± 0.2	3.9 ± 0.2

gating. Apart from the first peak in each run (which had not completely cleared the meniscus) all peaks were approximately Gaussian in shape for the duration of each run. Although there were some differences between the areas of the peaks calculated assuming the peak was Gaussian and those calculated by quadrature, the differences did not exhibit any consistent variation with concentration, pH or the progress of the run. On the whole the areas calculated using the Gaussian assumption were larger than those calculated by the tranpezoidal method.

The sedimentation coefficients were very similar at the different pHs especially at the same dilutions, indicating that the same system is present at each pH and that there was no radical change to the enzyme upon dialysis to the lower pH. Thus no evidence for dissociation at the lower pH was obtained.

Diffusion coefficients determined from the spreading of the boundary profile with time, were calculated by assuming that the peaks were Gaussian and also by quadrature. For each case the area of the peak divided by the maximum peak height all squared ($(A/H)^2$) was plotted against the time in seconds. If the assumption is made that the protein does not exhibit any variation in sedimentation coefficient with concentration (the concentration dependence was in any case much smaller than the value expected for a protein) then the slope of the plot should equal $4\pi D$. However with this assumption some variations in the diffusion coefficient with time of centrifugation will be observed.

The results of the two calculations are shown in Table 1 for pH 7.4 and Table 2 for pH 5.2. Within the uncertainties of the fits the diffusion coefficients determined by the two methods of measuring areas were the same, giving confidence that the original peaks were in fact Gaussian.

The molar mass of the enzyme was calculated by the sedimentation and diffusion method, which is based on the Svedberg equation, and also via the Stokes-Einstein equation, using the assumption that the enzyme was spherical. The values of sedimentation coefficient and diffusion coefficient used were those at the lowest concentration of enzyme. In the Svedberg equation the partial molar volume for catalase (0.73 mL/g) was substi-

Table 2. Diffusion coefficients of aldehyde dehydrogenase
at pH 5.2 from ultracentrifuge studies

Concentration (μM)	D (10^{-11} m^2 s^{-1})	D_G (10^{-11} m^2 s^{-1})
8.5	5.4 ± 0.4	5.8 ± 0.2
17.0	6.1 ± 0.2	6.1 ± 0.2
34.0	5.5 ± 0.1	7.7 ± 0.6
8.5*	5.0 ± 0.1	7.7 ± 0.4

*Denotes a sample that was respun a week later

Table 3. Molar mass calculated from the Svedberg and
Stokes-Einstein equations using ultracentrifuge data

pH	D $(10^{-11}$ m^2 s$^{-1})$	M_S	$M_{S\text{-}E}$
7.4	6.0	168000	172000
	5.5_G	184000	224000
5.2	5.4	178000	236000
	5.8_G	166000	191000
	5.0*	193000	298000

M_S is the molar mass obtained from the Svedberg equation
$M_{S\text{-}E}$ is themolar mass obtained via the Sstokes-Einstein equation
*Denotes a sample that was respun a week later

tuted for that of aldehyde dehydrogenase as this enzyme has a similar molar mass and
sedimentation coefficient to that of aldehyde dehydrogenase (Tanford, 1961) Molar
masses calculated using both the methods of estimating the area of a peak and both equations for calculating the molar mass are shown in Table 3. The molar masses were all the
same within experimental error to the expected molar mass of 212000 daltons for the un-
dissociated tetramer of aldehyde dehydrogenase. It should be noted that the Svedberg
method gives the molecular weight for the anhydrous molecule whilst the Stokes-Einstein
equation gives the hydrodynamic radius of the molecule.

In summary it is clear from the results that the tetrameric form of the enzyme with
molecular weight 212000 was present under all conditions used in the ultracentrifugation
studies. This was contrary to what was expected from the results of a gel chromatography
study (Buckley *et al.*, 1991) where evidence for dissociation of the enzyme at pH 5.2 was
observed. It seems unlikely that the associated form is favoured simply because of the
greater pressure experienced by the enzyme in the ultracentrifuge. Von Tigerstrom and
Razzell (1968) observed only a single peak with a molecular weight of 191000 for aldehyde dehydrogenase from *pseudomonas aeruginosa* but after dialysis against a pH 7.2
buffer containing low salt concentrations and incubation at 30°C for one hour they observed two major components because of partial dissociation of the enzyme. Similarly
Feldman and Weiner (1972) observed dissociation of the horse liver aldehyde dehydrogenase in the presence of magnesium ions. For yeast aldehyde dehydrogenase (Steinman
and Jakoby, 1967) and sheep liver mitochondrial aldehyde dehydrogenase (Hart and Dickinson, 1977) observed only single peaks in their ultracentrifuge studies.

3.2. Gel Filtration Studies

The gel filtration studies reported by Buckley *et al.* (1991) were repeated and similar
results were obtained. One major peak with a molecular mass corresponding to the tetrameric form of aldehyde dehydrogenase was observed at pH 7.4, while at pH 5.2 complete
dissociation occurred.

3.3. Laser Light Scattering Studies

3.3.1. Measurements at pH 7.4 and 25°C. Aldehyde dehydrogenase was expected to
have a diffusion coefficient in the vicinity of 5.x 10^{-11} m^2 s^{-1} at pH 7.4 based on a molecular weight of 212000 and with reasonable assumptions made about the shape and density
of the enzyme. Experimentally the diffusion coefficient obtained as an average from meas-

Table 4. Laser light scattering results from aldehyde
dehydrogenase at pH 7.4

Concentration (μM)	Sample time (μs)	D $(10^{-11}$ m^2 s$^{-1})$	Average D $(10^{-11}$ m^2 s$^{-1})$
15.5	50	2.4 ± 0.2	4.8 ± 0.3
	20	4.1 ± 0.2	
	15	7.3 ± 0.5	
	10	5.3 ± 0.2	
20.8	7	5.1 ± 0.3	5.4 ± 0.3
	4	5.7 ± 0.2	
52.0	3	5.0 ± 0.3	5.0 ± 0.2
	2	5.1 ± 0.3	
	1		

urements on three different pure samples was 5.1 x 10^{-11} m^2 s^{-1} at pH 7.4 (Table 4). These values followed the trend for a polydisperse sample of increasing diffusion coefficients with decreasing sample times.

When concentrations of magnesium ions of around 5 mM were added to the enzyme solution there was a decrease in the diffusion coefficients by a factor of up to two (Table 5). However some of the diffusion coefficients were comparable to those obtained earlier. The plots were curved and there was a trend towards larger values of the diffusion coefficient at shorter sample times. as expected from such curved plots. The higher value of the diffusion coefficient for the most concentrated sample was most likely due to the presence of more signal from the sample and measurements could then be taken at shorter sample times. There was no real difference between adding the magnesium ions an hour prior to measurement or adding them some days previously (Table 5). Some aggregation was a problem as the enzyme was cycled between refrigerator and laser beam but no evidence for dissociation was obtained.

Table 5. Summary of laser light scattering results from aldehyde dehydrogenase
at pH 7.4

Additive	Concentration(μM)	D$(10^{-11}$ m^2 s$^{-1})$	Average D$(10^{-11}$ m^2 s$^{-1})$
None	15.5	4.8 ± 0.3	5.1 ± 0.3
None	20.8	5.4 ± 0.3	
None	52	5.0 ± 0.2	
Mg^{2+} (5 mM)	13.9	3.0 ± 0.6	#
Mg^{2+} (5 mM)	21.0	2.1 ± 0.1	#
Mg^{2+} (5 mM)	52	5.0 ± 0.3	#
[3]Mg^{2+} (5 mM)	15.5	4.6 ± 0.3	4.6 ± 0.1
[1]Propanal	15.5	0.031 ± 0.001	0.036 ± 0.004
[1]Propanal	21.0	0.044 ± 0.001	
[1]Propanal	52	0.034 ± 0.01	
[2]Propanal	15.5	0.09 ± 0.01	0.68 ±0.04
[2]Propanal	21.0	0.15 ± 0.01	
[2]Propanal	52	1.8 ± 0.1	

[1]Longer sample times used (20 mM propanal).
[2]Shorter sample times used (20 mM propanal).
[3]Mg^{2+} added some days prior to light scattering.
#No average as sample exhibits apparent concentration dependence.

The addition of propanal (all measurements made in the absence of NAD⁺) had an immediate significant effect on light scattered from a solution of aldehyde dehydrogenase. Two decays were present. At short sample times the diffusion coefficient had decreased by varying factors signifying that varying degrees of aggregation had occurred (Table 5). **There appeared to be some concentration dependence with the most concentrated** sample of enzyme being much less affected by the addition of the propanal, even though at 20 mM the propanal concentration was far in excess of the enzyme concentration (54 µM). The main contribution to the biphasic decay had to be studied at longer sample times.

Clearly the propanal is binding to the enzyme. Presumably the reactive aldehyde is reacting non specifically forming for example Schiff's bases. Whether general multiple Schiff's base reactions are responsible for the aggregation or whether modification of a few specific sites triggers a conformation change in the enzyme which facilitates aggregation is not known.

It should be noted that the enzyme and the propanal remain mixed for much longer in these laser light scattering studies compared to the pre incubation times of only about fifteen minutes in studies carried out on the activity of the enzyme. A control sample of enzyme mixed with propanal does retain activity and so the aggregation effect may be kinetically invisible.

3.3.2. Measurements at pH 5.2 and 25°C. The enzyme was assayed at pH 7.4 and then dialyzed to pH 5.2. The active site concentration therefore reported is that at pH 7.4. At pH 5.2 the diffusion coefficient either decreased or remained similar to that at pH 7.4 (Table 6). A second decay process studied at very long sample times was of very low amplitude, indicating that there were few large aggregates of the enzyme present under these conditions. No evidence for extensive dissociation was obtained even at the lowest enzyme concentration used, a result which is in agreement with the results of the ultracentrifuge studies but is in opposition to what is observed on gel electrophoresis.

Samples treated with magnesium ions no more than 30 minutes before light scattering was started (Table 6) showed a decrease in the diffusion coefficient as compared to the magnesium ion free sample. An enzyme sample which had been treated with Mg^{2+} some days prior to light scattering exhibited two decays. The initial slope gave a diffusion coefficient less than half that of the untreated sample at similar sample times whilst the longer

Table 6. Summary of laser light scattering results for aldehyde dehydrogenase at pH 5.2

Additive	Concentration (µM)	D (10^{-11} m² s⁻¹)
None	6.8	2.7 ± 0.3
None	20.6	4.2 ± 0.3
After Centrifuge	15	0.065 ± 0.002
Mg^{2+} (5 mM)	6.8	2.0 ± 0.2
Mg^{2+} (5 mM)	20.6	3.0 ± 0.2
*Mg^{2+} (5 mM)	1.7	0.84 ± 0.02
Propanal (20 mM)	20.6	5.0 ± 0.2
Propanal (20 mM)	6.8	1.5 ± 0.2
Propanal (20 mM)	11	0.10 ± 0.001

Table 7. Average of the diffusion coefficients and
hydrodynamic radii at pH 7.4

Temperature (°C)	$D_{average}$ (10^{-11} m^2 s^{-1})	$R_{hydrodynamic}$ (10^{-9} m)
25	5.2	5.1
30	7.1	4.0
35	3.5	9.1
40	0.29	130
25	0.079	326
40	0.094	477

decay gave a smaller diffusion coefficient by a further factor of two. (Table 6) Aggregation seemed to have continued during the intervening period.

Unlike at pH 7.4, when 20 mM propanal was added to the enzyme at pH 5.2 (Table 6) there was no discernible change from the diffusion coefficients for the enzyme in the absence of propanal. The average value of 5.0×10^{-11} m^2 s^{-1} was very similar to the diffusion coefficient of the enzyme at pH 7.4 (Table 4). On long time scales a second decay could be detected indicating a very small amount of further aggregation. Whatever was causing the large scale aggregation in the presence of propanal at pH 7.4 was clearly not operating at the lower pH. This may be because of the protonation of amino groups prevents modification to form Stiff's bases at the low pH.

A sample that had been subject to ultracentrifugation was stored and studied by laser light scattering (Table 4) The data was dominated by the long decay which gave a diffusion coefficient indicative of very large aggregates.

3.3.3. Measurements at pH 7.4 and Elevated Temperatures. The diffusion coefficient measured at 25°C for the enzyme sample used in the temperature study (Table 7) was in agreement with the diffusion coefficients previously obtained (Table 4).

The diffusion coefficient at 30°C was slightly higher than that at 25°C, but as the temperature was raised further the diffusion coefficient dropped markedly. (Table 7). The hydrodynamic radius was also determined at each temperature (Table 7). After completion of the measurements at 40°C the sample was stored in the refrigerator over night and measurements were again made at 25°C. The marked changes that were observed in the diffusion coefficient were not reversible by simply lowering the temperature (Table 7). Indeed the general trend to decreasing diffusion coefficient and increasing hydrodynamic radius continued to be observed at 25°C.

The aim of the high temperature study was to observe thermal denaturation effects for aldehyde dehydrogenase under relatively mild conditions where precipitation of the protein did not occur. The 26 fold increase in the hydrodynamic radius however was unexpectedly large when compared to denaturation studies on other enzymes. For example the hydrodynamic radii of streptokinas and α-lactalbumin increased by 40 - 55% when denatured by guanidine hydochloride (as compared to 11 - 15 % when thermally unfolded (Gast et al., 1992)). The hydrodynamic radius as determined from light scattering studies of lysozyme at pH 4.2 increased up to 45% on addition of increasing concentrations of guanidine hydrochloride (Dubin et al., 1973). While an increase in the average radius of 18% was observed on the thermal denaturation of lysozyme at pH 1.45 (Nicoli and Benedek, (1976), the 2600% increase in radius at 40°C (and even the 200% increase observed at 35°C) for aldehyde dehydrogenase seemed too large to arise from the protein adopting some kind of molten globule conformation as was suggested for α-lactalbumin (Gast *et al.*, 1992).

Since after temperature cycling some enzyme eventually precipitated out of solution it seems likely that aldehyde dehydrogenase has adopted an unfolded conformation which perhaps exposed hydrophobic regions which then promoted extensive aggregation of the denatured enzyme. The effect seems too large to be accounted for by unfolding alone. This is consistent with the large increase in hydrodynamic radius without an intermediate value being observed. The activity (and hence the apparent active site concentration) of the control sample of enzyme increased as the temperature was raised to 30°C because of increases in k_{cat} with temperature. Both activity and active site concentration then decreased steadily throughout the high temperature measurements before finally precipitating out.

4. CONCLUSIONS

While the results of the ultracentrifuge studies, the laser light scattering studies and the gel filtration studies are in general agreement at pH 7.4, at pH 5.2 one significant difference is observed. Passing the enzyme through a gel filtration column at pH 5.2 results in substantial dissociation of tetrameric sheep liver aldehyde dehydrogenase. No such dissociation was observed with the other two techniques.

Ultracentrifuge studies with a range of enzyme concentrations gave no evidence for dissociation at either pH. The sedimentation and diffusion coefficients obtained for the enzyme at pH 7.4 were similar to those measured by von Tigerstrom and Razzell (1968) for an aldehyde dehydrogenase from *pseudomonas aeruginosa*. However using the ultracentrifuge von Tigerstrom and Razzell were able to observe dissociation of their enzyme after dialysis to a lower salt concentration. Clearly the technique was sensitive enough to detect dissociation of the sheep liver cytosolic enzyme at pH 5.2. Although the diffusion coefficients from the ultracentrifuge were determined assuming that the sedimentation coefficient did not vary with concentration the diffusion coefficients calculated from the data were in agreement with a protein with molecular weight of about 212000.

When making laser light scattering measurements two potential problems arise. If the sample is polydisperse the resultant curved plot may be difficult to resolve in to component decays. Although the greater the difference in size between the associated and dissociated forms the better the resolution of the plot into separate decays. The second potential problem is that the intensity of the light scattered is proportional to the square of the mass of the scattering object. Thus for aldehyde dehydrogenase, in a mixture composed of equal amounts of tetramer and monomer, the tetramer would scatter sixteen times more light than from the monomer. If only twenty per cent of the enzyme had dissociated, the tetramer would scatter sixty-four times more light than the monomer. However the situation is not as bleak as it might first appear because the data collected on the autocorrelator was inspected at shorter sample times for evidence of dissociation. In this way any significant amount of dissociation would have been detected.

Nevertheless one possible rationalization of the discrepancy is that dissociation at pH 5.2 was not detected by the laser light scattering experiment because of the potential problems referred to above and not observed in the ultracentrifuge because the high pressures present during ultracentrifugation favoured the tetrameric form of the enzyme. That both occurred seems unlikely.

When it is recognised that the concentration range of the enzyme used in the rate studies was an order of magnitude greater than the lowest concentration which could be conveniently studied with the ultracentrifuge and laser light scattering techniques, the failure to observe dissociation by these techniques does not preclude sufficient dissociation taking place in the reaction mixtures to account for the order of mixing effects at pH 5.2 (Buckley *et al.*, 1991).

It is of interest that the addition of propanal promoted large scale aggregation of the enzyme at pH 7.4, without great changes in activity. Laser light scattering is a particularly sensitive tool for studying denaturation because of the huge increase in size of the scattering species after unfolding has occurred. The effects were irreversible.

5. REFERENCES

Blackwell, L.F., Motion, R.L., MacGibbon, A.K.H., Hardman, M.J. and Buckley, P.D.: Evidence that the slow conformation change controlling NADH release from the enzyme is rate-limiting during the oxidation of propionaldehyde by aldehyde dehydrogenase. Biochem. J. 242 (1987) 803–808.

Buckley, P.D., Motion, R.L., Blackwell, L.F. and Hill, J.P.: pH effects on cytoplasmic aldehyde dehydrogenase from sheep liver. Adv. Exp. Med. Biol. 284 (1991) 31–41.

Daivis, P.J.: Polymer dynamics studied by NMR and light scattering methods, Ph. D. Thesis, Massey University 1989.

Dickinson, F.M., Hart, G.J. and Kitson, T.M.:The use of pH gradient ion-exchange chromatography to separate sheep liver cytoplasmic aldehyde dehydrogenase from mitochondrial contamination, and observations on the interaction between the pure cytoplasmic enzyme and disulfiram. Biochem. J. 199 (1981) 573–579.

Dubin, S.B., Feher, G. and Benedek, G.B.: Study of the chemical denaturation of lysozyme by optical mixing spectroscopy. Biochem.12 (1973) 714–719.

Feldman, R.I. and Weiner, H.: Horse liver aldehyde dehydrogenase I. Purification and characterization. J. Biol. Chem. 247 (1972) 260–266.

Gast, K., Damaschun, G., Damaschun, H. Misselwitz, R., Zirwer, D. and Bychkova, V.A.:Laser Light Scattering in Biochemistry. Royal Society of Chemistry, 1992 pp 209–224.

Hart, G.J. and Dickinson, F.M.: Some properties of aldehyde dehydrogenase from sheep liver mitochondria. Biochem. J. 163 (1977) 261–267.

MacGibbon, A.K.H., Blackwell, L.F. and Buckley, P.D.: Pre-steady-state kinetic studies on cytoplasmic sheep liver aldehyde dehydrogenase. Biochem. J. 167 (1977) 469–477.

Nicoli, D.F. and Benedek, G.B.: Study of thermal denaturation of lysozyme and other globular proteins by light-scattering spectroscopy. Biopolymers 15 (1976) 2421–2437.

Steinman, C.R. and Jakoby, W.B.: Yeast aldehyde dehydrogenase I. Purification and crystallization. J. Biol. Chem. 242 (1967) 5019–5023.

Tanford C.: Physical Chemistry of Macromolecules. John Wiles and Sons, 1991, pp 358,381.

Von Tigerstrom, R.G. and Razzell, W.E.: Aldehyde dehydrogenase II. Physical and molecular properties of the enzyme from *pseudomonas aeruginosa*. J. .Biol. Chem. 243 (1968) 6495–6503.

20

SUBUNIT INTERACTIONS IN MAMMALIAN LIVER ALDEHYDE DEHYDROGENASES

Henry Weiner, Saifuddin Sheikh, Jianzhong Zhou, and Xinping Wang

Department of Biochemistry
Purdue University
West Lafayette, Indiana 47907–1153

1. INTRODUCTION

Aldehyde dehydrogenases are multisubunit enzymes. The first ones purified were 500 amino acid homotetramers, cytosolic and mitochondrial in origin, and later called class 1 and 2, respectively. These two forms of the enzyme share 70% sequence identity with each other. A third class of enzyme characterized had 452 amino acids and was dimeric. The class 3 enzymes were microsomal, tumor specific or inducible in origin and shared less than 40% homology with the class 1 and 2 enzymes (Hempel et al., 1993). More is known about the enzymology of the class 1 and 2 enzymes, though the 3-dimensional structure of the class 3 enzyme has recently been determined (Liu, et al., 1997). Other classes of ALDHs are known but those enzymes have not been characterized.

The initial experiments performed to characterize the class 1 and 2 enzymes revealed that there is some type of interaction between the subunits. For example, we found that lag kinetics was observed with the horse liver mitochondrial enzyme. This was never studied in detail, but the lag time appeared to be related to the order of mixing of components in the cuvette. Though not proved, this could have been related to an association/dissociation phenomenon. That is, if the enzyme were capable of undergoing dimer/tetramer equilibrium and one form was more active than the other, a lag could have been observed as the less active, presumed tetramer, dissociated into the more active dimeric form. Studies with the horse liver enzyme revealed that the enzyme did dissociate to a pair of dimers in the presence of Mg^{2+} ions and the dimers were more active than was the tetramer (Takahashi and Weiner, 1980). We showed that the tetramer functioned with half-of-the-site reactivity while the dimer functioned with full-site reactivity (Takahashi et al., 1980).

We and other investigators could not find conditions which allowed enzyme from species other than horse to dissociate at pH 7. All mitochondrial class 2 ALDHs though, are activated by the presence of Mg^{2+} ions. What ever is the basis of the Mg^{2+} ion induced activation and increase in coenzyme binding stoichiometry, the data can best be interpreted as implying that there is subunit interaction which influences the properties of the enzyme so that the tetramer does not behave as four independent subunits.

Chemical modification studies also revealed that less than a stoichiometric number of modifiers covalently bound to enzyme (Pietruszko et al., 1985; Weiner et al., 1985). For example, a number of studies from Pietruszko's laboratory reported that binding of between 1 and 2 moles of covalent inhibitors led to complete inactivation of the human class 1 or 2 enzymes (Hempel and Pietruszko, 1981; Pietruszko et al., 1985). The chemical modification studies coupled with the above mentioned binding and kinetic studies all support the notion that there are interactions within the tetrameric enzyme which make it function with less than four active sites. The fact that in the presence of Mg^{2+} ions the enzyme appears to function with one active site per subunit dismisses the notion that it takes two subunits to make one active site.

To find an alternative way to study the possible interactions between subunits we choose to make a heterotetrameric enzyme composed of subunits with different properties. We took advantage of the one natural variant of the mitochondrial enzyme. This form is primarily found in Oriental people and was shown to be the result of lysine (K) residue being at position 487 rather than a glutamate (E) (Yoshida et al., 1984). People with the lysine (K)-enzyme have very little measurable ALDH activity in their mitochondria. Our laboratory recombinantly expressed the human and rat forms of the homotetrameric E_4 and K_4 enzymes. Compared to native enzyme the K_m for NAD with the K_4 mutant increased several fold while the k_{cat} decreased significantly (Farrés et al., 1994). In order to study possible subunit interactions the heterotetramers of the enzyme were produced to determine if this tetrameric form of the enzyme had properties of the individual subunits or possessed new hybrid properties.

2. EXPERIMENTAL PROCEDURES

The cloning, mutations and expression of the enzyme has been described in many of our publications (Zheng et al., 1993; Rout and Weiner, 1994; Farrés et al., 1994). Here we co-expressed on one plasmid the cDNA coding for the active E-subunit and the low active K-subunit. The plasmid is illustrated in Figure 1. After expression, the enzyme forms were purified and the kinetic properties determined. The details for the construction of the co-expressed plasmid are to be presented elsewhere (Wang et al., 1997). Kinetic assays were performed at pH 7.4 in 0.1 M phosphate with varied concentrations of NAD and propionaldehyde. Inhibition studies were performed at different fixed concentrations of inhibitors and varied concentrations of substrates.

3. RESULTS AND DISCUSSION

It was not possible to separate the three recombinantly expressed heterotetrameric forms of the human enzyme from each other. It was possible, though, to remove both homotetrameric forms of the enzyme from the mixture. Thus, the studies were performed with a mixture of EK_3, E_2K_2 and E_3K heterotetrameric forms of the enzyme. Assuming that there was no preference in assembly, then the mixture should be composed of 42.5% E_2K_2 and 28.7%, each EK_3 or E_3K.

Knowing the properties of the two homotetramers, it was possible to estimate what would have been the overall kinetic properties of the heterotetramers. When assayed with a concentration of NAD which would just saturate the E-subunits, the K-subunit would contribute very little activity, due to its very high K_m for NAD and low specific activity. It was found, though, that when assayed in the presence of 2.5 mM NAD the specific activ-

Figure 1. Plasmid to simultaneously coexpress human mitochondrial E487 and E487K subunits in *E. coli* BL21 cells. One T7 promoter (T7prom) and two ribose binding sites (RBS) were present on the pT7–7 plasmid.

ity of the mixture was just 13% of the native enzyme and not the 50% calculated. Thus, the presence of the K-subunit appeared to decrease the activity of the native E-subunit. Data suggesting that this indeed occurs was recently presented by Xiao et al. (1995) where it was shown that in HeLa cells the expression of the K-subunits decreased the activity of the E-subunits.

When the mixture of heterotetramers was assayed in the presence of a high concentration of NAD, enough to saturated the K-subunits, no significant increase in the activity was obtained. This leads us to suggest that the heterotetramers functioned with a new set of properties. Further proof that the E-subunit in the heterotetramer had properties that differed from those in the homotetramer was finding that the heterotetramer no longer pos-

Table 1. Properties of homo- and heterotetrameric liver mitochondrial aldehyde dehydrogenases

Enzyme form[a]	K_m(NAD) μM	k_{cat} min^{-1}
Rat		
Homo E	16	121
Homo K	845	45
Hetero E/K	-	58
Human		
Homo E	28	205
Homo K	7400	19
Hetero E/K	83	43

[a]Homo refers to homotetramers of either four subunits of E or K (E_4 or K_4) while hetero E/K refer to an unresolved mixture of heterotetramers possessing both the E- and the K-subunits (EK_3, E_2K_2 and E_3K). Saturating concentration of substrates were used to determine the activity of the enzyme (Data from Wang *et al.*, 1997).

sessed a pre-steady state burst of NADH formation. This is a property of the native enzyme (Wang and Weiner, 1995). Thus, the presence of the K-subunit altered the properties of the E-subunit.

It is still not known how the four subunits of tetrameric ALDH are spatially arranged. A simple model would be to have them in a tetrahedral arrangement. Here, each subunit could in theory be in contact with the other three subunits. If though, the subunits were not packed as the simple tetrahedral, one could envisage a packing such that the system functioned as a pair of dimers. We suggest that only one member of each pair is actually active. This is being proposed without corroborating structural data. Having the enzyme function as a pair of dimers would allow us to explain the burst magnitude and stoichiometry for coenzyme binding being just two for the E_4 homotetramer. We would need to postulate that in the E/K heterotetramers the pairing existed between an E- and a K-subunit. If the alternative existed, there would have been an E-E pair and a K-K pair. This arrangement would have produced an enzyme with a highly active dimeric pair which would have had half the activity of the E_4 homotetramer. Furthermore, the EE pair would have functioned identically to a pair in the homotetrameric form. No evidence for this was found. In the heterotetramer no presteady state burst of NADH formation was found and the specific activity was not that of an EE pair.

The physical basis for subunit interaction will not be clarified until the structure of the tetrameric enzyme is better understood. The enzyme, especially the mitochondrial isozyme, must have either inherent or coenzyme induced asymmetry to account for the stoichiometry of binding and pre-steady state burst magnitude. While investigating the binding of thyroxine analogs to ALDH (Zhou and Weiner, 1997) we found that 3,3',5-tri-iodo-L-thyronine and L-3,3',5-triiodothyroacetic acid were competitive inhibitors against NAD. Four moles of either thyroxine analogs bound to one mole of the tetrameric apo-enzyme, suggesting that the compounds bind to each subunit. It is not known if the inhibitor actually binds to the NAD site or just to the conformation of the enzyme which binds NAD. If the inhibitor interacts with the NAD binding site, it would indicate that there was the potential to form all four binding sites in the tetrameric enzyme.

A very different result was found when investigating the inhibition of ALDH by an isoflavone, prunetin. This class of compounds was recently shown to be excellent *in vitro* inhibitors against ALDH (Keung and Vallee, 1993). It is of possible interest to note that prunetin, found to be a competitive inhibitor against NAD and aldehyde, binds to the enzyme with a stoichiometry of 1 mole binding per mole of tetrameric enzyme (Sheikh and Weiner, 1997). The best interpretation of this finding is that the compound binds to a unique site, presumed to be an allosteric site, located perhaps at the junction where the subunits interact with each other. If there were an allosteric site on the enzyme then, chemical modification of the site might result in a diminution of the activity of the enzyme. It would be necessary to know that chemical modification of the enzyme which resulted in a loss of catalytic activity was not due to the modification of this site.

Finding that an additional potential allosteric site exists in the tetrameric enzyme and that its occupancy by an isoflavone affects enzyme activity can not be used to prove that the subunits of class 2 ALDH might not be totally independent of each other. The half-of-the-site-reactivity and the fact that the heterotetrameric form of the enzyme does not have properties of either of its subunits does supports the idea that there is some type of subunit interaction in the enzyme, especially as related to coenzyme binding. This implies that though the NAD binding site for the enzyme may ultimately be determined by x-ray crystallographic techniques, we still might not understand all aspects of the structural ramifications of coenzyme interaction with the enzyme.

4. REFERENCES

Farrés, J., Wang, X., Takahashi, K., Cunningham, S. J., Wang, T. T. and Weiner, H., 1994, Effect of changing glutamate to lysine in rat and human liver mitochondrial aldehyde dehydrogenase: A model to study human (oriental type) class 2 aldehyde dehydrogenase, *J. Biol. Chem.*, 269:13854–13860.

Hempel, J., Nicholas, H. and Lindahl, R., 1993, Aldehyde dehydrogenase: Widespread structural and functional diversity within a shared frame-work, *Protein. Sci.*, 2:1890–1900.

Hempel, J. D. and Pietruszko, R., 1981, Selective chemical modification of human liver aldehyde dehydrogenase E1 and E2 by iodoacetamide, *J. Biol. Chem.*, 256:10880–10896.

Keung, W.-M. and Vallee, B. L., 1993, Diadzin: A potent, selective inhibitor of human aldehyde dehydrogenase, *Proc. Natl. Acad. Sci. USA.*, 90:1247–1251.

Liu, Z-J., Hempel, J., Sun, J., Rose, J., Hsiao, D., Chang, W-R., Chung, Y-J., Kuo, I., Lindahl., R., and Wang, B-C., 1997, These Proceedings.

Pietruszko, R., Ferencz-Biro, K. and Mackerell, A. D. Jr., 1985, Chemical modification of aldehyde dehydrogenase, In: *Enzymology of Carbonyl Metabolism 2: Aldehyde Dehydrogenase, Aldo-Keto Reducatase and Alcohol Dehydrogenase* (Flynn TG and Weiner H Eds), pp 29–41, Alan R Liss, New York.

Rout, U. K. and Weiner, H, 1994, Involvement of serine 74 in the enzyme-coenzyme interaction of rat liver mitocondrial aldehyde dehydrogenase, *Biochemistry*, 33:8955.

Sheikh, S. and Weiner, H., 1997, Allosteric inhibition of human liver aldehyde dehydrogenase by the isoflavone prunetin, *Biochem. Pharmacol.* Submitted.

Takahashi, K. and Weiner, H., 1980, Magnesium stimulation of catalytic activity of horse liver aldehyde dehydrogenase, *J. Biol. Chem.*, 255:8206.

Takahashi, K., Weiner, H. and Hu, J. H. J., 1980, Increase in the stoichiometry of the functioning active sites of horse liver aldehyde dehydrogenase in the presence of magnesium ions, *Arch. Biochem. Biophys.*, 205:571–578.

Wang, X.-P. and Weiner, H., 1995, Investigation of the Role of Glutamate 268 in Human Liver Aldehyde Dehydrogenase, *Biochemistry*, 34:237–243.

Wang, X.-P., Sheikh, S., Saigal, D., Robinson, L. and Weiner, H., 1997, Heterotetramers of human liver mitochondrial (class 2) aldehyde dehydrogenase expressed in *E. coli.* A model to study the heterotetramers expected to be found in oriental people, *J. Biol. Chem.* In Press.

Weiner, H., Lin, F.-P. and Sanny, C. G., 1985, Chemical probes for the active site of aldehyde dehydrogenase, . In: *Enzymology of Carbonyl Metabolism 2: Aldehyde Dehydrogenase, Aldo-Keto Reducatase and Alcohol Dehydrogenase* (Flynn TG and Weiner H Eds), pp 57–70, Alan R Liss, New York.

Xiao, Q., Weiner, H., Johnston, T., and Crabb, D. W., 1995, The aldehyde dehydrogenase ALDH2*2 allele exhibits dominance over ALDH2*1 in transduced HeLa cells, *J. Clin. Invest.*. 96:2180–2186.

Yoshida, A., Huang, I.-Y. and Ikawa, M. (1984) Molecular abnormality of an inactive aldehyde dehydrogenase variant commonly found in Orientals, *Proc. Natl. Acad. Sci. USA* 81:258–261.

Zheng, C.-F., Wang, T. T. Y., and Weiner, H., 1993, Cloning and expression of the full length cDNAs encoding human liver class I and class II aldehyde dehydrogenase, *Alcoholism: Clin. Exp. Res.*, 17:828.

Zhou, J-Z and Weiner, H., 1997, Interactions of thyroxine analogs with human liver aldehyde dehydrogenases, *Eur. J.Biochem.*, submitted.

21

STUDIES ON THE DOMINANT NEGATIVE EFFECT OF THE ALDH2*2 ALLELE

Qing Xiao,[1] Henry Weiner,[2] and David Crabb[1]

[1]Departments of Medicine and Biochemistry and Molecular Biology
Indiana University School of Medicine
IB 424 975 West Walnut Street
Indianapolis, Indiana 46202–5121
[2]Department of Biochemistry
Purdue University
West Lafayette, Indiana 49707

1. INTRODUCTION

The alcohol flush reaction results from a dominantly inherited deficiency in the activity of mitochondrial aldehyde dehydrogenase (ALDH2) (Harada et al., 1982; Goedde et al., 1989). The active and inactive alleles are named *ALDH2*1* and *ALDH2*2*, respectively. Consumption of alcohol by individuals with the deficiency leads to dramatic increases in blood acetaldehyde levels. The aversive symptoms that they experience from this acetaldehyde accumulation reduces their consumption of alcohol and greatly reduces the risk that they will become alcoholics (Goedde et al., 1983; Harada et al., 1983; Thomasson et al., 1991). In fact, it is the most powerful genetic modifier of alcohol consumption yet discovered. This protection may be a two-edged sword; several recent studies have shown increased frequency of *ALDH2*2* heterozygotes among alcoholics with cirrhosis compared with alcoholics without evidence of liver disease (Chao et al., 1994; Yamauchi et al., 1995; Tanaka et al., 1996), although in both groups the frequency of heterozygotes is lower than in the non-alcoholic control population.

The mechanism by which the *ALDH2*2* allele exerts its dominant effect is incompletely understood. The active subunit synthesized from the gene is designated ALDH2E (for glutamate at position 487), and the product of the *ALDH2*2* allele is designated ALDH2K (for lysine at position 487) (Yoshida et al., 1984; Hempel et al., 1984). Weiner et al. showed that homotetramers of ALDH2K are active, albeit with a low V_{max} and very high K_m for NAD$^+$ compared with ALH2E tetramers (Farres et al., 1994). This enzyme is predicted to be virtually inactive under the conditions of the mitochondrial matrix space. The enzymology of pure heterotetramers has not yet been reported. We have therefore undertaken studies on a cell model of ALDH2 deficiency that has permitted us to postulate two mechanisms for the dominant negative effect of *ALDH2*2*.

2. METHODS

2.1. Cloning and Expression of ALDH2 cDNAs

ALDH2 cDNA was cloned from a human liver cDNA library using the rat cDNA as the probe. The 5' end of the cDNA was completed by polymerase chain reaction of genomic DNA. The cDNA contained the normal glutamate codon at position at 487. This was mutated to a lysine codon by site-directed mutagenesis. The cDNAs were expressed in HeLa cells using defective retroviruses as transducing agents. The vector pLNCX, conferring resistance to G418 (Miller and Rosman, 1989), was used for the *ALDH2*1* allele (Figure 1). To express both cDNAs in a single cell line, we constructed a second vector containing the resistance gene for hygromycin B (Gritz and Davies, 1983) in place of the neomycin resistance gene and the *ALDH2*2* cDNA with 3' untranslated sequences (named pLHCK3UT, Figure 1). A clone of cells expressing high levels of *ALDH2*1* (and approximately the same amount of ALDH2E protein as found in liver extracts) was transduced with the *ALDH2*2* expression vector, and individual clones were isolated after selelction with hygromycin B. Cells expressing active ALDH2 were named HeLa E cells, those expressing the inactive form only HeLa K cells, and cells expressing both alleles were named HeLa EK cells (Xiao et al., 1995).

2.2. Analysis of Expression of ALDH2 mRNA and Protein

Expression of the *ALDH2* alleles at the mRNA level was assessed by Northern blotting. Western blotting was performed with antibody generated against bovine liver mitochondrial ALDH2, using filters blotted from either SDS-PAGE or isoelectric focusing gels. Enzyme activity was determined in 50 mM sodium phosphate buffer (pH 8.8), containing 10 µM 4-methylpyrazole, using propionaldehyde as substrate at 15 µM and NAD$^+$

Figure 1. Construction of retroviral vectors for expression of ALDH2 alleles. The 5' LTR was derived from the long terminal repeat of the Moloney murine sarcoma virus; the 3' LTR was from the Moloney murine leukemia virus; MCS denotes a multiple cloning site; Neo denotes the neomycin/G418 resistance gene; Hyg denotes the hygromycin B resistance gene, and CMV denotes the cytomegalovirus immediate early promoter. The ALDH2E and ALDH2K cDNAs are shown driven by the CMV promoter in pLNCE and pLHCK3UT, respectively.

at 2 mM, and following the appearance of NADH spectrophotometrically. Gels were stained for ALDH2 activity using 1 mM NAD^+ and 1 mM propionaldehyde.

2.3. Determination of the Half-Life of ALDH2 Protein

Turnover of ALDH2 subunits was measured by two different methods. First, the cells were grown to confluence then puromycin was added at 100 μg/ml to inhibit the protein synthesis. The cells were harvested at varying times over 24 hours after addition of the drug. The cell extracts were prepared in sodium phosphate, pH 6.0 containing 1% Triton X100, 0.1 mM DTT, 1 mM EDTA, 1 mM PMSF, and 0.5 μg/ml leupeptin. The cells were disrupted by sonication on ice. Equal amounts of protein from the extracts were then subjected to SDS-PAGE and Western blotted with anti-ALDH2 antiserum. The antiserum had been shown to react equally well with both ALDH2E and ALDH2K subunits. The Western blots were then developed with radiolabelled protein A and the amount of ALDH2 protein was quantified by ß-scanning.

Pulse chase experiments were also performed to determine the rates of synthesis and degradation of the subunits. The cells were cultured in methionine- and cysteine-free medium for 30 minutes to deplete intracellular amino acids. They were then labelled for varying times up to 8 hours with Tran^{35}S-label, and were harvested at intervals to determine the rate of synthesis. The cells were then shifted to medium containing unlabelled amino acids and cultured for an additional 24 hours with harvesting at 0, 2, 4, 6, and 8 hours. The cells were trypsinized, washed in PBS, pelleted, and disrupted by sonication. Equivalent amounts of total protein in the clarified extracts were pre-incubated with pre-immune serum and then protein A Sepharose beads. ALDH2 protein was then immunoprecipitated with specific antibody and protein A Sepharose (Sastre et al., 1986). The immunoprecipicates were then fractionated by SDS-PAGE. The gels were dried and ß-scanned to quantify the amount of radiolabelled ALDH2 protein.

2.4. Genotyping of Human Autopsy Livers

Frozen human liver samples were stored frozen at -80°. DNA was isolated by proteinase K digestion, phenol-chloroform extraction, and spooling. *ALDH2* genotype was determined by the Molecular Biology Core of the Indiana Alcohol Research Center by polymerase chain reaction and hybridization to allele-specific oligonucleotides (Crabb et al., 1989).

3. RESULTS

3.1. Expression of *ALDH2* Alleles in HeLa Cells and Characterization of the Gene Products

The retroviral vectors used directed high level expression of the *ALDH2* alleles. RNA was easily detected using 10–20 μg of total cellular RNA on standard Northern blots. Western blots showed nearly identical amounts of ALDH2 protein in extracts of HeLa cells expressing either ALDH2E or ALDH2K and in human autopsy liver samples (Xiao et al., 1995). Since the viral constructs contained the mitochondrial leader sequence as well as the sequence for the ALDH2 coding region, it was important to determine if it was properly processed and transported into the mitochondria. Isoelectric focusing gels stained with propionaldehyde and NAD$^+$ showed a band of activity of the correct pI in the cells transduced with ALDH2E. In addition, Western blots of the isoelectric focusing gels showed that the pI of both ALDH2E and ALDH2K corresponded to those of the liver ALDH from individuals of known *ALDH2* genotype. These experiments indicated that the basically charged leader sequence was cleaved after mitochondrial import. Additional studies showed that the ALDH2 protein was present in highest concentration in the mito-chondrial fraction of the cells (prepared by differential centrifugation through sucrose gra-dients, not shown), and that ALDH2 protein was distributed in a pattern consistent with mitochondrial localization in immunohistochemical assays (M.J. Stewart, unpublished data). The enzyme present in the cells transduced with the ALDH2E cDNA expression vi-rus was active when assayed at 15 µM propionaldehyde. In cells transduced with ALDH2K cDNA-expressing virus, the enzyme was not active with 15 µM propionalde-hyde and 2 mM NAD$^+$, and was not detectable by staining for enzyme activity in isoelec-tric focusing gels. Thus, this strategy of retroviral transduction of ALDH2 alleles into HeLa cells generated good cellular models for ALDH2 expression.

3.2. Generation of Cell Lines Expressing Both *ALDH2*1* and *ALDH2*2*

A high expressing ALDH2E cell clone was selected and transduced with the ALDH2K viral construct conferring hygromycin resistance. Four clones of doubly transduced cells were selected by growth in hygromycin B containing medium and iso-lated as clones (referred to as HeLa EK cells). The presence of both *ALDH2*1* and *ALDH2*2* mRNAs in the cells was confirmed by Northern blotting, since the ALDH2K vector contained additional 3' untranslated sequences downstream of the coding region that resulted in a larger mRNA species being synthesized. Both subunits reached the same cell compartment and formed heterotetramers, as shown by a smear of protein with isoelectric points between that of the ALDH2E and ALDH2K homotetramers on isoelec-tric focusing/Western blots. The enzyme activity of these cells was in each case lower than that in the ALDH2E expressing cells from which they were derived (Figure 2). This supported the hypothesis that the ALDH2K is dominant over ALDH2E. The activity was reduced to about 30% of the parental cells. A model shown in Table 1, in which E4 tetra-mers have two active sites, E3K tetramers have one active site, and the remaining hetero-tetramers were essentially inactive was proposed to account for this degree of dominance (Xiao et al., 1995).

3.3. Metabolic Half-Lives of ALDH2E and ALDH2K Enzymes

Yoshida et al. had reported that ALDH2 deficient individuals had decreased im-munoreactive ALDH2 protein (Yoshida et al., 1983; Impraim et al., 1982). This was pur-sued by performing Western blotting of human autopsy liver samples. The amount of immunoreactive ALDH2 was apparently lower on Western blots in samples from *ALDH2*2* hetero- or homozygotes. This was confirmed by Western blotting the samples for another mitochondrial enzyme, 3-hydroxyisobutyrate dehydrogenase. This second en-zyme was used as an internal control for the possibility of post-mortem degradation of mi-tochondrial enzymes. There was little variation in the amount of 3-hydroxyisobutyrate dehydrogenase protein in the samples, regardless of ALDH2 genotype, while the amount of ALDH2 was about 50% lower in the liver of *ALDH2*2* heterozygotes (and the single *ALDH2*2* homozygote examined). This suggested that the ALDH2K subunit was either translated more slowly or degraded more rapidly.

Figure 2. Low K_m ALDH activity in Hela cells transduced with the ALDH2E and ALDH2K cDNAs. The ALDH activity with 15 µM propionaldehyde is shown on the vertical axis. Bars labelled EK2, EK3, EK12, and EK13 denote activities of individual cell clones expressing both ALDH2E and ALDH2K subunits. E indicates cells expressing only ALDH2E subunits, from which the doubly transduced cells were derived.

We therefore used the cell lines to estimate the synthetic rates and half-lives of the ALDH2 protein subunits. Puromycin decay curves indicated that the ALDH2E subunit was relatively stable (no apparent change in ALDH2 protein abundance over 24 hours), while the mutant enzymes had half-lives well under 24 hours. Pulse-chase labelling indicated no differences in rates of synthesis. The half-life of the ALDH2E subunit was about twice that of the ALDH2K subunits. The half-life of ALDH2 in heterozygotic cells was the same as that in the ALDH2K-expressing cells, indicating that the reduced half-life conferred by ALDH2K was also a dominant trait.

4. DISCUSSION

Since ALDH2 deficiency has such a pronounced effect on risk of alcoholism, and possibly on risk of alcoholic liver disease, it is important to understand the mechanism by which the variant *ALDH2*2* allele exerts its dominant effect. Moreover, the degree of dominance may vary between individuals. For instance, the blood level of acetaldehyde that results from drinking is highly variable among heterozygotes (Enomoto et al., 1991), with some experiencing mild and others intense flushing. Moreover, the prevalence of *ALDH2*2* heterozygotes among alcoholics in Japan has been increasing over the last 3 decades (Higuchi et al., 1994). These latter data suggest the possibility that environmental factors interact with the genetic deficiency to modulate the expression of ALDH2, the flushing phenotype, or the behavioral response to flushing.

Table 1. Models for the dominance of the *ALDH2*2* allele

Enzyme tetramer	Number of tetramers	Fraction of total	Active sites per tetramer	Uncorrected active sites	Correction for half-life	Corrected active sites
Hemizygous *ALDH2*1* cells						
E$_4$	n/4	16/16	2	n/2	100%	n/2
Heterozygous cells						
E$_4$	2n/4	1/16	2	n/16	100%	n/16
E$_3$K		4/16	1	2n/16	50%	n/16
Other inactive tetramers		11/16	0	0	-	0
				Sum: 3n/16		Sum: 2n/16
				37%*		25%*

The calculations are based on the model presented previously (Xiao et al., 1995) for ALDH2 heterotetramers. It assumes that association of the ALDH2E and ALDH2K subunits is random and therefore can be described by a binomial expansion. This modified model permits 2 active sites for E$_4$ and 1 active site for E$_3$K tetramers, and no active sites in the other tetramers (the low activity of the ALDH2K enzyme is ignored). The model assumes synthesis of n subunits per allele; thus, there will be n/4 tetramers in the parental hemizygous cells and 2n/4 (= n/2) tetramers in the heterozygous cells. This value is multiplied by the fraction of each tetramer and the number of active sites per tetramer to obtain the uncorrected number of active sites. The corrected number of active sites takes into account the instability of the E$_3$K tetramers. Without the correction for instability of the heterotetramer, the activity would be 37% of that of the parental cells (3n/16 divided by n/2), and with the correction, 25% of the activity of the parental cells (n/8 divided by n/2). This model is also the best fit of the data to predictions discussed elsewhere (Xiao et al., 1995).
*% of parent cell

The studies described here indicate that ALDH2 subunits can in fact form heterotetramers, and that these tetramers have reduced but not absent enzymatic activity. Complete dominance would have resulted in only the E4 tetramers having activity, with an estimated activity 12% of the ALDH2E expressing cells, while in the EK cell lines, activity was reduced to about 30% of that in the parental ALDH2E cells. The corresponding reduction in human liver ALDH2 activity would be twice as large since the heterozygotes loose an active *ALDH2*1* allele when they acquire an *ALDH2*2* allele. Thus, total dominance would result in only 6% ALDH2 activity in heterozygotes, while the partial dominance predicted by the model shown in Table 1 would result in about 15% activity. In fact, two reports indicate residual activity closer to the latter figure (Ferencz-Biro and Pietruszko, 1984; Enomoto et al., 1991). Understanding the precise effect of incorporation of ALDH2K subunits into the ALDH2 tetramer will await purification and characterization of the various heterotetramers, and possibly require crystal structures of the tetramers.

The apparent reduction in the half-life of ALDH2K-containing enzyme cannot yet be explained mechanistically; however, one possibility is that instability is related to the decreased binding of NAD$^+$ by ALDH2K (Farres et al., 1994). The K$_{ia}$ for NAD$^+$ (kinetically equivalent to the dissociation constant) was increased from 13 µM in the ALDH2E tetramers to 500 µM in ALDH2K tetramers. Although the K$_{ia}$ for NAD$^+$ of heterotetrameric enzyme has not been reported, decreased binding of NAD$^+$ could contribute to instability, as has been noted in many other genetically determined deficiencies of enzymes that have a coenzyme requirement. The estimated free NAD$^+$ concentration in the mitochondrion is about 6 mM (Tischler et al., 1977); thus, about 8% of the ALDH2K tetramers would be coenzyme-free at any particular time, and potentially more susceptible to degradation. Alternatively, binding of NAD$^+$ may be needed for folding and assembly of the nascent subunits, and unfolded subunits may be prematurely degraded. It will be interesting to determine if supplementation of the cells with high levels of nicotinamide stabilizes the ALDH2K-containing enzymes, and conversely, if nicotinamide depletion destabilizes ALDH2E enzymes.

5. ACKNOWLEDGMENTS

We appreciate the excellent technical assistance of Ruth Ann Ross, the genotyping servcies of the Alcohol Research Center, the gift of 3-hydroxyisobutyrate dehydrogenase antibody from Dr. Robert A. Harris, and the support of the NIAAA (AA 06434 and AA 10525 to DWC).

6. REFERENCES

Chao, Y.-C., Liou, S.-R., Chung, Y.-Y., Tang, H.-S., Hsu, C.-T., Li, T.-K. and Yin, S.-J. Polymorphism of alcohol and aldehyde dehydrogenase genes and alcoholic cirrhosis in Chinese patients. Hepatology 19 (1994) 360–366.

Crabb, D.W., Edenberg, H.J., Bosron, W.F. and Li, T-K. Genotypes for aldehyde dehydrogenase deficiency and alcohol sensitivity. The inactive ALDH2*2 allele is dominant. J Clin Invest 83 (1989) 314–316.

Enomoto, N., Takase, S., Yasuhara, M. and Takada, A. Acetaldehyde metabolism in different aldehyde dehydrogenase 2 genotypes. Alcoholism: Clin Exp Res 15 (1991) 141–144.

Farres, J., Wang, X., Takahashi, K., Cunningham, S.J., Wang, T.T. and Weiner, H. Effects of changing glutamate 487 to lysine in rat and human liver mitochondrial aldehyde dehydrogenase. J Biol Chem 269 (1994) 13854–13860.

Ferencz-Biro, K. and Pietruszko, R. Human aldehyde dehydrogenase: catalytic activity in Oriental liver. Biochem Biophys Res Comm 118 (1984) 97–102.

Goedde, H.W., Agarwal, D.P., Harada, S., Meier-Tackmann, D., Ruofo, D., Bienzle, U., Kroeger, A. and Hussein, L. Population genetic studies of aldehyde dehydrogenase isozyme deficiency and alcohol sensitivity. Am J Hum Genet 35 (1983) 769–772.

Goedde, H.W., Singh, S., Agarwal, D.P., Fritze, G., Stapel, K. and Paik, Y.K. Genotyping of mitochondrial aldehyde dehydrogenase in blood samples using allele-specific oligonucleotides: comparison with phenotyping in hair roots. Hum Genet 81 (1989) 305–307.

Gritz, L. and Davies, J. Plasmid encoded hygromycin B resistance: the sequence of hygromycin B phosphotransferase gene and its expression in Escherichia coli and Saccharomyces cerevisiae. Gene 25 (1983) 179–188.

Harada, S., Agarwal, D.P., Goedde, H.W., Tagaki, S. and Ishikawa, B. Possible protective role against alcoholism for aldehyde dehydrogenase isozyme deficiency in Japan. Lancet 2 (1982) 827.

Harada, S., Agarwal, D.P., Goedde, H.W. and Ishikawa, B. Aldehyde dehydrogenase isozyme variation and alcoholism in Japan. Pharm Biochem Behavior 18 (1983) 151–153.

Hempel, J., Kaiser, R. and Jornvall, H. Human liver mitochondrial aldehyde dehydrogenase: a C-terminal segment positions and defines the structure corresponding to the one reported to differ in the Oriental enzyme variant. FEBS Lett 173 (1984) 367–373.

Higuchi, S., Matsushita, S., Imazeki, H., Kinoshita, T., Takagki, S. and Kono, H. Aldehyde dehydrogenase genotypes in Japanese alcoholics. Lancet 343 (1994) 741–742.

Impraim, C., Wang, G. and Yoshida.A., Structural mutation in a major human aldehyde dehydrogenase gene results in loss of enzyme activity. Am J Hum Genet 34 (1982) 837–841.

Miller, A.D. and Rosman, G.J. Improved retroviral vectors for gene transfer and expression. Biotechniques 7 (1989) 980–990.

Sastre, L., Kishimoto, T.K., Gee, C., Roberts, T. and Springer, T.A. The mouse leukocyte adhesion proteins Mac-1 and LFA-1: Studies on mRNA translation and protein glycosylation with emphasis on Mac-1. J. Immunol. 137 (1986) 1060–1065.

Tanaka, F., Shiratori, Y., Yokosuka, O., Imazeki, F., Tsukada, Y. and Omata, M. High incidence of ADH2*1/ALDH2*1 genes among Japanese alcohol dependents and patients with alcoholic liver disease. Hepatology 23 (1996) 234–239.

Thomasson, H.R., Edenberg, H.J., Crabb, D.W., Mai, X-L., Jerome, R.E., Li, T-K., Wang, S-P., Lin, Y.-T., Lu, R.-B. and Yin, S.-J. Alcohol and aldehyde dehydrogenase genotypes and alcoholism in Chinese men. Am J Hum Genet 48 (1991) 677–681.

Tischler, M.E., Friedricks, D., Coll, K. and Williamson, J.R. Pyridine nucleotide distributions and enzyme mass action ratios in hepatocytes from fed and starved rats. Arch Biochem Biophys 184 (1977) 222–236.

Xiao, Q., Weiner, H., Johnston, T. and Crabb, D.W. The aldehyde dehydrogenase ALDH2*2 allele exhibits dominance over ALDH2*1 in transduced HeLa cells. J Clin Invest 96 (1995) 2180–2186.

Yamauchi, M., Maezawa, Y., Toda, G., Suzuki, H. and Sakurai, S. Association of a restriction fragment length polymorphism in the alcohol dehydrogenase 2 gene with Japanese alcoholic liver cirrhosis. J Hepatology 23 (1995) 519–523.

Yoshida, A., Huang, I-Y. and Ikawa, M. Molecular abnormality of an inactive aldehyde dehydrogenase variant commonly found in Orientals. Proc Nat Acad Sci USA 81 (1984) 258–261.

Yoshida, A., Wang, G. and Dave, V. Determination of genotypes of human liver aldehyde dehydrogenase ALDH2 locus. Am J Hum Genet 35 (1983) 1107–1116.

22

MUTATION OF THE CONSERVED AMINO ACIDS OF MITOCHONDRIA ALDEHYDE DEHYDROGENASE

Role of the Conserved Residues in the Mechanism of Reaction

Saifuddin Sheikh, Li Ni, and Henry Weiner

Department of Biochemistry
Purdue University
West Lafayette, Indiana 47907–1153

1. INTRODUCTION

During the last three decades much have been accomplished to help us understand the mechanism of aldehyde dehydrogenase. Earlier chemical modification studies suggested that C302 and E268 could be essential residues. We used site directed mutagenesis to elucidate the importance of these residues in catalysis. The enzyme was completely inactive when C302 was replaced by alanine. However, replacing C302 by serine did produce a mutant with some catalytic activity. This led us to conclude the C302 may function as a nucleophile (Farrés et al., 1995). Glutamate at position 268 was shown to act as a general base to activate the nucleophile that led to the initiation of the catalytic reaction (Wang and Weiner, 1995). There were other residues such as C49, S74, C162 and H235 which were thought to be essential based on the inhibition and protection experiments (Weiner et al., 1985; Tu and Weiner, 1988; Loomes et al., 1990). Later mutational analysis showed that replacing them with other residues did not render the enzyme completely inactive, hence, they might not be critical for the functioning of the enzyme (Zheng and Weiner, 1993; Rout and Weiner, 1994; Farrés et al., 1995).

The sequence alignment by Hempel et al. (1993) indicated that there are several amino acids which are completely conserved among the entire aldehyde dehydrogenase family of enzymes. The present mutational studies were carried out to understand the possible role of the conserved residues with reactive side chains. We found that mutation of conserved residues led to a significant loss of activity but only K192Q and S471(A/T) had impaired NAD binding properties. Out of the several residues mutated only K192 and E399 were found to have unique roles in the catalytic process.

Enzymology and Molecular Biology of Carbonyl Metabolism 6
edited by Weiner *et al.* Plenum Press, New York, 1996

2. EXPERIMENTAL PROCEDURES

All the point mutations were made according to the methods described elsewhere (Farrés et al., 1994). The enzymes were expressed in BL21 *E. coli* cells. The isolation and purification of enzymes was carried out according to the established methods used in our laboratory (Ghenbot and Weiner, 1992; Zheng et al., 1993; Farrés et al., 1994). The dehydrogenase activity was measured with a Aminco fluorescence spectrophotometer using similar procedures described by Takahashi and Weiner, (1980). The esterase reaction was measured spectroscopically with p-nitrophenylacetate as the substrate. NADH binding was determined by taking advantage of the increase in fluorescence of the bound NADH (Rout and Weiner, 1994).

3. RESULTS AND DISCUSSION

3.1. Kinetic Properties of the Mutant Enzymes

Table 1 lists the kinetic properties of the mutant enzymes obtained by the replacement of the conserved residues of liver mitochondrial aldehyde dehydrogenase. Mutational analysis showed that only C302 and E268 were found to be absolutely essential for the activity of the enzyme. C302 was found to be a nucleophile (Farrés et al., 1995) and that E268 functions as a general base (Wang and Weiner, 1995). Four mutants (S74A/T, K192Q, S471A/T and E487K) showed a very high increase of K_m for NAD and also a dramatic decrease in the specific activity of the enzyme. At a concentration where NAD saturates the native enzyme, the mutants with impaired binding for NAD showed negligible activity.

Table 1. Mutations of the conserved amino acids of mitochondrial aldehyde dehydrogenase. Kinetic constants of native and mutants of human and rat mitochondrial aldehyde dehydrogenases expressed in *E. coli* at pH 7.4

Enzyme	K_m (NAD), μM	K_m (prop)[b], μM	V^c_{max}
Native	28	0.5	818
Native[R]	4	0.08	190
S74A[a],[R]	2100	0.9	36
S74T[a,R]	2000	1.6	18
R84Q	32	1.3	260
K192Q	3600	3.5	168
E268Q	22	0.6	0.2
C302A[R]	-	-	0
T384A	160	0.9	50
E398K	137	0.1	295
E399Q	118	0.27	96
S471A	1465	0.2	127
E487K[a]	7400	1.4	86
E487Q[a]	90	0.2	400
K489E[a]	45	0.6	27
K489Q[a]	224	0.5	318

[R]Rat aldehyde dehydrogenase; [a]Not completely conserved.
[b]Propionaldehyde; [c]V_{max} activity is expressed in terms of nmoles/min/mg of protein.

The esterase activity was determined for some of the mutants. We previously reported that the C302 and E268 mutants did not have any measurable esterase activity (Farrés et al., 1995, Wang and Weiner, 1995). The mutants of the other conserved residues all possessed decreased esterase activity, ranging from 10 to 40% of the native enzyme (Sheikh et al., 1997).

3.2. Determination of the Coenzyme Binding Constants

Bisubstrate kinetic analysis was carried out in order to determine the dissociation constant (K_{ia}) of NAD for native and mutant enzymes (Farrés et al., 1994; Rout and Weiner, 1994). All the mutants with high K_m for NAD showed a higher dissociation constant for NAD. Mutation at either the N- (S74 and K192) or the C-terminal regions (S471 and E487) resulted in the enzyme with impaired NAD binding properties. It is difficult to envision that all of these residues are directly involved in the binding of NAD. However, it is possible that NAD binding can be effected by subtle changes in the conformation of the enzyme produced by the mutation.

The binding of NADH with the enzyme was measured by fluorescence enhancement of NADH in presence of enzyme at 450 nm (Takahashi et al., 1980). No significant increase in the K_d was obtained for the mutant with increase in K_{ia} for NAD. In addition, 2 moles of NADH per tetrameric enzyme molecule was still found for all the mutants as was observed for the native enzyme. Thus, it is concluded that a different enzyme form may interact with NAD and NADH, supporting the earlier notion that E-NAD and E-NADH have different conformations (Blackwell et al., 1987).

3.3. Effect of Mutation on the Mechanism of the Aldehyde Dehydrogenase Reaction

The mechanism of aldehyde dehydrogenase involves several steps including acylation (k_3), hydride transfer (k_5) and deacylation (k_7) as shown in Figure 1. The rate limiting step for the native enzyme was found to be deacylation. The criteria to determine the rate limiting step of the native or mutant enzyme was based on presteady state burst (Weiner et al., 1976) and comparison of the effect of electron donating and electron withdrawing groups in the substrates on the rate of the reaction (Feldman and Weiner, 1972). Mutation of several conserved amino acids resulted in an alteration of rate limiting step of the en-

Table 2. Binding of NAD and NADH to the native and mutants of human and rat mitochondrial aldehyde dehydrogenase

Enzyme	K_{ia} (µM) for NAD	K_d (µM) for NADH	Moles of NADH bound per mole enzyme
Native	11	3.2	2
Native[R]	2	1	2
S74A[R]	64	7	2
K192Q	590	9	2
S471A	282	17	2
E487K	548	N.D.	N.D.

K_{ia} represents the dissociation constant for NAD and was determined by the bisubstrate kinetic analysis. K_d represents the dissociation constant for NADH and was determined by the NADH binding assay.
N.D. not determined.

E_n = enzyme subunit; C = cysteine; E = glutamate; K = lysine

Figure 1. Model showing the various steps in the mechanism of the ALDH reaction. Deacylation is the rate limiting step for the native enzyme (k_7). Mutations at K192 and E399 produced enzymes where hydride transfer (k_5) became rate limiting. The proposed roles of various amino acids are indicated. Only C302 and E268 were found to be absolutely essential for catalysis.

zyme. The mutants S74A, R84Q, T384A, E398K, S471A and K489Q had the same rate limiting step as was found with the native enzyme. For K192Q and E399Q it changed from deacylation to hydride transfer whereas, acylation was found to be the rate limiting step for E487K enzyme.

4. CONCLUSIONS

Mutation of any residue totally or partially conserved in ALDH affected the properties of the enzyme. Even mutations made to residues conserved only in the mammalian enzymes affected the activity of the enzyme. Even mutations of S74', E487 and K489, conserved only in mammalian enzymes, decreased the specific activity of the enzyme.

Table 3. Rate-limiting step for mutant forms of human and rat live mitochondria aldehyde dehydrogenase

Enzyme	V^a (prop)[b]	V (chloro)[b]	Rate-limiting step
Native	818	3190	k_7
Native[R]	500	1000	k_7
S74A[R]	38	N.D.[c]	k_7
R84Q	260	948	k_7
K192Q	160	75	k_5
E268Q	0.2	N.D.	k_3
C302S[R]	4	48	k_3
T384A	51	128	k_7
E398K	293	966	k_7
E399Q	93	65	k_5
S471A	129	450	k_7
E487K[R]	214	428	k_3
K489Q	318	907	k_7

[a] The unit of the dehydrogenase activity is expressed in terms of nmoles/min/mg of protein.
[b] V (prop) and V (chloro) refers to dehydrogenase activity of the enzyme with propionaldehyde and chloroacetaldehyde as the substrates respectively.
[c] N.D. indicates not determined.

Mutations of two completely conserved residues, C302 and E268 produced an inactive enzyme. These two residues appear to be absolutely essential for the activity. Mutations of the other conserved residues which possessed side chains with reactive functional groups produced a variety of results. None caused a dramatic change in the K_m for aldehyde. There was an increase in the K_m for NAD in almost all of them. In some cases the increase in K_m was over a 100 fold while for others it was less than 10 fold. All the mutations caused a decrease in the specific activity of the enzyme, but the magnitude of the decrease did not correlate with the changes in the K_m values. It was also unexpected to find that mutations of two conserved residues not only decreased the kinetic properties of the enzyme but also changed the rate limiting step. For the native and many of the mutants this was deacylation, k_7 in Figure 1. Mutations at K192 and E399 produced an enzyme where hydride transfer became the rate limiting step (k_5).

Mutations affecting the affinity of the enzyme for NAD did not affect the binding of NADH. All of the mutated residues can not be in contact with the nicotinamide ring. Undoubtedly the mutations caused some conformational alteration in the enzyme which affected the binding of NAD. Why the K_d for NADH remained unaffected is not known. Perhaps it is related to a differential conformational affect on NAD binding to the enzyme compared to that caused by the binding of NADH.

We previously suggested that with native enzyme acylation, k_3, was rate limiting for esterase reaction while deacylation, k_7, was rate limiting for the dehydrogenase reaction. Even though the mutations altered steps other than acylation, all the mutants had decreased esterase activity (Sheikh et al., 1997). Even K192 and E399 mutations, which slowed the rate of hydride transfer, caused a decrease in the esterase reaction. Since this reaction is not involved in the ester hydrolysis, it can be concluded that the mutation caused a change to occur in many steps, though there may have been a pronounce effect on one particular step.

We have previously proposed that C302 functions as the nucleophile (Farrés et al., 1994) and E268 as the general base necessary to activate the nucleophilic residue cysteine 302 (Wang and Weiner, 1995). We now propose that K192 and E399 are involved in extracting the proton from the hemiacetal to facilitate hydride transfer. Obviously the precise role of these or any of the other amino acids will not be known until the detailed three dimensional structure of the enzyme is determined.

5. REFERENCES

Abroila, D. P., Fields, R., Stein, S., Mackerell, A. D. Jr., and Pietruzsko, R., 1987, Active site of aldehyde dehydrogenase, *Biochemistry*, 26:5679–5684.

Blackwell, L. F., Motion, R. L., MacGibbon, A. K. H., Hardman, M. J. and Buckley, P. D., 1987, Evidence for the slow conformational change controlling NADH release from the enzyme is rate limiting during the oxidation of propionaldehyde by aldehyde dehydrogenase, *Biochem. J.*, 175:753–756.

Farrés, J., Wang, X.-P., Takahashi, K., Cunningham, S. J., Wang, T. T. and Weiner, H., 1994, Effect of changing glutamate to lysine in rat and human liver mitochondrial aldehyde dehydrogenase: A model to study human (oriental type) class 2 aldehyde dehydrogenase, *J. Biol. Chem.* 269: 13854–13860.

Feldman, R. I. and Weiner, H., 1972, Horse liver aldehyde dehydrogenase: Kinetics and mechanism implications of dehydrogenase and esterase activity, *J. Biol. Chem.* 247:267–272.

Ghenbot, G. and Weiner, H., 1992, Purification of liver aldehyde dehydrogenase by *p*-hydroxyacetophenone-sepharose affinity matrix and the coelution of chloramphenicol acetyl transferase from the same matrix with recombinantly expressed aldehyde dehydrogenase, *Protein Exp. Purif.* 3:470–478.

Hempel, J., Nicholas, H. and Lindahl, R., 1993, Aldehyde dehydrogenase: Widespread structural and functional diversity within a shared frame-work, *Protein Sci.* 2:1890–1900.

Hempel, J., Pietruszko, R., Fietzek, P. and Jörnvall, H., 1982, Identification of segment containing reactive cysteine residue in human liver cytoplasmic aldehyde dehydrogenase (Isoenzyme E1), *Biochemistry,* 21:6834–6838.

Loomes, K. M., Midwinter, G. G., Blackwell, L. F. and Buckley, P. D., 1990, Evidence for the reactivity of serine 74 with *trans*-4-(N,N-dimethylamino)cinnamaldehyde during oxidation of cytoplasmic aldehyde dehydrogenase from sheep liver, *Biochemistry,* 29:2070–2075.

Rout, U. K. and Weiner, H., 1994, Involvement of serine 74 in the enzyme-coenzyme interaction of rat liver mitochondrial aldehyde dehydrogenase, *Biochemistry,* 33:8955–8961.

Sheikh, S., Ni, L. and Weiner, H., 1997, The potential role of conserved residues in aldehyde dehydrogenase, *Biochemistry* Submitted.

Takahashi, K., Weiner, H. and Hu, J. H. J., 1980, Increase in the stoichiometry of the functioning active sites of horse liver aldehyde dehydrogenase in the presence of magnesium ions, *Arch. Biochem. Biophys.,* 205: 571–578.

Tu, G.-C. and Weiner, H., 1988, Evidence for two distinct active sites on aldehyde dehydrogenase, *J. Biol. Chem.,* 263:1218–1222.

Wang, X. and Weiner, H., 1995, Involvement of glutamate 268 in the active site of human liver mitochondrial aldehyde dehydrogenase as probed by site directed mutagenesis, *Biochemistry,* 34:237–243.

Weiner, H., Hu, J. H. J. and Sanny, C.G., 1976, Rate limiting steps for esterase and dehydrogenase reaction catalyzed by horse liver aldehyde dehydrogenase, *J. Biol. Chem.,* 251:3853–3855.

Zheng, C.-F. and Weiner, H., 1993, Role of highly conserved residues in rat liver mitochondria aldehyde dehydrogenase as studied by site directed mutagenesis, *Arch. Biochem. Biophys.,* 305:460–466.

Zheng, C.-F., Wang, T. T. Y. and Weiner, H., 1993, Cloning and expression of full length cDNAS encoding human liver class 1 and class 2 aldehyde dehydrogenase, *Alcoholism: Clin. Exp. Res.,* 17:828–831.

23

THE ACTION OF CYTOSOLIC ALDEHYDE DEHYDROGENASE ON RESORUFIN ACETATE

Trevor M. Kitson and Kathryn E. Kitson

Departments of Chemistry and Biochemistry
Massey University
Palmerston North, New Zealand

1. INTRODUCTION

Aldehyde dehydrogenase is well known to have the ability to act as an esterase as well as to catalyse the oxidation of aldehydes by NAD^+. Some workers have been of the opinion that the twin activities of the enzyme are not closely related and occur at separate enzymic sites (see, for example, Motion et al., 1988), whereas we have thought for a number of years that the balance of evidence favours the simpler one-site model (see, for example, Kitson et al., 1991). Recently, the extensive studies of Klyosov et al. (1996) have rekindled the debate; these authors suggest that even within the dehydrogenase activity alone, separate sites are responsible for the oxidation of some substrates (such as acetaldehyde and 6-dimethylamino-2-naphthaldehyde) in human mitochondrial aldehyde dehydrogenase (though not in the cytosolic isozyme).

In the past, the esterase activity of aldehyde dehydrogenase has been exclusively studied with derivatives of *p*-nitrophenol; thus, for example, we have used various esters, carbonates, carbamates and lactones as substrates or modifiers, and all of these have had *p*-nitrophenoxide as the leaving group (Kitson, 1989a,1989b; Kitson and Kitson, 1996). It occurred to us that it might be instructive to probe the esterase ability of aldehyde dehydrogenase with substrates of markedly different structure, and to this end we report below the action of the cytosolic enzyme from sheep liver on resorufin acetate (7-acetoxy-3*H*-phenoxazin-3-one). Some derivatives of resorufin have been used as spectrophotometric and fluorimetric reagents for different forms of cytochrome P-450 (Burke et al., 1985) and for glycosidases (Tokutake et al., 1990). The acetate itself was suggested as a sensitive substrate for hydrolytic enzymes over thirty years ago (Kramer and Guilbault, 1964), but to the best of our knowledge it has received little attention until our recent work with chymotrypsin (Kitson, 1996) and the present study with aldehyde dehydrogenase. The structure of resorufin acetate and of its hydrolysis product, resorufin (shown as the symmetrical resonance-stabilised anion), are given in Figure 1.

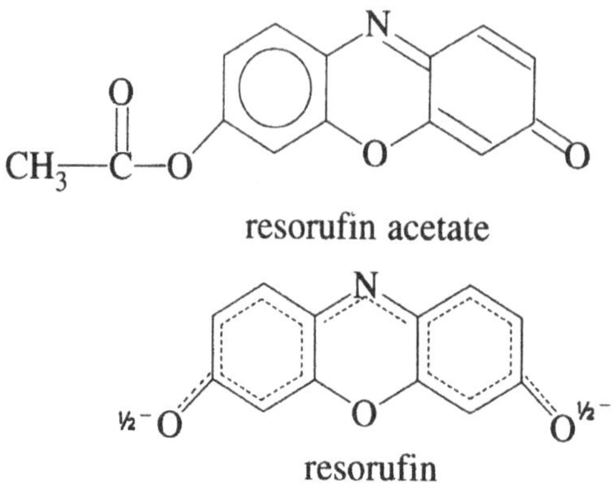

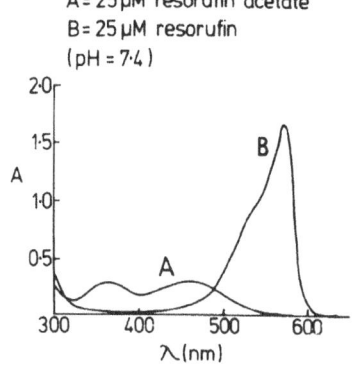

resorufin acetate

resorufin

Figure 1. Structures of resorufin acetate and the resorufin anion.

2. EXPERIMENTAL

Cytosolic aldehyde dehydrogenase was purified from sheep liver and its concentration was measured as reported previously (Kitson and Kitson, 1994). Resorufin acetate was synthesised from the sodium salt of resorufin (Aldrich) by the method of Nietzki et al. (1889). δ (CDCl$_3$): 2.36 (s, 3H), 6.33 (s, 1H), 6.84–6.88 (d, 1H), 7.11–7.15 (d, 1H), 7.15 (s, 1H, superimposed), 7.41–7.45 (d, 1H), 7.78–7.81 (d, 1H). C$_{14}$H$_9$NO$_4$ was confirmed by mass spectrometry (m/z = 255.052719). All spectra and all spectrophotometric assays of enzyme activity were recorded using a Varian Cary 1 spectrophotometer. Enzyme activity was measured in 50 mM sodium phosphate buffer, pH 7.4, 25 °C, at 571 nm for resorufin acetate and 399 nm for p-nitrophenyl acetate; kinetic data were corrected for the rate of spontaneous hydrolysis of the esters. In general all assays were repeated at least 2 or 3 times and the data were averaged.

Figure 2. UV/visible spectra of 25 μM solutions of resorufin acetate (A) and resorufin (B) in 50 mM sodium phosphate buffer, pH 7.4.

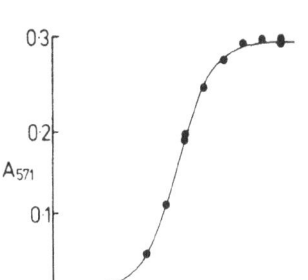

Figure 3. Ionisation profile of resorufin. The absorbance due to the anionic form of resorufin is plotted against pH.

3. RESULTS AND DISCUSSION

3.1. Spectrophotometric Properties of Resorufin and Its Acetate

The UV/visible spectra of equal concentrations of resorufin and resorufin acetate at pH 7.4 are shown in Figure 2. To the eye, the ester is pale orange in dilute solution whereas its hydrolysis product is a strikingly beautiful intense pink colour. The relatively long wavelength of the latter's absorption maximum (571 nm) immediately suggests that it could be conveniently monitored without any interference from protein, NADH or p-nitrophenoxide, for example. The extinction coefficient (molar absorptivity) of resorufin at this pH is 67,700, whereas that of p-nitrophenoxide is only 12,500; in other words, resorufin acetate is potentially a more sensitive chromophoric substrate than p-nitrophenyl acetate by a factor of 5.4. Figure 3 shows how the absorbance at 571 nm varies with pH. At low pH, the pink colour reverts to a pale orange as the resorufin anion becomes protonated to give the phenol form of the compound, and the pK_a of this phenol group is 5.8. The Figure clearly shows another potential advantage of the resorufin substrate; with it, the esterase activity of the enzyme could easily be assayed at relatively low pH (around 5), whereas this is not the case for p-nitrophenyl acetate. (The pK_a of p-nitrophenol is 7.1; Kitson and Freeman, 1993.)

3.2. Action of Cytosolic Aldehyde Dehydrogenase on Resorufin Acetate in the Absence of Cofactors

The spectrophotometer trace for the enzyme-catalysed hydrolysis of resorufin acetate (1 μM) is shown in Figure 4. The form of the trace (an essentially linear increase with a sudden tail-off as the substrate is exhausted) implies a very low value for the Michaelis constant; this conclusion is corroborated by the results shown in Figure 5, which give an approximate value for K_M of 0.15 μM. The Michaelis constant for p-nitrophenyl acetate is appreciably higher (in the range 2 to 3 μM), but the value for k_{cat} for this substrate is the same as that found from Figure 5 for resorufin acetate (0.23–0.27 s^{-1}). Previous work (Kitson and Kitson, 1996) has clearly shown that with p-nitrophenyl acetate in the absence of cofactors the rate of hydrolysis of the acyl-enzyme (i.e. the deacylation rate, see Figure 6)

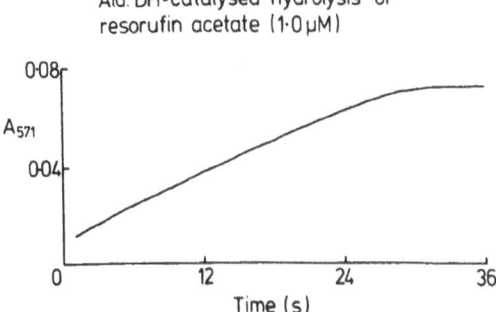

Figure 4. Copy of a spectrophotometric trace obtained with cytosolic aldehyde dehydrogenase (0.16 μM) and resorufin acetate (1 μM) at pH 7.4, 25 °C.

is rate-limiting, since there is a burst in the release of *p*-nitrophenoxide. If this is also true for resorufin acetate (and our preliminary results show that this substrate does indeed show a burst of product release under these conditions), then the identity of the k_{cat} values is of course to be expected, as both reactions involve the rate-limiting hydrolysis of the same *acetyl*-enzyme intermediate (Figure 6).

The very low Michaelis constant for resorufin acetate suggests that this relatively bulky, planar, hydrophobic, slab-shaped molecule may have an unusually high affinity for aldehyde dehydrogenase's binding site. In Figure 7 we give the structures and approximate dimensions of other molecules that seem to be particularly good at binding to or reacting with the active site of cytosolic aldehyde dehydrogenase. Retinal (another flat rigid molecule) is a substrate with a low K_M value (1.1 μM) in the dehydrogenase reaction (Klyosov (1996). [This author has recently determined far lower K_M values for some other hydrophobic substrates, such as decanal (about 3 nM).] Disulfiram (which is not constrained to be planar but can adopt such a conformation) is well known as a very potent inactivator of the cytosolic enzyme (see, for example, Kitson, 1987). 2-Bromo-4'-iodoacetanilide (which is essentially planar) is if anything an even more effective inactivator than disulfiram (see our other contribution to this volume). From these

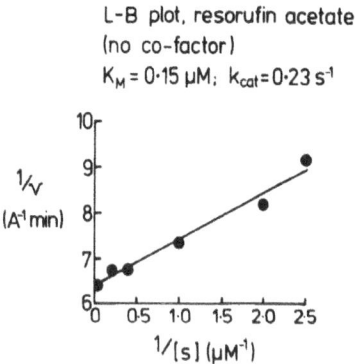

Figure 5. Lineweaver-Burk plot for the hydrolysis of resorufin acetate catalysed by cytosolic aldehyde dehydrogenase (0.16 μM) at pH 7.4, 25 °C, in the absence of cofactor.

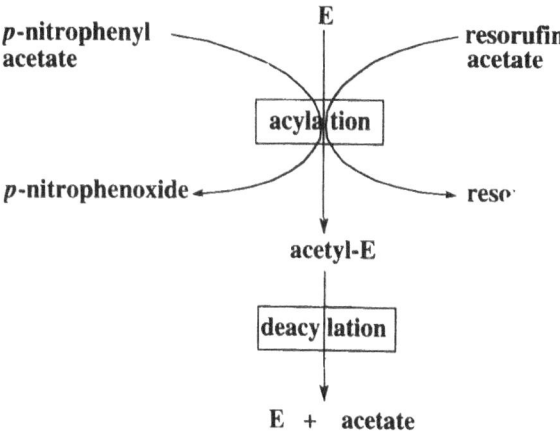

Figure 6. Pathway of the enzyme-catalysed hydrolysis of esters, showing the acylation and deacylation steps.

considerations we venture to suggest that when the tertiary structure of aldehyde dehydro-genase becomes elucidated, it will reveal a fairly large hydrophobic binding slot into which the molecules shown in Figure 7 can tightly fit. It will be interesting also to exam-ine resorufin acetate as a possible substrate for the mitochondrial form of aldehyde dehy-drogenase; the active site of this isozyme is evidently too small to accommodate disulfiram very easily (MacKerell et al., 1985), and retinal is a potent inhibitor of it rather than a substrate (Klyosov, 1996).

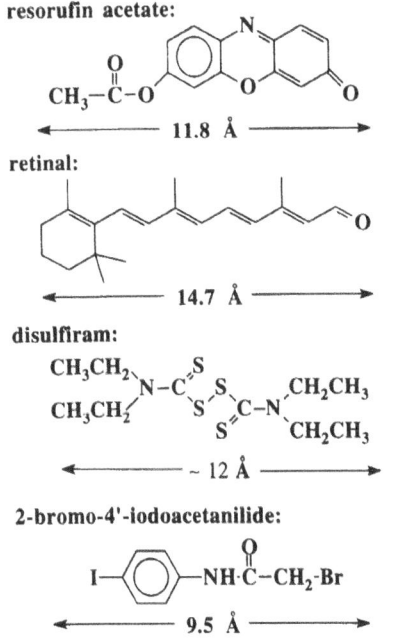

Figure 7. A comparison of the structures and dimensions of certain molecules that have a high affinity for cytoso-lic aldehyde dehydrogenase.

Table 1. Effect of cofactors (0.1 mM) on the hydrolysis of
p-nitrophenyl acetate or resorufin acetate (25 µM) catalysed
by cytosolic aldehyde dehydrogenase at pH 7.4, 25°C. For
each substrate the rate of reaction is expressed relative to the
rate in the absence of cofactor. For *p*-nitrophenyl acetate,
enzyme concentration was 0.32 µM; for resorufin acetate,
it was 0.08 µM

Substrate	Relative rate
p-nitrophenyl acetate	
No cofactor	1.00
+ NAD$^+$	0.49
+ NADH	0.58
Resorufin acetate	
No cofactor	1.00
+ NAD$^+$	21.8
+ NADH	24.7

3.3. Action of Cytosolic Aldehyde Dehydrogenase on Resorufin Acetate in the Presence of Cofactors

With *p*-nitrophenyl acetate, the presence of NAD$^+$ or NADH affects the activity of
aldehyde dehydrogenase in a way that depends on the relative concentrations of enzyme,
substrate and cofactor, resulting in either a modest activation or inhibition (Kitson and
Kitson, 1996). Under the conditions used in the present work, for example, the activity is
approximately halved, as seen in Table 1. By contrast, the activity of the enzyme towards
resorufin acetate under similar conditions is greatly enhanced, as also shown in the Table.
Because of these different effects, and because of the greater extinction coefficient of resorufin compared to *p*-nitrophenoxide, the measured rates with the two substrates in the
presence of 0.1 mM NAD$^+$ or NADH differ in terms of absorbance units per minute by a

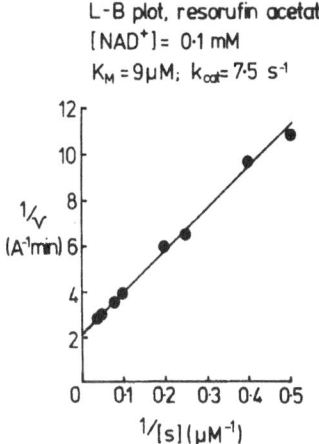

Figure 8. Lineweaver-Burk plot for the hydrolysis of resorufin acetate catalysed by cytosolic aldehyde dehydrogenase (0.016 µM) at pH 7.4, 25 °C, in the presence of NAD$^+$ (0.1 mM).

Table 2. Values of k_{cat} for ester hydrolysis catalysed by cytosolic aldehyde dehydrogenase in the presence and absence of cofactors. The data are as determined in the present work, with the exception of that marked by an asterisk, which is taken from Motion et al. (1988)

Substrate	$k_{cat}(s^{-1})$
p-nitrophenyl acetate	
No cofactor	0.25
+ NAD⁺	-
+ NADH	0.38*
Resorufin acetate	
No cofactor	0.23 - 0.27
+ NAD⁺	7.5
+ NADH	6.3

factor of at least 250 - a dramatic observation indeed. For resorufin acetate in the presence of 0.1 mM NAD⁺ the Michaelis constant is 9 µM and the k_{cat} value 7.5 s⁻¹, results calculated from Figure 8.

A major point put forward by Motion et al. (1988) in favour of their two-site model for the dehydrogenase and esterase activities of aldehyde dehydrogenase is that the value for k_{cat} for hydrolysis of p-nitrophenyl acetate in the presence of NADH is smaller by an order of magnitude than the rate constant for acyl-enzyme hydrolysis in the dehydrogenase action of the enzyme, which is 5–10 s⁻¹ (Blackwell et al., 1987) (see the data in Table 2). [The latter process takes place in the presence of NADH before it dissociates from the enzyme.] However, we have recently shown (Kitson and Kitson, 1996) that cofactors slow the rate of acylation of the enzyme with p-nitrophenyl acetate until this step becomes rate-limiting, and so the k_{cat} value shown in Table 2 (in the presence of NADH) is no longer a reflection of the rate of hydrolysis of the acyl enzyme and its comparison with the dehydrogenase deacylation rate is meaningless. With resorufin acetate, we have yet to establish the identity of the rate-determining step in the presence of cofactors, but at this stage we can definitely say that they accelerate the rate of acyl-enzyme hydrolysis, whilst not slowing the acylation rate to the major extent that happens with p-nitrophenyl acetate. Depending on whether acylation or deacylation is rate-limiting under these conditions, the rate constant for acyl-enzyme hydrolysis is either equal to the k_{cat} value of about 6–7 s⁻¹, or greater than this, but it cannot of course be less than this. Thus by choosing the right substrate, it becomes clear that the rate of deacetylation of the enzyme in the esterase mode is *not* too small to equate with that in the dehydrogenase mode, and this objection of Motion et al. (1988) to a single site for both activities disappears.

In conclusion, resorufin acetate has been shown to have a number of advantages over p-nitrophenyl acetate as a substrate for aldehyde dehydrogenase, both in terms of practical spectrophotometry and in terms of meaningful conclusions about the fundamental nature of the enzyme.

4. REFERENCES

Blackwell, L.F., Motion, R.L., MacGibbon, A.K.H., Hardman, M.J. and Buckley, P.D.: Evidence that the slow conformational change controlling NADH release from the enzyme is rate limiting during the oxidation of propionaldehyde by aldehyde dehydrogenase. Biochem. J. 242 (1987) 803–808.

Burke, M.D., Thompson, S., Elcombe, C.R., Halpert, J., Haaparanta, T. and Mayer, R.T.: Ethoxy-, pentoxy- and benzyloxyphenoxazones and homologues: a series of substrates to distinguish different induced cytochromes P-450. Biochem Pharmacol. 34 (1985) 3337–3345.

Kitson, T.M.: Effect of disulfiram on the pre-steady-state burst in the reactions of sheep liver cytoplasmic aldehyde dehydrogenase. Biochem. J. 248 (1987) 989–991.

Kitson, T.M.: Kinetics of p-nitrophenyl pivalate hydrolysis catalysed by cytoplasmic aldehyde dehydrogenase. Biochem. J. 257 (1989a) 573–578.

Kitson, T.M.: The action of cytoplasmic aldehyde dehydrogenase on methyl p-nitrophenyl carbonate and p-nitrophenyl dimethylcarbamate. Biochem. J. 257 (1989b) 579–584.

Kitson, T.M.: A comparison of resorufin acetate and p-nitrophenyl acetate as substrates for chymotrypsin. Bioorg. Chem. (1996) in the press.

Kitson, T.M. and Freeman, G.H.: 3,4-Dihydro-3-methyl-6-nitro-2H-1,3-benzoxazin-2-one, a reagent for labeling p-nitrophenyl esterases with a chromophoric reporter group - synthesis and reaction with chymotrypsin. Bioorg. Chem. 21 (1993) 354–365.

Kitson, T.M. and Kitson, K.E.: Probing the active site of cytoplasmic aldehyde dehydrogenase with a chromophoric reporter group. Biochem. J. 300 (1994) 25–30.

Kitson, T.M. and Kitson, K.E.: A comparison of nitrophenyl esters and lactones as substrates of cytosolic aldehyde dehydrogenase. Biochem. J. 316 (1996) 225–232.

Kitson, T.M., Hill, J.P. and Midwinter, G.G.: Identification of a catalytically essential residue in sheep liver cytoplasmic aldehyde dehydrogenase. Biochem. J. 275 (1991) 207–210.

Klyosov, A.A.: Kinetics and specificity of human liver aldehyde dehydrogenases toward aliphatic, aromatic and fused polycyclic aldehydes. Biochemistry 35 (1996) 4457–4467.

Klyosov, A.A., Rashkovetsky, L.G., Tahir, M.K. and Keung, W-M.: Possible role of liver cytosolic and mitochondrial aldehyde dehydrogenases in acetaldehyde metabolism. Biochemistry 35 (1996) 4445–4456.

Kramer, D.N. and Guilbault, G.G.: Resorufin acetate as a substrate for determination of hydrolytic enzymes at low enzyme and substrate concentrations. Anal Chem. 36 (1964) 1662–1663.

MacKerell, A.D., Vallari, R.C. and Pietruszko, R.: Human mitochondrial aldehyde dehydrogenase inhibition by diethyldithiocarbamic acid methanethiol mixed disulfide: a derivative of disulfiram. FEBS Lett. 179 (1985) 77–81.

Motion, R.L., Buckley, P.D., Bennett, A.F. and Blackwell, L.F.: Evidence that the cytoplasmic aldehyde dehydrogenase-catalysed oxidation of aldehydes involves a different active-site group from that which catalyses the hydrolysis of 4-nitrophenyl acetate. Biochem. J. 254 (1988) 903–906.

Nietzki, R., Dietze, A. and Mackler, H.: Ueber Weselsky's resorcinfarbstoffe. Chem. Ber. 22 (1889) 3020–3038.

Tokutake, S., Kasai, K., Tomikura, T., Yamaji, N. and Kato, M.: Glycosides having chromophores as substrates for sensitive enzyme analysis. II. Synthesis of phenolindophenyl-β-D-glucopyranosides having an electron-withdrawing substituent as substrates for β-glucosidase. Chem. Pharm. Bull. 38 (1990) 3466–3470.

24

INHIBITION OF AND INTERACTION WITH HUMAN RECOMBINANT MITOCHONDRIAL ALDEHYDE DEHYDROGENASE BY METHYL DIETHYLTHIOCARBAMATE SULFOXIDE

James J. Lipsky,[1] Dennis C. Mays,[1] Jennifer L. Holt,[1] Andy J. Tomlinson,[2] Kenneth L. Johnson,[2] Karen A. Veverka,[1,2] and Stephen Naylor[1,2]

[1]Clinical Pharmacology Unit
Department of Pharmacology
Mayo Clinic and Foundation
Rochester, Minnesota 55905
[2]Biomedical Mass Spectrometry Facility
Department of Biochemistry and Molecular Biology
Mayo Clinic and Foundation
Rochester, Minnesota 55905

1. INTRODUCTION

Disulfiram is used as aversion therapy in the treatment of alcoholism. Following the administration of disulfiram there is a decrease in the activity of hepatic aldehyde dehydrogenase. A consequence of this effect is the accumulation of acetaldehyde subsequent to ethanol ingestion. The increase in levels of acetaldehyde produce flushing, nausea, vomiting, and tachycardia which is referred to as the disulfiram-ethanol reaction (Hald, *et al.*, 1948; Kitson, 1977). *In vivo* disulfiram is rapidly reduced to *N,N*-diethylthiocarbamate (DDC) and it is therefore felt that the intact drug is not responsible for the inhibition of aldehyde dehydrogenase (ALDH) (Cobby, *et al.*, 1977). An active metabolite of disulfiram is felt to be responsible for the inhibition. One such candidate is methyl diethylthiocarbamate sulfoxide (MeDTC-sulfoxide) (Hart and Faiman, 1992). This compound has been shown to be an inhibitor of rat mitochondrial ALDH and has been detected in the plasma of rats given disulfiram (Hart and Faiman, 1992). Since the mitochondrial isoform of ALDH is felt to be important in the metabolism of acetaldehyde derived from ethanol metabolism (Svanas and Weiner, 1985), we examined the effect of MeDTC-sulfoxide upon the activity of human ALDH. We further investigated the interaction of the inhibitor with this enzyme by mass spectrometry.

Enzymology and Molecular Biology of Carbonyl Metabolism 6
edited by Weiner *et al.* Plenum Press, New York, 1996

2. METHODS

2.1. Materials

The recombinant human mitochondrial and cytosolic cDNA in pT7–7 were generous gifts from Dr. Henry Weiner (Purdue University, West Lafayette, IN); *S*-methyl *N,N*-diethylthiocarbamate sulfoxide (MeDTC-sulfoxide) was prepared by the method of Mays *et al.* (1996); acetaldehyde, ß-mercaptoethanol, isopropyl-ß-D-thiogalactopyranoside (IPTG), p-hydroxyacetophenone, protamine sulfate, sodium phosphate, sodium pyrophosphate, and epoxy-activated Sepharose 6B were obtained from Sigma (St. Louis, MO); ampicillin, nicotinamide adenine dinucleotide (NAD) and phenylmethylsulfonylfluoride (PMSF) were from Boehringer Mannheim (Germany); ethylenediaminetetraacetic acid (EDTA) and p-nitrophenylacetate were obtained from Aldrich (St. Louis, MO); sodium chloride was from Mallinckrodt (Kentucky); Terrific Broth was obtained from Bio 101 (La Jolla, CA); *Staphylococcus Aureus V8* (Glu C) endopeptidase was obtained from Boehringer Mannheim. BCA Protein Assay kit and Slide-A-Lyzer cassettes were obtained from Pierce; DEAE Sepharose Fast-Flow and FPLC System from Pharmacia; THERMOmax microplate reader and SOFTmax PRO software was from Molecular Devices Corporation (Sunnyvale, CA); Sigma Plot software was from Jandel.

2.2. Expression of Recombinant Human ALDH in *E. Coli*, and Purification

The enzyme was isolated using a procedure modified from the method published by Jeng and Weiner (1991). Cells carrying the plasmid containing the specific cDNA were grown in 2 liters of Terrific Broth supplemented with 100 µg/mL of ampicillin. When the culture reached an O.D. between 0.4 to 0.6, IPTG was added to a final concentration of 4 mM, and the culture was then allowed to grow overnight at room temperature (27°C) with shaking. The bacterial cells were harvested by centrifugation at 7,000 X g for 15 minutes, the pellet resuspended in Sonication Buffer (20 mM sodium phosphate, pH 7.4, 1 mM EDTA, 0.025% ß-mercaptoethanol and 10 µM PMSF) and lysed by sonication using a Branson sonifier at output 6 for 10 X 20-second bursts. The lysate was then centrifuged at 18,000 X g for 15 minutes to remove cell debris, and 125 mg of protamine sulfate per liter culture was added to the supernatant and mixed well. After incubating on ice for 10 minutes, the protamine sulfate precipitate was removed by centrifuging at 18,000 X g for 15 minutes, and the clarified lysate was dialyzed for 5 to 6 hours against 2 changes of 1 liter each of DEAE column equilibration buffer (20 mM sodium phosphate, pH 6.1, and 0.025% ß-mercaptoethanol). The sample was recentrifuged after dialysis and then loaded onto a DEAE Fast-Flow column. The protein of interest was eluted using a 0 to 200 mM NaCl gradient. Fractions containing ALDH activity were subsequently loaded onto a p-hydroxyacetophenone (p-HAP) affinity column and purified as described by Zheng *et al.* (1993). The p-HAP affinity resin was synthesized as described by Ghenbot and Weiner (1992). The purified enzyme was dialyzed extensively (4 changes of 1 liter of buffer each over 2 days) against p-HAP affinity column equilibration buffer (20 mM sodium phosphate, pH 7.4, 0.1 mM dithiothreitol, 1 mM EDTA, and 50 mM sodium chloride) and stored in 200 - 500 µL aliquots at -80°C. The final protein concentrations were from 0.5 to 8.6 mg/mL as determined by BCA Kinetic Protein Assay using bovine serum albumin as the standard.

2.3. ALDH Activity Assay

The microtiter-based assay was performed as described (Nelson and Lipsky, 1995), with the exception of the following modifications: purified recombinant human ALDH was used instead of solubilized rat mitochondria, rotenone and pyrazole were omitted. The final acetaldehyde concentration was 160 μM for the mitochondrial ALDH assays and 1 mM for the ctyosolic ALDH assays. Acetaldehyde and NAD were added together in a 25 μL volume.

Dose response curves were generated with inhibitor concentrations from 0.1 μM to 4 μM with a 10-minute incubation at 22°C between the addition of inhibitor and the addition of substrate to the enzyme.

The reversibility assay was performed by dialyzing inhibited enzyme against 2000 volumes of 50 mM sodium pyrophosphate buffer pH 8.8 (Buffer G) for a total of 30 minutes. The inhibitor (4 μM sulfoxide) was added to the enzyme and incubated for 10 minutes before 1.5 mL of the mixture was transferred to a dialysis cassette. The cassette was placed in a beaker with 150 mL of Buffer G and dialyzed for 15 minutes. Approximately 750 μL of the dialysate was removed using a 3-mL syringe, and the 750 μL remaining in the cassette was redialyzed for another 15 minutes in 150 mL of fresh Buffer G. ALDH activity was measured in samples taken before dialysis, after 15 and 30 minutes of dialysis, and of undialyzed enzyme incubated for the same length of time as the dialyzed sample.

2.4. Mass Spectrometry

Recombinant human mitochondrial ALDH was analyzed by HPLC-MS using a Finnigan MAT 900 mass spectrometer. A Finnigan MAT designed electrospray interface was used. A PLRP-S reversed phase, 300 Å pore size, 1.0 x 50 mm column (Michrom BioResources, CA) was used for the HPLC-MS analysis of intact proteins. A gradient using mobile phase A (98:2:0.1, v/v/v, water:acetonitrile:trifluoroacetic acid) and mobile phase B (80:10:10:0.1, v/v/v/v, acetonitrile:n-propanol:water:trifluoroacetic acid) from 20–70% mobile phase B, was run over 10 min at a flow rate of 75 μL/min. Twenty μL of ALDH solution (0.987 mg/ml in 20 mM Na phosphate pH 7.4, 100 μM DTT, 1 mM EDTA) was injected directly onto a protein trap inserted into the injector of a Michrom BioResources HPLC. Salts and other hydrophilic matrix components were removed by washing the trap with 300 μL of mobile phase A, prior to diversion of the mobile phase through the trap. A second aliquot of ALDH was analyzed after reaction with 40 μM MeDTC sulfoxide. The sheath gas was nitrogen at 3.5 bar with a flow rate of 1.58 L/min. Ions were detected by a PATRIC™ (Position and Time Resolved Ion Counter) scanning array detector. A mass window of 8% was used throughout. The estimated error margin associated with the mass measurements in this study was 0.01% of the molecular weight of the protein based upon daily measurement of reference protein standards. Peptide fragment analysis was performed by HPLC-MS using a Reliasil C-18, 300 Å pore size column, 1.0 x 150 mm. A gradient using mobile phase B from 10–70% was employed. The flow rate was 50 μL/min. Two μL of ALDH digest sample was injected directly onto a peptide trap inserted into the injector.

2.4.1. Digestion Conditions. Twenty μL of either native or MeDTC-S0 adducted ALDH was diluted with 40 μL of an aqueous solution containing 20 mM ammonium acetate and 1% acetic acid (pH 3.7) and incubated at 37° for two hours with 10% w/w *Staphy-*

Table 1. Effect of dialysis on inhibition of ALDH

Condition	% Remaining Activity
Pre-dialysis	12
Post-dialysis	15

ALDH was treated with MeDTC sulfoxide at a final concentration of 4 μM for 10 minutes prior to dialysis.

lococcus Aureus V8 (Glu C) endopeptidase. The resultant peptide mixtures were concentrated under vacuum to 20 μL and analyzed by HPLC-MS under the conditions detailed above.

3. RESULTS

The effect of MeDTC-sulfoxide upon the activity of human recombinant ALDH was examined. Results are shown in Figure 1. The concentration inhibition curve demonstrates that MeDTC-sulfoxide is a potent inhibitor of mitochondrial ALDH with a 50% inhibitory concentration of 1.5 μM. Since it is believed that *in vivo* the inhibition is irreversible (Deitrich and Erwin, 1971), we determined the effect of dialysis upon an inhibited enzyme solution. Following dialysis of the enzyme solution against 2000 volumes of buffer for a total of 30 min, the results shown in the Table indicate that enzyme activity could not be restored. Therefore, the enzyme appears to be irreversibly inhibited.

One explanation for irreversible inhibition of ALDH is that of a covalent modification of the enzyme by MEDTC-sulfoxide. Therefore we examined this possibility by mass spectrometry. We first determined the molecular weight of recombinant human mitochondrial ALDH and the results are shown in Figure 2. Our analysis demonstrated a good preparation of protein with a molecular weight of ~54432 Daltons. An aliquot of this solution of ALDH which had been reacted with 40 μM MeDTC-sulfoxide at room temperature for 25 min was subjected to HPLC-MS analysis. The mass spectrum shown in Figure 3, demonstrated a protein with a molecular weight of about 54535 Daltons. A shoulder peak

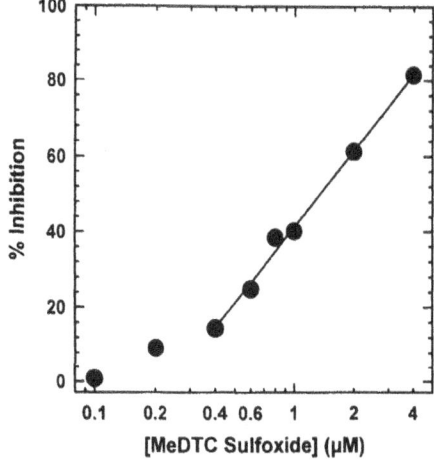

Figure 1. Inhibition of recombinant human mitochondrial ALDH by MeDTC sulfoxide.

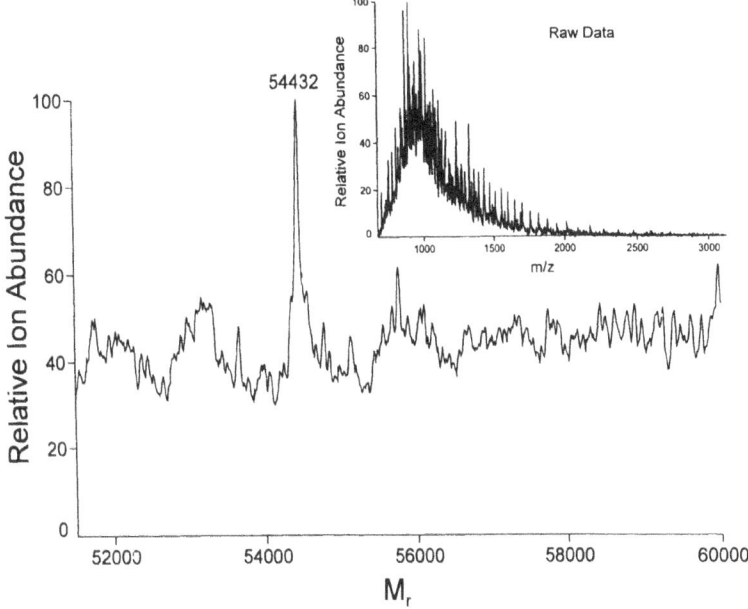

Figure 2. Deconvoluted mass spectrum of native recombinant human mitochondrial ALDH.

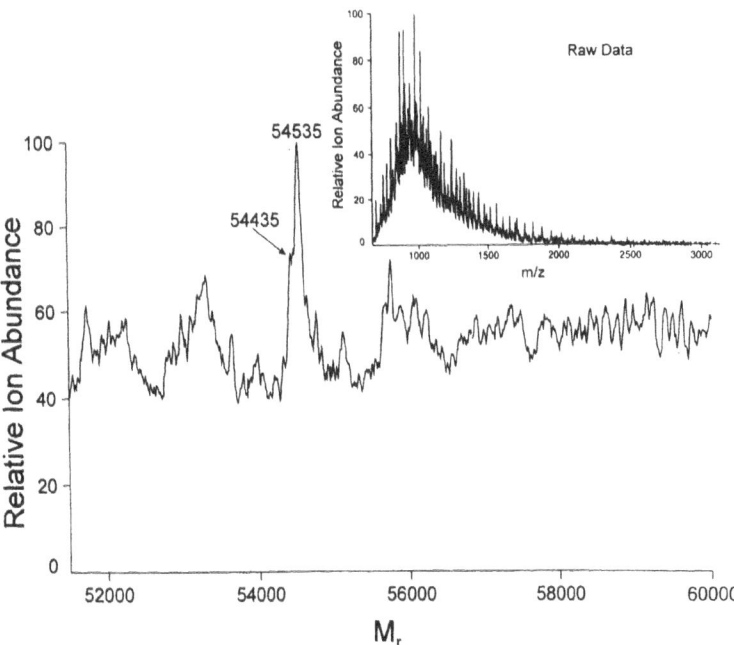

Figure 3. Mass spectrum of recombinant human mitochondrial ALDH which had been treated with MeDTC sulfoxide (40 µM).

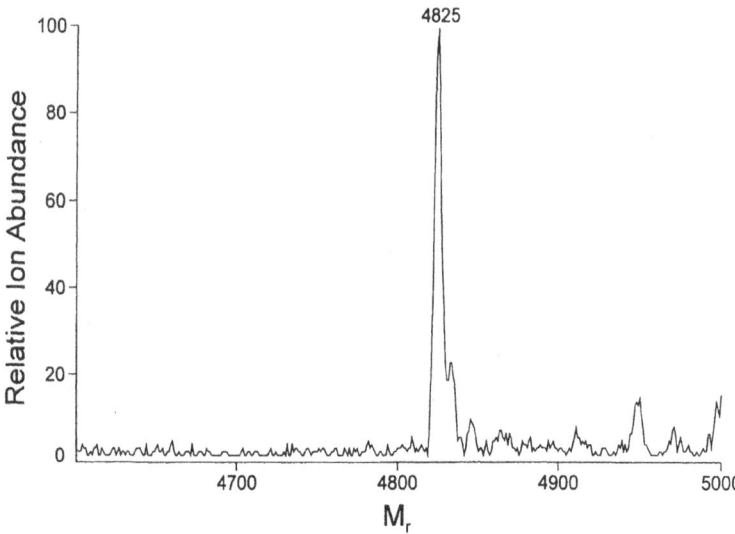

Figure 4. Mass spectrum of a peptide fragment consistent with Leu 269-Glu 312 obtained from a digestion of mitochondrial ALDH.

was also noted with a molecular weight of about 54435 Daltons. Native and MeDTC-sulfoxide adducted proteins were subjected to proteolytic digestion with endopeptidase Glu C and the resulting solution analyzed by HPLC-MS. Analysis indicated a complex mixture of peptides. In the native protein digestion, an incomplete digestion fragment was found which was consistant with a peptide which spanned Leu 269 to Glu 312 of ALDH. A deconvoluted spectrum of this peptide is shown in Figure 4 which demonstrates a molecular weight of 4825. A peptide fragment of molecule weight of 4924, spectrum shown in Figure 5, was found in the adducted recombinant human mitochondria ALDH preparation. This may represent a modification of the corresponding peptide from the unreacted molecule by approximately 100 Daltons.

4. DISCUSSION

Our results indicate that MeDTC-sulfoxide is a potent inhibitor of recombinant mitochondrial ALDH. The IC_{50} which we obtained of 1.5 μM is consistent with an IC_{50} of 0.93 μM using ALDH activity measured in solubilized rat liver mitochondria (Mays, *et al.*, 1996). Since activity in the present experiments could not be restored following dialysis of the enzyme, it appears that the inhibition is irreversible. This is also consistent with the results which we have obtained previously with MeDTC-sulfoxide and rat mitochondrial ALDH (Mays, *et al.*, 1996). Furthermore, it *in vivo* administration of disulfiram results in the irreversible inhibition of ALDH (Deitrich and Erwin, 1971). Since there are multiple pathways involved in the metabolism of disulfiram with many metabolites which may be capable of inhibiting ALDH, these results do not indicate whether MeDTC-sulfoxide is the actual metabolite responsible for all or part of the inhibition of this enzyme *in vivo*.

The data obtained from the mass spectrometry experiments indicate that ALDH is covalently modified by MeDTC-sulfoxide. The increase in approximately 100 mass units

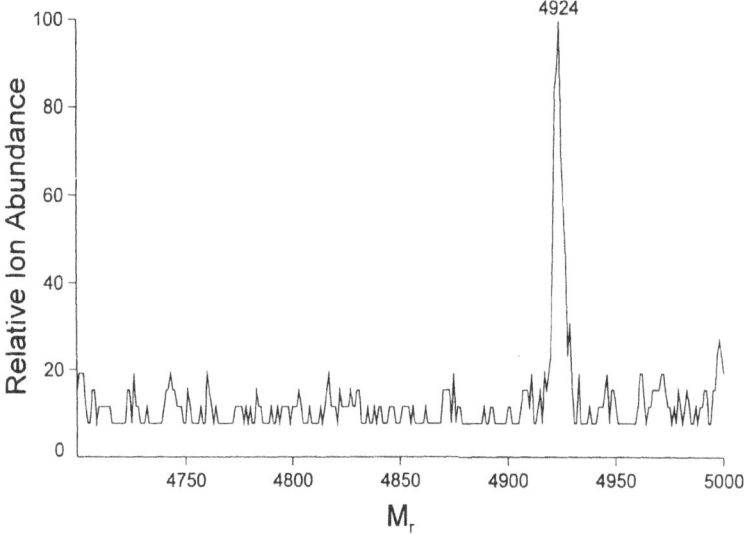

Figure 5. Mass spectrum of a peptide fragment obtained from mitochondrial ALDH which had been treated with 40 μM MeDTC sulfoxide. This fragment shows an increase in molecular weight of 100 as compared to a corresponding peptide obtained from a digestion of native mitochondrial ALDH shown in Figure 4.

of ALDH is consistent with carbamylation of this enzyme by MeDTC-sulfoxide. Our data are consistent with a single carbamylation per monomer of the enzyme. The analysis of the peptides following proteolytic digestion of the enzyme indicate that a peptide spanning the active site of the molecule is carbamylated. This peptide fragment contains three cysteines of which 302 is felt to be the amino acid directly involved in the mechanism of action of the enzyme (Pietruszko, *et al.*, 1993). Further analysis, including sequencing of this peptide, may indicate if indeed this amino acid is modified by MeDTC-sulfoxide. The relevance of our *in vitro* studies to *in vivo* administration of disulfiram and subsequent inhibition of ALDH will have to be determined by analysis of inhibited protein isolated from mammalian liver after treatment with disulfiram.

In conclusion, we have demonstrated in *in vitro* studies that a proposed active metabolite of disulfiram, MeDTC-sulfoxide, is capable of inhibiting human ALDH. The inhibition appears to be irreversible and involves covalent modification of the ALDH. Preliminary analysis of this covalent modification indicates that there is carbamylation in a peptide fragment which spans the active site of the enzyme. The nature of the modification of this peptide as well as its relation to *in vivo* inhibition of this enzyme is yet to be determined.

5. ACKNOWLEDGMENTS

This work was supported by grants NIH AA-09543 and FDA FDT-000–886. We wish to thank Dr. Henry Weiner for his advice and generous gift of cDNA's.

6. REFERENCES

Cobby, J., Mayersohn, M. and Selliah, S.: The rapid reduction of disulfiram in blood and plasma. J.Pharmacol.Exp.Ther. 202 (1977) 724–731.

Deitrich, R.A. and Erwin, V.G.: Mechanism of the inhibition of aldehyde dehydrogenase *in vivo* by disulfiram and diethyldithiocarbamate. Mol.Pharmacol. 7 (1971) 301–307.

Ghenbot, G. and Weiner, H.: Purification of liver aldehyde dehydrogenase by *p*-hydroxyacetophenone-sepharose affinity matrix and the coelution of chloramphenicol acetyl transferase from the same matrix with recombinantly expressed aldehyde dehydrogenase. Protein.Expr.Purif. 3 (1992) 470–478.

Hald, J., Jacobsen, E. and Larsen, V.: The sensitizing effect of tetraethylthiuramdisulphide (Antabuse) to ethylalcohol. Acta.Pharmacol. 4 (1948) 285–296.

Hart, B.W. and Faiman, M.D.: In vitro and in vivo inhibition of rat liver aldehyde dehydrogenase by S-methyl N,N-diethylthiolcarbamate sulfoxide, a new metabolite of disulfiram. Biochem.Pharmacol. 43 (1992) 403–406.

Jeng, J. and Weiner, H.: Purification and characterization of catalytically active precursor of rat liver mitochondrial aldehyde dehydrogenase expressed in *Escherichia coli*. Arch.Biochem.Biophys. 289 (1991) 214–222.

Kitson, T.M.: The disulfiram-ethanol reaction. J.Stud.Alcohol 38 (1977) 96–113.

Mays, D.C., Nelson, A.N., Lam-Holt, J.P., Fauq, A.H. and Lipsky, J.J.: S-Methyl-*N,N*-diethylthiocarbamate sulfoxide and S-methyl-*N-N*-diethylthiocarbamate sulfone, two candidates for the active metabolite of disulfiram. Alcohol.Clin.Exp.Res. 20 (1996) 595–600.

Nelson, A.N. and Lipsky, J.J.: Microtiter plate-based determination of multiple concentration-inhibition relationships. Anal.Biochem. 231 (1995) 437–439.

Pietruszko, R., Abriola, D.P., Blatter, E.E. and Mukerjee, N.: Aldehyde dehydrogenase: Aldehyde dehydrogenation and ester hydrolysis. In Weiner, H., Crabb, D.W. and Flynn, T.G. (Eds.), Enzymology and Molecular Biology of Carbonyl Metabolism 4. Plenum Press, New York, NY, 1993, pp. 221–231.

Svanas, G.W. and Weiner, H.: Aldehyde dehydrogenase activity as the rate-limiting factor for acetaldehyde metabolism in rat liver. Arch.Biochem.Biophys. 236 (1985) 36–46.

Zheng, C.-F., Wang, T.T.Y. and Weiner, H.: Cloning and expression of the full-length cDNAs encoding human liver class 1 and class 2 aldehyde dehydrogenase. Alcohol.Clin.Exp.Res. 17 (1993) 828–831.

25

INHIBITION OF MOUSE AND HUMAN CLASS 1 ALDEHYDE DEHYDROGENASE BY 4-(N,N-DIALKYLAMINO)BENZALDEHYDE COMPOUNDS

James E. Russo

Department of Chemistry
Whitman College
Walla Walla, Washington 99362

1. INTRODUCTION

The pharmacologic inhibition of the oxidation of aldehydes by aldehyde dehydrogenase (ALDH) has been pursued for over forty years. Disulfiram was first used clinically in an attempt to deter people from ethanol consumption (Hald and Jacobsen, 1948). Disulfiram inhibits the ALDH-catalyzed oxidation of acetaldehyde to acetate, resulting in the accumulation of the more toxic aldehyde metabolite. Unfortunately, disulfiram not only inhibits ALDH enzymes in the liver and erythrocytes, but it also interacts with reactive cysteine thiol side chains of other enzymes. Although identification of the active metabolite responsible for enzyme inhibition in vivo has proven difficult, bioactivation of disulfiram appears to be essential. While strong evidence has been presented that S-methyl N,N-diethylthiocarbamate sulfoxide is the disulfiram metabolite which acts as the in vivo inhibitor of rat liver mitochondrial ALDH (Hart and Faiman, 1992), the sulfone product has also been identified as a potent, irreversible inhibitor and potential in vivo metabolite of disulfiram (Mays et al, 1995).

Non-thiol containing molecules such as chloral hydrate and cyanamide, which block the oxidation of broad classes of aldehydic substrates, have been used to demonstrate the presence or absence of ALDH activity in the oxidative metabolism of putative aldehyde substrates. A series of structurally diverse compounds has more recently been identified as inhibitors or inactivators of the rat and human class 2 (mitochondrial) ALDH. These compounds include nitroxyl analogs (Nagasawa et al, 1993), cinnamic acid analogs (Poole et al, 1993), isosorbide dinitrate (Mukerjee and Pietruszko, 1994), and the natural products citral (Boyer and Petersen, 1991) and daidzin (Keung and Vallee, 1993a). Kinetic studies of the different ALDH enzymes have determined several mechanisms of inhibition, including competitive or noncompetitive inhibition of aldehyde or NAD binding and enzyme inactivation involving covalent modification of essential cysteine thiol residues on the enzyme. Although inhibition of acetaldehyde oxidation has been the major target of

Enzymology and Molecular Biology of Carbonyl Metabolism 6
edited by Weiner *et al.* Plenum Press, New York, 1996

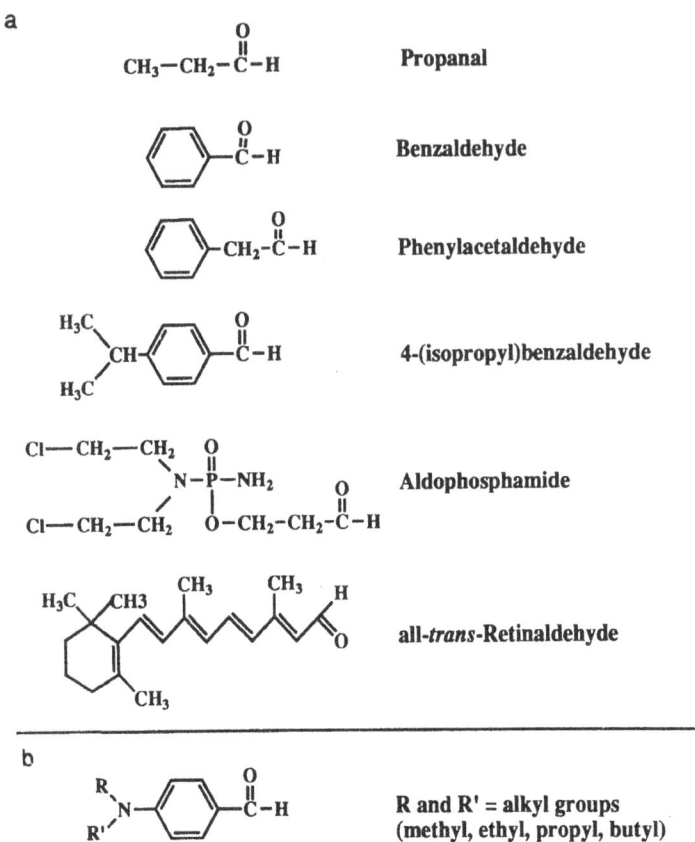

Figure 1. a) Aldehyde substrates and b) general structure for 4-(N,N-Dialkylamino)benzaldehyde inhibitors of class 1 ALDH for class 1 ALDH.

inhibitor design, the identification of inhibitors for specific classes of ALDH or for individual aldehyde substrates has acquired both pharmacologic and mechanistic interest. The ALDH enzymes which oxidize pro-hormones such as retinal, reactive aldehydes generated by lipid peroxidation, the metabolite aldophosphamide generated from the oxazaphosphorine class of antitumor alkylating agents (e.g. cyclophosphamide), and other drug metabolites have been identified. With this characterization of the specific class of enzyme for oxidation of a given aldehyde substrate, it becomes possible to modulate the bioactivation or detoxification of a particular aldehyde substrate (Fig. 1a).

2. RESULTS AND DISCUSSION

2.1. In Vitro and in Vivo Studies Identifying 4-(N,N-Diethylamino) Benzaldehyde (DEAB) as an Inhibitor of Mouse Class 1 Aldehyde Dehydrogenase

Our initial identification of DEAB as an inhibitor of class 1 ALDH was serendipitous. In the search for a molecule which would undergo bioactivation to a reactive alkylating agent following oxidation by class 1 ALDH, we synthesized the nitrogen mustard derivative 4-(N,N bischloroethylamino) benzaldehyde. However, when tested with mouse class 1 ALDH, this compound inhibited the oxidation of propanal and did not appear to undergo oxidation itself. Removal of the chlorine atoms to form DEAB resulted in a molecule with more potent inhibition of ALDH. We proceeded to screen a series of non-thiol, aromatic compounds with dialkylamino and carbonyl substituents for ALDH inhibition (Fig. 1b). Of the analogs tested, DEAB exhibited the greatest inhibition. DEAB strongly inhibits propanal and aldophosphamide oxidation by class 1 ALDH from mouse liver and a mouse leukemic cell line (L1210/CPA) which is resistant to cyclophosphamide by virtue of its overexpression of class 1 ALDH (Russo et al, 1988). Treatment of L1210/CPA cells in vitro with 50 μM DEAB abolished the tumor cells' resistance to 4-hydroperoxycyclophosphamide, an activated form of cyclophosphamide. The potency of class 1 ALDH inhibition increased as the alkyl chain length increased from methyl in 4-(N,N dimethylamino)benzaldehyde (DMAB) to ethyl in DEAB. Likewise, the inhibition potency increased as the alkyl chain length increased from methyl in tetramethylthiuram disulfide to ethyl in disulfiram. The 4-(N,N-dialkylamino)benzaldehyde compounds are structurally similar to disulfiram or the putative active metabolite by virtue of the common dialkylamino group. However, inhibition by a disulfiram metabolite results from an interaction between thiol groups of inhibitor and enzyme. Since the dialkylamino benzaldehyde compounds contain no thiol groups, they must inhibit ALDH via a different mechanism.

DEAB has also been an effective class 1 ALDH inhibitor in vivo. Studies in mice showed that in the presence of DEAB, cyclophosphamide became profoundly toxic to intestinal crypt cells (Russo et al, 1989), while no toxicity was observed in mice receiving up to 20 mg/kg DEAB alone via intraperitoneal injection. In doses of 50 mg/kg, DEAB inhibited ethanol metabolism in mice, as measured by an increased blood acetaldehyde concentration (Mahmoud et al, 1993).

2.2. Structure-Function Studies Identifying 4-(N,N-Dipropylamino) Benzaldehyde (DPAB) as the Most Potent Member of the 4-(N,N-Dialkylamino)Benzaldehyde Series

In order to obtain a better understanding of the structural requirements for ALDH inhibition by DEAB, we investigated the steric and electronic properties of the para substituent (Russo et al, 1995). We synthesized analogs of DEAB in which we varied the dialkyl chain lengths, mixed the chain length in the dialkyl substitution, or eliminated the amino group. The alkyl chain length of the 4-(N,N dialkylamino)benzaldehyde compounds plays a profound role in the reversible inhibition of mouse class 1 ALDH. Inhibition potency increases (K_i decreases 10-fold) as the alkyl chain length increases from ethyl in DEAB to propyl in DPAB (Fig. 2). However, introduction of butyl groups in DBAB results in a significant drop in inhibition potency, as seen by the large increase in the K_i value. The presence of a single propyl group in 4-(N-methyl, N-propyl-amino)benzaldehyde (MPAB) retains inhibitory properties for this analog, although the potency was less than that observed in DEAB or DPAB. We also investigated the necessity of the nitrogen atom at the 4-position on the aromatic ring for inhibition with the compound 4-(iso-propyl)benzaldehyde (IPB). IPB is a substrate, rather than an inhibitor, for class 1 ALDH. This suggests that the dialkyl groups are necessary, but not sufficient to confer inhibitory properties to these analogs. Although DPAB contains the benzaldehyde moiety, DPAB

Inhibitor		K$_i$ (µM)
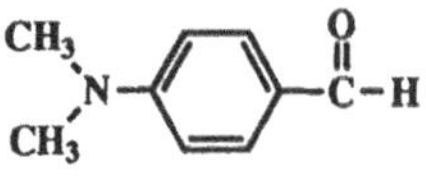 DMAB		> 10
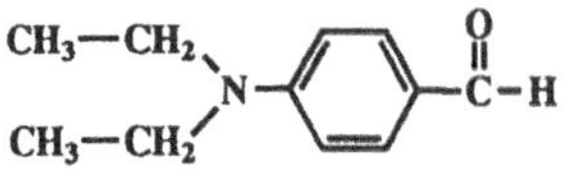 DEAB		0.12
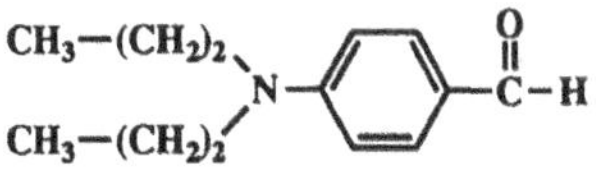 DPAB		0.010
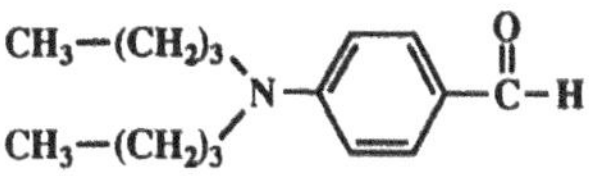 DBAB		> 10

ALDH inhibition assays were performed as described in Russo et al, 1995 with 1 mM NAD and varying concentrations of propanal at pH 7.4 and 37°C.

Figure 2. Inhibition constants (Ki) for the inhibition of mouse class 1 ALDH by 4-(N,N-dialkylamino)benzaldehyde compounds.

does not undergo detectable oxidation to the corresponding acid over a 30 minute incubation with enzyme. However, benzaldehyde is readily oxidized to benzoic acid by class 1 ALDH with NAD as cofactor. Therefore, we examined DPAB inhibition with respect to several aliphatic and aromatic ring-containing aldehydes. DPAB selectively inhibits the oxidation of the aliphatic aldehydes propanal and aldophosphamide, as compared with the aromatic (benzaldehyde) and aromatic ring-containing (phenylacetaldehyde) aldehydes. At 0.5 µM DPAB, propanal oxidation remains completely inhibited (>99%) and aldophosphamide oxidation is inhibited 79%, whereas inhibition of benzaldehyde (32%) and phenylacetaldehyde (19%) is markedly reduced (Russo et al, 1995).

Since DPAB exhibited the lowest Ki value of the 4-(N,N dialkylamino) benzaldehyde compounds tested, we investigated further the kinetics of inhibition by DPAB of class 1 ALDH from mouse liver with respect to aldehyde substrate and NAD cofactor. Lineweaver-Burk and Dixon plots were used to determine inhibition type and to estimate Ki values. Linear mixed-type inhibition is observed for DPAB with respect to propanal and phenylacetaldehyde oxidation by mouse class 1 ALDH. Uncompetitive inhibition is observed for DPAB inhibition with respect to the NAD cofactor.

The high affinity of DPAB for mouse class 1 ALDH is evident by the remarkably low Ki/Km ratio (0.0004) with propanal as substrate (Table 1). DPAB inhibits human (erythrocyte) class I ALDH with a slightly greater potency than mouse ALDH. However, inhibition of phenylacetaldehyde oxidation is markedly reduced, as shown by the higher Ki values (0.077 and 0.070 for mouse and human class I ALDH, respectively). In addition,

Table 1. Selectivity for DPAB inhibition of mouse and human class 1 ALDH

Substrate	K_i (μM)	K_m (μM)	K_i / K_m
Mouse ALDH			
Propanal	0.010	25	0.0004
Phenylacetaldehyde	0.077	5.4	0.014
Human ALDH			
Propanal	0.003	16	0.0002
Phenylacetaldehyde	0.070	5.7	0.012

ALDH assays were performed as described in Russo et al, 1995 with 1 mM NAD. DPAB concentrations were varied from 0.05 - 0.5 μM. K_i and K_m values are the mean of at least four trials. Reproduced from *Biochemical Pharmacology* by permission.

the K_i /K_m ratio of phenylacetaldehyde oxidation increases 35-fold for mouse ALDH and 60-fold for human ALDH, when compared to the K_i /K_m ratio of propanal oxidation obtained with DPAB (Russo et al, 1995).

2.3. Interaction of DPAB with ALDH

Our results indicate that the three carbon propyl group present in DPAB provides the optimal alkyl chain length for inhibition of propanal oxidation by class 1 ALDH in the series of dialkylamino-substituted benzaldehyde compounds studied. Although a single propyl chain can confer significant inhibition to MPAB, the dipropyl group provides the highest inhibition potency. The lack of inhibition observed with IPB indicates that the nitrogen contributes to the binding of these inhibitors. One possibility is that the amine nitrogen of DPAB is protonated, and thus capable of forming an ionic interaction with a carboxylate anion of a Glu or Asp residue essential for catalytic activity.

The mixed type inhibition observed with respect to aldehyde substrates and uncompetitive inhibition with respect to NAD suggests that DPAB does not interfere with the binding of NAD, but the dialkylamino group interferes with hydride ion transfer from aldehyde to NAD. DPAB may bind subsequently to NAD, similar to ordered binding observed for aldehyde substrates. Alternatively, DPAB may bind at a distinct site from either cofactor or aldehyde. The specific binding interactions of DPAB with the enzyme active site are unknown. Due to its apparent high affinity for binding to the class 1 ALDH enzyme, DPAB may be a good candidate as a ligand for x-ray crystallographic studies of class 1 ALDH.

The K_i of DPAB for class 1 ALDH is more than 100-fold lower than values reported for the inhibition of class 1 ALDH by the reversible, inhibitor chloral hydrate (Dockham et al, 1992; Crow et al, 1974) and 10-fold lower than the Ki reported for the inhibition of class 2 ALDH by the terpene citral (Boyer and Petersen, 1991). Both DPAB and citral exhibit linear-mixed type inhibition with respect to aldehyde but do not interfere with binding of NAD to enzyme.

2.4. Is Class-Specific or Substrate-Selective Inhibition of ALDH Feasible?

DPAB exhibits at least two desirable features for a clinically useful pharmacologic agent: high affinity for the target enzyme (Ki =10 nM) and reversible inhibition. Previous in vivo studies with DEAB in mice demonstrated a rapid onset for the inhibition of the oxidation of aldophosphamide (Russo et al, 1989) and acetaldehyde, but also a rapid clearance of DEAB from the blood (Mahmoud et al, 1993). Daidzin, a natural product isoflavone recently identified as a competitive inhibitor of ALDH, also has a high affinity (Ki 40 nM) and is a reversible inhibitor (Keung and Vallee, 1993a). In addition, daidzin is highly selective in vitro for the inhibition of acetaldehyde oxidation by human mitochondrial (class 2) ALDH compared with cytosolic (class 1) enzyme. Also, daidzin has been shown to suppress ethanol intake in Syrian Golden hamsters (Keung and Vallee, 1993b). Thus, daidzin appears to be an effective agent for blocking acetaldehyde metabolism via the specific inhibition of class 2 ALDH. It is possible that doses of daidzin and DPAB used in combination could provide an effective alcohol aversion drug therapy without inhibiting other key physiologic aldehyde oxidations.

In kinetic studies with human breast and colon carcinoma cell lines, DEAB is a substrate (estimated Km = 50 μM) for class 3 ALDH (Rekha et al, 1994). In MCF-7 cells whose class 3 ALDH activity has been transiently induced with 3-methylcholanthrene, the cellular resistance observed to mafosfamide, an activated oxazaphosphorine alkylating agent, is abolished by co-exposure of cells to mafosfamide and DEAB. Also, human class 1 and class 3 ALDH cDNAs have been transfected into hamster V79 cells, resulting in up to 20-fold and 12-fold resistance to mafosfamide following expression of the class 1 and 3 ALDH protein, respectively (Bunting and Townsend, 1996a,b). While a linear relationship between expression of class 1 ALDH and resistance was observed, a non-linear response was seen between expression of class 3 ALDH and resistance. Cell lines expressing either class 1 or class 3 ALDH could be sensitized to mafosfamide by co-exposure to 25 μM DEAB. At this concentration of DEAB, the mechanistic distinction of DEAB as a reversible inhibitor versus a competitive substrate may be unimportant. However, when considering the greater than 500-fold concentration ratio between the Km of DEAB for class 3 ALDH and the Ki for inhibition of class 1 ALDH, the possibility for identifying a DEAB or DPAB concentration which could selectively inhibit class 1 ALDH activity, but not compete for endogenous substrates of class 3 ALDH is tantalizing. Since increased expression of both class 1 and class 3 ALDH has been detected in leukemic and breast tumor cells selected for resistance by culturing in the presence of an activated oxazaphosphorine (4-hydroperoxycyclophosphamide or mafosfamide) the utility of using a class-specific ALDH inhibitor depends on the ability to identify the specific enzymatic determinant of cellular resistance in a given tumor cell population. The identification of a class 3-specific ALDH inhibitor would provide an important tool for further studies addressing the relative contribution of class 1 and class 3 ALDH in tumor cell resistance to alkylating agents.

Cellular levels of ALDH expression can play a critical role in the toxicity observed following exposure to a particular aldehyde-containing compound and in the essential bioactivation of endogenous aldehyde ligands. The oxidation of aldophosphamide by

ALDH has been identified as a key determinant in the survival of murine (Kohn and Sladek, 1985; Sahovic et al, 1988) and human (Kohn et al, 1987) hematopoietic progenitor cells following exposure to 4-hydroperoxy cyclophosphamide and mafosfamide. Although class 1 ALDH has been detected in human hematopoietic stem cells by flow cytometry (Kastan et al, 1990; Jones et al, 1995) the mouse homolog (AHD-2) does not appear to confer resistance to mafosfamide in murine progenitor cells (Maki and Sladek, 1993). In addition, since class 1 ALDH catalyzes the oxidation of retinaldehyde to retinoic acid, this oxidation may serve as an essential metabolic step in the proliferation and differentiation signalling pathways of the hematopoietic stem cells. If class 1 ALDH provides the major activity for aldophosphamide oxidation in hematopoietic or other stem cells, then DPAB would presumably have little efficacy. Concentrations of DPAB necessary to sensitize tumor cells, which are resistant to oxazaphosphorines due to expression of high levels of class 1 ALDH levels, would likely sensitize the proliferating hematopoietic stem cells. However, if the tumor cells were expressing class 3 ALDH, then a class 3-specific ALDH inhibitor may be an effective pharmacologic agent.

Finally, if the hematopoietic stem cells were oxidizing aldophosphamide via class 3 ALDH and the tumor cells were resistant due to expression of class 1 ALDH, then the efficacy of DPAB may depend on the substrate specificity for class 1 ALDH inhibition. DPAB inhibition of both aldophosphamide and retinaldehyde oxidation by class 1 ALDH may sensitize tumor cells, but would also abolish the retinoid differentiation signal of the stem cells. The above scenarios describing the contributions of the ALDH enzyme in aldophosphamide metabolism in tumor cells and hematopoietic stem cells, along with retinaldehyde metabolism in stem cells, illustrate some of the possibilities and caveats associated with attempts at selective modulation of aldehyde metabolism by inhibitors. In the latter scenario, efficacious use of DPAB would require the specific inhibiton of class 1 ALDH and the selective inhibition of the aldophosphamide substrate, but not retinaldehyde. To address the substrate specificity of inhibition, we are currently investigating the effect of DPAB on retinaldehyde oxidation by class 1 ALDH.

3. ACKNOWLEDGMENTS

Supported by Bristol-Myers Squibb Award of Research Corporation and the M.J. Murdock College Science Research Program.

4. REFERENCES

Boyer, C.S., and Petersen, D.R. The Metabolism of 3,7-dimethyl-2,6-octadienal (citral) in rat hepatic mitochondrial and cytosolic fractions. Drug. Metab. Dispos. 19: 81–86, 1991.

Bunting, K.D. and Townsend, A.J. De Novo Expression of Transfected Human Class 1 Aldehyde Dehydrogenase (ALDH) Causes Resistance to Oxazaphosphorine Anti- cancer Alkylating Agents in Hamster V79 Cell Lines. J. Biol. Chem. 271:11884- 11890, 1996a.

Bunting, K.D. and Townsend, A.J. Protection by Transfected Rat or Human Class 3 Aldehyde Dehydrogenasse against the Cytotoxic Effect of Oxazaphosphorine Alkylating Agents in Hamster V79 Cell Lines. J. Biol. Chem. 271:11892–11896, 1996b.

Crow, K.E., Kitson, T.M, MacGibbon, A.K.H., and Batt, R.D. Intracellular localization and properties of aldehyde dehydrogenases from sheep liver. Biochim. et Biophys. Acta 350: 121–128, 1974.

Dockham, P.A., Lee, M.O., and Sladek, N.E. Identification of Human Liver Aldehyde Dehydrogenases that Catalyze the Oxidation of Aldophosphamide and Retinaldehyde. Biochem. Pharmacol. 43: 2453–2469, 1992.

Hald, J. and Jacobsen, E. A drug sensitizing the organism to ethyl alcohol. Lancet 2: 1001–1004, 1948.

Hart, B.W. and Faiman, M.D. In vitro and in vivo inhibition of rat liver aldehyde dehydrogenase by S-methyl N,N-diethylthiolcarbamate sulfoxide, a new metabolite of disulfiram. Biochem. Pharmacol. 43:403–406, 1992.

Jones, R.J., Barber, J.P., Vala, M.S., Collector, M.I., Kaufmann, S.H., Ludeman, S.M., Colvin, O.M., and Hilton, J. Assessment of Aldehyde Dehydrogenase in Viable Cells. Blood 85:2742–2746, 1995.

Kastan, M.B., Schlaffer, E., Russo, J.E., Colvin, O.M., Civin, C.I., and Hilton, J. Direct Demonstration of Elevated Aldehyde Dehydrogenase in Human Hematopoietic Progenitor Cells. Blood 75:1947–1950, 1990.

Keung, W.M., and Vallee, B.L. Daidzin: A potent, selective inhibitor of human mitochondrial aldehyde dehydrogenase. Proc. Natl. Acad. Sci. USA 90: 1247- 1251, 1993a.

Keung, W.M. and Vallee, B.L. Daidzin and daidzein suppress free-choice ethanol intake by Syrian Golden Hamsters. Proc. Natl. Acad. Sci. USA 90: 10008–10012, 1993b.

Kohn, F.R., Landkamer, G.J., Manthey, C.L., Ramsay, N.K., and Sladek, N.E. Effect of aldehyde dehydrogenase inhibitors on the ex vivo sensitivity of human multipotent and committed hematopoietic progenitor cells and malignant blood cells to oxazaphosphorines. Cancer. Res. 47:3180–3185, 1987.

Kohn, F.R. and Sladek, N.E. Aldehyde dehydrogenase activity as the basis for the relative insensitivity of murine pluripotent hematopoietic stem cells to oxazaphosphorines. Biochem. Pharmacol. 34:3465–3471, 1985.

Maki, P.A. and Sladek, N.E. Sensitivity of aldehyde dehydrogenases in murine tumor and hematopoietic progenitor cells to inhibition by chloral hydrate as determined by the ability of chloral hydrate to potentiate the cytotoxic action of mafosfamide. Biochem. Pharmacol. 45:231–239, 1993.

Mahmoud, M.I.E., Potter, J.J., Colvin, O.M., Hilton, J., and Mezey, E. Effect of 4- (diethylamino)benzaldehyde on ethanol metabolism in mice. Alcohol Clin. Exp. Res. 17: 1223–1227, 1993.

Mays, D.C., Nelson, A.N., Fauq, A.H., Shriver, Z.H., Veverka, K.A., Naylor, S., and Lipsky, J.J. S-Methyl N,N-Diethylthiocarbamate Sulfone, A potential Metabolite of Disulfiram and Potent Inhibitor of Low Km Mitochondrial Aldehyde Dehydrogenase. Biochem. Pharmacol. 49: 693–700, 1995.

Mukerjee, N. and Pietruszko, R. Inactivation of Human Aldehyde Dehydrogenase by Isosorbide Dinitrate. J. Biol. Chem. 269:21664–21669, 1994.

Nagasawa, H.T., Yost, Y., Elberling, J.A., Shirota, F.N., and Demaster, E.G. Nitroxyl analogs as inhibitors of aldehyde dehydrogenase. Biochem. Pharmacol. 45: 2129- 2134, 1993.

Poole, R.C., Bowden, N.J., and Halestrap, A.P. Derivatives of cinnamic acid interact with the nucleotide binding site of mitochondrial aldehyde dehydrogenase. Biochem. Pharmacol. 45: 1621–1630, 1993.

Rekha, G.K., Sreerama, L., and Sladek, N.E. Intrinsic Cellular Resistance to Oxazaphosphorines Exhibited by a Human Colon Carcinoma Cell Line Expressing Relatively Large Amounts of a Class-3 Aldehyde Dehydrogenase. Biochem. Pharmacol. 48:1943–1952, 1994.

Russo, J.E., Hauquitz, D., and Hilton, J. Inhibition of mouse cytosolic aldehyde dehydrogenase by 4-(diethylamino)benzaldehyde. Biochem. Pharmacol. 3 7:1639–1642, 1988.

Russo, J.E., Hilton, J., and Colvin, O.M. The Role of Aldehyde Dehydrogenase Isozymes in Cellular Resistance to the Alkylating Agent Cyclophosphamide. In: Enzymology and Molecular Biology of Carbonyl Metabolism 2. Flynn, T.G. and Weiner, H,) pp. 65–79. Alan R. Liss, New York 1989.

Russo, J., Chung, S., Contreras, K., Lian, B., Lorenz, J., Stevens, D., and Trousdell, W. Identification of 4-(N,N-dipropylamino)benzaldehyde as a potent, reversible inhibitor of mouse and human class I aldehyde dehydrogenase. Biochem.Pharmacol. 50: 399–406, 1995.

Sahovic, E.A., Colvin, M., Hilton, J., and Ogawa, M. Role for aldehyde dehydrogenase in survival of progenitors for murine blast cell colonies after treatment with 4- hydroperoxycyclophosphamide in vitro. Cancer Res. 48: 1223–1226, 1988.

Sladek, N.E., Sreerama, L., and Rekha, G.K. Constitutive and Overexpressed Human Cytosolic Class-3 Aldehyde Dehydrogenases in Normal and Neoplastic Cells/Secretions. In Enzymology and Molecular Biology of Carbonyl Metabolism 5; Weiner, H., Holmes, R.S., and Wermuth, B., eds. Plenum Press, N.Y. pp. 103–113, 1995.

26

PHENETHYL ISOTHIOCYANATE AS AN INHIBITOR OF ALDEHYDE DEHYDROGENASES

Martti Koivusalo and Kai O. Lindros

Institute of Biomedicine
Department of Medical Chemistry
University of Helsinki and National Public Health Institute
Department of Alcohol Research
Helsinki, Finland

1. INTRODUCTION

There has been increasing interest on organic isothiocyanates especially on phenethylisothiocyanate (PEITC) as chemoprotective agents against several types of chemically induced tumors (Hecht 1995). These isothiocyanates are present in several cruciferous edible plants as complex thioglucosides called glucosinolates (Fenwick et al., 1983). They are released from the glucosinolates by the action of the hydrolytic enzyme myrosinase (thioglucoside glucohydrolase EC 3.2.3.1) which is present as several isoenzymes in the same plants and is released from the plants when the cells are crushed like in mastication. PEITC is found as its glucosinolate gluconasturtin in e.g. cabbage, broccoli, cauliflower and watercress, which are widely consumed as food. The chemoprotective action of isothiocyanates against carcinogens is thought to be due to their inhibitory effects against several Phase I detoxifying enzymes (cytochrome P-450s) which activate procarcinogens and their induction of Phase II detoxifying enzymes like glutathione S-transferase and quinone reductase (Zhang and Talalay, 1994) which aid in their excretion.

We have recently shown that when rats were treated with PEITC the aldehyde dehydrogenase activity was significantly decreased in their livers with no effect on alcohol dehydrogenase activity (Lindros et al., 1995). When both PEITC and ethanol were given to rats there was a significant increase in blood acetaldehyde levels. In the present paper we have further studied the effects of PEITC on rat liver aldehyde dehydrogenase activities and especially the inhibition kinetics by PEITC of purified class 3 cytosolic rat liver aldehyde dehydrogenase induced by 20-methylcholantrene.

2. EXPERIMENTAL

2.1. Materials

Class 3 aldehyde dehydrogenase was purified to homogeneity from livers of rats treated with 20-methylcholantrene by a combination of affinity chromatography on AMP-Sepharose and chromatofocusing (Koivusalo and Rautoma, 1985). Female rats of Wistar strain weighing about 200 g were used for the preparation of subcellular fractions from liver as described earlier (Koivula and Koivusalo, 1975). The mitochondrial and microsomal fractions were solubilized with Triton X-100 , final concentration 1% (w/v). All fractions were filtrated through 0.45 mm HAWP 025 00 ultrafilters (Millipore Filter Corp.) before assays. In the in vivo experiments PEITC (1 mM/kg) was administered i.p. to male Wistar rats as 1% solution in corn oil. The rats were killed 24 hours later and their oesophagi and stomachs were excised and the mucosal fractions were prepared by scraping. These were stored frozen at -80 Co. The soluble fractions were prepared after thawing by homogenization in a Sorvall Omnimixer in 0.1 M Na-phosphate buffer pH 7.4 and centrifuging for 1 hr at 100 000 g. Phenethyl isothiocyanate was purchased from Aldrich Chemical Co. (Milwaukee, WI). 20-Methylcholantrene, 4-methylumbelliferyl acetate, NAD and NADP were obtained from Sigma Chemical Company (St.Louis, MO).

2.2. Assay Methods

Aldehyde dehydrogenase activity was assayed spectrophotometrically at 25o as described earlier (Koivula and Koivusalo, 1975) using as a standard assay 9 mM propionaldehyde, 1.2 mM NAD and 3.0 mM pyrazole in 0.1 M sodium pyrophosphate buffer pH 8.0. In the kinetic experiments with NAD the formation of NADH was assayed fluorometrically. If not otherwise stated, the enzyme sample and PEITC were first incubated in buffer for 2 min and the reaction then started by adding aldehyde and NAD. The class 3 aldehyde dehydrogenase activity in the soluble fractions from oesophagi and stomachs was assayed with 0.4 mM benzaldehyde and 1.2 mM NADP in 0.1 M sodium pyrophosphate buffer pH 8.0. The esterase activity was assayed spectrophotometrically in 0.1 M sodium pyrophosphate buffer pH 8.0 with p-nitrophenylacetate or 4-methylumbelliferyl acetate as substrates at wavelengths of 405 nm or 340 nm, respectively. Spectrophotometric measurements were performed on either a Shimadzu UV-240 Graphicord spectrophotometer or on a vertical nine-channel FP-9 Analyzing System (Labsystems Oy). Fluorescence measurements were performed on a Hitachi F-2000 Fluorescence Spectrophotometer.

3. RESULTS

In in vivo experiments rats were treated i.p. with PEITC (1 mmol/kg) and 24 h later the constitutional class 3 aldehyde dehydrogenase activity was determined in the mucosal fractions of their oesophagi and stomachs using the NADP-benzaldehyde assay system as described under -Methods+. In the oesophagi the activity was decreased 80% when compared with the controls (control activity 56.7 nmoles per min per mg of protein). In the stomachs the decrease was 66% and the control value was 38.4 nmoles per min per mg of protein.

In the subcellular fractions of rat liver all ALDH activities were inhibited in vitro by low micromolar concentrations of PEITC (Table 1). The inhibitions were rapid and in-

Table 1. Inhibition of ALDH in rat liver subcellular fractions. Assay conditions as described under "Methods". 50 mM or 9 mM propionaldehyde were used as substrates. Means ± S.D. for three experiments are given

	IC50 mM	
Liver fraction	50 mM propionaldehyde	9 mM propionaldehyde
Mitochondrial	1.1 ± 0.4	5.8 ± 1.2
Cytosolic	0.9 ± 0.4	1.2 ± 0.6
Microsomal		4.1 ± 1.0

creased only little with time within 10 min. They were reversed by dilution and by a short incubation with 10 mM mercaptoethanol. Addition of NAD before PEITC gave a partial protection from the inhibition but addition of NAD after PEITC had no effect.

In the experiments using purified class 3 ALDH the enzyme was rapidly inhibited if PEITC was added before NAD and the inhibition did not increase with time. The inhibition was rapidly reversed by addition of 10 mM mercaptoethanol or DTT. It was also reversed by dilution or gel filtration. When the fractional inhibition was plotted against the concentration of the inhibitor a hyperbolic plot was obtained (Fig.1). An inhibition of 50 % was obtained with 23 nM PEITC and this was not dependent on the substrate concentration used. When the results obtained with different substrate and inhibitor concentrations were plotted as double reciprocal plots a noncompetitive pattern was obtained with respect to aldehyde. The K_i value calculated from the plots was 23 nM.When plotted wIth respect to NAD the pattern was competitive. This was, however, not due to direct reversible competition between PEITC and NAD but was a protective effect of binding of NAD. The inhibition was not reversed by addition of NAD after PEITC. Addition of NAD before PEITC protected the enzyme effectively from the rapid reversible inhibition (Fig.2.). No

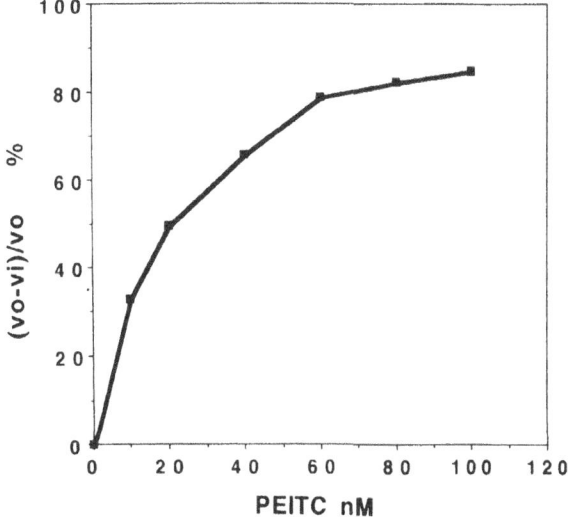

Figure 1. Inhibition of class 3 ALDH by PEITC in the absence of NAD.

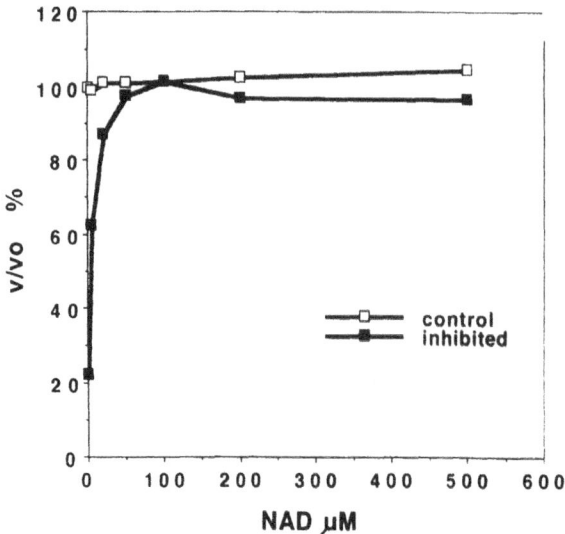

Figure 2. Protection by NAD of class 3 ALDH from the inhibition by PEITC. 60 nM PEITC was used as inhibitor and the indicated amounts of NAD were added before PEITC. The residual activity with no NAD added was 20 %.

such inhibition was seen when the enzyme was saturated with NAD before addition of PEITC.

The esterase activity of the purified enzyme with 4-methylumbelliferyl acetate as substrate was also inhibited by PEITC in the absence of NAD (Fig.3). An inhibition of 50 % was obtained with 27 nM PEITC. The inhibition was reversed by dilution or gel filtration. Similar results were obtained with p-nitrophenylacetate as substrate.

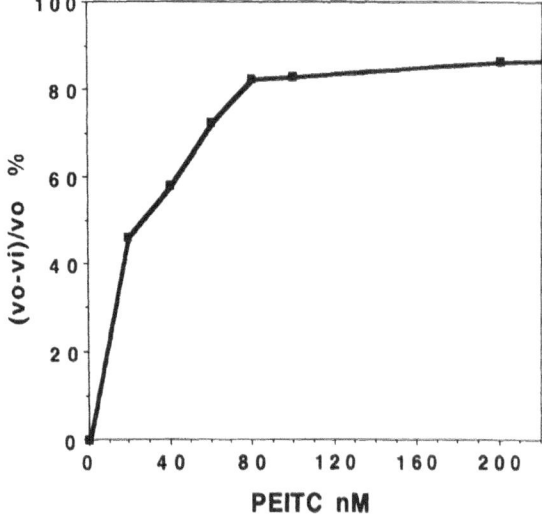

Figure 3. Inhibition of the esterase activity of class 3 ALDH by PEITC. Substrate was 0.4 mM 4-methylumbelliferyl acetate.

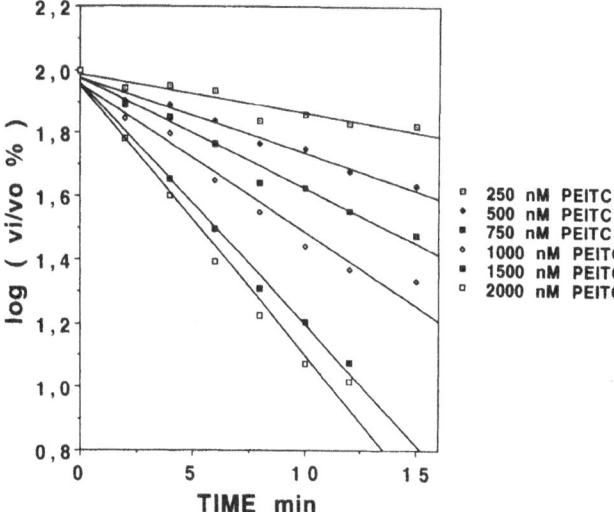

Figure 4. Inhibition of class 3 ALDH by PEITC in the presence of NAD. 1.2 mM NAD was added before PEITC.

In the presence of saturating concentrations of NAD the enzyme was less sensitive to PEITC. There was no rapid reversible inhibition but a slower time-dependent inhibition which was not reversed by dilution or gel filtration. When the logarithm of residual activity was plotted against time linear plots were obtained indicating a pseudo-first order reaction (Fig.4). When the values for the apparent rate constants kobs were plotted against the concentration of PEITC used a hyperbolic plot was obtained indicating a saturating effect. The double reciprocal plot of 1/kobs against 1/PEITC was linear and had an intercept on the y-axis (Fig. 5). This indicates a two-step process, where a reversible enzyme-inhibitor complex is first formed before the irreversible inactivation (Kitz and Wilson 1962, Fahrney and Gold 1963, Mares-Guia and Shaw 1967). The enzyme inactivated by PEITC in the presence of NAD was slowly reactivated by incubation with 20 mM mercaptoethanol (Fig. 6) and only a partial reactivation was obtained with an incubation for 30 min.

4. DISCUSSION

PEITC is a new naturally occurring inhibitor of aldehyde dehydrogenases. It inhibits all aldehyde dehydrogenase forms both *in vivo* and *in vitro* and does not inhibit alcohol dehydrogenase at these concentrations. The concentrations giving *in vitro* 50 % inhibition were 0.9–6.0 μM for the activities in the different subcellular fractions of rat liver.The active free PEITC concentrations have apparently been lower in these experiments due to binding to other proteins in the solubilized fractions so these values are upper estimates.

Class 3 aldehyde dehydrogenase is cytosolic and is either constitutively produced or inducible depending on the tissue (For a comprehensive review see Lindahl 1992). It prefers NAD as coenzyme but can use NADP in vitro. Its preferential substrates are aromatic aldehydes and medium chain length aliphatic aldehydes. It is not normally present in liver but can be induced by several carcinogenic and non-carcinogenic xenobiotics. It is also expressed in several chemically induced tumors appearing early during the tumorigenesis.

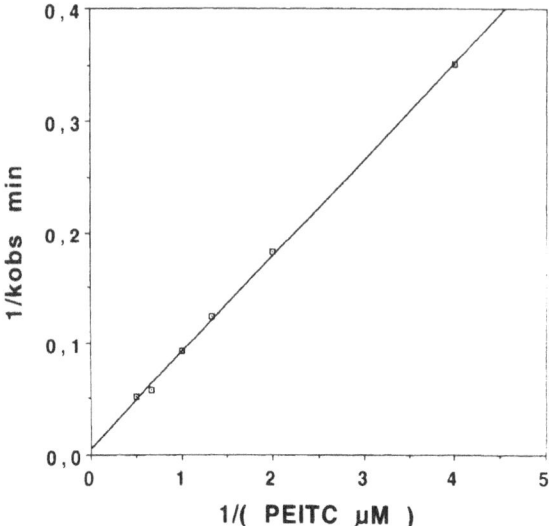

Figure 5. Dependence of kobs on the concentration of PEITC. The values were calculated from the results presented in Fig. 4. Double reciprocal plot.

In the present study the purified 20-methylcholantrene-induced enzyme was found to be effectively inhibited by PEITC. When PEITC was added before NAD there was a reversible noncompetitive inhibition with a low K_i value of 23 nM. If NAD was added before PEITC it effectively protected from the rapid time-independent inhibition. When the enzyme was saturated with NAD before addition of PEITC there was instead an irreversible time-dependent inhibition which followed pseudo-first order kinetics. Apparently when

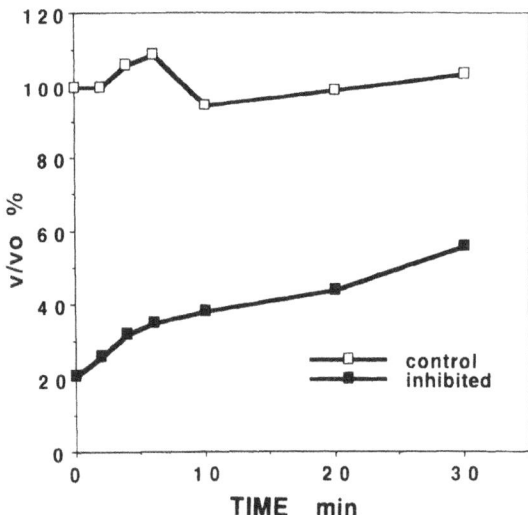

Figure 6. Effect of incubation with mercaptoethanol on the class 3 ALDH inactivated by PEITC in the presence of NAD. The enzyme was incubated first with 5 μM PEITC for 5 min in the presence of 1.2 mM NAD before addition of 20 mM mercaptoethanol. The residual activity before addition of mercaptoethanol was 20%.

NAD binds to the enzyme there is a conformational change which blocks earlier target groups and brings out new ones with different reactivities. When the k_{obs} values were calculated from the first-order plots and plotted against the concentration of PEITC used the plot was not linear but hyperbolic showing a saturation effect. This indicates, that there is first reversible formation of a complex between the enzyme and the inhibitor and this gives then the irreversibly inactivated enzyme-inhibitor complex. Class 3 ALDH is inhibited also by disulfiram but only in the absence of NAD (Koivusalo and Rautoma 1987). NAD protects partially also the mitochondrial low-K_m ALDH from inhibition by disulfiram but the inhibition is irreversible both in the presence and in the absence of NAD (Marchner and Tottmar 1983) in contrast to the inhibition of class 3 ALDH by PEITC.

PEITC inhibits the class 3 ALDH also *in vivo* .We have earlier reported that the ALDH activities in the livers of rats treated with PEITC (1 mmol/kg) is decreased 30% after 24 h (Lindros et al. 1995). We have now determined in similar experiments the constitutive class 3 ALDH activities in the mucosal fractions of oesophagi and stomachs and found them to be decreased by 80% and 66%, respectively. Class 3 ADH seems thus to be the ALDH form most sensitive to the effects of PEITC *in vivo*.

Organic isothiocyanates (R-N=C=S) are very reactive compounds due to the strong electrophilicity of the central carbon atom which reacts rapidly with several nucleophiles like thiols, cysteine and histidine (Zhang and Talalay 1994). The reactions with thiols and cysteine give rise to dithiocarbamylated compounds and reaction with essential cysteines is likely the mechanism by which PEITC inhibits the aldehyde dehydrogenases. Alkyl isocyanates (R-N=C=O) have also been reported to be inhibitors of aldehyde dehydrogenases presumably by carbamylation of essential cysteine groups (Nagasawa et al. 1994). Isothiocyanates can also react with glutathione either non-enzymically or catalyzed by glutathione transferase (Zhang et al. 1995) and N-acetylcysteine conjugates appear to be their major excretion products in urine (Chung et al. 1992). The conjugation with glutathione is freely reversible.

The induced class 3 ALDH has esterase activity with p-nitrophenylacetate or 4-methylumbelliferyl acetate as substrates (Koivusalo and Rautoma 1987). This esterase activity differs from that of class 1 or 2 aldehyde dehydrogenases in it that it is effectively inhibited by NAD(P)(H). In the absence of NAD the esterase activity was inhibited by PEITC with similar K_i than the aldehyde dehydrogenase activity. The group(s) with esterase activity are apparently in the active site associated with the binding of NAD and it seems likely that the acyl-enzyme intermediates in the aldehyde dehydrogenase reaction may form with different nucleophilic groups available in the NAD-enzyme complex.

We have shown in the present report that in addition to Class 1 and Class 2 aldehyde dehydrogenases PEITC inhibited the Class 3 enzymes both in vivo and in vitro. The effective inhibition of ALDH activities should be taken into account when the chemopreventive actions of PEITC against carcinogenesis are studied.

5. ACKNOWLEDGMENTS

We thank Mrs. Eija Haasanen for skillful technical assistance.

6. REFERENCES

Chung, F.-L., Morse, M.A., Eklind, K.I. and Lewis, J.: Quantitation of human uptake of the anticarcinogen phenethyl isothiocyanate after a watercress meal. Cancer Epidemiol.Biomark.Prevent. 1 (1992) 383–388.

Fahrney, D.E. and Gold, A.M.: Sulfonyl fluorides as inhibitors of esterases. I. Rates of reaction with acetylcholi-nesterase, a-chymotrypsin, and trypsin. J.Am.Chem.Soc. (1963) 997–1000.

Fenwick, G.R., Heaney, R.K. and Mullin, W.J.: Glucosinolates and their breakdown products in food and food plants. CRC Crit.Rev.Food Sci.Nutr. 18 (1983) 123–201.

Hecht, S.S.: Chemoprevention by isothiocyanates. J.Cell.Biochem. Suppl.22 (1995) 195–209.

Kitz, R. and Wilson, I.B.: Esters of methanesulfonic acid as irreversible inhibitors of acetylcholinesterase. J.Biol.Chem. 237 (1962) 3245–3249.

Koivula, T. and Koivusalo, M.: Different forms of rat liver aldehyde dehydrogenase and their subcellular distribu-tion. Bioch.Biophys.Acta 397 (1975) 9–23.

Koivusalo, M. and Rautoma, P.: Induction of cytoplasmic aldehyde dehydrogenases in rat tissues by carcinogenic and non-carcinogenic xenobiotics. Prog.Clin.Biol.Res. 174 (1985) 101–112.

Koivusalo, M. and Rautoma, P.: The cytoplasmic aldehyde dehydrogenase induced in rat liver by xenobiotics. Ki-netic studies and studies on esterase activity. Prog.Clin.Biol.Res. 232 (1987) 135–147.

Lindahl, R.: Aldehyde dehydrogenases and their role in carcinogenesis. CRC Crit.Rev.Biochem.Mol.Biol. 27 (1992) 283–335.

Lindros, K.O., Badger, T., Ronis, M., Ingelman-Sundberg, M. and Koivusalo, M.: Phenethyl isothiocyanate, a new dietary liver aldehyde dehydrogenase inhibitor. J.Pharmacol.Exp.Ther. 275 (1995) 79–83.

Marchner, H. and Tottmar, O.: Studies on the inactivation of mitochondrial rat-liver aldehyde dehydrogenase by the alcohol-sensitizing compounds cyanamide, 1-aminocyclopropanol and disulfiram. Biochem.Pharmacol. 32 (1983) 2181–2188.

Mares-Guia, M. and Shaw, E.: The specific inactivation of trypsin by ethyl p-guanidinobenzoate. J.Biol.Chem. 242 (1967) 5782–5788.

Nagasawa, H.T., Eberling, J.A., Goon, D.J.W. and Shirota F.N.: Latent alkyl isocyanates as inhibitors of aldehyde dehydrogenase in vivo. J.Med.Chem. (1994) 4222–4226.

Zhang, Y., Kolm, R.H., Mannervik, B. and Talalay, P.: Reversible conjugation of isothiocyanates with glutathione catalyzed by human glutathione transferase. Biochem.Biophys.Res.Comm. 206 (1995) 748–755.

Zhang, Y. and Talalay, P.: Anticarcinogenic activities of organic isothiocyanates: Chemistry and mechanism. Can-cer Res.(Suppl.) 54 (1994) 1976s-1981s.

27

EFFECT OF SUBSTITUTED BROMOACETANILIDES ON CYTOSOLIC ALDEHYDE DEHYDROGENASE

Susan E. Euston, Graham H. Freeman, Kathryn E. Kitson, and Trevor M. Kitson

Departments of Chemistry and Biochemistry
Massey University
Palmerston North, New Zealand

1. INTRODUCTION

The work described herein involves the covalent modification of cytosolic aldehyde dehydrogenase from sheep liver by various derivatives of bromoacetanilide (2-bromo-*N*-phenylethanamide). Previous studies have shown that aldehyde dehydrogenase is inactivated by haloacetyl compounds, including iodoacetamide (iodoethanamide, which labels Cys-302; Hempel et al., 1985) and bromoacetophenone (2-bromo-1-phenylethanone, which modifies both Glu-268 and Cys-302; Abriola et al., 1990). The aims of our work here are twofold: first, to investigate iodine-bearing derivatives of bromoacetanilide as potential vehicles for the specific incorporation of heavy atoms at the active site of the enzyme, with a view to facilitating the solution of its tertiary structure by X-ray crystallography, and second, to label the enzyme with nitrophenol-bearing derivatives of bromoacetamide in order to use the resulting covalently-linked 'reporter groups' as probes of the environment of the active site. The structures of the compounds investigated here are shown in Figure 1.

2. EXPERIMENTAL

Cytosolic aldehyde dehydrogenase was purified from sheep liver and its concentration was measured as reported previously (Kitson and Kitson, 1994). 2-Bromo-2'-hydroxy-5'-nitroacetanilide was purchased from Aldrich. 2-Bromo-4'-hydroxy-3'-nitroacetanilide was prepared as follows. 4-Amino-2-nitrophenol (2 g) and an equimolar amount of *N,N*-dimethylaniline were dissolved in acetone (15 ml) and cooled on ice. An equimolar amount of bromoacetyl bromide (dissolved in 10 ml of acetone) was added slowly with cooling. After 30 min, the acetone was evaporated giving a thick red-black oil. This was dispersed

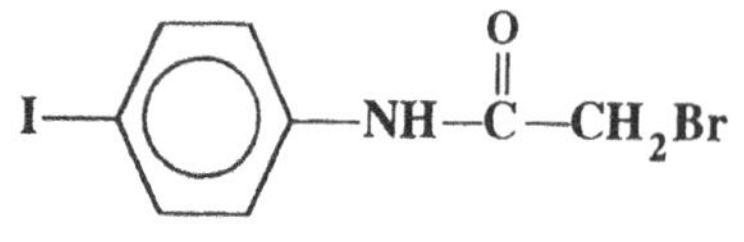

2-bromo-4'-iodoacetanilide

(BIA)

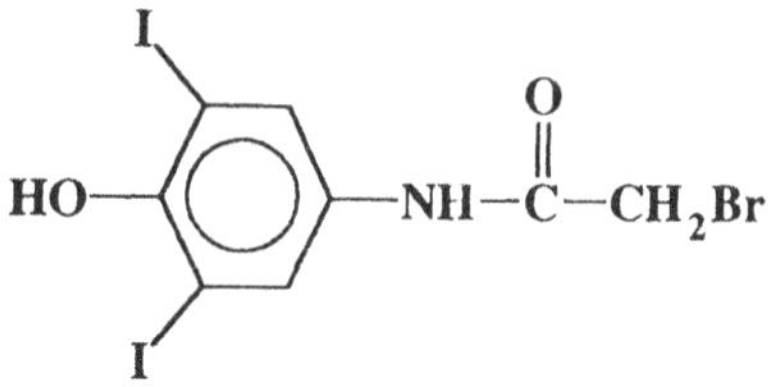

2-bromo-4'-hydroxy-3',5'-diiodoacetanilide

(BHDA)

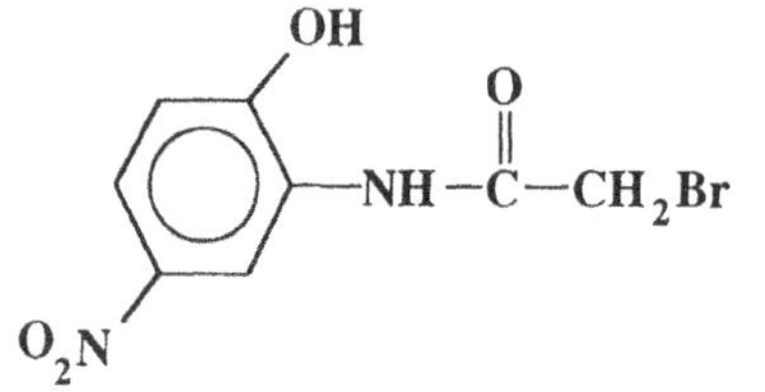

2-bromo-2'-hydroxy-5'-nitroacetanilide

(BHN-1)

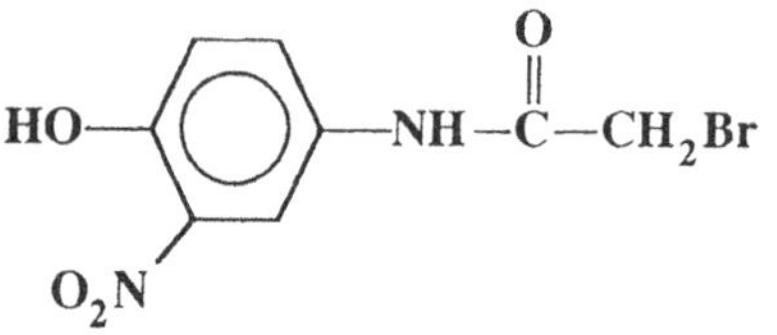

2-bromo-4'-hydroxy-3'-nitroacetanilide

(BHN-2)

Figure 1. Structures, names and abbreviations of the bromoacetanilides used in this work.

in 15 ml of hot chloroform and applied to a column of silica gel. Elution with chloroform yielded 200 ml of golden-yellow solution which was evaporated to dryness. The solid was dissolved in hot chloroform; after treatment with decolorising charcoal, filtration, and cooling, the product crystallised in the form of orange platelets. δ (CDCl$_3$): 4.05 (s, 2H), 7.16–7.20 (d, 1H), 7.76–7.79 (d, 1H), 8.21 (s, broad, 1H), 8.33 (s, 1H), 10.47 (s, 1H). C$_8$H$_7$N$_2$O$_4$Br was confirmed by mass spectrometry (m/z = 273.958740 [^{79}Br], 275.956833 [^{81}Br]. 2-Bromo-4'-iodoacetanilide was prepared as follows. 4-Iodoaniline (2.2 g) was dissolved in 75 ml of dry ether and cooled in ice. Separate equimolar amounts of bromoacetyl bromide and *N,N*-dimethylaniline were added slowly. A sticky grey solid separated; this was ground up with a glass rod and the mixture was left overnight at room temperature. The ether was evaporated and the solid residue was extracted twice by grinding with 25 ml of water. After recrystallising twice from ethanol the product was obtained as pale grey crystals. δ (CDCl$_3$): 4.02 (s, 2H), 7.41–7.44 (s, 2H), 7.64–7.68 (s, 2H). C$_8$H$_7$NOBrI was confirmed by mass spectrometry (m/z = 338.874832 [^{79}Br], 340.871078 [^{81}Br]. 2-Bromo-4'-hydroxy-3',5'-diiodoacetanilide was prepared as follows. 4-Nitrophenol (5 g) and sodium carbonate (15 g) were dissolved in hot water (100 ml). A solution of iodine (18.2 g) and potassium iodide (20 g) in water (100 ml) was added, giving a deep orange solution. To the warm solution was added sufficient sodium hydrosulphite to eliminate all the orange colour; the cream-coloured solid (4-amino-2,6-diiodophenol) was filtered off, washed well with water, and dried over phosphorus(V) oxide. A solution of 3 g of this and an equimolar amount of *N,N*-dimethylaniline in acetone was cooled on ice. An equimolar amount of bromoacetyl bromide was added slowly. The solution was evaporated to dryness and the residue was suspended in 15 ml of hot chloroform and applied to a column of silica gel. The colourless chloroform eluate was evaporated and the product was obtained from chloroform/ethanol as a white crystalline solid. δ (CD$_3$COCD$_3$): 4.01 (s, 2H), 8.12 (s, 2H), 9.48 (s, low integral). C$_8$H$_6$NO$_2$BrI$_2$ was confirmed by mass spectrometry

(m/z = 480.766296 [^{71}Br], 482.765533 [^{81}Br]. p-Nitrophenyl bromoacetate was prepared as follows. 4-Nitrophenol (3 g) was dissolved in 50 ml of dry chloroform containing an equimolar amount of triethylamine and cooled on ice. An equimolar amount of bromoacetyl bromide was added slowly. After leaving overnight, the solution was washed with water, dried and evaporated. The product was obtained from cyclohexane as white crystals. d (CDCl$_3$): 4.18 (s, 2H), 7.35–7.48 (s, 2H), 8.27–8.43 (s, 2H).

Assays of aldehyde dehydrogenase were carried out in 50 mM sodium phosphate buffer, pH 7.4, at 25 °C using a Varian Cary 1 spectrophotometer. Dehydrogenase activity was determined with acetaldehyde and NAD$^+$, both at 1 mM. Esterase activity was determined with p-nitrophenyl acetate (0.1 mM); all assays were corrected for the rate of spontaneous hydrolysis. In general all assays were repeated at least 2 or 3 times and the data were averaged. Modifiers were added in a small volume of acetone or ethanol (which was also added to the appropriate control assays). Two orders of mixing were investigated: (A) the enzyme and modifier were incubated for a set period of time before substrates were added to initiate the assay of remaining activity, and (B) the modifier was added to the enzyme already in the presence of its substrate(s) and the activity was measured after a set time. In labelling the enzyme with 2-bromo-2'-hydroxy-5'-nitroacetanilide, equimolar amounts of enzyme and modifier (0.114 mM) were incubated in 3 ml of 50 mM sodium phosphate buffer, pH 7.4, at 25 °C for 20 minutes. The modified enzyme was isolated by gel filtration through a small column of Bio-Gel P6, eluting with 10 mM sodium phosphate buffer, pH 7.4. Portions of the solution so obtained were added to various different buffers; the UV/visible spectrum was scanned and the pH was measured. The absorbance of the anionic form of the covalently linked nitrophenol group was plotted against pH and 'Enzfitter' was used to compute the best theoretical ionisation curve (Leatherbarrow, 1987).

3. RESULTS AND DISCUSSION

3.1. Interaction of Cytosolic Aldehyde Dehydrogenase with Iodine-Bearing Derivatives of Bromoacetanilide

The two iodine-containing compounds studied here, namely 2-bromo-4'-iodoacetanilide (BIA) and 2-bromo-4'-hydroxy-3',5'-diiodoacetanilide (BHDA), are both very potent inactivators of cytosolic aldehyde dehydrogenase, with the former being the more effective. The results in Table 1 show that in the absence of substrates (either NAD$^+$/acetaldehyde or p-nitrophenyl acetate), a very low concentration of the modifiers causes almost complete enzyme inactivation within one minute. This contrasts with the 5 to 10 hours needed for iodoacetamide to achieve its maximum inactivatory effect (Hempel and Pietruszko, 1981); clearly the enzyme has a very high affinity for the aromatic moiety of the modifiers used here, a point also addressed in our other contribution to this volume. [A similar difference can be noted between acetophenone, which is an inhibitor of the enzyme, and acetone, which is not (Weiner et al., 1982).] Although the modifiers react at different rates, the close similarity of the reactive groups of iodoacetamide, BIA and BHDA lead us to assume that the latter two compounds, like iodoacetamide, specifically modify Cys-302, which is implicated by much other evidence to be the enzyme's catalytically essential nucleophile (see, for example, Kitson et al., 1991).

Bromoacetophenone has been reported to modify both Glu-268 and Cys-302, but reaction with the thiol group apparently only happens after reaction with the carboxyl (Abri-

Table 1. Effect of BIA and BHDA on the esterase and dehydrogenase activity of cytosolic aldehyde dehydrogenase at pH 7.4, 25 °C. Order of mixing (A): enzyme (approx. 0.3 µM) and modifier were mixed and after 1 minute the addition of substrate(s) was made; (B) enzyme and substrate(s) were mixed, modifier was then added, and enzyme activity was recorded after a further 1 minute

| Order of mixing | Modifier and concentration | Residual activity (% of control) | |
		Esterase	Dehydrogenase
A	BIA, 10 µM	0.09	-
A	BIA, 1 µM	1.7	0.52
B	BIA, 1 µM	3.1	3.1
A	BHDA, 10 µM	1.6	3.1
A	BHDA, 1 µM	6.9	8.0
B	BHDA, 10 µM	22	2.7
B	BHDA, 1 µM	90	37

ola et al., 1990). We are a little doubtful about this result as enzyme and bromoacetophenone were routinely incubated for the extremely long (and surely unnecessary) time of 16 to 20 hours before the position of the label was investigated, and yet as Abriola et al. (1990) correctly remark bromoacetophenone is a "highly reactive compound which can form a covalent bond with any nucleophile in its vicinity". Is it possible that there could be equilibration of the label from one residue to the other during the long incubation period? It would have been interesting if these workers had found out if bromoacetophenone actually reacts very quickly with the enzyme (as we find BIA and BHDA to do) and if so, with which enzymic residue.

Other results in Table 1, and also Figures 2 and 3, show the effect of adding BIA or BHDA to an on-going assay, thus illustrating how the presence of substrates does or does not protect the enzyme from inactivation. It is striking that BIA at a concentration of only 1 µM is still a potent inactivator in the presence of a 100-fold excess of *p*-nitrophenyl acetate; Figure 2(a) shows that a considerable loss of activity has occurred within the time of mixing and placing the cuvette in the spectrophotometer and the rate declines further over the next minute, with the enzyme becoming about 97% inactive. Clearly, BIA must have an extremely high affinity for the active site of cytosolic aldehyde dehydrogenase, more so even than disulfiram has, since inactivation by disulfiram is protected against quite effectively by *p*-nitrophenyl acetate (Kitson, 1982). On the other hand, BHDA at 1 µM is not a very efficient inactivator (although better at 10 µM) in the presence of the ester substrate, as seen in Figure 2(b,c). The dehydrogenase substrates (NAD^+ and acetaldehyde, both at 1 mM) do not protect the enzyme at all against 1 µM BIA or 10 µM BHDA, although with 1 µM BHDA it seems that it takes a minute or so for the inactivation to occur under these conditions, as shown in Figure 3(a,b,c). The lack of protection by dehydrogenase substrates against inactivation has been observed before in the case of disulfiram, where it was concluded that the modifier reacts with the enzyme-NADH complex (the dissociation of which is rate-limiting), meaning that the presence of NAD^+ or acetaldehyde is immaterial (Kitson, 1982).

The somewhat lesser effectiveness of BHDA as an inactivator compared to BIA may be because the former's phenol group is likely to be partly ionised at pH 7.4 (the inductive effect of the two closely-situated halogen substituents will greatly reduce the pK_a of the phenol), and it is known that negatively-charged molecules do not bind well to aldehyde dehydrogenase's active site. (Iodoacetamide, for example, is an inactivator, but io-

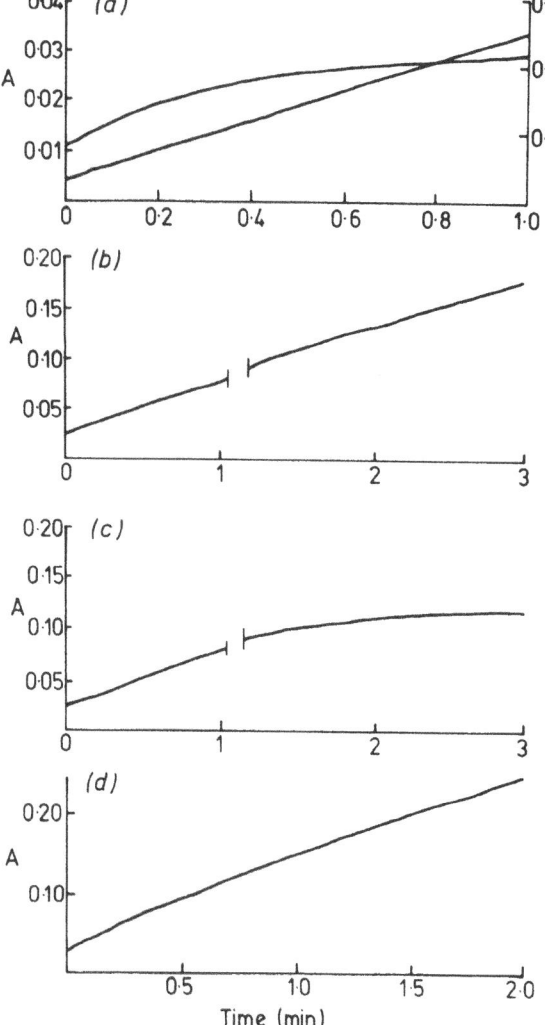

Figure 2. Effect of various bromoacetyl compounds on the esterase activity of cytosolic aldehyde dehydrogenase (approx. 0.3 μM) at pH 7.4, 25 °C, when added to the enzyme already in the presence of substrate (0.1 mM p-nitrophenyl acetate). (a) The linear trace is the control in the absence of modifier, obtained with the absorbance scale on the right. The other trace, with the scale on the left, was obtained with BIA (1 μM), recording the activity as soon as possible after the addition of modifier. In the other diagrams, the break in the trace shows the point at which modifier was added to an on-going assay. (b) BHDA (1 μM). (c) BHDA (10 μM). (d) p-nitrophenyl bromoacetate (10 μM).

doacetate is not; Hempel and Pietruszko, 1981.) Alternatively, the bulkier BHDA may not fit into the active site quite as easily as BIA.

Since the results discussed above are consistent with the idea that BIA and BHDA rapidly and specifically label the enzyme at the active site, it seems likely that they will prove useful in providing heavy-atom derivatives for helping with the X-ray crystallographic solution of aldehyde dehydrogenase's three-dimensional structure. Preliminary studies (Baker, 1996) show that enzyme modified by BIA or BDHA does indeed give very encouraging crystals.

3.2. Interaction between Aldehyde Dehydrogenase and p-Nitrophenyl Bromoacetate

During our work with derivatives of bromoacetanilide, we thought it would be interesting to investigate the interaction between aldehyde dehydrogenase and p-nitrophenyl

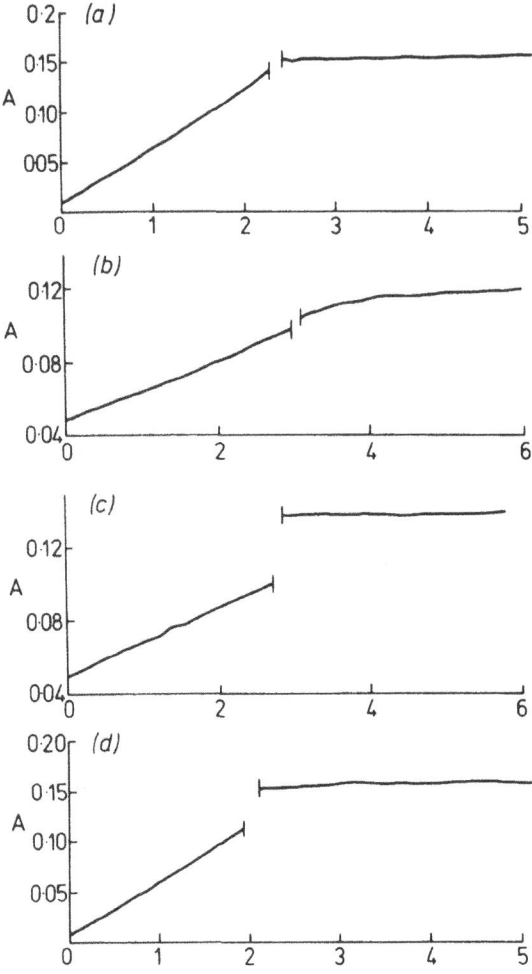

Fig. 3. Effect of various bromoacetyl compounds on the dehydrogenase activity of cytosolic aldehyde dehydrogenase (approx. 0.3 μM) at pH 7.4, 25 °C, when added to the enzyme already in the presence of substrates (1 mM NAD⁺, 1 mM acetaldehyde). The break in the traces shows the point at which modifier was added. (a) BIA (1 μM). (b) BHDA (1 μM). (c) BHDA (10 μM). (d) *p*-nitrophenyl bromoacetate (10 μM).

bromoacetate, since this compound might be expected to be an inactivator (like the structurally related BIA and BHDA), or to be a substrate (like *p*-nitrophenyl acetate), or possibly both. In fact we found that with 10 μM *p*-nitrophenyl bromoacetate there was no detectable enzyme-catalysed hydrolysis whatsoever. (A concentration of 100 μM gave an inconveniently high rate of spontaneous hydrolysis.) However, after 2 minutes' incubation of enzyme and 10 μM *p*-nitrophenyl bromoacetate, the addition of 100 μM *p*-nitrophenyl acetate to assay enzyme activity showed that 97 to 98 % inactivation had occurred. Adding substrate before modifier gave the result shown in Figure 2(d); the curvature of the assay trace shows the substrate partially protects the enzyme against inactivation. Dehydrogenase substrates give little or no protection, as expected; Figure 3(d). We conclude from this investigation that the methylene group in *p*-nitrophenyl bromoacetate must be considerably more electrophilic than the adjacent carbonyl group. It would be interesting to take a similar look at bromoacetaldehyde as a substrate or inactivator.

Table 2. Effect of BHN-1 and BHN-2 on the dehydrogenase activity of cytosolic aldehyde dehydrogenase at pH 7.4, 25 °C. Enzyme (0.37 μM) and modifier were incubated for the time shown and then substrates (NAD^+ and acetaldehyde, both 1 mM) were added and the enzyme activity was recorded

	Residual activity (% of control)	
Modifier	1 minute	10 minutes
[BHN-1], μM		
0.1	96.4	93.0
1	99.2	88.7
10	23.3	4.7
100	16.6	5.1
[BHN-2], μM		
0.1	102	91.8
1	4.3	4.7
10	5.9	6.2
100	5.1	5.1

3.3. Interaction of Cytosolic Aldehyde Dehydrogenase with Nitrophenol-Bearing Derivatives of Bromoacetamide

Just like BIA and BHDA, 2-bromo-2'-hydroxy-5'-nitroacetanilide (BHN-1) and 2-bromo-4'-hydroxy-3'-nitroacetanilide (BHN-2) were found to be rapid, potent inactivators of sheep liver cytosolic aldehyde dehydrogenase. Table 2 shows that incubation with a low concentration of modifier for a short time produces up to about 95% inactivation. (As with disulfiram, there seems to be a low level of resistant activity; Kitson, 1982.) With BHN-1, this is achieved with 10 μM modifier within 10 minutes, whereas with BHN-2 even only 1 μM modifier for 1 minute is sufficient.

BHN-1 and BHN-2 were some of the first reagents designed to supply proteins with covalently-linked reporter-groups. The idea of using a physical technique (such as spectrophotometry, fluorimetry, NMR or ESR spectroscopy, etc.) to probe the environment of a group bound to a macromolecule was originated by Burr and Koshland (1964). BHN-1 (but not BHN-2) has the disadvantage that in solutions of pH above the pK_a of its phenol group it can undergo an intramolecular cyclisation with the loss of bromide ion (Hille and Koshland, 1967) but the rate at which this happens is appreciably slower than the rate at which we find the compound reacts with aldehyde dehydrogenase. BHN-2, on the other hand, has the disadvantage of having a chromophoric group with a much smaller extinction coefficient than BHN-1; as a simple comparison we show in Figure 4 the UV/visible spectra of o-nitrophenol and p-nitrophenol in the ionised and unionised forms. Thus although BHN-2 has been found to be a potent inactivator of the enzyme, our attempts to monitor the absorbance spectrum of the modified enzyme have so far proved disappointing. Much clearer results, however, have been obtained with BHN-1, as will now be described.

After incubation of equimolar amounts of enzyme and BHN-1 at pH 7.4, the modified enzyme was isolated by gel filtration and samples of it were scanned spectrophotometrically over a range of pH, as shown in Figure 5. The pK_a of the covalently linked p-nitrophenol group is 8.2. (The λ_{max} of the ionised form is 420 nm.) Also shown in the Figure are the results of a similar experiment with enzyme modified by 3,4-dihydro-3-

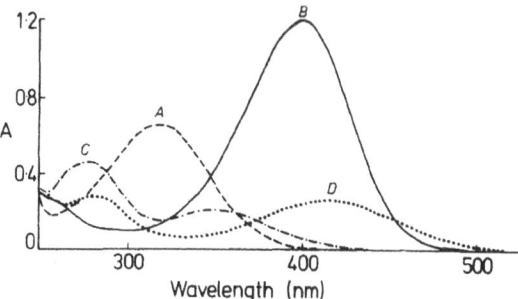

Figure 4. UV/visible spectra of *p*-nitrophenol (65 µM) at pH 3.5 (A) and pH 10.5 (B), and of *o*-nitrophenol (65 µM) at pH 3.5 (C) and pH 10.5 (D).

methyl-6-nitro-2*H*-1,3-benzoxazin-2-one (DMNB); these data are taken from Kitson and Kitson (1994), and show a pK_a of the reporter group of 9.8. (The λ_{max} of the ionised form is 433 nm.) A further difference is that in the presence of NAD^+ the pK_a of the group supplied by BHN-1 remains virtually unchanged (8.1), whereas we observed a dramatic shift in the case of DMNB to a pK_a of about 5.4. With DMNB, the results were discussed in terms of a binding site that is either very non-polar or possibly contains a negatively-charged group (perhaps Glu-268), and the suggestion was made that the reporter group may interact directly with the positively-charged nicotinamide ring of NAD^+ (Kitson and Kitson, 1994). We had expected more closely agreeing results from DMNB and BHN-1, since as shown in Figure 6 the structure of the covalently-linked reporter-groups are very similar. In the absence of any direct evidence, we are reluctant to suggest the unlikely possibility that the two compounds react at different sites on the enzyme. Instead it may be that reporter groups such as these are extremely sensitive to differences in their microenvironment (such as hydrophobicity or the proximity of charged amino acid sidechains), and although the number of atoms in the chain linking the nitrophenol group to the enzyme is the same for both reporter-groups, nevertheless these linkers do have different geometrical structures and may hold the nitrophenol moiety in a significantly different orientation within the active site. As an alternative explanation for the differences observed, we note that DMNB is neutral whereas BHN-1 is likely to be substantially negatively-charged at

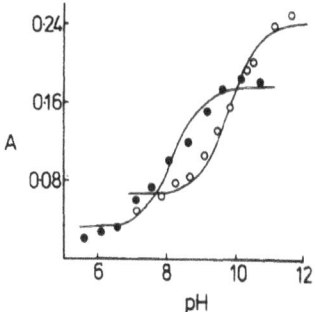

Figure 5. Ionisation curve of the nitrophenol reporter group attached to cytosolic aldehyde dehydrogenase by reaction with BHN-1 (filled circles). Also shown is the ionisation curve of the reporter group attached to the enzyme by reaction with DMNB (open circles) (data from Kitson and Kitson, 1994).

Enzyme modified by DMNB

Figure 6. A comparison of the structures of the reporter groups supplied to cytosolic aldehyde dehydrogenase by DMNB and by BHN-1. It is assumed that the groups are attached to the enzyme through the sulphur atom of Cys-302.

Enzyme modified by BHN-1

pH 7.4. (The pK_a of p-nitrophenol is 7.1; Kitson and Freeman, 1993.) Thus the former reagent may bind and produce its reporter-group directly within the putative hydrophobic pocket that the enzyme may possess (and which we suggest has a great affinity for planar, hydrophobic molecules such as retinal and BIA), whereas BHN-1 may tend to approach the active site 'bromine-first', with its nitrophenol moiety hanging away from the enzyme, in the surrounding aqueous solution. A similar picture could be imagined if the active site contains a negatively-charged group such as Glu-268. This scenario would maintain the idea of both reagents reacting with Cys-302, whilst explaining why only in the case of DMNB is the pK_a perturbed to such a high value. It will be interesting to test this idea in due course when the enzyme's tertiary structure becomes known; indeed, DMNB-modified enzyme may prove useful in the solution of the structure as noted previously (Baker et al., 1994).

4. REFERENCES

Abriola, D.P., MacKerell, A.D. and Pietruszko, R.: Correlation of loss of activity of human aldehyde dehydrogenase with reaction of bromoacetophenone with glutamic acid-268 and cysteine-302 residues. Biochem. J. 266 (1990) 179–187.

Baker, H.M. (1996) personal communication.

Baker, H.M., Blackwell, L.F., Brown, R.L., Buckley, P.D., Dobbs, A.J., Hardman, M.J., Hill, J.P., Kitson, K.E., Kitson, T.M. and Baker, E.N.: Crystallization and preliminary X-ray diffraction studies on cytosolic (class 1) aldehyde dehydrogenase from sheep liver. J. Mol. Biol. 241 (1994) 263–264.

Burr M. and Koshland, D.E.: Use of "reporter groups" in structure-function studies of proteins. Proc. Natl. Acad. Sci. (USA) 52 (1964) 1017–1024.

Hempel, J.D. and Pietruszko, R.: Selective chemical modification of human liver aldehyde dehydrogenases E_1 and E_2 by iodoacetamide. J. Biol. Chem. 256 (1981) 10889–10896.

Hempel, J., Kaiser, R. and Jornvall, H.: Mitochondrial aldehyde dehydrogenase from human liver; primary structure, differences in relation to the cytosolic enzyme, and functional correlations. Eur. J. Biochem. 153 (1985) 13–28.

Hille, M.B. and Koshland, D.E.: The environment of a reporter group at the active site of chymotrypsin. J. Amer. Chem. Soc. 89 (1967) 5945–5951.

Kitson, T.M.: Further studies of the action of disulfiram and 2,2'-dithiodipyridine on the dehydrogenase and esterase activities of sheep liver cytoplasmic aldehyde dehydrogenase. Biochem. J. 203 (1982) 743–754.

Kitson, T.M. and Kitson, K.E.: Probing the active site of cytoplasmic aldehyde dehydrogenase with a chromophoric reporter group. Biochem. J. 300 (1994) 25–30.

Kitson, T.M., Hill, J.P. and Midwinter, G.G.: Identification of a catalytically essential nucleophilic residue in sheep liver cytoplasmic aldehyde dehydrogenase. Biochem. J. 275 (1991) 207–210.

Leatherbarrow, R.J.: Enzfitter Manual, Elsevier Science Publishers, Amsterdam (1987).

Weiner, H., Freytag, S., Fox, J.M. and Hu, J.H.J.: Reversible inhibitors of aldehyde dehydrogenase. Enzymology of Carbonyl Metabolism; Aldehyde Dehydrogenase and Aldo/Keto Reductase (Weiner, H. and Wermuth, B., eds.) Alan R. Liss, New York (1982) 91–102.

28

HUMAN ALDEHYDE DEHYDROGENASE E3

Further Characterization

Regina Pietruszko, Alexandra Kikonyogo, Ming-Kai Chern, and
Gonzalo Izaguirre

Center of Alcohol Studies and Department of Molecular Biology and
 Biochemistry
Rutgers University
Piscataway, New Jersey 08855–0969

1. INTRODUCTION

The human "low Km" aldehyde dehydrogenases, E1, E2 and E3 share many common catalytic features. These include NAD as coenzyme, broad substrate specificity, low micromolar Km values for short chain aliphatic aldehydes, and similar Km values and maximal velocities with substrates such as imidazole acetaldehyde and the metabolites of monoamines like serotonin, dopamine and norepinephrine (Ambroziak and Pietruszko, 1991). All three isozymes also catalyze the dehydrogenation of aminoaldehydes; however, the Km values of the E3 isozyme for these compounds are considerably lower (5 µM for γ-aminobutyraldehyde, Ambroziak and Pietruszko, 1991) than those of E1 and E2 isozymes (Km values for γ-aminobutyraldehyde, 800 µM and 500 µM, respectively, Ambroziak and Pietruszko, 1987). Work on enzymes like E1 and E2 from other species has demonstrated that both E1-like (class 1) and E2-like (class 2) enzymes can catalyze the dehydrogenation of retinal to retinoic acid (Chen et al., 1994). There is good evidence for the involvement of the class 1 enzyme in mammalian eye development (McCaffery et al., 1991; Godbout et al., 1996). This enzyme may also be involved in prostaglandin metabolism (Westlund et al., 1994).

One of the postulated physiological roles of the E3 isozyme is in the metabolism of putrescine to γ-aminobutyric acid, which is the principal inhibitory neurotransmitter in the central nervous system. It can be formed from the diamine, putrescine, by two pathways. In the first pathway, occurring mainly in peripheral tissues, putrescine is converted into γ-aminobutyraldehyde by diamine oxidase, then oxidized to γ-aminobutyric acid by aldehyde dehydrogenase. The second pathway, thought to occur in brain, involves the acetylation of putrescine to N-acetylputrescine, which is converted to N-acetyl-γ-aminobutyraldehyde by monoamine oxidase, and then to N-acetyl γ-aminobutyric acid followed by deacetylation.

Our recent discovery that betaine aldehyde is a substrate of the E3 isozyme (Chern and Pietruszko, 1995) suggests the possibility of another metabolic function for the enzyme. In this paper the importance of this new substrate, its application to tissue distribution, as well as the properties of E3 isozyme from human brain (Kikonyogo and Pietruszko, 1996) are described. The catalytic properties of E3 isozyme are compared with those of the rat liver betaine aldehyde dehydrogenase (Rothschild and Barron, 1954; Goldberg and McCaman, 1968) and the γ-aminobutyraldehyde dehydrogenases from rat (Abe et al., 1990) and bovine (Lee and Cho, 1992) brain and from rat liver (Testore et al., 1995). By assembling all known information about betaine aldehyde dehydrogenases and γ-aminobutyraldehyde dehydrogenases an attempt is being made to gain some insight into the physiological function of the E3 isozyme.

2. MATERIALS AND METHODS

2.1. Aldehyde Substrates

γ-Aminobutyraldehyde was prepared by the acid hydrolysis of fractionally redistilled γ-aminobutyraldehyde diethyl acetal (Aldrich) as described previously (Ambroziak and Pietruszko, 1987). N-acetyl-γ-aminobutyraldehyde was generated from N-acetylputrescine (Sigma) by the action of monoamine oxidase (Kikonyogo and Pietruszko, 1996). Betaine aldehyde was from Sigma.

2.2. Enzyme Assays

Activity with γ-aminobutyraldehyde as substrate was assayed in 100 mM phosphate buffer, pH 7.4, containing 500 μM NAD, 1 mM EDTA and 100 μM γ-aminobutyraldehyde. Sodium pyrophosphate buffer, 100 mM, pH 9.0, containing 500 μM NAD and 1mM EDTA was used for assay with propionaldehyde (1mM) or betaine aldehyde (1mM). All assays carried out spectrophotometrically at 340 nm and 25°C were initiated by addition of enzyme. An extinction coefficient of 6.22 mM^{-1} cm^{-1} for NADH was used for the calculation of reaction rates. Km values were determined by the method of Lineweaver and Burk (1934) or by the single reaction curve method of Yun and Suelter (1977). Protein concentration was determined as described by Goa (1953).

2.3. Cloning and Sequencing of cDNA

Cloning and sequencing of liver and brain cDNA as well as Northern blot analyses are described in detail by Kurys et al, (1993) and Kikonyogo and Pietruszko (1996).

2.4. Enzyme Purification

Details of enzyme purification from human brain are described by Kikonyogo and Pietruszko (1996) and from liver by Chern and Pietruszko (1995).

2.5. Tissue Distribution

Human tissues were obtained from NDRI, Philadelphia, PA and human fetal brains were from the Anatomic Gift Foundation, Woodbine, GA. Tissue distribution of the E3 isozyme was studied by employing an enzyme assay with betaine aldehyde as substrate. A weighed amount of tissue was suspended in a volume equal to 5 x tissue weight of cold 50 mM phosphate buffer, pH 7.0, containing 1mM EDTA and 0.1% v/v 2-mercaptoethanol and homogenized employing Brinkman Polytron Homogenizer. Following extraction for 1h on ice, the cell debris were sedimented by centrifugation. The supernatant was collected and assayed for enzyme activity. Under these conditions, 1g of tissue enzyme activity was represented by 6 ml of centrifuged supernatant.

3. RESULTS AND DISCUSSION

3.1. Activity with Betaine Aldehyde and Content of the E3 Isozyme in Human Liver

The cloning of E3 isozyme cDNA (Kurys et al., 1993) revealed a primary structure which had ca. 40% positional identity with those of the E1 and E2 isozymes. The closest relative was betaine aldehyde dehydrogenase from *E. coli* with ca. 52% positional identity. For this reason it was not altogether surprising that betaine aldehyde was a substrate for the E3 isozyme. The activity with 1 mM betaine aldehyde at pH 9.0 of 8 μmol of NADH formed/min/mg was high when compared with that with 1 mM propionaldehyde (0.6 μmol/min/mg). γ-Aminobutyraldehyde activity (at pH 7.4} of 1.7 μmol/min/mg was also considerably lower than that with betaine aldehyde. Both E1 and E2 isozymes were inactive with betaine aldehyde. Total activity of Caucasian liver E3 isozyme, in terms of NADH formed, when assayed with betaine aldehyde constituted 30% of total liver propionaldehyde activity (Table 1). Thus, activity with betaine aldehyde is large when compared to total aldehyde dehydrogenase activity. Activity of the E3 isozyme with propionaldehyde as substrate (back calculated from the ratio of propionaldehyde to betaine aldehyde activity) corresponded to only ca 70 nmol/g liver/min (Table 1) and constituted only 2.5% of total liver propionaldehyde activity. The calculated E3 protein represented only about 4% of total E1 + E2 protein but has high catalytic activity.

3.2. Metabolic Importance of Choline

Choline is considered by many to be an essential ingredient in the human and animal diet for it cannot be readily made except indirectly (Zeisel, 1981; Ghoshal, 1995). Choline deficiency is rare because choline occurs in abundance in the diet as a component of the phospholipid, lecithin. Once taken up by the cell choline is immediately phosphorylated to

Table 1. Content of aldehyde dehydrogenase E3 in human liver

Enzyme	Activity (nmol NADH/min/g Liver)		Enzyme Protein mg/g Liver
	Propanal	Betaine Aldehyde	
E3	67	800	0.1
E1 + E2 +E3	2660	-	2.5

Assay as described in Materials and Methods, data represent average values from four human livers. Specific activity of E3 isozyme with betaine aldehyde is 8.0 μmol/min/mg; with propionaldehyde is 0.6 μmol/min/mg; and with γ-aminobutyraldehyde (100μM) 1.7μmol/min/mg.

phosphocholine. Some choline is acetylated to acetyl choline and functions in synaptic transmission. Choline is also oxidized to betaine via a betaine aldehyde intermediate. Liver and kidney are the organs where most of the choline metabolism occurs (Zeisel, 1981). In rats and other species a choline deficient diet causes the rapid accumulation of fats in the liver, followed by lipid peroxidation, cell death, necrosis and hepatocellular carcinoma (Ghoshal, 1995).

Betaine has been known for a long time to serve as methyl donor for the biosynthesis of methionine (Shapiro, 1965). Bacteria capable of accumulating high levels of betaine are able to survive in otherwise lethal salt concentrations; higher plants eg. spinach and barley synthesize betaine during osmotic stress (Yancey et al., 1982). It has been found recently that the response to osmotic stress of mammalian tissue culture cells and chick embryo fibroblasts is consistently modified by betaine (Petronini et al., 1992), suggesting that in all living organisms betaine may function as a defense against osmotic stress.

3.3. Tissue Distribution of E3 Isozyme

Since activity with betaine aldehyde in human liver appeared to be totally confined to the E3 isozyme (Chern and Pietruszko, 1995), betaine aldehyde was a substrate of choice to determine E3 isozyme distribution among the individual tissues using an enzyme assay. Tissue distribution of the E3 isozyme is shown in Table 2. The enzyme is present in all tissues tested with the largest amount in the liver, followed by the kidney and the brain, and the smallest amount in fetal brain. Mature human brain contains enzyme activity comparable to that present in the spleen and skeletal muscle. The distribution of betaine aldehyde dehydrogenase in rat tissues reported by Rothschild and Guzman Barron (1954) is in decreasing amounts: liver, kidney, small intestine, heart, spleen and brain, and is similar to that shown in Table 2. Thus, it appears, that tissue distribution of E3 isozyme is similar to the distribution of choline oxidation (Zeisel, 1981) system via which choline is converted to betaine. In the conversion of putrescine to γ-aminobutyric acid the first enzyme is diamine oxidase (EC 1.4.3.6). In the rat, the highest diamine oxidase content was found in the placenta, followed by small intestine (Shaff and Beaven, 1976) while in humans it was highest in the intestinal tract (Hesterberg et al., 1984).

Table 2. Tissue distribution of the E3 isozyme

Tissue	Specific Activity nmol NADH/g tissue/min		
	Mean	Range	Donors
Liver	802	685 - 921	3
Kidney	423	289 - 498	3
Pancreas	128	47 - 193	3
Spleen	91	61- 135	3
Brain	90	87 - 95	2
Muscle	76	55 - 96	2
Heart	54	31 - 77	2
Lung	23	19 - 26	2
Fetal Brain	15	11 - 19	3

Assay contained 1 mM betaine aldehyde, 500 μM NAD, 1 mM EDTA in 0.1M sodium pyrophosphate buffer, pH 9.0.

Table 3. Purification of the enzyme metabolizing g-aminobutyraldehyde, from 500 g of human brain

Purification Step	Total Protein (mg)	Total Activity (nmoles/min)	Specific Activity (nmoles/min/ mg)	Yield Yield (%)
$(NH_4)_2SO_4$ pptn. and dialysis	4600	6490	1.4	100
CM-Sephadex	1100	2660	2.4	40
DEAE-Sephadex	330	1590	4.8	24
5' AMP Sepharose	190	1000	5.3	15
Blue Sepharose	3.5	480	137	7.4
Mono P FPLC	-	-	1400	-

For enzyme activity measurement the assay used contained 0.1 M sodium phosphate buffer, pH 7.4, 1 mM EDTA, 500 µM NAD+ and 500 µM g-aminobutyraldehyde. Protein was determined by the method of Goa (1953).

3.4. Purification of E3 Isozyme from Brain

The presence of the E3 isozyme in the mature human brain has been confirmed by purification. The purification was attempted because of the possibility that the E3 isozyme in brain was different from that in liver. This is because putrescine metabolism in brain proceeds via a different metabolic pathway (see Introduction). The purification steps employed are shown in Table 3. The purification (Kikonyogo and Pietruszko, 1996) followed closely those previously described for rat (Abe et al., 1990) and bovine (Lee and Cho, 1992) brain γ-aminobutyraldehyde dehydrogenases and included ammonium sulfate fractionation. It was found that activity with γ-aminobutyraldehyde as substrate was precipitated by ammonium sulfate between 50–60% saturation. Pure enzyme was obtained after the MonoP FPLC step. The specific activity of the final product (1.4 µmol/min/mg) was similar to that obtained for the E3 isozyme from liver by Kurys et al. (1989). Kinetic properties and substrate specificity (Table 4) were also the same as those of the human liver E3 isozyme. During purification, however, the enzyme appeared more stable than the liver enzyme.

3.5. Cloning of E3 cDNA from Brain

The 2.0 Kb E3 cDNA clone isolated from a human cerebellar library (Stratagene) yielded an amino acid sequence that was, with one exception, identical to that of the liver

Table 4. Brain aldehyde dehydrogenase E3, kinetic constants for different substrates

Substrate	pH 7.4			pH 9.0			EC Number
	Km	V	V/Km	Km	V	V/Km	
Betaine Aldehyde	260	8,000	31	420	13,000	31	EC 1.2.1.8
g-Aminobutyraldehyde	14	2,000	143	-	-	-	EC 1.2.1.19
N-Acetyl-γ-Amino-butyraldehyde	100	-	-	-	-	-	
Acetaldehyde	50	280	5.6	68	470	6.9	EC 1.2.1.3
Propionaldehyde	8	310	39	9.6	560	58	"
Hexanal*	0.6	390	650				"
Glycolaldehyde	220	1,000	4.5				EC 1.2.1.21
NAD	4 - 14						

Km = µM; V = nmol/min/mg;Km for NAD determined in the presence of 300µM γ-aminobutyraldehyde, kinetic constants for other substrates determined in the presence of 500µM NAD. *Values for hexanal are at pH 7.0 and are taken from Ambroziak and Pietruszko (1991).

Figure 1. Sequence alignment of the part of amino terminal 36 residue tryptic peptide that contains serine to cysteine replacement. * Found in liver; **found in brain. The replaced residues are underlined and represent residue 128 (residue number as in E1 isozyme; Kurys et al., 1992).

E3 isozyme. Serine 128 was replaced by a cysteine in the brain sequence. It appears that serine 128 may be tissue specific. While both the tryptic peptide of the E3 isozyme shown in Figure 1 and the amino acid sequence derived from a 1.6 Kb liver E3 cDNA clone (Kurys et al., 1993) contain serine 128, cysteine 128 was found not only in the cerebellar cDNA derived sequence but also in the amino acid sequence derived from a fetal brain cDNA clone (Clontech).

3.6. Distribution of mRNA in Human Brain Areas

Two Multiple Tissue Northern Blots (Clontech) containing mRNA from various regions of the human brain were probed using a 1.6 kb liver E3 cDNA (Kurys et al., 1993). The probe hybridized with mRNA from all the brain regions (Figure 2). Only one band (2.9 Kb) was seen in all lanes demonstrating only one species of mRNA. The order of

Figure 2. Northern blot analysis of mRNA from various human brain regions. Shown are autoradiographs obtained following hybridization of the multiple tissue Northern blots (Clontech) with a radiolabelled 1.6 kb liver E3 cDNA and exposure for 18 h. The positions of the RNA molecular mass standards (in kb) are indicated in each blot. (Reproduced from Kikonyogo and Pietruszko, 1996, with permission).

mRNA distribution starting with the highest levels was: spinal cord, corpus callosum, subthalamic nucleus, medulla, thalamus, amygdala, cerebral cortex, substantia nigra, hippocampus, caudate nucleus, temporal lobe, cerebellum, putamen, frontal lobe, whole brain, occipital pole.

3.7. Some Catalytic Properties of the E3 Isozyme

The kinetic constants of the E3 isozyme with several substrates, some of them known metabolic intermediates, are listed in Table 4. Km values differ from 0.6 μM with hexanal to 260 μM with betaine aldehyde. There are also great differences in maximal velocity. Both Km values and maximal velocities increase at pH 9.0 as compared to pH 7.0. N-acetyl-γ-aminobutyraldehyde is also a substrate with a Km value about one order of magnitude larger than that for γ-aminobutyraldehyde. These constants are the same for the E3 isozyme from brain and liver. In addition to straight chain aliphatic aldehydes, betaine aldehyde and γ-aminobutyraldehyde, the enzyme also catalyses the dehydrogenation of glycolaldehyde. Thus, according to the Enzyme Commission Nomenclature the E3 isozyme represents several classes of specific aldehyde dehydrogenases. These enzyme classes are shown in the last column of Table 4. Glycolaldehyde is also a substrate of the E1 and E2 isozymes.

V/Km comparison shows that γ-aminobutyraldehyde is a better substrate than betaine aldehyde. Although acetaldehyde is a poor substrate (V/Km = 5.6) it also has the potential to inhibit betaine aldehyde metabolism because of its lower Km value. Hexanal, is the best substrate in this group with V/Km ratio of 650.

3.8. Comparison of Properties of E3 Isozyme with Those from Other Species

The properties of the E3 isozyme are compared with those of betaine aldehyde dehydrogenase from rat liver (Rothschild and Guzman Barron, 1954) and γ-aminobutyraldehyde dehydrogenases from rat liver (Testore et al., 1995), rat brain (Abe et al., 1990) and bovine brain (Lee and Cho, 1992) in Table 5. The Km value for the liver or brain E3 isozyme for betaine aldehyde agrees (within experimental error) with that reported by Rothschild and Guzman Barron (1954); there is a complete agreement with that reported by Goldberg and McCaman (1968) who confirmed Rothschild and Guzman Barron (1954) results. There is also an agreement about the instability of the enzyme during dialysis, its cytoplasmic subcellular localization and the narrow range of ammonium sulfate saturation that precipitates the enzyme. It was interesting that Rothschild and Guzman Barron (1954) observed that acetaldehyde was an inhibitor of betaine aldehyde activity. There is some disagreement between the properties of the E3 isozyme and those of other γ-aminobutyraldehyde dehydrogenases. The Km values of the E3 isozyme for γ-aminobutyraldehyde and NAD are lower (by an order of magnitude) than all reported for other γ-aminobutyraldehyde dehydrogenases. This may mean that the human enzyme is different from those from other species.

The Km value for NAD for the E3 isozyme, determined in the presence of several different substrates (acetaldehyde, betaine aldehyde, γ-aminobutyraldehyde) fell within the range shown in Table 5. These results are in disagreement with those of Testore et al., (1995) who found Km values for NAD to vary between 70 - 886 μM, depending on the constant substrate used during Km determination.

Table 5. Comparison of liver betaine aldehyde dehydrogenase and liver
γ-aminobutyraldehyde dehydrogenase with E3 isozyme and
rat and bovine brain γ-aminobutyraldehyde dehydrogenases

	Liver			Brain ABALDH	
Property	Rat BALDH	Rat ABALDH	Human E3	Rat	Bovine
K_m ABAL (µM)	-	110	5 - 14	151	154
K_m BAL (µM)	110	-	140 - 260	-	-
K_m Acetaldehyde (µM)	Inhibitor	111	50	20	625
K_m NAD (µM)	-	70 - 886	4 - 14	87	53
Subcellular Localization	C	C	C*	-	-
Stability on Dialysis	U	U	U	-	-
Precipitates with $(NH_4)_2SO_4$	50 - 60 %	-	50 - 60%	0 - 60%	40 - 75%
Native MW	-	115,000	220,000	210,000	230,000
Subunit MW	-	51,000	54,000	50,000	58,000

Rat BALDH (betaine aldehyde dehydrogenase) = Rothschild and Guzman Barron (1954); rat liver ABALDH (γ-ami-
nobutyraldehyde dehydrogenase) = Testore et al., (1995); rat brain ABALDH = Abe et al., (1990); bovine brain
ABALDH = Lee and Cho (1992); * Ambroziak et al., (1991). ABAL = γ-aminobutyraldehyde; BAL = betaine aldehyde;
C = cytoplasm; U = unstable; % with $(NH_4)_2SO_4$ = % saturation that precipitates the enzyme.

There is a general agreement about activity with acetaldehyde. There is also agreement among liver enzymes about instability to dialysis and cytoplasmic subcellular localization. The molecular properties (where determined) appear to be similar, with the
exception of rat liver γ-aminobutyraldehyde dehydrogenase which is a dimer instead of a
tetramer. Thus, it appears that more work may be necessary before conclusions can be
reached about the identity of the E3 isozyme with other γ-aminobutyraldehyde dehydrogenases.

3.9. Inhibition of Betaine Aldehyde Metabolism by Acetaldehyde

Inhibition of betaine aldehyde metabolism by acetaldehyde was first reported by
Rothschild and Guzman Barron (1954). Acetaldehyde is also a substrate with lower Km
value for the E3 isozyme than betaine aldehyde (Table 4) and an extremely low velocity
(only 3.2% of betaine aldehyde dehydrogenation velocity). Calculations based on the
Michaelis/Menten equation (McKerell et al., 1986) for three concentrations of betaine aldehyde being metabolized in the absence and presence of 100 µM acetaldehyde (the concentration of acetaldehyde in liver during ethanol metabolism, Erikson et al., 1984) are
shown in Table 6. The results show that betaine aldehyde metabolism would be inhibited
by 100 µM acetaldehyde to 34% of the uninhibited control, independent of the betaine aldehyde concentration between 5 - 100 µM. These calculations suggest that ethanol metabolism may have an inhibitory effect on betaine aldehyde metabolism.

3.10. Metabolic Role of E3 Isozyme

Although a vast majority of substrates are common to all three isozymes, betaine aldehyde appears to be a unique substrate of the E3 isozyme. Whether this activity is connected with its metabolic function is at present difficult to answer. One of the reasons is
that choline dehydrogenase, an enzyme that produces betaine aldehyde, is present at low
levels in the human liver mitochondria (Kensler and Langemann, 1954), suggesting that in
human liver choline oxidation would constitute only a minor pathway of choline metabo-

Table 6. Effect of 100 μM acetaldehyde on betaine aldehyde metabolism.
Calculations from Michaelis/Menten equation

Betaine Aldehyde (μM)	μmol NADH/min/mg		% Inhibition
	Betaine Aldehyde	Betaine Aldehyde + 100μM Acetaldehyde	
100	2.2	0.74	66
50	1.3	0.43	67
5*	0.15	0.05	67

*Concentration of betaine aldehyde in rat liver. Acetaldehyde (100μM) metabolism in the presence of 50μM betaine aldehyde would be 84 % of uninhibited rate and in the presence of 5μM betaine aldehyde would be 98%.

lism. Also, the choline oxidation pathway does not appear to be present in the brain or muscle of rats (Zeisel, 1981), which contain betaine aldehyde dehydrogenase (Rothschild and Guzman Barron, 1954). The E3 isozyme is also present in human brain and muscle (Table 2). Thus, it appears that E3 isozyme may perform several metabolic functions which may include a role in choline and putrescine metabolism.

Although some aldehyde dehydrogenase substrates include metabolic intermediates a large number of substrates are derived from the external environment, thus, making them enzymes of intermediary metabolism as well as detoxication. It is suggestive that an additional function for these enzymes is in the adaptation to the environment. This is especially true where endogenous substrates consist of metabolic intermediates of hormones and neurotransmitters. It is possible to visualize that compounds derived from the external environment would compete with natural substrates for metabolism. When natural substrates are hormone metabolites, their accumulation due to competition with an externally derived substrate might result in an altered message to the organism. Thus, for the E3 isozyme successful competition of externally derived substrates such as acetaldehyde with the metabolism of betaine aldehyde could occur. Substrates such as hexanal could easily and successfully compete with not only betaine aldehyde but also with γ-aminobutyraldehyde.

4. ACKNOWLEDGMENTS

This research was supported by PHS Grant 1R01 AA00186 from NIAAA and by Charles and Johanna Busch Memorial Fund.

5. REFERENCES

Abe, T., Takada, K., Okhawa, K., and Matsuda, M.: Purification and characterization of rat brain aldehyde dehydrogenase able to metabolize γ-aminobutyraldehyde to γ-aminobutyric acid. Biochem. J. 269 (1990) 25–29.

Ambroziak, W., and Pietruszko, R.: Human aldehyde dehydrogenase activity with aldehyde metabolites of monoamines, diamines and polyamines. J. Biol. Chem. 266 (1991) 13011–13018. Ambroziak, W., and Pietruszko, R.: Human aldehyde dehydrogenase: metabolism of putrescine and histamine. Alcohol. Clin. Expl. Res. 11 (1987) 528–532.

Ambroziak, W., Kurys, G., and Pietruszko, R.: Aldehyde dehydrogenase (EC 1.2.1.3) comparison of subcellular localization of the third isozyme that dehydrogenates 4-aminobutyraldehyde in rat, guinea pig and human liver. Comp. Biochem. Physiol. 100B (1991) 321–327.

Chen, M., Achkar, C., and Gudas, L.J.:Enzymatic conversion of retinaldehyde to retinoic acid by cloned murine cytosolic and mitochondrial aldehyde dehydrogenases. Molecular Pharmacology 46 (1994) 88–96.

Chern M-K., and Pietruszko, R.: Human aldehyde dehydrogenase E3 isozyme is a betaine aldehyde dehydrogenase. Biochem. Biophys. Res. Commun. 213 (1995) 561–568.

Erikson, C.J.P., Atkinson, N., Petersen, D.R., Deitrich, R.A.: Blood and liver acetaldehyde concentrations during ethanol oxidation in C57 and DBA mice. Biochem. Pharmacol. 33 (1984) 2213–2216.

Ghoshal, A.K.: New insight into the biochemical pathology of liver choline deficiency. Critical Revs. Biochem. Mol. Biol. 30 (1995) 263–273.

Goa, J.: A micro biuret method for protein determination. Determination of total protein in cerebrospinal fluid. Scand. J. Clin. Lab. Invest. 5 (1953) 218–222.

Godbout, R., Packer, M., Poppema, S., and Dabbagh, L.:Localization of the cytosolic aldehyde dehydrogenase in the developing chick retina: *in situ* hybridization and immunohistochemical analysis. Develop. Dynam. 205 (1996) 319–331.

Goldberg, A.M., and McCaman, R.E.: Betaine aldehyde dehydrogenase: assay and partial purification. Biochim. Biophys. Acta 167 (1968) 184–186.

Hesterberg, R., Sattler, J., Lorenz, W., Stahlknecht, C.D., Barth, H., Crombach, M., and Weber, D.: Distribution and metabolism of histamine, histamine content, diamine oxidase activity and histamine methyltransferase activity in human tissues: fact or fictions. Agents and Actions 14 (1984) 325–334.

Kensler, C.J., and Langemann, H.: Metabolism of choline and related compounds by hepatic tissue from several species including man. Proc. Soc. Expl. Biol. Med. 85 (1954) 364–367.

Kikonyogo, A., and Pietruszko, R.: Aldehyde dehydrogenase from adult human brain that dehydrogenates γ-aminobutyraldehyde: purification, characterization, cloning and distribution. Biochem. J. 313 (1996) 317–324.

Kurys, G., Ambroziak, W., and Pietruszko, R.: Human aldehyde dehydrogenase purification and characterization of the third isozyme with low Km for γ-aminobutyraldehyde. J. Biol. Chem. 264 (1989) 4715–4721.

Kurys, G., Shah, P., Kikonyogo, A., Reed, D., Ambroziak, W., and Pietruszko, R.: Human aldehyde dehydrogenase: cDNA cloning and primary structure of the enzyme that catalyzes dehydrogenation of γ-aminobutyraldehyde. Eur. J. Biochem. 218 (1993) 311–320.

Lee, J.E., and Cho, Y.D.:Purification and characterization of bovine brain γ-aminobutyraldehyde dehydrogenase. Biochem. Biophys. Res. Commun. 189 (1992) 450–454.

Lineweaver, H., and Burk, D.: The determination of enzyme dissociation constants. J. Amer. Chem. Soc. 56 (1934) 658–667.

MacKerell, A.D., Jr., Blatter, E.E., and Pietruszko, R.: Human aldehyde dehydrogenase: kinetic identification of the isozyme for which biogenic aldehydes and acetaldehyde compete. Alcohol. Clin. Expl. Res. 10 (1986) 266–270.

McCaffery, P., Tempst, P., Lara, G., and Drager, U.C.: Aldehyde dehydrogenase is a positional marker in the retina. Development 112 (1991) 693–702.

Petronini, P.G., De Angelis, E.M., Borghetti, P., and Borghetti, A.F.: Modulation by betaine of cellular responses to osmotic stress. Biochem. J. 282 (1992) 69–73.

Rothschild, H.A., and Guzman Barron, E.S.: The oxidation of betaine aldehyde by betaine aldehyde dehydrogenase. J. Biol. Chem. 209 (1954) 511–523.

Shaf, R.E., and Beaven, M.A.: Turnover and synthesis of diamine oxidase (DAO) in rat tissues, studies with heparin and cycloheximide. Biochem. Pharmacol. 25 (1976) 1057–1062.

Sessa, A., and Perin, A.: Diamine oxidase in relation to diamine and polyamine metabolism. Agents Actions 43 (1994) 69–77.

Testore, G., Colombato, S., Silvagno, F., and Bedino, S.: Purification and kinetic characterization of γ-aminobutyraldehyde dehydrogenase from rat liver. Int. J. Biochem. Cell Biol. 27 (1995) 1201–1210.

Tsugi, M., and Nakajima, T.: Studies on the formation of γ-aminobutyric acid from putrescine in rat organs and purification of its synthetic enzyme from rat intestine. J. Biochem. (Japan) 83 (1978) 1407–1412.

Westlund, P., Fylling, A.C., Cederlund, E., and Jornvall, H.: 11-Hydroxythromboxane B2 dehydrogenase is identical to cytosolic aldehyde dehydrogenase. FEBS Letters 345 (1994) 99–103.

Yancey, P.H., Clark, M.E., Hand, S.C., Bowlus, R.D., and Somero, G.N.: Living with water stress: evolution of osmolyte systems. Science 217 (1982) 1214–1222.

Yun, S-L., and Suelter, C.H.: A simple method for calculation of Km and V from a single enzyme reaction progress curve. Biochim. Biophys. Acta 480 (1977) 1–13.

Zeisel, S.H.: Dietary choline: biochemistry, physiology, and pharmacology. Ann. Rev. Nutr. 1 (1981) 95–121.

29

HUMAN SUCCINIC SEMIALDEHYDE DEHYDROGENASE

Molecular Cloning and Chromosomal Localization

Flavia Trettel,[1] Patrizia Malaspina,[1] Carla Jodice,[1] Andrea Novelletto,[1]
Clive A. Slaughter,[2] Deborah L. Caudle,[3] Debra D. Hinson,[3]
Ken L. Chambliss,[3] and K. Michael Gibson[3,4*]

[1]Department of Biology
University of Rome "Tor Vergata"
Rome, Italy
[2]Department of Biochemistry
Howard Hughes Medical Institute
University of Texas Southwestern Medical Center
Dallas, Texas
[3]Institute of Metabolic Disease
Baylor Research Institute and Baylor University Medical Center
Dallas, Texas
[4]Department of Neurology
University of Texas Southwestern Medical Center
Dallas, Texas

1. INTRODUCTION

4-Aminobutyric acid (GABA) is an important inhibitor of synaptic transmission (Tillakaratne, Medina-Kauwe and Gibson, 1995) in the mammalian central nervous system (CNS). Although bound forms of GABA contribute to the free GABA pool in CNS, most GABA is derived from glutamic acid in a reaction catalyzed by glutamic acid decarboxylase. The carbon skeleton of GABA eventually enters the Krebs cycle through the sequential action of two enzymes. GABA-transaminase converts GABA to succinic semialdehyde, by the stoichiometric conversion of 2-oxoglutarate to glutamic acid, which replenishes the main GABA precursor. Succinic semialdehyde is then oxidized by NAD^+-*dependent* succinic semialdehyde dehydrogenase (E.C. 1.2.1.24; SSADH) to form suc-

---

*Corresponding author: Baylor Research Institute, 3812 Elm Street, Dallas, Texas 75226. Telephone: 214–820–4749; FAX 214–820–4952.

Enzymology and Molecular Biology of Carbonyl Metabolism 6
edited by Weiner *et al.* Plenum Press, New York, 1996

cinic acid, thus playing a key role in maintenance of GABA homeostasis. In addition, as an inborn error in the metabolism of a neurotransmitter (McKusick 271980), SSADH deficiency represents an unusual neurometabolic disease. SSADH deficiency has been reported in approximately 100 patients ranging from 3 months to 25 years of age at the time of diagnosis (Scriver and Gibson, 1995).

Inherited SSADH deficiency results in intracellular accumulation of succinic semialdehyde, which is reduced by one or more 4-hydroxybutyrate dehydrogenases to 4-hydroxybutyric acid (4-HBA). Detection of elevated urinary 4-HBA, the metabolic marker for the disease, is the first step to achieving the correct differential diagnosis. Although not a neurotransmitter, 4-HBA is a compound that demonstrates unusual properties in the mammalian CNS (Snead, 1996). Hence, the pathophysiology of the SSADH deficiency may evolve from imbalances in two neurogenic compounds. A detailed discussion of SSADH deficiency is beyond the scope of this chapter, and the reader is referred to earlier reviews for detailed discussion (Scriver and Gibson, 1995; Jakobs, Jaeken and Gibson, 1993).

Since the first description of the disease in 1983, approximately 80 SSADH-deficient patients have been reported, and many other unreported cases are known (Gibson, Christensen and Jakobs et al, in press). Consanguinity in many families and equal gender distribution indicate an autosomal recessive inheritance. All patients manifest variable abnormalities of cerebral function, such as psychomotor or language delay, hypotonia, abnormal reflexes, seizures, or ataxia. Although our understanding of SSADH deficiency has continued to advance, diagnosis is generally hampered by the nonspecific clinical findings resulting in the provisional diagnosis of autism, cerebral palsy, Fragile X syndrome, or other forms of idiopathic mental retardation. The correct diagnosis will remain elusive without definitive urinary organic acid screening and diagnostic enzyme analysis.

A major goal of our laboratory is to elucidate the pathophysiology of SSADH deficiency through clinical, metabolic, enzymatic, and molecular genetic analysis. To understand the effect of mutant forms of SSADH, we must first identify the structure and function of the native SSADH gene and protein. In order to analyze the molecular genetics of SSADH deficiency, we isolated cDNAs encoding rat brain and human liver SSADH, and we verified a rat composite cDNA which encoded the mature protein by expression in bacteria (Chambliss, Caudle and Hinson et al, 1995).

This report extends our earlier cDNA and protein data by presenting additional cDNA sequence encoding human SSADH, the amino acid sequence of the mature polypeptide, and the chromosomal localization of the human SSADH gene.

2. METHODS

Unless otherwise stated, all molecular techniques were standard and previously reported (Chambliss, Caudle and Hinson et al, 1995). The construction of the contig covering human chromosome 6p22 has been presented (Malaspina, Roetto and Trettel et al, in press). The contig consisted of 134 yeast artificial chromosome (YAC) clones whose alignment is based upon the presence or absence of 52 DNA markers.

Initially, SSADH from rat and human brain tissue was purified in order to isolate cDNAs (Chambliss and Gibson, 1992). Polyclonal antibodies, raised in rabbit using these proteins as antigens, were employed to examine the subunit structure of SSADH in brain extracts from several mammalian species. Our results indicated a well conserved subunit M_r of approximately 54,000—58,000, consistent with most aldehyde dehydrogenase proteins (Hempel, Nicholas and Lindahl, 1993). Immunoblot analysis of human SSADH-deficient liver, kidney, and brain tissues confirmed a significant decrease in the amount of SSADH protein (Chambliss, Zhang and Rossier et al, 1995).

3. RESULTS

In earlier work, we reported a 1091 bp cDNA encoding human liver SSADH (Chambliss, Caudle and Hinson et al, 1995). This cDNA encoded approximately the C-terminal 2/3 of the human SSADH monomer. Repeated attempts to isolate additional 5_ sequence using PCR, PCR-RACE, and additional library screening were unsuccessful. Similar problems were encountered in isolation of cDNAs encoding 4-aminobutyraldehyde dehydrogenase (Kurys, Shah and Kikonygo et al, 1993). To circumvent difficulties with putative secondary structures in mRNA, we turned our focus to genomic cloning.

Sequence analysis of a YAC contig covering human chromosome 6p22 (in conjunction with PCR amplification) yielded a clone that contained a 244 bp insert, which shared identity with the 1091 bp cDNA encoding human SSADH. Using this clone as a probe against digested human genomic DNA revealed single bands, which was consistent with a single copy sequence covering more than 20 kb. Comparison of this 244 bp sequence with those in GenBank revealed three additional cDNA clones (whose identity was previously unknown) that had sequence homology with human SSADH (Fig. 1) (GenBank ID H46643; H06675; R20294; library constructed by B. Soares and M. Fatima Bonaldo; Wilson, R., the Washington University-Merck EST Project).

Figure 1. Schematic diagram of cDNA clones encoding human SSADH. For comparison purposes, a human liver cDNA (L34820; A) and rat brain cDNA (L34821; E) previously published (Chambliss, Caudle and Hinson et al, 1995) are depicted. Positions in parentheses refer to the original L34820 clone. The remaining positions refer to the composite sequence derived from the different cDNA clones. Position 1 corresponds to the first base in the inserts of both clones R20294 (B) and H46643 (C), and is the first base coding for Trp_{22}. Shaded regions indicate homologous sequences. Abbreviations include: infant brain R20294 cDNA, B; infant brain H06675 cDNA, C; human breast H46643 cDNA, D; TAG, stop codon; A(A/G)TAAA, polyadenylation signal; $(A)_n$, poly (A) tract. See text for details and insert sizes.

The infant brain R20294 cDNA (Fig. 1B) had an insert of approximately 1600 bp and contained the entire L34820 (Fig. 1A) sequence (the original 1091 bp cDNA encoding human SSADH) and two additional stretches: a 432 bp sequence upstream and a 162 bp sequence downstream. The sequences of R20294 and L34820 were identical with the exception of nucleotide 1463 (58 bp downstream of the stop codon), with a Ghe extreme 3' untranslated region revealed a second, probably rare, polyadenylation signal (AGTAAA) immediately followed by a poly (A) tail (Fig. 1C) (Wahle and Keller, 1992). The cDNA depicted in Fig. 1C indicated that this clone derived from the 6.0 kb mRNA encoding SSADH (Chambliss, Caudle and Hinson et al, 1995). Human breast H46643 cDNA (Fig. 1D) had an insert of 666 bp in length, and was identical to the 5' portion of clone R20294 (Fig. 1B), with the exception of position 334 where a Ctranslated regions of the human cDNA for SSADH.

The human cDNA sequences presented in Fig. 1 yielded a 2,323 bp composite clone with an open-reading frame encoding 467 amino acids (Fig. 2). The deduced amino acid sequence included a peptide with the sequence DRRALVLK (amino acids 138—145), which had previously been identified from sequencing purified human brain SSADH but was in a region of the protein for which no previous nucleotide sequence had been known (Chambliss, Caudle and Hinson et al, 1995). When aligned with the 24 N-terminal amino acids, known from sequencing human brain SSADH, the mature SSADH polypeptide contained 488 amino acids with a deduced M_r of 52,329 (rat deduced M_r 52,186) (Fig. 2).

The polypeptide presented in Fig. 2 is a composite of deduced amino acid sequence, amino acid sequences obtained from protein sequencing and amino acids inferred from the deduced rat SSADH polypeptide sequence (Chambliss, Caudle and Hinson et al, 1995). Amino acids deduced from cDNAs begin at residue 22 (Trp; Fig. 2), while the first 21 amino acids were obtained from sequencing of purified human brain SSADH. The nucleo-

AGRLAGLSAALLRTDSFVGGRWLPAAATFPVQDPASGAALGMVADCGVRE	50
ARAAVRAAYEAFCRWREVSAKERSSLLRKWYNLMIQNKDDLARIITAESG	100
KPLKEAHGEILYSAFFLEWFSEEARRVYGDIIHTPAKDRRALVLKQPIGV	150
AAVITPWNFPSAMITRKVGAALAAGCTVVVKPAEDTPFSALALAELASQA	200
GIPSGVYNVIPCSRKNAKEVGEAICTDPLVSKISFTGSTTTGKILLHHAA	250
NSVKRVSMELGGLAPFIVFDSANVDQAVAGAMASKFRNTGQTCVCSNQFL	300
VQRGIHDAFVKAFAEAMKKNLRVGNGFEEGTTQGPLINEKAVEKVEKQVN	350
DAVSKGATVVTGGKRHQLGKNFFEPTLLCNVTQDMLCTHEETFGPLAPVI	400
KFDTEEEAIAIANAADVGLAGYFYSQDPAQIWRVAEQLEVGMVGVNEGLI	450
SSVECPFGGVKQSGLGREGSKYGIDEYLELKYVCYGGL	488

Figure 2. Amino acid sequence of human succinic semialdehyde dehydrogenase. Bold residues represent invariant or highly conserved amino acids identified in 16 aldehyde dehydrogenase proteins (Hempel, Nicholas and Lindahl, 1993). Amino acids deduced from cDNA sequence begin at Trp_{22}, while the first 21 amino acid residues were known from amino acid sequencing of purified human brain SSADH. Gly_{19} and Gly_{20} were inferred from the corresponding rat SSADH amino acid sequence due to ambiguity in amino acid sequencing of purified human brain SSADH. Asterisked amino acids represent conversion of an invariant Gly_{212} to Cys_{212}, and conversion of an almost invariant Lys_{263} to Leu_{263} in human SSADH.

tide sequence encoding these amino acids, and the mitochondrial leader sequence, remain to be identified. Ambiguity in amino acid identification during sequence analysis resulted in Gly_{19} and Gly_{20} being inferred from the corresponding rat SSADH amino acid sequence (Fig. 2). Genomic sequencing will eventually clarify these sequences.

The subunit M_r for human SSADH was consistent with the estimated native M_r of 191,000 (homotetramer) for the human protein using molecular exclusion chromatography of the purified human brain SSADH. The M_r of the deduced polypeptide sequence, however, was somewhat lower than the expected subunit M_r of 58,000, which was determined from SDS—PAGE analysis. We had estimated that the rat SSADH subunit migrated with an estimated M_r of 54,000 (Chambliss and Gibson, 1992). Ryzlak and Pietruszko (1988) provided results which indicated that the human brain SSADH protein was a tetramer comprised of weight nonidentical subunits (M_r 61,000 and 63,000). Our results in rat and human, however, are consistent with a homotetrameric structure. Human and rat cDNAs, and their corresponding amino acid sequence alignment in overlapping regions, revealed 83% and 91% homology, respectively, between the two species.

Alignment of the human SSADH amino acid sequence with sequences from 16 aldehyde dehydrogenase proteins revealed that SSADH maintains the majority of conserved residues (Hempel, Nicholas and Lindahl, 1993). This alignment included strict consensus residues and those maintained at 87—93% identity. In Fig. 2, these residues are presented in bold, with alignment using the ALIGN program. As reported by Hempel and coworkers, 23 invariant residues, including eleven glycines and three prolines, have suggested evolutionary resistance to alteration of peptide chain-bending points. Human SSADH retained 22/23 invariant residues with one exception (Fig. 2). Gly_{212} was conserved in all aldehyde dehydrogenases, whereas in human SSADH this residue was a Cys. This Cys_{212} was maintained in rat SSADH, but the conserved Gly_{212} was detected in bacterial SSADH (Hempel, Nicholas and Lindahl, 1993). In addition, an almost invariant Lys_{263} was Leu in human and rat SSADH, and Asn in bacterial SSADH (see asterisked residues in Fig. 2). Invariant Cys and Glu residues, believed important in catalysis, were maintained in human SSADH at residues 259 and 293, respectively (Fig. 2). Human SSADH maintains an EiFGP sequence at residues 391—395 (Fig. 2), with Thr filling the 392 position (generally Ile in other aldehyde dehydrogenases) in bacterial, rat, and human SSADH.

The chromosomal location of the SSADH gene was determined by hybridizing the radiolabeled 1091 bp human liver cDNA to a Southern blot containing DNA purified from 20 different hamster-human somatic cell hybrids. Comparison of the content of human chromosomes in these hybrid cell lines against cell lines that did not hybridize to the radiolabeled probe, and analysis of concordance, revealed that human SSADH segregated with human chromosome 6 (Table 1). Partial sequencing of YACs and cosmids included in a contig encompassing human chromosome 6p22 verified the presence of SSADH coding sequences, and human SSADH was oriented from telomere to centromere.

4. FUTURE DIRECTIONS AND CONCLUDING REMARKS

A complete understanding of the genomic structure and differential tissue expression of SSADH transcripts will be valuable for mutation screening, in addition to understanding SSADH gene expression. Human and rat tissues differentially express 2.0 and 6.0 kb SSADH transcripts. Our data has revealed that the 6.0 kb message is polyadenylated ~ 0.8 kb downstream from the polyadenylation site of the 2.0 kb message; that both transcripts encode identical open reading frames; and that SSADH is a single copy gene. Therefore, the additional ~

Table 1. Segregation of the human SSADH gene with chromosome 6 in human-hamster somatic cell hybrids

Hybrid Clone	SSADH**	Human Chromosomes*																							
		1	2	3	4	5	6	7	8	9	10	11	12	13	14	15	16	17	18	19	20	21	22	X	Y
756	+	-	-	-	-	+	+	+	-	-	-	-	<	+	+	-	-	-	-	+	+	+	-	-	+
909	+	-	-	-	+	+	+	-	-	-	-	-	-	+	-	-	-	-	-	-	-	-	-	+	-
010	-	-	-	-	-	-	-	-	-	-	+	-	-	-	-	-	-	-	-	-	-	-	-	-	-
016	-	-	-	-	-	-	-	-	-	-	-	-	-	-	+	-	-	-	-	-	-	-	-	-	-
212	-	-	-	-	+	-	-	-	-	-	-	-	-	-	-	-	-	-	-	-	-	-	-	-	+
324	-	-	-	-	-	-	-	-	-	-	-	-	-	-	-	-	+	-	-	-	-	-	-	-	-
423	-	-	-	+	-	-	-	-	-	-	-	-	-	-	-	-	-	-	-	-	-	-	-	-	-
683	-	-	-	-	+	-	-	-	-	-	+	-	+	-	-	-	-	+	-	+	+	-	-		
734	-	-	-	-	+	-	-	+	-	-	-	-	-	-	-	-	-	-	-	-	-	-	-	-	-
750	-	-	-	-	+	-	-	-	-	-	-	+	+	+	-	-	-	+	-	-	-	-	-	-	-
803	-	-	-	<	+	-	+	-	-	-	-	-	-	-	-	-	-	-	-	-	+	+	-		
811	-	-	-	-	-	+	-	-	-	-	-	-	-	-	-	+	+	-	-	-	-	-	-	-	-
852	-	+	-	-	-	-	-	-	-	-	-	-	-	-	-	-	-	-	-	-	-	-	-	-	-
867	-	+	-	-	+	-	-	-	-	-	+	+	-	-	-	+	+	-	-	-	-	-		
937	-	+	-	-	+	-	-	-	-	-	-	+	<	-	<	-	-	-	+	-	-	-		
940	-	-	-	-	+	-	-	-	-	-	-	-	-	-	-	-	-	+	-	-	-	-	-	-	-
1006	-	-	-	+	+	-	+	-	-	-	-	+	-	+	-	-	-	+	-	+	-	-	-		
1049	-	-	-	-	+	-	-	-	-	+	-	-	-	-	-	-	-	+	-	+	-	-	<	
1079	-	-	+	-	+	-	-	-	-	-	-	-	-	-	-	-	-	-	-	-	-	-	-	-
1099	-	+	-	-	+	-	-	-	-	-	-	+	-	-	-	-	+	-	+	+	-	-		

* +, presence of the human chromosome in >30% of the cells; <, presence of the human chromosome in 5-30% of the cells; -, absence of the human chromosome

** +, presence of the SSADH DNA sequence; -, absence of the SSADH DNA sequence

3.2 kb difference between the two transcripts must lie in the, as yet, unisolated 5′ untranslated region (UTR) (Fig. 3). These two messages could have different transcription initiation sites (TIS) and potentially could be regulated by different promoter regions, or they may have a single TIS and promoter region and contain ~ 3.2 kb of differentially spliced non-coding exon(s) in the 5′ UTR between the TIS and coding region (Fig. 3).

The mechanism(s) responsible for the difference in 5′ UTR length between the transcripts is apparently coordinated with the selection of polyadenylation site at the 3′ end of the message. The short 5′ UTR is seen only when the most upstream polyadenylation site is used and the long 5′ UTR is observed only when the most downstream polyadenylation site is used. If the long 5′ UTR were combined with the upstream polyadenylation site, or the short 5′ UTR were combined wit,h the downstream polyadenylation site, the result would be transcripts of 6.0, 5.2, 2.8, and 2.0 kb, which has not been observed (Chambliss, Caudle and Hinson et al, 1995). Complete SSADH genomic sequence information will shed light on the possible mechanisms proposed in Fig. 3. Should genomic characterization provide evidence for different promoters, or we find an SSADH-deficient cell line with decreased or absent SSADH transcripts, an analysis of the promoter elements using a reporter gene would be of interest .

Current work focuses on the genomic structure of human SSADH and mutation screening in cultured cell lines derived from SSADH-deficient patients. SSADH deficiency, like many other inherited genetic diseases, presents with a broad range of phenotypic variability, from mildly to quite severely affected. One objective in determining the molecular basis of the disease is to determine if the nature and predicted severity of the mutation can be correlated with the phenotypic presentation of the disease. A comprehensive analysis of the molecular genetics of the disorder will eventually enable accurate car-

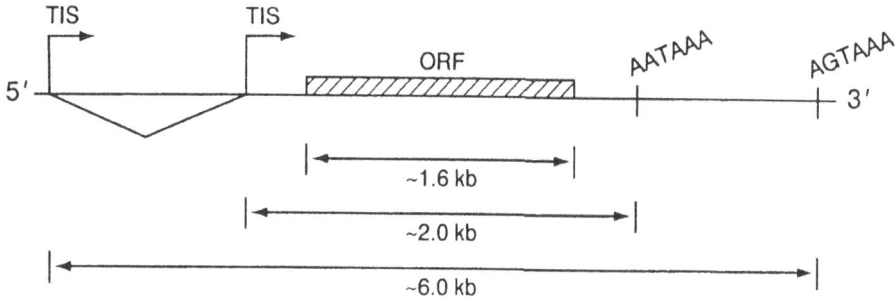

Figure 3. Hypothetical mechanisms controlling SSADH RNA processing. The most 3' TIS element will not exist if the 2.0 kb mRNA is simply the result of alternate splicing in the 5' untranslated region. In addition, more than one (or multiple) TIS elements may exist. The V-shaped line between TIS regions indicates sites of potential alternate splicing. Molecular weights in kb are approximations, and the figure is not drawn to scale. Abbreviations: TIS, transcription initiation site; ORF, open-reading frame (cross-hatched); A(A/G)TAAA, polyadenylation signals.

rier testing with the potential for genetic counseling, molecular diagnostics at the pre- and postnatal level, and increased awareness of the disease by primary care pediatricians, genetic specialists, and neurologists. The latter is important because the nonspecific clinical presentation of SSADH deficiency suggests that the disease is significantly under diagnosed (Gibson, Christensen and Jakobs et al, in press).

In terms of our basic understanding of aldehyde dehydrogenase, nature has provided us with an unfortunate, yet valuable, model system in which to analyze the relationship between structure and function. Molecular genetic analysis in SSADH deficiency will verify the amino acid residues critical to maintenance of viable SSADH protein. The effect of mutations on secondary structure may be modeled by computer analysis to determine predicted structural alterations. Recent research suggests that patients with type II hyperprolinemia may provide a similar model system. Type II hyperprolinemia, first described in 1974 as an inborn error of human metabolism (Valle, Phang and Goodman, 1974), is an autosomal-recessive inherited defect in the proline degradative pathway resulting from Δ^1-pyrroline-5-carboxylate dehydrogenase (P5CDHase) deficiency. P5CDHase is another member of the aldehyde dehydrogenase superfamily. The complete cDNA encoding human P5CDHase was recently published (Hu, Lin and Valle, 1996), which is the first step in beginning a molecular genetic analysis of the disease. Identification of the genetic defects in SSADH and P5CDHase deficiencies may provide novel insights into structure-function relationships in the aldehyde dehydrogenase superfamily of proteins.

5. ACKNOWLEDGMENTS

Supported by the GHB Research Fund, Baylor University Medical Center and Baylor Health Care System Foundation and E.U. Biomed GENE-CT93–0075 and GENE-CT93–0101.

6. REFERENCES

Chambliss, K.L. and Gibson, K.M.: Succinic semialdhyde dehydrogenase from mammalian brain: subunit analysis using polyclonal antiserum. Int.J.Biochem. 24 (1992) 1493—1499.

Chambliss, K.L., Caudle, D.L., Hinson, D.D., Moomaw, C.R., Slaughter, C.A., Jakobs, C. and Gibson, K.M.: Molecular cloning of the mature NAD$^+$-dependent succinic semialdehyde dehydrogenase from rat and human: cDNA isolation, evolutionary homology, and tissue expression. J.Biol.Chem. 270 (1995) 461—467.

Chambliss, K.L., Zhang, Y.-A., Rossier, E., Vollmer, B. and Gibson, K.M.: Enzymatic and immunologic identification of succinic semialdehyde dehydrogenase in rat and human neural and nonneural tissues. J.Neurochem. 65 (1995) 851—855.

Gibson, K.M., Christensen, E., Jakobs, C., Fowler, B., Clarke, M.A., Wallace, G., Hammersen, G., Raab, K., Kobori, J., Moosa, A., Vollmer, B., Rossier, E., Iafolla, A.K., Matern, D., Brouwer, O.F., Hofman, K., Aksu, F., Weber, H.-P., Bakkeren, J.A.J.M., Gabreels, F., Bluestone, D., Barron, T.F., Beauvais, P., Rabier, D., Santos, C., Umansky, R. and Lehnert, W.: The clinical phenotype of succinic semialdehyde dehydrogenase deficiency (4-hydroxybutyric aciduria): case reports of 23 new patients, in press.

Hempel J, Nicholas H. and Lindahl, R.: Aldehyde dehydrogenases: widespread structural and functional diversity within a shared framework. Protein Sci. 2 (1993) 1890—1900.

Hu, C.A., Lin W.-W. and Valle, D.: Cloning, characterization, and expression of cDNAs encloding human Δ^1-pyrroline-5-carboxylate dehydrogenase. J.Biol.Chem. 271 (1996) 9795—9800.

Jakobs, C., Jaeken, J. and Gibson, K.M.: Inherited disorders of GABA metabolism. J.Inher.Metab.Dis. 16 (1993) 704—715.

Kurys, G., Shah, P.C., Kikonygo, A., Reed, D., Ambroziak, W., Pietruszko, R.: Human aldehyde dehydrogenase: cDNA cloning and primary structure of the enzyme that catalyzes dehydrogenation of 4-aminobutyraldehyde. Eur.J.Biochem. 218 (1993) 311—320.

Malaspina, P., Roetto, A., Trettel, F., Jodice, C., Blasi, P., Frontali, M., Carella, M., Franco, B., Camaschella, C. and Novelletto, A.: Construction of YAC contig covering human chromosome 6p22. in press.

Ryzlak, M.T. and Pietruszko, R.: Human brain "high K$_m$" aldehyde dehydrogenase: purification, characterization, and identification as NAD$^+$-dependent succinic semialdehyde dehydrogenase. Arch.Biochem.Biophys. 266 (1988) 386—396.

Scriver, C.R. and Gibson, K.M.: Disorders of β- and γ- amino acids in free and peptide-linked forms. In Scriver, C.R., Beaudet, A.L., Sly, W.S. and Valle, D. (Eds.), The Metabolic and Molecular Basis of Inherited Disease. 7th ed. McGraw-Hill, N.Y., 1995, pp. 1349—1368.

Snead, O.C.: Antiabsence seizure activity of specific GABA (B) and gamma-hydroxybutyric acid receptor antagonists. Pharmacol.Biochem.Behav. 53 (1996) 73—79.

Tillakaratne, N.J.K., Medina-Kauwe, L. and Gibson, K.M.: Gamma-amniobutyric acid (GABA) metabolism in mammalian neural and nonneural tissues. Comp.Biochem.Physiol. 112A (1995) 247—263.

Valle, D.L., Phang, J.M., and Goodman, S.I.: Type 2 hyperprolinemia: absence of Δ^1-pyrroline-5-carboxylic acid dehydrogenase activity. Science 185 (1974) 1053—1054.

Wahle E., Keller W.: The biochemistry of 3'-end cleavage and polyadenylation of messenger RNA precursors. Annu.Rev.Biochem. 61 (1992) 419—440.

30

EFFECTS OF GLYCEROL ON THE KINETIC PROPERTIES OF BETAINE ALDEHYDE DEHYDROGENASE

Rosario A. Muñoz-Clares, Martina Vojtechová, Carlos Mújica-Jiménez, and Rogelio Rodríguez-Sotres

Department of Biochemistry
Faculty of Chemistry
National Autonomous University of Mexico
Mexico City, 04510, Mexico

1. INTRODUCTION

A general response to osmotic stress, observed in a wide range of organisms from bacteria to animals, is the synthesis and accumulation in the cytoplasm of neutral organic compounds known as compatible solutes or osmoprotectants (Yancey et al., 1982, Hanson and Hitz, 1982). These molecules prevent damage from cellular dehydration by balancing the osmotic potential of the cytoplasm with that of the environment and by stabilizing protein and membrane structures (Arakawa and Timasheff, 1985). In plants, the final, irreversible step in the biosynthetic pathway of the osmoprotectant glycine betaine is catalyzed in the chloroplasts by betaine aldehyde dehydrogenase (BADH, EC 1.2.1.8) (Weigel et al., 1986). The protein and activity levels of this enzyme increase in response to osmotic stress (Weretilnyk and Hanson, 1989, Arakawa et al., 1990, Arakawa et al., 1992, Valenzuela-Soto and Muñoz-Clares, 1994). We have recently found that the amaranth BADH also catalyzes the NAD$^+$-dependent oxidation of other small aldehydes with a positive charge, such as 3-dimethylsulfoniopropionaldehyde and γ-aminobutyraldehyde (Vojtechova et al., submitted). Therefore, this enzyme may be also involved in the synthesis of the osmoprotectants 3-dimethylsulfoniopropionate and γ-aminobutyrate.

The kinetics of amaranth BADH share some features with those of other aldehyde dehydrogenases, but present some interesting differences. Thus, the kinetic mechanism is ordered with NAD$^+$ as the leading substrate as it is in most aldehyde dehydrogenases, but amaranth BADH is the only one described to date to have an iso mechanism in which there are two forms of the free enzyme (Valenzuela-Soto and Muñoz-Clares, 1993). It is subjected to substrate inhibition by the aldehyde, which is also a common feature of aldehyde dehydrogenases (Jakoby, 1963), but exhibit substrate activation by the substrate nucleotide (Vojtechova et al., submitted) instead of the well known activation by the

Enzymology and Molecular Biology of Carbonyl Metabolism 6
edited by Weiner *et al.* Plenum Press, New York, 1996

aldehyde exhibited by other aldehyde dehydrogenases (MacGibbon et al., 1977, Green-field and Pietruszko, 1977, Hart and Dickinson, 1982). Finally, contrary to most aldehyde dehydrogenases, amaranth BADH is inhibited by the acid product of the reaction, glycine betaine (Muñoz-Clares et al., submitted).

Kinetic results obtained *in vitro* using highly diluted enzyme solutions has been subject of criticism, as these conditions are very far from the *in vivo* ones (Fulton, 1982). The total concentration of stromal proteins is estimated at 40–50% (w/w) (Ellis, 1979) and it is known that the kinetics properties of many enzymes change when in concentrated solutions due to homologous and heterologous protein-protein interactions (Bosca et al., 1985, Stamatakis et al., 1988). There are practical difficulties in assaying high concentrated enzymes, and because of that it is a common practice to perform the assays in the presence of high concentrations of glycerol and other polyhydric alcohols. These compounds are known to be excluded from the protein surface (Timasheff, 1993), thus favoring protein-protein interactions and mimicking the low water potential characteristic of the intracellular medium of a cell under osmotic stress. In this work, we examined the effect of high concentrations of glycerol on the steady-state kinetics of amaranth BADH with the aim of studying its kinetics properties under conditions closer to those prevailing *in vivo.*

2. EXPERIMENTAL PROCEDURES

Amaranth BADH was purified as described (Valenzuela-Soto and Muñoz-Clares, 1994). For steady-state studies, assays were conducted at 30 °C in HEPES-KOH buffer, pH 8.0, containing 1 mM EDTA and the indicated concentrations of substrates and products, in the absence or presence of 20% (v:v) glycerol. Absorbance increases at 340 nm were followed in a Beckman DU 7500 spectrophotometer. The pH studies were performed in a three buffer system consisting of CHES, HEPES, and MES, 33 mM each, titrated to the desired pHs. For the pH range from 9 to 11 a CHES-bicarbonate buffer, 50 mM each, was used. The molecular weight of the enzyme in the presence and absence of glycerol was estimated by HPLC (Waters) with a TSKG3000SW column. Kinetic data were analyzed using the nonlinear regression analysis program Microcal Origin.

3. RESULTS

3.1. Initial Velocity Patterns

In the presence of 20% (v:v) glycerol, the primary double reciprocal plots for the two substrates, in a concentration range from 25 to 500 μM, were linear and intersecting at the left of the $1/v$ axis, indicating a sequential mechanism as in the absence of glycerol (Valenzuela-Soto and Muñoz-Clares, 1993). The kinetic parameters obtained were similar to those found in the absence of the cosolute, except for $Vmax$ that was around 4-fold lower.

3.2. Substrate Inhibition

The addition of glycerol to the assay medium greatly affected the kinetics of saturation by NAD^+, decreasing the app$Vmax$ about 2-fold, and favoring substrate inhibition by the nucleotide (Fig. 1A). This inhibition was total, indicating that it is not caused by a

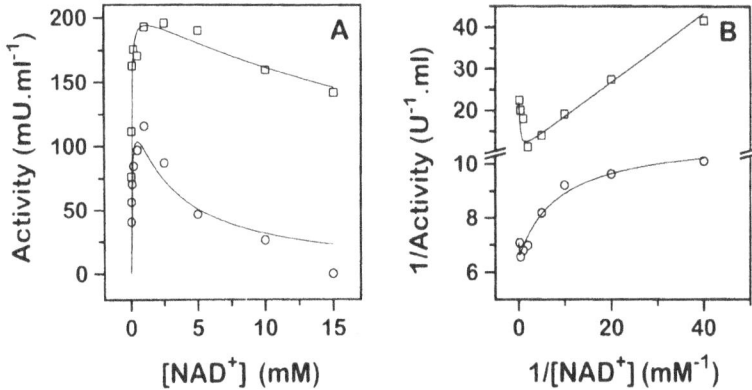

Figure 1. Saturation of amaranth BADH by NAD⁺. (A) Plot of initial velocity versus [NAD⁺] at 25 µM betaine aldehyde in the absence (□)or presence (O) of 20% (v:v) glycerol. The points are the experimental data, the lines are the best fit to the equation for total substrate inhibition. (B) Double reciprocal plots of $1/v$ versus $1/$[NAD⁺] at 0.025 (□) and 0.5 (O) mM betaine aldehyde. Assays were performed at 30 ºC, pH 8.0.

slower alternative route of reaction that takes place at high NAD⁺ concentrations. When the assays were performed at low concentration of betaine aldehyde the inhibition by NAD⁺ appeared at lower nucleotide concentrations (Fig. 1B). In these experiments downward curved double reciprocal plots of $1/v$ against $1/$[NAD⁺] were also observed, indicative of the alternative route of reaction that takes place at high betaine aldehyde and NAD⁺ concentrations (Vojtechova et al., submitted). It was found at only 0.1 mM betaine aldehyde and NAD⁺ concentrations above 0.1 mM, in contrast to the high concentrations of both substrates (above 1 mM) required to establish the alternative route of reaction in the absence of the cosolute (Vojtechova et al., submitted).

The kinetics of saturation by betaine aldehyde in the presence glycerol showed also the reduction in app$Vmax$, but the enzyme appears to be less sensitive to inhibition by high concentrations of betaine aldehyde (Fig. 2).

3.3. Product Inhibition

In the presence of glycerol, NADH produced a curvature in the double reciprocal plots of $1/v$ against $1/$[NAD⁺] (Fig. 3A), which suggests that NADH induced the establishment of an alternative reaction pathway at low concentrations of both substrates. These plots were totally linear when the same experiment was performed in the presence of 10 mM glycine betaine (Fig. 3B).

Glycine betaine inhibited the reaction by the same mechanism as in the absence of the cosolute. In both cases, the experimental data gave a good fit to the equation derived for product inhibition in a mechanism with random release of products (Muñoz-Clares et al., submitted). However, the partial reversion of inhibition that was consistently found in the studies in the absence of cosolute (Muñoz-Clares et al., submitted), was absent in its presence (Fig. 4A). Inhibition by added glycine betaine up to 0.1 mM with respect to NAD⁺ was partial uncompetitive (Fig. 4B), which means that the levels of the complex E'-glycine betaine are not kinetically significant to produce slope effects on the kinetics of

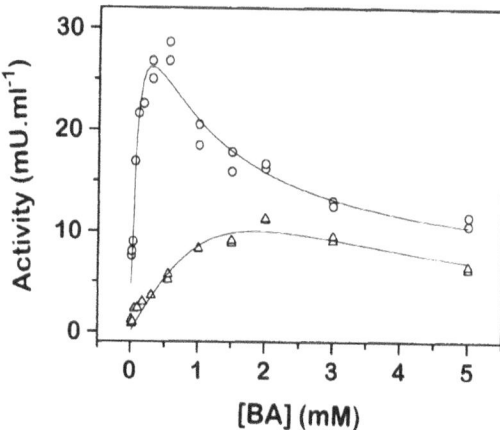

Figure 2. Saturation of amaranth BADH by betaine aldehyde in the absence (O) and presence (Δ) of 20% (v:v) glycerol. Plot of initial velocity versus [betaine aldehyde] at fixed 100 μM NAD$^+$. Assay conditions were as in Fig. 1. The points are the experimental data, the lines are the best fit to the equation for productive substrate inhibition.

saturation by NAD$^+$. Total noncompetitive inhibition was observed at higher glycine betaine concentrations (not shown).

3.4. pH Studies

Both, in the absence and presence of glycerol, we found "wave-shaped" profiles of *Vmax* versus pH, in which *Vmax* decreased to a constant lower value upon protonation of a residue of pK of 8.03 ± 0.11 and 9.49 ± 0.09, respectively (Fig. 5A). The pH of the assay also has a clear effect on the substrate inhibition by NAD$^+$, which was observed at lower concentrations of this substrate as the pH decreases (Fig. 5B).

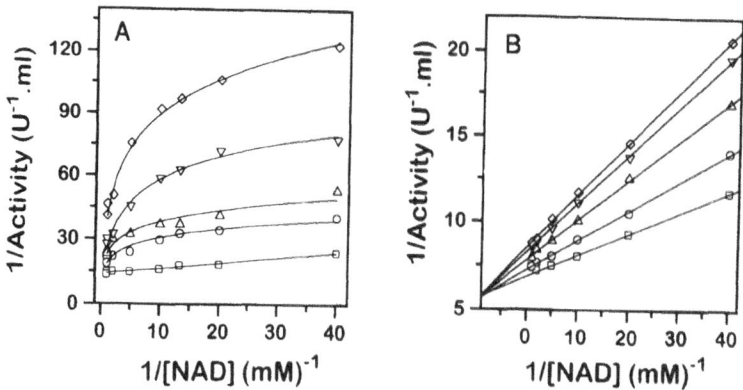

Figure 3. Inhibition of amaranth BADH by NADH in the presence of 20% (v:v) glycerol. (A) Double reciprocal plots of $1/v$ versus $1/NAD^+$ at 0 (□), 25 (O), 50 (Δ), 75 (▽) and 100 (◇) μM NADH. (B) Double reciprocal plots of $1/v$ versus $1/[NAD^+]$ at the same fixed NADH concentrations that in panel A, but in the presence of 10 mM glycine betaine. [Betaine aldehyde] was 0.1 mM. Others conditions as in Fig 1.

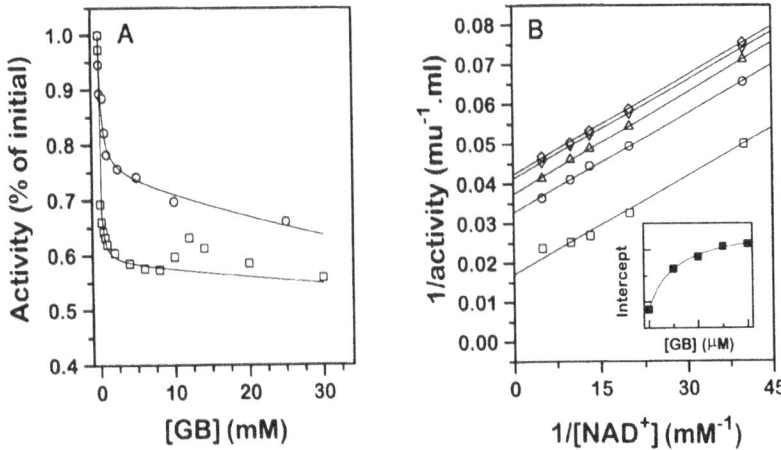

Figure 4. Inhibition of amaranth BADH by glycine betaine. (A) Plot of initial velocity versus [glycine betaine] at fixed 0.2 mM NAD⁺ and 0.1 mM betaine aldehyde in the absence (□) or presence (○) of 20% (v:v) glycerol. Assay conditions were as in Fig 1. The points are the experimental data, the lines are the best fit to $v = v_0 (1 + a\,[GB]) / (1 + b\,[GB] + c\,[GB]^2)$. (B) Double reciprocal plots of $1/v$ versus $1/[NAD^+]$ at 0 (□), 25 (○), 50 (△), 75 (▽), and 100 (◇) mM glycine betaine. Others conditions as in Fig 1.

3.5. Aggregation State

The dimeric nature of amaranth BADH (Valenzuela-Soto and Muñoz-Clares, 1993) was not changed by 20% glycerol, as assessed by exclusion chromatography (data not shown). Therefore, the observed effects of glycerol on the kinetics of the enzyme cannot be ascribed to a higher degree of association, as it was in other enzymes (Bosca et al., 1985, Stamatakis et al., 1988).

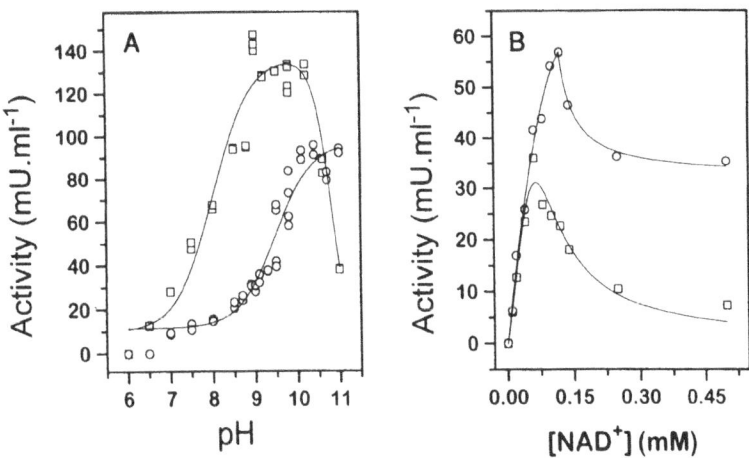

Figure 5. pH-Dependence of the effects of glycerol on the kinetics of amaranth BADH. (A) pH-profiles of $Vmax$ in the absence (□) and presence (○) of 20% (v:v) glycerol. The points are the experimental data, the lines are the best fit to $v = (v_L + v_H\,10^{(pKa1-pH)}) / (1 + 10^{(pKa1-pH)} + 10^{2\,(pH-pKa2)})$. Assay conditions were described in Experimental Procedures. (B) Substrate inhibition by NAD⁺ at pH 7.0 (□), and 9.0 (○). Others conditions as in Fig. 1A.

4. DISCUSSION

Glycerol induced important changes in the reaction catalyzed by amaranth BADH, the most significant being: (i) the appearance of substrate inhibition by the nucleotide at relatively low concentrations, (ii) the diversion of part of the reaction flux through an alternative reaction pathway at low concentrations of substrates and NADH, and (iii) the decrease in *Vmax*.

Substrate inhibition by the nucleotide of a reaction catalyzed by an aldehyde dehydrogenase is extremely rare. To our knowledge it has been only found in glutamic γ-semialdehyde dehydrogenase from human liver (Forte-McRobbie and Pietruszko, 1986). Given the iso mechanism with ordered addition of substrates followed by amaranth BADH (Valenzuela-Soto and Muñoz-Clares, 1993), NAD^+ can form two possible inhibitory complexes with the enzyme: E'-glycine betaine-NAD^+ and E'-NAD^+. Formation of the former requires kinetically significant levels of the complex E'-glycine betaine, which are not observed even when glycine betaine up to 0.1 mM is added to the reaction medium. Therefore, the most likely explanation for the NAD^+ inhibition is the combination of this substrate with E', the free enzyme form that normally binds NADH. It seems that betaine aldehyde cannot add to this inhibitory complex and so NAD^+ forms a true dead-end complex and produces total inhibition. The increased substrate inhibition by NAD^+ found in the presence of glycerol may be due either to increased steady-state levels of E' in relation to the other enzyme forms, to increased affinity of NAD^+ for E', or to both.

Curved double reciprocal plots are indicative of mechanisms with random addition of substrates (Cleland, 1963). In the case of amaranth BADH, we recently found that a slower alternative reaction pathway takes place at high betaine aldehyde concentrations by formation of the complex E'-NADH-betaine aldehyde, which leads to the E'-betaine aldehyde by release of NADH and finally to the central productive complex after combination with NAD^+ (Vojtechova et al., submitted). The establishment of the alternative route of substrate addition caused by NADH at much lower concentrations of NAD^+ than in the absence of glycerol (Fig. 3A) is also consistent with increased levels of E' and/or increased affinity of NAD^+ for the E' related forms, such as E'.BA. The finding that glycine betaine abolishes the NADH-induced nonlinearity of the double reciprocal plots support the conclusion that the operation of this alternative route of reaction requires the formation of the E'-NADH-betaine aldehyde complex, as in the absence of glycerol. Glycine betaine competes with betaine aldehyde for E'-NADH impeding the formation of the ternary complex with the aldehyde. The lesser sensitivity exhibited by the enzyme to inhibition by the aldehyde in the presence of glycerol may also be the result of the ready establishment of the alternative sequence of reaction, rather than of a lesser formation of the inhibitory complexes E'-NADH-betaine aldehyde and E'-betaine aldehyde.

Inhibition by both substrates can take place simultaneously. Thus, it can be observed in Fig. 1B that when the concentration of NAD^+ increases above 2.5 mM, inhibition by this substrate appeared in a reaction which was already occurring in part through the alternative reaction route and thus inhibited by the aldehyde.

The effect of glycerol on the pH profile of *Vmax* at first glance suggests that the cosolute markedly shifted the pK of the ionisable group involved in the rate-limiting step of the reaction, which may be due to a solvent perturbation effect (Cleland, 1990), to a cosolute-induced conformational change, or to both. Alternatively, glycerol may change the rate-limiting step of the reaction, and so the pK observed may correspond to a different ionisable group. The latter explanation is not unlikely, given that in the presence of this cosolute another reaction pathway may operate. Whatever the case, these results accounts for the decrease in *Vmax* observed in the presence of the cosolute. Glycerol also seems to affect the p*K* of the ionisable group(s) involved either in the step of isomerization of the free enzyme and/or in the binding of NAD^+ to the E' form, as indicated by the finding that substrate inhibition by NAD^+ appears at lower concentrations at pH 7.0 than 9.0. At pH 7.0 and low betaine aldehyde concentration inhibition by NAD^+ takes place at the very low concentration of 80 µM.

The pH-dependence of the observed effects of the cosolute on the kinetics of amaranth BADH described in this paper are of especial relevance, since osmotic stress decreases the stromal pH (Hartung et al., 1981) and under water stress the levels of NAD^+ in leaves increased up to 60 µM (Zagdanska, 1989), while those of NADH decreases. This, toghether wit the low intracellular levels of betaine aldehyde in leaves (2–20 nmol/g fresh weight, Lerma et al., 1991), may result in significant inhibition by NAD^+. Therefore, if the *in vivo* behavior of amaranth BADH resembles that observed in the presence of glycerol, the inhibition by NAD^+ may play an important role in the regulation of the activity of this enzyme.

5. ACKNOWLEDGMENTS

We are grateful to Consejo Nacional de Ciencia y Tecnología for financial support (grant CONACYT 1713-N9209).

6. REFERENCES

Arakawa, T. and Timasheff, S.N.: The stabilization of proteins by osmolytes. Biophys. J. 47 (1985) 411–414.

Arakawa, K., Katayama, M. and Takabe, T.: Levels of betaine and betaine aldehyde dehydrogenase activity in the green leaves , and etiolated leaves and roots of barley. Plant Cell Physiol. 31 (1990) 797–803.

Arakawa, K., Mizuno, K. Kishitani, S. and Takabe, T.: Immunological studies of betaine aldehyde dehydrogenase in barley. Plant Cell Physiol. 33 (1992) 833–840.

Blackwell, L.F., Motion, R.L., MacGibbon, A.K.H., Hardman, M.J. and Buckley, P.D.: Evidence that the slow conformation change controlling NADH release from the enzyme is rate-limiting during the oxidation of propionaldehyde by aldehyde dehydrogenase. Biochem. J. 242 (1987) 803–808.

Bosca, L., Aragon, J.J. and Sols, A.: Modulation of muscle phosphofructokinase at physiological concentration of enzymes. J. Biol. Chem. 260 (1985) 2100–2107.

Cleland, W.W.: The kinetics of enzyme-catalyzed reactions with two or more substrates or products. Biochim. Biophys. Acta 67 (1963) 104–137.

Cleland, W.W.: Steady-state kinetics. The enzymes 19 (1990) 99–158.

Ellis, R.J.: The most abundant protein in the world. Trends Biochem. Sci. 4 (1979) 241–244.

Forte-McRobbie, C.M. and Pietruszko, R.: Purification and characterization of human liver "high *Km*" aldehyde dehydrogenase and its identification as glutamic γ-semialdehyde dehydrogenase. J. Biol. Chem. 261 (1986) 2154–2163.

Fulton, A.B.: How crowded is the cytoplasm? Cell 30 (1982) 345–347.

Greenfield, N.J. and Pietruszko, R.: Two aldehyde dehydrogenases from human liver. Isolation via affinity chromatography and characterization of the isozymes. Biochim. Biophys. Acta 483 (1977) 35–45.

Hanson, A.D. and Hitz, W.D.: Metabolic responses of mesophytes to plant water deficits. Annu. Rev. Plant Physiol. 33 (1982) 163–203

Hart, G.J. and Dickinson, F.M.: Kinetic properties of highly purified preparations of sheep liver cytoplasmic aldehyde dehydrogenase. Biochem. J. 203 (1982) 617–627.

Hartung, W., Hilmann, B. and Gimpler, H.: Do chloroplasts play a role in Abscisic acid synthesis?. Plant Sci. Letters 22 (1981) 235–242.

Jakoby, W. B.: Aldehyde dehydrogenase. The Enzymes 7 (1963), 203–221.

Lerma, C., Rich, P.J., Ju, G.C., Yang, W.-Y., Hanson, A.D. and Rhodes D.: Betaine deficiency in maize. Complementation test and mtabolic basis. Plant Physiol. 95 (1991) 1113–1119.

MacGibbon, A.K.H., Blackwell, L.F. and Buckley, P.D.: Kinetics of sheep-liver cytoplasmic aldehyde dehydrogenase. Eur. J. Biochem. 77 (1977) 93–100.

Stamatakis, K., Gavalas, N.A. and Manetas Y.: Organic cosolutes increase the catalytic efficiency of phosphoenolpyruvate carboxylase from *Cynodon dactylon* (l.) Per., apparently through self-association of the enzymic protein. Aust. J. Plant Physiol. 15 (1988) 621–631.

Timasheff, S.N.: The control of protein stability and association by weak interactions with water: How do solvents affect these processes?. Annu. Rev. Biophys. Biomol. Struct. 22 (1993) 67–97.

Valenzuela-Soto, E.M. and Muñoz-Clares, R.A.: Betaine aldehyde dehydrogenase from leaves of *Amaranthus hypochondriacus* L. exhibits an iso ordered bi bi steady state mechanism J. Biol. Chem. 268 (1993) 23818–23823, and Additions and Corrections, J. Biol. Chem. 269, 4692.

Valenzuela-Soto, E.M. and Muñoz-Clares, R.A.: Purification and properties of betaine aldehyde dehydrogenase extracted from detached leaves of *Amaranthus hypochondriacus* L. subjected to water deficit. J. Plant Physiol. 143 (1994) 145–152.

Weigel, P., Weretilnyk, E.A. and Hanson, A.D.: Betaine aldehyde oxidation by spinach chloroplasts. Plant Physiol. 82 (1986) 753–759.

Weretilnyk, E.A. and Hanson, A.D.: Betaine aldehyde dehydrogenase from spinach leaves: Purification, *in vitro* translation of the mRNA, and regulation by salinity. Arch. Biochem. Biophys. 271 (1989) 56–63.

Yancey, P.H., Clark, M.E., Hand, S.C., Bowlus, R.D. and Somero, G.N.: Living with water stress: evolution of osmolyte systems. Science 217 (1982) 1214–1217.

Zagdanska, B.: Effect of water stress upon the pyridine nucleotide pool in wheat leaves. J. Plant Physiol. 134 (1989) 320–326.

31

STRUCTURAL AND FUNCTIONAL STUDIES OF A NADP$^+$-SPECIFIC ALDEHYDE DEHYDROGENASE FROM THE LUMINESCENT MARINE BACTERIUM *Vibrio harveyi*

Masoud Vedadi, Nathalie Croteau, Marc Delarge, Alice Vrielink, and Edward Meighen

Department of Biochemistry
McGill University
Montreal, PQ, H3G 1Y6, Canada

1. INTRODUCTION

Aldehyde dehydrogenase, isolated from the luminescent bacterium, *Vibrio harveyi*, catalyzes the oxidation of fatty aldehydes (Meighen et al.,1976). This enzyme has a preference for long chain aldehydes as reflected by a large decrease in the K_m for aldehydes on increasing the chain length from acetaldehyde to tetradecanal (Bognar & Meighen, 1978; Vedadi et al., 1995). As one of the substrates for luciferase is tetradecanal, the potential involvement of aldehyde dehydrogenase in the luminescent system has been proposed. Tetradecanal is produced in luminescent bacteria by a *lux*-specific fatty acid reductase complex which activates (plus ATP) tetradecanoic acid to fatty acyl-AMP; the acyl group is then transferred between the fatty acid reductase subunits before being reduced with NADPH (Riendeau et al., 1982; Rodriguez et al., 1983a, 1983b). The *V. harveyi* aldehyde dehydrogenase(Vh ALDH) may be present so as to maintain aldehyde at nontoxic levels under conditions where excess levels of aldehyde are produced by the fatty acid reductase complex. However, the possibility that the function of aldehyde dehydrogenase is unrelated to luminescence can certainly not be excluded as Vh ALDH is under different regulation than the *lux*-specific proteins and has not been detected in other luminescent bacteria (Byers et al.,1988; Vedadi, M. unpublished data).

Although Vh ALDH was originally discovered as a NAD$^+$-dependent enzyme (Meighen et al., 1976), later work demonstrated that it exhibited a high specificity for NADP$^+$ as reflected by a very low K_m for this cofactor (Byers & Meighen, 1984; Vedadi et al., 1995). In this regard it is quite unique with respect to its nucleotide specificity with a V_{max}/K_m about 40 times higher with NADP$^+$ than NAD$^+$ while most aldehyde dehydrogenases are NAD$^+$-specific and even those that function with both NAD$^+$ and NADP$^+$ (e.g. class 3 mammalian aldehyde dehydrogenases) do not have the low K_m or high specificity for NADP$^+$ (Lindahl, 1992).

Enzymology and Molecular Biology of Carbonyl Metabolism 6
edited by Weiner *et al.* Plenum Press, New York, 1996

The gene for the aldehyde dehydrogenase from *V. harveyi* (*Vh aldH*), which encodes a protein of 510 amino acids, has been cloned, sequenced, and expressed in *E. coli*. Its amino acid sequence differs to a large degree from most or all aldehyde dehydrogenases with sequence identity of <25%. Only 19 of 23 reported invariant residues and 12 of 28 almost invariant (94% conserved) residues in aldehyde dehydrogenases (Hempel, 1993) are conserved in Vh ALDH (Vedadi et al., 1995). Among the residues conserved in Vh ALDH are the cysteine nucleophile (Hempel & Pietruszko, 1981; Hempel et al., 1982; 1984; 1993; von Bahr-Lindstrom et al., 1985; Pietruzko et al., 1991; Farrés et al., 1995) and a glutamic acid residue proposed to function as a general base for activation of the cysteine residue (Abriola et al., 1987; Pietruszko et al., 1993; Wang & Weiner, 1995).

The unique preference and tight binding of NADP$^+$ by the Vh ALDH not only makes this an ideal protein to investigate the structural and functional basis for differences in nucleotide specificity, but is extremely advantageous for the rapid isolation and purification of the enzyme and its mutants for further characterization. Due to this high selectivity, the enzyme and most mutants can be purified from crude extracts to homogeneity in a single step. As a result, sufficient amounts of purified protein for crystallization studies have been produced and have facilitated the growth of crystals that diffract X-ray to better than 2.5Å resolution. In addition, a number of heavy atom derivatives have been produced which will enable the structure determination by the isomorphous replacement method. Elucidation of the structure of the Vh ALDH coupled with the present capability to rapidly generate and purify mutant enzymes, should make this an excellent system to obtain detailed insights into the structure-function relationships of aldehyde dehydrogenases in general.

2. METHODS

2.1. Expression and Purification

The *Vh aldH* gene was expressed in *E. coli* K38 in the pT7 plasmid using the T7 RNA polymerase/ promoter system for expression. Cells were grown in enriched medium up to an OD$_{660}$ of 0.9–1 at 30°C, incubated at 42 °C for 15 min and rifampicin added to specifically inhibit *E. coli* RNA ploymerase. Cells were then grown for an additional 2 hours at 30°C, centrifuged and resuspended in 10% of the original culture volume of ice cold 50 mM phosphate buffer, 10 mM ß-mercaptoethanol before being lysed by sonication (Vedadi et al., 1995). The lysate was centrifuged at 24000 g in a Sorvall RC 5B for 20 min, and the supernatant diluted to about 15 to 30 % of the original culture volume to give a solution of 1 to 2 mg/ml of protein. The sample was then directly loaded onto a 2',5'-ADP Sepharose (Pharmacia Biotech) column. The column was washed with 50–100 times the column volume of 50 mM phosphate, pH 7, 10 mM ß-mercaptoethanol and the pure Vh ALDH eluted with 100 μM NADP$^+$ in the same buffer. Pure enzyme in different fractions were pooled and concentrated followed by dialysis against 50 mM phosphate buffer, pH 7, 10 mM ß-mercaptoethanol. Aldehyde dehydrogenase assays were performed in 50 mM K/ phosphate buffer pH 8, 1 mM NADP$^+$.

2.2. Crystallization

Crystals of the enzyme were obtained by vapor diffusion using the hanging drop technique. Suitable crystals for X-ray diffraction studies were obtained from polyethylene glycol utilizing an iterative seeding procedure (Croteau et al,1996).

3. RESULTS AND DISCUSSION

3.1. Single Step Purification

The high affinity of *Vh* ALDH for NADP⁺ is reflected in its ability to tightly bind to 2', 5'-ADP Sepharose. Although initial studies used a two-step purification consisting of anion exchange chromatography followed by affinity chromatography on 2'5'-ADP Sepharose, it is possible to purify the recombinant enzyme directly from crude extracts of E. coli in a single step by chromatography on 2'5'-ADP Sepharose. In this purification, an *E. coli* centrifuged lysate(150 ml, 1.2 mg/ml) containing recombinant Vh ALDH in 50 mM phosphate, 10 mM β -mercaptoethanol was directly loaded onto a 1 x 10 cm column of 2',5'-ADP Sepharose. The column was washed extensively before elution of the enzyme with 100 μM NADP⁺. The enzyme was pure as judged by SDS-PAGE with increasing amounts(2.5 to 15 μg) of enzyme (Fig 1) as only a single polypeptide band of 54 kDa could be detected even on electrophoresis of 15 μg of protein. In this method, it was important to dilute the lysate to a protein concentration of < 2 mg/ml (1 liter of cells would be lysed and then diluted so that the final volume was ~150 ml) and wash the column with a relatively high amount of buffer(50–100 times column volume) to assure a high yield of homogeneous enzyme was obtained.

3.2. Relationship with Class 3 Aldehyde Dehydrogenases: Structural and Kinetic Properties

Although the amino acid sequence of Vh ALDH has only a low sequence homology with other aldehyde dehydrogenases (<25% identity), it does appear to be more closely related to the class 3 mammalian ALDHs which are N-terminally truncated and C-terminally extended (Hempel et al., 1992). This similarity is consistent with Vh ALDH having a dimeric structure as the longer N-terminal domains are implicated in the formation of tetramers (Loomes & Jörnvall, 1991).

A close similarity between Vh ALDH and class 3 ALDHs also appears to exist in some of their kinetic properties. As mentioned earlier, class 3 ALDHs as well as Vh ALDH can function with either NAD⁺ or NADP⁺ as cofactors. However, Vh ALDH differs in that it has a much stronger interaction with NADP⁺ than other aldehyde dehydrogenases.

A similar specificity for more hydrophobic aldehydes also appears to exist between V. harveyi and class 3 aldehyde dehydrogenases (Yin et al., 1993; 1995). Table 1 com-

Figure 1. SDS gel electrophoresis of Vh ALDH after a one step purification on 2',5' ADP Sepharose . Lanes 1 to 6 contain 2.5 to 15 μg respectively, of the sample eluted from the affinity column.

Table 1. Comparison of the kinetic properties of V. harveyi and human class 3 ALDHs[a]

	K_m(mM)		Vmax(%)		Vmax(%)/K_m	
	Vh	Hum-cl3	Vh	Hum-cl3	Vh	Hum-cl3
acetaldehyde	63	85	62	83	1	1
propanal	2.4	15	100	93	42	6
butanal	0.124	2.5	92	66	742	26
hexanal	0.044	0.058	84	73	1910	1260
decanal	-	0.0028	-	100	-	35700
dodecanal	0.003	-	84	-	28000	-

[a]Data for human class 3 ALDH modified from Yin et al., 1995; V. harveyi ALDH data from Vedadi et al., 1995.

pares the relative activity at saturating aldehyde concentration and the K_ms for aldehyde for *V. harveyi* and class 3 human aldehyde dehydrogenase (Yin et al., 1995) for aldehydes of different chain length. Both enzymes have a relatively high K_m for acetaldehyde with the K_m for aldehyde decreasing significantly as the chain length of the aldehyde is increased while the apparent V_{max} (given as percent activity) remains relatively constant with chain length. Comparison of the relative V_{max}/K_m (percent activity/Km) shows a preference of about 30000-fold for decanal/dodecanal over acetaldehyde indicating that both enzymes have a hydrophobic pocket at the active site that selects for the longer chain aldehyde.

3.3. Identification of Critical Residues by Site Specific Mutagenesis

Alignment of the amino acid sequence of Vh ALDH with different ALDHs (Vedadi et al., 1995) showed that 19 out of 23 conserved amino acids in 16 different ALDHs (Hempel et al., 1993) could be located in the corresponding position in Vh ALDH. Of these 19 residues, 11 are glycines and one each of eight other amino acids (K,R,E,N,F,P,C,T). The conserved cysteine residue at position 289 and a highly invariant glutamate residue at position 253 are located in the same relative positions as found in

Table 2. Comparison of the amino acid sequence of Vh ALDH flanking the proposed catalytic residues of E253 and C289

```
                          # *                              *# *
VhALDH     249   PFYGELGAINPTFIFPSAMRAKADLADQFVASMTMGCGQFCTKP
Cl3rt      205   PVTLELGGKSPCYVDKDC--DLDVACRRIAWGKFMNSGQTCVAP
Cl3mrt     203   PVTLELGGKSPCYIDRDC--DLDVⁱℂRRITWGKYMNCGQTCIAP
Cl1hu      264   RVTLELGGKSPCIVLADA--DLDNAVEFAHHGVFYHQGQCCIAA
Cl2hu      264   RVTLELGGKSPNIIMSDA--DMDWAVEQAHFALFFNQGQCCCAG
Colialdh   263   RVWLEAGGKSANIVFADC-PDLQQAASATAAGIFYNQGQVCIAG
```

* Conserved residues between all aldehyde dehydrogenases (shown in bold). # Nearly invariant residues(>94% conserved).Sequences are: Vh ALDH(Vedadi et al., 1995); Cl3rt from 2,3,7,8,- tetrachlorodibenzo-p-dioxin ⁱnduction of rat liver (class 3) (Hempel et al.,1989); Cl3mrt from rat liver microsomes (class3) (Miyauchi et al., 1991); Colialdh from *E. coli* (Heim & Strehler.,1991); Cl1hu from human liver cytosol(class 1) (Hempel et al.,1984); Cl2hu from human liver mitochondria(class 2) (Hempel et al., 1985).

Table 3. Mutants of V. harveyi ALDH

V.harveyi Residue	Corresponding residue in Human Liver ALDH	Proposed Function	Vh Mutant	Relative Activity (% wt)
C289	302	Nucleophile	C289A	0.0004
E253	268	General Base	E253A	< 0.01
G229	245	NAD(P)⁺ Binding	G229A	< 0.01
G234	250	NAD(P)⁺ Binding	G234A	~ 30
E377	399	Unknown, highly conserved		
			E377Q	< 1
			E377K	< 0.1

other aldehyde dehydrogenases (Table 2). Site specific mutagenesis has recently confirmed that cysteine 289 functions as the active site nucleophile (Vedadi, et al., 1995) and has the same apparent function as suggested for C302 of mammalian ALDH (Hempel et al., 1982; 1993; von Bahr-Lindstrom et al., 1985; Blatter et al., 1990; Pietruszko et al., 1991, 1993; Farrés et al., 1995). Moreover, preliminary experiments have shown that E253, which corresponds in location to E268 of human liver mitochondrial ALDH (Table 2) is important in Vh ALDH and the enzyme loses its activity on conversion of E253 to alanine(Table 3). This glutamic acid residue has been proposed to be in the active site of human cytosolic (class 1) and mitochondrial (class 2) ALDHs based on inactivation of the enzyme by chemical modification (Abriola et al., 1987; Pietruszko et al., 1991, 1993). Site directed mutagenesis has also supported the involvement of this residue in the catalytic mechanism and in particular to function as a general base in activation of the cysteine nucleophile (Wang & Weiner, 1995). The large drop of activity for the E253A mutant of Vh ALDH indicates that this residue may be involved in a similar role.

The initial results on mutagenesis of other residues in Vh ALDH and their corresponding locations in mammalian ALDHs are also compiled in Table 3. Although conversion of glycine 234 to alanine appears to have little affect on catalytic activity, changing glycine 229 to alanine has a large effect on the ability of the enzyme to function. The glycines in the corresponding positions in other ALDHs have been proposed to be important in nucleotide binding (Hempel et al., 1984, 1989, 1993; Loomes & Jörnvall, 1991) as they are part of a primary sequence that most closely resembles that expected for a consensus nucleotide binding site (Wierenga & Hol, 1983). However, before a more definitive conclusion can be reached, these mutants must first be purified and their molecular properties examined more closely . Mutation of E377, which is part of the most highly conserved region of ALDHs (eEiFGP) also has a large effect on aldehyde dehydrogenase activity with NAD⁺. This mutant has now been purified and its properties are being investigated in greater detail.

3.4. Elucidation of the Crystal Structure of *V. harveyi* ALDH

Recently, rapid progress has been made in the generation of crystals suitable for X-ray diffraction analysis(Croteau et al., 1996). Shown in Figure 2 are crystals of the recombinant wild type enzyme that diffract to better than 2.5A resolution. The crystals are monoclinic with two homodimers of 110 kDa in the asymmetric unit.

Generation of crystals with heavy atom derivatives has just recently been accomplished which should lead to the elucidation of the crystal structure within the near future.

Figure 2. Crystals of Vh ALDH. The crystals grow by the vapor diffusion method combined with iterative seeding techniques. They appear as diamond shaped plates. Typical crystal dimensions are 0.25 x 0.25 x 0.02 mm.

Of specific relevance for future analyses is that crystals have been obtained with bound NADP⁺ which will provide a superb opportunity to define the exact interactions of the NADP⁺ cofactor with the enzyme and should prove of general value for defining the nucleotide binding site for aldehyde dehydrogenases in general.

4. REFERENCES

Abriola, D.P., Fields, R., Stein, S., Mackerell, A.D., & Pietruszko, R., 1987, Active site of human liver aldehyde dehydrogenase, *Biochemistry* **26**, 5679–5684.

Aurich, U., Sorger, H., Bergman, R., & Lasch, J. (1987) *Biol. Chem. Hoppe.Seyler* **368**, 101–109.

Blatter, E.E., Tasayco, M.L., Prestwich, G., & Pietruszko, R., 1990, Chemical modification of aldehyde dehydrogenase by a vinyl ketone analogue of an insect pheromone, *Biochem. J.* **272**, 351-358.

Bognar, A., Meighen, E.A., 1978, An induced aliphatic aldehyde dehydrogenase from the bioluminescent bacterium *Beneckea harveyi*, Purification and Properties, *J. Biol. Chem.* **253**, 446- 450.

Byers, D., & Meighen, E.A., 1984, *V. harveyi* aldehyde dehydrogenase: Partial reversal of aldehyde oxidation and its possible role in the reduction of fatty acids for bioluminescent reaction, *J. Biol. Chem.* **259**, 7109–7114.

Byers, D., Bognar, A., & Meighen, E.A.,1988, Differential regulation of enzyme activities involved in aldehyde metabolism in luminescent bacterium *V. harveyi*, *J. Bacteriol.* **170**, 967–971.

Croteau, N.,Vedadi, M., Delarge, M., Meighen, E., Abu-Abed, M., Howell, L., Vrielink, A., 1996, Crystallization and preliminary X-ray analysis of aldehyde dehydrogenase from *V. harveyi* (manuscript submitted).

Farrés, J., Wang, T.T.Y., Cunningham, S.J., & Weiner, H.,1995, Investigation of the active site cysteine residue of rat liver mitochondrial aldehyde dehydrogenase by site-directed mutagenesis, *Biochemistry.* **34**, 2592–2598

Heim, R. & Strehler, E., 1991, Cloning an *Escherichia coli* gene encoding a protein remarkably similar to mammalian aldehyde dehydrogenases, *Gene* **99**, 15–23.

Hempel, J.D., & Pietruszko, R., 1981, Selective chemical modification of human liver aldehyde dehydrogenase E1 and E2 by iodoacetamide, *J. Biol. Chem.* **256**, 10889–10896

Hempel, J., Pietruszko, R., Fietzek, P., & Jörnvall, H.,1982, Identification of a segment containing a reactive, cysteine residue in human liver cytoplasmic aldehyde dehydrogenase (isoenzyme E1)., *Biochemistry* **21**, 6834–6838.

Hempel, J., von Bahr-Lindstrom, H., & Jörnvall, H., 1984, Aldehyde dehydrogenase from human liver. Primary structure of the cytoplasmic isoenzyme, *Eur. J. Biochem.* **141**, 21–35.

Hempel, J., Kaiser, R., Jörnvall, H., 1985, Mitochondrial aldehyde dehydrogenase from human liver: Primary structure, differences in relation to the cytosolic enzyme and functional correlations, *Eur.J.Biochem.***153**,13–28.

Hempel, J., Harper, K., & Lindahl, R., 1989, Inducible (class 3) aldehyde dehydrogenase from rat hepatocellular carcinoma and 2,3,7,8-tetrachlorodibenzo-p-dioxin-treated liver: Distant relationship to the class 1 and 2 enzymes from mammalian liver cytosol/mitochondria, *Biochemistry* **28**, 1160–1167.

Hempel, J., Nicholas, H., & Lindahl, R.,1993, Aldehyde dehydrogenases:Widespread structural and functional diversity within a shared framework, *Protein Science* **2**, 1892- 1900.

Lindahl, R., 1992, Aldehyde dehydrogenases and their role in carcinogenesis, *Crit. Rev. Biochem.* **27**, 283–335.

Loomes, K. & Jörnvall, H.,1991, Structural organization of aldehyde dehydrogenase probed by limited proteolysis, *Biochemistry* **30**, 8865–8870.

Meighen, E.A., Bogacki, I.G., Bognar, A., & Michaliszyn, G.A.,1976, Induction of fatty aldehyde dehydrogenase activity during the development of bioluminescence in *Beneckea harveyi*, *Biochem. Biophys. Res. Comm.* **69**, 423–430.

Miyauchi, K., Masaki, R., Taketani, S., Yamamoto, A., Akayama, M., & Tashiro, Y., 1991, Molecular cloning, sequencing and expression of cDNA for rat liver microsomal aldehyde dehydrogenase. *J. Biol. Chem.* **266**, 19536–19542.

Pietruszko, R., Blatter, E., Abriola, D.P., & Prestwich, G.,1991, Localization of cysteine 302 at the active site of aldehyde dehydrogenase. *Adv. Exp. Med. Biol.* **284**, 19–30.

Pietruszko, R., Abriola, D.P., Blatter, E.E., Mukerjee, N., 1993, Aldehyde dehydrogenase:Aldehyde dehydrogenation and ester hydrolysis, *Adv. Exp. Med. Biol.* **328**, 221–230.

Riendeau, D., Rodriguez, A., & Meighen, E.A.,1982, Resolution of the fatty acid reductase from *Photobacterium phosphoreum* into acyl-protein synthetase and acyl-CoA reductase activities. Evidence for an enzyme complex, *J. Biol. Chem.* **257**, 6908–6915.

Rodriguez, A., Riendeau,D., & Meighen, E.A.,1983a, Purification of the acyl-CoA reductase component from a complex responsible for the reduction of fatty acids in bioluminescent bacteria. Properties and acyl transferase activity, *J. Biol. Chem.* **258**, 5233–5237.

Rodriguez, A., Wall,L., Riendeau,D. & Meighen, E.A.,1983b, Fatty acid acylation of proteins in bioluminescent bacteria. *Biochemistry* **22**,5604–5611.

Vedadi, M., Szittner, R., Smillie, L., Meighen, E., 1995, Involvement of cysteine 289 in the catalytic activity of an NADP+-specific fatty aldehyde dehydrogenase from *V. harveyi*, *Biochemistry.* [34], 16725–16732.

von Bahr-Lindstrom, H., Jeck, R., Woenckhans, Christoph., Sohn, S., Hempel, J. & Jornvall, H.,1985, Characterization of the coenzyme binding site of liver aldehyde dehydrogenase: Differential reactivity of coenzyme analogues, *Biochemistry* **24**, 5847–5851.

Wang, X., & Weiner, H.,1995, Involvement of glutamate 268 in the active site of human liver mitochondrial (class2) aldehyde dehydrogenase as probed by site-directed mutagenesis, *Biochemistry* **34**, 237–243.

Wierenga, R.K., & Hol, W.G.J.,1983, Predicted nucleotide binding properties of p21 protein and cancer associated variant, *Nature* **302**, 842–844.

Yin, S.J., Wang, S.L., Liao, C.S., & Jörnvall, H. (1993) Human High-k_m Aldehyde dehydrogenase (ALDH3): Molecular, Kinetic and Structural Features., in "Enzymology and Molecular Biology of Carbonyl Metabolism 4. pp 87–98., H.Weiner, ed., Plenum Press, New York.

Yin, S.J., Wang, MF.,Han, CL.,Wang SL.,1995, Substrate binding pocket structure of human aldehyde dehydrogenases. A substrate specificity approach, *Adv. Exp. Med. Biol.* **372**, 9–16.

32

Saccharomyces cerevisiae ALDEHYDE DEHYDROGENASES

Identification and Expression

Xinping Wang, Yinglin Bai, Li Ni, and Henry Weiner

Department of Biochemistry
Purdue University
West Lafayette, Indiana 47907–1153

1. INTRODUCTION

Saccharomyces cerevisiae can grow aerobically on ethanol as a sole carbon and energy source. The oxidation of ethanol to acetate is catalyzed by the actions of alcohol dehydrogenase (ADH) and aldehyde dehydrogenase (ALDH). ADH catalyzes the formation of acetaldehyde from ethanol and ALDH oxidizes the acetaldehyde to acetate.

Four alcohol dehydrogenases have been found in *Saccharomyces cerevisiae* and were well studied (Drewke et al., 1990). ADHI was found to be constitutively expressed, located in cytosol and is involved the reduction of acetaldehyde to ethanol. ADHII also was found to be located in cytosol, but its expression was repressed by glucose. It had a low Km for NAD and ethanol and is thought to be involved in the oxidation of ethanol (Ganzhorn et al., 1987). ADHIII was located in mitochondria and presumably was involved the oxidation of ethanol in the mitochondrial matrix. The physiological role of ADHIV has not been clearly defined and it is missing in some yeast strains (Drewke and Ciriacy, 1988).

Saccharomyces cerevisiae aldehyde dehydrogenases have been studied for many years. It was found that mitochondrial ALDH was potassium-stimulated, repressed by glucose, and could use both NAD and NADP as a cofactor (Black 1951, Jacobson and Bernofsky, 1974). The cytosolic ALDH is a constitutive enzyme and use only NADP as a cofactor (Seegmiller 1953, Jacobson and Bernofsky, 1974).

2. RESULTS AND DISCUSSION

2.1. Gene

We screened a *Saccharomyces cerevisiae* genomic library and found the full length sequences for two yeast ALDHs, one cytosolic, the other mitochondrial. A 2774 bp en-

Table 1. Amino acid homology analysis of the *Saccharomyces cerevisiae*
cytosolic ALDH1 and mitochondrial ALDH2

Species	Homology (%)	
	ALDH1	ALDH2
Aspergillus niger	50	55
Emericella nidulans	50	52
Cladosporium herbarum	47	53
Caenorhabditis elegans	45	47
Alternaria alternata	45	51
Human, mitochondria	45	46
Human, cytosol	45	46
Rat, mitochondria	45	46
Rat, cytosol	43	45
Beef, mitochondria	45	44
Beef, cytosol	44	47

coded for one which we called ALDH1. The sequence contained 501 amino acids, including the initiation methionine. This ALDH was found to be located in cytosol by cell fractionation followed by activity staining and Western blot analysis. Another 2661 bp encoded a protein we called ALDH2. The sequence contained 519 amino acids of which the N-terminal 23 amino acids were the signal sequence. We found that the recombinantly expressed protein could be imported into mitochondria. The amino acid homology analysis (Table 1) showed that the cytosolic ALDH1 and the mitochondrial ALDH2 were about 45% and 46% identical to several mammalian class 1 and class 2 ALDHs, respectively. The homology was slightly higher to bacterial ALDHs.

After we cloned the two *Saccharomyces cerevisiae* ALDHs, the total genome of *Saccharomyces cerevisiae* strain, S288C, was published in Genebank. The cytosolic ALDH1 gene was found on chromosome XVI, and the mitochondrial ALDH2 gene was on chromosome V. It appears that five more ALDH genes were present in yeast. Amino acid sequence alignments of cytosolic and mitochondrial ALDHs along with these five other possible ALDHs are shown in Figure 1. The *Saccharomyces cerevisiae* cytosolic ALDH1 and mitochondrial ALDH2 contained all 23 conserved amino acids found in other ALDHs (Hempel et al., 1993). A gene on chromosome XV (ChromXV) also had all 23 conserved amino acids and had an N-terminal extension with four positive -charged amino acids. It is possible that this protein is another mitochondrial ALDH. Two ALDH genes located on chromosome XIII (Chrom13 and chrom13i) were adjacent each other and were almost identical (92%). It appears that they represent gene duplication. Those two ALDH genes coded for 506 amino acids, including the initiation methionine, and the protein would contain all the conserved amino acids except for glycine at position of 376 (corresponding to 370 of human class 1 ALDH), which was replaced by a serine. The N-terminal of these two duplicated proteins did not have properties associated with signal peptides which allow for the import of proteins into mitochondria. A gene found on chromosome VIII (cos8179) would code for a protein much longer in length than the other ALDHs: 644 amino acids. This protein would have a threonine at position of 310 (corresponding to 223 of human class 1 ALDH), instead of a glycine found in other ALDHs. Another sequence in chromosome XIII (2cos9718) would code for a protein with only 17 conserved amino acids within its 532 amino acids. It would have a long C-terminal extension. None of these five proteins have been identified and studied.

Figure 1. The amino acid sequence alignment of seven potential aldehyde dehydrogenase found in the *Saccharomyces cerevisiae* genome.

Yaldh1: The cytosolic ALDH1; Yaldh2: the mitochondrial ALDH2; Chromxv: a potential ALDH located on chromosome XV; Chrom13 and Chrom13i: two adjacent potential ALDHs located on chromosome XIII; Cos8179: a potential ALDH located on chromosome VIII. It would consist of 644 amino acids. Neither the first 41 amino acids, MSKVYLNSDMINHLNSTVQAYFNLWLEKQNAIMRSQPQIIQ, at the N-terminal nor the last 16 amino acids, TWQRIKSLFSLAKEAS, at the C-terminal were shown in this figure; 2cos9718: a potential ALDH also located on chromosome XIII and sequenced from cosmid 9718. It would contain 532 amino acids. The last 27 amino acids, NKWGLRQYFSLSAAVILISTIYAHCSS, at the C-terminal, were not shown in this figure. The underlined residues are those conserved in all known ALDHs.

2.2. Protein

How many aldehyde dehydrogenases are present in *Saccharomyces cerevisiae*? Even though seven potential ALDH genes were found in the *Saccharomyces cerevisiae* genome, we could find only two, the cytosolic ALDH1 and mitochondrial ALDH2, by activity staining and immunoreactions. Further investigations need to be performed to determine if the other five ALDHs are actually expressed in yeast.

We recombinantly expressed the two genes we cloned. The purified cytosolic ALDH1 had NADP preference and was not stimulated by potassium ions. It had a Km for NADP of 99 μM while the Km for NAD was 17400 μM. In contrast, the recombinantly expressed mitochondrial ALDH2 did not show cofactor preference (Km for NADP was 3470 μM and for NAD 6430 μM) and was stimulated by potassium ions . The role of the two ALDHs in the ethanol metabolism in yeast is being investigated.

3. REFERENCES

Black, S., 1951, Yeast aldehyde dehydrogenase, *Arch. Biochem. Biophys.*, 34:86–97.

Drewke, C. and Ciriacy, M., 1988, Overexpression, purification and properties of alcohol dehydrogenase IV from *Saccharomyces cerevisiae, Biochim. Biophys. Acta,* 950:54–60.

Drewke, C., Thielen, J., and Ciriacy, M., 1990, Ethanol formation in adh mutants reveals the existence of a noval acetaldehyde-reducing activity in *Saccharomyces cerevisiae, J. Bacteriol.*, 172:3909–3917.

Ganzhorn, A. J., Green, D. W., Hershey, A. D., Gould, R. M., and Plapp, B. V., 1987, Kinetic characterization of yeast Alcohol dehydrogenases: Amino acid residue 294 and substrate specificity, *J. Biol. Chem.*, 262:2754–3761.

Jacobson, M. K., and Bernofsky, C. (1974) Mitochondrial acetaldehyde dehydrogenase from *Saccharomyces cerevisiae, Biochim. Biophys. Acta,* 350:277–291.

Seegmiller, J. E. (1953) Triphosphopyridine nucleotide-linked aldehyde dehydrogenase from yeast, *J. Biol. Chem.*, 201:629–637.

33

ALCOHOL DEHYDROGENASE VARIABILITY

Evolutionary and Functional Conclusions from Characterization of Further Variants

Hans Jörnvall, Jawed Shafqat, Mustafa El-Ahmad, Lars Hjelmqvist,
Bengt Persson, and Olle Danielsson

Department of Medical Biochemistry and Biophysics
Karolinska Institutet
S-171 77 Stockholm, Sweden

1. INTRODUCTION

We have studied many alcohol dehydrogenases and related enzymes with the aim of defining functional properties, structural patterns, and evolutionary relationships. From this, four major conclusions have been drawn:

- The enzymes are clearly multiple and represent different protein families. Within the MDR family (medium-chain dehydrogenases/reductases), repeated duplications at different levels have produced the enzymes, classes, and isozymes that are now visible in human, mammalian, and other lines (Jörnvall *et al.*, 1987; Hjelmqvist *et al.*, 1995a).
- Of the alcohol dehydrogenases, class III, with its glutathione-dependent formaldehyde dehydrogenase activity (Koivusalo *et al.*, 1989), appears to be the parent form, locking much of the alcohol dehydrogenase family to cellular detoxication reactions (Danielsson and Jörnvall, 1992).
- Separate, internal molecular architectures are present (Danielsson *et al.*, 1994a). Class III has a protein-classical pattern, with a low variability overall like functionally constant enzymes in general, and with variable regions in non-functional segmens. The other classes are more variable, both overall and in their functional segments, in a protein-atypical manner, indicating evolution of new functions, or "enzymogenesis" (Danielsson and Jörnvall, 1992).
- Functional convergence toward ethanol activity has occurred in many lines. Thus, the ethanol-active enzymes in yeast, prokaryotes, plants, and animals all appear to have separate origins (Jörnvall, 1994).

We have now extended these studies. In particular, we have recently defined two further reptilian (Shafqat *et al.*, 1996a; Hjelmqvist *et al.*, 1996), one avian (Hjelmqvist *et al.*, 1995b), one bony fish (Danielsson *et al.*, 1996), and several other forms now in progress. We have purified a mushroom form of the enzyme, showing this to be a class III form. We have established that the variability patterns typical of the classes also apply to the isozymes within a class (Hjelmqvist *et al.*, 1996a). Hence, there is a continuity between late levels of isozyme duplications and the earlier levels of class duplications, with enzymogenesis being present and observable in a similar manner in both cases. Recently, we have also purified and characterized the pea class III form (Shafqat *et al.*, 1996b), which together with similar work on the *Arabidopsis* class III enzyme (Martinez *et al.*, 1996) establishes the separate nature of the plant and animal ethanol-active enzymes. Finally, we have crystallized the cod class I enzyme, leading to the crystallographic determination of that structure, showing localized differences toward previously known forms (Ramaswamy *et al.*, 1996) in agreement with the three regions of deviations predicted from comparisons and modelling (Danielsson *et al.*, 1992).

In spite of all this molecular knowledge available on alcohol dehydrogenases, questions have long remained on the origins of the classes, their distinct nature, and their separate functions. The novel structures give further answers on these points as outlined below.

2. ORIGINS

In this case, the major question has concerned the ancestral form and the origin(s) of the activity toward ethanol and low molecular weight alcohols. Thus, on the one hand, all observations on the alcohol dehydrogense structures suggest that the formaldehyde-active class III has an ancestral nature and that the ethanol-active class I has a later origin from class III through a duplication at early vertebrate times. Such a class I origin is consistent with three different sets of observation: with extrapolations from present class I species variants (Cederlund *et al.*, 1991), with the apparent absence of the class I enzyme in animals before the vertebrate emergence (Kaiser *et al.*, 1993; Danielsson *et al.*, 1994a,b), and with the presence of a class I alcohol dehydrogenase with mixed properties in that vertebrate line, bony fish, that originated at about the same time as the supposed class duplication (Danielsson *et al.*, 1992). All these three sets of observations suggest a fairly "late" origin, roughly with the vertebrates, of the ethanol activity of the "classical" zinc-containing alcohol dehydrogenase.

On the other hand, the enzyme branches of wider (i.e., older) separation, like the plant and animal ethanol-active forms, are clearly related (Dennis *et al.*, 1984; Sun and Plapp, 1992; Yokoyama and Harry, 1993), have the same quaternary structure (dimers), and have been correlated in three-dimensional structure (Brändén *et al.*, 1984). Hence, they look almost as related as the most different vertebrate class I forms are. Therefore, do the plant ethanol-active forms disturb the conclusion on class I origin in early vertebrates, and suggest the origin to be still earlier including also the plant forms? Or is the present estimate of distinct origins correct and the plant and vertebrate lines just similarly related from convergence or little divergence? Lack of a known class III structure in plants has long precluded a final answer on this point.

Similarly, in construction of phylogenetic trees, the class I form in bony fish falls outside all groupings, reflecting its "hybrid" structural properties between class I and class III (cf Danielsson and Jörnvall, 1992). Is this cod class I non-clustering with other forms expected and only a consequence of different evolutionary speeds at different times? Or does it suggest that the class origins are wrong in timing at the early vertebrate stage?

The recent finding of class III enzyme structures in plants gives novel answers to these questions:

- first, the plant class III enzymes as well as those from other kingdoms of living forms, prove that the class III structure is of a common nature, strictly conserved, and present in all kingdoms. Hence, the plant ethanol-active forms, although related to the vertebrate class I forms, reflect a far more distant relationship, not contradicting separate origins of these ethanol-active forms.
- second, construction of phylogenetic trees, involving both the ethanol-active plant/animal forms and corresponding formaldehyde-active forms (Shafqat *et al.*, 1996b; Martínez *et al.*, 1996), show separate branchings of the former (Fig. 1).

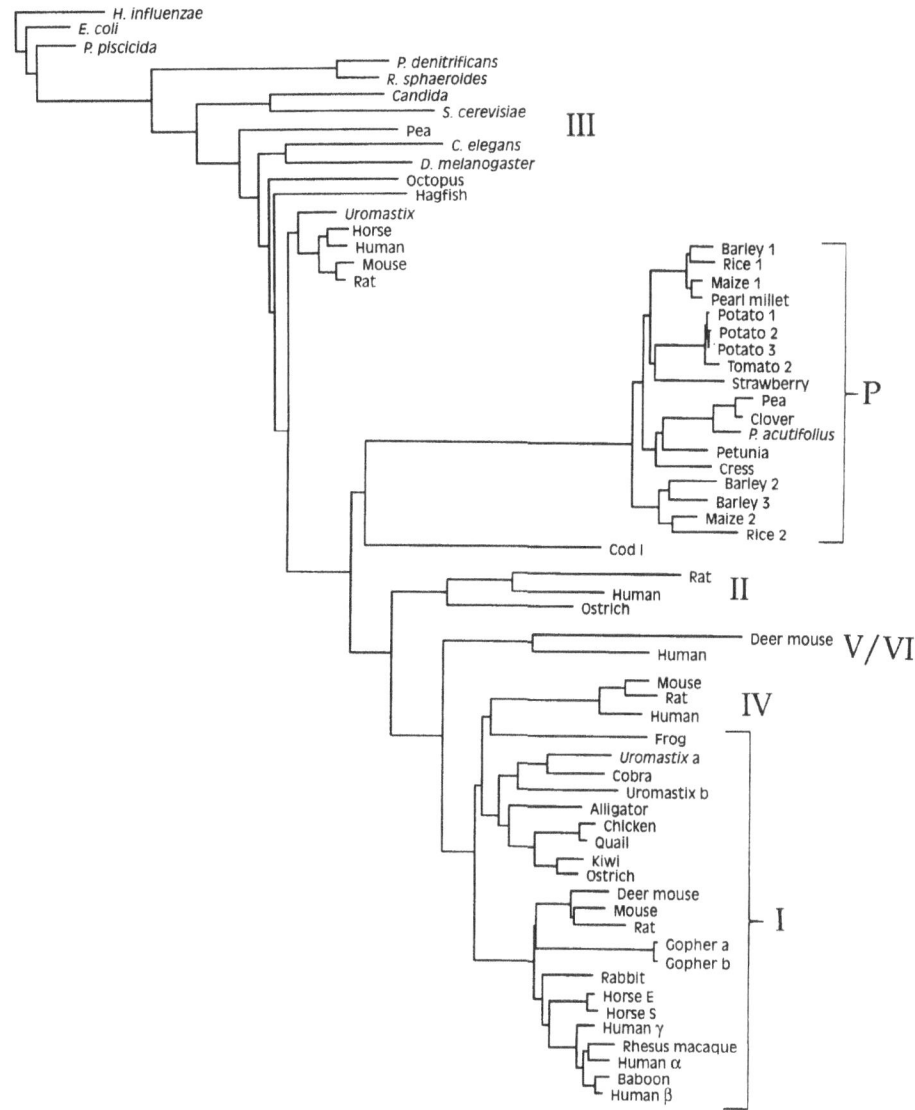

Figure 1. Phylogenetic tree of characterized animal and plant zinc-containing alcohol dehydrogenases. Amino acid sequences from data bases and covering those in Table I complemented with the plant class P and III forms.

They appear to support the concept of repeated origins of ethanol activity and hence separate origins of the ethanol-active forms. In fact, from these data, it is possible to distinguish the plant ethanol-active forms as a separate class, class P, distinct from the vertebrate ethanol-active forms (class I). In that sense, although the significance level of these tree branchings is still low, the presence of similar ethanol-active forms in plants and vertebrates, is no conceptual problem and does not contradict the previous conclusion of repeated origins of that activity but rather support it.

• third, the intra-class species variabilities are known to constitute distinct characteristics between the previously well-studied classes of the vertebrate enzymes as mentioned above, and when these patterns now are extended to the plant ethanol-active forms (Martinez *et al.*, 1996), they show a third pattern (Fig. 2). A segment of maximal variability is superficially positioned like in the basic fold of the class

Figure 2. Segment variability patterns of the universal class III (top), the vertebrate class I (middle), and the plant class P (bottom) forms, all adjusted into the fold of the human class I subunit crystallographically analyzed (Hurley et al., 1991). Light, segment of maximal intra-class species variability; dark, remaining parts. Class III has two maximally variable segments at superficial, non-functional regions in a manner typical for proteins in general, class I three segments of maximal variability in functional regions in a manner atypical for proteins with strictly conserved functions, while class P has a third pattern, with the most variable segment also superficial and non-functional like for class III but affecting a different segment. The two ethanol-active forms, I and P, differ fundamentally in internal variability architecture, compatible with separate duplicatory origins.

III form and distinct from the variations in functional segments typical of class I. Hence, the intra-class variability patterns also distinguish the ethanol-active plant and animal forms and do not contradict separate origins.

Therefore, the plant class III structures now defined, support a universal nature of the formaldehyde-active class III structure and help to distinguish the separate origins of the ethanol-active enzymes, with likely distinct forms in yeast (class Y), in plants (class P), and in vertebrates (class I and remaining mammalian enzyme classes).

The second of the previous complications regarding the origins, i.e. the extraordinary position of the cod class I enzyme outside that of all other branches, is also made less important than before by the novel structures. Thus, the most recent tree, based on 79 different Zn-containing alcohol dehydrogenase structures (Shafqat *et al.*, 1996b), shows that the cod class I structure, although still outside all other vertebrate enzyme branches, is insignificantly so (bootstrap value of only 243, which is insignificant). Hence, the divergent and "mixed-class" cod ethanol-active enzyme presents no problem for the origin of class I in general. It does not contradict present conclusions on the class I origin in early vertebrates, and may either reflect just another duplication (unique to fish), or is indeed a reflection of the original class I/III duplication in bony fish ancestors, but appearing deviant in phylogenetic trees because in standard tree calculations, constant evolutionary speeds are supposed. If instead, evolutionary speeds were greater at times immediately subsequent to the origin, before new restrictions have been set by development of novel functions, the patterns would be exactly as now found.

Thus, both the class P ethanol-active plant enzymes and the cod "hybrid" class I enzyme do not constitute great problems any longer. The tree constructions subgroup all classes, supporting the present evolutionary concept of multiple duplications, repeated creations of ethanol-active enzymes, and an hitherto apparently ancestral position for class III. Together these studies illustrate the evolution of the whole enzyme family, and suggest important but different functional roles for the classes.

3. FUNCTION

The recently determined structures also highlight functional conclusions in other manners.

On the one hand, a novel class II form (from ostrich, Hjelmqvist *et al.*, 1995c) also shows "hybrid" properties. Hence, enzymogenesis, like the one suggested for the cod class I/III separation, need not be restricted to just that class separation, but appears also to occur regarding the class II origin. Therefore, the concept of "hybrid" enzymes and evolution of new catalytic activities, "enzymogenesis", is apparently not limited to class I or bony fish at ~450 million years ago, but equally noticable in class II and the ancestral separation of the reptile/avian line from the mammalian line (at > 300 million years ago). Hence, class origins in general are early, also for the most variable class II enzyme, and reflect functional enzymogenesis in the creation of new activities. Consequently, each class is expected to have distinct functional roles, apart from their partly overlapping common activities.

On the other hand, the fairly large number of animal alcohol dehydrogenase structures now available show that intra-class variabilities are in all cases restricted (Table 1). Even class II, the one thus far most rapidly evolving, still exhibits only limited residue variability. Several positions, although some at first seemingly greatly or even randomly

Table 1. Amino acid residues at the functionally important substrate-binding and coenzyme-binding positions of animal alcohol dehydrogenases now characterized. Positions originally deduced from the horse class I enzyme crystallographically determined (cf Eklund and Brändén, 1987), amino acid sequences from data bases complemented with Danielsson et al. (1996) and Höög and Svensson (1996). Empty positions indicate non-analyzed parts, Δ denotes a gap

Class	Source	substrate-binding positions			coenzyme-binding positions	
		(inner)	(middle)	(outer)		
		48 67 93 140 141	57 115 116 294 318	110 306 309	47 48 51 178 203 223 224 228 269 271 369	
I	Human α	T H A F L	M D V V I	Y M L	G T H T V D I K I R R	
	Human β	T H F F L	L D L V V	Y M L	R T H T V D I K I R R	
	Human γ	S H F F V	L D L V I	Y M L	R S H T V D I K I R R	
	Rhesus α	T H A F L	M D V V V	Y M L	G T H T V D I K I R R	
	Baboon β	T H F F V	L D L V V	Y M L	R T H T V D I K I R R	
	Horse E	S H F F L	L D L V I	F M L	R S H T V D I K I R R	
	Horse S	S H F F L	L Δ L V I	L M L	R S H T V D I K I R R	
	Rabbit	S H F F I	I D L V I	F M L	R S H T V D I K I R R	
	G. b. major	S H F F L	I D L V V	L M L	R S H T V D I K I R R	
	G. knoxjonesi	S H F F L	I D L V V	L M L	R S H T V D I K I R R	
	Rat	S H F F L	L N L V I	L M L	R S H T V D I K I R R	
	Mouse	S H F F I	L D L V I	F M L	R S H T V D I K I R R	
	Deer mouse	S H F F I	L D L V I	L M L	R S H T V D I K I R R	
	Chicken	S H F F V	L D L V V	L M F	R S H T V D I K I R R	
	Quail	S H F F I	L D L V V	L M F	R S H T V D I K I R R	
	Mallard		V V	M F	V D I K I R R	
	Ostrich	S H F F A	L D I V V	L M F	R S H T V D I K I H R	
	Kiwi	S H F F I	L D I V V	L M F	R S H T V D I K I H R	
	Alligator	S H F F I	L D L V V	L M F	R S H T V D I K I R R	
	Uromastix a	S H F F I	F D L L V	L G F	R S H T V D I K I R R	
	Uromastix b	S H F F V	L E L V V	V L F	R S H T V D I K I R R	
	Cobra	S H F F I	F D L V V	L L F	R S H T V D I K I R R	
	Frog	S H F F I	L D I L V	L L L	R S H T V D L K I N R	
	Cod	T H F F L	H W A W M	Q I I	H T Y T V D L K V N R	
II	Human	T H Y F F	F N L V F	L E I	H T S T V D I K A G R	
	Rat	T H F F M	K N F A F	L V I	P T N S V D I K A T R	
	Rabbit 1	S H Y F F	F K G V N	F I I	R S H T V D I K A G R	
	Rabbit 2	S H Y F L	F E H V S	F V I	R S Y T V D I K A G R	
	Ostrich	T H Y F M	F K I V F	L M I	R T H T V D I K I N R	
III	Human	T H Y Y M	D R V V A	L F V	H T Y T V D I K I N R	
	Horse	T H Y Y M	D R T V A	L F V	H T Y T V D I K I N R	
	Rat	T H Y F M	D R V V A	L F V	H T Y T V D I K I N R	
	Mouse	T H Y F M	D R V V A	L F V	H T Y T V D I K I N R	
	Uromastix	T H Y F M	D R V V A	L F V	H T Y T V D L K I N R	
	Cod *l*	T H Y F M	D R V V A	L F V	H T Y T V D I K I N R	
	Cod *h*	T H Y F M	D R L V A	L F V	H T Y T L D V K V N R	
	Hagfish	T H Y F M	D R V V A	L F V	H T Y T V D R K V N R	
	Octopus	T H Y F M	D R V V A	L F V	H T Y T V D I K I N R	
	Drosophila	T H Y F M	D R L V A	L F V	H T F T V D I K I N R	
IV	Human	T H F F M	M D I V V	L M F	R T H T V D L K I H R	
	Rat	T H F F M	M D L A V	L M F	G T H T V D I K I R R	
	Mouse	T H F F M	M D L A V	L M F	R T H T V D I K I R R	
V	Human	T H F F G	H K Q V V	F Q F	G T K T V D V K I N Δ	
VI	Deer mouse	T H C Y I	K R L S V	I H F	G T K T V D T K T N R	

variable, are still class characteristic. In this context, positions with binding interactions at the active site appear most significant. For example, among the positions with hydrophobic residues, position 309, although variable between separate, mostly branched-chain residues, is restricted to be Ile in class II, Val in class III, and with the exception of the "class-mixed" cod enzyme, Leu or Phe, in other classes. Similarly, the variable position 318 is also characteristic, with Ala in all known animal class III forms, branched-chain residues in remaining classes except class II, where only other alternatives are found (Table 1). Therefore, the most variable class II form is also still subject to restrictions and apparently not free to evolve drastic changes. Combined, this finding gives extra emphasis on the functional importance of all the alcohol dehydrogenase forms. Each duplicatory origin appears to be unique and to have led to novel and distinct functions within the basic framework in common.

In this context, the recently determined tertiary structure of the class-mixed class I cod enzyme is of special interest (Ramaswamy *et al.*, 1996). It shows three segments with loops of different structure versus the classical class I enzymes (Eklund & Brändén, 1987; Hurley *et al.*, 1991), and these three segments are exactly those that were distinguished as different already from original alignments, comparisons and modelling (Danielsson *et al.*, 1992). Hence, apart from emphasizing the value of early comparisons and of distinction of the intra-class variable segments, the novel structure is likely to reflect the class-distinguishing properties and unique functional interactions in each case.

4. CONCLUSIONS

The alcohol dehydrogenase enzymes are unique within each class. All evidence is compatible with repeated origins of the ethanol-active forms, early but separate origins of the classes, an ancestral class III position, and consistent changes at critical residues with binding interactions at the active site. The derivation of other classes than just class I and III as of old origin is especially important, since class II has appeared highly variable and the function of class IV has been implicated in vertebrate regulation of embryogenesis and differentiation (Ang *et al.*, 1996). Minor shifts may still be detectable in the exact topology of the duplicatory orders, but origins are early, fairly uniform, and appear consistent, giving support to the evolution of novel function with the liver, and with the whole vertebrate system at large. In addition, the results also highlight the complexities in constructions of phylogenetic trees when the branches have not followed constant evolutionary speeds.

The multiplicity within the alcohol dehydrogenase family, and also, correspondingly, within the aldehyde dehydrogenase family, is impressive, like those for the cytochrome P450 family and the glutathione transferases. Hence, the complex nature of all these enzymes, with multiple forms and repeated duplications appear similar in pattern and suggest that they all participate in important basic cellular defense reactions, where each form has distinct functions although also exhibiting overlapping activites. The separate function is noticeable also in the phenotype of negative mutants. Thus, an inborn error of metabolism in humans, the Sjögren-Larsson syndrome, is caused by loss of just one aldehyde dehydrogenase (and in a few cases perhaps one alcohol dehydrogenase) in fatty alcohol metabolism, causing the disease affecting skin and the central nervous system (De Laurenzi *et al.*, 1996).

The multiplicity of the mammalian alcohol dehydrogenase family and its basic variability like those for aldehyde dehydrogenases, cytochrome P450s, and glutathione transferases, leading to the conclusions of functions in common, is also of historic interest. In part, we are now back to regard at least some forms of alcohol dehydrogenase as a defense enzyme toward endogenous alcohols, as apparent in its name and compatible with early suggestions (Krebs and Perkins, 1970), although many other substrates and functional roles have been tested and implicated in between. In addition, the unique nature of each class suggests special functional roles in each case. Hence, the structural data support an evolutionary pattern giving functional implications and emphasizing an importance of the entire ADH system.

5. REFERENCES

Ang, H.L., Deltour, L., Hayamizu, T.F., Zgomicknight, M. and Duester, G.: Retinoic acid synthesis in mouse embryos during gastrulation and craniofacial development linked to class IV alcohol dehydrogenase gene expression. J. Biol. Chem. 271 (1996) 9526–9534.

Brändén, C.-I., Eklund, H., Cambillau, C. and Pryor, A.J.: Correlation of exons with structural domains in alcohol dehydrogenase. EMBO J. 3 (1984) 1307–1310.

Cederlund, E., Peralba, J.M., Parés, X. and Jörnvall, H.: Amphibian alcohol dehydrogenase, the major frog liver enzyme. Relationships to other forms and assessment of an early gene duplication separating vertebrate class I and class III alcohol dehydrogenases. Biochemistry 30 (1991) 2811–2816.

Danielsson, O. and Jörnvall, H.: "Enzymogenesis": Classical liver alcohol dehydrogenase origin from the glutathione-dependent formaldehyde dehydrogenase line. Proc. Natl. Acad. Sci. USA 89 (1992) 9247–9251.

Danielsson, O., Eklund, H. and Jörnvall, H.: The major piscine liver alcohol dehydrogenase has class-mixed properties in relation to mammalian alcohol dehydrogenases of classes I and III. Biochemistry 31 (1992) 3751–3759.

Danielsson, O., Atrian, S., Luque, T., Hjelmqvist, L., Gonzàlez-Duarte, R. and Jörnvall, H.: Fundamental molecular differences between alcohol dehydrogenase classes. Proc. Natl. Acad. Sci. USA 91 (1994a) 4980–4984.

Danielsson, O., Shafqat, J., Estonius, M. and Jörnvall, H.: Alcohol dehydrogenase class III contrasted to class I. Characterization of the cyclostome enzyme, the existence of multiple forms as for the human enzyme, and distant cross-species hybridization. Eur. J. Biochem. 225 (1994b) 1081–1088.

Danielsson, O., Shafqat, J., Estonius, M., Jörnvall, H.: Isozyme multiplicity with anomalous dimer patterns in a class III alcohol dehydrogenase. Effects on activity and quaternary structure of residue exchanges at "nonfunctional" sites in a protein. Biochemistry (1996) submitted.

De Laurenzi, V., Rogers, G.R., Hamrock, D.J., Marekov, L.N., Steinert, P.M., Compton, J.G., Markova, N. and Rizzo, W.B.: Sjögren-Larsson syndrome is caused by mutations in the fatty aldehyde dehydrogenase gene. Nature Genetics 12 (1996) 52–57.

Dennis, E.S., Gerlach, W.L., Pryor, A.J., Bennetzen, J.L., Inglis, A, Llewellyn, D., Sachs, M.M., Ferl, R.J. and Peacock, W.J.: Molecular analysis of the alcohol dehydrogenase (Adh1) gene of maize. Nucleic Acids Res. 12 (1984) 3983–4000.

Eklund, H. and Brändén, C.-I.: Alcohol dehydrogenase. Biological Molecules and Assemblies 3 (1987) 73–142.

Hjelmqvist, L., Estonius, M. and Jörnvall, H.: Distinctive class relationships within vertebrate alcohol dehydrogenases. In Atassi, Z.M. and Appella, E. (Eds.), Methods in Protein Structure Analysis. Plenum, New York, 1995a, pp. 419–427.

Hjelmqvist, L., Metsis, M., Persson, H., Höög, J.-O., McLennan, J. and Jörnvall, H.: Alcohol dehydrogenase of class I: kiwi liver enzyme, parallel evolution in separate vertebrate lines, and correlation with 12S rRNA patterns. FEBS Lett. 367 (1995b) 306–310.

Hjelmqvist, L, Estonius, M. and Jörnvall, H.: The vertebrate alcohol dehydrogenase system: variable class II type form elucidates separate stages of enzymogenesis. Proc. Natl. Acad. Sci. USA 92 (1995c) 10904–10908.

Hjelmqvist, L., Shafqat, J., Siddiqi, A.R. and Jörnvall, H.: Linking of isozyme and class variability patterns in the emergence of novel alcohol dehydrogenase functions. Characterization of isozymes in Uromastix hardwickii. Eur. J. Biochem. 236 (1996) 563–570.

Höög, J.-O.: Cloning and characterization of a novel rat alcohol dehydrogenase of class II type. FEBS Lett. 368 (1995) 445–448.

Höög, J.-O. and Svensson, S.: Mammalian class II alcohol dehydrogense. A highly variable enzyme. In Weiner, H., Crabb, D. and Flynn, T.G. (Eds.), Enzymology and Molecular Biology of Carbonyl Metabolism 6, Plenum Press, New York, 1996, in press.

Hurley, T.D., Bosron, W.F., Hamilton, J.A. and Amzel, L.M.: Structure of human $\beta_1\beta_1$ alcohol dehydrogenase: catalytic effects of non-active-site substitutions. Proc. Natl. Acad. Sci. USA 88 (1991) 8149–8153.

Jörnvall, H., Höög, J.-O., von Bahr-Lindström, H. and Vallee, B.L.: Mammalian alcohol dehydrogenases of separate classes: intermediates between different enzymes and intraclass isozymes. Proc. Natl. Acad. Sci. USA 84 (1987) 2580–2584.

Jörnvall, H.: The alcohol dehydrogenase system. In Jansson, B., Jörnvall, H., Rydberg, U., Terenius, L. and Vallee, B.L. (Eds.), Toward a Molecular Basis of Alcohol Use and Abuse Birkhäuser, Basel, 1994, pp. 221–229.

Kaiser, R., Fernández, M.R., Parés, X. and Jörnvall, H.: Origin of the human alcohol dehydrogenase system. Implications from the structure and properties of the octopus protein. Proc. Natl. Acad. Sci. USA 90 (1993) 11222–11226.

Koivusalo, M., Baumann, M. and Uotila, L.: Evidence for the identity of glutathione-dependent formaldehyde dehydrogenase and class III alcohol dehydrogense. FEBS Lett. 257 (1989) 105–109.

Krebs, H.A. and Perkins, J.R.: The physiological role of liver alcohol dehydrogenase. Biochem. J. 118 (1970) 635–644.

Martínez, M.C., Achkor, H., Persson, B., Fernández, M.R., Shafqat, J., Farrés, J., Jörnvall, H. and Parés, X.: Arabidopsis formaldehyde dehydrogenase: molecular properties of plant class III alcohol dehydrogenase provide further insights into the origins, structure and function of plant class P and liver class I alcohol dehydrogenases. Eur. J. Biochem. (1996) in press.

Ramaswamy, S., El-Ahmad, M., Danielsson, O., Jörnvall, H. and Eklund, H.: Crystal structure of cod liver class I alcohol dehydrogenase: substrate pocket and structurally variable segments. Prot. Sci. 5 (1996) 663–671.

Shafqat, J., Hjelmqvist, L. and Jörnvall, H.: Liver class-I alcohol dehydrogenase isozyme relationships and constant patterns in a variable basic structure. Distinctions from characterization of an ethanol dehydrogenase in cobra, Naja naja. Eur. J. Biochem. 236 (1996a) 571–578.

Shafqat, J., El-Ahmad, M., Danielsson, O., Martinez, M.C., Persson, B., Parés, X. and Jörnvall, H.: Pea formaldehyde-active class III alcohol dehydrogenase: common derivation of the plant and animal forms but not of the corresponding ethanol-active forms (classes I and P). Proc. Natl. Acad. Sci. USA (1996b) 93, 5595–5599.

Sun, H.-W. and Plapp, B.V.: Progressive sequence alignment and molecular evolution of the Zn-containing alcohol dehydrogenase family. J. Mol. Evol. 34 (1992) 522–535.

Yokoyama, S. and Harry, D.E.: Molecular phylogeny and evolutionary rates of alcohol dehydrogenases in vertebrates and plants. Mol. Biol. Evol. 10 (1993) 1215–1226.

34

THREE-DIMENSIONAL STRUCTURES OF HUMAN ALCOHOL DEHYDROGENASE ISOENZYMES REVEAL THE MOLECULAR BASIS FOR THEIR FUNCTIONAL DIVERSITY

Thomas D. Hurley, Curtis G. Steinmetz, Peiguang Xie, and Zhong-Ning Yang

Department of Biochemistry and Molecular Biology
Indiana University School of Medicine
Indianapolis, Indiana 46202
E-mail: hurley@biochem6.iupui.edu

1. INTRODUCTION

Seven distinct genes for alcohol dehydrogenase (*ADH1-ADH7*) have been identified in humans (Edenberg and Bosron, 1996). The *ADH1-ADH5* and *ADH7* genes code for the α, β, γ, π, χ, and σ subunits, respectively. A protein product from the *ADH6* gene has not been identified *in vivo*. The individual subunits share between 58% and 93% sequence identity. The π, χ, and σ subunits are known to form only active homodimers *in vivo*, while the α, β, γ subunits can form active homo- and heterodimers (Edenberg and Bosron, 1996). Additional complexity is contributed by polymorphism at the *ADH2* gene locus giving rise to β_1, β_2, and β_3 subunits and at the *ADH3* gene locus giving rise to γ_1 and γ_2 subunits. The products of the seven different human ADH gene loci have been grouped into five classes based on their amino acid sequence and functional characteristics (Edenberg and Bosron, 1996). The human $\alpha\alpha$, $\beta\beta$, and $\gamma\gamma$ isoenzymes comprise Class I and are characterized by a relatively high V_{max}/K_M for ethanol, are primarily expressed in the liver, and are sensitive to inhibition by pyrazole and its 4-substituted derivatives in the low μM range. The Class I isoenzymes share approximately 93% sequence identity with each other. The human Class II, or $\pi\pi$, isoenzyme possesses a lower V_{max}/K_M for ethanol than the Class I enzymes, is primarily localized in the liver, and is less sensitive to inhibition by pyrazole. The Class II isoenzyme shares 58% sequence identity with the Class I isoenzymes. The human Class III, or $\chi\chi$, isoenzyme, is non-saturable with ethanol as a substrate, is constitutively expressed in all cell and tissue types, and is insensitive to inhibition by pyrazole. The human Class III isoenzyme shares 59% sequence identity to either the Class I or Class II isoenzymes and was shown to be identical to the glutathione-dependent formaldehyde dehydrogenase, which oxidizes the hydrated adduct of formaldehyde and glutathione, S-hydroxymethyl-glutathione, to S-formyl-glutathione (Koivusalo *et al.*, 1989, Kaiser, *et al.*, 1991). The human Class IV, or σσ, isoenzyme possesses the highest V_{max}/K_M for retinol oxidation of any human ADH isoenzyme (Yang *et al.*, 1994), is primarily expressed in epithelial tissue, and is sensitive to inhibition by pyrazole in the high µM range (Kedishvili *et al.*, 1995). The Class IV isoenzyme shares 69% sequence identity to the Class I isoenzymes and approximately 60% sequence identity to the Class II or Class III isoenzymes. The *ADH6* gene has been designated as a Class V isoenzyme based on its deduced amino acid sequence since it shares roughly 60% sequence identity with other human ADH isoenzymes (Jörnvall and Höög, 1996). The enzymatic properties of the *ADH6* gene product have not been well characterized.

Over the past 20 years, considerable effort has been devoted to examining the functional properties of the human ADH isoenzymes using various enzymes purified from tissue and, more recently, using recombinantly expressed enzyme. These studies have yielded the most information with regard to functional and structural properties of Class I isoenzymes. Our understanding of the properties of non-Class I isoenzymes has been hampered by the lack of an accurate structural model upon which to base our interpretations of the functional data. Structural models of these isoenzymes based on either the horse EE or human $\beta_1\beta_1$ Class I isoenzymes cannot predict many of the functional properties of the human Class II, III, or IV isoenzymes (Eklund *et al.*, 1990, Davis *et al*, 1994, Kedishvili *et al.*, 1995). However, these models have been useful in examining the properties of the human αα and $\gamma_1\gamma_1$ Class I isoenzymes (Eklund *et al.*, 1987, Stone *et al.*, 1989). Our understanding of the functional properties of the non-Class I isoenzymes would benefit greatly from direct structure determination of these distinct isoenzyme forms. Additionally, the ability to construct accurate models of new ADH enzyme Classes, based solely on sequence information, will benefit from examining the structural diversity among the ADH gene family. In an effort to achieve these goals, we have purified, crystallized, and determined the three-dimensional structures of the αα, $\beta_1\beta_1$, χχ, and σσ human ADH isoenzymes.

2. METHODS

2.1. Expression and Purification of the Recombinant Human ADH Isoenzymes

The human αα, $\beta_1\beta_1$, and σσ ADH isoenzymes were expressed in the JM105 strain of *E. coli* by subcloning the appropriate human ADH cDNA into the EcoRI site of the procaryotic expression vector pKK223–3 (Pharmacia Biotech). All enzymes were obtained from 20 liter cell cultures transformed with the appropriate expression vector system in TB media (Hurley *et al.*, 1990). The cultures were grown at 37°C until enzyme expression was induced by addition of 0.1 mM isopropyl-β-galactopyranoside (IPTG) and 10 µM ZnSO₄ once the cell cultures reached an optical density of 0.8 at 595 nm. Following induction of enzyme expression, the temperature of the incubator was reduced to 16°C and the cultures were incubated overnight. The methods for harvesting and lysis of the cell suspensions have been described (Stone, *et al.*, 1993). The purification of the recombinant human $\beta_1\beta_1$ isoenzyme has been described (Hurley, *et al.*, 1994).

The recombinant human αα isoenzyme was purified using modifications of the procedure for the purification of the human $\beta_1\beta_1$ isoenzyme (Hurley, *et al.*, 1994). Briefly, following batch chromatography over DE52 cellulose (Whatman International, Ltd.) at pH

8.8 (4°C) where the enzyme does not bind to the resin, the activity pool from the column flow-through was buffer exchanged into 10 mM HEPES, pH 7.2, 1 mM benzamidine, 2 mM DTT, and 10 μM ZnSO$_4$ using the Mini-tan tangential flow concentrator (Millipore, Corp.). The enzyme was then applied to a 5 x 5 cm column of S-Sepharose equilibrated with 10 mM HEPES, pH 7.2, 1 mM benzamidine, 2 mM DTT, and 10 μM ZnSO$_4$. The bound enzyme was washed and then step eluted with equilibration buffer containing 110 mM NaCl. The active fractions were pooled and dialyzed against 50 mM Tris, pH 7.5, 1 mM DTT, 10 μM ZnSO$_4$. The dialyzed pool of activity was applied to a 2.5 x 5 cm column of Affi-Gel Blue (Biorad Laboratories) equilibrated with 50 mM Tris, pH 7.5, 1 mM DTT, 10 μM ZnSO$_4$. The bound enzyme was washed and then step eluted with 800 mM NaCl in equilibration buffer. Prior to crystallization, the purified enzyme was dialyzed into 10 mM ACES, pH 7.2, 1 mM DTT and concentrated to 10 mg/ml.

The recombinant human σσ isoenzyme was purified using the following procedure. After batch chromatography over DE52 cellulose in 50 mM Tris, pH 8.8 (4°C), 1 mM benzamidine, and 2 mM DTT the flow-through fractions were buffer exchanged into 7 mM sodium phosphate, pH 6.4, 1 mM DTT using the Mini-tan tangential flow concentrator. The enzyme was then applied to a 5 x 5 cm column of S-Sepharose equilibrated with 7 mM sodium phosphate, pH 6.4, 1 mM DTT. The bound enzyme was washed and then eluted with a linear gradient of sodium phosphate from 7 to 65 mM. The active fractions were pooled and dialyzed against 10 mM sodium phosphate, pH 6.4, 1 mM DTT. Following dialysis the enzyme was applied to a 2.5 x 5 cm column of Affi-Gel Blue (Biorad Laboratories) and eluted with a linear gradient from 10 mM sodium phosphate, pH 6.4, 1 mM DTT to 100 mM Tris, pH 8.8, 1 mM DTT. Prior to crystallization, the purified enzyme was dialyzed into 10 mM HEPES, pH 7.0, 1 mM DTT and concentrated to 8 mg/ml.

The human Class III, χχ, isoenzyme was purified from autopsy liver tissue using modifications of published procedures (Wagner et al., 1984). Approximately 165 g of frozen liver was minced and divided into 4 parts. Each part was homogenized in a precooled blender containing 40 ml of cold 10 mM HEPES, pH 7.0, 2 mM DTT, 1 mM benzamidine, and 10 μM ZnSO$_4$. The four separate homogenates were pooled and centrifuged at 125,000 x g for 30 minutes. The supernatant was then dialyzed against two changes, 4 liters each, of homogenization buffer at pH 7.9. The dialyzed supernatant was the passed through a 400 ml bed of DE52 equilibrated with homogenization buffer and washed with 1 liter of homogenization buffer. The bound enzyme was eluted with 100 mM NaCl in 10 mM HEPES, pH 7.9, 1 mM DTT, 10 μM ZnSO$_4$. The active fractions were dialyzed against 10 mM HEPES, pH 7.9, 1 mM DTT, 10 μM ZnSO$_4$. The dialyzed activity pool was then applied to a 2.5 x 20 cm column of MIMETIC BLUE 2 (PIKSI) and washed with a linear gradient from 0–7.5 mM NaCl in 10 mM HEPES, pH 7.9, 1 mM DTT, 10 μM ZnSO$_4$. The enzyme was eluted with a linear gradient from 0–0.5 mM NADH in 10 mM HEPES, pH 7.9, 1 mM DTT, 7.5 mM NaCl, and 10 μM ZnSO$_4$. The active fractions were directly applied to a DE52 column equilibrated with BLUE 2 buffer. The bound enzyme was washed and eluted with a linear gradient from 0–70 mM NaCl in equilibration buffer. The active fraction were pooled and applied to an AcA-44 gel filtration column equilibrated with 10 mM HEPES, pH 7.9, 1 mM DTT, and 10 μM ZnSO$_4$. Prior to crystallization the enzyme was concentrated to 10 mg/ml in AcA-44 column buffer.

2.2. Crystallization and Collection of X-Ray Diffraction Data

Crystals of recombinant human β$_1$β$_1$ ADH were grown using the vapor diffusion method in the sitting drop configuration at a protein concentration of 15 mg/ml. The crystallization buffer contained 50 mM sodium phosphate, pH 7.5, 2 mM NAD+, 1 mM 4-io-dopyrazole, and 13.5% (w/v) polyethylene glycol molecular weight 8000 (PEG 8000). The $\beta_1\beta_1$ diffraction data were collected at room temperature to a resolution of 2.19 Å from two orientations of a single crystal at a crystal-to-detector distance of 11 cm.

Crystals of recombinant human $\alpha\alpha$ ADH were grown using the vapor diffusion method in the sitting drop configuration at a protein concentration of 10 mg/ml. The crystallization buffer contained 100 mM ACES, pH 7.2, 2 mM NAD+, 2 mM 4-iodopyrazole, and 14% (w/v) PEG 6000. The $\alpha\alpha$ diffraction data were collected at room temperature to 2.8 Å resolution from two crystals at a crystal-to-detector distance of 14.5 cm.

Crystals of recombinant human $\sigma\sigma$ ADH were grown using the vapor diffusion method in the sitting drop configuration at a protein concentration of 8 mg/ml. The crystallization buffer contained 100 mM sodium cacodylate, pH 6.5, 50–100 mM zinc acetate, 7.5 mM NAD+, and 18% (w/v) PEG 6000. The $\sigma\sigma$ diffraction data were collected at room temperature to a resolution of 3.0 Å from four crystals at a crystal-to-detector distance of 14.5 cm.

Crystals of liver purified human $\chi\chi$ ADH were grown using the vapor diffusion method in the sitting drop configuration at a protein concentration of 10 mg/ml. Large, moderately well diffracting crystals, could be obtained across a broad range of conditions (Hurley *et al.*, 1992). The crystallization drops contained 200–320 mM MES, pH 6.2–6.7, 1–4 mM NAD+, and 8–9% PEG 8000. The $\chi\chi$ diffraction data were collected at room temperature to a resolution of 2.7 Å from two crystals at a crystal-to-detector distance of 16 cm and a detector swing angle (2Θ position) of 3°.

All X-ray diffraction data measurements were collected on a Rigaku RU200-HB equipped with a Rigaku RAXIS IIC image plate area detector. The diffraction data were indexed, integrated, merged, and scaled using the RAXIS IIC data processing software (T. Hagashi, Rigaku, Corp., Japan).

2.3. Molecular Replacement Calculations and Structure Refinement

All structures were solved by molecular replacement using the coordinates of the human $\beta_1\beta_1$ ternary complex with NAD(H) and cyclohexanol (Hurley *et al.*, 1994) as the search model (PDB code 1HDX). The rotation and translation functions for each structure were calculated using the program AMORE (Navaza, 1994) and the data between 15 and 4 Å, with the exception of the human $\beta_1\beta_1$ ternary complex with NAD+ and 4-iodopyra-

Table 1. Crystal and data collection statistics

	$\beta_1\beta_1$	$\alpha\alpha$	$\sigma\sigma$	$\chi\chi$
Complex	NAD+	NAD+	NAD+	NAD+
	4-iodopyrazole	4-iodopyrazole	acetate	water
Space Group	P1	P2$_1$	P2$_1$	C222$_1$
Cell Dimensions	a=54.0 Å	a=55.8 Å	a=86.3 Å	a=141.3 Å
	b=44.5 Å	b=72.1 Å	b=94.7 Å	b=201.6 Å
	c=93.3 Å	c=97.3 Å	c=121.7 Å	c=69.6 Å
Cell Angles	α=92.6°	α=90.0°	α=90.0	α=90.0°
	β=103.3°	β=105.5°	β=100.0°	b=90.0°
	γ=68.8°	γ=90.0°	γ=90.0°	γ=90.0°
MaximalResolution	2.2 Å	2.8 Å	3.0 Å	2.7 Å
Completeness	83%	86%	93%	90%
Subunits per Asymm. Unit	2	2	4	2

zole. This new $\beta_1\beta_1$ data set was isomorphous with the 1HDX data, thus the molecule could be refined directly with the new data set after rigid-body refinement. All coordinate refinements were performed using XPLOR 3.1 (Brünger, 1988).

2.4. Coordinate Analysis and Active Site Volume Calculations

Alpha-carbon alignments of the different human ADH coordinate sets were performed using LSQKAB within the program package CCP4 (CCP4, 1994) and displayed using the graphics program QUANTA (Molecular Simulations, Inc.). Representations of the active site volumes were calculated using the SPHGEN subroutine within the program DOCK 3.0 (Kuntz et al.,1982). The surface representations were calculated and displayed using the MS program (Connolly, 1983), as implemented in QUANTA.

3. RESULTS

Crystals suitable for X-ray diffraction experiments have been produced for four of the seven known human ADH gene products. These enzymes crystallize in different space groups and with different intermolecular contacts within their respective lattices (Table 1). The three-dimensional structures of the human $\alpha\alpha$, $\beta_1\beta_1$, $\chi\chi$, and $\sigma\sigma$ isoenzyme have been solved by molecular replacement and refined to 2.8 Å, 2.2 Å, 2.7 Å, and 3.0 Å, respectively. The details of their molecular replacement calculations and refinements will be published elsewhere. Table 2 summarizes the current refinement results from these X-ray diffraction experiments. This manuscript will present only the general structural features of the different isoenzymes, the details of coenzyme and substrate binding will published elsewhere.

An alignment of the alpha-carbon atoms of the refined human ADH isoenzymes provides a means for examining the conserved and divergent structural features of this gene family (Figure 1). Not surprisingly, the amount structural divergence increases as the sequence identity between the isoenzymes decreases. The individual coenzyme and catalytic domains of the human $\alpha\alpha$ and $\beta_1\beta_1$ isoenzyme, which share 93% sequence identity, align with an average root-mean-square- difference (rmsd) of 0.35 Å, while all the alpha-carbon atoms align with an rmsd of 0.58 Å (Table 3). A small rotation of the catalytic domain (0.5°) in the $\alpha\alpha$ isoenzyme towards a more "closed" conformation relative to the $\beta_1\beta_1$ isoenzyme accounts for these alignment differences, similar to what we reported for the $\beta47G$ mutant enzyme. There are no areas of large structural difference between these two Class I isoenzymes. Despite their overall structural similarity, the active site volumes and topographies of the two isoenzymes are quite distinct (Figure 2).

Table 2. Refinement statistics of human ADH isoenzyme structures

	$\beta_1\beta_1$	$\alpha\alpha$	$\sigma\sigma$	$\chi\chi$
Data Included	8-2.2 Å	8-2.8 Å	8-3.0 Å	8-2.7 Å
Number of Reflections	32,940	15,205	32,781	23,718
R_{work}	18.0	19.4	22.4	20.6
R_{free}	26.0	28.9	31.6	28.3
Rmsd from Ideal Bond Lengths	0.009 Å	0.009 Å	0.01 Å	0.009 Å
Rmsd from Ideal Bond Angles	1.68°	1.79°	2.00°	1.52°
Luzatti estimate. of coord. error	0.21 Å	0.30 Å	0.33 Å	0.30 Å

Figure 1. Alpha-carbon alignments of the different human ADH isoenzymes with the $\beta_1\beta_1$ isoenzyme. (A) Alignment of human $\alpha\alpha$ with human $\beta_1\beta_1$ ADH. This alignment was produced by aligning only those residues in the coenzyme domains (residues 178 to 322). (B) Alignment of human $\sigma\sigma$ with $\beta_1\beta_1$ ADH. This alignment was produced by aligning residues 178 to 245 and residues 263 to 322 within their respective coenzyme binding domains. The positions of residues 116 and 260 are labelled in both subunits. (C) Alignment of human $\chi\chi$ and $\beta_1\beta_1$ ADH. This alignment was produced by aligning residues 178 to 322 in their respective coenzyme binding domains. The position of residue 116 within the new α-helix is labelled in both subunits.

Table 3. Average root-mean-square-difference in alpha-carbon positions between human $\beta_1\beta_1$ and other human ADH isoenzyme structures[1]

	$\alpha\alpha$	$\sigma\sigma$	$\chi\chi$
Common Cα atoms in Catalytic Domain	0.36 Å	0.51 Å	0.81 Å
Common Cα atoms in Coenzyme Domain	0.34 Å	0.65 Å	0.71 Å
Common Cα atoms in dimeric structure	0.58 Å	0.58 Å	1.8 Å

[1]The alignment procedure for the $\alpha\alpha$ isoenzyme utilized all alpha-carbon atoms in the two structures, while the procedure for aligning the $\sigma\sigma$ isoenzyme excluded residues 115-120 in the catalytic domain and residues 247-262 in the catalytic domain. The alignment procedure for the $\chi\chi$ isoenzyme excluded residues 56-61 and residues 113-120 in the catalytic domain. The excluded regions in each structure represent the areas of largest structural difference (eg., areas of amino acid insertions and/or deletions, or areas of unique secondary structure).

Figure 2. Active site volumes of different human ADH isoenzymes compared with the active site of the $\beta_1\beta_1$ isoenzyme. (A) The enclosed volume of the $\beta_1\beta_1$ isoenzyme is shown in grey (730 Å^3) and the enclosed volume of the $\alpha\alpha$ isoenzyme is shown in black (490 Å^3). The relative position of the active site zinc atom is indicated as Zn. Solvent access occurs from the extreme right of this representation. (B) The enclosed volume of the $\beta_1\beta_1$ isoenzyme is shown in grey (730 Å^3) and the enclosed volume of the $\sigma\sigma$ isoenzyme is shown in black (640 Å^3). The view in this representation is perpendicular to that shown in (A), solvent access still occurs from the right. (C) The enclosed volume of the $\beta_1\beta_1$ isoenzyme is shown in grey (730 Å^3) and the enclosed volume of the $\chi\chi$ isoenzyme is shown in black (1250 Å^3). The view shown here is identical to that in (A).

A similar comparison of the alpha-carbon positions in the $\beta_1\beta_1$ and $\sigma\sigma$ isoenzyme structures reveals an rmsd of 0.58 Å (Figure 1 and Table 3), similar results are obtained when individual domains are aligned (Table 3). Two areas of substantial structural variation are observed. The largest structural changes occur at residues 260 and 261 (numbering of the horse EE and human $\beta_1\beta_1$ sequences) where two Asn residues in the $\sigma\sigma$ structure replace two Gly residues in the $\beta_1\beta_1$ structure. A maximal displacement of 4.6 Å occurs at residue 261 when the $\sigma\sigma$ and $\beta_1\beta_1$ structures are aligned. The second substantial structural difference occurs within the loop comprised of residues 114 to 121 near the entrance to the substrate binding pocket, where residue 117 is deleted in the $\sigma\sigma$ isoenzyme. This deletion shortens the corresponding loop in the $\sigma\sigma$ isoenzyme and changes the position of residue 116 relative to its position in the $\beta_1\beta_1$ isoenzyme. The active site volume and topography of the $\sigma\sigma$ isoenzyme is more similar to the $\beta_1\beta_1$ isoenzyme than either the $\alpha\alpha$ or $\chi\chi$ active sites, but key differences exist near the catalytic zinc site and near the outer rim of the substrate binding site due to amino acid substitutions and the deletion of residue 117, respectively (Figure 2).

The structure of the human χχ isoenzyme shows the greatest structural divergence from the prototypical Class I structures (Figure 1), exemplified by the horse EE and human $\beta_1\beta_1$ isoenzymes. The rmsd between all alpha-carbon atoms in the χχ and $\beta_1\beta_1$ structures is 1.8 Å, while alignment of individual domains yields rmsd values less than half of this value (Table 3). These differences are due to the catalytic domains in the Class III isoenzyme adopting a more open conformation than the typical closed domain arrangement observed in previous ADH:NAD(H) complexes (Eklund *et al.*, 1981, Hurley *et al.*, 1994). Analysis of this domain motion reveals that the catalytic domain is 6° less closed than either human or horse Class I binary and ternary complex structures. The other major structural difference between the Class III isoenzyme and other known ADH structures is the presence of a novel α-helix comprised of residues 113 to 120 (Figure 1). In the Class I and Class IV structures, this stretch of amino acids adopts a loop structure which forms part of the entrance to the substrate binding pocket (Figure 1). Due the formation of an α-helix in the Class III isoenzyme, the position of these amino acids are much farther away from the catalytic zinc atom and amino acid residues involved in catalysis. The combination of this new α-helix and the domain motion creates a much different substrate binding site. The substrate binding site of the human χχ isoenzyme is roughly two times larger (Figure 2) and much more hydrophilic than any other ADH isoenzyme.

4. DISCUSSION

We have determined the three-dimensional structures of four of the seven known human ADH isoenzymes. These new structures include the previously determined human Class I $\beta_1\beta_1$ isoenzyme (Hurley *et al.*, 1991, Hurley *et al.*, 1994), now extended to 2.2 Å resolution (Davis, *et al.*, 1996), the Class III human χχ and Class IV σσ isoenzymes, as well as the Class I αα isoenzyme. The new structures have been solved by molecular replacement and refined to 2.8 Å, 3.0 Å, and 2.7 Å for the αα, σσ, and χχ isoenzymes, respectively. Each of the structures possess reasonable R-factors and stereochemistry for their respective resolution limits (Table 2). In this report we will focus on the general structural characteristics which are unique to each isoenzyme and which broadly influence their function.

The human αα isoenzyme exhibits the greatest overall structural similarity to the human $\beta_1\beta_1$ isoenzyme. This is not terribly surprising since both are Class I isoenzymes and share greater than 93% sequence identity. Interestingly, similar to what was observed when Gly was substituted for Arg47 in the $\beta_1\beta_1$ isoenzyme (Hurley *et al.*, 1994), the αα isoenzyme appears to be in a more "closed" conformation than the $\beta_1\beta_1$ isoenzyme. The αα isoenzyme, in addition to 22 other amino acid substitutions relative to the $\beta_1\beta_1$ isoenzyme, has a Gly at position 47. Since both structures are identical ternary complexes between enzyme, NAD+, and 4-iodopyrazole, it would appear that Gly at position 47 is both necessary and sufficient for a more closed catalytic domain conformation.

Although the $\beta_1\beta_1$ and αα isoenzymes differ by only 0.35 Å rmsd in their Cα positions within individual domains, their substrate binding pockets differ dramatically (Table 3 and Figure 2). The αα isoenzyme exhibits a more spacious active site near the catalytic zinc atom due to the substitution of Phe93, in the $\beta_1\beta_1$ isoenzyme, by Ala. The increased accessible volume near the catalytic zinc atom, due to this substitution, has been shown by modeling and mutagenesis studies to be responsible for this isoenzyme's ability to oxidize secondary alcohols with efficiencies approaching those of straight chain alcohol substrates (Stone *et al.*, 1989, Hurley and Bosron, 1992). Interestingly, the enclosed volume of the substrate binding pocket in the $\alpha\alpha$ isoenzyme is 35% smaller than that of the $\beta_1\beta_1$ isoenzyme. The middle region of the substrate binding pocket is narrowed by the substitutions of Met for Leu at positions 57 and 141, as well as the more closed conformation of the catalytic domain (Figures 1 and 2). This narrowing of the substrate binding pocket, in conjunction with the wider pocket near the catalytic zinc atom, appears to explain the dependence of V_{max}/K_m on the chain length of the substrate exhibited by the $\alpha\alpha$ isoenzyme (Stone et al., 1989). The $\beta_1\beta_1$ isoenzyme does not show a strong relationship between the V_{max}/K_m for an alcohol substrate and the chain length of the substrate. The active site of $\beta_1\beta_1$ is narrow and restrictive near the catalytic zinc atom and can bind small substrates in relatively few conformations, thus increasing the number of productive encounters between enzyme and small substrates. However, the active site of the $\beta_1\beta_1$ isoenzyme widens rapidly as the distance from the active site zinc increases, thus relative few specific interactions are gained with longer chain alcohol substrates (Figure 2). This property is reflected in K_m values which vary only four-fold for substrates from ethanol to hexanol (Hurley and Bosron, 1992). The active site of the $\alpha\alpha$ isoenzyme exhibits exactly the opposite characteristics. The $\alpha\alpha$ isoenzyme cannot restrict the binding modes of small substrates and more non-productive encounters between the enzyme and small substrates will occur, which will be reflected in higher K_m values (6 mM for ethanol). However, the active site narrows as the distance from the catalytic zinc atom increases. Thus, the enzyme gains productive interactions which act to increase the number of productive encounters between enzyme and substrate, resulting in lower K_m values (14 µM for 1-pentanol).

The overall structure of the Class IV human $\sigma\sigma$ isoenzyme is also similar the human $\beta_1\beta_1$ isoenzyme. The rmsd between common $C\alpha$ atoms in the two structures is only 0.58 Å (Table 3). There does not appear to be any difference in domain closure between these two isoenzymes. Two regions of substantial structural difference can be identified in this isoenzyme which shares 69% sequence identity with the $\beta_1\beta_1$ isoenzyme. The first region is due to the substitution of two Asn residues for two Gly residues at positions 260 and 261. In the $\beta_1\beta_1$ structure, the main chain conformations of Gly260 and Gly261 are not compatible with amino acids which possesses side chains. Thus, the main chain rearrangements at Asn260 and Asn261 are simply the result of the steric requirements of these new side chains. We do not expect that these changes influence the kinetic properties of the $\sigma\sigma$ isoenzyme to a great extent, but they can explain why Class IV and Class I heterodimers are not observed (Stone et al., 1993). The region around residues 260 and 261 form intersubunit contacts with residues in the structural zinc lobe of the opposite subunit (residues 101–109). Arg101 in the human Class I isoenzymes forms hydrogen bonds with the main chain atoms of residues 260 and 261. Due to the changes in main chain conformation at positions 260 and 261 and the substitution of Asn for Arg101 in the $\sigma\sigma$ isoenzyme, the Class I and Class IV subunits cannot form complementary interactions across this interface.

The substrate binding site of the $\sigma\sigma$ isoenzyme is more similar to the substrate binding site in $\beta_1\beta_1$, in both topology and enclosed volume, than is the substrate site in the $\alpha\alpha$ isoenzyme (Figure 2). Near the catalytic zinc atom the active site is differently shaped due to a rotation of the nicotinamide ring in the $\sigma\sigma$ isoenzyme by $3°$, relative to its position in the $\beta_1\beta_1$ or $\alpha\alpha$ isoenzymes. This rotation may explain the lower affinity for the inhibitor 4-methylpyrazole exhibited by the $\sigma\sigma$ isoenzyme. This positioning of the nicotinamide ring may distort the binding orientation of this transition state analog. Like the $\alpha\alpha$ isoenzyme, $\sigma\sigma$ also possesses Met57 and Met141, and like the $\alpha\alpha$ isoenzyme, the middle region of the substrate binding pocket is narrower than the corresponding region in the $\beta_1\beta_1$ isoenzyme (Figure 2). Thus, a similar explanation can be used to correlate the V_{max}/K_m for alcohols with their chain length for this isoenzyme. The σσ isoenzyme exhibits a K_m for ethanol of 28 mM, while its K_m for 1-hexanol is 0.14 mM (Kedishvili *et al.*, 1995). As we predicted in our initial modeling studies (Kedishvili *et al.*, 1995), the deletion at position 117 widens a key area near the entrance to the substrate binding site. This allows very long chain alcohols, such as retinol, to bind in an extended conformation. Thus, the high catalytic efficiency of the Class IV isoenzyme for retinol oxidation appears to be due to an enlarged active site which permits retinol to bind in a productive, low energy, conformation.

The Class III $\chi\chi$ isoenzyme exhibits the largest structural difference of any of these new structures when compared to the $\beta_1\beta_1$ isoenzyme structure. The Class III isoenzyme also shares the lowest sequence identity, 59%, to the $\beta_1\beta_1$ isoenzyme. The rmsd between all Cα atoms in the $\chi\chi$ and the $\beta_1\beta_1$ structures is 1.8 Å, although this reduces to an average of 0.75 Å if individual domains are aligned (Table 3). This large discrepancy between an alignment based on the dimer and alignments based on individual domains is due to a 6° rotation of the catalytic domain toward a more open conformation than observed in either the $\beta_1\beta_1$ or the σσ structures (Figure 1). The other substantial structural difference between the Class III isoenzyme and all other known ADH structures is the presence of a novel α-helix comprised of residues 113 to 120. It has been pointed out that the region between residues 114 and 120 is a highly variable region in ADH primary structures, with many substitutions, insertions, and deletions (Jörnvall *et al.*, 1995). Two examples of the structural consequences of these sequence differences are provided by these new structures. The Class IV enzyme exhibits a deletion at position 117 which shortens the loop between 114 and 120, widening the entrance to the substrate binding site, and permitting efficient binding of retinol (Kedishvili *et al.*, 1995). Similarly, the horse SS isoenzyme possesses a deletion at position 115 which renders the enzyme much more active toward steroids than is the EE isoenzyme (Park and Plapp, 1992). The Class III isoenzyme shows no insertions or deletions in this region but, as the consequence of many different amino acids in this region, the secondary structure of this stretch of residues changes from an extended loop to that of an α-helix. The position of this new helix and the rotation of the catalytic domain towards a more open conformation creates a substrate binding site which encloses almost twice the volume of the $\beta_1\beta_1$ substrate binding site.

An examination of the unique substrate binding site in the Class III structure provides an explanation for its substrate specificity. The inability to saturate this isoenzyme with ethanol as a substrate can be explained by the enormous size of this enzyme's substrate binding site (Figure 2). A substrate as small as ethanol cannot be productively bound in this substrate site, because there is literally nothing to hold it in place. Similar to what we observed in the σσ isoenzyme, the insensitivity of $\chi\chi$ ADH to inhibition by pyrazole derivatives appears to be due to a different active site geometry near the catalytic zinc atom and the nicotinamide ring of the coenzyme. Many studies have shown that the Class III ADH isoenzyme is identical to the glutathione-dependent formaldehyde dehydrogenase (Koivusalo et al., 1989, Kaiser et al., 1991). The enzyme oxidizes a hydrated adduct of formaldehyde and glutathione; S-hydroxymethyl-glutathione. The presence of an α-helix spanning residues 113 to 120 explains why modeling studies on Class I isoenzymes failed to explain the importance of Arg115 for binding the substrate hydroxymethyl-glutathione (Engeland *et al.*, 1993). Amino acid 115 is buried in the Class I structures and helps to stabilize the conformation of the loop structure in which it resides (Eklund *et al.*, 1976, Hurley *et al.*, 1994). Thus, when mutagenesis studies showed that Arg115 was critical for binding of substrate to this isoenzyme (Engeland *et al.*, 1993), it was difficult to rationalize this based on its position in the Class I structures. Consistent with its involvement in binding hydroxymethyl-glutathione, Arg115 is located on the solvent-exposed surface of the new helix and is in a perfect position to interact with the glycine carboxylate of the glutathione moiety. Asp57, also shown through mutagenesis to be critical for hydroxymethyl-glutathione binding (Estonius *et al.*, 1994), is located at the end of the α-helix comprised of residues 47 to 56 and can form a hydrogen bond with the α-amino group of the γ-glutamate in the glutathione moiety. Thus, this new structure can explain many of the novel features of this ADH isoenzyme.

In summary, we have determined the structures of four of the seven known human ADH isoenzymes. These structures include the $\alpha\alpha$, $\beta_1\beta_1$, $\sigma\sigma$, and $\chi\chi$ isoenzymes. These new structures provide the opportunity to investigate the functional properties of these isoenzymes at the molecular level. In addition, the structures of the $\sigma\sigma$ and $\chi\chi$ isoenzymes allow for the examination of the structural diversity which occurs among the different Classes of ADH isoenzymes. This new structural information will allow investigators to create a database of related, but not identical, structures which should permit better model-building of new ADH isoenzymes.

5. ACKNOWLEDGMENTS

The authors would like to thank Dave Vessell and Steve Parsons for technical assistance in the preparation of the recombinant human ADH isoenzymes. Supported by NIH grants K21-AA00150, P50-AA07611 and R29-AA10399.

6. REFERENCES

Brünger, A.: Crystallographic refinement by simulated annealing: application to a 2.8Å resolution structure of aspartate aminotransferase. J. Mol. Biol. 203 (1988) 803–816.

Collective Computational Project, 4.: The CCP4 suite of programs for protein crystallography. Acta Crystallogr., sect. D., 50 (1994) 760–763.

Connolly, M.L.: Solvent-accessible surfaces of proteins and nucleic acids. Science 221 (1983) 709–713.

Davis, G.J., Carr, L.G., Hurley, T.D., Li, T.-K., and Bosron, W.F.: Comparative roles of histidine 51 in human $\beta_1\beta_1$ and threonine 51 in πα alcohol dehydrogenases. Arch. Biochem.
Biophys. 311 (1994) 307–312.

Davis, G.J., Bosron, W.F., Stone, C.L., Owusu-Dekyi, K., and Hurley, T.D.: X-ray structure of human $\beta_3\beta_3$ alcohol dehydrogenase: the contribution of ionic interactions to coenzyme binding. J. Biol. Chem. 271 (1996) 17057–17061.

Edenberg, H.J. and Bosron, W.F.: Alcohol Dehydrogenases. In: Comprehensive Toxicology, Vol. 3 Biotransformation, (ed. Guengerich, F.P.) Pergamon Press, NY (in press).

Eklund, H., Nordström, B., Zeppezauer, E., Söderlund, G., Ohlsson, I., Boiwe, T., Söderberg, B.-O., Tapia, O., Brändén, C.-I., and Åkeson, Å.: Three-dimensional structure of horse liver alcohol dehydrogenase at 2.4 Å resolution. J. Mol. Biol. 102 (1976) 27–59.

Eklund, H., Samama, J.-P., Wallén, L., Brändén, C.-I., Åkeson, Å, and Jones, T.A.: Structure of a ternary complex of horse liver alcohol dehydrogenase at 2.9 Å resolution. J. Mol. Biol. 146 (1981) 561–587.

Eklund, H., Horjales, E., Vallee, B.L., and Jörnvall, H.: Computer-graphics interpretations of residue exchanges between the α, β, and γ subunits of human-liver alcohol dehydrogenase class I isozymes. Eur. J. Biochem. 167 (1987) 185–193.

Eklund, H., Muller-Wille, P., Horjales, E., Futer, O., Holmquist, B., Vallee, B.L., Höög, J.-O., Kaiser, R., and Jörnvall, H.: Comparison of three classes of human liver alcohol dehydrogenases. Emphasis on different substrate binding pockets. Eur. J. Biochem. 193 (1990) 303–310.

Engeland, K., Höög, J.-O., Holmquist, B., Estonius, M., Jörnvall, H., and Vallee, B.L.: Mutation of Arg-115 of human class III alcohol dehydrogenase: a binding site required for formaldehyde dehydrogenase activity and fatty acid activation. Proc. Nat. Acad. Sci. USA 90 (1993) 2491–2494.

Estonius, M., Höög, J.-O., Danielsson, O., and Jörnvall, H.: Residues specific for class III alcohol dehydrogenase. Site-directed mutagenesis of the human enzyme. Biochemistry 33 (1994) 15080–15085.

Hurley, T.D., Edenberg, H.J., and Bosron, W.F.: Expression and kinetic characterization of variants of human $\beta_1\beta_1$ alcohol dehydrogenase containing substitutions at position 47. J. Biol. Chem. 265 (1990) 16366–16372.

Hurley, T.D., Bosron, W.F., Hamilton, J.A., and Amzel, L.M.: Structure of human $\beta_1\beta_1$ alcohol dehydrogenase: catalytic effects of non-active site substitutions. Proc. Nat. Acad. Sci. USA 88 (1991) 8149–8153.

Hurley, T.D., Bosron, W.F.: Human alcohol dehydrogenase: dependence of secondary alcohol oxidation on the amino acids at positions 93 and 94. Biochem. Biophys. Res. Commun. 183 (1992) 93–99.

Hurley, T.D., Yang, Z.-N., Bosron, W.F., and Weiner, H.: Crystallization and preliminary X-ray analysis of bovine mitochondrial aldehyde dehydrogenase and human glutathione-dependent formaldehyde dehydrogenase. In: Enzymology and Molecular Biology of Carbonyl Metabolism 4, (Weiner, H., Crabb, D., and Flynn, G., eds) Plenum Press, New York, (1993) 245–250.

Hurley, T.D., Bosron, W.F., Stone, C.L., and Amzel, L.M.: Structures of three human β alcohol dehydrogenase variants: correlations with their functional differences. J. Mol. Biol. 239 (1994) 415–429.

Jörnvall, H., Danielsson, O., Hjelmqvist, L., Persson, B., and Shafqat, J.: The alcohol dehydrogenase system. In: Enzymology and Molecular Biology of Carbonyl Metabolism 5, (Weiner, H., Holmes, R., and Wermuth, B.) Plenum Press, New York, (1995) 281–294.

Jörnvall, H., Höög, J.-O.: Nomenclature of alcohol dehydrogenases. Alcohol and Alcoholism 30 (1995) 153–161.

Kaiser, R., Holmquist, B., Vallee, B.L., and Jörnvall, H.: Human class III alcohol dehydrogenase/glutathione-dependent formaldehyde dehydrogenase. J. Prot. Chem. 10 (1991) 69–73.

Kedishvili, N.Y., Bosron, W.F., Stone, C.L., Hurley, T.D., Peggs, C.F., Thomasson, H.R., Popov, K.M., Carr, L.G., Edenberg, H.J., and Li, T.-K.: Expression and kinetic characterization of recombinant human stomach alcohol dehydrogenase. J. Biol. Chem. 270 (1995) 3625–3630.

Koivusalo, M., Baumann, M., and Uotila, L.: Evidence for the identity of glutathione-dependent formaldehyde dehydrogenase and class III alcohol dehydrogenase. FEBS Letters 257 (1989) 105–109.

Kuntz, I.D., Blaney, J.M., Oatley, S.J., Langridge, R., Ferrin, T.E.: A geometric approach to macromolecular-ligand interactions. J. Mol. Biol. 161 (1982) 269–288.

Navaza, J.: AMoRe: An automated package for molecular replacement. Acta Cryst., sect. A, 50 (1994) 157–163.

Park, D.-H., Plapp, B.V.: Interconversion of E and S isoenzyme of horse liver alcohol dehydrogenase. J. Biol. Chem. 267 (1992) 5527–5533.

Stone, C.L., Li, T.-K., and Bosron, W.F.: Stereospecific oxidation of secondary alcohols by human alcohol dehydrogenases. J. Biol. Chem. 264 (1989) 11112–11116.

Stone, C.L., Bosron, W.F., and Dunn, M.F.: Amino acid substitutions at position 47 of human $\beta_1\beta_1$ and $\beta_2\beta_2$ alcohol dehydrogenases affect hydride transfer and coenzyme dissociation rate constants. J. Biol. Chem. 268 (1993) 892–899.

Stone, C.L., Thomasson, H.R., Bosron, W.F., and Li, T.-K.: Purification and partial amino acid sequence of a high-activity human stomach alcohol dehydrogenase. Alcoholism: Clin. Exptl. Res. 17 (1993) 911–918.

Wagner, F.W., Parés, X., Holmquist, B., and Vallee, B.L.: Physical and enzymatic properties of a class III isoenzyme of human liver alcohol dehydrogenase: χ-ADH. Biochemistry 23 (1984) 2193–2199.

Yang, Z.-N., Davis, G.J., Hurley, T.D., Stone, C.L., Li, T.-K., and Bosron, W.F.: Catalytic efficiency of human alcohol dehydrogenases for retinol oxidation and retinal reduction. Alcoholism: Clin. Exptl. Res. 18 (1994) 587–591.

35

MAMMALIAN CLASS II ALCOHOL DEHYDROGENASE

A Highly Variable Enzyme

Jan-Olov Höög and Stefan Svensson

Department of Medical Biochemistry and Biophysics
Karolinska institutet
S-171 77 Stockholm, Sweden

1. INTRODUCTION

The number of classes of mammalian alcohol dehydrogenase (ADH) has increased over the last years. Today six classes have been identified (Jörnvall and Höög, 1995), with class V and VI as the most novel forms. Class II ADH, one of the early characterized forms, (Li et al., 1977; Vallee and Bazzone, 1983) is poorly investigated, except for the human enzyme, in contrast to the classical ADH, class I, and the glutathione-dependent formaldehyde dehydrogenase, class III. In addition, the stomach/extrahepatic ADH, class IV, has been extensively investigated the last years.

The class II ADH with the π subunit was isolated as a high K_m enzyme from human liver and was at that time suggested to be a determinant for alcoholism (Li et al., 1977). The corresponding rat enzyme was confused with the stomach ADH (class IV) for a long time, but sequence analysis showed that the stomach ADH and class II were of separate origin (Parés et al., 1990; Höög, 1995). Class II ADHs have also been identified and to some extent characterized from different monkeys (Dafeldecker et al., 1981a; 1981b; 1985), horse (Seeley and Holmes, 1984) and ostrich (Hjelmqvist et al., 1995) (Table 1). The kinetic values differ widely between these species with K_m values for ethanol in the range of 0.7 mM - 120 mM (Hjelmqvist et al., 1995; Ditlow et al., 1984). The values from the different primates differ as well. The ostrich class II enzyme, also characterized structurally, shows kinetic constants similar to the class I ADHs, but the primary structure shows a class II enzyme (Hjelmqvist et al., 1995). The horse class II ADH has values very similar to the constants determined for the human class II enzyme (Seeley and Holmes, 1984). The human enzyme has been shown to participate in the metabolism of dopamine, norepinephrine, serotonin and 4-hydroxyalkenals (Mårdh and Vallee, 1985; Mårdh et al., 1986; Consalvi et al., 1986; Sellin et al., 1990). The class II ADH has further been postulated to be regulated by thyroid hormones (Mårdh et al., 1987) and to reduce p-nitrosophenol (Maskos and Winston, 1994). Recently, it was shown

Enzymology and Molecular Biology of Carbonyl Metabolism 6
edited by Weiner *et al.* Plenum Press, New York, 1996

Table 1. Characterized class II ADHs

	Enzymatical Analysis	Structural Analysis	References
Human	1977	1987	Li et al., 1977; Höög et al., 1987
Squirrel monkey	1981	--	Dafeldecker et al., 1981a
Rhesus monkey	1981	--	Dafeldecker et al., 1981b
Horse	1984	--	Seeley et al., 1984
Macaca nemestrina	1985	--	Dafeldecker et al., 1985
Ostrich	1995	1995	Hjelmqvist et al., 1995
Rat	1996*	1995	Höög, 1995; unpublished data
Rabbit	1996*	1996	unpublished data

* Isolated as recombinant protein

that the class II enzyme was the most active hepatic ADH in retinol metabolism, but less active than the extrahepatic class IV enzyme (Yang et al., 1994). From tissue distribution studies the rat class II enzyme shows limited distribution with an mRNA localization to the liver and the intestinal tract (Estonius et al., 1993). In human, the distribution is limited mainly to the liver with trace amounts detected in the small intestine (unpublished data). At the gene level the class II gene, *ADH4*, has the same organization as the other investigated ADHs with eight introns and nine exons, but the introns are larger as compared to other ADH genes (von Bahr-Lindström et al., 1991). From cloning work and protein characterization it is clear that different variants (allelic forms) occur of the class II ADH (Höög et al., 1987; 1988).

In addition, class II ADHs have now been characterized from rabbit and rat. All results taken together show that this enzyme is highly variable both at the structural and functional level.

2. MATERIALS AND METHODS

2.1. Isolation of cDNAs

5'extended cDNA-libraries in λgt11, with mRNA isolated from New Zealand female rabbits and Sprague-Dawley rats (Clontech) were screened with a 350-bp cDNA fragment coding for human class II ADH (Höög et al., 1987). Isolated phages were treated with restriction enzymes to liberate cDNA-inserts for size estimation by agarose gel electrophoresis and subsequent subcloning into plasmid vectors. In cases of non-full-length cDNAs, an adaptor-ligated cDNA library was created from poly-A enriched liver mRNA (Clontech) utilizing reverse transcription and second-strand synthesis with a mixture of *E. coli* DNA polymerase I, Rnase H and DNA ligase (Marathon cDNA Amplification, Clontech). The 5'-ends of the cDNAs were subsequently isolated by adapter-mediated PCR amplification using adapter and cDNA specific primers. DNA sequence analyses were performed on both strands of the entire cDNAs using T7 DNA polymerase (Pharmacia Biotech) in combination with [α-^{35}S]dATP (Amersham). All DNA sequences obtained were analyzed on a Decstation computer using the Wisconsin Package (Genetics Computer Group, Univ. of Wisconsin) and compared with EMBL data banks.

2.2. Expression of Recombinant Protein

cDNA coding for human class II ADH was subcloned into a pET3d expression vector (Novagen) and the corresponding rat cDNA was subcloned into pET15b. The cDNAs coding for rabbit class II ADHs were subcloned into pET3d (II-1) and pET12b (II-2). Prior to subcloning into the expression plasmids the coding parts of the cDNAs were PCR amplified with *Pfu* polymerase (Stratagene) to introduce suitable restriction enzyme sites.

Recombinant proteins were expressed from pET3d and pET12b in 2 l cultures using an *E. coli lac*I^q strain after bursts of 0.2 mM isopropyl-β-thiogalactopyranoside. After harvesting, the cells were disrupted in 1 mM dithiothreitol, 10 mM Tris-Cl, pH 8, sonicated intermittently and centrifuged at 48,000xg. The supernatants were applied on DEAE-cellulose columns (DE-52, Whatman). The void volumes, containing recombinant ADH, were applied to a HiTrap Sephablue column (Pharmacia Biotech), human and rabbit II-1, or to an AMP-Sepharose column (Pharmacia Biotech), rabbit II-2. Recombinant protein from pET15b was expressed in 100 ml cultures. After harvesting the cells were disrupted in loading buffer, 20 mM imidazole and 15 mM NaCl, pH 7.5. The supernatant, after sonication and centrifugation, was applied to a metal chelating column (Novagen) loaded with nickel-ions. The column was washed with loading buffer and the protein was eluted with 200 mM imidazole. In all cases a final purification step was performed on a Superose 12 column/FPLC (Pharmacia Biotech). Recombinant proteins were analyzed for purity by SDS-polyacrylamide gel electrophoresis. Protein concentrations were determined according to the method of Bradford (1970), standardized with bovine serum albumin.

Enzymatic activities were determined spectrophotometrically at 340 nm using a Pharmacia Biochrom 4060 or a Hitachi 3000 spectrophotometer to follow the NADH formation in 0.1 M glycine-NaOH, pH 10, 25°C or in sodium phosphate, pH 7.5, 25°C. The NAD$^+$ concentration used was 2.4 mM, and the absorption coefficient for NADH 6.22x10^3 M^{-1}cm^{-1} was used. Ethanol of analysis grade was used without further purification and NAD$^+$ grade III was from Sigma. A weighted non-linear regression analysis program, was used to fit all lines to data points and to calculate kinetic parameters.

3. RESULTS

Rat and rabbit liver cDNA libraries were screened with cDNA-fragments coding for human class II ADH. For rabbit class II ADH, one full-length clone and one clone of partial length were isolated, coding for class II-1 and II-2, respectively. The rat cDNA coding for class II was of partial length, but the entire 5′-region was isolated with reverse-transcription/PCR amplification (Höög, 1995). For rabbit class II-1 cDNA a poly-A-tail was isolated but no perfect poly-A-signal was identified. The rabbit class II-2 cDNA was not complete and the 5'part was isolated with reverse transcription and adapter-mediated PCR amplification to yield a full-length cDNA. The two forms of cDNA coding for rabbit class II ADH show 89.0% positional identity in the coding part.

The rabbit class II ADHs translate into 378 amino acid residues and show that mammalian class II ADHs differ to a large extent (Fig. 1). The rabbit class II enzymes both have two insertions, as compared to the class I ADHs, one close to position 60 and four residues close to position 120. One deletion is observed in the region around position 325. All these characteristics are found in the two variants of the rabbit class II ADH which form isozymes, class II-1 and II-2. These variants show 44 positions with altered amino acid residues which results in 88.4% positional identity at the protein level. Residues affected at the active site pocket are at positions 51, 117, 146 and 323 (corresponding to positions 51, 116, 141 and 318 in the standard class I numbering system). The N-terminal residues differ as well between the two forms, Gly in II-1 and Ser in II-2. The amino acid residue exchanges between the class II isozymes are clustered into three regions, the same regions affected by the insertions and the deletion.

Figure 1. Alignment of class II ADH amino acid sequences. From top to bottom: Rabbit isozyme 1, rabbit isozyme 2, human, rat and ostrich. Conserved residues are inverted in black. - denotes deletions. The numbering above each sequence refers to the individual numbering also indicating the difference in polypeptide lengths.

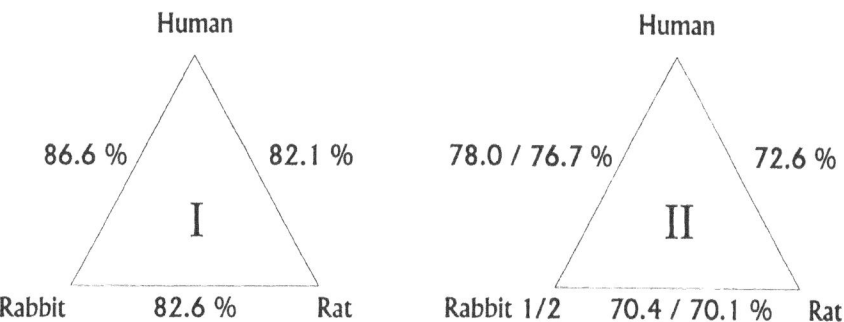

Figure 2. Species comparisons for class I and class II ADHs. The positional identities given are at the protein level. The double values given for rabbit refer to isozyme 1 and 2.

The rat structure shows an active site that differs at many positions as compared to other ADHs. Worthwhile to notice is Pro at position 47 (Fig. 1) and a deletion of two residues in the region around position 300. With the alignment used here (Fig. 1; cf. Höög, 1995) the rat class II shows Ala at position 298 (Fig. 1), a position showing Val in all analyzed mammalian ADHs.

Positional identities for rabbit class II ADHs to the corresponding enzymes from human and rat are 76.7–78.0% and 70.1–70.4%, respectively. The corresponding value for class I ADHs are 86.6% and 82.6%, respectively (Fig. 2). The now characterized rabbit enzymes consist of 378 amino acid residues, in contrast to the human and rat class II ADHs which consists of 379 and 376 residues, respectively (Fig 1).

All isolated cDNAs coding for mammalian class II ADH were expressed in *E. coli*. The recombinant proteins were purified to homogeneity with ion-exchange and affinity chromatography. The rat class II ADH was purified as a fused protein with a His-tag. Specific activities were determined in glycine-NaOH, pH 10, for the four different expressed enzymes. The obtained results are: 1.1 U/mg for human class II, 0.6 U/mg for rabbit 1, 1.8 U/mg for rabbit 2 and 30 U/mg for the rat enzyme. The kinetic constants determined at pH 7.5 differed as well between the species (Table 2). The difference in catalytic efficiency (k_{cat}/K_m) for ethanol is over 1000-fold, from 1.1 $\text{min}^{-1}\text{mM}^{-1}$ for the human enzyme to 1500 $\text{min}^{-1}\text{mM}^{-1}$ for the highly active rat enzyme.

Table 2. Comparison of kinetic constants for different class II ADHs. All values are given for ethanol as substrate with an NAD^+ concentration of 2.4 mM. The values for the human, rat and rabbit enzymes were determined in sodium phosphate, pH 7.5

Species	k_{cat} (min^{-1})	K_m (mM)	k_{cat}/K_m (mM^{-1}min^{-1})
Human	11.5	10.5	1.1
Rat	1200	0.8	1500
Rabbit -1	15	8.6	1.7
Rabbit -2	60	0.6	100*
Ostrich (pH 10)	36	0.71	51

* The values given for rabbit isozyme 2 are from one preparation only.

4. DISCUSSION

cDNA clones coding for class II ADHs have been isolated from human, rat and rabbit liver cDNA libraries. Human is the only source of these from which a class II enzyme has been isolated at the protein level (Höög et al., 1987), and this enzyme has been shown to exist in different variants that form allelozymes (Höög et al., 1988; unpublished). The rat enzyme has been cloned and the deduced amino acid sequence show a protein with many residue exchanges especially at the active site (Höög, 1995). Recently, two different clones coding for rabbit class II ADH were isolated. These latter clones differed at several positions where 44 positions altered amino acid residues. This clearly shows that isozymic forms occur in the rabbit with a positional identity of 88.4% between the isozymes. Furthermore, the rabbit enzymes are the first mammalian non-class I ADHs with a serine residue at position 48, a residue shown to be essential for steroid dehydrogenase activity and testosterone inhibition (Höög et al., 1992). Further prerequisites for steroid dehydrogenase activity and testosterone inhibition are small residues around position 115 which will enlarge the substrate binding pocket. In the horse S subunit of class I a deletion is found at position 115, which will enlarge the pocket for steroid binding (Park & Plapp, 1991). The rabbit class II isozymes differ in many respects as compared to other characterized class II ADHs. A comparison shows that one of the rabbit isozymes is a mixture of class I and class II ADHs at the active site. The residues after the first zinc ligand, Cys46, are almost identical to residues of class I ADHs, including His at position 51. The latter residue has been suggested to participate in a charge relay system (Eklund et al., 1982). However, in the rabbit class II-2 the residues are of non-class I type. Position 51, e.g., is occupied by Tyr in II-2, a residue typical of class III ADH. The residues around position 115 in the rabbit class II ADH are to some extent small, but in isozyme 1 position 117 is occupied by Lys which introduces a positive charge at the middle part of the substrate binding pocket. Rabbit class II-2, on the other hand, shows Ser116, Glu117 and His118. Ser116 is also found in the human enzyme, but Glu117 and His118 introduce charges at these positions (positions 116, 117 and 118 correspond to positions 115, 116 and 117 in class I and rat class II). The rat class II shows Arg at position 115, a residue which has been shown to be essential for glutathione-dependent formaldehyde dehydrogenase activity in the class III ADH (Engeland et al., 1993). However, the rat class II ADH has no such activity. At position 146 rabbit class II-1 shows Phe, but II-2 shows Leu (Fig. 1). Phe at this position is also found in the human enzyme and a large hydrophobic residue at the middle part of the active site pocket can be an explanation for the high K_m value (cf. Fig. and Table 2). At the outer part of the pocket in the rabbit enzyme Asn is identified at position 323 in II-1, but the corresponding residue is Ser in II-2. This position is occupied by hydrophobic residues in other characterized mammalian ADHs. The differences between rabbit class II-1 and II-2 suggest that the kinetic characteristics will differ between the two isozymes (see below and cf. Table 2). The human class II ADH was isolated as a high K_m enzyme (Li et al., 1977), but a high K_m for class II ADHs seems not to be of general validity (Table 2). Comparison of residues lining the active site pocket in the different class II ADHs suggest that the kinetic values will differ widely between the species. Position 47, shown to interact with the coenzyme, is occupied by His, Pro and Arg in the human, rat and rabbit enzymes, respectively. The rat class II ADH has several unique residues at the inner part of the pocket, e.g. Pro47, which probably is involved in the low class I type K_m value for ethanol (Table 2; unpublished data). Furthermore, the ostrich class II ADH has similar kinetic characteristics as the class I enzymes which is, in some respect, expected from the amino acid residues lining the substrate binding pocket (Table 2; Hjelmqvist et al., 1995).

If this is due to the typical His51 of class I is impossible to judge, but probably that residue has a large influence on the characteristics. In contrast, the rabbit class II-1, with His at position 51, has a K_m value similar to that of human class II ADH.

As a result of the highly divergent primary structures for the class II ADHs the catalytic efficiencies differ widely. The K_m values for the class II ADHs can be divided into two groups. One group has K_m values around 1 mM, like the class I ADHs, and another group has K_m values around 10 mM. The first group includes the rabbit isozyme 2, rat and ostrich ADHs and the other group includes the human enzyme and rabbit isozyme 1. Furthermore, these latter class II ADHs show almost identical catalytic efficiencies. The K_m value determined for the liver isolated human enzyme is 34 mM (Li et al., 1977), but lower values have been reported for recombinant enzymes (Davis et al., 1994). However, the now determined K_m value is lower yet, which maybe can be explained by different allelic variants. A more remarkable difference is observed for the rat enzyme with a very high k_{cat} value resulting in an over 1000-fold higher catalytic efficiency as compared to the human enzyme (Table 2).

From sequence alignments of class II ADHs a pattern is visible. Insertions, deletions and residue exchanges are clustered in three regions. These are around positions 60, 120 and 300 (Fig. 1). Insertions are found around position 60 (one residue) and around position 120 (four residues). Deletions are found around position 300 in the rat and around position 325 in the rabbit enzymes. These regions have been shown to be highly variable from alignment of different classes (Jörnvall et al., 1987; Persson et al., 1993). Hence, it is now evident that class II ADHs from different species show similar pattern as ADHs of different classes. In addition to the highly variable regions, class II ADHs also differ in polypeptide chain lengths. A span from 376 to 379 residues is observed. The total positional identities are low, 70% - 78%, as compared to other investigated classes of ADHs, showing that mammalian class II ADH is the most divergent form of the three main hepatic ADHs. The conserved structure of the class III ADHs is also shown at the functional level with almost identical kinetic values between the species. The class I ADH has been classified as an enzyme with a high evolutionary rate (Yin et al., 1991), and now the class II ADH is shown to have a very high evolutionary rate. The non hepatic ADH, class IV, has an evolutionary rate between class I and class III (Parés et al., 1994).

The tissue distributions show that class II alcohol dehydrogenase is liver specific with very low amounts identified in other tissues like kidney, stomach and the intestinal tract (Estonius et al., 1993; unpublished data). The very similar tissue distributions for human and rat class II ADHs are in contrast to their very different catalytic efficiencies, probably reflecting the overall high ethanol metabolic rate in the rat. The results obtained for the rabbit class II ADHs further strengthen the divergence of this ADH class (Höög, 1995) and show isozymic forms of rabbit class II ADH. For the human enzyme there is evidence for different forms of class II (Höög et al., 1988), but these forms are rather allelozymes than isozymes. However, the isozymic forms of rabbit class II ADH are conceivable with the pattern of the system of ADH with a split into different enzymes, classes, isozymes and finally allelic variants (Jörnvall et al., 1987). The entire ADH system is complex with isozymes now found in a further mammalian class of ADH.

5. ACKNOWLEDGMENTS

This work was supported by grants from the Swedish Medical Research Council and the Karolinska Institutet.

6. REFERENCES

Bradford, M.M.: A rapid and sensitive method for quantitation of microgram quantities of protein utilizing the principle of protein-dye binding. Anal. Biochem. 72 (1976) 248–254.

Consalvi, V., Mårdh, G. and Vallee, B.L.: Human alcohol dehydrogenases and serotonin metabolism. Biochem. Biophys. Res. Commun. 139 (1986) 1009–1016.

Dafeldecker, W.P., Parés, X., Vallee, B.L., Bosron, W.F. and Li, T.-K: Simian liver alcohol dehydrogenase: Isolation and characterization of isozymes from Saimiri sciureus. Biochemistry 20 (1981a) 856–861.

Dafeldecker, and Vallee, B.L.: Simian liver alcohol dehydrogenase: Isolation and characterization of isozymes from Macaca mulatta. Biochemistry 20 (1981b) 6729–6734.

Dafeldecker, W.P., Liang, S.-J. and Vallee, B.L.: Simian liver alcohol dehydrogenase: Isolation and characterization of isozymes from Macaca Nemestrina. Biochemistry 24 (1985) 6474–6479.

Davis, G.J., Carr, L.G., Hurley, T.D., Li, T.-K. and Bosron, W.F.: Comparative roles of histidine 51 in human $\beta_1\beta_1$ and threonine in $\pi\pi$ alcohol dehydrogenases. Arch. Biochem. Biophys. 311 (1994) 307–312.

Ditlow, C.C., Holmquist, B., Morelock, M.M. and Vallee, B.L.: Physical and enzymatic properties of a class II alcohol dehydrogenase isozyme of human liver: π-ADH. Biochemistry 23 (1984) 6363–6368.

Eklund, H., Plapp, B.V., Samama, J.-P. and Brändén, C.-I. (1982) Binding of substrate in a ternary complex of horse liver alcohol dehydrogenase. J. Biol. Chem., 257 (1982) 14349–14358.

Engeland, K., Höög, J.-O., Holmquist, B., Estonius, M., Jörnvall, H. and Vallee, B.L.: Mutation of Arg-115 of human class III alcohol dehydrogenase: A binding site required for formaldehyde dehydrogenase activity and fatty acid activation. Proc. Natl. Acad. Sci. USA 90 (1993) 2491–2494.

Estonius, M., Danielsson, O., Karlsson, C., Persson, H., Jörnvall, H. and Höög, J.-O.: Distribution of alcohol and sorbitol dehydrogenases: Assessment of mRNAs in rat tissues. Eur. J. Biochem. 215 (1993) 497–503.

Hjelmqvist, L., Estonius, M. and Jörnvall, H.: The vertebrate alcohol dehydrogenase system: Variable class II type form elucidates separate stages of enzymogenesis. Proc. Natl. Acad. Sci. USA 92 (1995) 10905–10909.

Höög, J.-O.: Cloning and characterization of a novel rat alcohol dehydrogenase of class II type. FEBS Lett. 368 (1995) 445–448.

Höög, J.-O., von Bahr-Lindström, H., Hedén, L.-O., Holmquist, B., Larsson, K., Hempel, J., Vallee, B.L. and Jörnvall, H.: Structure of the class II enzyme of human liver alcohol dehydrogenase. Combined cDNA and protein sequence determination of the π subunit. Biochemistry 26 (1987) 1926–1932.

Höög, J.-O., von Bahr-Lindström, H., Hedén, L.-O., Vallee, B.L. and Jörnvall, H.: Evidence for structural heterogeneity among subunits of class II human liver alcohol dehydrogenase. In: Biomedical and Social Aspects of Alcohol and Alcoholism (eds. Kuriyama, K., Takada, A. & Ishii, H.) (1988) Elsevier pp. 87–90.

Höög, J.-O., Eklund, H. and Jörnvall, H.: A single-residue exchange gives human recombinant $\beta\beta$ alcohol dehydrogenase $\gamma\gamma$ isozyme properties. Eur. J. Biochem. 205 (1992) 519–526.

Jörnvall, H. and Höög, J.-O.: Nomenclature of alcohol dehydrogenases. Alcohol Alcoholism 30 (1995) 153–161.

Jörnvall, H., Höög, J.-O., von Bahr-Lindström, H. and Vallee, B.L.: Mammalian alcohol dehydrogenases of separate classes. Intermediates between different enzymes and intraclass isozymes. Proc. Natl. Acad. Sci. USA 84 (1987) 2580–2584.

Li, T.-K., Bosron, W.F., Dafeldecker, W.P., Lange, L.G. and Vallee, B.L.: Isolation of Π-alcohol dehydrogenase of human liver: Is it a determinant of alcoholism? Proc. Natl. Acad. Sci. USA 74, (1977) 4378–4381.

Maskos, Z. and Winston, G.W.: Mechanism of p-nitrosophenol reduction catalyzed by horse liver and human π-alcohol dehydrogenase (ADH). J. Biol. Chem. 269 (1994) 31579–31584.

Mårdh, G. and Vallee, B.L.: Human class I alcohol dehydrogenases catalyze the interconversion of alcohols and aldehydes in the metabolism of dopamine. Biochemistry 25 (1986) 7279–7282.

Mårdh, G., Dingely, A.L., Auld, D.S. and Vallee, B.L.: Human class II (π) alcohol dehydrogenase has a redox-specific function in norepinephrine metabolism. Proc. Natl. Acad. Sci. USA 83 (1986) 8908–8912.

Mårdh, G., Auld, D.S. and Vallee, B.L.: Thyroid hormones selectively modulate human alcohol dehydrogenase isozyme catalyzed ethanol metabolism. Biochemistry 26 (1987) 7585–7588.

Parés, X., Moreno, A., Cederlund, E., Höög, J.-O. and Jörnvall, H.: Class IV mammalian alcohol dehydrogenase. Structural data of the stomach enzyme reveal a new class well separated from those already characterized. FEBS Lett. 277 (1990) 115–118.

Parés, X., Cederlund, E., Moreno, A., Hjelmqvist, L., Farrés, J. and Jörnvall, H.: Mammalian class IV alcohol dehydrogenase (stomach alcohol dehydrogenase): Structure, origin and correlation with enzymology. Proc. Natl. Acad. Sci. USA 91 (1994) 1893–1897.

Park, D.-H. & Plapp, B.V.: Isoenzymes of horse liver alcohol dehydrogenase active on ethanol and steroids. J. Biol. Chem. 266 (1991) 13296–13302.

Persson, B., Bergman, T., Keung, W.M., Waldenström, U., Holmquist, B., Vallee, B.L. and Jörnvall, H.: Basic features of class-I alcohol dehydrogenase: variable and constant segments coordinated by inter-class and intra-class variability. Conclusions from the characterization of the alligator enzyme. Eur. J. Biochem. 216 (1993) 49–56.

Seeley, T.-L. and Holmes, R.S.: Purification and molecular properties of a class II alcohol dehydrogenase (ADH-C2) from horse liver. Int. J. Biochem. 16 (1984) 1037–1042.

Sellin, S., Holmquist, B., Mannervik, B. and Vallee, B.L.: Oxidation and reduction of 4-hydroxyalkenals catalyzed by isozymes of human alcohol dehydrogenase. Biochemistry 30 (1990) 2514–2518.

von Bahr-Lindström, H., Jörnvall, H. and Höög, J.-O.: Cloning and characterization of the human *ADH4* gene. Gene 103 (1991) 269–274.

Vallee, B.L. and Bazzone, T.J.: Isozymes of human liver alcohol dehydrogenase. Isozyme 8 (1983) 219–244.

Yang, Z.N., Davies, G.J., Hurley, T.D., Stone, C.L., Li, T.-K. and Bosron, W.F.: Catalytic efficiency of human alcohol dehydrogenases for retinol oxidation and retinal reduction. Alcohol Clin. Exp. Res. 18 (1994) 587–591.

Yin, S.-J., Vagelopoulos, N., Wang, S.-L. and Jörnvall, H.: Structural features of aldehyde dehydrogenase distinguish dimeric aldehyde dehydrogenase as a variable "enzyme." "Variable" and "constant" enzymes within the alcohol and aldehyde dehydrogenase families. FEBS Lett. 283 (1991) 85–88.

36

ACTIVITY OF LIVER ALCOHOL DEHYDROGENASES ON STEROIDS

Françoise Strasser, Minh N. Huyng, and Bryce V. Plapp

Department of Biochemistry
The University of Iowa
Iowa City, Iowa 52242

1. INTRODUCTION

Four mammalian class I liver alcohol dehydrogenases catalyze stereospecific oxidoreduction at the 3-position of 5β-steroids: the horse S, human γ, rat and rabbit enzymes (Waller et al., 1965; McEvily et al., 1988; Reynier et al., 1975; Höög et al., 1993). These enzymes may provide a connection between the metabolism of ethanol and steroids, since alcohol oxidation raises the NADH/NAD$^+$ ratio in the liver and thus shifts the redox state in favor of the reduced steroid (Anderson et al., 1986). Evaluating the role of liver alcohol dehydrogenase in steroid metabolism requires more complete knowledge of substrate specificities, kinetic parameters and species variation.

The homologous horse E and the other human liver alcohol dehydrogenases (class I (α,β), class II (π) and class III (χ)) have no, or low, activity on steroids, but oxidize ethanol and a variety of other alcohols. The substrate binding pockets of the enzymes active on steroids must differ in size and shape to be able to accommodate bulky tetracyclic molecules.

Amino acid sequence comparisons (Sun and Plapp, 1992) show three main differences between the substrate binding pockets of subunits E and γ: Phe110, Ser117 and Leu141 in E are replaced by Tyr110, Gly117 and Val141 in γ (Table 1). Gly117 is also found in rabbit liver ADH, and Leu110 is found in rat liver ADH; both of these enzymes are active on steroids. Computer modeling was used to dock 5β-androstane-3β,17β-diol in the active site of the E subunit (solved at 2.1 Å by x-ray crystallography; Ramaswamy et al., 1994). The modeling (Figure 1) suggests that Ser117 and Phe110 are not directly involved in binding steroid as they are located at the opening of the substrate binding pocket. Leu141 (E subunit) appears to have a steric conflict with the steroid, and this would be relieved with Val. However, the horse S subunit has Leu and the rat and rabbit enzymes have Ile, so that the Leu to Val substitution does not appear to be critical. The side chain of Leu116 also exhibits steric hindrance with the C and D rings of the 5β-androstane-3β,17β-diol, but this residue is conserved in all four enzymes active on steroids

Enzymology and Molecular Biology of Carbonyl Metabolism 6
edited by Weiner *et al.* Plenum Press, New York, 1996

Table 1. Residues in the substrate binding pockets of liver alcohol dehydrogenases[a]

| Residue | Horse | | Human | | | Rat | Rabbit |
	E	S	α	β	γ		
48	Ser	Ser	Thr	Thr	Ser	Ser	Ser
57	Leu	Leu	Met	Leu	Leu	Leu	Ile
93	Phe	Phe	Ala	Phe	Phe	Phe	Phe
110	Phe	Leu	Tyr	Tyr	Tyr	Leu	Phe
115	Asp	Del	Asp	Asp	Asp	Asn	Asp
116	Leu	Leu	Val	Leu	Leu	Leu	Leu
117	Ser	Ser	Ser	Gly	Gly	Thr	Gly
141	Leu	Leu	Leu	Leu	Val	Ile	Ile
318	Ile	Ile	Ile	Val	Ile	Ile	Ile
Steroid activity	low	yes	no	no	yes	yes	yes

[a] Höög et al. (1993). In addition, Val294 and Met306 (furnished by the second subunit) are conserved.

and must be accommodated. Previous studies showed that deletion of Asp115 in the horse E enzyme resulted in an enzyme active on steroids (similar to horse S subunit). The deletion could move the adjacent Leu116 into a space provided by the substitution of Phe110 with Leu in the S subunit (Park and Plapp, 1992).

In order to elucidate the structural requirements of the γ enzyme for steroid recognition, we attempted to convert the active site of the well-characterized horse E subunit into a γ-like substrate binding pocket. Changing Ser117 into Gly and/or Phe110 into Leu should increase the size of the substrate pocket and allow for more flexibility in binding large molecules such as steroids. We prepared the S117G, F110L and F110L-S117G enzymes and studied their kinetic properties and activities on steroids. Reduction of 5β-cholanoic acid-3-one (negatively charged) and 5β-androstane-17β-ol-3-one (uncharged) was measured to evaluate steroid activity. In contrast to previous studies, we found that the

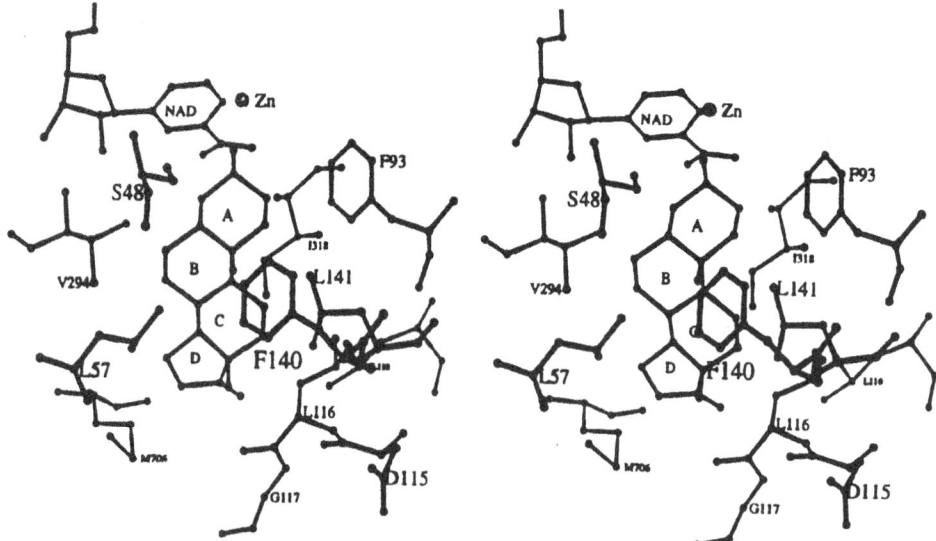

Figure 1. Modeling of 5β-androstane-3β,17β-diol binding to *Eq*ADH-E with the F110L and S117G mutations.

horse liver E isoenzyme has low, but measurable, activity on steroids. The mutations, however, had relatively small effects on the activities. It appears that other residues or indirect effects are responsible for activities on steroids.

2. EXPERIMENTAL PROCEDURES

Native horse liver ADH (EqADH), NAD$^+$ (grade I, free acid or Li salt), NADH (grade I, disodium salt) were purchased from Boehringer Mannheim; DEAE-Sepharose Fast Flow and SP-Sepharose Fast Flow were from Pharmacia. Steroids (5β-cholanoic acid-3-one and 5β-androstane-17β-ol-3-one) were purchased from Steraloids, Inc.

Recombinant EqADH was purified from $E.$ $coli$ XL1-Blue (Stratagene) transformed with the expression phagemid pBPE/EqADH (5.2-kb), which contains the coding region for horse liver ADH (1.5-kb) under the control of tac promoter (Park and Plapp, 1991). This construct also contains a replication origin for f1 bacteriophage that allows the phagemid to be rescued as single-stranded DNA with VCSM13 helper phage, and used as a template for site-directed mutagenesis. Three oligonucleotides of 21 bases each were synthesized to randomly change Phe110, Ser117 and Leu141. GAAGGCAACT(A/T)(A/T)TGCTTGAAA was prepared to modify the TTC codon of Phe110 into TTA for Leu, TAT for Tyr (or TTT for Phe and TAA for the ochre stop codon). AATGATCTGG(G/T)CATGCCTCGG was prepared to change AGC of Ser117 into GGC for Gly (or CTG for Val). CAC-CACTTC($G/T/A$)TTGGCACCAGC was prepared to change the CTT codon of Leu141 into GTT for Val, TTT for Phe and ATT for Ile. Mutagenesis was performed on the recombinant single-stranded DNA template with one, two or three oligonucleotides, using the Sculptor in vitro mutagenesis system (Amersham) based on the phosphorothioate technique (Nakamaye and Eckstein, 1986). Mutations were identified by sequencing the phagemids with the dideoxy chain termination method using the Sequenase Version 2.0 Sequencing Kit (United States Biochemical) and [α^{35}S]dATP (Amersham).

The mutated enzymes were purified essentially as described for wild-type enzyme (Park and Plapp, 1991), except that the protein eluted from the DEAE-Sepharose gel was precipitated with 25 % PEG4000 and resuspended in 5 mM sodium phosphate buffer, pH 7.5, containing 0.25 mM EDTA before being loaded onto the SP-Sepharose column. The concentration of active sites (normality, N) was determined by titration with NAD$^+$ in the presence of 10 mM pyrazole ($\Delta\varepsilon_{294\ nm}$ = 9,000 M^{-1} cm^{-1}, Shearer et al., 1993) using double difference spectroscopy (Theorell and Yonetani, 1963). ADH protein has an A$_{280\ nm}$ for 1 mg/ml of 0.455.

3. RESULTS

Random mutagenesis yielded three single mutants (F110L, F110Y, S117G), three double mutants (F110L-S117G, F110L-L141V, S117G-L141V), but no triple mutant. We studied the F110L-S117G and F110L-S117G enzymes. F110Y enzyme was not stable, preventing kinetic characterization of this form.

The expression yield of F110L enzyme was the same as for wild-type enzyme: 450 units in the lysate from 50 g of wet cells, harvested from 9 liters of culture. The expression yields for S117G and S117G-F110L enzymes were lower, with 300 units and 180 units, respectively. During purification, the mutated enzymes all behaved like wild-type ADH. The overall enzyme recovery of 10% after purification, typical of wild type enzyme, was

Table 2. Kinetic constants for wild-type and mutant alcohol dehydrogenases[a]

Kinetic constants	ADH-E	F110L	S117G	F110L-S117G	ADH-S[b]
K_a NAD$^+$ μM	4.5	7.9	10	3.9	11
K_b CH$_3$CH$_2$OH, mM	0.18	3.1	0.34	1.1	4.8
K_p CH$_3$CHO, mM	0.11	0.4	0.25	0.4	24
K_q, NADH μM	2.5	4.1	5.7	3.4	7.3
K_{ia}, NAD$^+$ μM	30	30	31	12	62
K_{iq} NADH, μM	0.5	0.9	0.7	0.7	1.5
V_1, s^{-1}	1.9	11	10	3.6	1.5
V_2, s^{-1}	20	82	67	53	73
V_1/K_b, mM^{-1}s^{-1}	10	3.5	31	3.1	0.3
V_2/K_p, mM^{-1}s^{-1}	180	190	270	130	3
TN, s^{-1}	2.4	2.9	2.4	1.9	1.0
K_{eq}, pM	16	6	25	12	24

[a] Initial velocities were studied at 25 °C in 33 mM sodium phosphate buffer, pH 8.0, with 0.25 mM EDTA, using a SLM fluorometer (Model 4800C). The change in NADH concentration was monitored with an excitation wavelength of 340 nm and an emission wavelength of 460 nm. Ethanol and acetaldehyde were redistilled before use; the concentrations of nucleotides were determined by absorbance. The initial velocity data were fitted to a straight line or parabola with the least-square analysis program provided with the instrument software. K_a, K_b, K_p and K_q are the Michaelis constants for NAD$^+$, ethanol, acetaldehyde and NADH, respectively. V_1 and V_2 are turnover numbers of ethanol oxidation and acetaldehyde reduction, respectively. These constants were determined by varying coenzyme and substrate concentrations in a 9-fold range around the K_m values for the enzyme. Data were fitted with the program SEQUEN (Cleland, 1979), and standard errors were ≤ 20 %. V_1/K_b and V_2/K_p are catalytic efficiencies for ethanol oxidation and acetaldehyde reduction, respectively. TN is the turnover number in the standard assay (Plapp, 1970) based on the active site titration. The equilibrium constant was calculated from $K_{eq} = (V_1 K_p K_{iq} [H^+]) / (V_2 K_b K_{ia})$. K_i values are product inhibition constants determined by varying coenzyme concentrations over a 4-fold range at a fixed concentration of substrate. Data were fitted with the equation for competitive inhibition. [b] Data from Park and Plapp (1992).

also obtained with F110L (45 mg of protein) and S117G enzymes (26 mg). The protein recovery for F110L-S117G was 2.3 mg after purification (corresponding to only 2%). The fraction of protein sites titratable with NAD$^+$ in the presence of pyrazole varied slightly: 70% for F110L, 75 % for S117G, and 82 % for F110L-S117G. The mutated proteins were at least 90% pure as determined by denaturing sodium dodecyl sulfate polyacrylamide gel and agarose gel electrophoresis, and migrated at the same position as wild-type enzyme.

The initial velocity patterns for the F110L, S117G and F110L-S117G enzymes followed a sequential bi bi mechanism for ethanol oxidation and acetaldehyde reduction, as described for wild-type enzyme. The turnover numbers for ethanol oxidation (V_1) and acetaldehyde reduction (V_2) by F110L and S117G enzymes were 3 to 5-fold higher than for wild-type enzyme (Table 2). Michaelis constants for F110L enzyme were also higher (17-fold for ethanol and 3.5-fold for acetaldehyde), but the K_m values for S117G enzyme were not significantly different from wild-type. Consequently, the catalytic efficiency for ethanol oxidation by F110L enzyme was 3-fold lower and the value for S117G was 3-fold higher, whereas catalytic efficiencies for acetaldehyde reduction were essentially unchanged for both enzymes. The turnover numbers (V_1 and V_2) for F110L-S117G enzyme were not significantly different from wild-type, and the Michaelis constants were higher (6-fold for ethanol, 3.5-fold for acetaldehyde). As with F110L enzyme, the catalytic efficiency for ethanol oxidation by F110L-S117G enzyme was 3-fold lower as compared to wild-type, whereas the catalytic efficiency for acetaldehyde reduction was unchanged. The dissociation constants of the oxidized and reduced coenzymes from enzyme-coenzyme complexes (K_{ia} and K_{iq}) were not significantly different for F110L, S117G, F110L-S117G and wild-type enzymes, showing that the mutations do not affect the coenzyme

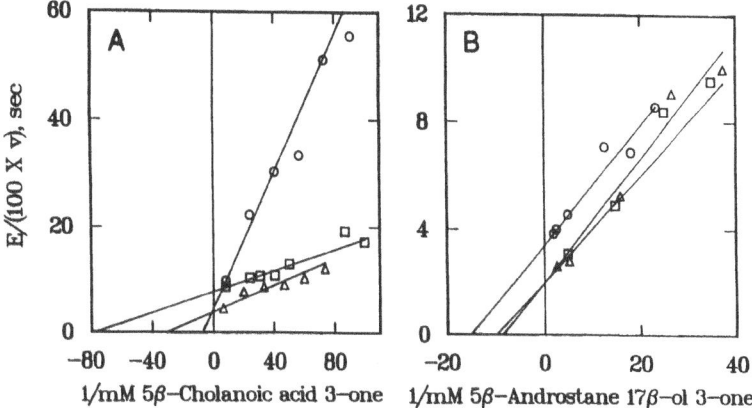

Figure 2. Kinetics of steroid reduction catalyzed by wild-type ADH-E and mutated enzymes. The experimental conditions are described in Table 3. The velocities are 100 times the turnover numbers. Wild-type (o), F110L (Δ) and S117G ($\square$) enzymes. (**A**) β-cholanoic acid-3-one. (**B**) 5β-androstane-17β-ol-3-one.

binding sites. The turnover numbers based on NAD^+ titration were not significantly different.

Activity in reduction of steroids by wild-type and mutated forms of ADH-E was tested with 5β-cholanoic acid 3-one and 5β-androstane-17β-ol-3-one and compared to S enzyme. Surprisingly, wild-type ADH-E was found to reduce both steroids with low rates and in a reproducible manner (Figure 2). Michaelis constants and catalytic efficiencies with both compounds were similar (Table 3). The rates of reduction of steroids with S117G, F110L and S117G-F110L enzymes were not significantly different from wild-type enzyme, but the Michaelis constant for 5β-cholanoic acid 3-one for S117G and F110L enzymes were decreased 10 and 4-fold respectively, resulting in a catalytic efficiency 6-fold higher than for wild-type. Kinetic constants for the reduction of 5β-androstane-17β-3-one by F110L and S117G enzymes were not different from wild-type ADH-E. ADH-E and F110L enzymes were not detectably active on 5α-androstane-17β-ol-3-one, consistent with the stereospecificity for 5β configuration described for human γ enzyme (McEvily et al., 1988).

4. DISCUSSION

The kinetic parameters of F110L, S117G and F110L-S117G enzymes with ethanol and acetaldehyde are very similar to those for the E enzyme. Catalytic efficiencies for the forward direction vary in a 3-fold range and are unchanged in the reverse direction. Replacing Phe110 by Leu and Ser117 by Gly in the E subunit results in a larger substrate pocket, but these mutations do not significantly affect reaction of small substrates, suggesting that the local structural changes are inconsequential. In contrast, when the size of substrate binding pocket of the E subunit was enlarged by site-directed mutagenesis to mimic the S enzyme, binding of short-chain alcohols became less efficient due to poorer interactions with the mutant enzymes (Park and Plapp, 1992).

The S117G single mutation increases the catalytic efficiency for 5β-cholanoic acid 3-one by about 6-fold, to a level that is still 3 orders of magnitude less than for horse S en-

Table 3. Reduction of steroids by wild-type and mutant alcohol dehydrogenases[a]

	5β-cholanoic acid 3-one			5β-androstane-17β-ol-3-one		
	K_m (μM)	V_2 (s^{-1})	V_2/K_m ($M^{-1}s^{-1}$)	K_m (μM)	V_2 (s^{-1})	V_2/K_m ($M^{-1}s^{-1}$)
ADH-E	140	0.0022	15	71	0.0029	41
S117G	13	0.0013	100	120	0.0054	45
F110L	32	0.0024	92	100	0.0051	51
F110L-S117G[b]	≤10	0.0020				
D115Δ[c]			1,000	60	1.2	20,000
ADH-S[c]	8	0.81	100,000	11	1.2	100,000

[a] Reduction of steroids was measured by following the disappearance of NADH at 340 nm and 25 °C. The assay mixture contained 33 mM sodium phosphate buffer, pH 8, with 0.25 mM EDTA, 0.2 mM NADH and enzyme (5.5 μN ADH-E, 1.4 μN F110L, 1.6 μN S117G for β-cholanoic acid-3-one; 1.8 μN ADH-E, 1.4 μN F110L, 1.8 μN S117G for 5β-androstane-17βol-3one). The reaction was initiated by the addition of 5β-androstane-17β-ol-3one after a 15 min equilibration (during which time contaminating aldehydes were reduced and the rate decreased to a blank value (between 1 and 3.5 x 10^{-4} $\Delta A_{340\,nm}$/min) or β-cholanoic acid-3-one after a 50 min equilibration time (blank value between 5 and 50 x 10^{-5} $\Delta A_{340\,nm}$/min). After the steroids were added, the velocities were at least 10 times faster than the blank rates. The steroids were dissolved in acetonitrile (final concentrations were 3.7 % for β-cholanoic acid-3-one and 2.5 % for 5β-androstane-17β-ol-3one) and buffer. All solutions were prepared with Millipore Q water, which significantly reduced the blank reactions. Data were fitted with HYPER (Cleland, 1979). [b] The K_m value of 5β-cholanoic acid 3-one for F110L-S117G enzyme could not be determined because saturation was reached at 10 μM substrate, and the low rates recorded below this concentration were indistinguishable from the blank values. [c] From Park and Plapp, 1992; data at pH 7.3, 30 °C; acetonitrile was not used.

zyme. Comparison of our results with the literature values for other mammalian enzymes active on steroids, in particular human γ, is more difficult, not only because these enzymes have been tested with different steroids and their isomers, but also because they were studied either in the oxidative or the reductive directions. Oxidation of 3β-hydroxy-5β-steroids by the γ subunit was reported by McEvily et al. (1988), with K_m values between 46 and 320 μM, and k_{cat} values between 0.12 and 1.2 s^{-1}. The Michaelis constants for 5β-cholanoic acid 3-one and 5β-androstane-17β-ol-3-one for S117G and F110L enzymes are within the same range (13 and 40 μM), but the catalytic rates are 100 times lower than the values for the γ enzyme (McEvily et al., 1988). Our results clearly show that the changes introduced in the E enzyme at positions 110, 117, or both combined, do not transform the substrate binding pocket into a site with high reactivity for steroids. Although sequence alignment and molecular modeling suggest that residues 110 and 117 could be involved, they apparently do not by themselves control the substrate specificity.

The small change of Leu141 in horse E enzyme into Val in human γ might be more important than anticipated. Substituting Leu141 with Val introduces a β-methyl group that appears to be too close to Leu116 or Phe93. It seems unlikely, however, that the single change of Leu into Val is sufficient to transform substrate specificity. It appears that other residues, or combination of residues, more distant from the substrate binding pocket, indirectly affect the active site, reflecting the flexible nature of the hydrophobic active site of ADH.

The roles of other residues in steroid recognition have been investigated by comparison of the structure of ADH γ with the other class I human enzymes (Eklund et al., 1987, 1990; Höög et al., 1992, 1993). The horse E subunit differs at about 64 residues from any of the subunits of the human class I, and two-thirds of these are surface substitutions that are compatible with a highly conserved three-dimensional structures. Only six of the substitutions among the human isoenzymes occur at the substrate binding site (Eklund et al., 1987). The γ subunit is most similar to the horse E subunit; the α subunit is the most different.

Position 48, at the top of the substrate barrel, was proposed to be an important residue for substrate specificity, since both α and β have Thr48, whereas γ has Ser48 (Eklund et al., 1987). When Thr48 was changed to Ser in ADH-β, the resulting enzyme was able to oxidize 3β-hydroxy-5β-androstane-17-one (Höög et al., 1992), although with only 3 % of the efficiency of ADH-γ. Since most ADHs have Thr or Ser at position 48, and ADH-E, which has Ser48, is not active on steroids, Ser48 cannot account by itself for the high activity on steroids.

The catalytic constants for reduction of steroids by the mutant enzymes are not very different from the values we obtained for wild-type horse E enzyme. The ability of the E subunit to reduce steroids is surprising, since it is currently accepted that the E subunit is not active on steroids, in contrast to the S subunit. Nevertheless, the catalytic efficiency for the E enzyme, as compared to the S enzyme, is 3 to 4 orders of magnitude lower and difficult to detect.

5. ACKNOWLEDGMENT

This work was supported by NSF Grants MCB 91–18657 and 95–06831 and NIH Grant AA00279.

6. REFERENCES

Anderson, S., Cronholm, T., and Sjövall, J.: Redox effects of ethanol on steroid metabolism. Alcoholism Clin. Exp. Res. 10 (1986) 555–635.

Cleland, W.W.: Statistical analysis of enzyme kinetic data. Methods Enzymol. 63, (1979) 103–138.

Eklund, H., Horjales, E., Vallee B.L., and Jörnvall, H.: Computer-graphics interpretations of residue exchanges between the α, β, and γ subunits of human-liver alcohol dehydrogenase class-I isoenzymes. Eur. J. Biochem. 167 (1987) 185–193.

Eklund, H., Müller-Wille, P., Horjales, E., Futer, O., Holmquist, B., Vallee, B.L., Höög, J.-O., Kaiser, R., and Jörnvall, H.: Comparison of three classes of human liver alcohol dehydrogenase. Eur. J. Biochem. 193 (1990) 303–310.

Höög, J.-O., Eklund, H., and Jörnvall, H.: A single residue exchange gives human recombinant ββ alcohol dehydrogenase γγ isoenzyme properties. Eur. J. Biochem. 205 (1992) 519–526.

Höög, J.-O., Vagelopoulos, N., Yip, P.-K., Keung, W.M., and Jörnvall, H.: Isoenzyme developments in mammalian class-I alcohol dehydrogenase. Eur. J. Biochem. 213 (1993) 31–38.

McEvily, A.J., Holmquist, B., Auld, D.S., and Vallee, B.L.: 3β-Hydroxy-5β-steroid dehydrogenase activity of human liver alcohol dehydrogenase is specific to γ-subunits. Biochemistry 27 (1988) 4284–4288.

Nakamaye, K., and Eckstein, F.: Inhibition of restriction endonuclease Nci I cleavage by phosphorothioate groups and its application to oligonucleotide-directed mutagenesis. Nucl. Acids Res. 14 (1986) 9679–9698.

Park, D.-H., and Plapp, B.V.: Isoenzymes of horse liver alcohol dehydrogenase active on ethanol and steroids. J. Biol. Chem. 266 (1991) 13296–13302.

Park, D.-H., and Plapp, B.V.: Interconversion of E and S isoenzymes of horse liver alcohol dehydrogenase. J. Biol. Chem. 267 (1992) 5527–5533.

Plapp, B.V.: Enhancement of the activity of horse liver alcohol dehydrogenase by modification of amino groups at the active sites. J. Biol. Chem. 245 (1970) 1727–1735.

Ramaswamy, S., Eklund, H., and Plapp, B.V.: Structure of horse liver alcohol dehydrogenase complexed with NAD⁺ and substituted benzyl alcohols. Biochemistry 33 (1994) 5230–5237.

Reynier, M., Theorell, H., and Sjövall, J.: Studies on the specificity of liver alcohol dehydrogenase for 3β-hydroxy-5β-steroids. Acta Chem. Scand. 23 (1969) 1130–1136.

Shearer, G.L., Kim, K., Lee, K.M., Wang, C.K., and Plapp, B.V.: Alternative pathways and reactions of benzyl alcohol and benzaldehyde with horse liver alcohol dehydrogenase. Biochemistry 32 (1993) 11186–11194.

Sun, H.W., and Plapp, B.V.: Progressive sequence alignment and molecular evolution of the Zn-containing alcohol dehydrogenase family. J. Mol. Evol. 34 (1992) 522–535.

Theorell, H., and Yonetani, T.: Liver alcohol dehydrogenase-DPN-pyrazole complex: A model of a ternary inter-
    mediate in the enzyme reaction. Biochem. Z. 338 (1963) 537–553.
Waller, G., Theorell, H., and Sjövall, J.: Liver alcohol dehydrogenase as a 3β-hydroxy-5β-cholanoic acid dehydro-
    genase. Arch. Biochem. Biophys. 111 (1965) 671–684.

37

ROLE OF ALCOHOL DEHYDROGENASES IN STEROID AND RETINOID METABOLISM

Natalia Y. Kedishvili,[1] Carol L. Stone,[1] Kirill M. Popov,[1] and
Ellen A. G. Chernoff[2]

[1]Department of Biochemistry and Molecular Biology
[2]Department of Biology
Indiana University School of Medicine
635 Barnhill Drive, MS 402
Indianapolis, Indiana 46202

1. INTRODUCTION

Retinoid and steroid hormones play an important role in the regulation of differentiation and maintenance of a wide range of animal tissues. These tissues include reproductive organs, liver, kidney, heart, brain, and skin of species from fish to humans. Several isozymes of cytosolic NAD^+-dependent 40 kDa subunit molecular weight alcohol dehydrogenases catalyze oxidation and reduction of retinoid and steroid substrates *in vitro*. The isozymes are grouped into classes based on the similarities in amino acid sequence and their substrate specificities. Currently, a total of six classes of mammalian ADHs are known (Jörnvall and Höög, 1995). Each class has a characteristic tissue-specific and developmental pattern of expression (Edenberg and Bosron, 1996). Class I ADHs are basic isozymes with a wide range of Km for ethanol (0.05–36 mM). In humans, class I is comprised of multiple molecular forms, $\beta_1\beta_1$, $\beta_2\beta_2$, $\beta_3\beta_3$, $\gamma_1\gamma_1$, $\gamma_2\gamma_2$, $\alpha\,\alpha$, and their heterodimers. During development, $\alpha\,\alpha$ is the first ADH isozyme detectable in fetal liver. β-ADH appears by mid-gestation, and γ-ADH is first detected about six month after birth. Human class II π-ADH has a relatively high K_M for ethanol (34 mM) and is found in fetal and adult liver. The ubiquitously expressed class III ADH, also known as glutathione-dependent formaldehyde dehydrogenase, is not saturable with ethanol and is not active with either steroid or retinoid alcohols. Human class IV σ-ADH exhibits high K_M for ethanol (28 mM) and is present in the adult stomach, esophagus and epithelium. In mice embryos, class IV ADH is detected on day 7.5 of development in the craniofacial region as well as trunk and forelimb bud mesenchyme (Ang, H.L. *et al.*, 1996). Little is known about the catalytic properties of human class V and deermouse class VI ADH isozymes.

Retinol is oxidized *in vitro* by multiple mammalian ADH isozymes with varying catalytic efficiency. In humans, class IV ADH is the most catalytically efficient retinol dehydrogenase followed by human class II and class I αα ADH (Yang *et al.*, 1994). Rat class IV ADH is less efficient for retinol oxidation than the human class IV ADH (Boleda *et al.*, 1993).

The ability to oxidize steroid alcohols has been limited to class I ADHs that have Ser in position 48 (Park and Plapp, 1991). The only human ADH active with steroid alcohols is class I γγ ADH (McEvily *et al.*, 1988). This isozyme is also active with retinol. Horse SS class I ADH exhibits a hundred times higher catalytic efficiency for oxidation of 3β,5β-hydroxysteroids than either human or rat class I ADHs (Cronholm *et al.*,1975). Thus, the physiological significance of these ADHs for steroid and retinoid metabolism may vary in different species.

The metabolism of retinoid and steroid hormones is important for embryonal development (Gudas, 1994; Domm and Ericson, 1972; Payne and Jaffe, 1972). Retinoic acid regulates morphogenesis of multiple tissues during embryogenesis. The chick embryo is widely accepted as a model for studies on vertebrate development. The retinoic acid-target tissues include limb buds and heart of the chick stage 21 (3 days old) embryo (Thaller and Eichele, 1987; Osmond *et al.*, 1991). It is not clear whether retinoic acid is synthesized in each target tissue from retinol (retinol⇌retinal→retinoic acid) or is delivered to them from the other sites of synthesis. Thus, retinol dehydrogenase may be present in limb buds and/or heart of the 3 days old chick embryo. ADH isozymes appear at different stages of embryogenesis. To evaluate the physiological role of these enzymes for steroid and retinoid metabolism of the fetus, it is important to determine which isozymes are expressed in the early embryo.

Multiple ADH forms are observed in the fetal and adult liver extracts of avian species (LeVine and Haley, 1975). Class I ADH isozymes have been purified and sequenced from chicken, ostrich, quail, and kiwi. They share over 70% identity and have similar substrate specificity. In addition, a class II isoform of ADH has been identified in the ostrich liver recently (Hjelmqvist *et al.*, 1995).

In the fetal chick and quail livers, two cathodal zonal regions of ADH activity are observed on starch gel electrophoretograms (LeVine and Haley, 1975). The slower moving bands become detectable in the 5 day old embryos. The faster moving bands are detected first on day 9. By day 13 of incubation, only the faster-migrating bands are detected. Methanol and 1-propanol are poor substrates, yielding little or no staining of any bands. Retinoid and steroid alcohols have not been not tested as substrates. To determine which forms of ADH are expressed in the chick 3-day old embryo, we isolated the total mRNA from the limb buds and hearts and analyzed them for the presence of messages encoding 40 kDa ADHs.

2. MATERIALS AND METHODS

Degenerate oligonucleotide primers were synthesized according to amino acid sequences conserved in animal ADHs (Sun and Plapp, 1992) (Fig. 1).

Total RNA was isolated from the pooled limb buds and hearts of 30 chick embryos and used for RT-PCR. The PCR products of the predicted size were purified, subcloned into M13mp19RF and sequenced. A chicken liver cDNA library was screened with the radiolabeled PCR product. The hybridizing λ phage was purified, the cDNA insert was released by *Eco*R I digestion, subcloned into M13 and sequenced. The coding region of ADH-F (chick fetal alcohol dehydrogenase) was amplified by PCR and subcloned into a prokaryotic expression vector. The cDNA construct was expressed in *E. coli* cells. Activ-

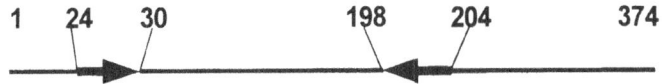

Figure 1. Localization of the conserved regions in the ADH primary structure. The numbering corresponds to the human β_1-ADH.

ity of the recombinant protein was determined with 100 mM ethanol, 2.5 mM NAD$^+$, in 0.1 M sodium phosphate, pH 7.4 at 25° C. Activity with all-*trans* retinol was determined by the production of all-*trans* retinal at 400 nm (Julia *et al.*,1986). The retinol and steroid stock solutions were prepared in methanol. The reaction mixtures with these substrates contained 0.02% Tween-80. Protein sequences of ADH isozymes were obtained from GenBank and aligned with ADH-F by progressive alignment method according to Feng and Doolittle, 1990. The phylogenetic tree was constructed using the TREE program.

```
        AAA GCT CAT GAA GTT CGC ATC AAG ATA GTT GCC ACT GCT CTC TGT CAC ACT GAT
Chi      K   A   H   E   V   R   I   K   I   V   A   T   A   L   C   H   T   D  -47
Hum                                          I               V

        GCC TAT ACT CTG AGT GGT GCT GAT CCT GAG GGA TGT TTC CCT GTG ATT CTG GGT
Chi      A   Y   T   L   S   G   A   D   P   E   G   C   F   P   V   I   L   G  -65
Hum

        CAT GAA GGA GCA GGA ATT GTA GAA AGT GTT GGG GAA GGA GTA ACA AAA GTA AAG
Chi      H   E   G   A   G   I   V   E   S   V   G   E   G   V   T   K   V   K  -83
Hum                                                                  L

        CCA GGG GAC ACC GTG ATC CCT CTG TAC ATC CCC CAG TGT GGC GAG TGC AAG TAC
Chi      P   G   D   T   V   I   P   L   Y   I   P   Q   C   G   E   C   K   Y -101
Hum      A                                                                   F

        TGC AAG AAT CCT AAA ACT AAT CTG TGC CAA AAG ATA AGA GTT ACT CAA GGG AAA
Chi      C   K   N   P   K   T   N   L   C   Q   K   I   R   V   T   Q   G   K -119
Hum          L

        GGA CTC ATG CCT GAT GGT ACG ATC AGA TTC ACC TGC AAA GGA AAG CAG ATT TAC
Chi      G   L   M   P   D   G   T   I   R   F   T   C   K   G   K   Q   I   Y -137
Hum                                  S                           T       L

        CAC TTC ATG GGG ACT AGC ACC TTC TCG GAG TAC ACA GTG GTG GCC GAT ATC TCA
Chi      H   F   M   G   T   S   T   F   S   E   Y   T   V   V   A   D   I   S -155
Hum          Y

        GTA GCT AAG ATA GAT CCT GCA GCA CCT TTC GAT AAA GTG TGC CTG CTG GGT TTG
Chi      V   A   K   I   D   P   A   A   P   F   D   K   V   C   L   L   G   C -173
Hum                          L           L

        GCA TCT CTA CAG GTT ATG GGG CTG CTG TTA ACA CTG CAA AGG TGG AAC CTG GCT
Chi      G   I   S   T   G   Y   G   A   A   V   N   T   A   K   V   E   P   G -191
Hum                                                          L

        CCA CGT GTG CAG TGT TCG GCT TGG GCG GCG
Chi      S   T   C   A   V   F   G   L   G   G-201
Hum          V
```

Figure 2. The partial nucleotide and deduced protein sequence of the chick class III ADH (Chi) aligned with the human class III χ-ADH (Hum). The amino acid numbers correspond to human χ-ADH.

Table 1. Percent identity of ADH-F with AD
isozymes from various species. Class V
is represented by a single human gene
ADH6, class VI is represented by the
deermouse *Adh-2* gene

Class	% Identity
I	61-69
II	54-62
III	62-64
IV	62-63
V	63
VI	57

3. RESULTS

Chick embryonic tissues were analyzed for the presence of ADH messages by RT-PCR with degenerate ADH-specific primers. The resulting PCR products were purified and subcloned into M13 vector for sequencing. Two ADH-related PCR products were identified. One product had 87% deduced protein sequence identity with human class III ADH (Fig. 2).

The second ADH-related product encoded a protein with the highest identity to class I ADH isozymes (Table 1).

The complete cDNA sequence encoding ADH-F was determined from the five cDNA clones purified from the chicken cDNA library. Two of the clones contained the entire coding region, while the other three lacked the N-terminus. The overlapping regions were identical in all clones. The schematic of the two complete cDNA clones is shown in Fig. 3.

The longest clone contains 76 bp of the 5' non-coding sequence and 206 bp of the 3' non-coding sequence. The coding region is 1128 bp long and encodes a 376 amino acid polypeptide including the initiator methionine.

The phylogenetic relationships of ADH-F with other ADH isozymes were analyzed by progressive alignment method (Fig. 4).

The phylogenetic tree shows that the precursor to ADH-F preceded the divergence to amphibians and the divergence of class IV ADH.

Figure 3. Schematic of the cDNA clones for chick ADH-F.

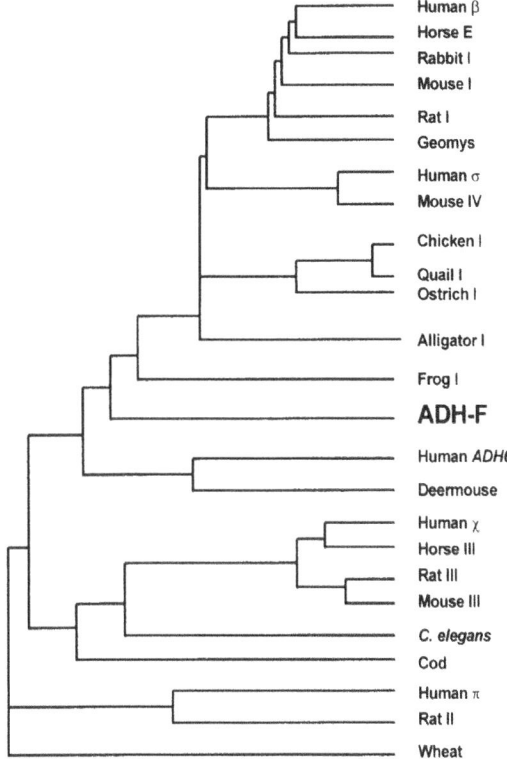

Figure 4. Phylogenetic tree of ADH isozymes.

The coding region of ADH-F cDNA was subcloned into the expression vector and expressed in *E. coli* cells. The purified ADH-F had a subunit molecular weight of 40,000 Da in the SDS-PAGE. The recombinant ADH-F was active with all-*trans* retinol, epiandrosterone, dehydroepiandrosterone and dihydrotestosterone.

4. DISCUSSION

Two ADH-related cDNAs were PCR-amplified from chick embryonal mRNA. Class III ADH was obtained from chick limb buds. This ADH is ubiquitously expressed in all adult and embryonal tissues. Since it is not active with steroid and retinoid alcohols, we did not pursue the isolation of a complete cDNA. The alignment of the partial deduced protein sequence of chick class III ADH with human χ-ADH confirmed the high degree of conservation (87%) of this gene throughout the evolution (Fig. 2).

A new ADH-encoding cDNA (ADH-F) was found in the chick embryonal heart. A hybridizing message was also detected in the adult chicken liver. ADH-F had the highest sequence identity with the avian class I ADH (69% with ostrich class I ADH) and was less similar to class II-VI ADH isozymes (about 63% identity) (Table I). The evolutionary tree suggested that the precursor to ADH-F diverged before avian class I ADHs and that ADH-F may, thus, represent a separate class VII of ADH isozymes (Fig. 4).

Figure 5. Synthesis of active steroids from dehydroepiandrosterone.

The coding region of ADH-F cDNA encodes a full-length 376 amino acid (~40 kDa) polypeptide. The cDNA expressed in *E. coli* translates into a functional recombinant protein with approximately neutral isoelectric point. The new ADH is active with all-*trans* retinol and steroid alcohols, suggesting that ADH-F may play a role in steroid and retinoid metabolism of lower species.

All ADH isozymes examined thus far are specific for oxidation and reduction of the 3β-hydroxy group of steroid substrates. The human γγ ADH is specific for the 3β,5β-hydroxysteroid substrates (McEvily *et al.*, 1988), while horse SS and rat class I ADHs oxidize the 3β,5α conformation also, although with a lesser turnover rate (Cronholm *et al.*, 1975). The new ADH-F prefers the 3β,5α-hydroxysteroids. The activity with 3β,5β-hydroxysteroids was not detected by the spectrophotometric assay. 3β-Hydroxysteroid dehydrogenase activity is required for the synthesis of biologically active steroids such as cortisol, progesterone, aldosterone, dihydrotetosterone and estradiol (Fig. 5 from Labrie *et al.*, 1992).

The membrane-associated 42 kDa 3β-hydroxysteroid dehydrogenases are responsible for this activity in the adrenal glands of male and female rats, testes and ovary. Two human isozymes and three rat isozymes have been characterized. Interestingly, a 42 kDa 3β-hydroxysteroid dehydrogenase found in male rat liver is the type III isozyme which functions in the reductive direction as a NADPH-prefering 3-ketosteroid reductase. Rat

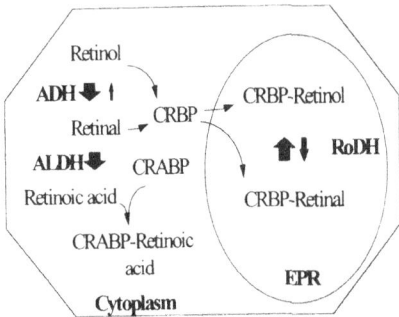

Figure 6. Metabolism of retinol to retinoic acid. EPR, endoplasmic reticulum; RoDH, microsomal retinol dehydrogenase.

liver cytosol contains 28 mU/g liver of dehydroepiandrosterone-oxidizing activity versus 4 mU/g liver in microsomal fraction (Ghraf *et al.*, 1975). The identity of the cytosolic 3β-hydroxysteroid dehydrogenases is not clear. This could be a yet unidentified 42 kDa steroid dehydrogenase type IV, and/or the 40 kDa class I /VII ADH.

The other potential physiological substrate of the 40 kDa ADHs, all-*trans* retinol, serves as a precursor for the synthesis of retinoic acid (Fig. 6).

Most of the retinol in the liver is bound to the cellular retinol-binding protein (CRBP) (Ross, 1993). The physiological role of this family of proteins that includes retinoic acid binding proteins, CRABP-I and II, is not clear. CRBPs and CRABPs are often expressed in different cell types within a tissue. Currently, it is not known how the presence of CRBP affects retinol oxidation by the 40 kDa ADH isozymes. Recently, a new family of short-chain retinol dehydrogenases/reductases has emerged that oxidize the CRBP-bound form of all-*trans* retinol (Chai *et al.*, 1995). The only NAD$^+$-dependent member of this family of enzymes known thus far is the membrane-bound retinol dehydrogenase that converts 11-*cis* retinol into 11-*cis* retinal in the pigment epithelium of the retina (Simon *et al.*, 1995). Three other microsomal isozymes prefer NADP$^+$. The role of these enzymes in retinoic acid synthesis is not clear, and their expression pattern in the embryo is yet to be established. However, in some cells, retinoic acid is synthesized in the absence of CRBP, suggesting that free retinol can serve as a substrate for retinoic acid synthesis. One of the enzymes that use free retinol is the recently discovered and cloned retinol dehydratase (Grün *et al.*, 1996). This enzyme converts retinol to anhydroretinol in *Spodoptera frugiperda* and has a K_M for free retinol in the low nanomolar range. It is possible that the K_M values for the free retinol for ADH isozymes have been overestimated as a result of the micelle formation in the presence of detergent added to aqueous solutions to increase retinol solubility.

By homology with 17β- and 11β-steroid dehydrogenases, which include both NAD$^+$ and NADP$^+$-dependent enzymes that work in the oxidative and reductive directions, respectively (Labrie *et al.*, 1995), it is likely that the family of the short-chain retinol dehydrogenases will expand to include a cytosolic NAD$^+$-dependent retinol dehydrogenase with a low nanomolar K_M for the free retinol.

Thus, the first non-mammalian ADH isoenzyme with activity towards steroid and retinoid alcohols has been cloned. Within the cell, the competing multiple substrates, inhibitors and availability of coenzymes will contribute to the factors that determine the physiological significance of the observed *in vitro* steroid and retinoid-oxidizing activities of ADH isoenzymes. A cell culture model may help to approach these *in vivo* conditions.

5. ACKNOWLEDGMENT

Supported by NIAAA grants AA00221–01, AA02342, and AA07117.

6. REFERENCES

Ang, H.L., Deltour, L., Hayamizu, T.F., Żgombić-Knight, M. and Duester, G.: Retinoic acid synthesis in mouse embryos during gastrulation and craniofacial development to class IV alcohol dehydrogenase gene expression. J. Biol. Chem. 271 (1996) 9526–9534.

Boleda, M.D., Saubi, N., Farrés, J. and Parés, X.: Physiological substrates for rat alcohol dehydrogenase classes: Aldehydes of lipid peroxidation, ω-hydroxyfatty acids, and retinoids. Arch. Biochem. Biophys. 307 (1993) 85–90.

Chai, X., Zhai, Y. Popescu, G. and Napoli, J.L.: Cloning of a cDNA for a second retinol dehydrogenase type II. J. Biol. Chem. 270 (1995) 28408–28412.

Cronholm, T., Larsén, C., Sjövall, J., Theorell and Åkeson, Å.: Steroid oxidoreductase activity of alcohol dehydrogenases from horse, rat, and human liver. Acta Chemica Scandinavica B 29 (1975) 571–576.

Domm, L.V. and Ericson, G.C.: 3β-Hydroxysteroid dehydrogenase activity in the adrenals of normal and hypophysectomized chick embryos. Proc. Soc. Exp. Biol. Med. 140 (1972) 1215–1220.

Edenberg, H.J. and Bosron, W.F.: Alcohol dehydrogenases. Comprehensive Toxicology (1996), in press.

Feng, D.F. and Doolittle R.F.: Progressive alignment and phylogenetic tree construction of protein sequences. Methods Enzymol. 183 (1990) 375–389.

Ghraf, R., Lax, E.R. and Schriefers, H.: The hypophysis in the regulation of androgen and oestrogen dependent enzyme activities of steroid hormone metabolism in rat liver cytosol. Hoppe-Seyler's Z. Physiol. Chem. 356 (1975) 127–134.

Grün, F., Noy, N., Hämmerling, U. and Buck, J.: Purification, cloning, and bacterial expression of retinol dehydratase from Spodoptera frugiperda. J. Biol. Chem. 271 (1996) 16135–16138.

Gudas, L.J.: Retinoids and vertebrate development. J. Biol. Chem. 269 (1994) 15399–15402.

Hjelmqvist, L., Estonius, M., Jörnvall, H.: The vertebrate alcohol dehydrogenase system: variable class II type form elucidates separate stages of enzymogenesis. Proc. Natl. Acad. Sci. USA 92 (1995) 10904–10908.

Jörnvall, H. and Höög, J.-O.: Nomenclature of alcohol dehydrogenases. Alcohol & Alcoholism 30 (1995) 153–161.

Julia, P., Farres, J., Pares, X.: Ocular alcohol dehydrogenase in the rat: Regional distribution and kinetics of ADH-1 isoenzyme with retinol and retinal. Exp. Eye Res. 42 (1986) 305–314.

Labrie, F., Bélanger A., Simard, J., Luu-The, V. and Labrie, C.: DHEA and peripheral androgen and estrogen formation: Intracrinology. Ann. N.-Y. Acad. Sci. 774 (1995) 17–28.

Labrie, F., Simard, J., Luu-The, V., Bélanger A. and Pelletier, G.: Structure, function and tissue-specific gene expression of 3β-hydroxysteroid dehydrogenase/5-ene-4-ene isomerase enzymes in classical and peripheral intracrine steroidogenic tissues. J. Steroid Biochem. Molec. Biol. 43 (1992) 805–826.

LeVine, J.P. and Haley, L.E.: Gene activation of alcohol dehydrogenase in Japanese quail and chicken-quail hybrid embryos. Biochemical Genetics 13 (1975) 435–446.

McEvily, A.J., Holmquist, B., Auld, D.S. and Vallee, B.L.: 3β-Hydroxy-5β-steroid dehydrogenase activity of human liver alcohol dehydrogenase is specific to γ-subunits. Biochemistry 27 (1988) 4284–4288.

Osmond, M.K., Butler, A.J., Voon, F.C.T. and Bellairs, R.: The effects of retinoic acid on heart formation in the early chick embryo. Development 113 (1991) 1405–1417.

Park, D.-H. and Plapp, B.V.: Isoenzymes of horse liver alcohol dehydrogenase active on ethanol and steroids. J. Biol. Chem. 266 (1991) 13296–13302.

Payne, A.H. and Jaffe, R.B.: Comparison of androgen synthesis in human fetal testis and adrenal: 3β-Hydroxysteroid dehydrogenase-isomerase and 17β-steroid dehydrogenase activities. Biochim. Biophys. Acta 279 (1972) 202–207.

Ross, A.C.: Cellular metabolism and activation of retinoids: Roles of cellular retinoid-binding proteins. FASEB J. 7 (1993) 317–327.

Simon, A., Hellman, U., Wernstedt, C. And Eriksson, U.: The retinal pigment epithelial-specific 11-*cis* retinol dehydrogenase belongs to the family of short chain alcohol dehydrogenases. J. Biol. Chem. 270 (1995) 1107–1112.

Sun, H.-W. and Plapp, B.V.: Progressive sequence alignment and molecular evolution of the Zn-containing alcohol dehydrogenase family. J. Mol. Evol. 34 (1992) 522–535.

Thaller, C. and Eichele, G.: Identification and spatial distribution of retinoids in the developing chick limb bud. Nature 327 (1987) 625–628.

Waller, G. and Theorell, H.: Liver alcohol dehydrogenase as a 3β-hydroxy-5β-cholanic acid dehydrogenase. Arch. Biochem. Biophys. 111 (1965) 671–684.

Yang, Z.-N., Davis, G.J., Hurley, T.D., Stone, C.L., Li, T.-K. and Bosron, W.F.: Catalytic efficiency of human alcohol dehydrogenases for retinol oxidation and retinal reduction. Alcohol. Clin. Exp. Res. 18 (1994) 587–591.

38

MECHANISTIC ENZYMOLOGY OF LIVER ALCOHOL DEHYDROGENASE

Kinetic and Stereochemical Characterization of Retinal Oxidation and Reduction

Y. Pocker and Hong Li

Department of Chemistry, Campus Box 351700
University of Washington
Seattle, Washington 98195–1700

1. INTRODUCTION

Vitamin A is vital for promotion of general growth, differentiation of epithelial tissues, visual function and reproduction. A connection between vitamin A and cancer was made 70 years ago when it was found that carcinomas of the stomach appeared in albino rats fed a vitamin A deficient diet (Fujimaki, 1926). The history of the carotenoids predates that of vitamin A. However, it was during the subsequent four years that the high vitamin A activity of carotene was demonstrated and its metabolism to vitamin A in the liver documented (for a review see Pawson, 1981). In 1930, Karrer presented a structural assignment for β-carotene and a year later, he proposed a structural formula for vitamin A alcohol (Karrer, 1938; Pawson, 1981).

A further demonstration of the importance of vitamin A derivatives came from the elegant studies of Wald when he carried out his research on the visual process (for a review see Wald, 1968).

Figure 1 shows the interrelationships of retinol, retinal, and retinoic acid and their biological roles.

Liver alcohol dehydrogenase (LADH, E.C.1.1.1.1) is an $NAD^+/NADH$ dependent zinc metallo-enzyme (Jörnvall, 1970; Brändén, et al., 1975; Klinman, 1981; Eklund, et al., 1982; Eklund and Brändén, 1983; Zeppezauer, 1986; Pocker, et al., 1986; Eklund and Brändén, 1987; Pettersson, 1987; Pocker,1989; Light, et al., 1992) with a broad substrate specificity (Pietruszko, 1979; Pocker and Raymond, 1985; Pocker, et al., 1986). In the presence of NAD^+, the enzyme oxidizes the alcohol retinol to the aldehyde retinal which is necessary for vision (Bliss, 1951; Mezey and Holt, 1971; Pocker et al., 1987). This is a reversible reaction as retinal is converted to retinol by the same enzyme in the presence

Enzymology and Molecular Biology of Carbonyl Metabolism 6
edited by Weiner *et al*. Plenum Press, New York, 1996

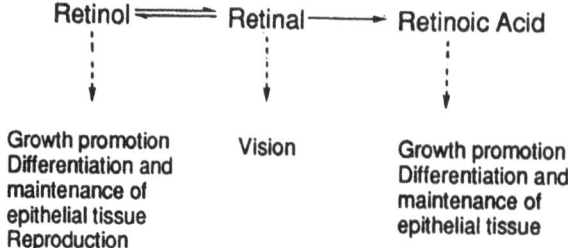

Figure 1. Biological roles of retinol, retinal and retinoic acid.

of reduced coenzyme NADH. Oxidation of retinal by liver alcohol dehydrogenase, LADH, in the presence of NAD$^+$ produces retinoic acid (Pocker *et al.*, 1987; Pocker and Li, 1993), which has also been shown to be a major metabolite of retinol (Emerick *et al.*, 1967). Retinol is required for growth, for differentiation and maintenance of epithelial tissues, and for reproduction (Bridges, 1984; Roberts and Sporn, 1984; Sporn, *et al.*, 1984a, 1984b; Moon and Itri, 1984; Packer, 1990a, 1990b). In vitamin A deficient animals, retinoic acid alone cannot substitute completely for retinol in maintaining the reproductive function.

Recent studies form the basis for the renaissance in the mechanistic enzymology of alcohol dehydrogenase which has occurred in the last few years. The present study focuses on kinetic discoveries, many of which were made with *cis*- and *trans*-retinoids that differ in structure through variations in the ring, the side chain and the terminal group of the molecule, Figure 2.

Liver alcohol dehydrogenase also catalyzes the oxidation of retinals to the corresponding retinoic acids (Futterman, 1962; Frolik, 1984; Pocker, *et al.*, 1987; Pocker and Li, 1993). One manifestation of this catalytic activity is the aldehyde dismutation reaction (Hinson and Neal, 1972; Oppenheimer and Henehan, 1995). This ability of alcohol dehydrogenase can sometimes, but not always, be rationalized by the fact that most aliphatic aldehydes exist in aqueous solution in equilibrium with their hydrated form where the *gem*-diol form acts as a structural analog of an alcohol and may be presumed to be the substrate in this reaction (Pocker and Li, 1991; Oppenheimer and Henehan, 1995).

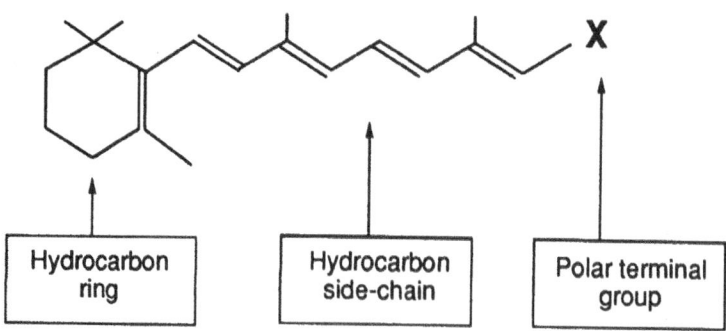

Figure 2. Sites for structural modifications in vitamin A. Chemical structures of all-*trans* retinol X=CH$_2$OH, all-*trans* retinal X=CHO and all-*trans* retinoic acid X=CO$_2$H.

2. MATERIALS AND EXPERIMENTAL PROCEDURES

2.1. Enzyme and Coenzymes

Horse liver alcohol dehydrogenase was obtained as a crystallized and lyophilized preparation from the Sigma Chemical Company. The oxidized and reduced forms of the coenzyme β-nicotinamide adenine dinucleotide, NAD$^+$ and NADH, were also purchased from Sigma.

2.2. Substrates and Other Reagents

All-*trans* retinol, all-*trans* retinal, all-*trans* retinoic acid, 13-*cis* retinol, 13-*cis* retinal and t-octylphenoxypolyethoxyethanol, TX-100, were purchased from Sigma and used as received. The alcohols 9-*cis* retinol and 11-*cis* retinol, the aldehydes 9-*cis* retinal and 11-*cis* retinal and the acids 9-*cis* retinoic acid, 11-*cis* retinoic acid and 13-*cis* retinoic acid were generous gifts from Hoffman-LaRoche Inc., from the National Eye Institute, or were synthesized from generously donated materials. All chemicals appeared to be either re-agent or spectral grade (Moon and Itri, 1984; Roberts and Sporn, 1984; Bridges, 1984; Dawson and Hobbs, 1990; Pocker and Li, 1993).

2.3. Buffers

Unless otherwise specified, aqueous phosphate buffers were used for preparation of all solutions of enzyme, coenzyme, substrate and inhibitor. Phosphate buffers were prepared by dissolving weighed amounts of KH_2PO_4 and K_2HPO_4 in distilled deionized water. The phosphate concentration was 30 mM and the ionic strength was maintained at 0.1 by the addition of Na_2SO_4 (Pocker and Li, 1993).

2.4. Enzyme and Coenzyme Solutions

Enzyme concentration was determined by titration of the active sites with NAD$^+$ and pyrazole (Theorell and Yonetani, 1963). Enzyme solutions prepared in the manner de-scribed by Pocker and Page (1990, 1993) and stored at 0°C lost no activity within a 6 to 8 hour time period. Coenzyme solutions were made by dissolving weighed amounts of NAD$^+$ and NADH in buffer and determining the concentration spectrophotometrically (Pocker and Li, 1993). Fresh solutions were prepared for each day's experiments.

2.5. Retinol, Retinal and Retinoate Solutions

Due to their low solubility in water retinol, retinal and retinoate stock solutions were prepared by the method described in Pocker and Li (1993). The final concentration of sub-strate was ascertained spectrophotometrically using the experimentally determined extinc-tion coefficients. Stock solutions of the various retinoids were stored under nitrogen in the dark below 5°C.

2.6. Instrumentation

The pH measurements were conducted on a PHM84 Research pH Meter (Radiome-ter Copenhagen) equipped with a glass electrode. Spectroscopic and kinetic measurements were performed on a Varian Cary 210 UV-visible spectrophotometer interfaced to an Apple II/e microcomputer and thermostated at 25.00 ± 0.05°C with a circulating water bath controlled by a Forma-Temp Jr. Model 2095 thermonitor.

2.7. Stopped-Flow Apparatus

Fast reactions were monitored on an extensively modified Durrum-Gibson stopped flow spectrophotometer with the temperature kept at 25.0 ± 0.05°C by a Forma-Temp Model 2095–2 thermonitor (Pocker and Bjorkquist, 1977a, 1977b; Pocker and Fong, 1980; Pocker and Janji, 1988). The output from the photomultiplier was filtered and amplified by a specially constructed amplification board. The signals were collected as described earlier (Pocker and Li, 1993).

2.8. Molecular Modeling Studies

Calculations were performed on Silicon Graphic Indigo 2 computers using Insight II and Discover software packages (Molecular Simulations, Inc., Release 950). The structure of LADH as determined by Eklund, Samama, and Jones (1984) and given in the Protein Data Bank was used. The structures of the retinoids investigated possessed an s-cis conformation about the C6-C7 bond as described by Dawson and Hobbs (1990). Parallel calculations were also performed on the corresponding s-trans conformers. The system $LADH \cdot NAD^+$ + 9-cis retinol exhibits unfavorable interactions with Phe 110, Phe 93 and Ile 318. The system $LADH \cdot NAD^+$ + 11-cis retinol exhibits unfavorable interactions with Phe 110, Phe 93 and Ile 318. The system $LADH \cdot NAD^+$ + 13-cis retinol exhibits unfavorable interactions with Thr 94, Phe 93 and Ile 318.

3. RESULTS

Our kinetic studies indicate that the retinol-retinal interconversion follows an ordered bi-bi mechanism (Pocker and Raymond, 1980, 1985). The kinetic parameters (Haldane, 1930; Wratten and Cleland, 1963; Segel, 1975) were determined from the initial rate equation for an ordered bi-bi mechanism using double reciprocal plots. Table 1 summarizes the apparent k_{cat} and K_m values for reactions involving all-trans, 9-cis, 11-cis and 13-cis retinol-retinal interconversions. It also contains data on the oxidation of all-trans, 9-cis, 11-cis and 13-cis retinals to the corresponding retinoic acids (actually retinoate anions).

The apparent K_m values indicate that the interaction between the long chain carbon skeleton of the retinoid and the hydrophobic pocket of the enzyme provides a major driving force for the binding process (substrate recognition and docking). However, the apparent k_{cat} values for the oxidation of cis-retinols to cis-retinals and for the reduction of cis-retinals to cis-retinols are significantly slower (over 100 fold slower) than those corresponding to the interconversion of all-trans isomers.

Furthermore, in this report we show that the enzyme can also oxidize cis-retinals to the corresponding acids. An interesting feature of these oxidations is that the initial rates appear to be roughly 100 fold slower than the initial rates of reduction of cis-retinals to the corresponding cis-retinols.

Table 1. Apparent kinetic parameters for the reactions of LADH with retinols and retinals[a]

Reaction	k_{cat} (s^{-1})	$K_m^{retinol}$ μM
all-*trans* retinol + NAD$^+$	1.55 ± 0.10	59 ± 4
9-*cis* retinol + NAD$^+$	0.025 ± 0.005	300 ± 30
11-*cis* retinol + NAD$^+$	0.022 ± 0.005	280 ± 30
13-*cis* retinol + NAD$^+$	0.018 ± 0.005	257 ± 30
Reaction	k_{cat} (s^{-1})	$K_m^{retinol}$ μM
all-*trans* retinal + NADH	2.50 ± 0.10	340 ± 40
9-*cis* retinal + NADH	0.023 ± 0.005	242 ± 20
11-*cis* retinal + NADH	0.021 ± 0.005	200 ± 20
13-*cis* retinal + NADH	0.018 ± 0.005	137 ± 10
Reaction	k_{cat} (s^{-1})	$K_m^{retinol}$ μM
all-*trans* retinal + NAD$^+$	0.022 ± 0.012	260 ± 120
11-*cis* retinal + NAD$^+$	0.00020 ± 0.00005[b]	160 ± 60
13-*cis* retinal + NAD$^+$	0.00017 ± 0.00005[b]	110 ± 30

[a] Reactions were carried out in 30 mM phosphate buffer containing 0.1% TX-100 at pH 7.00 and 25.0°C

[b] Preliminary values.

4. DISCUSSION

The k_{cat} values for all-*trans* retinol oxidation and all-*trans* retinal reduction have the same order of magnitude, with reduction proceeding about 1.5–2 times faster. Similarly, the reduction of 9-*cis*, 11-*cis* and 13-*cis* retinals exhibits the same order of magnitude as the oxidation of the corresponding *cis*-retinols. This is in contrast to the interconversion ethanol-acetaldehyde and other short-chain alcohol-aldehyde pairs where reduction proceeds many times faster than oxidation.

Comparing the kinetic constants of all-*trans* retinal reduction with those of 9-*cis*, 11-*cis* and 13-*cis* retinal reductions, it is noted that there exist large differences in k_{cat} values while K_m values are more nearly similar. Apparently, the all-*trans* form can achieve a conformation in the ternary complex that facilitates hydride transfer. On the other hand, it is important to note that all the retinals so far studied possess reasonably similar dissociation constants. Indeed, the binding of the coenzyme induces a conformational change in the enzyme which increases the lipophilicity of the active site that in turn stabilizes the binding of substrates having long hydrophobic chains (Pettersson, 1987; Pocker, 1989). Indeed, previous studies have shown that K_m decreases as the hydrophobic carbon chain increases in length (Sund and Theorell, 1963).

Our molecular modeling calculations indicate that when 9-*cis*, 11-*cis* and 13-*cis* retinols are in a correct juxtaposition for hydride transfer, unfavorable steric interactions between enzyme•NAD$^+$ complex and *cis*-substrate occur in the active site. These steric interactions prevent 9-*cis*, 11-*cis* and 13-*cis* retinols from achieving a favorable conformation for hydride transfer in the ternary complex. Similar modeling also indicates that when 9-*cis*, 11-*cis* and 13-*cis* retinals are in a correct juxtaposition for hydride transfer, unfavorable steric interactions between enzyme•NADH complex and *cis*-substrate occur in the active site. These steric interactions prevent 9-*cis*, 11-*cis* and 13-*cis* retinals from achieving a favorable conformation for hydride transfer in the ternary complex. The net result is that the rate of hydride transfer is greatly diminished in comparison to that of the all-*trans* isomers, compounds that can assume the proper molecular orientation for hydride transfer in the active site. It is not surprising therefore to find that the all-*trans* alcohol and aldehyde exhibit the highest reactivity.

Understanding the mechanisms by which alcohol dehydrogenases recognize and bind retinols, retinals and retinoic acids poses important intellectual, experimental and modeling challenges. No one single approach or strategy has so far produced a complete picture. The multi-faceted research program we have undertaken has generated some new insights. We are focusing more attention on the metal ion as well as on the protein component of the system by using mutagenesis approaches to test our hypotheses concerning the mechanisms of retinoid recognition and transformation.

5. ACKNOWLEDGMENTS

We wish to thank Professor Clarita C. Bhat for helpful kinetic measurements, Dr. Greg T. Spyridis for effective comments and Mr. Thillainatarajan Sivakumaran for valuable literature surveys.

6. REFERENCES

Bliss, A. F., 1951, The equilibrium between vitamin A alcohol and aldehyde in the presence of alcohol dehydrogenase, *Arch. Biochem. Biophys.*, *31*, 197.

Brändén, C.-L., Jörnvall, H., Eklund, H. and Furugren, B., 1975, Alcohol dehydrogenases *in:* "The Enzymes", (Boyer, P. D. ed.) vol. 11, Academic Press, New York, pp 103–190.

Bridges, C. D. B., 1984, Retinoids in photosensitive systems, *in:* "The Retinoids", (Sporn, M. B., Roberts, A. B. and Goodman, D. S. eds.) vol. 2, Academic Press, New York, pp 90–124.

Dawson, M. I. and Hobbs, P. D., 1990, Synthetic retinoic acid analogs: handling and characterization, *in:* "Methods in Enzymology", (Packer, L. ed.) vol. 189 Academic Press, San Diego. pp 13–43.

Eklund, H. and Brändén, C.-L., 1983, The role of zinc in alcohol dehydrogenases *in:* "Zinc Enzymes", (Spiro, T. G. ed.), John Wiley and Sons, New York, pp 123–152.

Eklund, H. and Brändén, C.-L., 1987, Alcohol dehydrogenase, *in:* "Biological Macromolecules and Assemblies", (Jurnak, F. and McPherson, A. eds.), vol. 3, John Wiley and Sons, New York, pp 73–142.

Eklund, H., Plapp, B. V., Samama, J.-P. and Brändén, C.-L., 1982, Binding of substrate in a ternary complex of horse liver alcohol dehydrogenase, *J. Biol. Chem.*, *257*, 14349.

Eklund, H., Samama, J.-P. and Jones, T.A., 1984, Crystallographic investigations of nicotinamide adenine dinucleotide binding to horse liver alcohol dehydrogenase, *Biochemistry, 23*, 5982–5996.

Emerick, R. J., Zile, M., DeLuca, H. F., 1967, Formation of retinoic acid from retinol in the rat, *Biochem. J. 102*, 606–611.

Frolik, C. A., 1984, Metabolism of retinoids, *in:* "The Retinoids", (Sporn, M. B., Roberts, A. B. and Goodman, D. S. eds.) vol. 2, Academic Press, New York, pp 177–208.

Fujimaki, Y., 1926, Formation of carcinoma in albino rats fed on deficient diets, *J. Cancer Res., 10*, 469–447.

Futterman, S., 1962, Enzymatic oxidation of vitamin A aldehyde to vitamin A acid, *J. Biol. Chem. 237*, 667–680.

Haldane, J. B. S., 1930, "Enzymes", Longmans Green, London.

Hinson, J. A. and Neal, R. A., 1972, An examination of the oxidation of aldehydes by horse liver alcohol dehydrogenase, *J. Biol. Chem., 247*, 7106.

Jörnvall, H., 1970, Horse liver alcohol dehydrogenase, *Eur. J. Biochem., 16*, 25.

Karrer, P., 1938, "Organic Chemistry", Elsevier, New York, pp 656–657 and 672–673.

Klinman, J. P., 1981, Probes and mechanism and transition state structure in the alcohol dehydrogenase reaction, *CRC Crit. Rev. Biochem., 10*, 39.

Light, D. R., Dennis, M. S., Forsythe, I. J., Liu, C. C., Green, D. W., Kratzer, D. A. and Plapp, B. V., 1992, α-Isoenzyme of alcohol dehydrogenase from monkey liver: cloning, expression, mechanism, coenzyme, and substrate specificity, *J. Biol. Chem., 267*, 12592.

Mezey, E. and Holt, P. R., 1971, The inhibitory effect of ethanol on retinol oxidation by human liver and cattle retina, *Experimental and Molecular Pathology, 15*, 148.

Moon, R. C. and Itri, L. M., 1984, Retinoids and cancer, in: "The Retinoids", (Sporn, M. B., Roberts, A. B. and Goodman, D. S. eds.) vol. 2, Academic Press, New York, pp 327–372.

Oppenheimer, N. J. and Nenehan, G. T. M., 1995, Horse liver alcohol dehydrogenase-catalyzed aldehyde oxidation, in: "Enzymology and Molecular Biology of Carbonyl Metabolism 5", (Weiner, H., Holmes, R. S. and Wermuth, G. eds.) Plenum Press, New York, pp 407–415.

Packer, L. ed., 1990a, "Methods in Enzymology" vol. 189 Academic Press, San Diego.

Packer, L. ed., 1990b, "Methods in Enzymology" vol. 190 Academic Press, San Diego.

Pawson, B. A., 1981, A historical introduction to the chemistry of vitamin A and its analogs (retinoids) Ann. New York Acad. Sci., 359, 1–8.

Pettersson, G., 1987, Liver alcohol dehydrogenase, CRC Crit. Rev. Biochem., 21, 349.

Pietruszko, R., 1979, Nonethanol substrates of alcohol dehydrogenase, in: "Biochemistry and Pharmacology of Ethanol", vol. 1 (Majchrowicz, E. and Noble, E. P., eds.) Plenum Press, New York, pp 87–106.

Pocker, Y., 1989, Alcohol dehydrogenase: structure catalysis and site directed mutagenesis, in: "Metal Ions in Biological Systems", (Sigel, H., ed.) Marcel Dekker, New York, pp 335–358.

Pocker, Y. and Bjorkquist, D. W., 1977a, Stopped-flow studies of carbon dioxide hydration and bicarbonate dehydration in H_2O and D_2O. Acid-base and metal ion catalysis, J. Am. Chem. Soc. 99, 6537.

Pocker, Y. and Bjorkquist, D. W., 1977b, Comparative studies of bovine carbonic anhydrase in H_2O and D_2O. Stopped-flow studies of the kinetics of interconversion of carbon dioxide and bicarbonate, Biochemistry 16, 5698.

Pocker, Y. and Fong, C. T. O., 1980, Kinetics of inactivation of erythrocyte carbonic anhydrase by sodium 2,6-pyridinedicarboxylate, Biochemistry 19, 2045.

Pocker, Y. and Janji, N., 1988, Differential modification of specificity in carbonic anhydrase, J. Biol. Chem. 263, 6169.

Pocker, Y. and Li, H., 1991, Kinetics and mechanism of methanol and formaldehyde interconversion and formaldehyde oxidation catalyzed by liver alcohol dehydrogenase, in: "Enzymology and Molecular Biology of Carbonyl Metabolism 3", (Weiner, H., Wermuth, B. and Crabb, D. W. eds.) Plenum Press, New York, pp 315–325.

Pocker, Y. and Li, H., 1993, The catalytic specificity of liver alcohol dehydrogenase: vitamin A alcohol and vitamin A aldehyde activities, in: "Enzymology and Molecular Biology of Carbonyl Metabolism 4", (Weiner, H., Crabb, D. W. and Flynn, T. G. eds.) Plenum Press, New York, pp 411–418.

Pocker, Y., Li, H. and Page, J. D., 1987, Liver alcohol dehydrogenase: metabolic and energetic aspects, in: "Advances in Biomedical Alcohol Research", (Lindros, K. O., Ylikahri, R. and Kiianmaa, K. eds.) Pergamon Press, London, pp 181–185.

Pocker, Y. and Page, J. D., 1990, Zinc-activated alcohols in ternary complexes of liver alcohol dehydrogenase, J. Biol. Chem., 265, 22101–22108.

Pocker, Y. and Page, J. D., 1993, Fluorescence studies of ternary complexes of liver alcohol dehydrogenase, in: "Enzymology and Molecular Biology of Carbonyl Metabolism 4" (Weiner, H., Crabb, D. W. and Flynn, T. G. eds.) Plenum Press, New York, pp 513–521.

Pocker, Y. and Raymond, K. W., 1980, Kinetics and mechanistic studies of vitamin A alcohol to vitamin A aldehyde by liver alcohol dehydrogenase. The inhibition by ethanol and pyrazole, in: "Alcohol and Aldehyde Metabolizing Systems-IV", (Thurman, R. G. ed.) Plenum Press, New York, pp 137–150.

Pocker, Y. and Raymond, K. W., 1985, Liver alcohol dehydrogenase: substrate inhibition and competition between substrates, Alcohol 2, 3.

Pocker, Y. Raymond, K. W. and Thompson III, W. H., 1986, Liver alcohol dehydrogenase: kinetic and mechanistic studies, in: "Zinc Enzymes", (Bertini, I., Luchinat, C., Maret, W. and Zeppezauer, M. eds.) Birkhäuser, Boston, pp 435–449.

Roberts, A. B. and Sporn, M. B., 1984, Cellular biology and biochemistry of the retinoids in: "The Retinoids", (Sporn, M. B., Roberts, A. B. and Goodman, D. S. eds.) vol. 2, Academic Press, New York, pp 209–286.

Segel, I. H., 1975, "Enzyme Kinetics", John Wiley and Sons, New York.

Sporn, M. B., Roberts, A. B. and Goodman, D. S. eds., 1984a, "The Retinoids", vol. 1, Academic Press, New York.

Sporn, M. B., Roberts, A. B. and Goodman, D. S. eds., 1984b, "The Retinoids", vol. 2, Academic Press, New York.

Sund, H. and Thoerell, H., 1963, Alcohol dehydrogenases, in: "The Enzymes", (Boyer, P. D., Lardy, H. and Myrback, K. eds.) vol. 7, Academic Press, New York, pp 25–83.

Theorell, H. and Yonetani, T., 1963, Liver alcohol dehydrogenase-DPN-pyrazole complex: a model of a ternary intermediate in the enzyme reaction, Biochemische Zeitschrife, 338, 537.

Wald, G., 1968, Molecular basis of visual excitation, Science, 162, 230–239.

Wratten, C. C. and Cleland, W. W., 1963, Product inhibition studies on yeast and liver alcohol dehydrogenases, *Biochemistry*, 2, 935.

Zeppezauer, M., 1986, The metal environment of alcohol dehydrogenase: aspects of chemical speciation and catalytic efficiency in a biological catalyst, *in:* "Zinc Enzymes", (Bertini, I., Luchinat, C., Maret, W. and Zeppezauer, M. eds.) Birkhäuser, Boston, pp 417–434.

39

REGULATION OF THE SEVEN HUMAN ALCOHOL DEHYDROGENASE GENES

Howard J. Edenberg,[*] Celeste J. Brown,[†] Man-Wook Hur,[‡] Shailaja Kotagiri, Mei Li, Lu Zhang, and Xin Zhi

Department of Biochemistry and Molecular Biology and of Medical and
  Molecular Genetics
Indiana University School of Medicine
635 Barnhill Drive, MS 418
Indianapolis, Indiana 46202–5122

1. INTRODUCTION

The human alcohol dehydrogenase gene family consists of seven genes with distinct but overlapping patterns of expression. ADH enzymes metabolize a wide range of alcohols. Some of the ADH enzymes catalyze the first, rate-limiting step in the oxidative metabolism of ethanol. This pathway generates acetaldehyde as an intermediate, which in turn is oxidized by aldehyde dehydrogenase to acetate. Allelic differences in *ADH2* and *ADH3* genes encode forms of the enzymes that differ in their kinetic properties toward ethanol. These differences have been shown to affect the risk for alcoholism in Japanese and Chinese individuals (Thomasson et al. 1991; Thomasson et al. 1994; Thomasson et al. 1993; Crabb et al. 1995; Tanaka et al. 1996; Yamauchi et al. 1995; Maezawa et al. 1995; Tu and Israel, 1995). We hypothesize that differences in the expression of these genes could also affect the metabolism of ethanol, and thereby modify the risk for alcoholism in other populations.

The *ADHs* have evolved from a common ancestral gene, as evidenced by their common genomic structure (Edenberg and Brown, 1992; Zgombic-Knight et al. 1995). The *ADH* genes are expressed in different but overlapping sets of tissues. Therefore, this gene family provides a window into the evolution of tissue-specificity on two time-scales: significant divergence among genes of the different classes, and more recent divergence among the class I genes.

---

[*] Phone (317) 274–2353. FAX (317) 274–4686. EMAIL: edenberg@iupui.edu
[†] Current address: Department of Biological Science, University of Idaho, Moscow, ID.
[‡] Current address: Department of Biochemistry and Molecular Biology, Yonsei Univ. College of Medicine, Seodaemoon-Ku, Shinchon-Dong 134, Seoul, Korea.

We are examining the regulation of the *ADH* genes, determining the *cis*-acting promoter elements and the *trans*-acting transcription factors involved in controlling their expression. Six of the *ADH* genes are expressed in restricted sets of tissues; these genes have TATA boxes, as is common for tissue-specific genes. *ADH5* is the exception: it encodes the class III ADH/formaldehyde dehydrogenase that is expressed in all tissues. The *ADH5* proximal promoter lacks a TATA box and is G+C-rich, as is common for "housekeeping" genes. This report discusses some common themes in the regulation of these genes, including important common *cis*-acting elements.

2. CLASS I ADH GENES: *ADH1*, *ADH2*, AND *ADH3*

The class I genes, *ADH1*, *ADH2* and *ADH3*, metabolize ethanol at low concentrations. These three genes are very closely related, encoding proteins that are about 94% identical in amino acid sequence and form homo- and hetero-dimers. They evolved recently in the primate lineage (Cheung et al. 1995). All are expressed in adult liver. Their patterns of expression outside liver differ (reviewed in Edenberg and Brown, 1992). For example, only *ADH2* is expressed in lung, and both *ADH2* and *ADH3* are expressed in kidney. None is expressed in heart or brain. Their promoters share 80–85% pairwise identity for approximately 800 bp upstream of the transcriptional start site (Stewart et al. 1990a; Edenberg and Brown, 1992; Edenberg et al. 1995).

The proximal 210 bp of the *ADH2* promoter is bound at 7 distinct sites by transcription factors found in liver extracts (Brown et al. 1994). Proteins from other tissues bind at different subsets of these sites. There are a pair of C/EBP sites that flank the TATA-box (Stewart et al. 1990b; Brown et al. 1994). A promoter extending just through these C/EBP sites is weak but functional (Brown et al. 1994). Just upstream of these sites are a USF site and the G3T site. The G3T site can bind Sp1 (Brown et al. 1992). Deletion analyses demonstrates that the region containing the USF and G3T sites is a strong positive element; the promoter fragment extending just through these sites gives the most expression of any construct tested (Brown et al. 1994). *In vitro* transcription assays show similar results, and competitions with oligonucleotides covering each site demonstrate that both are independent strong positive *cis*-acting elements (Carr and Edenberg, 1990). Just upstream of these is a CTF/NFI-related site that functions as a negative element (Edenberg et al. 1993). A site whose transcription factor is still not identified lies further upstream, followed by an HNF-1 site. This proximal *ADH2* promoter functions well in a rat hepatoma cell line (H4IIE-C3) that expresses its own homologous *ADH* gene (Brown et al. 1994). Surprisingly, it also functions in HeLa cells, a cell line that does not express its endogenous *ADH*. The promoters are not functional in all cells, since expression in CV-1 cells is negligible (Brown et al. 1994).

The *ADH1* and *ADH3* promoters are very similar, although small sequence differences lead to differences in the binding of transcription factors (Brown et al. 1996). For example, one of the two C/EBP sites and the USF site in *ADH1* are bound less well than the corresponding sites in the other genes. The G3T site in *ADH3* is bound less well than the related sites in *ADH1* and *ADH2*. Under our standard conditions the G3T site in *ADH3* is not footprinted (Brown et al. 1996), but binding can be demonstrated in gel mobility shift experiments (Brown et al. 1992). The apparent differences in affinity may be on the order of 3 to 5-fold, as judged by gel shift competition experiments. Differences of this magnitude may be important in establishing the ability of the genes to be transcribed in cells that differ in the amount of specific transcription factors.

There is an interesting difference among the genes in a CTF/NFI-related site (Edenberg et al. 1993). All three genes bind to a CTF/NFI-related protein somewhere within a 50 bp region, but the location of the binding site within that region differs among the genes despite the great overall similarity of sequence. The site is bound close to the G3T site in *ADH2*, and at a non-overlapping site just upstream in *ADH1* and *ADH3* (with the latter having a much larger footprint (Edenberg et al. 1993; Brown et al. 1996). This CTF/NFI-related site, unlike most, acts as a negative element, reducing transcription (Edenberg et al. 1993; Brown et al. 1996).

The *ADH1* promoter functions at least as well as the *ADH2* promoter in H4IIE-C3 cells and HeLa cells; the *ADH3* promoter does not work well in either cell line (Brown et al. 1996). None of the three promoters works well in CV-1 cells. Coexpression of C/EBPα (by cotransfection with an expression plasmid) changes the cell-type specificity of *ADH1* and *ADH2* and allows expression in CV-1 cells (Brown et al. 1996). The strong response of the *ADH1* promoter to C/EBPα in these cells contrasts with an earlier report that *ADH1* responded poorly in HepG2 cells (Stewart et al. 1991). *ADH2* promoters of all lengths tested are stimulated by coexpression of C/EBPα (Brown et al. 1994). The shortest *ADH2* promoter tested, which contains the two C/EBP sites but no other upstream *cis*-acting elements, functions only weakly when transfected alone, but is greatly stimulated in all three cells (Brown et al. 1994). This demonstrates that the C/EBP sites can act as powerful transcriptional stimulators in the presence of sufficient C/EBPα. These results illustrate the importance of the C/EBP sites in gene expression, as well as the ability of C/EBPα to work with the other transcription factors present in CV-1 cells.

These results, taken together, emphasize that subtle differences in the sequence of *cis*-acting elements can affect the affinity of transcription factors for those sites. Even small differences in the binding of transcription factors can affect the expression of genes in different cells, particularly if the level of an important transcription factor is low. This suggests that small differences in the promoter sequence in different individuals could also have significant effects on the level (and perhaps even the tissue distribution) of gene expression. Conversely, one should be cautious in interpreting the results of transfection studies in which a very high level of a transcription factor is coexpressed; although such experiments provide useful information about the potential role of that factor in gene expression, by altering the balance among factors in the cell they may obscure interactions among competing and cooperating factors.

Table 1. ADH promoters and primary sites of expression

Gene	TATA	C/EBP	G±C rich	Site(s) of expression[1]
ADH1	Yes	Yes	No	Liver
ADH2	Yes	Yes	No	Liver, kidney
ADH3	Yes	Yes	No	Liver, kidney
ADH4	Yes	Yes	No	Liver, kidney, stomach
ADH5	**No**	(by seq)[2]	**Yes**	**Ubiquitous** (high in liver)
ADH6	Yes	Yes	No	[Liver, stomach][3]
ADH7	Yes	Yes	No	Stomach, esophagus

[1] Primary sites.
[2] A possible consensus sequence was found in a footprinted area.
[3] RNA detected.

3. OTHER TISSUE-SPECIFIC ADH GENES: *ADH4, ADH6, ADH7*

ADH4 is expressed in human liver, with lower levels in kidney and stomach. In the 400 bp extending upstream from the transcriptional start site there are four sites at which transcription factors bind (unpublished data). The proximal *ADH4* promoter functions in all three cell lines in which it has been tested: hepatoma cells, HeLa cells and CV-1 cells (unpublished data). Although the relative activity is greater in the hepatoma cells, the degree of cell-specificity is not nearly what one would expect. Thus important elements that restrict expression to only a few tissues are not present in the proximal promoter region. As in the class I *ADH* genes, there is a C/EBP site that is important in gene expression (unpublished data).

ADH6 was discovered by homology to a conserved exon of *ADH2* (Yasunami et al. 1991). Its mRNA was detected in liver and stomach (Yasunami et al. 1991; and unpublished data), but the protein has not yet convincingly been demonstrated in human tissues. Therefore, its physiological role is still unclear. We have examined its promoter region, and shown that the *ADH6* promoter can function in several different cell types. These include hepatoma cells and CV-1 cells (unpublished data). As with *ADH4*, the proximal promoter shows little cell-specificity. The *ADH6* promoter also includes an apparent C/EBP site (unpublished data).

In adults, σ-ADH (also called μ-ADH) activity is highest in the esophagus, stomach and epithelial tissues (Yin et al. 1990; Yin et al. 1993; Moreno and Pares, 1991; Stone et al. 1993). It is the ADH most efficient in oxidation of retinol to retinal (Kedishvili et al. 1995a; Yang et al. 1994). Retinoids are important in many physiological processes, from vision to development and gene regulation. σ-ADH is encoded by *ADH7* (Satre et al. 1994; Kedishvili et al. 1995b; Zgombic-Knight et al. 1995; Yokoyama et al. 1994). We have shown that the *ADH7* proximal promoter functions in HeLa cells and in CV-1 cells (unpublished data). We have mapped *cis*-acting elements in this region, and examined by deletion analysis and site-directed mutagenesis their roles in promoter function. It is interesting that a C/EBP site and an AP1 site are important in promoter function (unpublished data).

The promoters of the different classes of *ADH* genes have diverged greatly, leaving little homology among them. The six tissue-specific *ADH* genes share TATA-boxes. It is interesting that all six have apparent C/EBP sites, suggesting that this family of regulatory proteins plays an important role in their expression. This also partly explains the expression of five of the six genes (*ADH1, ADH2, ADH3, ADH4, ADH6*) in liver, a tissue with high levels of C/EBPα. The proximal promoters of these genes, while showing some cell-specificity, clearly do not show the extent of specificity seen among tissues of an intact organism. There must be other important *cis*-acting elements that stimulate or restrict expression in other cell types. One phase of our ongoing research is to pursue these elements.

4. CLASS III ADH GENE: *ADH5/FDH*

ADH5 is expressed in all tissues, but to different extents. Evolutionary analyses of the coding region suggests that *ADH5* is the ancestral *ADH* (Jornvall et al. 1995; and see article by Jornvall et al. in this volume). It is surprising that comparisons of the promoter regions of the genes in this family show that it is least like the other members. The G+C-rich promoter of *ADH5* lacks a TATA-box (Hur and Edenberg, 1992), and in both of these aspects it is very different from all of the other *ADH* promoters.

There are two strong Sp1 sites flanking the primary transcriptional start site of *ADH5*. A very small DNA fragment extending to -34 bp and containing these Sp1 sites is a strong promoter (Hur and Edenberg, 1995). Although this might lead one to expect a simple promoter, the *ADH5* promoter is surprisingly complex. Upstream of these Sp1-sites there are many *cis*-acting elements that are bound to different extents in different cells and tissues (Hur and Edenberg, 1995). These include *cis*-acting elements that function as positive transcriptional elements in some cells and as negative or neutral elements in other cells. Thus the level of expression of *ADH5* is controlled by a different set of transcription factors in different cells, rather than just by different levels of a single transcription factor. This might well be true of other "housekeeping genes."

ADH5 is also unusual in having upstream ATGs in the 5' non-translated region that might play a role in gene regulation. Two upstream ATGs start overlapping open reading frames of 10 and 20 codons. We have shown that in transient transfection assays, differences in the length of the upstream open reading frames defined by these ATGs affect the level of gene expression about two-fold (Hur and Edenberg, 1995). Thus the regulation of this ubiquitously expressed gene is quite complex.

5. SUMMARY

Our comparative studies of the expression of all seven human *ADH* genes (and of the homologous mouse genes) have shown common features and individual diversity. Proximal promoters of the six tissue-specific genes show some cell-type differences in expression, but none shows the degree of specificity found *in vivo*. Despite major differences among their promoters, these six genes have both TATA-boxes and C/EBP sites as important *cis*-acting elements in common. We are continuing our studies to locate other *cis*-acting elements that contribute to tissue specificity, and to identify the transcription factors that bind to and regulate them. We are also examining the surprisingly complex set of *cis*-acting elements that regulates the ubiquitously expressed *ADH5* gene and determines its level of expression in different tissues. Studies on the members of the *ADH* gene family will illuminate the evolution of tissue-specificity.

6. ACKNOWLEDGMENTS

We thank Ronald E. Jerome and Jinghua Zhao for excellent technical assistance. This research was supported by PHS grant R01 AA06460 from the National Institute of Alcohol Abuse and Alcoholism.

7. REFERENCES

Brown, C. J., Baltz, K. A. and Edenberg, H. J.: Expression of the human *ADH2* gene: an unusual Sp1-binding site in the promoter of a gene expressed at high levels in liver. Gene 121 (1992) 313–320.

Brown, C. J., Zhang, L. and Edenberg, H. J.: Tissue-specific differences in the expression of the human *ADH2* alcohol dehydrogenase gene and in binding of factors to *cis*-acting elements in its promoter. DNA Cell Biol. 13 (1994) 235–247.

Brown, C. J., Zhang, L. and Edenberg, H. J.: Gene expression in a young multigene family: tissue-specific differences in the expression of the human alcohol dehydrogenase genes ADH1, ADH2 and ADH3. DNA Cell Biol. 15 (1996) 187–196.

Carr, L. G. and Edenberg, H. J.: *Cis*-acting sequences involved in protein binding and in vitro transcription of the human alcohol dehydrogenase gene *ADH2*. J. Biol. Chem. 265 (1990) 1658–1664.

Cheung, B., Holmes, R. S. and Beacham, I. R.: Molecular evolution of class I alcohol dehydrogenases in primates: Models for gene evolution and comparisons of 3' untranslated regions of cDNAs. Adv. Exp. Med. Biol. 372 (1995) 315–320.

Crabb, D. W., Edenberg, H. J., Thomasson, H. R. and Li, T. -K. in *The Genetics of Alcoholism*, H. Begleiter and B. Kissin, Eds. (Oxford Univ. Press, New York, 1995), p. 202.

Edenberg, H. J. and Brown, C. J.: Regulation of human alcohol dehydrogenase genes. Pharmacogenet. 2 (1992) 185–196.

Edenberg, H. J., Brown, C. J. and Zhang, L.: Regulation of the human alcohol dehydrogenase genes *ADH1*, *ADH2* and *ADH3:* Differences in *cis*-acting sequences at CTF/NF-I sites. Adv. Exp. Med. Biol. 328 (1993) 561–570.

Edenberg, H. J., Brown, C. J. and Zhang, L.: Alcohol dehydrogenases: molecular biology and gene regulation. Alcohol. Alcohol. Suppl.2 (1995) 123–127.

Hur, M. -W. and Edenberg, H. J.: Cloning and characterization of the *ADH5* gene encoding human alcohol dehydrogenase 5, formaldehyde dehydrogenase. Gene 121 (1992) 305–311.

Hur, M. -W. and Edenberg, H. J.: Cell-specific function of cis-acting elements in the regulation of human alcohol dehydrogenase 5 gene expression and effect of the 5'-nontranslated region. J. Biol. Chem. 270 (1995) 9002–9009.

Jornvall, H., Danielsson, O., Hjelmqvist, L., Persson, B. and Shafqat, J.: The alcohol dehydrogenase system. Adv. Exp. Med. Biol. 372 (1995) 281–294.

Kedishvili, N. Y., Bosron, W. F., Stone, C. L., Hurley, T. D., Peggs, C. F., Thomasson, H. R., Popov, K. M., Carr, L. G., Edenberg, H. J. and Li, T. -K.: Expression and kinetic characterization of recombinant human stomach alcohol dehydrogenase: Active site amino acid sequence explains substrate specificity compared with liver isozymes. J. Biol. Chem. 270 (1995a) 3625–3630.

Kedishvili, N. Y., Bosron, W. F., Stone, C. L., Peggs, C. F., Thomasson, H. R., Popov, K. M., Carr, L. G., Hurley, T. D., Edenberg, H. J. and Li, T. -K.: Cloning and expression of a human stomach alcohol dehydrogenase isozyme. Adv. Exp. Med. Biol. 372 (1995b) 341–347.

Maezawa, Y., Yamauchi, M., Toda, G., Suzuki, H. and Sakurai, S.: Alcohol-metabolizing enzyme polymorphisms and alcoholism in Japan. Alcohol. Clin. Exp. Res. 19 (1995) 951–954.

Moreno, A. and Pares, X.: Purification and characterization of a new alcohol dehydrogenase from human stomach. J. Biol. Chem. 266 (1991) 1128–1133.

Satre, M. A., Zgombic-Knight, M. and Duester, G.: The complete structure of the human class IV alcohol dehydrogenase (retinol dehydrogenase) determined from the ADH7 gene. J. Biol. Chem. 269 (1994) 15606–15612.

Stewart, M. J., McBride, M. S., Winter, L. A. and Duester, G.: Promoters for the human alcohol dehydrogenase genes *ADH1*, *ADH2*, and *ADH3:* interaction of CCAAT/enhancer-binding protein with elements flanking the *ADH2* TATA box. Gene. 90 (1990a) 271–279.

Stewart, M. J., Shean, M. L. and Duester, G.: Trans activation of human alcohol dehydrogenase gene expression in hepatoma cells by C/EBP molecules bound in a novel arrangement just 5' and 3' to the TATA box. Mol. Cell. Biol. 10 (1990b) 5007–5010.

Stewart, M. J., Shean, M. L., Paeper, B. W. and Duester, G.: The role of CCAAT/enhancer-binding protein in the differential transcriptional regulation of a family of human liver alcohol dehydrogenase genes. J. Biol. Chem. 266 (1991) 11594–11603.

Stone, C. L., Thomasson, H. R., Bosron, W. F. and Li, T. -K.: Purification and partial amino acid sequence of a high-activity human stomach alcohol dehydrogenase. Alcohol. Clin. Exp. Res. 17 (1993) 911–918.

Tanaka, F., Shiratori, Y., Yokosuka, O., Imazeki, F., Tsukada, Y. and Omata, M.: High incidence of ADH2*1/ALDH2*1 genes among Japanese alcohol dependents and patients with alcoholic liver disease. Hepatol. 23 (1996) 234–239.

Thomasson, H. R., Edenberg, H. J., Crabb, D. W., Mai, X. L., Jerome, R. E., Li, T. K., Wang, S. P., Lin, Y. T., Lu, R. B. and Yin, S. J.: Alcohol and aldehyde dehydrogenase genotypes and alcoholism in Chinese men. Am. J. Hum. Genet. 48 (1991) 677–681.

Thomasson, H. R., Crabb, D. W., Edenberg, H. J. and Li, T. -K.: Alcohol and aldehyde dehydrogenase polymorphisms and alcoholism. Behavior Genetics 23 (1993) 131–136.

Thomasson, H. R., Crabb, D. W., Edenberg, H. J., Li, T. -K., Hwu, H. -G., Chen, C. C., Yeh, E. K. and Yin, S. -J.: Low frequency of the ADH2*2 allele among Atayal natives of Taiwan with alcohol use disorders. Alcohol. Clin. Exp. Res. 18 (1994) 640–643.

Tu, G. -C. and Israel, Y.: Alcohol consumption by Orientals in North America is predicted largely by a single gene. Behavior Genetics 25 (1995) 59–65.

Yamauchi, M., Maezawa, Y., Toda, G., Suzuki, H. and Sakurai, S.: Association of a restriction fragment length polymorphism in the alcohol dehydrogenase 2 gene with Japanese alcoholic liver cirrhosis. J. Hepatology 23 (1995) 519–523.

Yang, Z. -N., Davis, G. J., Hurley, T. D., Stone, C. L., Li, T. -K. and Bosron, W. F.: Catalytic efficiency of human alcohol dehydrogenases for retinol oxidation and retinal reduction. Alcohol. Clin. Exp. Res. 18 (1994) 587–591.

Yasunami, M., Chen, C. S. and Yoshida, A.: A human alcohol dehydrogenase gene (ADH6) encoding an additional class of isozyme. Proc. Natl. Acad. Sci. USA 88 (1991) 7610–7614.

Yin, S. -J., Wang, M. -F., Liao, C. -S., Chen, C. -M. and Wu, C. -W.: Identification of a human stomach alcohol dehydrogenase with distinctive kinetic properties. Biochem. Int. 22 (1990) 829–835.

Yin, S. -J., Chou, F. -J., Chao, S. -F., Tsai, S. -F., Liao, C. -S., Wang, S. -L., Wu, C. -W. and Lee, S. -C.: Alcohol and aldehyde dehydrogenases in human esophagus: comparison with the stomach enzyme activities. Alcohol. Clin. Exp. Res. 17 (1993) 376–381.

Yokoyama, S., Matsuo, Y., Ramsbotham, R. and Yokoyama, R.: Molecular characterization of a class IV human alcohol dehydrogenase gene (ADH7). FEBS Lett. 351 (1994) 411–415.

Zgombic-Knight, M., Foglio, M. H. and Duester, G.: Genomic structure and expression of the ADH7 gene encoding human class IV alcohol dehydrogenase, the form most efficient for retinol metabolism in vitro. J. Biol. Chem. 270 (1995) 4305–4311.

40

EXPRESSION, ACTIVITIES, AND KINETIC MECHANISM OF HUMAN STOMACH ALCOHOL DEHYDROGENASE

Inference for First-Pass Metabolism of Ethanol in Mammals

Shih-Jiun Yin,[1] Chih-Li Han,[1] Chin-Shya Liao,[1] and Chew-Wun Wu[2]

[1]Department of Biochemistry
National Defense Medical Center
Taipei, Taiwan, Republic of China
[2]Department of Surgery
Veterans General Hospital
Taipei, Taiwan, Republic of China

1. INTRODUCTION

Alcohol dehydrogenase (ADH) is a dimeric molecule containing two zinc atoms per subunit, one for catalytic and the other for structural functions (Eklund and Branden, 1987; Hurley et al., 1994). Human ADH constitutes a complex family. Seven genes have thus far been identified. ADH_1 through ADH_5 and ADH_7 encode α, β, γ, π, χ, and μ (or denoted σ) polypeptides, respectively (Smith, 1986; Yoshida et al., 1991; Jornvall and Hoog, 1995). The ADH_6-encoding subunit has not yet been designated a Greek letter (Yasunami et al., 1991). Both ADH_2 and ADH_3 exhibit polymorphism. The allelic variants, ADH_2^1, ADH_2^2, ADH_2^3, and ADH_3^1, ADH_3^2 code for β_1, β_2, β_3, and γ_1, γ_2, respectively. These polymorphic genes display different allelic distribution among racial populations (Agarwal and Goedde, 1992; Bosron et al., 1993; Yin, 1994). Based on the kinetic and structural features, human ADH can be grouped into five classes (Vallee and Bazzone, 1983; Jornvall and Hoog, 1995). Class I ADH comprises homo- and heterodimers with α, β or γ subunits. Class II, III, IV enzymes are $\pi\pi$, $\chi\chi$, and $\mu\mu$, respectively. The intraclass and interclass sequence homologies at the amino acid/nucleotide level are approximately 90% and 60%, respectively (Yoshida et al., 1991; Jornvall and Hoog, 1995). Kinetic properties of these five class ADHs vary strikingly (Vallee and Bazzone, 1983; Bosron et al., 1993; Yin, 1994). K_m values for ethanol for class I isoenzymes are generally lower than 5 mM, suggesting that they are active after low alcohol consumption. Class II, IV and V enzymes exhibit K_m about 30 mM, hence active at moderate to high ethanol concentrations. Class III enzyme does not seem to significantly contribute to ethanol metabolism due to its extremely high K_m, > 3 M.

ADH is the principal enzyme responsible for ethanol metabolism in humans (Li, 1977) and exhibits tissue-specific expression (Smith, 1986; Yoshida et al., 1991; Yin et al., 1988, 1990, 1992, 1993, 1994; Dong et al., 1996). Class I $\alpha\alpha$, $\beta\beta$, and $\gamma\gamma$ ADHs are expressed in the liver. $\beta\beta$ ADH is found in the lung, kidney, spleen, pancreas, and the muscle layer of digestive tract. $\gamma\gamma$ ADH distributes in the mucosal layer of stomach and of the lower digestive tract. Class II $\pi\pi$ appears solely in the liver. Class III $\chi\chi$ appears ubiquitously expressed. Class IV $\mu\mu$ displays a limited distribution to the mucosa of upper digestive tract including gingiva, tongue, pharynx, esophagus, stomach, and also in the cornea (Holmes, 1988). Ethnic variability of the expression of gastric $\mu\mu$ has been reported (Baraona et al., 1991; Yin et al., 1993). It is detectable by electrophoresis in all the Caucasian and African subjects examined whereas only in 20–30% of the Orientals studied. Tissue distribution of the enzyme activity of class V ADH has not been defined, although its mRNA transcripts were detected in the liver and stomach (Yasunami et al., 1991).

First-pass metabolism of ethanol, i.e., the difference between the quantity of alcohol that reaches the systemic circulation by the intravenous route and the quantity that entered by the oral dose, has recently been an issue of extensive interest and debate because of its medical and legal implications (Gentry et al., 1994a,b; Levitt, 1993, 1994; Smith and Levitt, 1995). Three major questions remain controversial: (1) What is the extent of first-pass metabolism? (2) What is the relative contribution of liver, stomach and small intestine to this metabolism? (3) Is there a gender difference in first-pass metabolism? In attempt to partly answer these complex questions, we have elucidated the steady-state kinetic mechanism of human stomach $\mu\mu$ ADH, determined the ethanol-oxidizing activity in gastric muscosa in relation to sex, genetic polymorphism at the *ADH₃* locus and the expressed type of $\mu\mu$, and assessed the capacity of first-pass metabolism of ethanol in the stomach and liver based on the total organ activity in conjunction with the expression pattern and kinetic constants of ADH.

2. EXPERIMENTAL PROCEDURES

$\mu\mu$ ADH was isolated from pooled surgical gastric mucosae via SP-Sephadex, Q-Sepharose, AMP-Sepharose, and Blue Sepharose chromatographic steps to apparent homogeneity. The subunit molecular weight was estimated by SDS-polyacrylamide gel electrophoresis to be 40,000 daltons. The protein concentration was determined according to Bradford (1976), using bovine serum albumin as the standard. The specific activity of $\mu\mu$, 65 μmol/min/mg measured in 100 mM glycine/NaOH (pH 10.0) at 25°C containing 100 mM ethanol and 2.4 mM NAD, was similar to that reported by Stone et al. (1993). Kinetic assay was performed with a Cary 3E spectrophotometer in 100 mM sodium phosphate, pH 7.5, at 25°C. The kinetic data were evaluated with HYPER, COMP, NONCOMP, or UNCOMP statistical programs (Cleland, 1979). The coefficients of variation for the kinetic constants were less than 15%.

Stomach and liver specimens from adult Chinese were obtained during routine operations for therapeutic treatment. None of the patients had a history of heavy alcohol consumption. The normal portion was dissected off and stored at -70°C until use. The tissue (0.2–0.3 g) was homogenized in 2 volumes (v/w) of ice-cold 10 mM sodium phosphate, pH 7.5, with a Polytron homogenizer. The homogenate was centrifuged at 100,000 g for 1 hr at 4°C. Multiple ADH forms were identified by an agarose isoelectric focusing procedure using a mixture of Ampholine, pH 3.5–10, 7–9, and 9–11, and stained for enzyme ac-

Table 1. Product and dead-end inhibition patterns of
human stomach μμ ADH

Inhibitor	Varied substrate	Pattern[a]
NADH	NAD	C
NAD	NADH	C
acetaldehyde	ethanol	N
ethanol	acetaldehyde	N
4-methylpyrazole	ethanol	C
trifluoroethanol	ethanol	C
	NAD	U
isobutyramide	acetaldehyde	C
	NADH	U

[a]C, competitive; N, noncompetitive; U, uncompetitive.

tivity at 120 mM ethanol (Yin et al., 1984a). ADH activity was determined at 30°C with 33 or 500 mM ethanol, 2.4 mM NAD, and 1 mM semicarbazide in 100 mM sodium phosphate, pH 7.5. A 5-min assay in the absence of ethanol was subtracted as blank.

3. RESULTS AND DISCUSSION

Reaction mechanism of human stomach μμ ADH was studied by steady-state kinetic approaches. The pattern of covarying NAD and ethanol as well as product inhibition patterns for forward or reverse reaction (Table 1) suggest that μμ follows an ordered bi bi sequential mechanism with pyridine nucleotide cofactor being binding first and leaving last in the catalytic cycle. The proposed mechanism was further supported by the patterns of dead-end inhibitors with respect to ethanol or acetaldehyde (Table 1). The K_m values for cofactors and substrates, i.e., K_a, K_b, K_p, and K_q, were determined to be 0.13, 27, 16, and 0.11 mM, respectively. The k_{cat} values for forward and reverse reactions, V_f and V_r, were determined to be 1,500 and 24,000 min⁻¹, respectively. The data of K_m and k_{cat} for ethanol oxidation for μμ are similar to those reported for σσ isolated from Caucasian stomach (Stone et al., 1993; Farres et al., 1994) or determined from the recombinant enzyme (Kedishvili et al., 1995). The rate constant for NADH dissociation, k_{-4}, was calculated to be 1,300 min⁻¹ from the kinetic constants (Cleland, 1963). Since k_{-4} of μμ is nearly equal to its V_r, NADH dissociation appears to be rate limiting in catalysis. The same rate-limiting step has been demonstrated for the horse liver EE ADH (Dalziel and Dickinson, 1966; Wratten and Cleland, 1963) and the human liver αα, $β_1β_1$, and $β_2β_2$ (Bosron et al., 1983; Yin et al., 1984b).

Expression pattern of ADH in the mucosa of the body of stomach was identified by agarose isoelectric focusing (Fig. 1). Two genetic phenotypes for the class I γγ were observed, homozygous ADH₃ 1–1 (exhibiting $γ_1γ_1$) and heterozygous ADH₃ 1–2 (exhibiting $γ_1γ_1$, $γ_1γ_2$, and $γ_2γ_2$). Two phenotypes for the class IV μμ were also found, the expressed type μ-ADH(+) and the absent type μ-ADH(−). Of the 209 mucosal samples studied, the frequencies of the homozygosity for $γ_1γ_1$ and heterozygosity were 81% and 19%, respectively. No homozygous ADH₃ 2–2 phenotypic samples were found, indicating the alleles ADH_3^2 (encoding $γ_2$ subunit) was much less common in the Chinese population in Taiwan. The frequencies of the alleles ADH_3^1 and ADH_3^2 were calculated to be 0.90 and 0.10, respectively. This allele frequency, similar to that determined from leukocyte DNA by PCR

Figure 1. Agarose isoelectric focusing of ADHs from surgical gastric mucosae. Gels were stained for enzyme activity. Lanes 2–4 and 5–7 are ADH_3 1–1 and ADH_3 1–2 phenotypes, respectively. Lanes 3, 4 and 6 are mm express phenotype. For comparison, lane 1 was from a surgical liver with ADH_2 2–2 and ADH_3 1–1 phenotype.

genotyping method (Yin, 1994), fits the Hardy-Weinberg equilibrium distribution. It is interesting to note that the $ADH_3{}^1$ and $ADH_3{}^2$ occur about equal frequencies among Caucasians and American Indians (Smith, 1986; Bosron et al., 1993). Thirty-one percent (65/209) of the gastric samples displayed weak to moderately intense activity bands of class IV μμ (Fig. 1). This frequency is similar to that reported for the Japanese (Baraona et al., 1991). Absence of the gastric μμ activity was not found in Caucasians and African Americans. This ethnic variation of the expression of stomach μμ may be regulated at the transcriptional or translational level rather than due to a point mutation or deletion of the gene (Yin et al., 1993).

Ethanol oxidizing activities in the gastric mucosa with different phenotypes of class I and IV ADHs are shown in Table 2. The homozygous ADH_3 1–1 and the μ-ADH expressed type exhibited significantly greater activity than the heterozygous ADH_3 1–2 and the μ-ADH absent type, respectively, at both high and low ethanol concentrations. This finding is consistent with the reported two-fold higher maximal activity of $\gamma_1\gamma_1$ compared to $\gamma_2\gamma_2$ (Bosron et al., 1983). At 500 mM ethanol, the activities of the μ-ADH(+) mucosae showed about 30% greater than that of the μ-ADH(–) ones. This is also compatible with the high K_m value (27 mM) of μμ. As expected, the activity difference between these two μ-ADH phenotypes became smaller when measured at 33 mM ethanol (Table 2). In the μ-ADH(–) gastric mucosae, the ethanol oxidizing activity at 500 mM appeared to be about

Table 2. ADH activities in surgical human gastric mucosa with different phenotypes

Class	Phenotype	N	Specific activity (nmol/min/g tissue)	
			33 mM	500 mM
I	ADH_3 1–1	169	238 ± 7	216 ± 6
	ADH_3 1–2	40	166 ± 13[a]	176 ± 17[a]
IV	μ-ADH (+)	65	245 ± 11	254 ± 12
	μ-ADH (–)	144	216 ± 8[a]	188 ± 6[a]

[a]$p < 0.05$, compared to the corresponding phenotype.

Table 3. Expression pattern and kinetic constants for ethanol oxidation of mammalian stomach and liver ADHs

Species	Organ	Class	Form	K_m (mM)	k_{cat} (min^{-1})	k_{cat}/K_m (mM^{-1}·min^{-1})
Human	Liver	I	$\alpha\alpha$	4.2	27	6.4
			$\beta\beta_1$	0.049	9.2	190
			$\beta_2\beta_2$	0.94	400	430
			$\beta_3\beta_3$	36	320	8.9
			$\gamma_1\gamma_1$	1.0	87	87
			$\gamma_2\gamma_2$	0.63	35	56
		II	$\pi\pi$	34	20	0.59
	Stomach	I	$\gamma_1\gamma_1\gamma_2\gamma_2$			
		IV	$\mu\mu$	27	1,500	56
Baboon	Liver	I	ADH-2	0.25	45	180
	Stomach	IV	ADH-3	385	12,400	32
Rat	Liver	I	ADH-3	1.4	39	28
	Stomach	IV	ADH-1	5,000	1,000	0.20

[a]Data for human class I and II, baboon, and rat enzymes are from Bosron et al. (1993), Algar et al. (1992), and Julia et al. (1987), respectively.

10% lower than that at 33 mM. This is in agreement with the previously reported inhibition of $\gamma\gamma$ activity at high ethanol concentrations (Buhler and von Wartburg, 1982). No inhibition at such high ethanol level was found in the gastric mucosae expressing intense $\mu\mu$ activity, compatible with the observed saturation kinetics with high ethanol for $\mu\mu$. When further compared the activities of the gastric samples according to their ADH$_3$ 1–1, μ-ADH(+) or μ-ADH(−) phenotype, no significant sex and age effects on the ADH activity at both high and low ethanol concentrations were found (data not shown). This result is in conflict with the reports by Frezza et al. (1990) and Seitz et al. (1993), but in general agreement with the findings by Thuluvath et al. (1994), Salmela et al. (1994), Moreno et al. (1994) and Brown et al. (1995). Our recent studies with a total of 115 endoscopic stomach biopsies have confirmed the findings of no significant correlation of ADH activity with sex and age (Yin et al., 1996). This is also in agreement with our previous report of no age and gender differences in human colon ADH activity (Yin et al., 1994).

To evaluate the ethanol oxidizing capacitity in mammalian stomach and liver, expression pattern and kinetic properties of their ADHs are compared (Table 3). K_m values for human $\gamma_1\gamma_1$ and $\gamma_2\gamma_2$ are actually $[S]_{0.5}$, which reflect negative cooperativity for ethanol oxidation (Bosron et al., 1983). The kinetic constants were determined in 100 mM sodium phosphate, pH 7.5 at 25°C, for human and rat ADHs (Bosron et al., 1993; Julia et al., 1987; and this study), and in 50 mM sodium phosphate, pH 7.4 at 30°C for baboon enzymes. Unlike the rat and baboon, human liver exhibits high-K_m class II $\pi\pi$ in addition to the complex low-K_m class I enzymes as well as both the class I and IV ADHs are significantly expressed in the human stomach. K_m for human stomach class IV $\mu\mu$ is at least one order of magnitude lower than that of the baboon ADH-3. Rat stomach ADH-1 exhibits an extremely high K_m, 5,000 mM, making it the least efficient class IV enzyme compared with those from the human and baboon. The species distinctions in the organ expression pattern and kinetic constants of ADH are important in the comparison of the capacity of first-pass metabolism of ethanol in mammals.

We take the average human liver weight as 1,500 g and assume one third of the average stomach weight of 150 g as mucosa, i.e., 50 g. Based on the ADH activities determined from surgical human liver with homozygous ADH$_2$ 1–1 (a less common phenotype

Table 4. Total ADH activities in mammalian stomach and liver

Species	Ethanol concentration	ADH activity (μmol·min^{-1} organ^{-1})		Stomach/Liver ratio (%)
		Liver	Stomach	
Human	33 mM	3,500	12	0.34
	500 mM	—[b]	13	
Baboon[a]	50 mM	108	1.5	1.4
	500 mM	79	6.9	8.7
	1,000 mM	77	8.4	11
Rat[a]	33 mM	3.5	0.020	0.57
	1,000 mM	—[b]	0.22	

[a]Data for baboon and rat are from Algar et al. (1992) and Boleda et al. (1989), respectively.
[b]Not determined.

in Orientals but predominant in Caucasians), 2.3 μmol/min/g tissue (Yin, S.-J., unpublished observation) and from human stomach mucosa with $\mu\mu$ express type (Table 2), we can estimate the ethanol oxidizing capacities of whole liver and stomach (Table 4). Humans exhibit a stomach-to-liver organ activity ratio, 0.34%, similar to that of the rat, 0.57%. It is worth noting that the human stomach displayed similar activities at 33 and 500 mM ethanol. However, the activities of rat and baboon stomach increased considerably when measured at higher ethanol concentrations (Table 4). This is in agreement with the high K_m values for the stomach enzyme in rat and baboon (Table 3).

Gastric and hepatic contributions to the first-pass metabolism of ethanol in human and rat can be estimated based on the alcohol concentration differences between the systemic blood and portal vein or gastric lumen, Michaelis constants of the liver or stomach ADH, and the total organ activities (Table 5). Since the peripheral blood ethanol concentrations reached 2–6 mM in human during the first 2-hr period following ingestion of alcohol with a dose of 0.3 g/kg at fed state (Frezza et al., 1990), we assume that the average blood ethanol concentrations in the systemic circulation and portal vein are 3 and 33 mM, respectively, and in the gastric lumen, 500 mM. We also assume that in human liver, 30% (as suggested by the extent of 4-methylpyrazole inhibition) of the total organ activity at 33 mM ethanol is contributed by the class II $\pi\pi$ with a K_m of 34 mM, and assume 50% of total stomach activity at 500 mM ethanol is contributed by the class IV $\mu\mu$ with a K_m of 27 mM. The capacity of the first-pass metabolism in the liver and stomach can then be calculated by using the Michaelis-Menten equation. The human stomach-to-liver FPM ratio was estimated to be 0.68% (Table 5), indicating that the gastric first-pass metabolism in human may be quantitatively negligible. By contrast, this ratio in rat was estimated to be much greater, 22%. Hence, first-pass metabolism of ethanol in human can not be adequately interpreted in terms of rat data. Human hepatic first-pass metabolism of ethanol by class II $\pi\pi$ for the first 1-hr period was estimated, 870 μmol/min/organ × 60 min = 52 mmol (2.4 g), equivalent to 12% of total ethanol intake at 0.3 g/kg for a 70 kg person. This estimated value is comparable to the reported 10–30% of first-pass metabolism of ethanol in humans at the same dose (Frezza et al., 1990; Gentry et al., 1994a). It should be stressed that the crucial point is to assess the magnitude of ADH activity increase in the liver or stomach corresponding to the ethanol concentration difference in the organ and systemic blood during the period of first-pass metabolism, which reflects the effective potential to metabolize the first-passed ethanol. It should also be pointed out that alcohol elimination follows Michaelis-Menten kinetics and the comparison of the areas under the alcohol concentration-time curve for first-pass metabolism is valid only when the metabolic clearance follows first-order kinetics.

Table 5. Estimations of gastric and hepatic contribution to mammalian first-pass metabolism of ethanol

Species	Assumed ethanol concentration			FPM capacity (μmol $\times$ min^{-1} $\times$ organ^{-1})		FPM capacity stomach/liver ratio (%)
	Systemic blood	Portal vein	Gastric lumen	Liver	Stomach	
Human	3 mM	33 mM	500 mM	870	5.9	0.68
Rat[a]	3 mM	33 mM	1,000 mM	1.0	0.22	22

[a]Data were calculated using Michaelis-Menten equation based on the measured total organ activity (Table 4) and the K_m values for ethanol for the liver and stomach ADHs, 1.4 and 5,000 mM, respectively (Table 3).

4. CONCLUSIONS

Human gastric mucosa exhibits significant ethanol oxidizing activities which can be correlated with the polymorphism at the class I ADH_3 locus and also with the expression of class IV $\mu\mu$. Stomach ADH activities appear not to differ significantly between men and women with respect to age and polymorphisms. Steady-state kinetic studies suggest that $\mu\mu$ follows ordered bi bi sequential mechanism with dissociation of the enzyme-NADH complex being rate-limiting.

Differential expression of class I and IV ADHs in the stomach suggests that different vulnerability to ethanol-induced mucosal injury may exist among racial populations. The results of tissue expression pattern, kinetic features of the ADH family and relative organ ethanol-oxidizing activity suggest that stomach may play a very minor role in the first-pass metabolism of alcohol in humans. In rats, the relative contributions to the first-pass metabolism of hepatic and gastric origins may vary depending on the ethanol dose and delivery rate.

5. ACKNOWLEDGMENTS

This research was supported by grants from the Department of Health (DOH 85-HR-304), the National Science Council (NSC 84–2331-B016–62), and Academia Sinica, Republic of China.

6. REFERENCES

Agarwal, D. P. and Goedde, H. W.: Pharmacogenetics of alcohol metabolism and alcoholism. Pharmacogenetics 2 (1992) 48–62.

Algar, E. M., VandeBerg, J. L. and Holmes, R. S.: A gastric alcohol dehydrogenase in the baboon: purification and properties of a high K_m enzyme, consistent with a role in first pass alcohol metabolism. Alcohol.Clin.Exp.Res. 16 (1992) 922–927.

Baraona, E., Yokoyama. A., Ishii, H., Hernandez-Munoz, R., Takagi, T., Tsuchiya, M. and Lieber, C. S.: Lack of alcohol dehydrogenase isoenzyme activities in the stomach of Japanese subjects. Life Sci. 49 (1991) 1929–1934.

Boleda, M. D., Julia, P., Moreno, A. and Pares, X.: Role of extrahepatic alcohol dehydrogenase in rat ethanol metabolism. Arch.Biochem.Biophys. 274 (1989) 74–81.

Bosron, W. F., Magnes, L. J. and Li, T.-K.: Kinetic and electrophoretic properties of native and recombined isoenzymes of human liver alcohol dehydrogenase. Biochemistry 22 (1983) 1852–1857.

Bosron, W. F., Ehrig, T. and Li, T.-K.: Genetic factors in alcohol metabolism and alcoholism. Semin.Liver Dis. 13 (1993) 126–135.

Bradford, M. M.: A rapid and sensitive method for the quantitation of microgram quantities of protein utilizing the principle of protein-dye binding. Anal.Biochenm. 72 (1976) 248–254.

Brown, A. St. J. M., Fiatarone, J. R., Wood, P., Bennett, M. K., Kelly, P. J., Rawlins, M. D., Day, C. P. and James, O. P. W.: The effect of gastritis on human gastric alcohol dehydrogenase activity and ethanol metabolism. Aliment.Pharmacol.Ther. 9 (1995) 57–61.

Buhler, R. and von Wartburg, J.-P.: Purification and substrate specificities of three human liver alcohol dehydrogenase isoenzymes. FEBS Lett. 144 (1982) 135–139.

Cleland, W. W.: The kinetics of enzyme-catalyzed reactions with two or more substrates or products. Biochem.Biophys.Acta 67 (1963) 104–196.

Cleland, W. W.: Statistical analysis of enzyme kinetic data. Methods Enzymol. 63A (1979) 103–138.

Dalziel, K. and Dickinson, F. M.: The kinetics and mechanism of liver alcohol dehydrogenase with primary and secondary alcohols as substrates. Biochem.J. 100 (1966) 34–46.

Dong, Y.-J., Peng, T.-K. and Yin, S.-J.: Expression and activities of class IV alcohol dehydrogenase and class III aldehyde dehydrogenase in human mouth. Alcohol 13 (1996) 257–262.

Eklund, H. and Branden, C.-I.: Alcohol dehydrogenase. In Jurnak, F. A. and McPherson, A. (Eds.), Biological Macromolecules and Assemblies. Vol. 3, Wiley, New York, 1987, pp. 73–142.

Farres, J., Moreno, A., Crosas, B., Peralba, J. M., Allali-Hassani, A., Hjelmqvist, L., Jornvall, H. and Pares, X.: Alcohol dehydrogenase of class IV ($\sigma\sigma$-ADH) from human stomach. cDNA sequence and structure/function relationships. Eur.J.Biochem. 224 (1994) 549–557.

Frezza, M., DiPadova, C., Prozzato, G., Terpin, M., Baraona, E. and Lieber, C. S.: High blood alcohol levels in women. The role of decreased gastric alcohol dehydrogenase activity and first pass metabolism. N.Engl.J.Med. 322 (1990) 95–99.

Gentry, R. T., Baraona, E. and Lieber, C. S.: Agonist: gastric first pass metabolism of alcohol. J.Lab.Clin.Med. 123 (1994a) 21–26.

Gentry, R. T., Baraona, E. and Lieber, C. S.: Rebuttal to antagonist. J.Lab.Clin.Med. 123 (1994b) 32–33.

Holmes, R. S.: Alcohol dehydrogenases and aldehyde dehydrogenases of anterior eye tissues from humans and other mammals. In Kuriyama, K., Takada, A. and Ishii, H. (Eds.), Biomedical and Social Aspects of Alcohol and Alcoholism. Elsevier, Amsterdam, 1988, pp. 51–57.

Hurley, T. D., Bosron, W. F., Stone, C. L. and Amzel, L. M.: Structures of three human β alcohol dehydrogenase variants. Correlations with their functional differences. J.Mol.Biol. 239 (1994) 415–429.

Jornvall, H. and Hoog, J.-O.: Nomenclature of alcohol dehydrogenases. Alcohol Alcohol. 30 (1995) 153–161.

Julia, P., Farres, J. and Pares, X.: Characterization of three isoenzyme of rat alcohol dehydrogenase. Tissue distribution and physical and enzymatic properties. Eur.J.Biochem. 162 (1987) 179–189.

Kedishvili, N. Y., Bosron, W. F., Stone, C. L., Hurley, T. D., Peggs, C. F., Thomasson, H. R., Popov, K. M., Carr, L. G., Edenberg, H. J. and Li, T.-K.: Expression and kinetic characterization of recombinant human stomach alcohol dehydrogenase. Active-site amino acid sequence explains substrate specificity compared with liver isozymes. J.Biol.Chem. 270 (1995) 3625–3630.

Levitt, M. D.: Review article: lack of clinical significance of the interaction between H$_2$-receptor antagonists and ethanol. Aliment.Pharmacol.Ther. 7 (1993) 131–138.

Levitt, M. D.: Antagonist: the case against first-pass metabolism of ethanol in the stomach. J.Lab.Clin.Med. 123 (1994) 28–31.

Li, T.-K.: Enzymology of human alcohol metabolism. Adv.Enzymol. 45 (1977) 427–483.

Moreno, A. and Pares, X.: Purification and characterization of a new alcohol dehydrogenase from human stomach. J.Biol.Chem. 266 (1991) 1128–1133.

Moreno, A., Pares, A., Ortiz, J., Enriquez, J. and Pares, X.: Alcohol dehydrogenase from human stomach: variability in normal mucosa and effect of age, gender, ADH$_3$ genotype and gastric region. Alcohol Alcohol. 29 (1994) 663–671.

Salmela, K. S., Salaspuro, M., Gentry, R. T., Methuen, T., Hook-Nikanne, J., Kosunen, T. U. and Roine, R. P.: *Helicobacter* infection and gastric ethanol metabolism. Alcohol.Clin.Exp.Res. 18 (1994) 1294–1299.

Seitz, H. K., Egerer, G., Simanowski, U. A., Waldherr, R., Eckey, R., Agarwal, D. P., Goedde, H. W. and von Wartburg, J.-P.: Human gastric alcohol dehydrogenase activity: effect of age, sex, and alcoholism. Gut 34 (1993) 1433–1437.

Smith, M.: Genetics of human alcohol and aldehyde dehydrogenases. Adv.Hum.Genet. 15 (1986) 249–290.

Smith, T. and Levitt, M. D.: Ethanol concentration and gastric alcohol dehydrogenase activity. Gastroenterology 109 (1995) 663–664.

Stone, C. L., Thomasson, H. R., Bosron, W. F. and Li, T.-K.: Purification and partial amino acid sequence of a high-activity human stomach alcohol dehydrogenase. Alcohol.Clin.Exp.Res. 17 (1993) 911–918.

Thuluvath, P., Wojno, K. J., Yardley, J. H. and Mezey, E.: Effects of *Helicobacter pylori* infection and gastritis on gastric alcohol dehydrogenase activity. Alcohol.Clin.Exp.Res. 18 (1994) 795–798.

Vallee, B. L. and Bazzone, T. J.: Isoenzymes of human liver alcohol dehydrogenase. Isozymes Curr. Top.Biol.Med.Res. 8 (1983) 219–244.

Wratten, C. C. and Cleland, W. W.: Product inhibition studies on yeast and liver alcohol dehydrogenases. Biochemistry 2 (1963) 935–941.

Yin, S.-J., Bosron, W. F., Li, T.-K., Ohnishi, K., Okuda, K., Ishii, H. and Tsuchiya, M.: Polymorphism of human liver alcohol dehydrogenase: identification of ADH_2 2–1 and ADH_2 2–2 phenotypes in the Japanese by isoelectric focusing. Biochem.Genet. 22 (1984a) 169–180.

Yin, S.-J., Bosron, W. F., Magnes, L. J. and Li, T.-K.: Human liver alcohol dehydrogenase: purification and kinetic characterization of the $\beta_2\beta_2$, $\beta_2\beta_1$, $\alpha\beta_2$, and $\beta_2\gamma_1$ "Oriental" isoenzymes. Biochemistry 23 (1984b) 5847–5853.

Yin, S.-J., Chang, T.-C., Chang C.-P., Chen, Y.-J., Chao, Y.-C., Tang, H.-S., Chang, T.-M. and Wu, C.-W.: Human stomach alcohol and aldehyde dehydrogenase (ALDH): A genetic model proposed for ALDH III isoenzymes. Biochem.Genet. 26 (1988) 343–360.

Yin, S.-J., Wang. M,-F., Liao, C.-S., Chen, C.-M. and Wu, C.-W.: Identification of a human stomach alcohol dehydrogenase with distinctive kinetic properties. Biochem.Int. 22 (1990) 829–835.

Yin, S.-J., Liao, C.-S., Chen, C.-M., Fan, F.-T. and Lee, S.-C.: Genetic polymorphism and activities of human lung alcohol and aldehyde dehydrogenases: implications for ethanol metabolism and cytotoxicity. Biochem.Genet. 30 (1992) 203–215.

Yin, S.-J., Chou, F.-J., Chao, S.-F., Tsai, S.-F., Liao, C.-S., Wang, S.-L., Wu, C.-W. and Lee, S.-C.: Alcohol and aldehyde dehydrogenases in human esophagus: comparison with the stomach enzyme activities. Alcohol.Clin.Exp.Res. 17 (1993) 376–381.

Yin, S.-J.: Alcohol dehydrogenase: enzymology and metabolism. Alcohol Alcohol. 29(Suppl.2) (1994) 113–119.

Yin, S.-J., Liao, C-S., Lee, Y.-C., Wu, C.-W. and Jao, S.-W.: Genetic polymorphism and activities of human colon alcohol and aldehyde dehydrogenases: no gender and age differences. Alcohol.Clin.Exp.Res. 18 (1994) 1256–1260.

Yin, S.-J., Liao, C.-S., Lai, C.-L. and Chao, Y.-C.: Human stomach alcohol and aldehyde dehydrogenase activities: no gender and age difference. Alcohol.Clin.Exp.Res. 20(Suppl.) (1996) 35A.

Yasunami, M., Chen, C.-S. and Yoshida, A.: A human alcohol dehydrogenase gene (*ADH6*) encoding an additional class of isozyme. Proc.Natl.Acad.Sci.USA 88 (1991) 7610–7614.

Yoshida, A., Hsu, L. C. and Yasunami, M.: Genetics of human alcohol-metabolizing enzymes. Prog.Nucl.Acid Res.Mol.Biol. 40 (1991) 255–287.

41

EVIDENCE THAT CLASS IV ALCOHOL DEHYDROGENASE MAY FUNCTION IN EMBRYONIC RETINOIC ACID SYNTHESIS

Gregg Duester, Louise Deltour, and Hwee Luan Ang

The Burnham Institute
10901 North Torrey Pines Road
La Jolla, California 92037

1. INTRODUCTION

The signaling effects of the vitamin A derivative retinoic acid are mediated by the retinoic acid receptor family (Kastner et al., 1994; Mangelsdorf et al., 1994). Recent studies on mice carrying retinoic acid receptor mutations have demonstrated that retinoic acid is essential for development of several embryonic tissues including the craniofacial region, eye, heart, and other organs (Lohnes et al., 1994; Mendelsohn et al., 1994). However, little is known about the enzymatic mechanism involved in the correct spatial and temporal regulation of retinoic acid synthesis from retinol (vitamin A) during embryonic development.

Retinoids are ultimately derived from oxidative cleavage of dietary β-carotene which produces retinal, a reactive aldehyde that is primarily reduced to retinol to form retinyl esters for storage. Since the retinol/retinal interconversion is reversible, retinol can undergo oxidation to produce retinal, and a further oxidation step produces retinoic acid which serves as the ligand for nuclear retinoic acid receptors (Fig. 1). The rate-limiting step in retinoic acid synthesis from retinol is the oxidation of retinol, a reaction which can be catalyzed *in vitro* by several classes of cytosolic alcohol dehydrogenase (ADH) which use NAD as coenzyme (Connor and Smit, 1987; Boleda et al., 1993; Yang et al., 1994; Kedishvili et al., 1995). A microsomal retinol dehydrogenase which also catalyzes retinol oxidation *in vitro* has been identified as a member of the short-chain dehydrogenase/reductase (SDR) enzyme family (Chai et al., 1995). However, since microsomal retinol dehydrogenase prefers NADP(H) as a coenzyme, it may function primarily as a retinal reductase due to the high ratio of NADPH to NADP in cells. The oxidation of retinal to retinoic acid can be catalyzed *in vitro* by either class I aldehyde dehydrogenase (ALDH) which uses NAD as coenzyme (Lee et al., 1991; Dockham et al., 1992; Bhat et al., 1995) or cytochrome P450 1A1 which utilizes molecular oxygen and NADPH as coenzyme (Raner et al., 1996). These aspects of retinoid metabolism are depicted in Fig. 1.

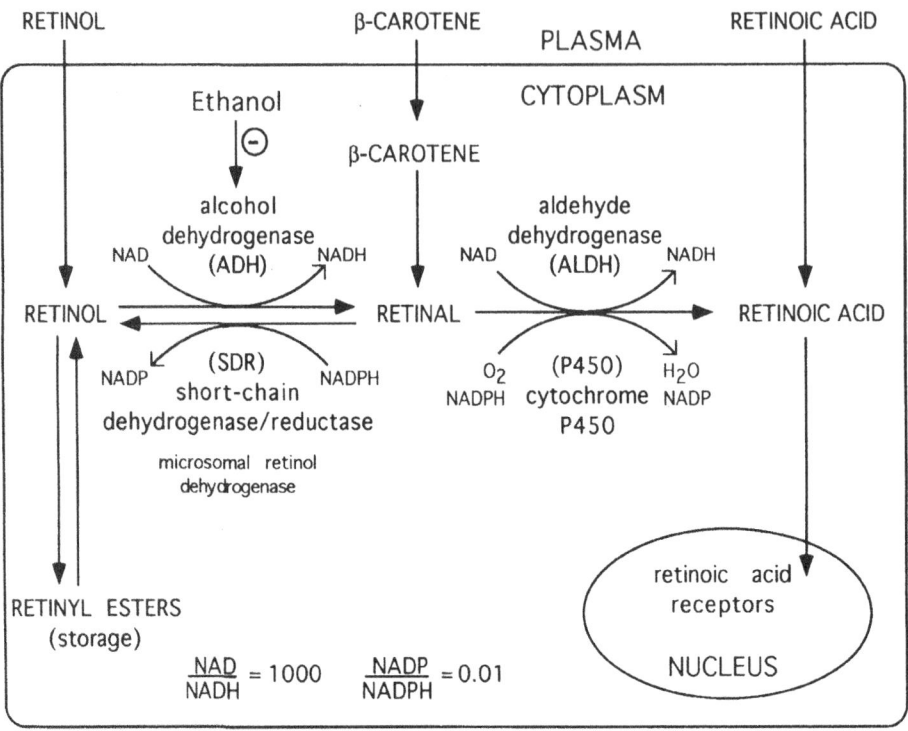

Figure 1. Enzymes involved in retinoic acid synthesis. The retinol/retinal interconversion is catalyzed *in vitro* by some forms of ADH and SDR, with the preference for NAD(H) or NADP(H) likely having an influence on the direction of the reaction. Oxidation of retinol by ADH is competitively inhibited by ethanol. Oxidation of retinal to retinoic acid is catalyzed *in vitro* by some forms of ALDH and P450.

Protein purification and gene cloning studies have identified seven classes of vertebrate ADH, although not all classes have been identified in each species analyzed (Fig. 2). Purified class I ADH, class II ADH, class IV ADH, and class VII ADH can each function as retinol dehydrogenases *in vitro* (Connor and Smit, 1987; Boleda et al., 1993; Yang et al., 1994; Kedishvili et al., 1995; Kedishvili et al., this volume). Ethanol is also a substrate for these classes of ADH, and high levels of ethanol competitively inhibit ADH-catalyzed retinol oxidation *in vitro* (Mezey and Holt, 1971; Van Thiel et al., 1974; Julià et al., 1986; Kedishvili et al., this volume; Yin et al., this volume). Ethanol treatment of cultured mouse embryos has been shown to reduce retinoic acid levels, implicating an ADH in the enzymatic pathway (Deltour et al., 1996). Studies on the expression patterns of the three known mouse ADH genes have shown that class I and class IV ADHs are expressed in patterns which correlate with sites of retinoic acid production in late embryos and adults (Zgombic-Knight et al., 1995; Ang et al., 1996b). However, only class IV ADH gene expression correlates spatially and temporally with early embryonic retinoic acid production (Ang et al., 1996a). These findings suggest that class IV ADH is involved in early embryonic retinoic acid synthesis, and that ethanol-induced craniofacial defects observed in fetal alcohol syndrome may be caused by ethanol inhibition of retinoic acid synthesis catalyzed by class IV ADH (Deltour et al., 1996).

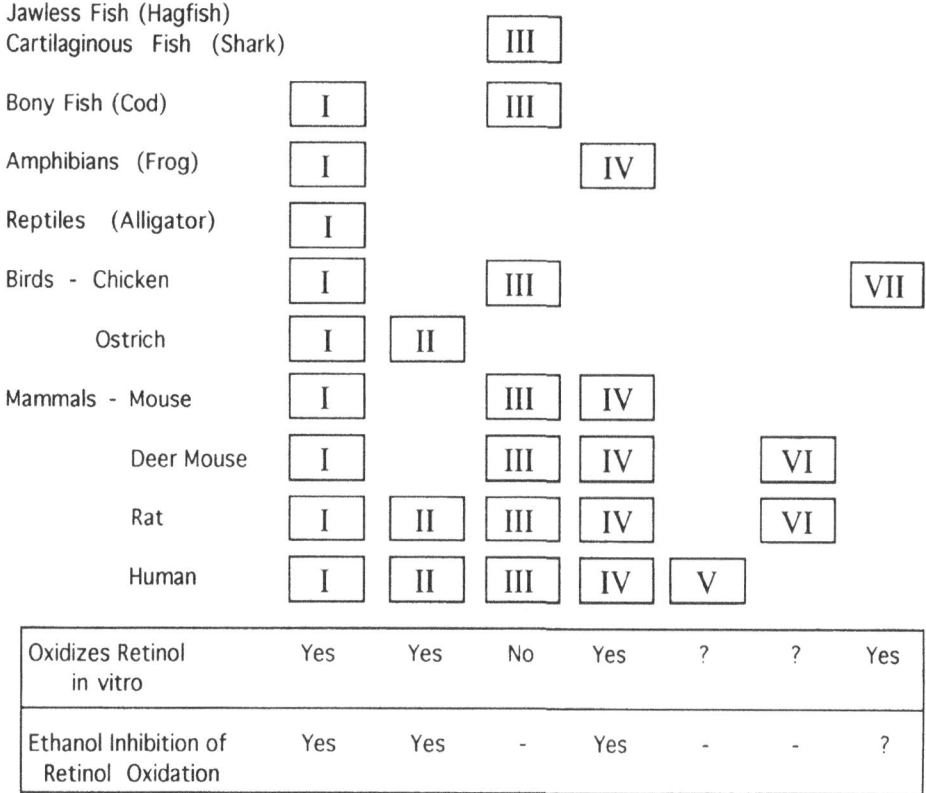

Figure 2. Classes of ADH which catalyze retinol oxidation *in vitro*. Seven classes of ADH have been identified in the various vertebrate species analyzed. Classes I-VI were first detected in mammals, with some forms also found in other vertebrates (Jörnvall et al, 1995; Duester et al., 1995). Class VII ADH was recently discovered in the bird lineage (Kedishvili et al., this volume). Within a species the interclass amino acid sequence identity is in the 60% range. Classes I, II, IV, and VII are known to oxidize retinol, and ethanol can inhibit this reaction in those enzymes tested.

2. ETHANOL INHIBITION OF RETINOIC ACID PRODUCTION IN MOUSE EMBRYOS

A bioassay has been developed to examine individual mouse embryos for the presence of endogenous retinoic acid, and for the effect of ethanol on retinoic acid levels. A monolayer of retinoic acid reporter cells from the cell line F9-RARE-*lacZ* serves as a bioassay to monitor the diffusion of retinoic acid from embryos cultured as explants on top of the reporter cells (Wagner et al., 1992). This cell line harbors a transgene in which the *E. coli lacZ* gene encoding β-galactosidase is controlled by a retinoic acid-inducible promoter which is stimulated by exogenous retinoic acid derived from the co-cultured embryo. Embryo dissection, culturing of tissue explants with or without ethanol, and detection of β-galactosidase activity has been described previously (Ang et al., 1996a; Deltour et al., 1996).

Retinoic acid is not detected in embryos undergoing the early phase of gastrulation on gestation day 6.5 (E6.5; gestation time for the mouse is approximately 20 days with

Table 1. Effect of ethanol on retinoic acid (RA) detection in
cultured mouse embryos

Mouse Strain and Treatment	Total No. of embryos	No. embryos with RA detection	No. embryos without RA detection	% embryos with RA detection
FVB/N (E6.5)				
Control	25	0	25	0
FVB/N (E7.5)				
Control	40	35	5	88
100 mM Ethanol	41	22	19	54[a]
FVB/N (E8.5)				
Control	25	24	1	96
100 mM Ethanol	7	7	0	100
C57BL/6J (E7.5)				
Control	47	26	21	55
10 mM Ethanol	93	38	55	41[b]
100 mM Ethanol	50	13	37	26[c]

[a] $P = 0.001$, $\chi^2 = 11.12$, ethanol-treated embryos versus control embryos.

[b] $P = 0.106$, $\chi^2 = 2.61$ (not statistically significant).

[c] $P = 0.003$, $\chi^2 = 8.66$, ethanol-treated embryos versus control embryos.

noon on the day of conception designated as E0.5). However, retinoic acid is detected in a majority of E7.5 embryos undergoing late gastrulation, and in essentially all E8.5 embryos undergoing neurulation. Treatment of cultured E7.5 embryos with 100 mM ethanol leads to a significant decrease in retinoic acid detection, with no significant effect of ethanol on embryonic survival in culture (Table 1). This effect of 100 mM ethanol is seen in two strains of mice (FVB/N and C57BL/6J), and is dosage dependent since no statistically significant effect is seen with 10 mM ethanol treatment of C57BL/6J embryos. There was no noticeable effect of ethanol on retinoic acid detection in E8.5 embryos, possibly due to a significantly higher baseline level of retinoic acid by this stage.

Overall, these results suggest that mouse embryos first begin to synthesize retinoic acid at E7.5, and that this synthesis may involve retinol oxidation catalyzed by an ADH since it is inhibited by ethanol.

3. EXPRESSION OF ADH GENE FAMILY DURING MOUSE DEVELOPMENT

The expression patterns of all known mouse ADH genes (i.e. the genes encoding ADH classes I, III, and IV) have been examined by Northern blot analysis (Zgombic-Knight et al., 1995) and *in situ* hybridization. Mouse embryos from E6.5-E9.5 were subjected to whole-mount *in situ* hybridization to detect class I, III, and IV ADH mRNAs as described previously using digoxigenin-labelled antisense riboprobes and detection with an antidigoxigenin antibody coupled to alkaline phosphatase (Ang et al., 1996a). Adult mouse tissues and embryos from E10.5-E16.5 were embedded in paraffin, sectioned, and processed for *in situ* hybridization to detect class I, III, and IV ADH mRNAs using ^{35}S-labelled antisense riboprobes as described (Ang et al., 1996b).

Detection of class IV ADH mRNA, but not class I or class III ADH mRNAs, coincides with the onset of retinoic acid synthesis; i.e. class IV ADH mRNA is absent in E6.5 embryos but present in posterior mesoderm of E7.5 embryos. During cranial neural tube development (E8.0–9.5), class IV ADH mRNA is localized in craniofacial mesenchyme adjacent to the neural tube rather than in the neuroepithelium itself. Class I ADH mRNA is not detected from E6.5-E8.5, but is detected at E9.5 in the mesonephros, a progenitor of the genitourinary tract. Class III ADH mRNA was observed ubiquitously from E6.5-E9.5.

In older embryos, class IV ADH mRNA is undetectable from E10.5-E12.5, but is detectable by E14.5-E16.5 in the adrenal gland and nasal epithelium. In adults class IV ADH mRNA is detected in the epithelia of the esophagus, stomach, testis, epididymis, epidermis, and adrenal cortex. In E10.5-E16.5 embryos, class I ADH mRNA is observed in the genitourinary tract, intestinal tract, adrenal gland, liver, conjunctival sac, epidermis, nasal epithelium, and lung. By adulthood, class I ADH mRNA is detected in the epithelia of the testis, epididymis, uterus, kidney, intestine, adrenal cortex, and liver. Class III ADH mRNA is ubiquitously present at low levels in late embryonic and adult tissues, with a little higher level in the liver.

The enzymatic properties and expression patterns of class I, III, and IV ADHs suggest that only class IV ADH (known to function as a retinol dehydrogenase) can be linked to early embryonic retinoic acid synthesis which begins at E7.5. Class I ADH (also a retinol dehydrogenase) is expressed too late to participate in retinoic acid synthesis during E7.5-E8.5, but may participate by E9.5 when its gene is first expressed. Both class I and class IV ADHs may participate in retinoic acid synthesis needed for the establishment and maintenance of specialized epithelia during late embryonic development and adulthood. Class III ADH is expressed throughout early and late embryogenesis as well as adulthood, but cannot participate in retinoic acid synthesis since it is unable to oxidize retinol (Boleda et al., 1993; Yang et al., 1994). Instead, class III ADH is known to function as a glutathione-dependent formaldehyde dehydrogenase needed to eliminate toxic formaldehyde (Koivusalo et al., 1989; Holmquist and Vallee, 1991). The low ubiquitous expression pattern for class III ADH supports the view that this enzyme performs a basic housekeeping function. The expression patterns of the three mouse ADH genes are summarized (Fig. 3).

4. CONCLUSIONS

The ability of some classes of ADH to function as retinol dehydrogenases presents the possibility that some ADH isozymes may participate physiologically in retinoic acid

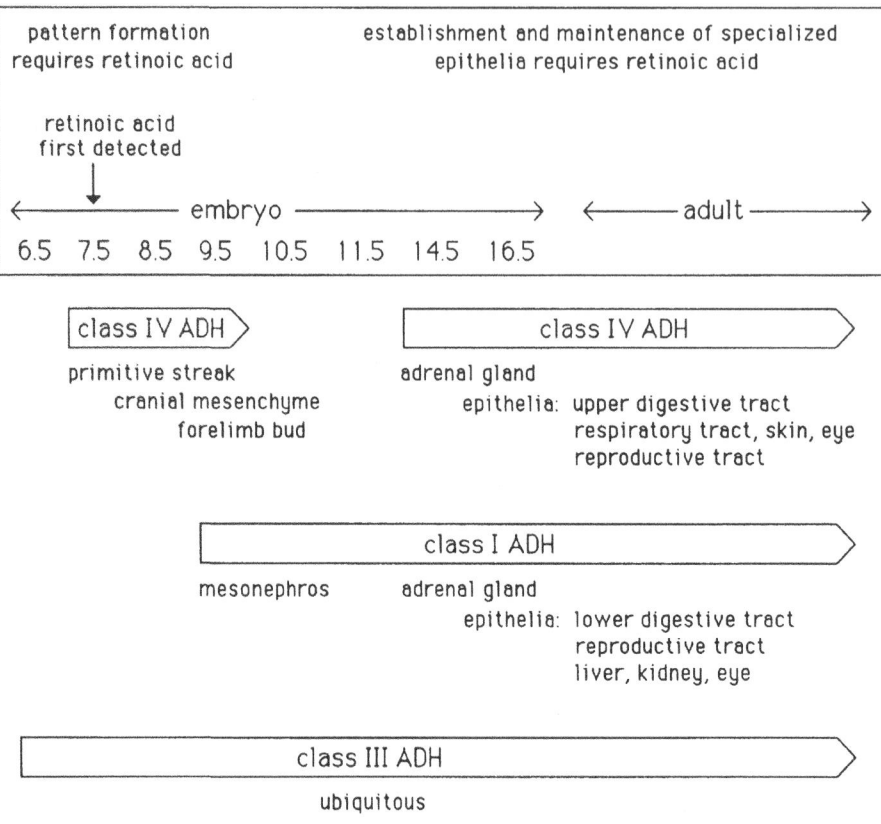

Figure 3. Expression of the mouse ADH gene family during development. Expression of the gene encoding class IV ADH (an enzyme able to catalyze retinol oxidation) is first noticed in embryonic development at E7.5, the stage when endogenous retinoic acid is first detected and first believed to play a role in embryonic pattern formation. Later in development and into adulthood, class I and class IV ADHs (both known to function as retinol dehydrogenases) are expressed in several epithelia known to require retinoic acid for their establishment and maintenance. In contrast, class III ADH expression is low and ubiquitous correlating with its proposed role as a housekeeping enzyme to remove formaldehyde, and with its inability to oxidize retinol.

synthesis. So far, the best evidence supporting this contention comes from studies on class IV ADH as summarized below:

1. Class IV ADH functions as a retinol dehydrogenase *in vitro* (Connor and Smit, 1987; Boleda et al., 1993; Yang et al., 1994; Kedishvili et al., 1995). This enzyme has a high catalytic efficiency for retinol oxidation, but a very low catalytic efficiency for ethanol oxidation.
2. Class IV ADH-catalyzed retinol oxidation *in vitro* is competitively inhibited by high levels of ethanol (Julià et al., 1986; Kedishvili et al., this volume; Yin et al., this volume).
3. Treatment of E7.5 mouse embryos with high levels of ethanol (100 mM), but not lower levels (10 mM), leads to a significant reduction in endogenous retinoic acid levels (Deltour et al., 1996). This suggests that an ADH with a low catalytic efficiency for ethanol (such as class IV ADH) catalyzes retinol oxidation for early embryonic retinoic acid synthesis and is sensitive to competitive inhibition by high levels of ethanol.

4. Class IV ADH mRNA is first detected in mouse embryos at E7.5, the stage when endogenous retinoic acid is first detectable and the stage most sensitive to ethanol-induced teratogenesis in the mouse model of fetal alcohol syndrome (Ang et al., 1996a). During E8.0-E9.5 class IV ADH mRNA is detected in craniofacial mesenchyme, a tissue known to require retinoic acid for proper development, and known to be quite sensitive to ethanol teratogenesis.

5. By adulthood, class IV ADH mRNA is detected in epithelia that are known to require retinoic acid for proper differentiation including the testis, epididymis, and epidermis (Ang et al., 1996b).

5. ACKNOWLEDGMENTS

Thanks to M. Foglio, R.J. Haselbeck, and I. Hoffmann for stimulating discussions. This work was supported by NIH grants AA07261 and AA09731.

6. REFERENCES

Ang, H.L., Deltour, L., Hayamizu, T.F., Zgombic-Knight, M., and Duester, G., 1996a, Retinoic acid synthesis in mouse embryos during gastrulation and craniofacial development linked to class IV alcohol dehydrogenase gene expression, *J. Biol. Chem.* 271: 9526–9534.

Ang, H.L., Deltour, L., Zgombic-Knight, M., Wagner, M.A., and Duester, G., 1996b, Expression patterns of class I and class IV alcohol dehydrogenase genes in developing epithelia suggest a role for ADH in local retinoic acid synthesis, *Alcohol. Clin. Exp. Res.* 20: in press.

Bhat, P.V., Labrecque, J., Boutin, J.M., Lacroix, A., and Yoshida, A., 1995, Cloning of a cDNA encoding rat aldehyde dehydrogenase with high activity for retinal oxidation, *Gene* 166: 303–306.

Boleda, M.D., Saubi, N., Farrés, J., and Parés, X., 1993, Physiological substrates for rat alcohol dehydrogenase classes: Aldehydes of lipid peroxidation, omega-hydroxyfatty acids, and retinoids, *Arch. Biochem. Biophys.* 307: 85–90.

Chai, X., Boerman, M.H.E.M., Zhai, Y., and Napoli, J.L., 1995, Cloning of a cDNA for liver microsomal retinol dehydrogenase. A tissue-specific, short-chain alcohol dehydrogenase, *J. Biol. Chem.* 270: 3900–3904.

Connor, M.J. and Smit, M.H., 1987, Terminal-group oxidation of retinol by mouse epidermis: Inhibition *in vitro* and *in vivo*, *Biochem. J.* 244: 489–492.

Deltour, L., Ang, H.L., and Duester, G., 1996, Ethanol inhibition of retinoic acid synthesis as a potential mechanism for fetal alcohol syndrome, *FASEB J.* 10: in press.

Dockham, P.A., Lee, M.-O., and Sladek, N.E., 1992, Identification of human liver aldehyde dehydrogenases that catalyze the oxidation of aldophosphamide and retinaldehyde, *Biochem. Pharmacol.* 43: 2453–2469.

Duester, G., Ang, H.L., Deltour, L., Foglio, M.H., Hayamizu, T.F., and Zgombic-Knight, M., 1995, Class I and class IV alcohol dehydrogenase (retinol dehydrogenase) gene expression in mouse embryos, *Adv. Exp. Med. Biol.* 372: 301–313.

Holmquist, B. and Vallee, B.L., 1991, Human liver class III alcohol and glutathione dependent formaldehyde dehydrogenase are the same enzyme, *Biochem. Biophys. Res. Commun.* 178: 1371–1377.

Jörnvall, H., Danielsson, O., Hjelmqvist, L., Persson, B., and Shafqat, J., 1995, The alcohol dehydrogenase system, *Adv. Exp. Med. Biol.* 372: 281–294.

Julià, P., Farrés, J., and Parés, X., 1986, Ocular alcohol dehydrogenase in the rat: Regional distribution and kinetics of the ADH-1 isoenzyme with retinol and retinal, *Exp. Eye Res.* 42: 305–314.

Kastner, P., Chambon, P., and Leid, M., 1994, Role of nuclear retinoic acid receptors in the regulation of gene expression, in *"Vitamin A in Health and Disease,"* R. Blomhoff, ed.,Marcel Dekker, Inc., New York, p. 189.

Kedishvili, N.Y., Bosron, W.F., Stone, C.L., Hurley, T.D., Peggs, C.F., Thomasson, H.R., Popov, K.M., Carr, L.G., Edenberg, H.J., and Li, T.-K., 1995, Expression and kinetic characterization of recombinant human stomach alcohol dehydrogenase. Active-site amino acid sequence explains substrate specificity compared with liver isozymes, *J. Biol. Chem.* 270: 3625–3630.

Koivusalo, M., Baumann, M., and Uotila, L., 1989, Evidence for the identity of glutathione-dependent formaldehyde dehydrogenase and class III alcohol dehydrogenase, *FEBS Lett.* 257: 105–109.

Lee, M.-O., Manthey, C.L., and Sladek, N.E., 1991, Identification of mouse liver aldehyde dehydrogenases that catalyze the oxidation of retinaldehyde to retinoic acid, *Biochem. Pharmacol.* 42: 1279–1285.

Lohnes, D., Mark, M., Mendelsohn, C., Dollé, P., Dierich, A., Gorry, P., Gansmuller, A., and Chambon, P., 1994, Function of the retinoic acid receptors (RARs) during development. (I) Craniofacial and skeletal abnormalities in RAR double mutants, *Development* 120: 2723–2748.

Mangelsdorf, D.J., Umesono, K., and Evans, R.M., 1994, The retinoid receptors, in *"The Retinoids: Biology, Chemistry, and Medicine, 2nd Edition,"* M.B. Sporn, A.B. Roberts, and D.S. Goodman, eds., Raven Press, Ltd., New York, p. 319.

Mendelsohn, C., Lohnes, D., Décimo, D., Lufkin, T., LeMeur, M., Chambon, P., and Mark, M., 1994, Function of the retinoic acid receptors (RARs) during development. (II) Multiple abnormalities at various stages of organogenesis in RAR double mutants, *Development* 120: 2749–2771.

Mezey, E. and Holt, P.R., 1971, The inhibitory effect of ethanol on retinol oxidation by human liver and cattle retina, *Exp. Mol. Pathol.* 15: 148–156.

Raner, G.M., Vaz, A.D.N., and Coon, M.J., 1996, Metabolism of all-*trans*, 9-*cis*, and 13-*cis* isomers of retinal by purified isozymes of microsomal cytochrome P450 and mechanism-based inhibition of retinoid oxidation by citral, *Mol. Pharmacol.* 49: 515–522.

Van Thiel, D.H., Gavaler, J., and Lester, R., 1974, Ethanol inhibition of vitamin A metabolism in the testes: Possible mechanism for sterility in alcoholics, *Science* 186: 941–942.

Wagner, M., Han, B., and Jessell, T.M., 1992, Regional differences in retinoid release from embryonic neural tissue detected by an in vitro reporter assay, *Development* 116: 55–66.

Yang, Z.-N., Davis, G.J., Hurley, T.D., Stone, C.L., Li, T.-K., and Bosron, W.F., 1994, Catalytic efficiency of human alcohol dehydrogenases for retinol oxidation and retinal reduction, *Alcohol. Clin. Exp. Res.* 18: 587–591.

Zgombic-Knight, M., Ang, H.L., Foglio, M.H., and Duester, G., 1995, Cloning of the mouse class IV alcohol dehydrogenase (retinol dehydrogenase) cDNA and tissue-specific expression patterns of the murine ADH gene family, *J. Biol. Chem.* 270: 10868–10877.

42

EXPRESSION OF FORMALDEHYDE DEHYDROGENASE AND S-FORMYLGLUTATHIONE HYDROLASE ACTIVITIES IN DIFFERENT RAT TISSUES

Lasse Uotila and Martti Koivusalo

Department of Clinical Chemistry and Institute of Biomedicine
Department of Medical Chemistry
University of Helsinki, Helsinki, Finland

1. INTRODUCTION

Formaldehyde is oxidized in animal cells into formate in two consecutive reactions catalyzed by the specific enzymes formaldehyde dehydrogenase (EC 1.2.1.1) and S-formylglutathione hydrolase (EC 3.1.2.12) (Uotila and Koivusalo, 1974a; 1974b). Formaldehyde reacts nonenzymically with glutathione (GSH) to form the adduct S-hydroxymethylglutathione. Formaldehyde dehydrogenase catalyzes the NAD-dependent oxidation of this adduct to S-formylglutathione. S-Formylglutathione hydrolase catalyzes the hydrolysis of S-formylglutathione to formate and GSH.

Formaldehyde dehydrogenase is an identical enzyme with class III alcohol dehydrogenase (Koivusalo et al., 1989). Both formaldehyde dehydrogenase (Uotila and Koivusalo, 1974a; 1989) and class III alcohol dehydrogenase (Julià et al.,1989; Boleda et al., 1989; Estonius et al., 1993; Höög et al., 1994) have been reported to have a wide, perhaps ubiquitous tissue distribution, and S-formylglutathione hydrolase also occurs widely (Uotila and Koivusalo, 1974b; Uotila, 1989; Koivusalo et al., 1995). However, quantitative data on the relative activities of these enzymes in different tissues are largely lacking. Especially few studies have been performed on S-formylglutathione hydrolase.

We have determined in this work the activities of these enzymes in the cytosolic fraction of sixteen different tissues of the rat. Because by the ordinary assay some tissues contained low activities of formaldehyde dehydrogenase or the enzyme was apparently lacking, and nonspecific aldehyde dehydrogenase(s) interefered in the assay in the crude tissue extracts used by also reacting with formaldehyde, we devised an additional more sensitive assay method for formaldehyde dehydrogenase. Because the hydrolysis of S-formylglutathione is also catalyzed in crude tissue extracts by glyoxalase II (S-2-hydroxyacylglutathione hydrolase, EC 3.1.2.6) (Uotila, 1973a; 1973b; 1979; 1989), we also determined the activities of the latter enzyme in the tissues studied. These data were used to estimate the relative hydrolytic activities for S-formylglutathione in these tissues of the specific hydrolase and of glyoxalase II.

2. METHODS

2.1. Animals and Preparation of the Tissue Extracts

The rat tissues used were obtained from female rats of the Wistar strain weighing 200–220 g, except testis tissues which were obtained from male animals of the same strain and age. The tissues were homogenized in a Potter homogenizer in seven-fold volume of 0.25 M sucrose also containing 10 mM sodium phosphate pH 7.5 and 0.5 mM dithiothreitol. The homogenates were centrifuged at 10,000 g for 30 min, and the supernatants then centrifuged at 105,000 g for 60 min. The supernatants thus obtained were applied to Sephadex G-25 columns, and the protein fraction, free from glutathione and other low molecular weight compounds, was collected and used for the enzyme activity and protein assays.

The assays were performed with nine tissues for four separately prepared pools of two individual rats. For these tissues the results are reported as mean ± SEM (n= 4). With the rest of the tissues, only one tissue pool from two individual rats was analyzed. For these tissues only the result obtained from closely agreeing duplicate or triplicate assays is reported. At the concentrations of the tissue extracts used all the enzymic reaction rates analyzed were linearly dependent on the extract concentration.

2.2. Enzyme Activity Assays

2.2.1. Formaldehyde Dehydrogenase and Nonspecific Aldehyde Dehydrogenase(S). Method 1. The blank assay mixture contained 80 mM sodium pyrophosphate buffer pH 8.0, 3 mM pyrazole, 1.2 mM NAD, a suitable amount of the tissue extract and water to 1.0 ml. The assay mixture for aldehyde dehydrogenase activities contained all the compounds in the blank assay and in addition 1.8 mM formaldehyde. The assay mixture for formaldehyde dehydrogenase contained all the compounds in the blank assay and in addition 1.8 mM formaldehyde and 1.0 mM GSH. The absorbance increase was recorded in a spectrophotometer at 340 nm and 25 °C. The difference in the initial rates observed with and without formaldehyde represented the nonspecific aldehyde dehydrogenase activities. The difference in the initial rates observed, in the presence of formaldehyde, with and without GSH represented the activity for the glutathione-dependent formaldehyde dehydrogenase. The molar absorbance used for NADH was 6 220 $M^{-1}cm^{-1}$.

Method 2. Similar assay mixtures to those in Method 1 were used except that NAD was replaced by 0.8 mM 3-acetylpyridine adenine dinucleotide (APAD) and the reaction rates were recorded at 363 nm. The molar absorbance used for the reduced APAD at this wavelength was 9 100 $M^{-1}cm^{-1}$.

2.2.2. S-Formylglutathione Hydrolase. The assay mixture contained 90 mM sodium phosphate buffer pH 7.0, 0.5 mM S-formylglutathione (prepared according to Uotila, 1981), a suitable dilution of the tissue extract, and water to 1.0 ml. A blank without the extract was always used and subtracted from the total rate. The decreasing absorbance was recorded at 240 nm and 25 °C. The molar absorbance used for S-formylglutathione was 3 300 $M^{-1}cm^{-1}$ (Uotila, 1973a).

2.2.3. Glyoxalase II. The assay mixture contained 70 mM MES buffer pH 6.5, 0.2 mM 4,4′-dithiodipyridine, 0.5 mM *S*-lactoylglutathione (Uotila, 1981), a suitable amount of the tissue extract and water to 1.0 ml. A blank without the extract was always included and subtracted from the total rate. The absorbance increase was recorded at 324 nm. The molar absorbance used for the reduced form of 4,4′-dithiodipyridine was 19 800 $M^{-1}cm^{-1}$.

2.3. Protein Assays

These were performed by the biuret method with bovine serum albumin as the standard. The enzyme activities were calculated as milliunits (mU) (nmol of substrate used or product formed per min) and the specific activities as mU/mg of protein.

3. RESULTS AND DISCUSSION

3.1. Formaldehyde Dehydrogenase and Aldehyde Dehydrogenase Activities in Different Tissues

Pyrazole was included in the assay mixtures to inhibit the NADH-dependent reduction of formaldehyde by class I alcohol dehydrogenase which otherwise could occur at least in the liver preparations. Most of the tissues investigated gave some continuous absorbance increase in the assay used at 340 nm even when no formaldehyde was added to the assay mixture. Additional activity obtained by formaldehyde represented aldehyde dehydrogenase activities. The further activity obtained with GSH represented formaldehyde dehydrogenase; nonspecific aldehyde dehydrogenase activities thus formed the blank reaction for formaldehyde dehydrogenase. High blank rates made accurate assays of both formaldehyde dehydrogenase and aldehyde dehydrogenase difficult with the extracts of several tissues.

Purified formaldehyde dehydrogenases from both human liver (Uotila and Mannervik, 1979) and rat liver (Uotila and Koivusalo, 1983) react with APAD at a rate more than 20-fold higher than that with NAD. We thus also performed modified assays in which NAD was replaced with APAD. In some tissues like liver the GSH-dependent formaldehyde dehydrogenase activity was even with NAD considerably higher than the GSH-independent activity, but in the assay with APAD the ratio of these activities was markedly further increased since only formaldehyde dehydrogenase was activated. In other tissues like heart the majority of the activity observed with NAD apparently represented aldehyde dehydrogenase(s) because only a small activity increase was observed with GSH. Even with the heart preparations the use of APAD to replace NAD gave a high ratio of the rates with and without GSH, thus making the formaldehyde dehydrogenase assay much more reliable.

With the assays performed with either NAD or APAD, liver gave the highest specific activity for formaldehyde dehydrogenase. Colon, kidney, stomach, small intestine and brain also were among the richest sources for the activity (Tables 1 and 2). With the assay with NAD, brown fat tissue contained only questionable formaldehyde dehydrogenase activity, and rectum, skin, esophagus and white fat tissue contained no activity (Table 1). However, with the APAD assay all of these sources unequivocally contained formaldehyde dehydrogenase activity (Table 2). All tissues for which the assays could either be made with NAD or APAD, gave approximately 15-fold higher formaldehyde dehydrogenase activity with APAD. For brown fat tissue the activation appeared to be much

Table 1. Specific activities with NAD as the coenzyme of formaldehyde dehydrogenase (FaldDH) and of aldehyde dehydrogenase (with formaldehyde as the substrate) (AldDH) in various tissues of the rat. The assays were performed at 340 nm as detailed in Methods. u.a., undetectable activity

Tissue	Formaldehyde dehydrogenase nmol/min/mg prot.	Aldehyde dehydrogenase nmol/min/mg prot.	FaldDH/AldDH activity ratio
Liver	5.08 ± 0.54	2.71 ± 0.34	1.9
Colon	1.94	0.29	6.6
Kidney	1.89 ± 0.05	1.69 ± 0.29	1.1
Stomach	1.76 ± 0.29	2.04 ± 0.60	0.9
Small intestine	1.55 ± 0.29	1.41 ± 0.11	1.1
Brain	1.11 ± 0.20	0.27 ± 0.14	4.1
Spleen	0.92 ± 0.08	0.20 ± 0.09	4.6
Heart	0.89 ± 0.23	2.65 ± 0.59	0.3
Muscle	0.85 ± 0.03	0.27 ± 0.06	3.2
Lung	0.72 ± 0.16	0.45 ± 0.08	1.6
Testis	0.46	0.61	0.8
Brown fat tissue	0.11	4.13	0.03
Rectum	u.a.	1.61	-
Skin	u.a.	0.90	-
Esophagus	u.a.	u.a.	-
White fat tissue	u.a.	u.a.	-

Table 2. Specific activities with APAD as the coenzyme of formaldehyde dehydrogenase (FaldDH) and of aldehyde dehydrogenase (with formaldehyde as the substrate) (AldDH) in various tissues of the rat. The assays were performed at 363 nm as described in Methods. u.a., undetectable activity

Tissue	Formaldehyde dehydrogenase nmol/min/mg prot.	Aldehyde dehydrogenase nmol/min/mg prot.	FaldDH/AldDH activity ratio
Liver	73.5 ± 1.59	2.41 ± 0.39	30.4
Colon	28.7	1.13	25.3
Kidney	26.6 ± 0.68	2.31 ± 0.64	11.5
Stomach	21.6 ± 1.86	2.58 ± 1.69	8.4
Small intestine	18.6 ± 1.25	1.45 ± 0.94	12.8
Brain	17.9 ± 0.81	3.42 ± 1.60	5.2
Spleen	13.9 ± 1.57	0.814 ± 0.34	17.1
Heart	16.0 ± 0.76	1.75 ± 0.91	9.1
Muscle	7.94 ± 1.80	1.15 ± 0.49	6.9
Lung	10.4 ± 1.73	1.00 ± 0.33	10.4
Testis	6.73	u.a.	-
Brown fat tissue	8.55	4.06	2.1
Rectum	7.63	0.81	9.4
Skin	5.62	u.a.	-
Esophagus	5.44	u.a.	-
White fat tissue	2.59	1.96	1.3

higher, probably because of an underestimation of the activity with NAD, which was negligible in comparison to the glutathione-independent rate due to aldehyde dehydrogenases.

Thus formaldehyde dehydrogenase occurred in all the tissues investigated but the specific activity values observed for the enzyme varied nearly 30-fold (Table 2). This variation is higher than was reported on the basis of some earlier qualitative or semiquantitative studies (Höög et al., 1994).

The cytosolic aldehyde dehydrogenase activity (with formaldehyde) was highest in brown fat tissue, liver, heart, stomach, kidney and with APAD only, in brain. In the other tissues mentioned APAD did not increase the activity over that observed with NAD. In brain and in some other tissues apparently containing only minor aldehyde dehydrogenase activity with NAD, significant activation by APAD of the aldehyde dehydrogenase seemed to occur (Tables 1 and 2). However, this probably resulted largely from difficulties in measuring the true aldehyde dehydrogenase activity with NAD since the assays were in these tissues complicated by marked blank absorbance increases at 340 nm.

3.2. S-Formylglutathione Hydrolase and Glyoxalase II Activities in Different Tissues

Purified *S*-formylglutathione hydrolase is a highly active enzyme (Uotila and Koivusalo, 1974b; Uotila, 1989; Koivusalo et al., 1995). All the tissues investigated in this work contained a high level of the hydrolase activity (Table 3). Kidney and liver were the most active tissues with the specific activities in the crude extract of as high as 3 to 4 U/mg at 25 °C. Several other tissues (stomach, colon, heart, intestine, brain, muscle) gave specific activities of over or nearly 1 U/mg, and even the tissue with the lowest activity, skin, gave a specific activity of over 0.3 U/mg. The activities in different tissues of *S*-for-

Table 3. Specific activities of *S*-formylglutathione (F-SG) hydrolase and of glyoxalase II (*S*-2-hydroxyacylglutathione hydrolase) in various tissues of the rat. The assays were performed as described in Methods. For the calculation of the activity ratios in the last column, the activities of formaldehyde dehydrogenase (FaldDH) with the natural coenzyme NAD (Table 1) were used

Tissue	S-Formylgluta-thione hydrolase nmol/min/mg prot.	Glyoxalase II nmol/min/mg prot.	F-SG hydrolase/ glyoxalase II	F-SG hydrolase/ FaldDH
Kidney	4,020 ± 270	333 ± 19	12	2,130
Liver	3,350 ± 322	304 ± 1	11	660
Stomach	1,610 ± 68	75.9 ± 8	21	910
Colon	1,600	47.7	34	830
Heart	1,250 ± 76	50.6 ± 5	25	1,410
Small intestine	1,150 ± 115	54.3 ± 9	21	740
Brain	987 ± 18	180 ± 8	6	890
Muscle	944 ± 58	28.3 ± 3	33	1,110
Spleen	918 ± 103	51.7 ± 3	18	1,000
Brown fat tissue	899	119	8	8,100
White fat tissue	536	29	19	-
Esophagus	529	46	12	-
Rectum	494	17	30	-
Lung	483 ± 48	25 ± 1	19	670
Testis	435	543	0.8	940
Skin	322	25	13	-

mylglutathione hydrolase correlated relatively well with those of formaldehyde dehydrogenase, but the hydrolase activity was generally 600 to 2000-fold higher in comparison to the dehydrogenase (Table 3).

Glyoxalase II, tested with S-lactoylglutathione as the substrate, was also present in all the tissues investigated. Testis, one of the tissues with the lowest level of S-formylglutathione hydrolase activity, gave among all tissues the highest specific activity for glyoxalase II (Table 3). Kidney and liver, the richest sources of S-formylglutathione hydrolase, contained the next highest specific activities of glyoxalase II, followed by brain. All other tissues contained considerably less of the glyoxalase II activity. Jerzykowski et al. (1978) found in accordance with this study the highest glyoxalase II activity in rat kidney, liver and brain in that order, with less activity in spleen, heart and muscle. The specific activities reported by them under similar conditions were, however, only 30 to 50 % of those observed in this work.

In most of the tissues investigated the specific activities for S-formylglutathione hydrolase were 10 to 30-fold higher in comparison to those for glyoxalase II, although in brain this ratio was only 5.5. As a remarkable exception, testis gave a higher specific activity for glyoxalase II than for S-formylglutathione hydrolase (Table 3). Glyoxalase II is also able to catalyze the hydrolysis of S-formylglutathione (Uotila, 1973b; 1979; 1989). Therefore it was necessary to estimate the proportion of glyoxalase II in the hydrolysis of S-formylglutathione in different tissues, since the specific activities given (Table 3) represented the total activities in the tissue extracts. We have found that partially purified rat liver glyoxalase II gives with S-formylglutathione as the substrate 40 % of the rate with S-lactoylglutathione, but due to partial inhibition of glyoxalase II by phosphate, the activities for S-formylglutathione due to glyoxalase II were only 20 % of those observed with S-lactoylglutathione as the substrate (Table 3) under the conditions used for S-formylglutathione hydrolase assay (unpublished results). Supposing a similar substrate specificity for glyoxalase II in the other tissues to the enzyme in liver, in most tissues only 0.6 to 1.8 % of the total hydrolytic rate for S-formylglutathione was due to glyoxalase II. Only a minor correction would thus be required for the numbers reported in Table 3 to obtain the specific activities for the true S-formylglutathione hydrolase. A slightly higher proportion of the hydrolytic activity for S-formylglutathione was due to glyoxalase II in brain and brown fat (3.7 % and 2.7 %, respectively). However, in testis extracts as much as 25 % of the hydrolytic rate found for S-formylglutathione appeared to be due to glyoxalase II.

4. CONCLUSIONS

Formaldehyde dehydrogenase, S-formylglutathione hydrolase and glyoxalase II were present in all the sixteen rat tissues investigated and may be ubiquitous enzymes in animal cells. The specific activities of the enzymes varied 12 to 30-fold in the tissues investigated. Liver, kidney, stomach, colon and small intestine were among the richest sources for these enzymes. Due to interfering enzyme activities, formaldehyde dehydrogenase could be reliably assayed in some tissues only when the natural coenzyme NAD was replaced by 3-acetylpyridine adenine dinucleotide. This gave 15-fold activities for formaldehyde dehydrogenase compared to those with NAD, but did not significantly increase nonspecific aldehyde dehydrogenase activities. S-Formylglutathione hydrolase is a highly active enzyme giving in most tissues 10 to 30-fold higher activities in comparison to glyoxalase II, and 600 to 2000-fold higher activities in comparison to formaldehyde dehydrogenase. Testis contained, however, a higher activity of glyoxalase II than of S-formylglutathione hydrolase.

5. ACKNOWLEDGMENTS

Ms. Eija Haasanen provided skilful technical assistance in the experiments performed.

6. REFERENCES

Boleda, M.D., Julià, P., Moreno, A., and Parés, X.: Role of extrahepatic alcohol dehydrogenase in rat ethanol metabolism. Arch. Biochem. Biophys. 274 (1989) 74–81.

Estonius, M., Danielsson, O., Karlsson, C., Persson, H., Jörnvall, H., and Höög, J.-O.: Distribution of alcohol and sorbitol dehydrogenases. Assessment of mRNA species in mammalian tissues. Eur. J. Biochem. 215 (1993) 497–503.

Höög, J.-O., Estonius, M., and Danielsson, O.: Site-directed mutagenesis and enzyme properties of mammalian alcohol dehydrogenases correlated with their tissue distribution. In Jansson, B., Jörnvall, H., Rydberg, U., Terenius, L. and Vallee, B.L. (Eds.), Toward a Molecular Basis of Alcohol Use and Abuse. Birkhäuser Verlag, Basel, 1994, pp. 301–309.

Jerzykowski, T., Winter, R., Matszewski, W., and Piskorska, D: A re-evaluation of studies on the distribution of glyoxalases in animal and tumor tissues. Int. J. Biochem. 9 (1978) 853–860.

Julià, P., Farrés, J., and Parés, X.: Characterization of three isoenzymes of rat alcohol dehydrogenase. Tissue distribution and physical and enzymatic properties: Eur. J. Biochem. 162 (1989) 179–189.

Koivusalo, M., Baumann, M., and Uotila, L.: Evidence for the identity of glutathione-dependent formaldehyde dehydrogenase and class III alcohol dehydrogenase. FEBS Lett. 257 (1989) 105–109.

Koivusalo, M., Lapatto, R., and Uotila, L.: Purification and characterization of *S*-formylglutathione hydrolase from human, rat and fish tissues. Adv. Exp. Med. Biol. 372 (1995) 427–433.

Uotila, L.: Preparation and assay of glutathione thiol esters. Survey of human liver glutathione thiol esterases. Biochemistry 12 (1973a) 3938–3943.

Uotila, L.: Purification and characterization of *S*-2-hydroxyacylglutathione hydrolase (glyoxalase II) from human liver. Biochemistry 12 (1973b) 3944–3951.

Uotila, L.: Glutathione thiol esterases of human red blood cells. Fractionation by gel electrophoresis and isoelectric focusing. Biochim. Biophys. Acta 580 (1979) 277–288.

Uotila, L.: Thioesters of glutathione. Meth. Enzymol. 77 (1981) 424–430.

Uotila, L.: Glutathione thiol esterases. In Dolphin, D., Poulson, R. and Avramovic, O. (Eds.), Coenzymes and Cofactors, vol. III, Glutathione. Chemical, Biochemical and Medical Aspects, Part A. John Wiley & Sons, Inc., New York, 1989, pp. 767–804.

Uotila, L., and Koivusalo, M.: Formaldehyde dehydrogenase from human liver. Purification, properties and evidence for the formation of glutathione thiol esters by the enzyme. J. Biol. Chem. 249 (1974a) 7653–7663.

Uotila, L., and Koivusalo, M.: Purification and properties of *S*-formylglutathione hydrolase from human liver, J. Biol. Chem. 249 (1974b) 7664–7672.

Uotila, L., and Koivusalo, M.: Formaldehyde dehydrogenase. In Larsson, A., Orrenius, S., Holmgren, A. and Mannervik, B. (Eds.), Functions of Glutathione. Biochemical, Physiological, Toxicological and Clinical Aspects. Raven Press, New York, 1983, pp. 175–186.

Uotila, L., and Koivusalo, M.: Glutathione-dependent oxidoreductases: Formaldehyde dehydrogenase. In Dolphin, D., Poulson, R. and Avramovic, O. (Eds.), Coenzymes and Cofactors, vol. III, Glutathione. Chemical, Biochemical and Medical Aspects, Part A. John Wiley & Sons, Inc., New York, 1989, pp. 517–551.

Uotila, L., and Mannervik, B.: A steady-state kinetic model for formaldehyde dehydrogenase from human liver. A mechanism involving NAD and the hemimercaptal adduct of glutathione and formaldehyde as substrates and free glutathione as an allosteric activator of the enzyme. Biochem. J. 177 (1979) 869–878.

43

FORMALDEHYDE DEHYDROGENASE FROM YEAST AND PLANT

Implications for the General Functional and Structural Significance of Class III Alcohol Dehydrogenase

M. Rosario Fernández, Josep A. Biosca, M. Carmen Martínez, Hakima Achkor, Jaume Farrés, and Xavier Parés

Department of Biochemistry and Molecular Biology
Faculty of Sciences
Universitat Autónoma de Barcelona
08193 Bellaterra, Barcelona, Spain

1. INTRODUCTION

Glutathione-dependent formaldehyde dehydrogenase, also known as class III alcohol dehydrogenase (class III ADH), is widely distributed in animals, plants and microorganisms (Uotila and Koivusalo, 1989). It is the main enzymatic system responsible for the formaldehyde elimination, although a role in the oxidation of long chain alcohols and ω-hydroxyfatty acids has also been proposed (Boleda *et al.* 1993). Class III ADH shows a well conserved structure and function throughout evolution, with 74% identities between the human and the octopus enzymes (Kaiser *et al.*, 1993). However, differences exist between the catalytic constants of the enzymes isolated from different organisms (Uotila and Koivusalo, 1989; Fernández *et al.*, 1995). In the present work we have studied the structure and function of class III ADH from two plants, *Arabidopsis thaliana* and pea (*Pisum sativum*), and from yeast (*Saccharomyces cerevisiae*), with the aim to characterize the kinetics of the enzymes and to relate their specific features with structural differences. The plant class III enzyme have been also compared with the classical alcohol dehydrogenase from plants, that we call class P (Shafqat *et al*, 1996), to establish functional and structural relationships between the two plant enzymes. Finally, the use of yeast has allowed the study of the physiological importance of the enzyme, by preparing a strain defective in class III ADH.

2. EXPERIMENTAL

2.1. Enzyme Characterization

Class III ADH was purified from baker's yeast, pea and *Arabidopsis* by column chromatography as previously described (Fernández *et al.*, 1995; Shafqat *et al.*, 1996; Martínez *et al.*, 1996). Activity was determined spectrophotometrically at 340 nm, using 0.1M glycine/NaOH, pH 10.0 for alcohols and 0.1 M sodium pyrophosphate, pH 8.0, for S-hydroxymethylglutathione (Koivusalo *et al.* 1989), with 4 mM NAD^+. Kinetic constants were calculated with the program Enzfitter. Formaldehyde concentration was determined as previously reported (Fernández *et al.* 1995).

2.2. DNA and Protein Sequences

Yeast class III ADH structure was obtained by cloning and sequencing the genomic DNA from X2180 *Saccharomyces cerevisiae* (Wehner *et al.*, 1993; Fernández *et al.*, 1995). Structure of the pea enzyme was determined by protein sequencing (Shafqat *et al.*, 1996) while the *Arabidopsis* enzyme structure was obtained from the cloned cDNA (Martínez *et al.*, 1996).

2.3. Gene Deletion

A linear fragment containing URA3, plus two flanking regions homologous to those flanking the class III ADH gene coding sequence, were used to transform the diploid strain of *S. cerevisiae* W303D. The new heterozygous strain was induced to sporulate and spores were dissected and grown on rich medium. The strain with the deleted gene was selected after activity and Southern blotting analyses.

2.4. Expression of Glutathione-Dependent Formaldehyde Dehydrogenase

For high level expression of class III ADH in *S. cerevisiae*, the *NcoI/ClaI* genomic fragment was subcloned into the episomal plasmid Yeplac181 (Gietz and Sugino, 1988) and transformed into the deleted haploid strain.

3. RESULTS AND DISCUSSION

A *S. cerevisiae* haploid strain was constructed such that the gene for class III ADH was replaced by the URA3 gene. This resulted in a deletion of the gene, and, therefore, in the absence of class III activity, the strain, that was named Fdh⁻. However, both Fdh⁻ and Fdh⁺ strains could grow equally well in rich medium, confirming that the deletion is not lethal, and therefore, the class III ADH gene is not essential, at least in rich medium and under laboratory conditions, as previously reported (Mack *et al.*, 1988). The physiological importance of the enzyme is, however, demonstrated in Fig. 1. The Fdh⁻ strain could not grow in the presence of 0.6 mM formaldehyde in the culture medium. In contrast, the Fdh⁺ strain could grow under these conditions, although after a long lag phase. Fdh⁻ could, however, grow in the presence of 0.3 mM formaldehyde, suggesting the existence of addi-

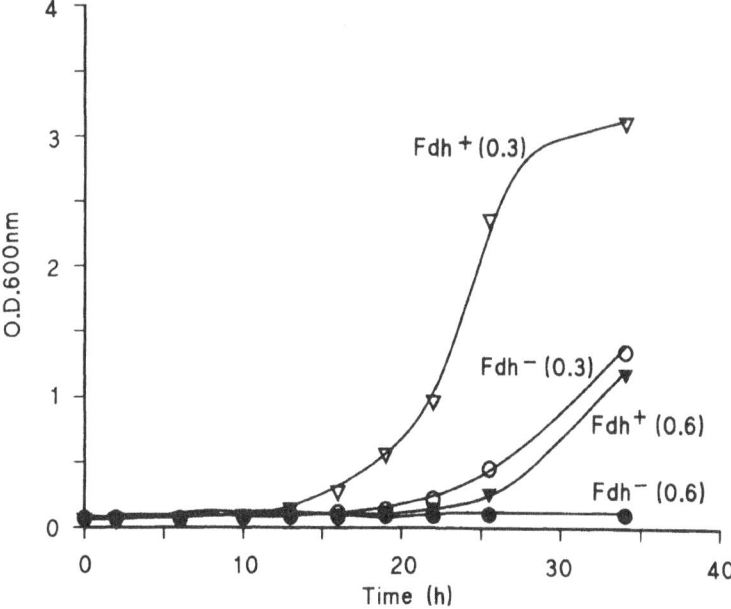

Figure 1. Growth of *S. cerevisiae* strains in the presence of formaldehyde. Fdh⁺ is the wild strain. Fdh⁻ is the strain with the deletion of the class III ADH gene. Formaldehyde concentrations are indicated in mM within parenthesis.

tional metabolic systems that can eliminate formaldehyde in yeast besides class III ADH. At 0.3 mM formaldehyde, the growth of the Fdh⁺ strain is much faster than that of Fdh⁻, demonstrating that class III ADH is the most effective enzymatic system for formaldehyde detoxication in *Saccharomyces*.

Fig. 2 confirms the essential role of class III ADH for yeast growth in the presence of environmental formaldehyde. Both Fdh⁻ and Fdh⁺ cannot grow in the presence of 1 mM formaldehyde. However the strain that contains an episomal plasmid overexpressing class III (Fdh⁺⁺ strain) grows with a similar generation time as in the absence of formaldehyde, although with a longer lag phase. The growth of the Fdh⁺⁺ strain is accompanied by a decrease of the formaldehyde concentration in the medium, demonstrating an active metabolism of formaldehyde by the strain. After an initial lag phase, also observed in the absence of formaldehyde, the Fdh⁺⁺ strain starts growing very slowly but the initial small amount of cells effectively eliminate formaldehyde from the medium (Fig. 2, bottom). The exponential growth starts when formaldehyde levels are about 0.6 mM, indicating that normal growth can take place as soon as the amount of the toxic aldehyde decreases below this threshold.

Table 1 shows the kinetic constants of the purified yeast and pea class III alcohol dehydrogenases. *Arabidopsis* class III exhibits constants similar to those of the pea enzyme, with a Km for S-hydroxymethylglutathione of 1.4 μM (Martinez *et al.*, 1996). The best substrate, in terms of both Km and kcat, is S-hydroxymethylglutathione, a common feature with the enzyme from animals, demonstrating that this is the physiological substrate of the enzyme also in plants. However, it is clear from Table 1 that yeast class III kinetic constants are significantly different from those of the plant enzyme. The yeast enzyme exhibits higher Km and higher kcat for all substrates, including NAD⁺.

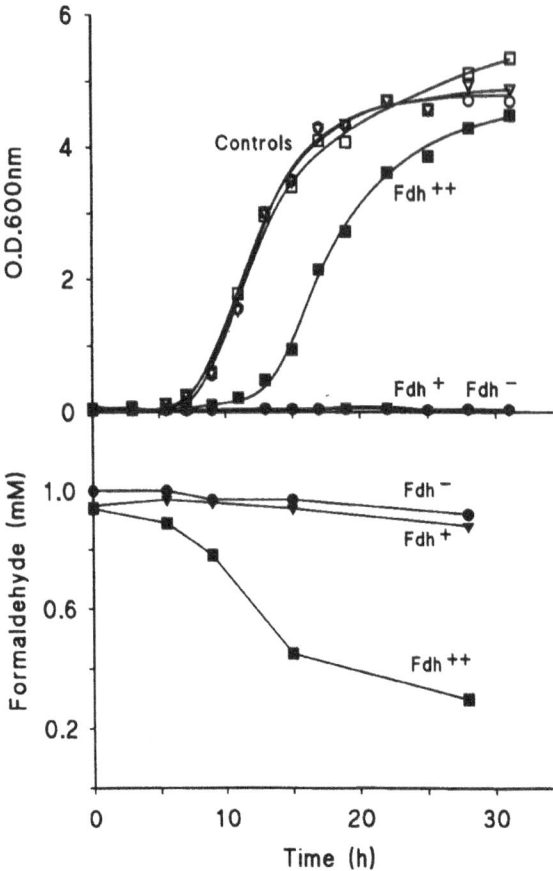

Figure 2. Growth of *S.cerevisiae* strains in the presence of 1 mM formaldehyde (top) and formaldehyde concentration in the culture medium (bottom). Fdh⁺ and Fdh⁻ are as indicated in Fig.1. Fdh⁺⁺ is the strain with a plasmid that overexpresses class III ADH. Open symbols indicate the growth of the strains without formaldehyde (controls).

The kinetic constants of class III ADH from different species, using the physiological substrate S-hydroxymethylglutathione, are indicated in Table 2. Two class III ADH types are clearly distinguished: the animal and plant type, with very low Km and relatively high kcat, and the microorganism type with high Km and very high kcat. kcat/Km values are similarly high in all cases, leading to a value of about 200,000 mM^{-1}.min^{-1}. Constants reported for other yeast species (*Pichia pastoris, Candida boidini, Kloeckera* sp.) (Allais *et al.* 1983), although determined at different experimental conditions and not directly

Table 1. Enzymatic properties of plant and yeast class III alcohol dehydrogenase

	Pea class III		*S. cerevisiae* class III	
Substrate	Km (µM)	kcat (min-1)	Km (µM)	kcat (min-1)
Ethanol	N.S.	--	N.S.	--
Octanol	840	190	2400	1100
12-OH-dodecanoic ac.	180	110	1800	1800
S-OH-methyl glutathione	2	380	40	5300
NAD⁺	6.5		140	

Km for NAD⁺ was determined with S-hydroxymethylglutathione as a substrate. N.S., nonsaturable. Taken from Shafqat *et al.*, 1996 and Fernández *et al.*, 1995.

Table 2. Enzymatic properties of class III alcohol dehydrogenases from different species with S-hydroxymethylglutathione

Species	Km (μM)	kcat (min⁻¹)	k_{cat}/Km (mM⁻¹ min⁻¹)
Human	4	200	50,000
Rat	0.92	216	235,000
Octopus	1.5	300	200,000
D. melanogaster	6	960	160,000
Pea	2	380	190,000
S. cerevisiae	40	5,300	134,000
E. coli	94	9,350	100,000

Determined at pH 8.0 (Fernández et al., 1995; Shafqat et al., 1996).

comparable with the present data, are also much higher than those for the animal species, supporting the two class III type distinction.

The two types of kinetics, distinct for multicellular and unicellular organisms, suggest an adaptation to the formaldehyde levels in their environment. These are low in the plant and animal cells, where formaldehyde is produced by the endogenous normal metabolism of amino acids and other compounds (Uotila and Koivusalo, 1989). Hence, it appears significant that the animal and plant enzymes have very low Km values, adapted to these conditions. In contrast, microorganisms are exposed to hazardous environments, with potentially higher content of formaldehyde. At a higher formaldehyde concentration, an enzyme with both higher Km and higher kcat values will be more efficient (Fersht, 1985). Consistent with this fact, methanol- and formaldehyde-utilizing yeast, which are naturally exposed to high formaldehyde concentrations, exhibit the highest Km of all known class III enzymes (Uotila and Koivusalo, 1989).

Overall these data support the hypothesis of a close correlation between class III catalytic constant values and the need for an efficient formaldehyde elimination in the corresponding organism, maintaining an optimal kcat/Km ratio. Depending on the formaldehyde concentration in their respective environments, the Km and kcat values would change in an evolutionary adaptation to acquire the highest catalytic rate with the maximal Km value. Class III is, therefore, a good example for the accepted theory of the evolution of the kinetic constants (Fersht, 1985; Pettersson, 1989). Evolutionary pressure leads to the highest kcat/Km values, keeping both kcat and Km high, but Km cannot be much higher than the physiological substrate concentrations. The correspondence between the kinetic constants and the physiological needs indicates a strong evolutionary restriction on the catalytic constants, and emphasizes the central role of class III ADH in formaldehyde detoxication.

The primary structures of the plant (*Arabidopsis*), yeast (*Saccharomyces*) and animal (human) class III enzymes were aligned and the positional identities calculated (Table 3). The class III structures were also compared with the alcohol dehydrogenases active with ethanol from human liver (class I) and plant (*Arabidopsis* class P). The high percentage of identities between the three class III proteins supports the notion of a high conservation between the class III enzymes throughout evolution. Interestingly, the class III ADHs from plant and animal exhibit 69% identities, whereas the corresponding ethanol dehydrogenases appear much more distant lines with only 51% identities.

Residues at functionally important positions of the yeast, plant and human class III enzymes have been compared (Table 4). As expected, the active site residues are very

Table 3. Sequence identity (%) between ADH classes III, P, and I. Amino acid sequences were compared using the program Clustal W (Thompson *et al.*, 1994)

	Saccharomyces III	*Arabidopsis* III	Human III	*Arabidopsis* P
Arabidopsis III	60			
Human III	63	69		
Arabidopsis P	43	59	54	
Human I	47	53	62	51

conserved between the four enzymes. Only 3 out of the 23 positions compared present differences. The most interesting substitution is at position 269, at the coenzyme-binding site, were *Saccharomyces* class III, an enzyme of the microorganism type, contains Thr (also present in the *Candida maltosa* class III structure, Sasnauskas *et al.*, 1992) while all enzymes of the animal-plant type show a big hydrophobic residue, Leu or Ile. Class I ADH also exhibits an Ile269, while the substitution by a Ser by site-directed mutagenesis results in an increase of kcat by 26-fold due to faster release of NADH (Fan and Plapp, 1995). The exchange to Thr in yeast class III may also produce a similar effect, which could contribute to the increased kcat for the microorganism type of class III ADH.

In contrast to the highly conserved active site structure within the class III ADH family, class I and class P show a surprisingly large number of differences in the functionally important residues. Thus, 13 of 23 residues exhibit substitutions (Table 4, bottom). The inner part of the substrate binding cleft is similar in the animal and plant enzymes, explaining the common activity versus ethanol. The middle part is very different with important exchanges such as Leu/Gly57 and Asp/Arg115. This part is involved in the binding of medium and long chain substrates and the changes may explain the different specificity of the two enzymes. Activity is usually higher with long chain alcohols than with ethanol in class I enzymes, but lower for class P. This can be related with a more specific physiological role for plant class P in the last step of alcoholic fermentation, in contrast to class I, involved in detoxication of a broad variety of structures. The coenzyme binding site exhibits many differences suggesting different coenzyme binding properties.

Comparison between the classes III, I and P active site residues (Table 4) indicates more conservation between the class III and P lines (11 residues conserved) than between the classes III and I (7 residues conserved). Notably Arg115 is present in all class P and all class III enzymes but not in class I, where an Asp is found instead. It has been demonstrated that Arg115 has an important role in class III specificity (Holmquist *et al.*, 1993; Engeland *et al.* 1993). We tested the possibility that this residue has a similar role in class P that would confer some class III properties to class P. However, class P ADH isolated from *Pisum sativum* did not show any activity with S-hydroxymethylglutathione nor activation by medium chain fatty acids (Martínez *et al.* 1996). This indicates that additional residues, like Tyr93 and Asp57, typical of class III and not present in class P, are essential for class III specificity.

4. CONCLUSIONS

Class III alcohol dehydrogenase is not essential for *Saccharomyces cerevisiae* under laboratory conditions. However a Fdh⁻ mutant strain cannot grow in 0.6 mM formaldehyde, in contrast to the wild type strain that can grow under these conditions. Overexpression of class III provides resistance to formaldehyde.

Table 4. Residues at functionally important positions of ADH classes III, P and I. Boxes indicate residue differences within each III and P/I enzyme groups. Modified from Martínez et al. (1996)

						Substrate-interacting residues									Coenzyme-interacting residues										
	48	67	93	140	141	57	115	116	294	318	110	306	309	47	48	51	178	203	223	224	228	269	271	369
Yeast III	T	H	Y	F	M	D	R	A	V	A	L	F	V	H	T	Y	T	V	D	I	K	T	N	R
Ara III	T	H	Y	F	M	D	R	S	V	A	L	F	V	H	T	Y	T	V	D	I	K	–	N	R
Pea III	T	H	Y	F	M	D	R	A	V	A	L	F	V	H	T	Y	T	V	D	I	K	L	N	R
Human III	T	H	Y	Y	M	D	R	V	V	A	L	F	V	H	T	Y	T	V	D	I	K	–	N	R
Ara P	T	H	F	F	L	G	R	I	V	F	M	M	L	H	T	Y	T	V	D	F	R	T	S	R
Human I	S	H	F	F	V	L	D	L	V	I	Y	M	L	R	S	H	T	V	D	T	K	I	R	R

Within the class III line two types of enzymes can be distinguished: the microorganism (bacteria and yeast) type, with high Km and kcat, and the plant-animal type with low Km and kcat. The two types probably represent adaptations to different physiological needs for formaldehyde elimination. The change Ile269 (plant-animal type) to Thr269 (microorganism type) is probably involved in the functional differences between the two class III types.

Classes I and P exhibit many changes in the active site residues, compatible with the observed kinetic properties and distinct physiological functions. While class III represents a single structural and functional line common to microorganisms, plants and animals, classes P and I represent separate lines.

5. ACKNOWLEDGMENTS

Supported by the Spanish Dirección General de Investigación Científica y Técnica (PB92–0624) and the European Commission (BMH1-CT93–1601).

6. REFERENCES

Allais, J.J., Louktibi, A. and Baratti, J.: Oxidation of methanol by the yeast *Pichia pastoris*. Purification and properties of the formaldehyde dehydrogenase. Agric. Biol. Chem. 47 (1983) 1509–1516.

Boleda, M.D., Saubi, N., Farrés, J. and Parés, X.: Physiological substrates for rat alcohol dehydrogenase classes: Aldehydes of lipid peroxidation, ω-hydroxyfatty acids, and retinoids. Arch. Biochem. Biophys. 307 (1993) 85–90.

Engeland, K., Höög, J.O., Holmquist, B., Estonius, M., Jörnvall, H. and Vallee, B.L.: Mutation of Arg-115 of human class III alcohol dehydrogenase: A binding site required for formaldehyde dehydrogenase activity and fatty acid activation. Proc. Natl. Acad. Sci. USA. 90 (1993) 2491–2494.

Fan, F. and Plapp, B.V.: Substitutions of isoleucine residues at the adenine binding site activate horse liver alcohol dehydrogenase. Biochemistry 34 (1995) 4709–4713

Fernández, M.R., Biosca, J.A., Norin, A., Jörnvall, H. and Parés, X.: Class III alcohol dehydrogenase from *Saccharomyces cerevisiae*: Structural and enzymatic features differ toward the human/mammalian forms in a manner consistent with functional needs in formaldehyde detoxication. FEBS Lett. 370 (1995) 23–26.

Fersht, A.: Enzyme Structure and Mechanism, 2nd Edn., Freeman, New York (1985).

Gietz, R.D., and Sugino, A.: New yeast-*Escherichia coli* shuttle vectors constructed with *in vitro* mutagenized yeast genes lacking six-base pair restriction sites. Gene 74 (1988) 527–534.

Holmquist, B., Moulis, J.M., Engeland, K. and Vallee, B.L.: Role of arginine 115 in fatty acid activation and formaldehyde dehydrogenase activity of human class III alcohol dehydrogenase. Biochemistry 32 (1993) 5139–5144.

Kaiser, R., Fernández, M.R., Parés, X. and Jörnvall, H.: Origen of the human alcohol dehydrogenase system: Implications from the structure and properties of the octopus protein. Proc. Natl. Acad. Sci. USA. 90 (1993) 11222–11226.

Koivusalo, M., Baumann, M. and Uotila, L.: Evidence for the identity of glutathione-dependent formaldehyde dehydrogenase and class III alcohol dehydrogenase. FEBS Lett. 257 (1989) 105–109.

Mack, M., Gömpel-Klein, P., Haase, E., Hietkamp, J., Ruhland, A. and Brendel, M.: Genetic characterization of hyperresistance to formaldehyde and 4-nitroqunoline-N-oxide in the yeast *Saccharomyces cerevisiae*. Mol. Gen. Genet. 211 (1988) 260–265.

Martínez, M.C., Achkor, H., Persson, B., Fernández, M.R., Shafqat, J., Farrés, J. Jörnvall, H. and Parés, X. *Arabidopsis* formaldehyde dehydrogenase: Molecular properties of plant class III alcohol dehydrogenase provide further insights into the origins, structure and function of plant class P and liver class I alcohol dehydrogenase. Eur. J. Biochem. In press.

Petersson, G.: Effect of evolution on the properties of enzymes. Eur. J. Biochem. 184 (1989) 561–566.

Sasnauskas, K., Jomantiené, R., Januska, A., Lebediené, E., Lebedys, J. and Janulaitis, A.: Cloning and analysis of *Candida maltosa* gene wich confers resistance to formaldehyde in *Saccharomyces cerevisiae*. Gene 122 (1992) 207–211.

Shafqat, J., El-Ahmad, M., Danielsson, O., Martinez, M.C., Persson, B., Parés, X. and Jörnvall, H.: Pea class III alcohol dehydrogenase: Common derivation of the plant and animal formaldehyde-active forms (class III alcohol dehydrogenase) but not of the corresponding ethanol-active forms (classes I and P). Proc. Natl. Acad. Sci. USA. 93 (1996) 5595–5599.

Thompson, J.D., Higgins, D.G. and Gibson, T.J.: CLUSTAL W: Improving the sensitivity of progressive multiple sequence alignment through sequence weighting, position-specific gap penalties and weight matrix choice. Nucleic. Acids. Res. 22 (1994) 4673–4680.

Uotila, L. and Koivusalo, M.: Glutathione-dependent Oxidoreductases: Formaldehyde dehydrogenase. In Coenzymes and Cofactors, Vol III. D. Dolphin, R Poulson and O. Avramovic. Eds. John Wiley & Sons, New York (1989) 517–551.

Wehner, E.P., Rao, E. and Brendel, M.: Molecular structure and genetic regulation of SFA, a gene responsible for resistance to formaldehyde in *Saccharomyces cerevisiae*, and characterization of its protein product. Mol. Gen. Genet. 237 (1993) 351–358.

44

HUMAN SORBITOL DEHYDROGENASE – A SECONDARY ALCOHOL DEHYDROGENASE WITH DISTINCT PATHOPHYSIOLOGICAL ROLES

pH-Dependent Kinetic Studies

Wolfgang Maret

Center for Biochemical and Biophysical Sciences and Medicine
Harvard Medical School
250 Longwood Avenue
Boston, Massachusetts 02115

1. INTRODUCTION

1.1. The Polyol Pathway and Diabetic Complications

Aldose reductase and sorbitol dehydrogenase (SDH, L-iditol:NAD$^+$ 2-oxidoreductase, EC 1.1.1.14) metabolically link and glucose and fructose in the polyol pathway:

$$\text{D-Glucose} \underset{\text{NADPH/NADP}^+}{\xleftarrow{\hspace{2cm}}} \text{Sorbitol} \underset{\text{NAD}^+/\text{NADH}}{\xleftarrow{\hspace{2cm}}} \text{S-Fructose}$$

The significance of this pathway under normal physiological conditions is not clear. Comparatively more studies have addressed its pathophysiological role in diabetes mellitus. In this disease, inappropriate elevation of blood glucose leads to severe medical problems that affect tissues where glucose uptake is not dependent on insulin, such as the eye (retina - retinopathy/lens - "sugar" cataracts), nerves (neuropathy), kidney (nephropathy), and blood vessels (vasculopathy). Diabetic pathology is thought to arise from (i) the glycation of proteins which changes their functional and structural properties, (ii) accumulation of sorbitol and other polyols which impairs osmoregulation, and (iii) metabolic imbalances which resemble hypoxia and oxidative/reductive stress. The SDH-catalyzed oxidation of sorbitol to fructose is involved in all three of these areas, and, thus, has a central position in the development of diabetic complications. First, in hyperglycemia SDH raises the tissue levels of fructose. The increased reactivity of this ketose in comparison to glucose poses a relatively higher risk for advanced glycation end products through Maillard chemistry. Second, SDH is responsible for the removal or control of polyols in many tissues. Third, it has now been shown that the metabolic imbalances in diabetes are primarily the result of increased substrate flux through the polyol pathway and a changed NAD/NADH redox status determined by SDH (Tilton et al., 1995; van den Enden et al., 1995). Accordingly, vascular and neural dysfunction in diabetic individuals are believed to be closely linked to the action of SDH (Williamson et al., 1993). SDH has been studied much less than aldose reductase which has received inordinate attention as the exclusive therapeutic target for drugs designed to counteract the development of diabetic complications. Our approach, therefore, has been to study basic aspects of the mechanism of action of SDH. The strength of this approach has been proven already by demonstrating that sorbitol concentrations as high as those in diabetic tissue completely inhibit the enzyme (Maret, 1991). This report briefly summarizes our previous work on human SDH (Maret and Auld, 1988; Maret, 1989; Maret, 1991) and presents our recent studies on the specificity of SDH toward ketoses and on the pH dependence of SDH, followed by a short discussion of the possible physiological role of SDH.

1.2. Human Sorbitol Dehydrogenase

SDH is a zinc-dependent tetrameric alcohol dehydrogenase (Jeffery and Jörnvall, 1988). While clearly related to the "classical" alcohol dehydrogenases in terms of primary structure (24% residue identity compared to class I alcohol dehydrogenases), mammalian SDH differs from these in zinc content (one instead of two zinc atoms per subunit), quaternary structure (tetramer instead of dimer), and subunit size (356 versus 374 residues). It is the only known acyclic polyol dehydrogenase in mammals, which is somewhat surprising because it would assign SDH a central role in the metabolism of all acyclic polyols. Human SDH is a strict secondary alcohol dehydrogenase (Maret and Auld, 1988). It has never been shown to oxidize primary alcohols, nor has it been reported to reduce aldehyde functions. SDH exhibits a relatively high degree of stereospecificity (Maret and Auld, 1988). Importantly, this specificity holds through all classes of chemically different alcohol substrates, even for the smallest substrates, 2,3-butanediols, where there are fewer elements of molecular recognition, and for aromatic secondary alcohols, both of which we have identified as new substrate classes (Maret and Auld, 1988; Maret, 1991). There is considerable evidence for multiple SDH forms in non-human species (Jeffery and Jörnvall, 1988). For human SDH, genetic polymorphism was detected only recently (Karlsson et al., 1989; Carr and Markham, 1995). Its extent, functional significance and relation to reported different forms of the protein are unknown. SDH is expressed almost ubiquitously, with an absence noted only for the intestinal tract of the rat (Estonius et al., 1993). The pattern of SDH gene expression in humans (Iwata et al., 1995; Carr and Markham, 1995) is strikingly different from that of rodents. In this regard, the high levels of expression in tissues affected by diabetic complications in humans (kidney, lens) is particularly noteworthy, and by itself draws attention to the pathophysiological role of human SDH and the need to study the human enzyme.

2. EXPERIMENTAL PROCEDURES

2.1. Materials

L-erythrulose was purchased from Aldrich Chemical Co. (Milwaukee, WI), D-xylulose and other ketoses from Sigma Chemical Co. (St. Louis, MO). The sources of all other chemicals and the preparation of the enzyme were described previously (Maret and Auld, 1988).

2.2. Steady-State Kinetic Studies

Initial velocity data were collected spectrophotometrically by following the increase or decrease in NADH absorbance at 25 °C. Ionization constants were evaluated from Dixon-Webb plots (Dixon and Webb, 1964) by visually fitting the data to theoretical titration curves corresponding to a single ionization.

2.3. Equilibrium Dialysis

SDH was dialyzed exhaustively in a DiaCell™ system (InstruMed Inc., Union Bridge, MD) at 4 °C against metal-free 0.1 M sodium acetate (pH 4–6) and 0.1 M 2-(4-morpholino)ethanesulfonic acid (pH 6–7) buffers, and then its specific enzymatic activity measured and its zinc content determined by atomic absorption spectrophotometry.

3. RESULTS AND DISCUSSION

3.1. Ketose Substrates

Virtually nothing seems to be known about the recognition of ketoses by SDH. When testing the entire series of C-4, C-5, and C-6 D-ketoses and a few L-ketoses as possible substrates for SDH (Table 1), it was found that the specificity observed for polyol substrates, i.e. sorbitol>xylitol>L-threitol, is not paralleled by a corresponding trend of K_m values of their respective products, i.e. D-fructose, D-xylulose, L-erythrulose. D-sorbose, D-tagatose, and D-psicose showed only 6%, 14%, and 41% of the activity of L-sorbose at identical substrate concentrations of 140 µM. High apparent K_m values and low rates with these latter substrates precluded the determination of their kinetic parameters with the spectrophotometric assay. The millimolar K_m value for D-xylulose is particularly noteworthy (Table 1). This substrate exists between 8 and 22% in the free ketose/chain form (Hayward and Angyal, 1977). In contrast, a solution of D-fructose at equilibrium is a complex mixture of α- and β-furanose forms, α- and β-pyranose forms, and the free ketose, the contribution of which is about 0.8% (Goux, 1985). It, therefore, appears that the true substrate is the chain form of the sugar which is only a small percentage of the total. As a re-

Table 1. Ketose substrates for human liver sorbitol dehydrogenase[a]

Ketose	$k_{cat}(s^{-1})$	$K_m (M)$	$k_{cat}/K_m (M^{-1} s^{-1})$
C-4			
L-erythrulose	34	2.5×10^{-2}	1.3×10^3
C-5			
D-xylulose[b]	7	1×10^{-3}	7×10^3
D-ribulose	12	3.4×10^{-2}	3.5×10^2
C-6			
L-sorbose	42	0.8	5.2×10^1
D-fructose	45	0.14	3.2×10^2

[a]Determined in 10 mM phosphate buffer, pH 7.4, containing 230 µM NADH
[b]The comparatively low k_{cat} value for D-xylulose is due to substrate inhibition which becomes noticeable at concentrations above K_m.

sult, ketoses have very high apparent K_m values and variation of their structure is not reflected in large changes of kinetic parameters. If one takes into account the percentage of substrate being in the chain form, the corrected K_m value for D-fructose would be 1.12 mM instead of the apparent K_m of 0.14 M (Table 1). Two additional observations further support the postulate that SDH acts on the chain form of the sugar. Polyol substrates do not cyclize. Hence, the chemical step produces D-fructose in the chain form. Equilibrium constants for the substrate pairs L-threitol/L-erythrulose, xylitol/D-xylulose and sorbitol/D-fructose of 6.9×10^{-12}, 4.2×10^{-11}, and 4.2×10^{-10} (Hollmann, 1959) are thought to reflect the tendency of the ketoses to form a ring structure with increasing chain lengths. For the other enzyme in the polyol pathway, aldose reductase, recent results suggest that the chain form of the aldose (0.02%) is the true substrate (Grimshaw, 1986), lowering the apparent millimolar K_m to the micromolar range. Any metabolic flux consideration of glucose and fructose through the polyol pathway should be based on those corrected K_m values.

In the context of these data it should be recalled that the spectrophotometric assay of SDH is based on NADH formation or consumption and, therefore, can neither establish which polyols are formed from a particular ketose nor can it identify the product(s) of those polyols that can undergo oxidation at either end of the molecule. At least two different mechanisms lead to the simultaneous formation of multiple products from a single substrate (scheme 1). First, flipping of substrates that lack rotational symmetry results in different products. For example, L-iditol is an unambiguous substrate. It is converted to L-sorbose regardless of whether oxidation occurs at the C-2 or C-5 position. In contrast, L-arabinitol is an ambiguous substrate and gives rise to L-ribulose (95%) and L-xylulose (about 5%) (Smith, 1962). Sorbitol is also an ambiguous substrate and can form both D-fructose and L-sorbose (box with dashed lines in scheme). Thus, reaction products are not clear even for the most common substrate. Second, human, sheep, and rat SDH exhibit significant activity with D-mannitol as a substrate (Jeffery and Jörnvall, 1988; Maret and Auld, 1988; Lindstad et al., 1992). It is thought that D-fructose is the product of oxidation of this polyol (Jeffery and Jörnvall, 1988). The identification of 1-deoxyfructose as the product of 1-deoxymannitol corroborates this postulate (Dills and Meyer, 1976). Thus, D-fructose is formed from either sorbitol or mannitol (box with dotted lines in scheme). In this case, substrate flipping cannot be the cause for the formation of D-fructose, because mannitol is an unambiguous substrate with rotational symmetry. The product can be obtained only if hydrogen abstraction at C-2/C-5 is not highly stereospecific. This stereochemical permissiveness, i.e. addition of the "wrong" hydride ion to a certain degree, yields different products from one sugar in the reverse reaction (D-fructose→D-mannitol/sorbitol). Further studies are needed to address this peculiar regiochemistry of SDH.

3.2. pH Dependence of Steady-State Kinetic Parameters

In order to establish the pH range in which SDH is stable for kinetic studies, the enzyme was incubated at different pH values and then the relative enzymatic activity measured (Figure 1, closed squares).

Above pH 10.5 and below pH 4.5 the half-life of the enzyme is shorter than the time required for performing enzymatic assays, thus precluding data collection at these extremes of pH. Equilibrium dialysis of SDH at various pH values and subsequent zinc analyses demonstrated loss of zinc from SDH already below pH 6 (Figure 1, open squares) and a concomitant loss of enzymatic activity (data not shown). Thus, dissociation of zinc must be so slow in the pH range 5–6 that it does not affect enzymatic activity in the time required for the assay.

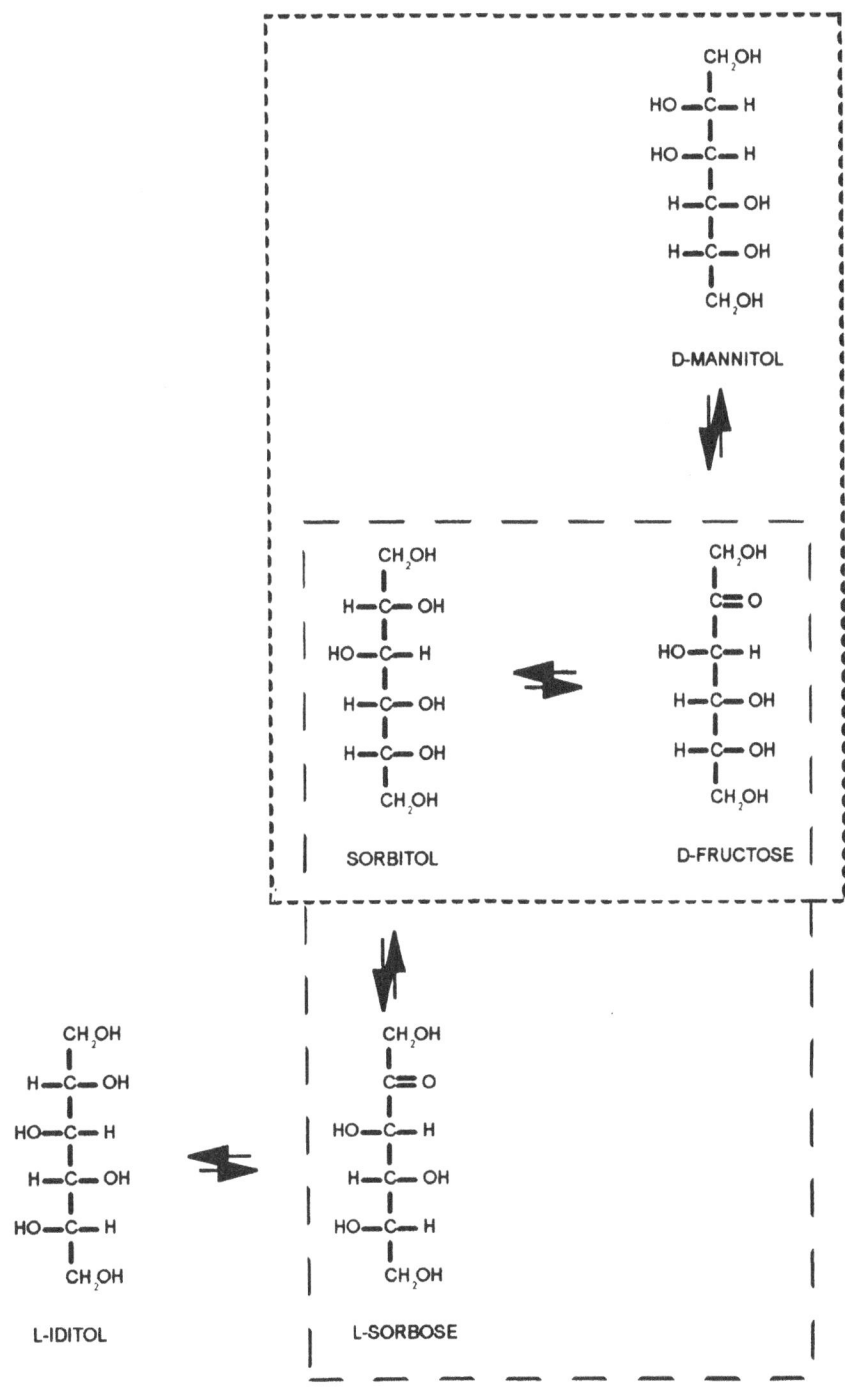

Scheme 1

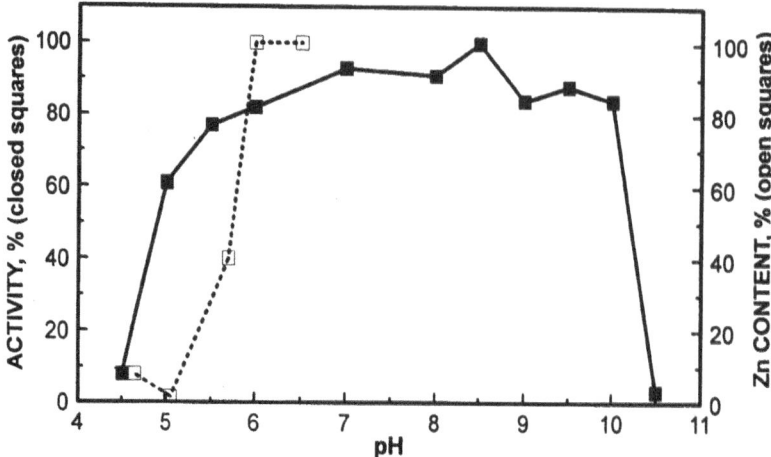

Figure 1. SDH activity and zinc content as a function of pH. The stability of the enzyme as a function of pH was studied by incubating 5 μL of the enzyme (5.5 μM) in 95 μL of buffer for 5 min at 4 °C and then assaying SDH activity at pH 10.0 and 25 °C. Zinc content of SDH was determined by atomic absorption spectroscopy after dialysis of the enzyme at corresponding pH values.

The pH profile of $k_{cat}/K_{m(sorbitol)}$ for sorbitol oxidation is controlled by a pK_a value of 7.1 with enzymatic activity decreasing toward acidic pH in the two buffer systems employed (Figure 2).

In the reverse reaction, i.e. fructose reduction, it was not possible to saturate SDH with fructose at high pH values, because the K_m value increases with pH and becomes greater than 1 M above pH 8. Hence, the kinetic behavior of the enzyme can be described only at low substrate concentrations (k_{cat}/K_m conditions). In this case, $k_{cat}/K_{m(fructose)}$ is controlled by a pK_a value of 8 with the enzymatic activity decreasing toward basic pH (Figure 3).

In contrast to the k_{cat}/K_m profile, the k_{cat} profile for sorbitol oxidation differs in the two buffers systems and appears to be controlled by pK_a values of 4.5 and 4.9, respectively (Figure 4). However, in this pH range SDH is no longer stable (Figure 1). This ionization, therefore, controls the stability of the enzyme rather than its kinetic behavior, and it will not be discussed here.

Several buffers and anions influence the binding of substrate and coenzyme in alcohol dehydrogenases, resulting in enzyme activation or inhibition. In order to detect possible buffer effects on the pH profiles, data were collected in two different buffer systems, in particular since it was noted that varying glycine from 0 to 0.5 M in the assay buffer (bis(tris)propane, pH 9.5) increased the rate of sorbitol oxidation up to 5-fold. Therefore, the increase in k_{cat} at high pH (Figure 4, bracketed open squares) is thought not to be due to an ionization but rather to an effect of glycine on the enzyme. Similarly, the decrease in k_{cat}/K_m at pH 9.7 (Figure 3, bracketed open square) is also likely due to a buffer effect. In the pH profile of k_{cat}, the wave in the region where phosphate was used as a buffer probably also reflects a buffer effect rather than an ionization, because the corresponding pH profile using bis(tris)propane buffer follows a single ionization. Alternatively, the pH profile may be sigmoidal and controlled by two ionizations with a relatively small difference in activity between the alkaline and acidic form of the enzyme (Lindstad and McKinley-

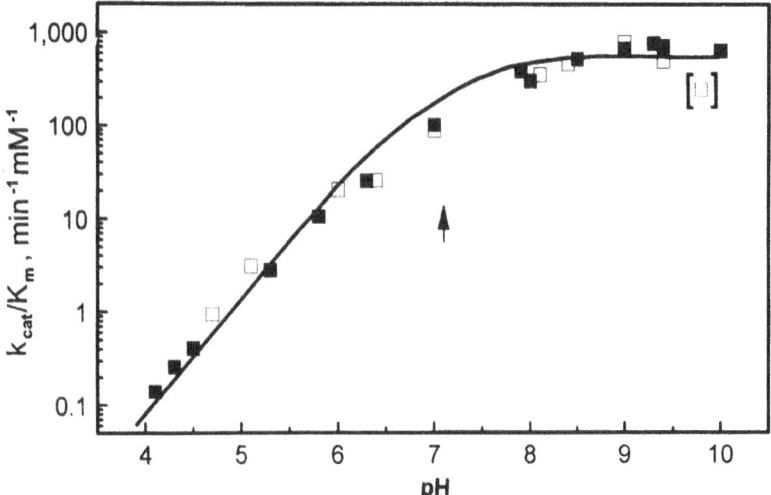

Figure 2. pH Dependence of k_{cat}/K_m for sorbitol oxidation. Buffers used were: sodium acetate (below pH 5), phosphate (pH 5–8), pyrophosphate (pH8–9.5), and carbonate (above pH 9.5) at an ionic strength of 0.05 (closed squares); and sodium acetate (below pH 6), 1,3-bis[tris(hydroxymethyl)methylamino]propane (pH 6–9), and glycine (above pH 9) at an ionic strength of 0.1 (open squares). Concentrations of coenzymes were kept at saturating levels, i.e., 5 mM NAD+ and 250 µM NADH. The arrow indicates the estimated pKa value.

McKee, 1995), an interpretation that seems less likely since the wave is seen only in one of the two buffer systems. Chloride inhibits at low substrate concentration (K_I = 0.1 M) and activates at high substrate concentration. Inhibition by pyrophosphate which has been observed for sheep liver SDH and attributed to chelation of the active site zinc ion (Reiersen et al., 1994) was not noted at the relatively low pyrophosphate concentrations employed here and during the time required for the enzymatic assays.

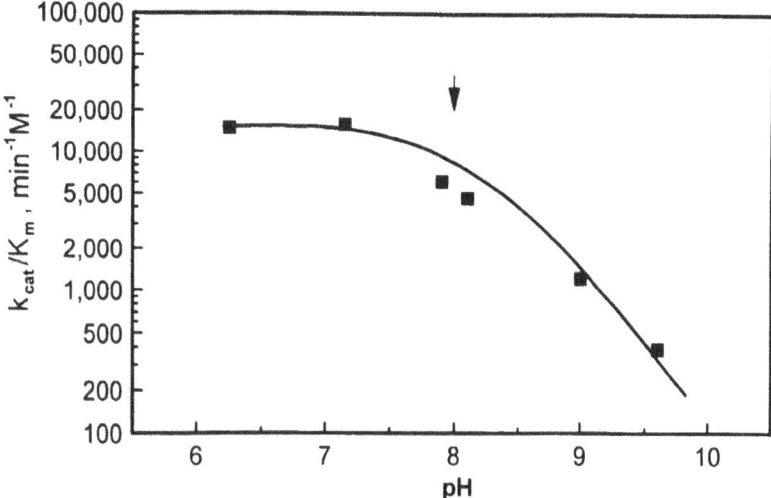

Figure 3. pH Dependence of k_{cat}/K_m for fructose reduction. Buffers used were phosphate (pH 6–8) and pyrophosphate (pH 8–9.5) at an ionic strength of 0.05. Other conditions are described in the legend of Figure 2.

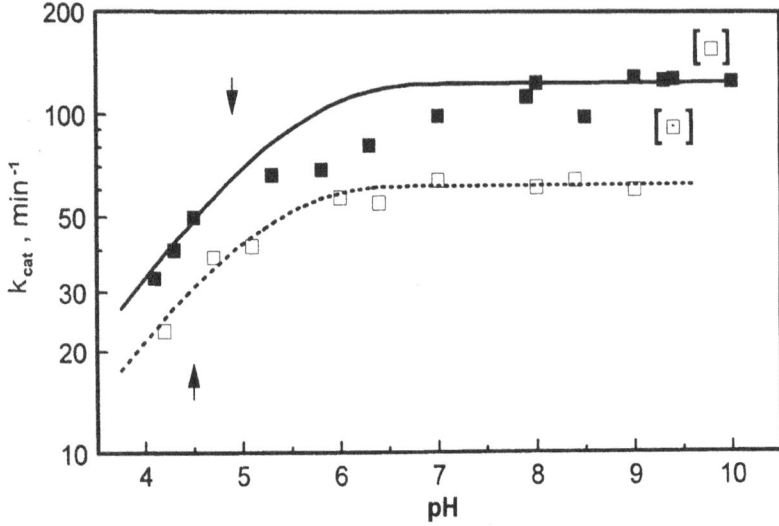

Figure 4. pH Dependence of k_{cat} for sorbitol oxidation. The constant difference between k_{cat} values in the two buffers systems is in part due to the slightly different specific enzymatic activities of the two different enzyme preparation used. Other conditions are described in the legend of Figure 2.

The pH dependence of steady-state kinetic parameters can reveal ionizations of the free reactants (enzyme and substrate) in k_{cat}/K_m profiles and of enzyme/substrate complexes in k_{cat} profiles (Brocklehurst, 1994). Catalytically essential groups in SDH have not yet been firmly assigned. A compulsory-order mechanism holds for sheep SDH in the pH range 7–9.9 (Lindstad et al., 1992). In such an ordered mechanism with coenzyme binding first, the pH profiles of k_{cat}/K_m reflect ionizations of the binary enzyme/coenzyme complexes. The pK_a value of 8.0 in fructose reduction for human SDH (Figure 3) is higher than that of 7.7 for ovine SDH (Lindstad and McKinley-McKee, 1995), while the pK_a value of 7.1 in sorbitol oxidation (Figure 2) is identical in both enzymes. In the latter study, a slightly different approach was employed. Steady-state kinetic data were analyzed in terms of the pH dependence of individual rate constants for a six-velocity-constant Theorell-Chance mechanism. For the discussion of what group(s) might be responsible for the observed pK_a values, we turn to alcohol dehydrogenase (ADH, specifically the class I enzyme from horse liver), but not without emphasizing the potential of investigations on SDH for elucidating specific aspects of the catalytic mechanism of ADH that have remained conjectural. Such a comparative approach is commended since those amino acids that have been invoked in the mechanism of ADH are all conserved in SDH. An ionization with a pK_a value of 7.5 in ADH, which corresponds to the ionization in the k_{cat}/K_m profile of SDH, has been assigned to a the side chain of His-51 (Cook and Cleland, 1981; Maret and Makinen, 1991), a residue that has been implicated in proton transfer (Eklund and Brändén, 1987). This interpretation is in agreement with the data on human SDH and ovine SDH (Lindstad and McKinley-McKee, 1995), but it is fundamentally different from those of other proposals that have linked an ionization with a pK_a value of 7.6 to the zinc-bound water molecule in the ADH/NAD$^+$ complex (Pettersson, 1987). The latter assignment would seem to be problematic since water dissociates from zinc in the proposed mechanism, and thus is not an essential group in the enzyme in the sense that it is required for enzymatic catalysis.

A second ionization in the bell-shaped pH profile of k_{cat}/K_m in ADH with a pK_a value of 8.9 decreases the enzymatic activity at high pH and has been assigned to the zinc-bound water molecule (Maret and Makinen, 1991). This ionization is not observed in SDH (Figure 2). Intuitively, it could be argued that the substitution of one of the cysteine metal ligands in ADH by a glutamate in SDH (Eklund et al., 1985) increases the negative charge at the zinc atom and raises the pK_a value of the zinc-bound water molecule to a pH range where it is no longer detected. Again, the previous assignment in ADH is consistent with the data for SDH.

4. THE POSSIBLE PHYSIOLOGICAL ROLE OF SDH

Sorbitol is probably the most important endogenous acyclic sugar alcohol in human physiology. It has a specific transporter (Napathorn and Spring, 1994) and undergoes further metabolic changes to sorbitol 3-P (Lal et al., 1993); it serves as an osmolyte (Burg and Kador, 1988), an energy source, and a possible regulator of gluconeogenesis in specific tissues. For example, sorbitol and SDH have a specific function in glycogen metabolism in insects. In *Bombyx mori* eggs, SDH is cold-inducible and converts sorbitol to fructose for the formation of glycogen (Niimi et al., 1993). SDH may exert some control in the formation of glycogen in mammalian liver as well and, thus, have a more important role in energy metabolism than currently appreciated. First, sorbitol is so far the most potent sugar metabolite in the stimulation of glucokinase translocation (Agius, 1994). Second, sorbitol 6-P inhibits the binding of the regulatory protein to glucokinase (Vandercammen et al., 1992). Third, when the SDH inhibitor S-0773 was tested in rats, a potentially very important finding was made (Tilton et al., 1995): "In view of the unknown physiological function of the sorbitol pathway, the paradoxical observation that S-0773-treated control and diabetic rats consumed considerably more food (but gained less weight) than did untreated rats is of particular interest. This observation is consistent with the possibility that increased flux of glucose via the sorbitol pathway during periods of hyperglycemia may attenuate flux of glucose to lactate. The increased ratio of $NADH:NAD^+$, resulting from increased oxidation of sorbitol to fructose, will favor utilization of glycolytic intermediates (including those derived from fructose) for glycogen and lipid synthesis rather than for lactate production." This new awareness of cellular functions of sorbitol stresses the significance of SDH, the major enzyme responsible for its metabolism, and also indicates that knowledge regarding SDH has not yet been fully integrated into intermediary metabolism.

5. ACKNOWLEDGMENTS

This work was supported by a grant from the Samuel Bronfman Foundation, Inc. to the Endowment for Research in Human Biology, Inc., with funds provided by Joseph E. Seagram and Sons, Inc., and a grant from the Milton Fund of the Harvard Medical School to WM. I thank Drs. Bert L. Vallee and James F. Riordan for advice and encouragement, and Dr. Douglas S. Auld for assistance.

6. REFERENCES

Agius, L.: Control of Glucokinase Translocation in Rat Hepatocytes by Sorbitol and the Cytosolic Redox State. Biochem. J. 298 (1994) 237–243.

Burg, M.B. and Kador, P.F.: Sorbitol, Osmoregulation, and the Complications of Diabetes. J. Clin. Invest. 81 (1988) 635–640.

Brocklehurst, K.: A Sound Basis for pH-Dependent Kinetic Studies on Enzymes. Protein Eng. 7 (1994) 291–299.

Carr, I.M. and Markham, A.F.: Molecular Genetic Analysis of the Human Sorbitol Dehydrogenase Gene. Mamm. Genome 6 (1995) 645–652.

Cook, P.F. and Cleland, W.W.: pH Variation of Isotope Effects in Enzyme-Catalyzed Reactions. 2. Isotope-Dependent Step Not pH Dependent. Kinetic Mechanism of Alcohol Dehydrogenase. Biochemistry 20 (1981) 1805–1816.

Dills Jr, W.L. and Meyer, W.L.: Studies on 1-Deoxy-D-fructose, 1-Deoxy-D-glucitol, and 1-Deoxy-D- mannitol as Antimetabolites. Biochemistry 15 (1976) 4506–4512.

Dixon, M. and Webb, E.C.: In Enzymes (2nd Ed.). Academic Press, New York, 1964, pp. 116–166.

Eklund, H. and Brändén, C.-I.: Alcohol Dehydrogenase. In Jurnak, F.A. and McPherson, A. (Eds.), Biological Macromolecules and Assemblies, Vol. 3. John Wiley & Sons, New York, 1987, pp. 73- 142.

Eklund, H., Horjales, E., Jörnvall, H., Brändén, C.-I. and Jeffery, J.: Molecular Aspects of Functional Differences between Alcohol and Sorbitol Dehydrogenases. Biochemistry 24 (1985) 8005–8012.

Estonius, M., Danielsson, O., Karlsson, C., Persson, H., Jörnvall, H. and Höög, J.-O.: Distribution of Alcohol and Sorbitol dehydrogenases. Eur. J. Biochem. 215 (1993) 497–503.

Goux, W. J.: Complex Isomerizations of Ketoses: A 13C NMR Study of the Base-catalyzed Ring-opening and Ring-closing Rates of D-Fructose Isomers in Aqueous Solution. J. Am. Chem. Soc. 107 (1985) 4320–4327.

Grimshaw, C.E.: Direct Measurement of the Rate of Ring Opening of D-Glucose by Enzyme Catalyzed Reduction. Carbohydr. Res. 148 (1986) 345–348.

Hayward, L. D. and Angyal, S. J.: A Symmetry Rule for the Circular Dichroism of Reducing Sugars and the Proportion of Carbonyl Forms in Aqueous Solutions Thereof. Carbohydr. Res. 53 (1977) 13- 20.

Hollmann, S.: Trennung, Reinigung und Eigenschaften der mitochondrialen Xylit-Dehydrogenasen der Meerschweinchenleber. Hoppe-Seyler's Z. Physiol. Chem. 317 (1959) 193–215.

Iwata, T., Popescu, N.C., Zimonjic, D.B., Karlsson, C., Höög, J.-O., Vaca, G., Rodriguez, I.R. and Carper, D.: Structural Organization of the Human Sorbitol Dehydrogenase Gene (SORD). Genomics 26 (1995) 55–62.

Jeffery, J. and Jörnvall, H.: Sorbitol Dehydrogenase. Adv. Enzymol. Relat. Areas Mol. Biol. 61 (1988) 47–106.

Karlsson, C., Maret, W., Auld, D.S., Höög, J.-O. and Jörnvall, H.: Variability within Mammalian Sorbitol Dehydrogenases. The Primary Structure of the Human Liver Enzyme. Eur. J. Biochem. 186 (1989) 543–550.

Lal, S., Szwergold, B.S., Kappler, F. and Brown ,T.: Detection of Fructose-3-Phosphokinase Activitiy in Intact Mammalian Lenses by ^{31}P NMR Spectroscopy. J. Biol. Chem. 268 (1993) 7763–7767.

Lindstad, R.I., Hermansen, L.F. and McKinley-McKee, J.S.: The Kinetic Mechanism of Sheep Liver Sorbitol Dehydrogenase. Eur. J. Biochem. 210 (1992) 641–647.

Lindstad, R.I. and McKinley-McKee, J.S.: Effect of pH on Sheep Liver Sorbitol Dehydrogenase Steady- State Kinetics. Eur. J. Biochem. 233 (1995) 891–898.

Maret, W. and Auld, D.S.: Purification and Characterization of Human Liver Sorbitol Dehydrogenase. Biochemistry 27 (1988) 1622–1628.

Maret, W.: Novel Substrates and Inhibitors of Human Liver Sorbitol Dehydrogenase. In Weiner, H., Wermuth, B. and Crabb, D.W. (Eds.), Enzymology and Molecular Biology of Carbonyl Metabolism 3. Plenum Press, New York, 1991, pp. 327–336.

Maret, W. and Makinen, M.W.: The pH Variation of Steady-state Kinetic Parameters of Site-specific Co^{2+}- reconstituted Liver Alcohol Dehydrogenase. J. Biol. Chem. 266 (1991) 20636–20644.

Maret, W.: Cobalt(II)-Substituted Class III Alcohol and Sorbitol Dehydrogenase from Human Liver. Biochemistry 28 (1989) 9944–9949.

Napathorn, S. and Spring, K.R.: Further Characterization of the Sorbitol Permease in PAP-HT-25 Cells. Am. J. Physiol. 267 (1994) C514–519.

Niimi, T., Yamashita, O. and Yaginuma, T.: A Cold-inducible *Bombyx* Gene Endoding a Protein Similar to Mammalian Sorbitol Dehydrogenase. Eur. J. Biochem. 213 (1993) 1125–1131.

Pettersson, G.: Liver Alcohol Dehydrogenase. Crit. Rev. Biochem. 21 (1987) 349–389.

Reiersen, H., Lindstad, R.I. and McKinley-McKee, J.S.: The Inactivation of Sheep Liver Sorbitol Dehydrogenase by Pyrophosphate and Some Analogous Metal Chelators. Arch. Biochem. Biophys. 311 (1994) 450–456.

Smith, M.G.: Polyol dehydrogenases. 4. Crystallization of the L-Iditol Dehydrogenase of Sheep Liver. Biochem. J. 83 (1962) 135–144.

Tilton, R.G., Chang, K., Nyengaard, J.R., Van den Enden, M., Ido, Y. and Williamson, J.R.: Inhibition of Sorbitol Dehydrogenase. Effects on Vascular and Neural Dysfunction in Streptozocin-induced Diabetic Rats. Diabetes 44 (1995) 234–242.

Van den Enden, M.K., Nyengaard, J.R., Ostrow, E., Burgan, J.H. and Williamson, J.R.: Elevated Glucose Levels Increase Retinal Glycolysis and Sorbitol Pathway Metabolism. Invest. Opthalmol. Vis. Sci. 36 (1995) 1675–1685.

Vandercammen, A., Deux, M. and Van Schaftingen, E.: Binding of Sorbitol 6-Pi and of Fructose 6-Pi to the Regulatory Protein of Liver Glucokinase. Biochem. J. 286 (1992) 253–256.

Williamson, J.R., Chang, K., Frangos, M., Hasan, K.S., Ido, Y., Kawamura, T., Nyengaard, J.R., Van den Enden, M., Kilo, C. and Tilton, R.G.: Hyperglycemic Pseudohypoxia and Diabetic Complications. Diabetes 42 (1993) 801–813.

45

STRUCTURAL AND MECHANISTIC ASPECTS OF A NEW FAMILY OF DEHYDROGENASES, THE β-HYDROXYACID DEHYDROGENASES

John W. Hawes, Edwin T. Harper, David W. Crabb, and Robert A. Harris

Department of Biochemistry and Molecular Biology and Department of
  Medicine
Indiana University School of Medicine
Indianapolis, Indiana 46202–5122

1. INTRODUCTION

The catabolism of valine, unlike that of other branched-chain amino acids, occurs with the formation of a free branched-chain acid, (S)-β-hydroxyisobutyrate or HIBA, whereas other branched-chain amino acids are metabolized solely as coenzyme A thioesters. Because it exists as a free acid, HIBA can be released into the blood stream by specific tissues and is cleared by the liver where it can serve as a substrate for gluconeogenesis (Letto et al., 1986). During the past decade there has been significant interest in the metabolism and interorgan trafficking of the R- and S- enantiomers of HIBA. HIBA is oxidized in mitochondria to methylmalonate semialdehyde by a highly specific, NAD^+-dependent dehydrogenase (HIBADH or 3-hydroxy-2-methyl-propionate: NAD^+ oxidoreductase, EC 1.1.1.31). Previous studies of rat HIBADH tentatively placed the enzyme in the now well-established short-chain alcohol dehydrogenase family (Crabb et al., 1993; Hawes et al., 1995). This assignment was based on amino acid sequence homology, enzymatic properties such as the lack of a metal requirement for catalysis, and effects of tyrosine-specific chemical modification. However, site-directed mutagenesis studies indicated that HIBADH differs in mechanism from the short-chain dehydrogenases studied to date, such as *Drosophila* alcohol dehydrogenase (Hawes et al., 1995). Furthermore, the short-chain dehydrogenases mostly prefer secondary alcohols as optimal substrates whereas HIBADH is only active with primary alcohol substrates. HIBADH, therefore, is most likely not closely related to the short-chain dehydrogenases. More recent studies showed that HIBADH shares better amino acid sequence homology and enzymatic properties with a separate, previously unrecognized family of enzymes that includes D-phenylserine dehydrogenase from *Pseudomonas syringae*, 6-phosphogluconate dehydrogenase from numerous species, and numerous hypothetical proteins of microbial origin (Hawes et al., 1996). Like the short-chain dehydrogenases, these enzymes have no requirement for divalent

Enzymology and Molecular Biology of Carbonyl Metabolism 6
edited by Weiner *et al.* Plenum Press, New York, 1996

metal ions, and contain a dinucleotide cofactor-binding domain located at the N-terminus. The amino acid sequence motif representative of the dinucleotide binding sites of dehydrogenases and present at the N-terminus in members of this new family is highly conserved, but differs somewhat from that of other dehydrogenases such as the short-chain dehydrogenases. The X-ray crystal structure of native sheep 6-phosphogluconate dehydrogenase was recently reported (Adams et al., 1994). Many of the amino acid residues determined to be important for substrate binding and catalysis from the crystallographic data are well conserved in all of the present sequences, including specific lysine and asparagine residues proposed to be of importance in the catalytic mechanism of 6-phosphogluconate dehydrogenase. We have used lysine-specific chemical modification and site-directed mutagenesis to address the possibility that HIBADH and 6-phosphogluconate dehydrogenase may share a common mechanism as β-hydroxyacid dehydrogenases. The present study suggests that 6-phosphogluconate dehydrogenase and HIBADH may indeed share a common evolutionary origin and enzymatic mechanism.

2. AMINO ACID SEQUENCE HOMOLOGY

Previous comparisons of the amino acid sequence of rat HIBADH with those of other dehydrogenases suggested that HIBADH was most closely related to the short-chain dehydrogenases / reductases (SDRs) (Crabb et al., 1993; Hawes, et al., 1995). HIBADH, like the SDRs, has a conserved dinucleotide cofactor-binding domain present at its N-terminus. However, the amino acid sequence of HIBADH is approximately 50 residues longer than those of most of the SDRS, and lacks a highly conserved lysine residue which has been established as having an important role in catalysis by the SDRs (Jornvall et al., 1995). Further searching of amino acid sequences present in published databases has recently revealed a separate group of enzymes with homology to HIBADH (Hawes et al.,

Figure 1. Amino acid sequence comparison of HIBADH with D-phenylserine dehydrogenase, and 6-phosphogluconate dehydrogenase. All sequences were obtained from NCBI databases and alignments were performed using the NCBI BLAST network service. Residues conserved between HIBADH and at least 50% of the other sequences are shown with reversed text. PSDH (PS) indicates D-phenylserine dehydrogenase from *Pseudomonas syringae*. HIBADH (PA) and HIBADH (RN) indicate 3-hydroxyisobutyrate dehydrogenase from *Pseudomonas aeruginosa* and *Rattus norvegicus*, respectively. PGDH (EC), PGDH (SH), and PGDH (SC) indicate 6-phosphogluconate dehydrogenase from *Escherichia coli*, sheep, and *Saccharomyces cerevisiae*, respectively.

1996) which includes 6-phosphogluconate dehydrogenase (PGDH) and a specific D-phenylserine dehydrogenase. The HIBADHs of both rat and *Pseudomonas aeruginosa* display significant homology to PGDH throughout their entire amino acid sequences. Figure 1 shows an alignment of the HIBADH sequences with those of PGDH from sheep, *S. cerevisiae*, and *E. coli*. Also shown is the N-terminal sequence of the D-phenylserine dehydrogenase purified from *Pseudomonas syringae* (Packdibamrung et al., 1993). Pairwise alignments indicate approximately 21% identity between HIBADH and these three PGDHs. Eight hypothetical proteins encoded by bacterial genes were also found to display the same pattern of conserved residues found in HIBADH and PGDH (Figure 2). HIBADH shares somewhat higher homology, overall, with the bacterial hypothetical proteins as compared to the PGDHs, with pairwise alignments indicating a range from 30% to over 70% identity with different combinations of these sequences. However, the level of homology between the bacterial proteins alone suggests that these sequences most likely do not represent a single bacterial enzyme such as a bacterial HIBADH. Although these bacterial sequences represent unidentified proteins, one sequence (32782 of *Haemophilus influenzae*) was reported to be present on an operon specifically inducible by glyoxylate, and one of the enzymes common to the glyoxylate cycle of plants and bacteria, tartronate semialdehyde reductase, catalyzes a reaction very similar to that catalyzed by HIBADH (Hawes et al., 1996):

$$CH_2OHCH_2OHCOO^- + NAD^+ \xrightleftharpoons[]{NAD^+/NADH} CHOCH_2OHCOO^- + NADH + H^+$$

Many of the most highly conserved residues in all of these sequences are alanines and glycines which may indicate a common structural framework among these proteins. However, several hydrophobic and polar residues, including two lysines and one glutamate, are also completely conserved. The N-terminus comprises one of the most highly conserved regions in all of these enzymes and shows a strict consensus sequence, **(A/G)XXGL(A/G)XMGX$_5$NX$_4$G**. This sequence is typical of the dinucleotide cofactor-binding fold of many dehydrogenases, and this region of PGDH is known to comprise the cofactor-binding domain from X-ray crystallographic data (Adams et al., 1994). However, this sequence differs slightly from the consensus sequences found in other types of dehydrogenases. Several residues which have been implicated in substrate binding and catalysis in PGDH also appear to be strictly conserved in this entire family of enzymes. These residues include aspartate 33, valine 123, serine 124, glycine 125, glycine 126, lysine 173, and asparagine 177 (numbered according to rat HIBADH). In the published structures of sheep PGDH complexed with substrates and cofactors, the residue corresponding to D33 is involved in cofactor binding and provides specificity for NADPH. The residue corresponding to serine 124 of HIBADH is involved in binding the carboxylate moiety of the substrate, whereas the residues corresponding to lysine 173 and asparagine 177 provide specific hydrogen bonds to the β-hydroxyl moiety of the substrate. Like 6-phosphogluconate, HIBA is a β-hydroxy-carboxylic acid, and it may be expected that similar binding sites would be required in HIBADH as compared to PGDH.

3. SITE-DIRECTED MUTAGENESIS

One of the most crucial residues in the cofactor-binding domain of dehydrogenases in general is a charged residue located approximately 18–20 residues C-terminal to the

Figure 2. Amino acid sequence comparison of HIBADH and eight hypothetical bacterial proteins. All sequences were obtained from NCBI databases and alignments were performed using the NCBI BLAST network service. Residues concerved in 50% or more of the sequences are shown with reversed text. Hypothetical proteins are identified by Genbank accession numbers. EC, *Escherichia coli*; BS, *Bacillus subtilis*; HI, *Haemophilis influenzae*; MS, *Mycobacterium smegmatis*; SC, *Streptomyces coelicolor*; SS, *Synechocystis* sp.; RN, *Rattus norvegicus*.

Table 1. Kinetic parameters of wild-type and mutant HIBADH[a]

Enzyme	V_{max} units/mg	K_a μM	k_{cat}/K_a mM^{-1}s^{-1}	K_b μM	k_{cat}/K_b mM^{-1}s-1
NAD$^+$					
wild-type	11.0 ± 0.8	22 ± 6	625 ± 110	56 ± 7	240 ± 30
D33R	0.7 ± 0.1	1240 ± 110	0.70 ± .08	112 ± 9	7.9 ± 0.7
NADP$^+$					
wild-type	3.0 ± 0.4	960 ± 90	4 ± 1	104 ± 14	38 ± 5
D33R	2.0 ±0.2	240 ± 60	10 ± 2	110 ± 12	22 ± 3

[a]Values are mean ± S.D. for three determinations. K_a and K_b represent the K_m values for dinucleotide cofactor and substrate, respectively.

conserved GXGXXG motif which is preceded by several conserved hydrophobic residues. A number of investigators have established that an acidic residue such as aspartate at this position usually provides stringent specificity for NAD$^+$ whereas a basic residue such as arginine endows specificity for NADP$^+$. All of the published sequences for 6-phosphogluconate dehydrogenase, an NADP$^+$-specific enzyme, show an asparagine and an adjacent arginine at this position, and the crystal structure shows that this arginine is involved in binding NADP$^+$ (Adams et al., 1994). HIBADH has an aspartate residue at this position, and is specific for NAD$^+$. To unambiguously identify this N-terminal sequence as the cofactor-binding domain of HIBADH, aspartate 33 was substituted with arginine. This mutation produced a 56-fold increase in K_m (NAD$^+$) and a 900 fold decrease in k_{cat}/K_m (NAD$^+$), whereas K_m (NADP$^+$) actually decreased by 4-fold compared to wild-type (Table 1). Therefore the D33R mutation produced an enzyme with a grossly decreased catalytic efficiency with NAD$^+$, but with a reversal of the preference for NAD$^+$ vs. NADP$^+$.

In the published PGDH-substrate cocrystal structure two residues corresponding to lysine 173 and asparagine 177 in HIBADH provided hydrogen bonds to the β-hydroxyl group of the substrate. A mechanism was proposed whereby the asparagine residue acts as the donor in the hydrogen bond to the β-hydroxyl, and the β-hydroxyl acts as the donor in the hydrogen bond to the lysine residue. This lysine residue was, therefore, proposed to act as a base catalyst. To test this mechanistic proposal, lysine 173 of HIBADH was substituted with alanine, arginine, asparagine, and histidine. Asparagine 177 was substituted with glutamine. Each of the lysine 173 mutants was well expressed and purified without apparent proteolysis. However, each of these mutations produced inactive enzymes, even when assayed at very high protein concentrations, high concentrations of substrate, and various conditions of pH. Each of the lysine 173 mutant enzymes displayed circular dichroism spectra identical to that of wild-type enzyme, suggesting that no large changes in secondary structure or assembly of the enzymes resulted from these mutations. Thus lysine 173 appears to have a crucial role in catalysis by HIBADH. It is, therefore, of considerable interest that this residue is completely conserved in all of the presently reported sequences. Asparagine 177 of HIBADH is also highly conserved in the present sequences, differing only by the presence of a glutamine residue in several of the bacterial sequences (Figure 1 and Figure 2). Substitution of asparagine 177 of HIBADH with glutamine pro-

Table 2. Kinetic parameters of wild-type and mutant HIBADH[a]

Enzyme	V_{max} units/mg	K_a μM	K_b μM
wild-type	11.0 ± 0.8	22 ± 6	56 ± 7
C215A	-	-	-
C215D	-	-	-
C215S	1.8 ± 0.2	28 ± 5	55 ± 8
N177Q	0.40 ± 0.03	55 ± 6	890 ± 70

[a]Values are mean $\pm$ S.D. for three determinations. Dashes indicate no measurable activity or that the parameter could not be determined. K_a and K_b represent the Km values for dinucleotide factor and substrate, respectively.

duced an enzyme which retained catalytic activity, but displayed a 27-fold decrease in V_{max}, and a 16-fold increase in K_m (HIBA) (Table 2), but no change in K_m (NAD$^+$), consistent with the previously proposed catalytic role for this conserved residue involving a specific interaction with the β-hydroxyl group of the substrate.

4. CHEMICAL MODIFICATION

Previous studies utilizing chemical modification of HIBADH showed the enzyme to be sensitive to chemical modifiers of cysteine and tyrosine (Hawes et al., 1995). Inactivation by tyrosine specific modification was shown to be in large part due to modification of tyrosine 32, present near the N-terminal cofactor-binding domain (Hawes et al., 1995). HIBADH is readily inactivated by a number of cysteine modifiers such as iodoacetate, however, the identity of any specific residue leading to this inactivation was not known before the use of site-directed mutagenesis. In order to determine if any specific cysteine residues play important structural or catalytic roles in this enzyme, each of the six cysteine residues were substituted with alanine residues. Each of the cysteine sulfhydryl groups were found to be dispensable, with the exception of cysteine 215 (Hawes et al., 1995). Alanine and aspartate substitutions of this residue each produced enzymes with no detectable catalytic activity, but with CD spectra identical to that of wild-type enzyme (Table 2; Hawes et al., 1995). However, a serine substitution at position 215 led to an active enzyme with a five-fold decrease in V_{max}, but no change in the K_m values for HIBA or NAD$^+$ (Table 2). Furthermore, the C215S mutant HIBADH was completely insensitive to treatment with iodoacetate, identifying this residue as responsible for inactivation by this reagent (Hawes et al., 1995). Although the exact role of cysteine 215 in catalysis by HIBADH is not known, it is perhaps important that this residue is highly conserved as a serine residue in the presently reported hypothetical, bacterial proteins (Figure 2), and could possibly play a similar functional role in these unidentified enzymes. We recently reported that recombinant rat HIBADH is also readily inactivated by treatment with pyridoxal phosphate, a lysine-specific chemical modification reagent. In the presence of 5 mM pyridoxal phosphate and 2.5 mM $NaBH_3CN$, at 25 °C, inactivation occurs with a first order rate constant of 0.05 min^{-1}, and is irreversible upon extended dialysis suggesting the formation of a stable Schiff base with a lysine residue. Treatment with pyridoxal phosphate in the presence and absence of cofactor and substrates showed significant protection against inactivation in the presence of saturating concentrations of either R- or S-HIBA, both of which are active substrates, but virtually no protection in the presence of NAD^+. In light of this finding it is interesting to note that the crystal structures of 6-phosphogluconate dehydrogenase in the presence and absence of bound cofactor indicated a rigid structure with no change in conformation upon cofactor binding. The present data are consistent with the presence of a lysine residue, such as lysine 173, in the active site of rat HIBADH which may be involved in substrate binding although other lysine residues may be modified and could possibly contribute to inactivation of the enzyme.

5. SUBSTRATE SPECIFICITY

β-Hydroxyisobutyrate, 6-phosphogluconate, and phenylserine have in common only the β-hydroxyacid functionality, suggesting that this may be a common feature of the substrate specificity of the presently proposed family of enzymes. Previously, rabbit and rat HIBADH were reported to be specific for S-HIBA and R-HIBA although with approximately 20-fold increased K_m for the R enantiomer. No activity is reported with other short-chain or branched-chain substrates such as lactate, 4-hydroxybutyrate, malate, or malonate. Further analysis of the substrate specificity of rat HIBADH indicated that the enzyme is active with a series of β-hydroxyacid substrates including S- and R-HIBA, L-glycerate, and L-serine. The V_{max} with most of these substrates was approximately the same, however, the K_m increased with increasing polarity of the substituent on the α-carbon (Table 3). The enzyme was inactive with similar compounds containing β-hydroxy-carboxylic acid functionalities but with carbon backbones of four or five carbons in length (Table 3). Therefore the optimal substrate for this enzyme can be defined as an L or S-β-hydroxyacid with a three carbon backbone and a nonpolar substituent such as a methyl group on the α-carbon. Substrate specificity studies of the bacterial D-phenylserine dehy-

Table 3. β-Hydroxyacid substrate specificity of HIBADH[a]

Substrate	K_m	V_{max} (units/mg)	k_{cat}/K_m (mM^{-1}s^{-1})
S-HIBA	59 ± 6 μM	10.3 ± 0.9	215 ± 22
R-HIBA	1.5 ± 0.1 mM	0.9 ± 0.1	0.89 ± 0.06
L-glycerate	9.8 ± 0.8 mM	8.0 ± 1.0	1.06 ± 0.09
L-serine	21 ± 1 mM	10.5 ± 1.2	0.64 ± 0.10
S-3-hydroxybutyrate	–	–	–
L-mevalonate	–	–	–

[a]Values are mean ± S.D. for three determinations. Dashes indicate no measurable activity or that the parameter could not be determined.

drogenase also indicated that only β-hydroxyacids (in this case, D-β-hydroxyaminoacids) could serve as active substrates for this enzyme (Packdibamrung et al., 1993). Thus, HIBADH, PGDH, and phenylserine dehydrogenase appear to be similar in chemistry, as specific β-hydroxyacid dehydrogenases, as well as in primary structure.

6. CONCLUSIONS

HIBADH shares a limited degree of amino acid sequence homology with the short-chain alcohol dehydrogenases but differs in mechanism from this family of enzymes. A previously unrecognized family of enzymes exists which includes HIBADH, 6-phospho-gluconate dehydrogenase, and D-phenylserine dehydrogenase. These enzymes each cata-lyze the oxidation of specific β-hydroxy-carboxylic acid substrates and, therefore, are similar in chemistry as well as primary structure. Site-directed mutagenesis and chemical modification studies indicate that the reaction catalyzed by HIBADH may be mechanisti-cally similar to that catalyzed by PGDH. Although eight of the sequences now identified as members of this enzyme family remain as unidentified open reading frames, we pro-pose that these homologous proteins represent a distinct family of enzymes most likely consisting of β-hydroxyacid dehydrogenases. This family of enzymes appears to be most prevalent in bacterial genomes, however, further research will be required to determine if they play specific and important roles in bacterial metabolism.

7. ACKNOWLEDGMENTS

The authors wish to acknowledge Dr. Leodis Davis (Department of Chemistry, The University of Iowa) for suggesting the use of L-serine as a β-hydroxyacid substrate. This work was supported in part by grants from the US Public Health Services (NIH DK 40441), the Grace M. Showalter Residuary Trust, and a postdoctoral fellowship from the American Heart Association, Indiana Affiliate (to J.W.H.).

8. REFERENCES

Adams, M.J., Ellis, G.H., Gover, S., Naylor, C.E., and Phillips, C.: Crystallographic study of coenzyme, coenzyme analogue, and substrate binding in 6-phosphogluconate dehydrogenase: implications for NADP specificity and the enzyme mechanism. Structure. 2 (1994) 651–668.

Crabb, D.W., Kedishvili, N.Y., Popov, K.M., Rougraff, P.M., Zhao, Y., and Harris, R.A.: Evolutionary relationships of branched-chain and non-specific alcohol and aldehyde dehydrogenases. In Weiner, H., Crabb, D.W., and Flynn, T.G. (Eds), Enzymology and Molecular Biology of Carbonyl Metabolism. Plenum Press, N. Y., 1993, pp. 523–531.

Hawes, J.W.H., Crabb, D.W., Chan, R.M., Rougraff, P.M., and Harris, R.A.: Chemical modification and site-di-rected mutagenesis studies of rat 3-hydroxyisobutyrate dehydrogenase. Biochemistry 34 (1995) 4231–4237.

Hawes, J.W.H., Harper, E.T., Crabb, D.W., and Harris, R.A.: Structural and mechanistic similarities of 6-phospho-gluconate and 3-hydroxyisobutyrate dehydrogenases reveal a new enzyme family, the 3-hydroxyacid dehy-drogenases. FEBS Lett. 3899 (1996) 263–267.

Jornvall, H., Persson, B., Krook, M., Atrian, S., Gonzalez-Duarte, R., Jeffery, J., and Ghosh, D.: Short-chain dehy-drogenases / reductases (SDR). Biochemistry 34 (1995) 6003–6013.

Letto, J., Brosnan, M.E., and Brosnan, J.T.: Valine metabolism. Gluconeogenesis from 3-hydroxyisobutyrate, Bio-chem. J. 240 (1986) 909–912.

Packdibamrung, K., Misono, H., Harada, M., Nagata, S., and Nagasaki, S.: An inducible NADP-dependent D-phenylserine dehydrogenase from Pseudomonas syringae NK-15: purification and biochemical charac-terization. J. Biochem. 114 (1993) 930–935.

STRUCTURE–FUNCTION RELATIONSHIPS OF SDR HYDROXYSTEROID DEHYDROGENASES[*]

Udo C. T. Oppermann, Bengt Persson, Charlotta Filling, and Hans Jörnvall

Department of Medical Biochemistry and Biophysics
Karolinska Institutet
S-171 77 Stockholm, Sweden

1. INTRODUCTION

1.1. The SDR Superfamily

Since its discovery and subsequent establishment as a superfamily, rapid progress on the knowledge of short chain dehydrogenases/reductases (SDR) has been achieved (Jörnvall *et al.*, 1981; Persson *et al.*, 1991; Jörnvall *et al.*, 1995). Based on conserved sequence characteristics, over 100 different enzymes in the databases now belong to the SDR family. The conserved residues are restricted to certain sequence segments and include the coenzyme binding site and the catalytic center (Krook *et al.*, 1990; Persson *et al.*, 1991; Krozowski, 1994; Ghosh *et al.*, 1994; Jörnvall *et al.*, 1995). Despite a residue identity level of 20–30% between different SDR members, the 3D structures thus far analyzed (Ghosh *et al.*, 1991, 1994; Varughese *et al.*, 1992; Ghosh *et al.*, 1995; Tanaka *et al.*, 1996a,b; Benach *et al.*, 1996) reveal a highly similar architecture with a one-domain α/β folding pattern. In particular, most of the conserved residues are found at positions 10–40 (comprising strands βA and βB, helix αB and the joining turns), 80–90 (strand βD), 110 (in helix αE), 130–180 (strands βE and βF, and helix αF) and 183–184 (at the end of strand βF) in the $3\beta/17\beta$-hydroxysteroid dehydrogenase numbering system. Table 1 lists these segments with residues conserved in more than 80% of all SDR structures and relates them to the secondary structure elements. Sequence comparisons, chemical modifications, site-directed mutageneses and crystallographic analyses reveal that most of these parts form the coenzyme binding and catalytic sites of SDR proteins, thus establishing the secondary structure elements βA to αD as parts of the coenzyme binding site with a typical "Rossmann fold" and a highly conserved Y-X-X-X-K segment (residues 150–154, helix αF) as part of the catalytic center (Persson *et al.*, 1991; Ghosh *et al.*, 1994; Jörnvall *et al.*, 1995).

[*] *Comamonas testosteroni* is used instead of *Pseudomonas testosteroni* throughout the text (Marcus & Talalay, 1956; Tamaoka *et al.*, 1987).

Enzymology and Molecular Biology of Carbonyl Metabolism 6
edited by Weiner *et al.* Plenum Press, New York, 1996

Table 1. Conserved sequence elements of SDR proteins

Sequence	Position	Secondary structure
TGxxxGxG	12–19	turn between βA and αB
NNAG	85–88	end of βD
N	110	αE
GxxxxxxS	130–137	βE
YxxxK	150–154	αF
PG	183–184	turn after αF

The highly similar overall fold and the conserved Tyr-X-X-X-Lys motif found in close to all SDR structures thus far determined, suggest a similar reaction mechanism for the individual proteins. The enzymatic reaction follows an ordered Bi-Bi mechanism with binding of the coenzyme occuring first, followed by the substrate, enzymatic reaction and leaving of products, with the reacted coenzyme leaving last and being the rate limiting step in the overall reaction (Schultz *et al.*, 1977; de Pouplana and Fothergill-Gilmore, 1994).

A reaction mechanism has been proposed on the basis of these findings with the conserved Tyr acting as a catalytic base, facilitated by the adjacent protonated Lys residue, which lowers the pK_a of the phenolic Tyr-OH group (McKinley-McKee *et al.*, 1991; Ghosh *et al.*, 1994; Jörnvall *et al.*, 1995). However, the exact roles of the active site residues are not yet clarified, especially not concerning the function of the conserved Ser137, which is located in the vicinity of the conserved Tyr and Lys residues, indicating an essential role in catalysis for Ser137 in SDR proteins.

1.2. Hydroxysteroid Dehydrogenases (HSDs)

Thus far, hydroxysteroid dehydrogenases (HSDs) belong to either the SDR family containing the majority of the available primary structures of these enzymes or to the monomeric aldo–keto reductase family (Bohren *et al.*, 1989). HSDs play key roles in the cellular action of steroid hormones and other compounds both in prokaryotes and eukaryotes.

1.2.1. Role of HSDs in Vertebrates. In higher vertebrates, HSDs fulfill several roles in cellular physiology. First, they are involved at three crucial points in the synthesis of all classes of vertebrate steroid hormones, achieved by the 3β-HSD/Δ4–5 isomerase, the 17β-HSD and 11β-HSD gene families, all belonging to the SDR superfamily. The importance of these genes is highlighted by the occurrence of several inborn errors of steroid synthesis and metabolism, manifested in adreno-genital syndromes as pseudohermaphroditism or vascular dysregulation like the "apparent mineralocorticoid excess syndrome", AME (New and White, 1995). Furthermore, the local activation of steroid hormone precursors, a process termed "intracrinology" (Labrie, 1991) requires the involvement of HSDs, thus establishing these enzymes as prime targets for pharmacotherapy in prostate and breast cancer treatment (Labrie, 1991). Regulation of intracellular levels of "active" steroid hormones or "protection" of steroid receptors by HSDs is an accepted concept, at least in the action of 11β-HSD type II in mineralocorticoid tissues such as the renal distal tubules and other epithelial tissues where transcellular electrolyte transport takes place (Edwards *et al.*, 1996).

Second, the role of HSDs in bile acid metabolism and hepatocellular transport and binding has yet not been clearly defined, but appears to be important. This hepatocellular binding of bile acids is effected by several HSD forms of the monomeric aldo-keto reductase superfamily (Stolz et al., 1995). However, the intestinal conversion to the corresponding iso-bile acids seems to be achieved by microbial HSDs, some of which belong to the SDR family as in the case of 3β/17β–HSD from Comamonas testosteroni.

Third, the ability of several HSDs to metabolize xenobiotic carbonyl compounds points to a general involvement of HSDs in the phase I metabolism of exogenous substances including drugs, pesticides and carcinogens (Felsted and Bachur, 1980; Maser, 1995).

1.2.2. Role of HSDs in Prokaryotes and Lower eukaryotes. The occurrence of several HSDs in bacteria and fungi is early documented (Marcus and Talalay, 1956). Some of these enzyme activities appear to be constitutive, some are steroid-inducible. The role of HSDs in steroid mediated actions in these organisms is far from clear. It can be anticipated that they are involved in the catabolism of exogenous steroid compounds, which at least in some cases can be the sole source of carbon (Marcus and Talalay, 1956). Steroid induction with the concomitant increase in HSD activities can also be the reason for the enhanced resistance toward certain steroid antibiotics and the elevated and different metabolism of insecticides in Gram-negative bacteria such as Comamonas testosteroni (Oppermann et al., 1996a). In analogy to the mammalian HSDs, the bacterial enzymes can reduce xenobiotic carbonyl compounds, including drugs, pesticides and carcinogens (Oppermann and Maser, 1996). The occurrence of some of these bacteria in the intestine of mammals, suggests also a role for bacterial HSDs in the human metabolism of foreign compounds (Barbaro et al., 1987; Oppermann et al., 1996a; Oppermann and Maser, 1996). Some of the characterized HSDs can act as quinone reductases (cf. Lanisnik-Rizner, this volume) and therefore might be involved in the metabolism of cytotoxic semiquinone compounds (Storz et al., 1990).

2. SCOPE OF THE STUDY

Our current research interests focus on structure-function relationships of SDR-hydroxysteroid dehydrogenases. In order to understand the general role of highly conserved amino acid residues lining the catalytic cavity in SDR proteins (Ghosh et al., 1994; Jörnvall et al., 1995) we exchanged these residues by site-directed mutagenesis and investigated the corresponding effects on the enzymatic activity, protein conformation and stability (Oppermann et al., 1996b). For this purpose we chose 3β/17β-hydroxysteroid dehydrogenase from the Gram-negative bacterium Comamonas testosteroni (Yin et al., 1991) as a model system. The successful crystallization (Benach et al., 1996) and structure determination of this protein will further promote the interpretation of the mutagenesis studies. Finally, we compare and relate structural and topological features of the major mammalian hydroxysteroid dehydrogenases, i.e. 3β-HSD/Δ4–5 isomerase, 11β-HSDs and 17β-HSDs.

3. EXPERIMENTAL

3.1. Molecular Cloning and Mutagenetic Replacements

Molecular cloning of 3β/17β-HSD (EC 1.1.1.51) from Comamonas testosteroni ATCC 11996 was carried out by generation of a PCR product from genomic DNA, using

Vent DNA polymerase with sequence specific primers encompassing the complete gene (Abalain *et al.*, 1993) and containing compatible restriction sites for further cloning. The resulting PCR product was cloned into the NdeI/BamHI restriction sites of the pET 29a vector (Novagen), containing an F′ site, allowing single-stranded DNA rescue by superinfection with helper phage R 408. These constructs were used for mutagenetic replacements. Site directed mutagenesis of the (+) strand was carried out based on the phosphorothioate technique. Mutated inserts were cloned into the NdeI/BamHI restriction sites of the pET 15b vector, resulting in the overexpressed recombinant protein with an N-terminal His-tag sequence that could be cleaved by thrombin.

3.2. Overexpression and Purification of Wild-Type and Mutant 3β/17β-HSD Proteins

Recombinant proteins were overexpressed in BL21 DE3 cells grown in LB medium by addition of 1 mM IPTG. After 2.5 h, the cells were harvested and lysed using a French Press or sonication. Overexpressed proteins containing the N-terminal His-tag were further purified by metal-chelate chromatography on a His-Bind resin. After buffer exchange by gelfiltration, the proteins were cleaved by thrombin thus releasing the His-tag sequence. Cleavage of the His-tag and efficient purification was confirmed by amino acid analysis after separation on 10% SDS/PAGE using a tricine-based buffer system. To exclude the possibility that the proteins were uncorrectly folded or expressed, CD spectroscopy and Western Blot were performed. Judged from these experiments no difference in folding, size or immunogenic reactivity between the wild-type and the mutant proteins could be observed.

3.3. Analysis of Enzyme Activity and Determination of Kinetic Constants

Enzyme activities were measured as NAD(H) dependent 3β-oxidoreductase activities from the absorbance at 340 nm. Confirmation of steroid products was achieved by thin-layer chromatography on silica plates after extraction in diethylether. Reactions were performed in 1.0 ml at 25°C under the following conditions. Dehydrogenase reactions: 20 mM Tris/Cl, pH 8.5, 250 μM NAD^+, varying the amount of steroid. Reductase reactions: 20 mM Tris/Cl, pH 7.0, 100 μM NADH, varying the amount of steroid. Steroids used as substrates for the activity measurements were 3β,12α-dihydroxy-5β-cholanoic acid (3β-HSD), 5α-dihydrotestosterone (3-oxo-reductase). Cleavage of the His-tag did not alter the enzymatic properties compared to those with the uncleaved recombinant protein.

3.4. Prediction of Transmembrane Segments

Prediction of membrane spanning segments was performed by use of the TMAP program (Persson and Argos, 1994; available via http://www.embl-heidelberg.de/tmap/tmap_info.html) using multiple sequence alignments.

4. RESULTS AND DISCUSSION

4.1. Role of Conserved Residues in SDR Hydroxysteroid Dehydrogenases

To determine the role of highly conserved residues found in SDR-HSDs we performed site-directed mutagenetic replacements of the residues and positions indicated in

Table 2. Mutagenetic replacements in 3β/17β-HSD

Position/Mutation	3β-HSD		3-oxo-reductase	
	K_m	rel. k_{cat}/K_m	K_m	rel. k_{cat}/K_m
Wild-type	22.3	100%	16.1	100%
12 Thr → Ala	nd	0%*	15.8	90%
12 Thr → Ser	28.3	97.2%	13.9	78%
16 Ser → Ala	27.6	88.3%	23.8	90%
86 Asn →Ala	28.8	16.5%*	18.6	120%
137 Ser→ Ala	nd	<0.04%*	nd	0 %*
137 Ser→ Thr	18.8	126%	14.2	78.0%

Table 2 in 3β/17β-HSD from *Comamonas testosteroni* and analyzed the influence of these mutations on protein conformation and enzymatic activity.

Depending on the amino acid substitution, several effects on enzymatic constants were observed. A replacement of Ser137 with Ala results in a close to complete loss of all enzyme activities (<0.04% residual activity compared to wild-type activities). However, substitution of Ser137 with Thr yields a mutant enzyme, which is fully active. Enzyme constants with this mutant protein are not significantly different from those with the wild-type enzyme, indicating that a side-chain hydroxyl at position 137 is essential for catalytic activity. The data obtained strongly support the role of Ser137 as part of a catalytically important "triad" (Ghosh *et al.*, 1994) of residues involving Ser137, Tyr150 and Lys154. The importance of hydrogen bonds at this position is clearly shown by the Ala and Thr exchanges. Together with recent crystallographic findings (Tanaka *et al.*, 1996a,b), the role of Ser137 can now be interpreted as to form hydrogen bonds to the carbonyl substrate in order to stabilize and/or polarize the substrate group for subsequent proton and hydrid transfer from the conserved Tyr-OH and the coenzyme, respectively. The deduced catalytic mechanism of SDR proteins thus far known is presented in Figure 1.

The mutations carried out with Thr12 and Asn86 reveal differential effects on the reaction directions in the SDR protein. A Thr→Ala exchange results in a complete loss of the dehydrogenase reaction, whereas the reduction direction remains unaffected. A Thr→Ser substitution yields a protein with indistinguishable properties from the wild-type, showing the importance of the hydroxyl side-chain at position 12 to form hydrogen bonds. A similar effect is observed by an Asn86→Ala replacement. The dehydrogenase activity is decreased by >80%, again with unchanged reductase properties. A determina-

Figure 1. Putative reaction mechanism of SDR steroid dehydrogenases. The interactions of the three conserved residues Ser137, Tyr150 and Lys154 with the steroid substrate and the coenzyme NADH are shown. The dotted arrow from Lys154 indicates the NH_3^+ influence in lowering the pK_a of Tyr150.

Figure 2. Detail of the three-dimensional structure of 3β/20β-hydroxysteroid dehydrogenase (Ghosh *et al.*, 1991, 1994) showing the orientation of Thr12, Asn86 and NAD. The 2.0 Å hydrogen bond between Thr12 and Asn86 is shown by the dotted line. The picture was created using the program ICM (version 2.5, Molsoft LLC, Metuchen, NJ, USA 1996).

tion of K_m values with NAD$^+$ for the wild-type and the active mutants revealed no significant difference. Therefore, the binding of the reduced and oxidized forms of the coenzyme to the protein must be differentially influenced by introduced mutations. Our interpretation of the available 3α/20β-HSD structure from *Streptomyces hydrogenans* (Ghosh *et al.*, 1994) suggests that the important determinant in this crucial process is the ability of the Thr12 side-chain to form hydrogen bond interactions with the main-chain of Asn86 in strand βD, thus positioning this strand for correct coenzyme binding. This new structural aspect is illustrated in Figure 2. The Asn86 does not form direct interaction with the coenzyme, however, hydrogen bonds *via* a hydrated coenzyme are possible.

   In SDR enzymes, position 16 is mostly a basic or polar residue. In NAD(H) dependent enzymes like 3β/17β-HSD it can be replaced by a nonpolar residue like Ala without a major change in activity, indicating the non-essential nature of these residues for catalysis. In NADP(H) dependent enzymes, however, a basic residue at position 16 in combination with a basic residue at 38 confers the specificity for NADPH, by performing electrostatic interactions with the 2′phosphate group (Tanaka *et al.*, 1996a; Baker *et al.*, 1992; Chen *et al.*, 1991). In NAD(H) enzymes, the coenzyme specificity is obtained by an Asp residue at position 37 which performs electrostatic repulsion of the 2′phosphate group of NADPH. Table 3 summarizes the effects and functions of conserved residues in SDR proteins as determined in this and other studies.

Table 3. Role of conserved residues in SDR proteins

Position	Residue	Mode	Function	Reference
12	T, S	H-bond to backbone amide N86	coenzyme positioning	this study
13	G	βA/αB positioning	coenzyme binding	Chen et al., 1990
16	S, R, K	interaction with 2′phosphate	coenzyme specificity	Tanaka et al., 1996, this study
17	G	βA/αB positioning	coenzyme binding	Chen et al., 1990
19	G	βA/αB positioning	coenzyme binding	Chen et al., 1990
37	D	interaction with 2′phosphate	coenzyme specificity	Tanaka et al., 1996
38	R,K	interaction with 2′phosphate	coenzyme specificity	Tanaka et al., 1996
86	N	possible H-binding to coenzyme	coenzyme positioning	this study
110	N	unknown	possibly oligomerization	
137	S	H-bonds to coenzyme, Tyr-OH	catalytic triad	this study, Tanaka et al., 1996
150	Y	catalytic base	catalytic triad	Obeid and White, 1991; Ghosh et al., 1991
154	K	interactions with Tyr-OH, coenzyme, substrate	catalytic triad	Ensor and Tai, 1991, Ghosh et al., 1994,1995; Tanaka et al., 1996
183–184	PG	end of βF	loop closure mechanism	de Pouplana and Fothergill-Gilmore, 1994; Hara et al., this volume

4.2. Summary of the Catalytic Mechanism in SDR Proteins

The conserved sequence elements described above are obviously important to maintain the functions described in Table 3 with the catalytically important triad of residues – Ser137, Tyr150 and Lys154 – acting as the catalytic center (Figure 1; Krook et al., 1990; Ghosh et al., 1994; Jörnvall et al., 1995). Interestingly, the reaction mechanism of SDR proteins with catalytically active Ser, Tyr and Lys residues appears to be similar to the reaction mechanism, found in the monomeric aldo–keto reductase superfamily, where a conserved His residue fulfills the role of the SDR Ser residue (Barski et al., 1995, this volume). Notably, also partial similarities in molecular architecture between these two superfamilies possibly reflect basic patterns in common (Jörnvall et al., 1995). Binding and positioning of the coenzyme is effected by the "Rossmann fold" with secondary structure elements βA–αE outlined above. Specificity for NAD(H) or NADP(H) is mainly achieved through electrostatic interactions obtained by basic and acidic residues placed at crucial positions, pos. 16, 37, 38 (Tanaka et al., 1996). A dynamic behavior of the protein following coenzyme binding, similar to that found in lactate dehydrogenase and other MDR enzymes and involving a loop closure mechanism, seems to be possible for the SDR enzymes (de Pouplana and Fothergill-Gilmore, 1994; Krook et al., 1992; Hara et al., this volume) and might be achieved by a flexible loop, located in the C-terminal part of the molecule including the conserved PG motif (residues 183–184). Substrate specificity seems to be also determined by the C-termini of the enzymes and thus gives an explanation for these highly variable parts of the respective sequences. Among the sequence elements described above only the TGxxxGxG (pos. 12–19), the GxxxxxxS (pos. 130–137) and the YxxxK (pos.150–154) motifs are strictly conserved among the hydroxysteroid dehydrogenases from pro- and eukaryotic sources, which are compared in Figure 3.

Figure 3. 11β-HSD2: human 11β-HSD type 2 (Krozowski *et al.*, 1995); 17β-HSD2: human 17β-HSD type 2 (Wu *et al.*, 1993); 3α/20β-HSD: 3α/20β-HSD from *Streptomyces* (Marekov *et al.*, 1990); 3β/17β-HSD: 3β/17β-HSD from *Comamonas* (Yin *et al.*, 1991); Ke6: possible HSD involved in polycystic kidney disease (Aziz *et al.*, 1994); NucHSD: nuclear HSD from hepatoblastoma (Gabrielli *et al.*, 1995); 11β-HSD1: human 11β-HSD type 1A (Tannin *et al.*, 1991). Conserved residues (3 or more out of 7) are marked black on grey, completely conserved are marked white on black. Horizontal lines above the alignment indicate the highly conserved sequence motifs in the SDR family, arrows indicate the positions of site-directed mutagenesis performed in this study, diamonds mark the active-site Y/K residues.

The complete NNAG motif is not found in the Ke6, NucHSD and 11β-HSD type 1 proteins, although similar residues might maintain the assumed function of correct coenzyme positioning. A highly conserved Asn residue at position 110 in helix αE seems to have a structural role, since this helix is involved in oligomerization of subunits. This Asn residue produces a slight kink in the helix as judged from the available crystallographic data, the function of this twist still remains obscure. The PG motif is not found in 3β/17β-HSD and 11β-HSD1, but at least one residue of this motif remains in the sequences. A conserved Thr residue three residues downchain is obvious and might have a special role in HSDs. Because of its distant relationship, the 3β-HSD/Δ4–5 isomerase sequence was omitted from the alignment, but the N-terminal TGxGxxxG and the active site YxxxK motifs are well conserved also in this distantly related SDR subgroup.

4.3. Membrane Insertion, Posttranslational Modifications and Topological Signals in Vertebrate Hydroxysteroid Dehydrogenases

In addition to the general features described above, several other factors influence the activity and intracellular distribution of HSDs. The majority of the main mammalian SDR-HSDs, namely the different 3β-HSD/Δ4–5 isomerases, 17β-HSDs and 11β-HSDs appear to be located in or associated with either endoplasmatic reticulum (ER) membranes or specialized vesicles.

4.3.1. 3β-HSD/Δ4–5 isomerase. The characterized forms (I–V) of the 3β-HSD/Δ4–5 isomerase gene family are located in the microsomal fractions of the cells and appear to be anchored by a C-terminal transmembrane segment. The bulk of the protein extends to the cytoplasmic side of the ER (Alvarez *et al.*, 1994).

4.3.2. 11β-HSD. The membrane environment of the ER seems to be of critical importance for 11β-HSD type 1 (Obeid *et al.*, 1993), yet it is unknown whether membrane insertion is required for protein folding or whether the lipid interactions are necessary for activity. Furthermore, posttranslational events like glycosylation also influence the biological activity. Mutations of N-linked glycosylation sites, inhibition of glycosylation and truncation of potential membrane spanning domains (11β-HSD 1B, Obeid *et al.*, 1993) result in altered or inactive enzymes as in the case of 11β-HSD1 (Obeid *et al.*, 1993; Agarwal *et al.*, 1995). However, these events are not completely understood and currently under further investigation. The 11β-HSD 1A form, which contains the N-terminal transmembrane segment, might be located in the lumen of the ER (Ozols, 1995), contrary to the type 2 isoform which contains cytoplasmic localization signals (Krozowski *et al.*, 1995). Two possible N-terminal segments are identified by use of the TMAP transmembrane prediction program. Separate isoforms derived from differential splicing and transcription start of the 11β-HSD type 1 were detected (Obeid *et al.*, 1993; Yang *et al.*, 1995) but thus far, no enzymatic activity of these forms could be proven. Different type 1A isoforms are detected but not further characterized in rabbit liver (Ozols, 1995). A signal peptide cleavage signal is present in the transmembrane segment of the 1A form but not further processed (Krozowski, 1992; Oppermann *et al.*, 1995).

4.3.3. 17β-HSD. The 17β-HSD isoforms of the SDR type appear to be localized in different cellular compartments. Type 1 is a cytosolic enzyme, types 2 and 3 are ER bound and type 4 is covalently linked to actin and associated with the peroxisomal fraction (Peltoketo *et al.*, 1988; Wu *et al.*, 1993; Andersson *et al.*, 1995; Markus *et al.*, 1995). Type 2

Table 4. Topological factors of mammalian hydroxysteroid dehydrogenases. + and − denote presence and absence, respectively, of transmembrane segment. (+) for 17β-HSD type 2 indicates that the second transmembrane segment is putative

HSD	Class	Transmembrane segment	Location	Signals/Remarks	Reference
3β-HSD	I-V	+	C-terminal, ER	Cytosolic orientation	Simard *et al.*, 1995; Mason, 1993; Alvarez *et al.*, 1994
11β-HSD	1A	+	N-terminal, ER	N-glycosylation, signal peptide cleavage?	Krozowski, 1992
	1B	−	·	No membrane segment, inactive enzyme	Obeid *et al.*, 1993
	1C	?	?	Differential splicing product, missing exon 5	Yang *et al.*, 1995
	2	+, (+)	N-terminal, ER	N-glycosylation, cytosolic orientation signal	Krozowski *et al.*, 1995
17β-HSD	1	−	·	Cytosolic localization	Peltoketo *et al.*, 1988
	2	+	N-terminal, ER	N-glycosylation, ER retention signal	Wu *et al.*, 1993
	3	+	N-terminal, ER	N-glycosylation	Andersson *et al.*, 1995
	4	−	peroxisomes	Covalent linkage to actin, peroxisome signal	Markus *et al.*, 1995

might contain an additional, second N-terminal transmembrane segment, indicated by parentheses in Table 4. Corresponding subcellular localization signals can be found in the type 2 and 4 isoforms.

4.3.4. Other Possible HSDs. Included in the sequence alignment of Figure 3 are two mammalian proteins, which possibly are hydroxysteroid dehydrogenases but such a function has still to be proven. The Ke6 protein is a gene product identified in the pathogenesis of polycystic kidney disease, PKD (Aziz *et al.*, 1994) and displays significant similarities to SDR-HSDs. Steroid hormones might play a crucial role in the development of PKD, therefore a HSD could play a central role in this disease. NucHSD abbreviates for an SDR protein, localized to the nucleus and accumulating upon growth arrest in human HepG2 hepatoblastoma cells (Gabrielli *et al.*, 1995). According to the criteria raised by Tanaka *et al.* (1996), this SDR protein will be a NADP(H) dependent enzyme. It contains a bipartite signal like the p53 protein or the thyroid and estrogen receptors (Robbins *et al.*, 1991). This motif might be responsible for the nuclear localization of this protein.

5. CONCLUSIONS

The enzymes of the SDR family represent a wide variety of substrate specificities. Several of these enzymes are involved in steroid metabolism and degradation of toxic compounds. Amino acid sequence comparisons have revealed conserved residues, and the importance of these residues has been shown in site-directed mutagenesis experiments.

Further mutagenesis studies, combined with heterologous expression of particular HSD forms, will allow the determination of other factors important for biological activity of this class of SDR proteins, which play central roles in steroid hormone synthesis, steroid hormone-dependent cancer forms, regulation of blood pressure and phase I metabolism of hormones, mediators and xenobiotics including carcinogens.

6. ACKNOWLEDGMENTS

This study was supported by grants from the Swedish Medical Research Council (projects 13X-3532 and 13X-11210). U.O. is a recipient of a research scholarship from the DFG (Deutsche Forschungsgemeinschaft).

7. REFERENCES

Abalain, J.H., Di Stefano, S., Amet, Y., Quemener, E., Abalain-Colloc, M.L. and Floch, H.H.: Cloning, DNA sequencing and expression of (3–17)(β-hydroxysteroid dehydrogenase from *Pseudomonas testosteroni*. J. Steroid Biochem. Mol. Biol. 44 (1993) 133–139.

Agarwal, A.K., Mune, T., Monder, C. and White, P.C.: Mutations in putative glycosylation sites of rat 11β-hydroxysteroid dehydrogenase affect enzymatic activity. Biochem. Biophys. Acta (1995) 1248, 70–74.

Alvarez, C.I., Genti-Raimondi, S., Patrito, L.C. and Flury, A.: Topography of human placental 3b-hydroxysteroid dehydrogenase/D5–4 isomerase in microsomal membranes. Biochem. Biophys. Acta 1207 (1994) 102–108.

Andersson, S.: Molecular genetics of androgenic 17β-hydroxysteroid dehydrogenases. J. Steroid Biochem. Mol. Biol. 55 (1995) 533–534.

Aziz, N., Maxwell, M.M. and Brenner, B.M.: Coordinate regulation of 11β-HSD and Ke6 genes in cpk mouse: implications for steroid metabolic defect in PKD. Am. J. Physiol. 267 (1994) F791–F797.

Barbaro, D.J., Mackowiak, P.A:, Barth, S.S. and Southern, P.M.J.: *Pseudomonas testosteroni* infections. Rev. Infect. Dis. 9 (1987) 124–129.

Baker, M.E.: Sequence analysis of steroid and prostaglandin metabolizing enzymes: application to undertsanding catalysis. Steroids 59 (1994) 248–258.

Barski, O.A., Gabbay, K-H., Grimshaw, C.E. and Bohren, K.M.: Mechanism of human aldehyde reductase: characterization of the active site pocket. Biochemistry 34 (1995) 11264–11275.

Benach, J., Knapp, S., Oppermann, U.C.T., Hägglund, O., Jörnvall, H. and Ladenstein, R.: Crystallization and crystal packing of recombinant 3 (or 17) (β-hydroxysteroid dehydrogenase from *Comamonas testosteroni* ATCC 11996. Eur. J. Biochem. 236 (1996) 144–148.

Bohren, K., Bullock, B., Wermuth, B. and Gabbay, K.H.: The aldo-keto reductase superfamily. J. Biol. Chem. 264 (1989) 9547–9551.

Chen, Z., Lu, L., Shirley, M., Lee, W.R: and Chang, S.H.: Site-directed mutagenesis of Glycine 14 and two critical cysteinul residues in Drossophila alcohol dehydrogenase. Biochemistry 29 (1990) 1112–1118.

Chenevert, S.W., Fossett, N.G., Chang, S.H., Tsigelny, I., Baker, M.E. and Lee, W.R.: Amino acids important in enzyme activity and dimer stability for Drosophila alcohol dehydrogenase. Biochem. J. 308 (1995) 419–423.

de Pouplana, L.R. and Fothergill-Gilmore, L.A.: The active site architecture of a short-chain dehydrogenase defined by site-directed mutagenesis and structure modeling. Biochemistry (1994) 33, 7047–7055.

Edwards, C.R.W., Benediktsson, R., Lindsay, R.S. and Seckl, J.R.: 11β-Hydroxysteroid dehydrogenases: Key enzymes in determining tissue-specific glucocorticoid effects. Steroids (1996) 61, 263–269.

Ensor, M.E. and Tai, H.H.: Bacterial expression and site-directed mutagenesis of two critical residues (tyrosine 151 and lysine 155) of human placental NAD dependent 15-hydroxyprostaglandin dehydrogenase. Biochem. Biophys. Acta (1994) 1208, 151–156.

Felsted, R.L. and Bachur, N.R.: Mammalian carbonyl reductase. Drug Metab. Rev. 11 (1980) 1–60.

Gabrielli, F., Donadel, G., Bensi, G., Heguy, A. and Melli, M.: A nuclear protein, synthesized in growth-arrested human hepatoblastoma cells, is a novel member of the short-chain alcohol dehydrogenase family. Eur. J. Biochem. 232 (1995) 473–477.

Ghosh, D., Weeks, C.M., Grochulski, P., Duax, W.L., Erman, M., Rimsay, R.L. and Orr, J.C.: Three-dimensional structure of holo 3α,20β-hydroxysteroid dehydrogenase: a member of a short-chain dehydrogenase family. Proc. Natl. Acad. Sci. USA. 88 (1991) 10064–10068.

Ghosh, D., Wawrzak, Z., Weeks, C.M., Duax, W.L. and Erman, M.: The refined three-dimensional structure of 3α/20β-hydroxysteroid dehydrogenase and possible roles of the residues conserved in short-chain dehydrogenase. Structure 2 (1994) 629–640.

Ghosh, D., Pletnev, V.Z., Zhu, D.W., Wawrzak, Z., Duax, W.L., Pangborn, W., Labrie, F. and Lin, S.X.: Structure of human estrogenic 17b-hydroxysteroid dehydrogenase at 2.20 Å resolution. Structure 3 (1995) 503–513.

Jörnvall, H., Persson, M. and Jeffery, J.: Alcohol and polyol dehydrogenases are both divided into two protein types and structural properties cross-relate the different enzyme activities within each type. Proc. Natl. Acad. Sci. USA, 78 (1981) 4226–4230.

Jörnvall, H., Persson, B., Krook, M., Atrian, S., Gonzalez-Duarte, R., Jeffery, J. and Ghosh, D.: Short-chain dehydrogenase/reductases (SDR). Biochemistry 34 (1995) 6003–6013.

Krook, M., Marekov, L. and Jörnvall, H.: Purification and structural characterization of placental NAD⁺-linked 15-hydroxyprostaglandin dehydrogenase. The primary structure reveals the enzyme to belong to the short-chain alcohol dehydrogenase family. Biochemistry 29 (1990) 738–743.

Krook, M., Prozorowski, V., Atrian, S., Gonzalez-Duarte, R. and Jörnvall, H.: Short-chain dehydrogenases. Eur. J. Biochem. 209 (1992) 233–239.

Krozowski, Z.: 11β-HSD and the short-chain alcohol dehydrogenase (SCAD) family. (1992) Mol. Cell. Endocrinol. 84, C25–C31.

Krozowski, Z.: The short-chain dehydrogenase superfamily. J.Steroid Biochem. Mol. Biol. 51 (1994) 125–130.

Krozowski, Z., Albiston, A.L., Obeyesekere, V.R., Andrews, R.K. and Smith, R.E.: The human 11β-hydroxysteroid dehydrogenase type II enzyme: Comparisons with other species and localization to the distal nephron. J. Steroid Biochem. Mol. Biol. 55 (1995) 457–464.

Labrie, F.: Intracrinology. Mol. Cell. Endocrinol. 78 (1991) C113–C118.

Marcus, P.I and Talalay, P.: Induction and purification of α- and β-hydroxysteroid dehydrogenases. J. Biol. Chem. 218 (1956) 661–674.

Marcus, M., Husen, B. and Adamski, J.: The subcellular localization of 17β-HSD type 4 and its interaction with actin. J. Steroid Biochem. Mol. Biol. 55 (1995) 617–621.

Marekov, L., Krook, M. and Jörnvall, H.: Prokaryotic 20β-hydroxysteroid dehydrogenase is an enzyme of the short-chain, non metalloenzyme alcohol dehydrogenase type. FEBS Lett. 266 (1990) 51–54.

Maser, E.: Xenobiotic carbonyl reduction and physiological steroid oxidoreduction. Biochem. Pharmacol. (1995) 49, 421–440.

Mason, J.I.: The 3β-Hydroxysteroid Dehydrogenase gene family of enzymes. Trends Endocrinol. Met. 4 (1993) 199–203.

McKinley-McKee, J.S., Winberg, J.-O. and Pettersson, G.: Mechanism of action of *Drosophila melanogaster* alcohol dehydrogenase. Biochem. Int. 25 (1991) 879–885.

New, M.I. and White, P.C.: Genetic disorders of steroid hormone synthesis and metabolism. Baillieres Clinical Endocrinol. Metabol. 9 (1995) 525–554.

Obeid, J. and White, P.C.: Tyr 179 and Lys 183 are essential for enzymatic activity of 11β-hydroxysteroid dehydrogenase. Biochem. Biophys. Res. Comm. 188 (1992) 222–227.

Obeid, J., Curnow, K.M., Aisenberg, J. and White, P.C.: Transcripts originating in intron 1 of the HSD11 (11β-hydroxysteroid dehydrogenase) gene encode a truncated polypeptide that is enzymatically inactive. Mol. Endocrinol. 7 (1993) 154–160.

Oppermann, U.C.T. and Maser, E.: Characterization of a 3α-hydroxysteroid dehydrogenase/carbonyl reductase from the Gram-negative *bacterium Comamonas testosteroni*. Eur. J. Biochem. (1996), submitted.

Oppermann, U.C.T., Netter, K.J. and Maser, E.: Cloning and primary structure of murine 11β-hydroxysteroid dehydrogenase/microsomal carbonyl reductase. Eur. J. Biochem. 227 (1995) 202–208.

Oppermann, U.C.T., Belai, I. and Maser, E.: Antibiotic resistance and enhanced insecticide catabolism as consequences of steroid induction in the Gram-negative bacterium *Comamonas testosteroni*. J. Steroid Biochem. Mol. Biol. (1996a), in press.

Oppermann, U.C.T., Filling, C., Berndt, K.D., Persson, B., Benach, J., Ladenstein, R. and Jörnvall, H.: Active site directed mutagenesis of 3β/17β-hydroxysteroid dehydrogenase establishes differential effects on SDR reactions. Biochemistry (1996b), submitted.

Ozols, J.: Lumenal orientation and post-translational modifications of the liver microsomal 11β-hydroxysteroid dehydrogenase. J. Biol. Chem. 270 (1995) 2305–2312.

Peltoketo, H., Isomaa, V., Mäentausta, O. and Vihko, R.: Complete amino acid sequence of human placental 17β-hydroxysteroid dehydrogenase deduced from cDNA. FEBS Lett.239 (1988) 73–77.

Persson, B., Krook, M. and Jörnvall, H.: Characteristics of short-chain alcohol dehydrogenases and related enzymes. (1991) Eur. J. Biochem. 200, 537–543.

Persson, B. and Argos, P.: Prediction of transmembrane segments in proteins utilising multiple sequence alignments. J. Mol. Biol. 237 (1994) 182–192.

Robbins, J., Dilworth, S.M., Laskey, R.A: and Dingwall, C.: Two interdependent basic domains in nucleoplasmin nuclear targeting sequence: identification of a class of bipartite nuclear targeting sequence. Cell 64 (1991) 615–623.

Schultz, R.M., Groman, E.V. and Engel, L.L.: 3(17)β-Hydroxysteroid dehydrogenase of *Pseudomonas testosteroni*. J. Biol. Chem. 252 (1977) 3784–3790.

Simard, J., Sanchez, R:, Durocher, F., Rheaume, E., Turgeon, C., Labrie, Y., Luu-The, V., Mebarki, F., Morel, Y., de Launoit, Y. and Labrie, F.: Stucture-function relationships and molecular genetics of the 3β-hydroxysteroid dehydrogenase gene family. J. Steroid Biochem. Mol. Biol. 55 (1995) 489–505.

Stolz, A:, Hammond, L. and Lou, H.: Rat and human bile acid binders are members of the monomeric reductase gene family. In Weiner, H., Holmes, R.S. and Wermuth, B. (Eds), Enzymology and Molecular Biology of Carbonyl Metabolism 5, Plenum Press, New York, 1995, pp. 269–280.

Storz, G., Tartaglia, L.A., Farr, S.B. and Ames, B.N.: Bacterial defenses against oxidative stress. Trends Genet. 6 (1990) 363–368.

Tamaoka, J., Ha, D.M.and Komagata, K.: Reclassification *of Pseudomonas acidovorans* den Dooren de Jong 1926 and *Pseudomonas testosteroni* Marcus and Talalay 1956 as *Comamonas acidovorans* comb.nov. and *Comamonas testosteroni* comb. nov., with an amended description of the genus Comamonas. (1987) Int. J. Syst. Bacteriol. 37, 52–59.

Tanaka, N., Nonaka, T., Nakanishi, M., Deyashiki, Y., Hara, A. and Mitsui, Y.: Crystal structure of the ternary complex of mouse lung carbonyl reductase at 1.8 Å resolution: the structural origin of coenzyme specificity in the short-chain dehydrogenase/reductase family. Structure 4 (1996a), 33–45.

Tanaka, N., Nonaka, T., Tanabe, T., Yoshimoto, T., Tsuru, D., Mitsui, Y.: Crystal structures of the binary and ternary complexes of 7α-hydroxysteroid dehydrogenase from *Escherichia coli*. Biochemistry 35 (1996b) 7715–7730.

Tannin, G.M., Agarwal, A.K., Monder, C., New, M.I. and White, P.C.: The human gene for 11β-hydroxysteroid dehydrogenase. J. Biol. Chem. 266 (1991) 16653–16658.

Varughese, K.I., Skinner, M.M., Whiteley, J.M., Matthews, D.A. and Xuong, N.H. : Crystal structure of rat liver dihydropteridine reductase. Proc. Natl. Acad. Sci. USA. 89 (1992) 6080–6084.

Wu, L., Einstein, M., Geissler, W.M., Chan, H.K., Elliston, K.O. and Andersson, S.: Expression cloning and characterization of human 17β-HSD type 2, a microsomal enzyme possessing 20α-HSD activity. J.Biol. Chem. 268 (1993) 12964–12969.

Yang, K., Yu, M. and Han, V.K.M.: Identification and tissue distribution of a novel variant of 11β-hydroxysteroid dehydrogenase 1 transcript. J. Steroid Biochem. Mol. Biol. 55 (1995) 247–253.

Yin, S.J., Vagelopoulos, N., Lundquist, G. and Jörnvall, H.: Pseudomonas 3β-hydroxysteroid dehydrogenase Eur. J. Biochem. 197 (1991) 359–365.

47

P. putida FORMALDEHYDE DEHYDROGENASE

An Alcohol Dehydrogenase Masquerading as an Aldehyde Dehydrogenase

Norman J. Oppenheimer,[1] Gary T. M. Henehan,[1] Jorge A. Huete-Pérez,[1] and Kiyoshi Ito[2]

[1]Department of Pharmaceutical Chemistry, S-926
University of California, San Francisco
San Francisco, California 94143
[2]School of Pharmaceutical Sciences
Nagasaki University
1–14 Bunkyo-machi
Nagasaki 852, Japan

1. INTRODUCTION

The *P. putida* formaldehyde dehydrogenase (P-FDH) [EC 1.2.1.46] (Ogushi, *et al.*, 1984a) meets the operational criteria of an aldehyde dehydrogenase. When P-FDH is incubated with NAD^+ and either formaldehyde or acetaldehyde it catalyzes the production of NADH as monitored by an increase in A_{340}, whereas no increase in A_{340} is observed with propanal or longer chain aldehydes; thus, it is specific for short chain aldehydes. Furthermore, no decrease in A_{340} is observed when aldehydes and NADH are incubated with P-FDH, nor is there any NADH generated with short chain primary alcohols and NAD^+ (Ogushi, *et al.*, 1984b; Ito, *et al.*, 1994). P-FDH does show a weak, but measurable ability to dehydrogenate long chain primary alcohols, but only at high pH (Ogushi, *et al.*, 1984b). While P-FDH nominally functions as an aldehyde dehydrogenase, it is a class III alcohol dehydrogenase (Kaiser, *et al.*, 1988; Holmquist & Vallee, 1991; Danielsson, *et al.*, 1994; Estonius, *et al.*, 1994) based on sequence analysis, with an overall 26% sequence identity (42% similarity) to the human χ-ADH (Ito, *et al.*, 1994), a class of ADHs that shows sequence similarities to the human class I ADHs (Danielsson, *et al.*, 1994).

Recently we demonstrated that the reported ability of *Drosophila* alcohol dehydrogenase (D-ADH) to oxidize aldehydes consisted of a small burst of NADH that was a function of its catalysis of a rapid dismutation of aldehydes (Henehan, *et al.*, 1995). At the time we speculated that if an alcohol dehydrogenase could catalyze a sufficiently large burst, then it could be mistaken for an aldehyde dehydrogenase. We report herein that P-

FDH is such an enzyme. Rather than being a short chain aldehyde dehydrogenase, we find P-FDH to be a functional alcohol dehydrogenase that conducts the efficient dismutation of a wide range of aldehydes where NADH production represents a pH-dependent burst.

2. MATERIALS AND METHODS

NAD$^+$ was obtained from the Sigma Chemical Company and aldehyde substrates were obtained from Aldrich Chemical Company. Aldehydes were distilled under a stream of oxygen-free nitrogen, diluted 1:10 with double distilled water, and stored frozen in aliquots at -20 °C until used. Buffers were of the highest grade commercially available. P-FDH was isolated according to the procedure of Ito, *et al.*, 1994 and a stock solution was prepared in 0.1 M sodium phosphate buffer pH 7.5 before use. NAD$^+$ was standardized by absorbance measurement at 260 nm using an extinction coefficient of 1.8×10^4 l mol^{-1} cm^{-1} in distilled water. P-FDH concentration was standardized using a spectrophotometric assay conducted at 37 °C in 0.1 M sodium phosphate buffer, pH 7.5 containing 2 mM NAD$^+$ and 5 mM acetaldehyde. Specific activities were also calibrated using the original assay conditions which contain glycine and found to be identical to those previously reported (Ito, *et al.*, 1994).

Proton nuclear magnetic resonance data were acquired on a General Electric GN-500 instrument operating at 500 MHz using a probe temperature of 37°C and 5 mm NMR tubes. Each time point consisted of 64 scans acquired with a spectral width of ±3000 Hz, using 16K data points. The pulse width was set to correspond to a 45° tip angle and the time between acquisitions provided equilibrium intensities of the resonances. The reactions were carried out in aqueous buffers with 10% D$_2$O to provide a lock signal. Suppression of the water resonance was achieved by presaturation with a decoupler pulse. Spectrophotometric assays were conducted on a Hewlett-Packard 8542A spectrophotometer with diode array detection using 1 cm pathlength cells in a thermostatted cell holder.

Aldehyde oxidation was monitored spectrophotometrically and by NMR. A typical reaction mixture contained 0.1 M sodium phosphate buffer pH 7.5, 2.0 mM NAD$^+$, 10% D$_2$O, and 9 mM aldehyde for a total volume of 2 mL. For NMR spectra, 480 µL of the assay mix, without P-FDH, was incubated for at least 10 min to achieve temperature equilibration before a control spectrum was taken. The reaction was initiated by addition of P-FDH to give a final volume of 500 µL. Spectra were taken every 8.05 min. The resonances monitored to determine changes in concentration of assay components were as follows: ethanol was monitored by measuring the disappearance of its methyl group, a 3H triplet at 1.17 ppm.; acetaldehyde was monitored by measuring the appearance of the aldehydic proton, a 1H quartet at 9.67 ppm; acetate was monitored by following the appearance of its methyl group at 1.90 ppm.; butanol was monitored by following the disappearance of the methylene protons, a 2H triplet at 3.65 ppm; butanal was monitored by following the appearance of the aldehydic proton, a 1H quartet at 9.71 ppm; and butyrate was monitored by following the appearance of the α-CH$_2$ group at 2.15 ppm. The integrated areas of these resonances were converted to concentrations by comparison with the integrated areas of the combined adenine ring protons of NAD$^+$ and NADH whose concentration was determined spectrophotometrically. The reported aldehyde concentrations represent the sum of free and hydrated species (Henehan & Oppenheimer, 1993; Henehan, *et al.*, 1995).

For spectrophotometer assays, 960 µL of the corresponding assay mix was incubated in a 1 mL cuvette for ten minutes to establish a base line then 40 µL of P-FDH was added

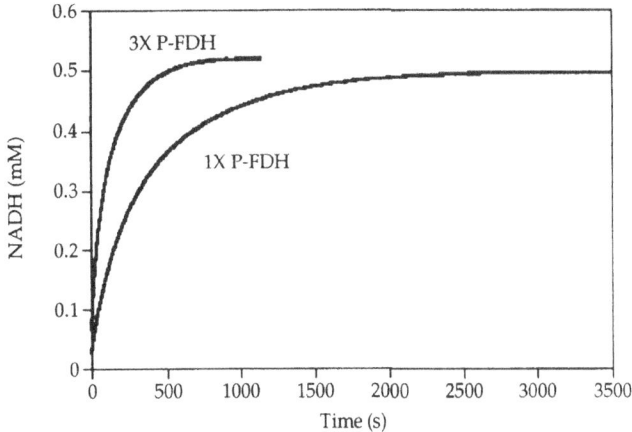

Figure 1. Enzyme-dependent increase in NADH from incubation with two concentrations of P-FDH. Note that while there is a linear dependence on the concentration of enzyme for the rate of increase in NADH, the limiting concentration of NADH remains unchanged. Assay conditions, 0.1 M sodium phosphate buffer, pH 7.5 with 2 mM NAD^+ and 9 mM acetaldehyde.

and the reaction monitored at 340 nm. For acetaldehyde, where the A_{340} quickly exceeded 2 OD, the reaction was also monitored at 370, 384 and 500 nm and the total concentration of NADH was calculated from the extinction coefficients at these other wavelengths using the absorbance at 500 nm as an internal blank. Spectrophotometric assays were also conducted at high pH using 0.1 M sodium pyrophosphate buffer at pH 8.6 and 9.5 (no NMR spectra were obtained for the reaction at higher pH).

3. RESULTS

Incubation of NAD^+ and acetaldehyde with P-FDH at neutral pH leads to a rapid increase in A_{340} and kinetic values based on measurements of initial rates are consistent with previous studies (Ogushi, *et al.*, 1984b; Ito, *et al.*, 1994). If the full reaction progress is monitored at 340 nm, however, rather than just an initial rate measurement, the increase in A_{340} plateaus before all the NAD^+ has been converted to NADH; i.e., as shown in Figure 1 NADH production is a burst. Product inhibition can be excluded as a cause of the burst because addition of acetate does not alter the final concentration of NADH. Likewise enzyme inactivation can be dismissed because P-FDH retains full activity throughout the course of the assay (note, 1X and 3X concentrations of P-FDH yield the same final concentration of NADH). Furthermore, propanal, an aldehyde that does not give any increase in A_{340} at neutral pH, shows a pH-dependent A_{340} burst above pH 7.5 (Fig. 2). Similarly we find that the magnitude of the burst of A_{340} in the oxidation of acetaldehyde increases with increasing pH (Fig. 3). No increase in A_{340}, even at high pH, is observed with butanal and longer chain aldehydes, indicating that the magnitude of the burst for these substrates remains too small to be observed at any achievable pH.

Given the similarity of the spectrophotometric progress curve to that previously observed for D-ADH (Henehan, *et al.*, 1995) we monitored the reaction by ^{1}H NMR in parallel experiments, incubating P-FDH with NAD^+ and acetaldehyde or butanal. Plots of the

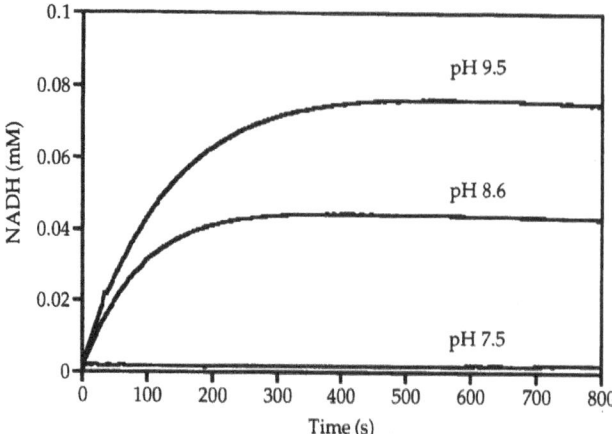

Figure 2. Dependence of the burst in NADH production on pH for the reaction of propanal catalyzed by P-FDH. At pH 7.5 there is no significant increase in A_{340}, but above that pH, a pH-dependent burst is observed. Assay conditions, 0.1 M sodium phosphate buffer, pH 7.5 with 2 mM NAD^+ and 9 mM propanal and 0.1 M sodium pyrophosphate buffer, pH 8.6 and pH 9.5 with 2 mM NAD^+ and 9 mM propanal.

time-dependent changes in concentrations of the reaction components for acetaldehyde and butanal in an assay mix at neutral pH are shown in Figures 4 and 5. As can be seen with both substrates, there is a rapid decrease in aldehyde with a concomitant increase in acid and alcohol. Therefore even though no absorption at 340 nm is observed in the spectrophotometric assay with butanal, the ${}^1H$ NMR assay shows that P-FDH catalyzes its rapid dismutation. Note that while the dismutation of butanal results in equimolar concentrations of butanol and butyrate (within the limits of detection), the dismutation of acetaldehyde leads to more acetate than ethanol by approximately 0.5 mM, an amount corresponding to the concentration of NADH measured spectrophotometrically at that pH (Fig. 4).

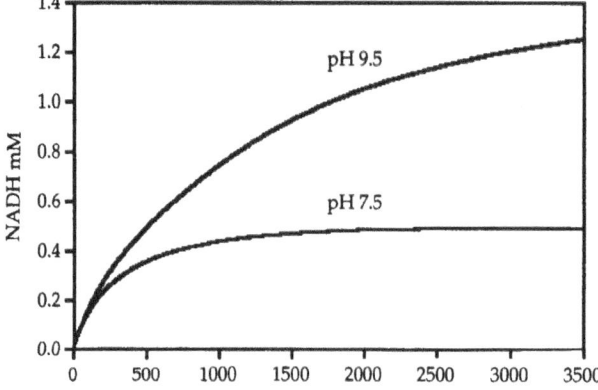

Figure 3. The pH-dependence of the burst in NADH production for the reaction of acetaldehyde catalyzed by P-FDH. Assay conditions, 0.1 M sodium phosphate buffer, pH 7.5 with 2 mM NAD^+ and 9 mM acetaldehyde and 0.1 M sodium pyrophosphate buffer, pH 9.5 with 2 mM NAD^+ and 9 mM acetaldehyde.

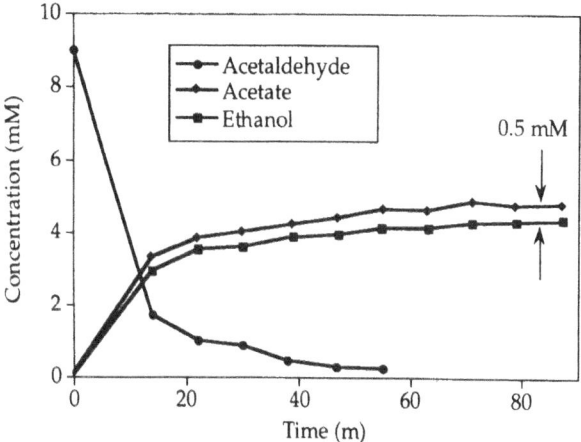

Figure 4. P-FDH-catalyzed dismutation of acetaldehyde monitored by ¹H NMR to a mixture of ethanol and acetate. Acetate is in excess of ethanol by approximately 0.5 mM, identical to the net production of NADH observed spectrophotometrically. Assay conditions, 0.1 M sodium phosphate buffer, pH 7.5 with 2 mM NAD⁺ and 9 mM acetaldehyde.

4. DISCUSSION

The results for P-FDH clearly parallel those we have reported for D-ADH (Henehan, *et al.*, 1995). As argued previously, the magnitude of the burst is a function of the concentration and dissociation constant of the E•NADH complex, while the rate of build-up of NADH is governed by a partitioning between NADH dissociation from E•NADH and aldehyde reduction by E•NADH. As a consequence, the increase in A_{340} is not a direct meas-

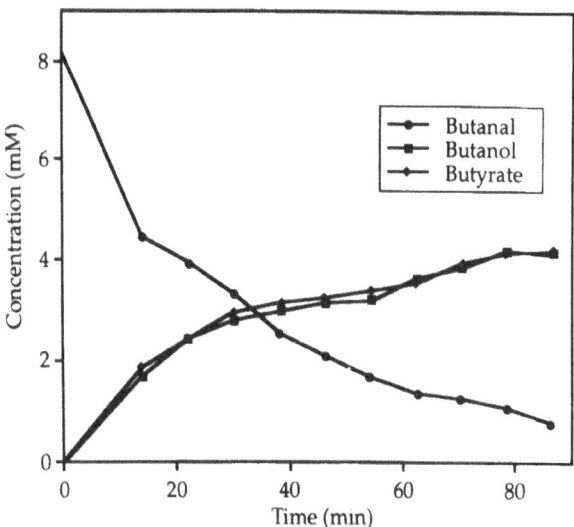

Figure 5. P-FDH-catalyzed dismutation of butanal monitored by ¹H NMR. No increase in A_{340} is observed and butanol and butyrate are equimolar to within the limits of detection. Assay conditions, 0.1 M sodium phosphate buffer, pH 7.5 with 2 mM NAD⁺ and 8 mM butanal.

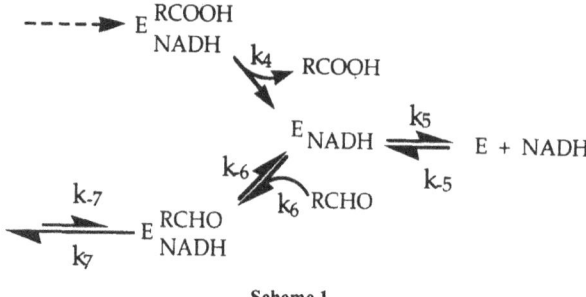

Scheme 1

ure of aldehyde oxidation and the kinetic values derived from the ΔA_{340} cannot be used for comparisons with kinetic measurements for other dehydrogenases.

The apparent specificity of P-FDH for short chain aldehydes based on the spectrophotometric assay is now readily explained. For short chain aldehydes, the rate of aldehyde reduction is slow; thus, some NADH dissociates to give the observed burst. As the chain length of the aldehyde increases, the rate of aldehyde reduction also increases, thus E•NADH is successfully intercepted by aldehyde and thus no NADH is released; i.e., the reaction becomes a pure dismutation. Increasing the pH for propanal slows aldehyde reduction enough to allow observation of a burst of NADH, but for longer aldehydes, pH alone is insufficient to slow aldehyde reduction. Therefore, far from being a short chain aldehyde dehydrogenase, P-FDH functions as a broad spectrum aldehyde dismutase.

In summary, we have demonstrated that aldehyde dismutation can mimic the properties of aldehyde dehydrogenation when the sole criteria are spectrophotometric assays based on changes in A_{340}. This raises the disturbing question of how many other "aldehyde" dehydrogenases are really alcohol dehydrogenases conducting the dismutation of aldehydes with a burst of NADH production and how many assays of "aldehyde dehydrogenase" activity in crude extracts include contributions from alcohol dehydrogenases? As we have discussed elsewhere (Henehan, *et al.*, 1995), the ability of many alcohol dehydrogenases to dismutate aldehydes requires a reinvestigation of their metabolic functions, now "aldehyde" dehydrogenases may also need to be scrutinized.

5. ACKNOWLEDGMENTS

This research was supported in part by PHS grant GM-52529. The UCSF Magnetic Resonance Laboratory is in part supported by grants from the National Science Foundation (DMB-8406826) and the National Institutes of Health (RR-01668).

6. REFERENCES

Danielsson, O., Shafqat, J., Estonius, M. & Jornvall, H. (1994). Alcohol dehydrogenase class III contrasted to class I. Characterization of the cyclostome enzyme, the existence of multiple forms as for the human enzyme, and distant cross-species hybridization. Eur. J. Biochem. 225, 1081–8.

Estonius, M., Hoog, J. O., Danielsson, O. & Jornvall, H. (1994). Residues specific for class III alcohol dehydrogenase. Site-directed mutagenesis of the human enzyme. Biochemistry 33, 15080–5.

Henehan, G. T. M., Chang, S. H. & Oppenheimer, N. J. (1995). Aldehyde dehydrogenase activity of *Drosophila melanogaster* alcohol dehydrogenase: Burst kinetics at high ph and aldehyde dismutase activity at physiological pH. Biochemistry 34, 12294–12301.

Henehan, G. T. M. & Oppenheimer, N. J. (1993). Horse liver alcohol dehydrogenase-catalyzed oxidation of aldehydes: Dismutation precedes net production of reduced nicotinamide adenine dinucleotide. Biochemistry 32, 735–738.

Holmquist, B. & Vallee, B. L. (1991). Human liver class III alcohol and glutathione dependent formaldehyde dehydrogenase are the same enzyme. Biochem. Biophys. Res. Commun. 178, 1371–1377.

Ito, K., Takahashi, M., Yoshimoto, T. & Tsuru, D. (1994). Cloning and high-level expression of the glutathione-independent formaldehyde dehydrogenase gene from *Pseudomonas putida*. J. Bacteriol. 176, 2483–2491.

Kaiser, R., Holmquist, B., Hempel, J., Vallee, B. L. & Jornvall, H. (1988). Class III human liver alcohol dehydrogenase: a novel structural type equidistantly related to the class I and class II enzymes. Biochemistry 27, 1132–1140.

Ogushi, S., Ando, M. and Tsuru, D. (1984a). Formaldehyde dehydrogenase from *Pseudomonas putida*: a zinc metalloenzyme. J. Biochem. (Tokyo) 96, 1587–1591.

Ogushi, S., Ando, M., and Tsuru, D. (1984b). Substrate specificity of formaldehyde dehydrogenase from *Pseudomonas putida*. Agric. Biol. Chem. 48, 597–601.

48

NICOTINOPROTEIN ALCOHOL/ALDEHYDE OXIDOREDUCTASES

Enzymes with Bound NAD(P) as Cofactor

Sander R. Piersma, Simon de Vries, and Johannis A. Duine

Department of Microbiology and Enzymology
Delft University of Technology
Julianalaan 67 2628 BC Delft, The Netherlands

1. INTRODUCTION

Enzymes with tightly bound NAD(P), acting as cofactor, have been described in the past. Well known examples include UDP-galactose 4-epimerase from *E. coli* (Wilson and Hogness, 1964) and lactate-oxaloacetate transhydrogenase from *V. alcalescens* (Allen, 1966). Recently enzymes have been discovered in which the bound NAD(P) acts as cofactor in the oxidation/reduction of alcohols/aldehydes and the reducing equivalents do not exchange with the cytosolic NAD(P) pool but are probably transferred to the respiratory chain (Arfman, 1991, Bystrykh, 1993, van Ophem, 1993). These enzymes operate in the same way as, e.g., flavoprotein or quinoprotein alcohol dehydrogenases and are clearly distinct from the NAD(P)(H) dependent alcohol/aldehyde oxidoreductases in a physiological sense. In the latter case nicotinamide pyridine dinucleotide acts as coenzyme (i.e. as co-substrate) and forms part of the cytosolic NAD(H) pool. We have introduced the name nicotinoproteins for this new group of NAD(P)(H) containing enzymes (van Ophem, 1993). Nicotinoproteins are proteins which contain NAD(P)(H) as tightly bound, redox active cofactor. The pyridine dinucleotide in these enzymes is strongly but not covalently bound and does not exchange with the cytosolic NAD(P) pool. In Table 1 enzymological data of some of the nicotinoproteins known presently is summarised.

The reaction types catalysed by this group of enzymes all involve two-electron oxidation-reduction steps. The overall reactions catalysed by nicotinoproteins like epimerisation, oxidoreduction, and dismutation are redox neutral (see Figure 1). In the epimerisation and dismutation reaction the enzymes only require one substrate for catalysis, both for oxidation and reduction of the cofactor. For the oxidoreduction or transhydrogenase reaction two substrates are needed for catalysis, yielding ping-pong kinetics (Allen, 1966, Bystrykh, 1993, Kato, 1986), which is in contrast with the ternary complex kinetics observed for NAD$^+$-dependent dehydrogenases. The enzyme-bound NAD(P)$^+$ co-

Enzymology and Molecular Biology of Carbonyl Metabolism 6
edited by Weiner *et al.* Plenum Press, New York, 1996

Table 1. Overview of enzymological data on nicotinoproteins

Enzyme	Source	Cofactor	MW*	Metals
UDP-galactose 4-epimerase[a]	*E. coli*	NAD	39 (2)	-
Glucose-fructose oxidoreductase[b]	*Z. mobilis*	NADP	40 (4)	-
Lactate-oxaloacetate transhydrogenase[c]	*V. alcalescens*	NAD(H)	35 (2)	-
Formaldehyde dismutase[d]	*P. putida F61*	NAD(H)	44 (4)	2 Zn^{2+}
Methanol:NDMA oxidoreductase[e]	*A. methanolica*	NADPH	50 (10)	Zn^{2+},Mg^{2+}
Alcohol:NDMA oxidoreductase[f]	*A. methanolica*	NADH	39 (3-4)	Zn^{2+}

*Monomeric molecular weight in kDa and in between brackets the multimer composition (2 indicating a dimer, etc).
[a](Vanhooke, 1994), [b](Zachariou, 1986), [c](Allen, 1966), [d](Kato, 1986), [e](Bystrykh, 1993), [f](van Ophem, 1993).

factor of the nicotinoprotein is reduced by one substrate to bound NAD(P)H and subsequently oxidised by the second substrate.

A number of spectroscopic techniques have been used to study the (reduced) enzyme-bound pyridine dinucleotide including absorbance, fluorescence, circular dichroism and ^{31}P nuclear magnetic resonance spectroscopy (Allen, 1966, Vanhooke, 1994, Wong, 1977, Wong, 1978). Also rapid kinetic measurements monitoring the oxidation and reduction of the enzyme-bound pyridine dinucleotide have been performed to gain insight in the oxidative and reductive half-reactions (Hardman, 1988, Hardman, 1992).

Figure 1. Different types of reactions catalysed by nicotinoproteins. 1: oxidoreduction by glucose fructose oxidoreductase; 2: formaldehyde dismutation by formaldehyde dismutase; 3: epimerisation by UDP-galactose 4-epimerase.

It is an interesting question to find out whether there are common structural features among the group of nicotinoproteins and if so, which of them are responsible for the tightly-bound pyridine dinucleotide cofactor. Until now only one high-resolution crystal structure of a nicotinoprotein is available, being that of UDP-galactose 4-epimerase (both in the E:NAD$^+$ and E:NADH form). Another crystal structure, that of catalase, is available in which a bound NADH was found but it is not clear whether this pyridine dinucleotide is actually involved in catalysis and therefore we do not -as yet- classify this enzyme as a nicotinoprotein.

In the first part of this paper we present an overview on the nicotinoproteins presently known (UDP-galactose 4-epimerase from *E. coli*, glucose-fructose oxidoreductase from *Z. mobilis*, lactate-oxaloacetate transhydrogenase from *V. alcalescens* and formaldehyde dismutase from *P. putida F61*) and in the second part we will focus on some of our recent results obtained with the nicotinoprotein alcohol/aldehyde oxidoreductases (methanol:NDMA oxidoreductase (MNO) and alcohol:NDMA oxidoreductase (NDMA-ADH)).

2. NICOTINOPROTEINS

2.1. Udp-galactose 4-epimerase

UDP-galactose 4-epimerase catalyses the conversion of UDP-galactose to UDP-glucose through a mechanism involving the transient reduction of enzyme-bound NAD$^+$. UDP-galactose binding causes a conformational change which activates the cofactor toward reduction. In general hydride transfer involving NAD$^+$ dependent dehydrogenases is highly stereospecific, this, however, is not the case with UDP-galactose 4-epimerase. The epimerase catalyses the nonstereospecific hydride transfer between enzyme-bound NAD$^+$ and the chiral substrate. The mechanism of catalysis involves two steps. First the alcohol moiety of the sugar is oxidised to a ketone and then the hydride is reintroduced to either side of the carbonyl group. From fluorescence and circular dichroism studies on the reduced enzyme it was concluded that the pyridine dinucleotide is fully immobilised in the active site (Wong, 1977, Wong, 1978). *In vivo* the reduced form of the enzyme is inactive in catalysis as expected from the catalytic mechanism (Fig. 1). Recently the high resolution 3D structures of both the reduced and the oxidised UDP-galactose 4-epimerase(1.8 Å) have been solved. Analysis shows that a greater number of protein-dinucleotide interactions, such as van der Waals and electrostatic interactions, being 35 in total for UDP-galactose 4-epimerase compared to typical NAD$^+$ dependent dehydrogenases like malate dehydrogenase (22), horse liver alcohol dehydrogenase (25) or glyceraldehyde-3-phosphate dehydrogenase (27) accounts for the non-dissociable nature of the enzyme-bound pyridine dinucleotide (Thoden, 1996). In the UDP-galactose 4-epimerase structure fewer of these contacts are contributed by ordered solvent molecules compared to glyeraldehyde-3-phosphate (7 vs. 15). The number of protein-cofactor interactions is higher throughout the whole pyridine dinucleotide cofactor but especially in the adenine part of the molecule. In horse liver ADH the adenine part has few specific interactions with the protein; in the epimerase, however, the adenine moiety is bound with 5 hydrogen bonds. Another striking feature of the epimerase is the conformation of the nicotinamide ring which is anti (with respect to the ribose) in the reduced enzyme and is syn in the oxidised enzyme. This change in conformation can give an explanation for the nonstereospecific hydride transfer by the epimerase. The enzyme is capable to accommodate the dinucleotide in two conformations equally strong.

2.2. Glucose Fructose Oxidoreductase (GFOR)

GFOR is the enzyme responsible for sorbitol production in the ethanologenic bacterium *Zymomonas mobilis*. The enzyme transfers reducing equivalents between the substrates glucose and fructose: the respective reaction products are gluconolactone and sorbitol. The physiological substrates have been identified to be glucose and fructose, although the enzyme has a low affinity for these compounds requiring upto 1 M fructose and 50–100 mM glucose for maximum activity (Hardman, 1988, Zachariou and Scopes, 1986). The kinetics of enzyme-bound cofactor oxidation and reduction with the respective substrates was studied extensively (Hardman, 1988, Hardman, 1992). Rate controlling steps for the overall reaction are probably dissociation of gluconolactone in the forward reaction and hydride transfer from sorbitol to enzyme-NADP$^+$ in the reverse reaction. The GFOR gene has been cloned and sequenced (Kanagasundaram, 1992): however, no significant homology with other enzymes was found from the primary sequence. The enzyme is localised in the periplasm of *Z. mobilis* and is produced in a precursor form which includes a leader sequence of 52 amino-acids and (Loos, 1991, Loos, 1993). GFOR is present in the organism up to 0.5% of the soluble protein.

2.3. Lactate-Oxaloacetate Transhydrogenase

The reaction catalysed by lactate-oxaloacetate transhydrogenase is comparable to the glucose-fructose oxidoreductase, i.e. two substrates are used *in vivo*, L-lactate and oxaloacetate. The enzyme catalyses the reversible oxidation of lactate to pyruvate and reduction of oxaloacetate to L-malate and does not require the further addition of coenzymes (Allen, 1982). *V. alcalescens* does not contain a lactate dehydrogenase; the transhydrogenation reaction is coupled with an NAD$^+$-dependent malate dehydrogenase. The organism uses the two enzymes to oxidise lactate in a two step process. The net reaction is L-lactate+NAD$^+$ → pyruvate+NADH+H$^+$. The cofactor of the transhydrogenase can only be removed with methods which destroy the protein matrix (perchloric acid, guanidine HCL, urea) (Allen, 1966). In the absence of substrates the enzyme-bound pyridine dinucleotide equilibrates to a mixture of 40 % reduced and 60% oxidised cofactor, irrespective of the starting ratio.

2.4. Formaldehyde Dismutase

The formaldehyde tolerant bacterium *Pseudomonas putida* F61 produces a formaldehyde dismutase (FDM), an enzyme which catalyses the dismutation of aldehydes (including formaldehyde) and which can also catalyse alcohol:aldehyde oxidoreduction in the absence of an electron acceptor. FDM can also oxidise alcohols with concomitant reduction of the artificial electron acceptor N,N-dimethyl-p-nitrosoaniline (NDMA) via a ping-pong mechanism. The K_m for alcohol oxidation (with NDMA) ranged from 40 mM for ethanol to 1.4 mM for 1-pentanol; the enzyme is inactive with methanol and hardly active (0.15% of ethanol activity) with benzyl alcohol (Kato, 1986). Because of similar sensitivity of both formaldehyde dismutase and alcohol dehydrogenase activities to pyrazole (a class I ADH specific inhibitor), it was concluded that both activities occurred at the same active site.

The whole organism as well as the purified enzyme are able to catalyse alcohol:aldehyde oxidoreduction reactions. A variety of aldehydes and ketones can be reduced to their respective primary and secondary alcohols with the concomitant oxidation of alcohol. Instead of an alcohol as reductant, also formaldehyde can be used. In this aldehyde:formaldehyde dismutation reaction aldehyde is reduced and formaldehyde is oxidised to formic acid. The FDM gene has been cloned and sequenced (Yanase, 1995). The primary sequence revealed the classical GXGXXG (G193–198) motif most likely involved in binding the pyridine dinucleotide. Other conserved NAD⁺-dependent alcohol dehydrogenase residues include the catalytic zinc binding domain (C45, C178, H66) and structural zinc binding domain (C96, 99, 102, 110). Moreover, the proton relay network residues S47 and H50 are also present. The overall sequence identity with horse liver ADH was, however, low (15%). The role of FDM *in vivo* is presumed to be part of the formaldehyde detoxification system of *P. putida* F61.

3. NICOTINOPROTEIN ALCOHOLDEHYDROGENASES

3.1 Methanol:NDMA Oxidoreductase

From the facultative methylotroph *A. methanolica* two distinct nicotinoprotein dehydrogenases were isolated. Both enzymes are reactive with the artificial electron acceptor NDMA and primary alcohols. The most abundant enzyme is the methanol:NDMA oxidoreductase (MNO) an enzyme which is proposed to be involved in the oxidation of methanol in a multienzyme complex upon growth on methanol as sole carbon source (Bystrykh, 1993) and personal communication L. Bystrykh). In the multienzyme complex, termed TD-ADH (tetrazolium dye-dependent alcohol dehydrogenase), MNO is the first enzyme in the oxidation of methanol by *A. methanolica*. Reducing equivalents are transported via other components of the complex, presumably to the respiratory chain. MNO can oxidise methanol (specific activity 57 mU/mg, K_m 2.7 mM with NDMA) as well as longer primary and secondary alcohols and it can also oxidise aldehydes displaying ping-pong kinetics. In addition to these activities it is an effective formaldehyde dismutase in the absence of NDMA with a specific activity of 85000 mU/mg and a K_m of 19 mM. The enzyme can also reduce aldehydes with NADH (formaldehyde: specific activity 14 mU/mg, K_m 0.7 mM) (Bystrykh, 1993). The enzyme is a homodecamer with five-fold symmetry, as observed with electron microscopy (Bystrykh, 1993). The same type of enzyme is also isolated from the other gram positive bacteria *R. erythropolis* (Nagy, 1995) and *M. gastri* (Bystrykh, 1993). The genes of *A. methanolica* MNO (personal communication H. Hektor) and *R. erythropolis* MNO (Nagy, 1995) have been cloned and sequenced. MNO has limited sequence homology with class III ADH. The *R. erythropolis* MNO has highest sequence identity with the methanol dehydrogenase from the thermotolerant bacterium *B. methanolicus* (37%), an enzyme related to the MNO types but capable of transferring reducing equivalents from methanol to the cytosolic NADH pool without losing its firmly-bound pyridine dinucleotide (being NADH, not NADPH as for MNO). The tightly bound pyridine dinucleotide in this enzyme is not exchanged with NAD(H) from solution, but during turnover the enzyme-bound nucleotide is oxidised by NAD⁺ from solution. Methanol dehydrogenase is also a homodecamer in solution and needs an activator protein in addition to Mg^{2+} ions for maximal activity (Arfman, 1991).

3.2. NDMA:Alcohol Oxidoreductase

The second nicotinoprotein ADH from *A. methanolica* is the alcohol:NDMA oxidoreductase (NDMA-ADH) (van Ophem, 1993). The enzyme does not react with exoge-

Table 2. Inhibitors of NDMA-ADH

Inhibitor	K_i (µM)
Pyrazole	1.6
Isobutyramide	33
Trifluoroethanol	1.2

nous NAD(P)(H). NDMA-ADH catalyses the oxidation of primary alcohols with the artificial electron acceptor NDMA. The enzyme is, however, not reactive with methanol and cannot oxidise or dismutate aldehydes. From N-terminal sequencing of the first 76 amino-acids (personal communication A. Norin and H. Jörnvall) a striking homology (identical residues) of 50% was found with horse liver ADH. Homology of microbial class I NAD⁺-dependent ADH's with horse liver ADH is normally less than 28% (Reid, 1994). The N-terminal part includes the tentative proton relay network residues H51 and S48 and the active site zinc ligands C46 and H67 (numbering horse liver ADH).

Class I ADH inhibitors like pyrazole, isobutyramide and trifluoroethanol are also potent inhibitors of NDMA-ADH (Table 2). Pyrazole forms a covalent adduct with NAD⁺ bound in the active site of an NAD⁺-dependent alcohol dehydrogenase. Pyrazole is also a competitive inhibitor of NDMA-ADH with respect to ethanol (van Ophem, 1993). Isobutyramide on the other hand is a compound which binds to class I ADH complexed with NADH. Isobutyramide is a competitive inhibitor of NDMA with respect to the electron acceptor NDMA.

NDMA-ADH can use NDMA as artificial electron acceptor (specific activity for ethanol 5900 mU/mg, K_m 82 µM). Acetaldehyde was first reported to act as inhibitor of the enzyme (van Ophem, 1993). Upon reinvestigation, however, acetaldehyde turned out to be a competing substrate in the reduction reaction (competing with NDMA). Later it became clear that NDMA-ADH is a true alcohol:aldehyde oxidoreductase, since the enzyme can reduce aldehydes with the concomitant oxidation of alcohols, in the absence of NDMA. Currently the reactions involving ethanol, acetaldehyde, benzyl alcohol, and benzaldehyde in various alcohol/aldehyde combinations are under investigation. Alcohol oxidation coupled to benzaldehyde reduction displays ping-pong kinetics and substrate inhibition by ethanol and product inhibition by benzyl alcohol are also observed. The apparent k_{cat}/K_m for benzaldehyde reduction (at fixed concentration of ethanol) is dependent on pH and can be described with a single protonatable group with a pKa of 8.5 (unpublished observations S. Piersma). No specific residue has been assigned to this ionisable group yet. Pre-steady-state experiments involving the oxidation of enzyme-bound NADH, were monitored with stopped-flow fluorescence spectroscopy. The kinetics observed are comparable to those observed for glucose-fructose oxidoreductase displaying a rapid equilibrium binding step followed by hydride transfer. This type of kinetics is characterised by a K_d for substrate binding, a k_{max} for forward hydride transfer and a K_{off} for the reverse hydride transfer (Hardman, 1992). The kinetic parameters for nucleotide oxidation by benzaldehyde at pH 7 are 149 µM for the K_d, 328 s⁻¹ for the k_{max} and 26 s⁻¹ for K_{off} (unpublished observations S. Piersma).

NAD⁺-dependent alcohol dehydrogenases are known for their high stereospecificity with respect to hydrogen transfer from the nicotinamide moiety and also with respect to the alcohol and aldehyde substrates (Loewus, 1953, Rétey, 1982). Usually the stereochemistry of alcohol oxidation and aldehyde reduction by dehydrogenases is established using isotopically labelled pyridine dinucleotides. This approach is, however, not applicable to nicotinoproteins since there is no exchange of NADH with bulk solution. The isotope label has to be introduced at the substrate level. Class I ADH transfers a hydrogen from the *Re* side of C4 of the nicotinamide ring (the A-side) to the *Re* side of substrate aldehyde

Table 3. Optical properties of NDMA-ADH

Spectroscopic method	λ Maximum (nm)
UV/vis Absorbance	325
Fluorescence, emission	422
Fluorescence, excitation	326
UV/vis Circular Dichroism	323

and for the reverse reaction from the *R* hydrogen of the alcohol to the *Re* side of nicotinamide ring C4. Inspired by the classical work of Vennesland, Westheimer and co-workers (Loewus, 1953) we used deuterated acetaldehyde (CD_3CDO), NADH and yeast ADH to couple the production of isotopically labelled ethanol (CD_3CH_RDOH), after separation, with the oxidation of this alcohol by NDMA-ADH. Reaction products were analysed by mass spectrometry. NDMA-ADH is specific (>95%) for the *R* hydrogen of ethanol and for the *Re* side of acetaldehyde (unpublished observations S. Piersma).

The reduced pyridine dinucleotide was studied with various optical techniques including steady-state and time-resolved fluorescence spectroscopy and absorbance and circular dichroism spectroscopy. From CD and fluorescence anisotropy decay experiments it can be concluded that the NADH is rigidly bound in the enzyme active site, as no mobility could be observed. Absorbance and fluorescence excitation and emission maxima (see Table 3) are blue-shifted compared to NADH bound to NAD^+-dependent dehydrogenases, but are comparable to those observed for horse liver ADH. The fluorescence emission maximum is even blue-shifted compared to that of horse liver ADH (430 nm). The blue-shift indicates that the NADH binding environment is more compact and hydrophobic compared to NAD^+-dependent dehydrogenases (Baumgarten, 1988, Fischer, 1988). Time-resolved fluorescence shows a very fast excited state fluorescence decay, much faster than observed for horse liver ADH (Gafni, 1976, Ladokhin, 1995). The active site environment affords little stabilisation to the more polar excited-state (Visser, 1987) of NADH in NDMA-ADH indicating slow dipole-dipole interactions. Inhibition by isobutyramide indicates that this aldehyde analogue binds to the active site of NDMA-ADH. Bound to NAD^+-dependent ADH's, isobutyramide enhances NADH fluorescence by forming a ternary complex. The same is observed for NDMA-ADH, isobutyramide causes a maximum fluorescence enhancement of a factor 3.

Until recently NDMA-ADH was found only in the organism *A. methanolica*. In our lab we have been able to isolate a NDMA-ADH type enzyme from *R. erythropolis*. This latter organism has already been shown to contain a MNO type nicotinoprotein upon induction by herbicides (Nagy, 1995). Also NDMA-dependent alcohol dehydrogenase activity in crude extracts of the organism had already been reported in literature (van Ophem, 1993). The NDMA-ADH enzyme from *R. erythropolis* is induced under different conditions than MNO (personal communication, P. Schenkels), this being the probable reason why the NDMA-ADH type was not discovered previously. The enzymological characteristics, including molecular mass, subunit composition and N-terminal sequence, are very similar to those from the *A. methanolica* enzyme. The physiological roles of these NDMA-ADH enzymes remains to be elucidated.

4. CONCLUSIONS

The small diverse group of nicotinoproteins that have pyridine dinucleotides as cofactor are all involved in oxidation/reduction reactions. The known physiological roles of these en-

Table 4. Redox states (%) and UV/vis absorbance maxima (nm) of pyridine dinucleotides bound to nicotinoproteins, as isolated and in the absence of substrates

Enzyme	NAD(P)$^+$	NAD(P)H	λ Maximum
UDP-galactose 4-epimerase	>95	<5	345
Glucose-fructose oxidoreductase	±85	±15	344
Lactate-oxaloacetate transhydrogenase	60	40	345
Formaldehyde dismutase	71	29	325
Methanol:NDMA oxidoreductase	±10	±90	330
Alcohol:NDMA oxidoreductase	<5	>95	325

zymes include detoxification (FDM, MNO) and metabolic conversions (epimerase, GFOR, transhydrogenase, and MNO). The extraction of the cofactor, even under mild conditions, often results in denaturation of the enzymes (Allen, 1966, Thoden, 1996). This finding indicates that the enzyme-bound dinucleotide is, in addition to catalysis, also important to maintain structural integrity. This may be the reason why so far no successful cofactor substitution experiments have been reported in literature. The redox state of the enzyme-bound pyridine dinucleotide that is stabilised in the different nicotinoproteins (see Table 4) encompasses the whole range, spanning from fully oxidised (UDP-galactose 4-epimerase) via mixtures of the oxidised and reduced form (FDM, lactate-oxaloacetate transhydrogenase) to the fully reduced form (NDMA-ADH). It appears that the redox state of tightly bound pyridine dinucleotide is tuned by the protein matrix to stabilise either one of the redox states, or a mixture of the two. It is an interesting question which type of binding interactions of the protein with the nicotinamide moiety govern which redox state of the bound pyridine dinucleotide is stabilised. Preferential stabilisation of NAD$^+$ in UDP-galactose 4-epimerasecan be rationalised from the catalytic mechanism (Fig 1). A ratio of 1 for oxidised and reduced cofactor seems optimal for dismutation, in fact this ratio is observed upon addition of formaldehyde to FDM. An interesting correlation can be made between the λ_{max} of the enzyme-bound NAD(P)H and the stabilised redox state. If λ_{max} is higher than 340 nm then the nicotinoprotein stabilises the oxidised cofactor and vice versa (although with FDM this is not the case). A blue-shift (from 340 nm) in NAD(P)H absorbance indicates a more hydrophobic active site environment, favouring binding of the NAD(P)H form over the NAD(P)$^+$ form (as observed for NDMA-ADH and MNO). No direct measurements of the redox potentials of the enzyme-bound pyridine dinucleotides of nicotinoproteins, however, have appeared in literature yet (to our knowledge).

From the 3-D structural information on UDP-galactose 4-epimerase and (limited) sequence information on FDM and NDMA-ADH it appears that the nucleotide binding motifs are comparable to those observed for NAD$^+$-dependent ADH's. The number of protein-cofactor interactions, however, is increased in the nicotinoproteins. The case is not so clear for the GFOR and MNO enzymes where limited sequence homology with other nicotinoproteins and ADH's is observed. However, more 3-D structures are required to establish the nature of the tight binding of pyridine dinucleotide by nicotinoproteins. If there are specific protein-cofactor interactions characteristic for nicotinoproteins, this structural information can be used to establish nicotinoproteins as a group, separate from NAD$^+$-dependent dehydrogenases.

REFERENCES

Allen, S. H.: The isolation and characterisation of malate-lactate transhydrogenase from *Micrococcus lactilyticus*. J. Biol. Chem. 241 (1966) 5266–5275.

Allen, S. H. G.: Lactate-oxaloacetate transhydrogenase from *Veillonella alcalescens*. Meth. Enzymol. 89 (1982) 367–376.

Arfman, N., Van Beumen, J., De Vries, G.E., Harder, W., Dijkhuizen, L.: Purification and characterization of an activator protein for methanol dehydrogenase from thermotolerant Bacillus spp. J. Biol. Chem. 266 (1991) 3955–3960.

Baumgarten, B., Nönes, J.: Spectroscopic investigation of dihydronicotinamides-II. dihydronicotinamide adenine dinucleotide complexes with dehydrogenases. Photochem. Photobiol. 47 (1988) 201–205.

Bystrykh, L. V., Vonck, J., Van Bruggen, E.F.J.,Van Beumen, J., Samyn, B., Govorkhina, N. I., Duine, J.A., Dijkhuizen, L.: Electron microscopic analysis and structural characterisation of novel NADP(H)-containing methanol:*N,N*-dimethyl-4-nitrosoaniline oxidoreductases from the gram positive methylotrophic bacteria *Amycolatopsis methanolica* and *Mycobacterium gastri* MB19. J. Bacteriol. 175 (1993) 1814–1822.

Bystrykh, L. V., Govorkhina, N. I., Van Ophem, P.W., Hektor, H.J., Dijkhuizen, L., Duine, J.A.: Formaldehyde dismutase activities in gram-positive bacteria oxidising methanol. J. Gen. Microbiol. 139 (1993) 1979–1983.

Fischer, P., Fleckenstein, J., Hönes, J.: Spectroscopic investigation of dihydronicotinamides-I: Conformation, absorption and fluorescence. Photochem. Photobiol. 47 (1988) 193–199.

Gafni, A., Brand, L.: Fluorescence decay studies of reduced nicotinamide adenine dinucleotide in solution and bound to liver alcohol dehydrogenase. Biochemistry. 15 (1976) 3165–3171.

Hardman, M. J., Scopes, R. K.: The kinetics of glucose-fructose oxidoreductase from *Zymomonas mobilis*. Eur. J. Biochem. 173 (1988) 203–209.

Hardman, M. J., Tsao, M.,Scopes,R. K.: Changes in the fluorescence of bound nucleotide during the reaction catalysed by glucose-fructose oxidoreductase from *Zymomonas mobilis*. Eur. J. Biochem. 205 (1992) 715–720.

Kanagasundaram, V., Scopes, R. K.: Cloning, sequencing and expression of the structural gene encoding glucose-fructose oxidoreductase from *Zymomonas mobilis*. J. Bact. 174 (1992) 1439–1447.

Kato, N., Yomagami, T., Shimao, M., Sakazawa, C.: Formaldehyde dismutase, a novel NAD-binding oxidoreductase from *Pseudomonas putida* F61. Eur. J. Biochem. 156 (1986) 59–64.

Ladokhin, A. S., Brand, L.: Evidence for an excited-state reaction contributing to NADH fluorescence. J. of Fluorescence. 5 (1995) 99–106.

Loewus, F. A., Westheimer, F.H., Vennesland, B.: Enzymatic synthesis of the enantiomorphs of ethanol-1-d. J. Am. Chem. Soc. 75 (1953) 5018–5023.

Loos, H., Völler, M., Rehr, B., Stierhof, Y. -D., Sahm, H., Sprenger, G. A.: Localisation of the glucose-fructose oxidoreductase in wild type and overproducing strains of *Zymomonas mobilis*. FEMS Lett. 84 (1991) 211–216.

Loos, H., Sahm, H., Sprenger, G. A.: Glucose-fructose oxidoreductase, a periplasmatic enzyme of *Zymomonas mobilis*, is active in its precursor form. FEMS Micr. Biol. Lett. 107 (1993) 293–298.

Nagy, I., Verheijen, S., De Schrijver, A., Van Damme, J., Proost, P., Schoofs, G., Vanderleyden, J., De Mot, R.: Characterization of the *Rhodococcus* sp NI86/21 gene encoding alcohol:*N,N*'-dimethyl-4-nitrosoaniline oxidoreductase inducible by atrazine and thiocarbamate herbicides. Arch. Microbiol. 163 (1995) 439–446.

Reid, M. F., Fewson, C. A.: Molecular characterisation of microbial alcohol dehydrogenases. Crit. Rev. Microbiol. 20 (1994) 13–56.

Rétey, J., Robinson, J.A.: Stereospecificity in organic chemistry and enzymology, Monographs in Modern Chemistry 13 (1982) ., Verlag Chemie, Weinheim.

Thoden, J. B., Frey, P. A., Holden, H. M.: Crystal structures of the oxidized and reduced forms of UDP-galactose 4-epimerase isolated from *Escherichia coli*. Biochemistry. 35 (1996) 2557–2566.

van Ophem, P. W., Duine, J.A.: Microbial alcohol, aldehyde and formate ester oxidoreductases, in H. Weiner, D. W. Crabb and T. G. Flynn, (Eds.), Enzymology and molecular biology of carbonyl metabolism 4 , Plenum press, New York, 1993, pp 605–620

van Ophem, P. W., van Beumen, J., Duine, J. A.: Nicotinoprotein (NAD(P)-containing) alcohol/aldehyde oxidoreductases. purification and characterisation of a novel type from *Amycolatopsis methanolica*. Eur. J. Biochem. 156 (1993) 819–826.

Vanhooke, J. L., Frey, P.: Characterisation and activation of naturally occurring abortive complexes of UDP-galactose 4-epimerase from *Escherichia coli*. J. Biol. Chem. 269 (1994) 31496–31504.

Visser, A. J. W. G.: Excited states of pyridine nucleotide coenzymes: fluorescence, and phosphorescence, in D. Dolphin, Poulson, R., Avramovic, O., (Eds.), Pyridine Nucleotide Coenzymes , John Wiley, New York, 1987, pp 163–183.

Wilson, D. B. and Hogness, D. S.: The enzymes of the galactose operon in *Escherichia coli*. I. Purification and characterisation of uridine diphosphogalactose 4-epimerase. J. Biol. Chem. 239 (1964) 2469–2481.

Wong, S. S., Frey, P.: Fluorescence and nucleotide binding properties of *Escherichia coli* uridine diphosphate galactose 4-epimerase: support for a model for nonstereospecific action. Biochemistry. 16 (1977) 298–305.

Wong, S. S., Cassim, J. Y., Frey, P. A.: *Escherichia coli* uridine diphosphate galactose 4-epimerase: circular dichroism of the protein and the protein bound dihidronicotinamide adenine dinucleotide. Biochemistry. 17 (1978) 516–520.

Yanase, H., Noda, H, Aoki, K., Kita, K., Kato, N.: Cloning, sequence analysis, and expression of the gene encoding formaldehyde dismutase from *Pseudomonas putida* F61. Biosci. Biotech. Biochem. 59 (1995) 197–202.

Zachariou, M. and Scopes, R. K.: Glucose-fructose oxidoreductase, a new enzyme isolated from *Zymomonas mobilis*, that is responsible for sorbitol production. J. Bacteriol. 167 (1986) 863–869.

49

STRUCTURAL STUDIES OF ALDO-KETO REDUCTASE INHIBITION

David K. Wilson,[1] Takayuki Nakano,[2] J. Mark Petrash,[2] and
Florante A. Quiocho[1]

[1]Howard Hughes Medical Institute and Baylor College of Medicine
One Baylor Plaza
Houston, Texas 77030
[2]Department of Ophthalmology and Visual Sciences
Washington University School of Medicine
660 South Euclid Avenue
St. Louis, Missouri 63110

1. INTRODUCTION

Aldose reductase (ALR2) has long been a target for drug design to combat complications which arise in diabetes (Dvornik *et al.*, 1973). The enzyme's ability to reduce in an NADPH-dependent manner, glucose in its carbonyl-containing, open chain form to sorbitol has been linked to a number of these complications affecting a number of tissues. A large number of ALR2 inhibitors have been developed (reviewed in Sarges & Oates, 1993) but most have not proven to be clinically effective. These disappointing results may be attributed to the inhibition of other members of the aldo-keto reductase family of proteins. This set of enzymes shares considerable sequence homology and consequently has overlapping substrate and inhibitor specificity.

In order to understand the mechanism and to gain insights into the inhibition of the aldo-keto reductases, crystallographic structures have been determined of ALR2 from pig (Rondeau *et al.*, 1992) and human (Borhani *et al.*,1992; Wilson *et al.*, 1992). Structures of other members of the family have also been determined including aldehyde reductase from human and pig (El-Kabbani *et al.*, 1994, 1995) and rat 3α-hydroxysteroid dehydrogenase (Hoog *et al.*, 1994). All of the structures to date show that these proteins fold into a $(\beta/\alpha)_8$ barrel which occurs in a large number of enzymes with unrelated sequences, functions and cofactors. Although the $(\beta/\alpha)_8$ fold appears to be the most common seen in protein structures (Farber & Petsko, 1990), ALR2 was the first NAD(P)H-dependent enzyme to adopt this motif. All of the structures of the holoenzyme have shown that the cofactor is bound in an extended conformation with the nicotinamide at the bottom of the active site which is centered in the middle of the barrel (Figure 1).

Figure 1. A stereo view of the C_α trace of the ALR2 holoenzyme. The NADPH cofactor (shown as a ball and stick model) extends from the active site in the middle of the barrel across to the right.

The chemical mechanism for the aldo-keto reductase family involves a hydride transfer to or from the carbonyl/alcoholic carbon in the reduction and oxidation reaction respectively. A proton is abstracted in either a concerted action or a following step from a general acid. Conclusions derived from several different crystallographic determinations (Wilson *et al.*, 1992; Hoog *et al.*, 1994; El-Kabbani *et al.*, 1994, 1995) indicate that the phenolic proton from tyrosine 48 in ALR2 (or its analog in the other enzymes) serves as the general acid. Further studies of site-directed mutants of ALR2 (Tarle *et al.*, 1993; Bohren *et al.*, 1994) support this hypothesis. Recently, in conflict with these experimental findings, theoretical calculations have suggested that histidine 110 in ALR2 may serve as the general acid (Lee *et al.*, 1996).

Despite the work being done on the enzyme's mechanism, there can be little doubt that the most interesting structural facet of ALR2 is in its clinically relevant inhibition. Since the substrate binding site of the enzyme is studded with hydrophobic side chains and the vast majority of known ALR2 inhibitors have a large hydrophobic component, it was logical to assume that the inhibitors bound in this same pocket although there were suggestions that it might bind elsewhere (Kador *et al.*, 1995). To examine this possibility and to determine the interactions which mediate drug binding, we solved the structure of ALR2, NADPH and zopolrestat (an ALR2 inhibitor developed by Pfizer, Inc.) (Mylari *et al.*, 1991) in a ternary complex. Finally, to understand what interactions confer specificity to zopolrestat binding to ALR2, the structure of a similar ternary complex of a related protein, murine fibroblast growth factor induced protein 1 (FR-1) (Donohue *et al.*, 1994) was determined and compared with the ALR2 structure.

Table 1. Crystal parameters, data collection and refinement statistics for
ALR2/NADPH/zopolrestat and FR-1/NADPH/zopolrestat structures

	ALR2/NADPH/zopolrestat	FR-1/NADPH/zopolrestat
Space group	P1	$P2_12_12_1$
Unit cell	a = 47.64 Å a = 67.47°	a = 59.45 Å
	b = 48.04 Å β = 76.77°	b = 65.38 Å a = β = γ = 90°
	c = 40.48 Å γ = 76.07°	c = 90.29 Å
Observations/reflections	53,037 / 26,145	213,331 / 43,337
R_{merge}	3.4 %	4.6 %
Reflections used (F ≥ σ(F))	25,891	36,797
R_{cryst}	0.18	0.16
r.m.s. deviation from idealbonds/angles	0.006 Å / 1.7°	0.008 Å / 1.8°
Protein atoms refined	2,517	2,537
NADPH atoms refined	48	48
Zopolrestat atoms refined	29	29
Ordered waters refined	175	448

2. MATERIALS AND METHODS

Aldose reductase was produced and purified as described previously (Petrash *et al.*, 1992). The purification for recombinant FR-1 has also been reported (Wilson *et al.*, 1995). Crystallization conditions using hanging drop vapor diffusion were identical for both ALR2- and FR-1-NADPH-zopolrestat complexes. Drops containing 7 mg/ml holoenzyme, 8.5 % (w/v) polyethylene glycol 6000, 200 µM zopolrestat, 3.5 mM β-mercaptoethanol, 25 mM citrate, pH 5.0 were suspended over wells containing 17 % (w/v) polyethylene glycol 6000, 7 mM β-mercaptoethanol, 50 mM citrate (pH 5.0). Seeding was used to improve crystal size and quality in both cases.

Data collection and space group determination were done on a Rigaku R-Axis area detector for the ALR2 complex and SDMS multiwire area detector for the FR-1 complex. Crystal parameters and data collection statistics are summarized in Table 1.

Both crystals were not isomorphous with the ALR2/NADPH structure so molecular replacement was carried out using the 1.4 Å structure of the ALR2 holoenzyme (D.K.W., J.M.P. & F.A.Q., unpublished data) as a search model. The highest peak in the ALR2 rotation map was subjected to a Patterson correlation refinement which yielded Euler angles of (θ_1 = 48°, θ_2 = 84°, θ_3 = 90°). A translation peak corresponding to this rotation solution was found at 17.78 Å, 28.09 Å and 11.73 Å and applied to the rotated molecule to give the initial structure. The rotation map obtained using the FR-1 data showed a peak at Euler angles of (θ_1 = 234°, θ_2 = 72°, θ_3 = 147°) which were used without further refinement. No translation function was needed due to the lack of crystallographic origin in space group P1. "Mutation" of the ALR2 side chains was done to agree with the FR-1 sequence and $|F_o - F_c|$ maps were used to position these atoms when necessary. In the case of ALR2 and of FR-1, $|F_o - F_c|$ and of $|2F_o - F_c|$ maps revealed the location of the zopolrestat, ordered waters and the nature of conformational changes which took place upon inhibitor binding. Results of the final round of refinement for both structures is shown in Table 1. All molecular replacement and refinement calculations were performed using the XPLOR set of programs (Bruenger, 1992).

Figure 2. A stereo view of the C$_\alpha$ trace of the ALR2 complexed with the NADPH cofactor (drawn in lines) and the inhibitor zopolrestat (shown in ball and stick). The zopolrestat is perched atop the nicotinamide ring of the NADPH and extends down underneath the carboxyl-terminal segment of the enzyme which shifts upon inhibitor binding. This figure and Fig. 1 were produced using the program MOLSCRIPT (Kraulis, 1991).

3. RESULTS AND DISCUSSION

A central element in the design of ALR2 inhibitors with increased binding affinity is the structural understanding of how the enzyme is inhibited. Before the structure of an ALR2/NADPH/inhibitor complex was available, many attempts at manually modeling known inhibitors into the active site were made with disappointing results. Although the hydrophobicity of these compounds complemented the largely aromatic character of the binding site, the preferred orientation was impossible to determine.

Efforts to computationally dock large numbers of compounds from libraries which contained known ALR2 inhibitors using the DOCK program (Shoichet & Kuntz, 1993) also met with failure. This procedure evaluated each compound's optimum steric fit in the binding site objectively but did not consider potential polar interactions that could be made. Furthermore, the protein was considered to be rigid and the putative inhibitor was not allowed to have any conformational freedom. Any or all of these shortcomings may have contributed to the program's inability to identify previously known ALR2 inhibitors from the database.

The determination of the ALR2 holoenzyme structure complexed with zopolrestat (Figure 2) demonstrated conclusively that the inhibitor bound to the active site of the enzyme, atop the nicotinamide ring of the NADPH (Wilson *et al.*, 1993). The majority of the hydrophobic surfaces associated with the ring systems come in contact with the numerous hydrophobic side chains in the binding site (Table 2) which also provide excellent steric complementarity. Many of the polar atoms in zopolrestat made favorable interactions with

Table 2. Interactions ≤ 4.0 Å seen in crystal structures between the zopolrestat inhibitor and ALR2 or FR-1

Residue (ALR2/FR-1 if different)	# contacts (aldose reductase)	# contacts (FR-1)
Trp 20	16	19
Tyr 48	5	5
Trp 79	1	1
His 110	8	8
Trp 111	38	40
Thr / Gln 113	6	2
Phe / Leu 115	2	-
Phe 122	6	9
Trp 219	6	-
Cys 298	1	2
Ala / Leu 299	3	-
Leu 300	5	10
Cys / Thr 303	3	6
Met 306	-	1
Tyr 309	4	1
Pro 310	5	1

the enzyme (Figure 3). The most notable of these is the salt bridge between the carboxylate of the inhibitor and the positively-charged side chain of His-110. This interactions carries the implication that most or all of the carboxylate-containing inhibitors, which constitute a large fraction of ALR2 inhibitors, would also bind similarly.

The ALR2/NADPH/zopolrestat structure provided a major reason as to why manual modeling and computational docking to the holoenzyme model were not successful in determining how ALR2 inhibitors bound to the enzyme. The structure appears similar in most respects to the holoenzyme structure with one large exception: two loops in the pro-

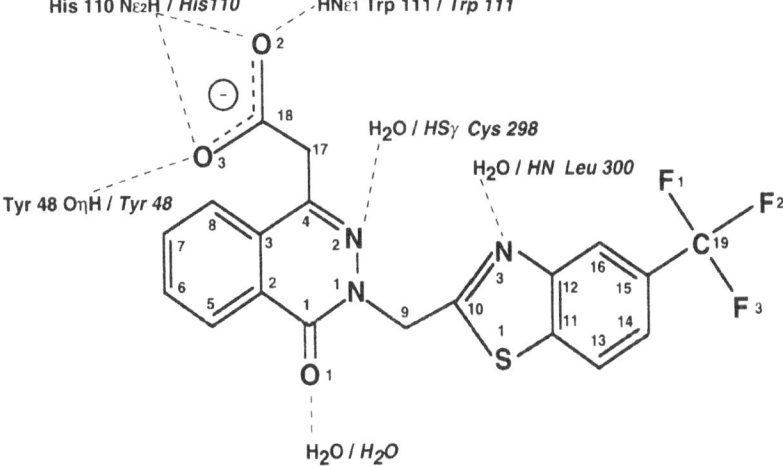

Figure 3. A schematic of the polar interactions made between the zopolrestat inhibitor and ALR2 or FR-1. Interactions seen in the FR-1 structure are listed first followed by those from the ALR2 structure.

tein, consisting of residues 121–135 and 298–303 have had their positions perturbed as a result of zopolrestat binding. While one loop (121–135) collapsed toward the nicoti-namide so that Phe-115 and Phe-122 could interact with the inhibitor, the other loop was pried away from the active site. These two movements had the effect of completely chang-ing the shape of the enzyme's binding site. The holoenzyme possesses a hydrophobic groove which formed the presumptive inhibitor binding site. It extends away from the catalytic site towards the carboxyl-terminal loop consisting of residues from approx. 298 through the terminal 315. Rather than binding entirely within this groove, zopolrestat bur-rows beneath this loop creating an unforeseen binding site of entirely different shape.

The large number of complementary interactions between zopolrestat and ALR2 ac-count for the high affinity (IC_{50} = 3 nM, Mylari *et al.*, 1991) of the enzyme for the inhibi-tor. The possibility exists that ALR2 inhibitors could suffer not from lack of affinity but from lack of specificity owing to the number of enzymes in the aldo-keto reductase family with sequence homology and therefore, presumably structural similarity in the substrate specificity pocket. In order to understand what structural features may be involved in such specificity and lack of specificity, the structure of the related (70% sequence identity with human ALR2) murine fibroblast growth factor-induced protein 1 (FR-1) was determined at 1.7 Å resolution. Although nothing was known about the protein's ability to bind NADPH, to possess oxidoreductase activity or to be inhibited by zopolrestat, the high de-gree of sequence conservation strongly suggested that it bore these properties so NADPH and zopolrestat were included in the crystallization.

Unsurprisingly, the structure was very similar to the ALR2 ternary complex with the r.m.s. deviations between C_α's of only 0.82 Å. The cofactor and zopolrestat were clearly seen interacting with the protein in a manner which was almost identical to ALR2 and were found to have dissociation constants of 0.45 µM and 30 nM respectively. All resi-dues involved in catalysis were in conserved conformations and the enzyme was shown to utilize DL-glyceraldehyde as a substrate (K_m = 0.92 mM).

The similarity between the structure of the inhibitor binding site seen in ALR2 and FR-1 is illustrated by the number of van der Waals interactions made with zopolrestat by each enzyme (Table 2). A small number of differences are noted when comparing the binding site of each enzyme. The binding site of FR-1 is somewhat larger and accommo-dates 4 additional water molecules. The extra volume is due to the slight shift outward of the enzyme's carboxyl-terminal loop. It is likely that this difference is due to the diver-gence in sequence of the enzymes in this region. It is also possible that the structure and flexibility of this loop, which contains the majority of sequence divergence among the aldo-keto reductases, may be a significant factor in the determination of substrate and in-hibitor specificity.

4. CONCLUSIONS

The failed attempts at predicting the structure of the ALR2/NADPH/zopolrestat ter-nary complex using the structure of the complex with NADPH illustrates the dangers of using theoretical approaches to this sort of problem without considering all of the parame-ters associated with this type of system. These include allowing for the flexibility of the protein, evaluating all energies associated with binding and allowing for many possible conformations of the candidate inhibitor. A rigorous examination of all of these parame-ters in ALR2 is computationally prohibitive, probably by orders of magnitude, particularly if many compounds are to be surveyed. The experimental results provided by the crystal structure show in an unbiased manner, the unpredictable change in the shape of the binding site caused by the zopolrestat-induced conformational change.

The problem associated with the therapeutic inhibition of ALR2 may be complicated by the fact that compounds which bind to it will be likely to bind to related enzymes. In contrast to the structure of zopolrestat bound to its target enzyme ALR2, the structure of the FR-1/NADPH/zopolrestat complex constitutes an example of the drug bound to such a potential "anti-target". Structures of zopolrestat and other compounds bound to other anti-targets may play a key role in improving not the affinity but the specificity of such compounds. The large amount of similarity between the ALR2 and FR-1 binding sites indicates that this may not be an easy task, however.

5. REFERENCES

Bohren KM, Grimshaw CE, Lai CJ, Harrison DH, Ringe D, Petsko GA, Gabbay KH. (1994) Tyrosine-48 is the proton donor and histidine-110 directs substrate stereochemical selectivity in the reduction reaction of human aldose reductase: enzyme kinetics and crystal structure of the Y48H mutant enzyme. *Biochemistry* 33:2021–3032.

Borhani DW, Harter TM, Petrash JM. (1992) The crystal structure of the aldose reductase - NADPH binary complex. *J. Biol. Chem.* **267**:24841–24847.

Bruenger AT. (1992) *XPLOR: A system for crystallography and NMR.* Version 3.1 Manual, Yale University Press, New Haven CT.

Donohue PJ, Alberts GF, Hampton BS, Winkles JA. (1994) A delayed-early gene activated by fibroblast growth factor-1 encodes a protein related to aldose reductase. *J. Biol. Chem.* **269**:8604–8609.

Dvornick D, Simard-Duguesne N, Kraml M, Sestanj K, Gabbay KH, Kinoshita JH, Varma DS, Merola LO. Polyol accumulation in galactosaemic and diabetic rats: control by an aldose reductase inhibitor. *Science* **182**:1146–1148.

El-Kabbani O, Green NC, Lin G, Carson M, Narayanam SVL, Moore K, Flynn TG, DeLucas LJ. (1994) Structures of human and porcine aldehyde reductase: an enzyme implicated in diabetic complications. *Acta Crystallogr.* **D50**:859–868.

El-Kabbani O, Judge K, Ginell SL, Myles D, DeLucas LJ, Flynn TG. (1995) Structure of porcine aldehyde reductase holoenzyme. *Nat. Struct. Biol.* **2**:687–692.

Farber GK, Petsko GA. (1990) The evolution of α/β barrel enzymes. *Trends Biochem. Sci.* **15**:228–234.

Hoog SS, Pawlowski JE, Alzari PM, Penning TM, Lewis M. (1994) Three-dimensional structure of rat liver 3α-hydroxysteroid/dihydrodiol dehydrogenase: a member of the aldo-keto reductase superfamily. *Proc. Natl. Acad. Sci. U.S.A.* **91**:2517–2521.

Kador PF, Lee YS, Rodriguez L, Sato S, Bartoszko-Malik A, Abdel-Ghany YS, Miller DD. (1995) Identification of an aldose reductase inhibitor site by affinity labeling. *Bioorg. Med. Chem.* **3**:1313–1324.

Kraulis PJ. (1991) MOLSCRIPT: a program to produce both detailed and schematic plots of protein structures. *J. Appl. Crystall.* **24**:946–950.

Mylari BL, Larson ER, Beyer TA, Zembrowski WJ, Aldinger CE, Dee MF, Siegel TW. Singleton DH. (1991) Novel, potent aldose reductase inhibitors: 3,4-dihydro-4-oxo-3-[[5-(trifluoromethyl)-2-benzothiazolyl]methyl]-1-phthalazine-acetic acid (zopolrestat) and congeners. *J. Med. Chem.* **35**:457–465.

Petrash JM, Harter TM, Devine CS, Olins PO, Bhatnagar A, Liu S, Srivastava SK. (1992) Involvement of cysteine residues in catalysis and inhibition of human aldose reductase: site directed mutagenesis of cys-80, cys-298 and cys-303. *J. Biol. Chem.* **267**:24833–24840.

Rondeau JM, Tete-Favier F, Podjarny A, Reymann JM, Barth P, Biellmann JF, Moras D. (1992) Novel NADPH-binding domain revealed by the crystal structure of aldose reductase. *Nature* **355**:469–472.

Sarges R, Oates P. (1993) Aldose reductase inhibitors: recent developments. *Prog. Drug Res.* **40**:99–161.

Shoichet BK, Kuntz ID. (1993) Matching chemistry and shape in molecular docking. *Prot. Eng.* **6**:723–732.

Tarle I, Borhani DW, Wilson DK, Quiocho FA, Petrash JM. (1993) Probing the active site of human aldose reductase: site directed mutagenesis of asp-43, tyr-48, lys-77 and his-110. *J. Biol. Chem.* **268**:25687–25693.

Wilson DK, Bohren KM, Gabbay KH, Quiocho FA. (1992) An unlikely sugar substrate site in the 1.65 Å structure of the human aldose reductase holoenzyme implicated in diabetic complications. *Science* **257**:81–84.

Wilson DK, Tarle I, Petrash JM, Quiocho FA. (1993) Refined 1.8 Å structure of human aldose reductase complexed with the potent inhibitor zopolrestat. *Proc. Natl. Acad. Sci. U.S.A.* **90**:9847–9851.

Wilson DK, Nakano T, Petrash JM, Quiocho FA. (1995) 1.7 Å structure of FR-1, a fibroblast growth factor-in-duced member of the aldo-keto reductase family complexed with coenzyme and inhibitor. *Biochemistry* **34**:14323–14330.

Lee YS, Hodoscek M, Brooks B, Kador PF (1996) Investigation of the geometry of asp-43, lys-77, tyr-48 and his-110 in the active site of human aldose reductase by the QM/MM potential. in the 8th International Work-shop for Enzymology and Molecular Biology of Carbonyl Metabolism. Deadwood, South Dakota, June 29 - July 3, 1996.

50

ALDEHYDE REDUCTASE

Catalytic Mechanism and Substrate Recognition

Oleg A. Barski,[1] Kenneth H. Gabbay,[2] and Kurt M. Bohren[1]

[1]Department of Pediatrics, Molecular Diabetes and Metabolism Section, and
[2]Department of Cell Biology
Baylor College of Medicine
Houston, Texas 77030

1. INTRODUCTION

Aldehyde reductase is a member of a superfamily of NADPH-dependent aldo-keto reductases, which comprises well over 30 different proteins in seven subfamilies. The enzymes of this superfamily are monomeric proteins with the molecular weight between 30 and 40 kDa. They are characterized by a broad substrate specificity and a great preference of the reduction reaction. They share many common structural and functional characteristics, namely an α/β barrel tertiary structure, a NADPH cofactor which is enfolded by a mobile loop that varies among the different members, and an active site located at the C-terminus of the barrel (Wilson et al., 1992, Harrison et al., 1994, El-Kabbani et al., 1995). The physiological role(s) of these enzymes have not yet been completely established but many are thought to be involved in general detoxification of reactive aldehydes (Bachur, 1976). More specific functions were found for some members, e.g., osmoregulation within the renal tubular cells (aldose reductase; Garcia-Perez and Burg, 1991) or bile acids metabolism (bile acid binders; Stolz et al., 1995). The most studied member of this protein family, aldose reductase, is implicated in the pathogenesis of certain diabetic complications (Gabbay, 1973; Dvornik, 1987; Kador, 1988). Aldehyde and aldose reductase differ significantly in their substrate specificity and tissue distribution (Davidson et al., 1977; Wermuth, 1982), and despite similar kinetic mechanisms, there are differences in the function of the two enzymes that are not yet understood.

Aldehyde reductase was first purified from rat liver and characterized by Mano et al. (1961) under the name of TPN L-hexonate dehydrogenase. Later on it was also classified as glucuronate reductase, mevaldate reductase, "high Km aldehyde reductase," ALR1 and aldehyde reductase. The first enzyme from a human source was purified from liver in 1977 (Wermuth et al., 1977). Aldehyde reductase is widely distributed among all mammalian species and birds; it was also detected in reptiles, amphibia, fish, insects, flowering plants, and fungi (Davidson et al., 1978). The highest levels of aldehyde reductase in mammals are present in kidney cortex and liver. Some level of activity is detectable in virtually every tissue.

Human liver aldehyde reductase is a cytosolic protein consisting of 325 amino acid residues with a calculated molecular weight of 36,578 (Wermuth et al., 1987; Bohren et al., 1989). It has a relatively acidic isoelectric point of 5.3. The rate of reduction is maximal between pH 6 and 7, while the pH optimum for alcohol oxidation is at pH 9–9.5 (Wermuth et al., 1977). During catalysis the pro-4R hydrogen atom of NADPH attacks the re-face of the aldehyde (Flynn et al., 1975). The enzyme can also utilize NADH as a cofactor, but the activity is less than 5% of that observed with NADPH (Wermuth et al., 1977).

As with the other aldo-keto reductases, aldehyde reductase reduces a broad array of both biogenic and xenobiotic aldehydes. The best substrates are isocorticosteroids, containing the 17β -aldol side chain (Wermuth and Monder, 1983). These substrates have low values of K_m (<1µM) and the values of k_{cat} are in the range of 10 s^{-1}. Steroids are followed in catalytic effectiveness by aromatic aldehydes, and further by sugar aldehydes and aliphatic aldehydes (Table 1). In contrast to aldose reductase, aldehyde reductase is notable for its preference for substrates having a negatively charged carboxyl group such as glucuronate, succinic semialdehyde, and p-carboxybenzaldehyde (Branlant & Biellmann, 1980, Wermuth, 1985). This paper compares aldehyde reductase with aldose reductase to identify the common and unique features in terms of kinetics, catalytic mechanism, inhibition and substrate specificity.

2. KINETIC MECHANISM

Aldehyde reductase, as well as aldose reductase, follow a sequential ordered kinetic mechanism (Boghosian, R.A., & McGuinness, E.T., 1981; Davidson, W. S. & Flynn T. G., 1979). NADPH binds to the enzyme first, followed by an isomerization of the enzyme. This isomerization involves the folding of a loop that holds the nucleotide firmly in place and allows the active site to assume a conformation suitable for substrate binding and catalysis. After the alcohol product leaves the enzyme, the loop opens in order to release the oxidized cofactor and replace it with a fresh molecule of reduced cofactor. The opening of

Table 1. Kinetic constants of wild type and mutant aldehyde reductase

Substrate	Wild Type			H112Q		
	k_{cat}	K_m	k_{cat}/K_m	k_{cat}	K_m	k_{cat}/K_m
20α -isocorticosterone[b]	11	0.004	2750000	-	-	-
p-carboxybenzaldehyde	3.7	0.038	97534	0.054	2.8	19
p-nitrobenzaldehyde	5.1	0.16	32025	0.056	5.9	9.5
succinic semialdehyde	3.7	0.17	22186	0.28	18	16
DL-glyceraldehyde	1.3	1.7	786	0.21	76	2.8
D-glucuronate	2.7	4.2	637	0.97	48	20
butyraldehyde[b]	0.67	1.3	515	-	-	-
D-glucose[b]	0.08	400	0.2	-	-	-

[a]k_{cat} is expressed in s^{-1}, K_m is in mM, and k_{cat}/K_m is in s^{-1}M^{-1}.
[b]Data from Wermuth et al., 1977.

Table 2. Primary deuterium and solvent isotope effects for DL-Glyceraldehyde reduction

	Wild type aldose reductase	Wild type aldehyde reductase	Aldehyde reductase H112Q
$^D(k_{cat})$	0.99±0.02	1.48±0.02	1.79±0.21
$^D(k_{cat}/K_m)$	1.82±0.08	2.28±0.05	2.30±0.15
$^{D2O}(k_{cat})$	1.04±0.06	1.06±0.03	2.06±0.09
$^{D2O}(k_{cat}/K_m)$	4.73±0.23	2.02±0.06	0.96±0.11

the loop is the sole rate-limiting step in the reduction reaction catalyzed by aldose reductase (Grimshaw et al., 1995).

Primary deuterium and solvent isotope effects (Table 2) indicate that, unlike aldose reductase, the chemistry of the reaction is partially rate-limiting in the overall catalytic cycle of wild-type aldehyde reductase. The primary deuterium isotope effect measured with NADPD is significantly different from 1 both for k_{cat} and k_{cat}/K_m indicating that hydride transfer is partially rate-limiting. The effect on k_{cat} is smaller than on $k_{cat}/K_{m(aldehyde)}$, a term which comprises all the steps in the catalytic cycle from the binding of aldehyde to the E:NADPH complex through the release of the alcohol product. The fact that the isotope effect on k_{cat}/K_m is larger than on k_{cat} suggests that steps outside this part of the reaction sequence, i.e., nucleotide exchange and the associated enzyme isomerization, are partially rate limiting for overall turnover. There is no solvent isotope effect on k_{cat} although a significant effect is observed on k_{cat}/K_m (2.02 ± 0.06). If we use a value of 6.5 for ^Dk, the intrinsic isotope effect on hydride transfer based on that determined for aldose reductase (Grimshaw et al., 1995), and the value of 8 for ^{D2O}k determined for the solvent isotope effect on k_{cat}/K_m of the Y14F mutant of Δ^5–3-ketosteroid isomerase (Xue et al., 1991), then a simple calculation of the fractional rates (Northrop, 1975) would suggest about a 23% rate-limitation of k_{cat}/K_m by hydride transfer and about 15% rate-limitation by proton transfer, with the remainder contributed by other steps comprising k_{cat}/K_m. A similar calculation using the isotope effect on k_{cat} indicates that hydride transfer accounts for ~9% limitation of the overall reaction velocity.

The apparent "acceleration" in the rate of opening of the nucleotide enfolding loop in aldehyde reductase (in comparison to aldose reductase) has broad kinetic implications. First, it is most probably responsible for the generally one order of magnitude higher turnover numbers characteristic of this enzyme. Second, the values of K_d for NADPH and $NADP^+$ are approximately one order of magnitude higher in aldehyde (70 and 360 nM, respectively; Barski et al., submitted for publication), than in aldose reductase (10 and 6 nM, respectively; Ehrig et al., 1994). As a consequence of this turnover acceleration, the chemical steps appear more rate-limiting in the overall turnover rate of aldehyde reductase than in aldose reductase. Thus, the kinetic mechanism of aldehyde reductase is more balanced in a sense that the rate-limitation is dissipated along the whole sequence of steps instead of being concentrated in one step, as it is in aldose reductase.

3. CATALYTIC MECHANISM

The mechanism of aldehyde reduction catalyzed by aldehyde reductase and homologous enzymes consists of the transfer of the pro-4R hydrogen atom of NADPH in the form of a hydride onto an aldehyde carbon and the obligate addition of a proton to form the hydroxyl group of the alcohol product. Thus, to successfully perform the catalysis, the enzyme has to do the following:

a. position the substrate and the nicotinamide close to each other and in a right orientation for the hydride ion to transfer,

b. polarize the nicotinamide and aldehyde to facilitate hydride transfer,

c. provide the necessary proton.

The problem of which residue acts as a general acid catalyst has been already addressed in aldose reductase, where we determined that Tyr48 is the proton donor in the aldose reductase catalyzed reaction (Bohren et al., 1994). Our studies showed that a Y48F mutant is totally inactive, while studies of a catalytically active Y48H mutant enzyme showed that a water molecule had replaced the phenolic hydroxyl of Tyr48 as the source of the required proton. The crystal structure of the Y48H mutant revealed the location of the water molecule precisely in the space previously occupied by the hydroxyl of Tyr48 in the wild-type enzyme. Tyr48 is conserved throughout the aldo-keto reductase superfamily and corresponds to Tyr49 in aldehyde reductase. Mutation of this residue to phenylalanine, which conserves the hydrophobic properties and spatial fit of a tyrosine residue, yet eliminates the possibility of general acid catalysis by removing the tyrosyl hydroxyl group, leads to a completely inactive enzyme. A histidine substitution for tyrosine 49 in aldehyde reductase (Y49H) also yields an inactive enzyme. Thus, unlike human aldose reductase, the geometry of the aldehyde reductase active site does not permit substitution for the Tyr49 hydroxyl function by some other element such as the water molecule detected in the crystal structure of the Y48H aldose reductase mutant. Alternatively, if such a substitution does take place, it diminishes the catalytic efficiency for all the substrates to a greater extent than in aldose reductase making it virtually impossible to detect any enzymatic activity.

The normal pKa of a tyrosine side chain is 10.5, and we previously suggested that the proton donor function of Tyr48 in human aldose reductase is consistent with a downward shift in its pK_a to 8.4 imparted by a hydrogen-bonding/salt bridge network formed by Asp43$^-$/Lys77$^+$ and Tyr48. A K77M mutation in aldose reductase inactivates the enzyme. The crystal structure of the porcine aldehyde reductase shows that a similar network exists in aldehyde reductase (El-Kabbani et al., 1995). Mutation of Lys 79 to methionine also yields an inactive enzyme, indicating that this residue plays a similarly critical role in aldehyde reductase catalysis. Our previous studies (Barski et al., 1995) of the pH profiles indicated the ionization of a catalytically active group with an apparent pK_a of 9.4. This pK_a was ascribed to Tyr49, which apparently experiences a 1 log unit acidic shift in its pK_a due to its interaction with Lys79.

His112 residue is also conserved throughout the aldo-keto reductase superfamily and plays an important role in catalysis. In aldose reductase, the homologous His110 was shown to direct the stereochemical orientation of substrates in the active site pocket (Bohren et al., 1994). Based on molecular modeling calculations, De Winter and von Itzstein (1995) concluded that His110 cannot be excluded as a proton donor in aldose reductase. However, they assumed NADPH as the crystal bound cofactor when in fact it is NADP$^+$. This assumption might have flawed their calculations. Recent pH studies by Grimshaw et al. (1995) clearly identify the pK_a of Tyr48 and its preferred ionization state for catalysis and inhibition. The mutation of His 112 to glutamine has a drastic effect on enzymatic activity with the catalytic constant, k_{cat}, decreasing 6 to 100-fold while the K_m increases 10 to 100-fold depending on the substrate. These changes are reflected in severe decreases in catalytic efficiencies (k_{cat}/K_m) of 30 to 5000-fold, with aromatic substrates like p-nitrobenzaldehyde and p-carboxybenzaldehyde most drastically affected (3400 and 5100-fold respectively), while polar substrates such as DL-glyceraldehyde and D-glucuronate are much less affected (277 and 32-fold respectively). These findings suggest that His112 is important for the accommodation and chemistry of interaction with aromatic and polar substrates in the active site pocket.

The porcine aldehyde reductase crystal structure indicates that the Nϵ2 of His113 (corresponding to His 112 in the human enzyme) is hydrogen-bonded to the amide oxygen of the nicotinamide ring of NADPH (El-Kabbani et al., 1995). This interaction is not observed in the aldose reductase crystal structure. It was suggested that this interaction helps to polarize the nicotinamide ring of NADPH and thus, facilitates the hydride transfer (Harrison, 1995).

The important role of His112 for the catalytic mechanism is reflected in the isotope effects (Table 2). While in the wild type enzyme we observe an effect of 2.02 on k_{cat}/K_m, but no effect (1.06) on k_{cat}, the H112Q mutant exhibits an effect of 2.06 on k_{cat}, but no effect on k_{cat}/K_m. This inversion of the isotope effects is an indication of the major change in the mechanism brought about by the mutation of histidine. These isotope effect data are consistent with a two step mechanism, where hydride transfer occurs first and irreversibly, followed by the proton transfer (Barski et al., 1995). It differs from the wild type mechanism where hydride and proton transfers occur in a reversible and probably concerted mode.

This hypothesis is consistent with a major role for the His112 residue in catalysis, and more specifically, hydride transfer. In the H112Q mutant, due to the key histidine substitution, the location of the substrate is less favorable for proton transfer from Tyr49. Proton transfer to the carbonyl oxygen of the aldehyde substrate is impaired and slower than the hydride transfer at the moment when the chemical reaction is about to start. On the other hand, glutamine is still capable of polarizing the nicotinamide ring through interactions with its amide group, thus facilitating hydride transfer. The process of hydride transfer develops the alcoholate anion and the proton from Tyr49 transfers to the alcoholate anion instead of the carbonyl group, a process which is now favored due to the positive (proton) and negative (alcoholate) charges involved. This process can go fast despite the unfavorable location of Tyr49. Thus, His112 in aldehyde reductase may facilitate catalysis in two ways: positioning the substrate in the active site pocket in an orientation favorable for hydride and proton transfer; and promoting the resonance form of NADPH with a negative charge on the amide oxygen and a positive charge on the ring nitrogen, which allows the hydride to transfer more readily.

4. MECHANISM OF INHIBITION

The implication of aldose reductase in the etiology of diabetic complications sparked great interest in the development of inhibitors for this enzyme. Carboxylic acids, dicarboxylic acids, spyrohydantoins, barbiturates and other heterocyclic compounds inhibit aldose reductase. These same compounds are also potent inhibitors of aldehyde reductase.

All known inhibitors of aldehyde and aldose reductase are un- or non-competitive in the reduction and competitive in the oxidation reaction. As an example, the inhibition pattern of glutaric acid with aldehyde reductase in the forward and the reverse directions is presented in fig. 1. This inhibition pattern is consistent with the inhibitor binding preferentially to the enzyme:NADP$^+$ complex. Direct binding studies of the most potent aldehyde reductase inhibitor AL1576 indicated that, indeed, the inhibitor binds strongly to the E:NADP$^+$ complex, very weakly to the E:NADPH complex, and not at all to the apoenzyme (Barski et al., 1995). Similar results were obtained with aldose reductase binding to alrestatin, another aldose reductase inhibitor (Ehrig et al., 1994). The reason lies in the fact that all potent aldose and aldehyde reductase inhibitors have at least one negative charge. Crystallographic studies of aldose reduc-

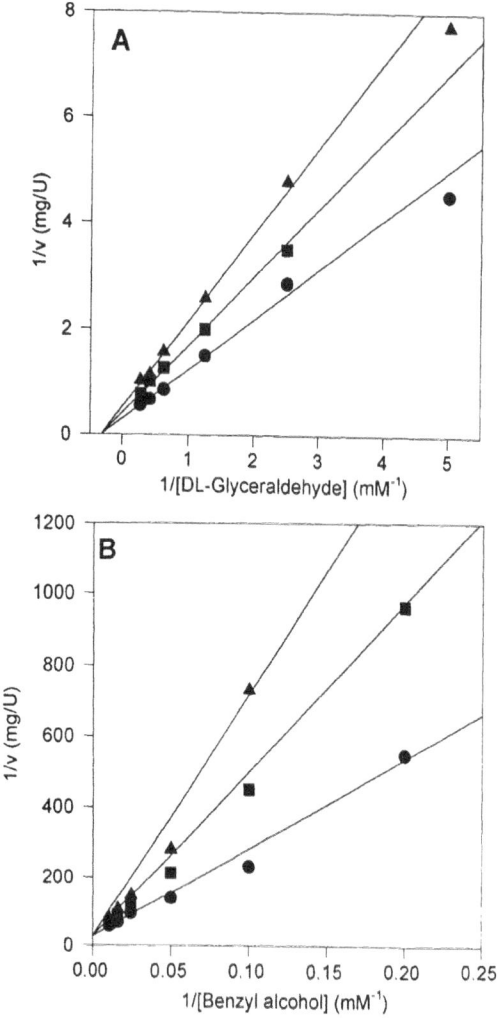

Figure 1. Inhibition of aldehyde reductase by glutaric acid in forward (A) and reverse (B) reactions. Glutaric acid concentrations are: 0 (circles), 1 mM (squares) and 2 mM (triangles).

tase complexes with a number of inhibitors showed that the negative charge of the inhibitor binds in an anion binding site at the bottom of the active site pocket, delineated by the C4N of nicotinamide, OH of Tyr48, and the Nε of His110 (Harrison et al., 1994). This site bears a positive charge when the bound cofactor is $NADP^+$ and provides the basis for the preference of the $E:NADP^+$ complex by the negatively charged inhibitors. A similar anion binding site is present in aldehyde reductase.

5. SUBSTRATE AND INHIBITOR SPECIFICITY (RECOGNITION)

Inefficiency in clinical trials of very potent commercially developed aldose reductase inhibitors demonstrated the necessity to strive for specificity. Most of the tested com-

```
                         295         305           315
Aldehyde Reductase    NKNWRYIVPMLTVDGKRVPRDAGHPLYPFNDPY
Aldose Reductase      -R---VCALLSCTSH-D.........---HEEF
Chlordecone Reductase -R-Y--V-MDFLM-HPD.........---S-E-
Bile Acid Binder      -R-V--LTLDIFAGPPN.........---S-E-
```

Figure 2. C-terminus sequences of human aldo-keto reductases.

pounds are nonspecific for aldose reductase and inhibit other enzymes of the aldo-keto reductase superfamily as well. Efforts are being directed to develop inhibitors specific for each enzyme of the superfamily.

It has long been known that dicarboxylic acids, bearing two negative charges, are better inhibitors of aldehyde than aldose reductase (Branlant 1982a,b). Also, aldehyde reductase is notable for its preference of the carboxyl-containing negatively charged substrates. If we compare catalytic efficiencies (Table 1), glucuronate is a much better substrate than glucose, succinic semialdehyde is better than butyraldehyde, p-carboxybenzaldehyde is better than p-nitrobenzaldehyde, although the carboxy group is less electron-withdrawing than the nitro group. Study of the Hammet relationship in para-substituted benzaldehydes indicated that p-carboxybenzaldehyde had specific interactions of the carboxyl substituent of the substrate with the enzyme (Bhatnagar et al., 1991). These facts suggest the presence of an anion recognition site, that is absent in other reductases, in the active site of aldehyde reductase.

Despite extensive homology throughout the whole sequence, the C-terminal loops of the aldo-keto reductases are unique for each member and differ drastically in length and amino acid composition (fig. 2). Studies in aldose and aldehyde reductase have suggested that the C-terminal loop is critical for catalytic effectiveness, substrate and inhibitor specificity (Bohren et al., 1992). Recently we identified Arg311, present in the C-terminal loop, as the residue that binds the negatively charged ω-carboxyl group of substrates and is therefore the basis for the preference of aldehyde reductase for carboxyl-containing substrates (Barski et al., submitted for publication). This residue also interacts with the negatively charged carboxyl group of dicarboxylic acid inhibitors. The mode of binding of dicarboxylic inhibitors to aldehyde reductase is illustrated in fig. 3. Fig. 3 also provides the basis for the long known fact that the inhibitor potency depends on the distance between the two carboxyl groups, since this distance should correspond to the distance between Arg311 and the anion binding site at the bottom of the active site pocket. Thus, the C-terminal loop of aldehyde reductase directly interacts with substrates and inhibitors.

6. CONCLUSION

This chapter summarizes the kinetics, inhibition and amino acid residues involved in the catalytic mechanism and substrate recognition of aldehyde reductase. The chemical mechanism of aldehyde reductase catalysis appears to be quite similar to that of aldose reductase. A major difference in the kinetic mechanism is the faster rate of cofactor exchange in aldehyde reductase probably due to an accelerated opening of the nucleotide enfolding loop, making the chemical steps of the reaction appear to be partially rate-limiting. The structure of the active site pocket in aldehyde reductase is significantly different from that of aldose reductase, which mainly result from participation of the C-terminal

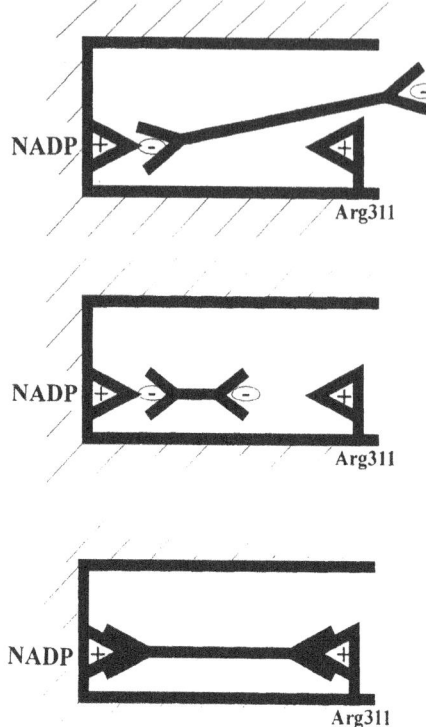

Figure 3. Cartoon showing the mode of binding of dicarboxylic acid inhibitors to the aldehyde reductase active site. One carboxyl group of such inhibitors binds to the anion binding site at the bottom of the active site pocket outlined by NADP$^+$ on this drawing. The second carboxyl group should bind to the arginine311. The cartoon shows the inhibitor that is too long (top), too short (middle) and perfect (bottom).

loop in the formation of the active site pocket. Observed similarities and differences have broad implications for future design of pharmaceutical effectors specifically targeted at the various members of the aldo-keto reductase superfamily.

7. REFERENCES

Bachur, N.R.: Cytoplasmic aldo-keto reductases: a class of drug metabolizing enzymes. Science 193 (1976) 595–597.

Barski, O.A., Gabbay, K.H., and Bohren, K.M.: The C-terminal loop of aldehyde reductase determines substrate and inhibitor specificity. Submitted for publication.

Barski, O.A., Gabbay, K.H., Grimshaw, C.E., Bohren, K.M.: Mechanism of human aldehyde reductase: characterization of the active site pocket. Biochemistry 34 (1995) 11264–11275.

Bhatnagar, A., Liu, S-Q., and Srivastava, S.K.: Structure-activity correlations in human kidney aldehyde reductase-catalyzed reduction of para-substituted benzaldehyde by 3-acetyl pyridine adenine dinucleotide phosphate. Biochim. et Biophys. Acta 1077 (1991) 180–186.

Boghosian, R.A., & McGuinness, E.T.: Pig brain aldose reductase: a kinetic study using the centrifugal fast analyzer. Int. J. Biochem. 13 (1981) 909–914.

Bohren K.M., Grimshaw, C.E., and Gabbay, K.H.: Catalytic effectiveness of human aldose reductase. J. Biol. Chem. 267 (1992) 20965–20970.

Bohren, K. M., Grimshaw, C. E., Lai, C.-J., Harrison, D. H., Ringe, D., Petsko, G. A., & Gabbay, K. H.: Tyrosine-48 is the proton donor and histidine-110 directs substrate stereochemical selectivity in the reduction reaction of human aldose reductase: enzyme kinetics and crystal structure of the Y48H mutant enzyme. Biochemistry 33 (1994) 2021–2032.

Bohren, K.M., Bullock, B., Wermuth, B., & Gabbay, K.H.: The aldo-keto reductase superfamily. J. Biol. Chem. 264 (1989) 9547–9551.

Branlant, G., and Biellmann, J-F.: Purification and some properties of aldehyde reductases from pig liver. Eur. J. Biochem. 105 (1980) 611–621.

Branlant, G.: Properties of an aldose reductase from pig lens. Eur. J. Biochem. 129 (1982b) 99–104.

Branlant, G.: The substrate binding site of aldehyde reductase from pig liver. Eur. J. Biochem. 121 (1982a) 407–411.

Davidson, W. S. & Flynn T. G.: Kinetics and mechanism of action of aldehyde reductase from pig kidney. Biochem. J. 177 (1979) 595–601.

Davidson, W. S., Walton, D. J., & Flynn, T. G.: A comparative study of the tissue and species distribution of NADPH-dependent aldehyde reductase. Comp. Biochem. Physiol. 60B (1978) 309–315.

De Winter H.L., and von Itzstein M.: Aldose reductase as a target for drug design: molecular modeling calculations on the binding of acyclic sugar substrates to the enzyme. Biochemistry 34 (1995) 8299–308.

Dvornik, D (1987) Aldose reductase inhibition. An approach to the prevention of diabetic complications. (Port, D. ed. Biomedical Information Corporation, McGraw-Hill New York)

Ehrig, T., Bohren, K.M., Prendergast, F.G., & Gabbay, K.H.: Mechanism of aldose reductase inhibition: binding of NADP$^+$/NADPH and alrestatin-like inhibitors. Biochemistry 33 (1994) 7157–7165.

El-Kabbani, O., Judge, K., Ginell, S.L., Myles, D.A.A., Delucas, L.J., and Flynn, T.G.: Structure of porcine aldehyde reductase holoenzyme. Nature Structural Biology 2 (1995) 687–692.

Flynn, T.G., Shires, J., and Walton, D.J.: Properties of the NADP-dependent aldehyde reductase from pig kidney. J. Biol. Chem. 250 (1975) 2933–2940.

Gabbay, K.H.: The sorbitol pathway and the complications of diabetes. N. Engl. J. Med. 288 (1973) 831–836.

Garcia-Perez, A., & Burg M.B.: Renal medullary organic osmolytes. Physiol. Rev. 71 (1991) 1081–1115.

Grimshaw C.E., Bohren K.M., Lai C.J., and Gabbay K.H.: Human aldose reductase: pK of tyrosine 48 reveals the preferred ionization state for catalysis and inhibition. Biochemistry 34 (1995) 14374–84.

Grimshaw, C.E., Bohren, K.M., Lai, C-J., and Gabbay, K.H.: Human aldose reductase: rate constants for a mechanism including interconversion of ternary complexes by recombinant wild-type enzyme. Biochemistry 34 (1995) 14356–14365.

Harrison, D. H., Bohren, K. M., Ringe, D., Petsko, G. A. & Gabbay, K. H.: An anion binding site in human aldose reductase: mechanistic implications for the binding of citrate, cacodilate, and glucose 6-phosphate. Biochemistry 33 (1994) 2011–2020.

Harrison, D.H.T.: All in the family. Nature Struct. Biol. 2 (1995) 719–720.

Kador, P.F.: The role of aldose reductase in the development of diabetic complications. Med. Res. Rev. 8 (1988) 325–352.

Mano, Y., Suzuki, K., Yamada, K., and Shimazono, N.: Enzymic studies on TPN L-hexonate dehydrogenase from rat liver. J. Biochem. 49 (1961) 618–634.

Northrop, D.B.: Steady-state analysis of kinetic isotope effects in enzymic reactions. Biochemistry 14 (1975) 2644–2650.

Stolz, A., Hammond, L., and Lou H.: Rat and human bile acid binders are members of the monomeric reductase gene family. Advances in Experimental Medicine & Biology. 372 (1995) 269–280.

Wermuth, B. and Monder, C.: Aldose and aldehyde reductase exhibit isocorticosteroid reductase activity. Eur. J. Biochem. 131 (1983) 423–426.

Wermuth, B., Münch, J.D.B., and von Wartburg, J-P.: Purification and properties of NADPH-dependent aldehyde reductase from human liver. J. Biol. Chem. 252 (1977) 3821–3828.

Wermuth, B., Omar, A., Forster, A., di Francesco, C., Wolf, M., von Wartburg, J.P., Bullock, B., and Gabbay, K.H.: Primary structure of aldehyde reductase from human liver. Progress in Clinical & Biological Research. 232 (1987) 297–307.

Wermuth, B.: Aldo-keto reductases. in Enzymology of Carbonyl Metabolism 2: Aldehyde Dehydrogenase, Aldo/Keto Reductase, and Alcohol Dehydrogenase (1985) pp 209–230, Alan R. Liss, Inc., New York.

Wermuth, B.: Human carbonyl reductases. in Enzymology of Carbonyl Metabolism: Aldehyde Dehydrogenase and Aldo/Keto Reductase (1982) pp 261–274, Alan R. Liss, Inc., New York.

Wilson, D.K., Bohren, K.M., Gabbay, K.H., & Quiocho, F.A.: An unlikely sugar substrate site in the 1.65 Å structure of the human aldose reductase holoenzyme implicated in diabetic complications. Science 257 (1992) 81–84.

Xue, L.A., Talalay, P., & Mildvan, A.S.: Studies of the catalytic mechanism of an active-site mutant (Y14F) of delta 5–3-ketosteroid isomerase by kinetic deuterium isotope effects. Biochemistry 30 (1991) 10858–10865.

51

STUDY OF NON-COVALENT ENZYME-INHIBITOR COMPLEXES OF ALDOSE REDUCTASE BY ELECTROSPRAY MASS SPECTROMETRY

Noelle Potier,[1] Patrick Barth,[2] Denis Tritsch,[2] and Jean-François Biellmann,[2] and Alain Van Dorsselaer[1]

[1]Laboratoire de Spectrometrie de Masse Bioorganique and
[2]Laboratoire de Chimie Organique Biologique
URA31, Faculte de Chimie, Universite Louis Pasteur
1 rue Blaise Pascal, F-67008 Strasbourg, France

1. INTRODUCTION

Electrospray mass spectrometry (ESMS) of biological macromoleculeshas benefited from recent technical advances. Now it is feasible to observe the species injected at physiological pH and thus the mass determined has a direct relevance to these species present in solution (Ganem, et al., 1991, Cheng, et al., 1995). Intra- and intermolecular weak non-covalent interactions can now be determined (Smith and Light-Wahl, 1993).

2. METHODS

Aldose reductase was purified from pig lens according to the published proceedure (Reymann, et al., 1992). All the mass spectrometric studies were performed on a VG Bio-Q triple quadrupole mass spectrometer manufactured by Fisons (Manchester, UK) and upgraded so that the second quadrupole had an m/z range up to 8000 Thomson.

3. RESULTS AND DISCUSSION

Aldose reductase apoenzyme (AR) crystallizes as a dimer of a dimer (Rondeau, et al., 1992). In solution dimer of aldose reductase had been detected and on addition of the NADP+, only the monomer existed (Rondeau, et al., 1991). With ESMS, the 10 microM solution of AR in 25 mM ammonium acetate buffer pH 7 gives signals corresponding to

Enzymology and Molecular Biology of Carbonyl Metabolism 6
edited by Weiner *et al*. Plenum Press, New York, 1996

the dimer (71549 ± 9 Da, with mainly 18 and 19 charges) and to the monomer (35772 ± 5 Da, calculated from the sequence : 35780 Da, with mainly 11 and 12 charges). The fact that the number of charges of the dimer was reduced, agreed with a specific interaction. On addition of the NADP+ in stoechiometric amount, the dimer disappeared and only the monomer with the cofactor bound was detected (36512 ± 5 Da, calculated monomer plus cofactor 36522 Da).

The inhibition of aldose reductase is of therapeutic interest and a number of inhibitors of this enzyme have been and are being prepared (Dvornik, 1987). The rather low turnover of this enzyme requires rather high concentration of the protein in kinetic studies and does not allow the determination of the type of inhibition with the strong inhibitors. By ESMS, in presence of sorbinil or tolrestat, no binding of the inhibitor on AR apoenzyme (as above) was detected. And on addition of the cofactor, the ternary complex : AR-NADP+ - inhibitor was quantitatively titrated. Thus the inhibitor binds with a great affinity to the holoenzyme and not to the apoenzyme, even if the strong binding site of the inhibitors and the binding site of the cofactor are different (Wilson, et al., 1993).

On addition of 10 molar equivalents of inhibitors to AR - NADP+, multiple complexes were detected. With tolrestat, species corresponding to the binding of one, two and up to seven molecules of tolrestat per molecule of holoenzyme were detected. However sorbinil gave mainly the 1 to1 complex [(holoenzyme)$_1$ - (sorbinil)$_1$] beside a small amount of a 1 to 2 complex [(holoenzyme)$_1$ - (sorbinil)$_2$]. Thus multiple binding of inhibitor to the enzyme was detected with ESMS. Is this multiple binding of the inhibitor related to the lack of inhibition specificity ? Is this multiple binding a protection from catabolism for the inhibitor? Is this multiple binding a more general phenomena for other active substances? These remain open questions. The multiple binding has been detected by computer modeling and has been used to design a very potent inhibitor for acetylcholinesterase (Pang, et al., 1996).

4. REFERENCES

Cheng, X., Chen, R., Bruce, J.E., Schwarz, B.L., Anderson, G.A., Hofstadler, S.A., Gale, D.C., Smith, R.D., Gao, J., Sigal, G.B., Mammen,M. and Whitesides, G.M., Using Electrospray Ionisation FTICR Mass Spectrometry to Study Competive Binding of Inhibitors to Carbonic Anhydrase. (1995) J. Am. Chem. Soc. 117, 8859–8860.

Dvornik, D., (1987) Aldose Reductase Inhibition, Mc Graw-Hill, New York.

Ganem, B., Li, Y.T. and Henion, J.D., Detection of Noncovalent Receptor-Ligand Complexes by Mass Spectrometry. (1991) J. Am. Chem. Soc. 113, 6294–6296.

Pang Y.-P.,Quiram P., Jelacic T. and Brimijoin, S., A Novel Strategy for Developing Potent and Selective Acetylcholinesterase Inhibitors. (1996) Jour. Biol. Chem., in press.

Reymann, J.M., Rondeau, J.M., Barth, P., Jaquinod, M., Van Dorsselaer, A. and Biellmann, J.F., Purification and electrospray mass spectrometry of aldose reductase from pig lens. (1992) Biochim. Biophys. Acta. 1122, 1–5.

Rondeau, J.M., Moras D.,Tete, F., Podjarny, A., Dorsselaer, A. V., Reymann, J.-M., Barth,P.and Biellmann, J.F., Structural Studies of Pig Lens Aldose Reductase: Reversible Dimerisation of the Enzyme. (1991) in Enzymology and Molecular Biology of Carbonyl Metabolism 3, Weiner, H., Wermuth, B., Crabb, D. W., Ed. Plenum Press, New York, p 113–118.

Rondeau, J.M., Tête-Favier, F., Podjarny, A., Reymann, J.M., Barth, P., Biellmann, J.F. and Moras, D., Novel NADPH-binding domain revealed by the crystal structure of aldose reductase. (1992) Nature, 355, 469–472.

Smith, R.D. and Light-Wahl, K.J., The observation of non-covalent interaction in solution by electrospray ionization mass spectrometry promise, pitfalls and prognosis. (1993) Biol. Mass Spectrom. 22, 493–501.

Wilson, D.K., Tarle, I., Petrash, J.M. and Quiocho, F.A., Refined 1.8 Angstrom structure of human aldose reductase complexed with the potent inhibitor zopolrestat. (1993) Proc. Natl.Sci.Acad.U.S.A., 90, 9847–9851.

52

CHARACTERIZATION OF A NOVEL MURINE ALDO-KETO REDUCTASE

Kurt M. Bohren,[1] Oleg A. Barski,[1] and Kenneth H. Gabbay[1,2]

[1]Department of Pediatrics and
[2]Department of Cell Biology
Baylor College of Medicine
Houston, Texas 77030
kbohren@mbcr.bcm.tmc.edu

1. INTRODUCTION

The aldo-keto reductase superfamily (Bohren et al., 1989) consists of many reductases that differ in their primary structure, substrate specficities and catalytic properties. Many subfamilies have been recognized, including the aldose reductase, aldehyde reductase, 3α-hydroxysteroid dehydrogenase, androgen regulated protein from the mouse vas deferens, and many more. In the course of cloning murine liver aldo-keto reductases, we discovered yet another aldo-keto reductase with some unique properties. We hereby describe the cloning, sequencing, over-expression and characterization of this novel murine aldo-keto reductase which appears to be uniquely different from any hitherto described member of the superfamily.

2. MATERIALS AND METHODS

The methodologies used in this study were described in our previous papers (Bohren et al., 1991, 1994, Ehrig et al. 1994, Grimshaw et al 1995).

2.1. Isolation of a Novel Aldo-Keto Reductase

A mouse (strain: SV129) liver cDNA library constructed in Lambda ZAPII vector (a gift from Dr. A. Beaudet, Baylor College of Medicine), was used to screen about 0.5 million clones using a ^{32}P-labeled full length human aldose reductase cDNA probe. Colony hybridization was carried out on duplicate nylon membranes (0.45 μM, MSI, Westboro, MA) at 50 °C overnight at a final ratio of 1x10^6 cpm/2 ml of solution/membrane; hybridization solution was 6xSSC (1xSSC = 150 mM NaCl, 15 mM sodium citrate, pH 7.0), mM EDTA, and 0.5 % dry milk. Membranes were washed for three 30-min intervals in 2 x SSC containing 0.1 % SDS at 60 °C, with a final wash in 1 x SSC containing 0.05 % SDS at 62°C. After a third screening the Lambda ZAP II vectors containing various inserts were excised with helper phage and recircularized to generate subclones in the pBluescript SK-phagemid vector as outlined by the manufacturer (Stratagene).

2.2. Expression in *E. coli* and Enzyme Purification

The clone containing the 1700 bp cDNA insert was partially digested with NcoI and a 1.1 kb piece containing the encoded novel aldo-keto-reductase was ligated into the expression plasmid pET11d (Novagen, Madison, WI). The recombinant plasmids were over-expressed and the resulting proteins purified, with slight modifications noted in the results section, as previously described for human aldose and aldehyde reductase in detail elsewhere (Bohren et al, 1991).

3. RESULTS AND DISCUSSION

3.1. Library Screening and Sequence Analysis of the Novel Aldo-Keto Reductase

Restriction analysis and DNA sequencing revealed several different clones,[*] two of which could not be identified with any known sequence in the genebank using the sequence analysis GCG Package software (Genetics Computer, Inc.) although high similarity scores were obtained as described below. Digestion with endonucleases EcoRI and XhoI released two different sized inserts, i.e. 1200 and 1700 bp. Sequencing of both inserts showed that the smaller insert is completely contained in the larger one which extends 26 bp on the 5' and 351 bp on the 3' end. Interestingly, both inserts exhibit a poly(A) tail. Fig.1 shows the cDNA sequence of both clones and the open reading frame corresponding to a protein with 301 amino acids and a calculated M_r of 34,482. Putative polyadenylation signals are found at position 1272 (AATTAAAA) and at position 1623 (AATATAAA) of the two sequences.

Comparison of the deduced amino acid sequence of the new mouse cDNA with members of the aldo-keto reductase superfamily revealed extensive similarities between aldose reductases from several species, human aldehyde reductase (Bohren et al, 1989), bovine prostaglandin F synthase, (Watanabe et al, 1988), 3α-hydroxysteroid dehydrogenase (Cheng et al, 1991), mouse androgen-dependent protein from vas deferens (Pailhoux et al 1990), a mouse protein whose gene is activated by fibroblast growth factor-1 (Donohue et al , 1994), and many others. The overall similarity between the novel mouse protein and murine aldose reductase, excluding one gap consisting of six residues, is 60 % and increases to 72 % if conservative substitutions according to Dayhoff (1978) are considered. The same numbers are obtained when the comparison is made with human aldose reductase although the similarities are slightly different than the ones between the two mouse proteins. The most striking difference besides the absence of the first nine residues in the deduced sequence of the new mouse cDNA and aldose reductase is the gap compris-

---

[*] The nucleotide sequence reported in this paper has been submitted to the GenBank™/EMBL Data Bank with accession number U 68535.

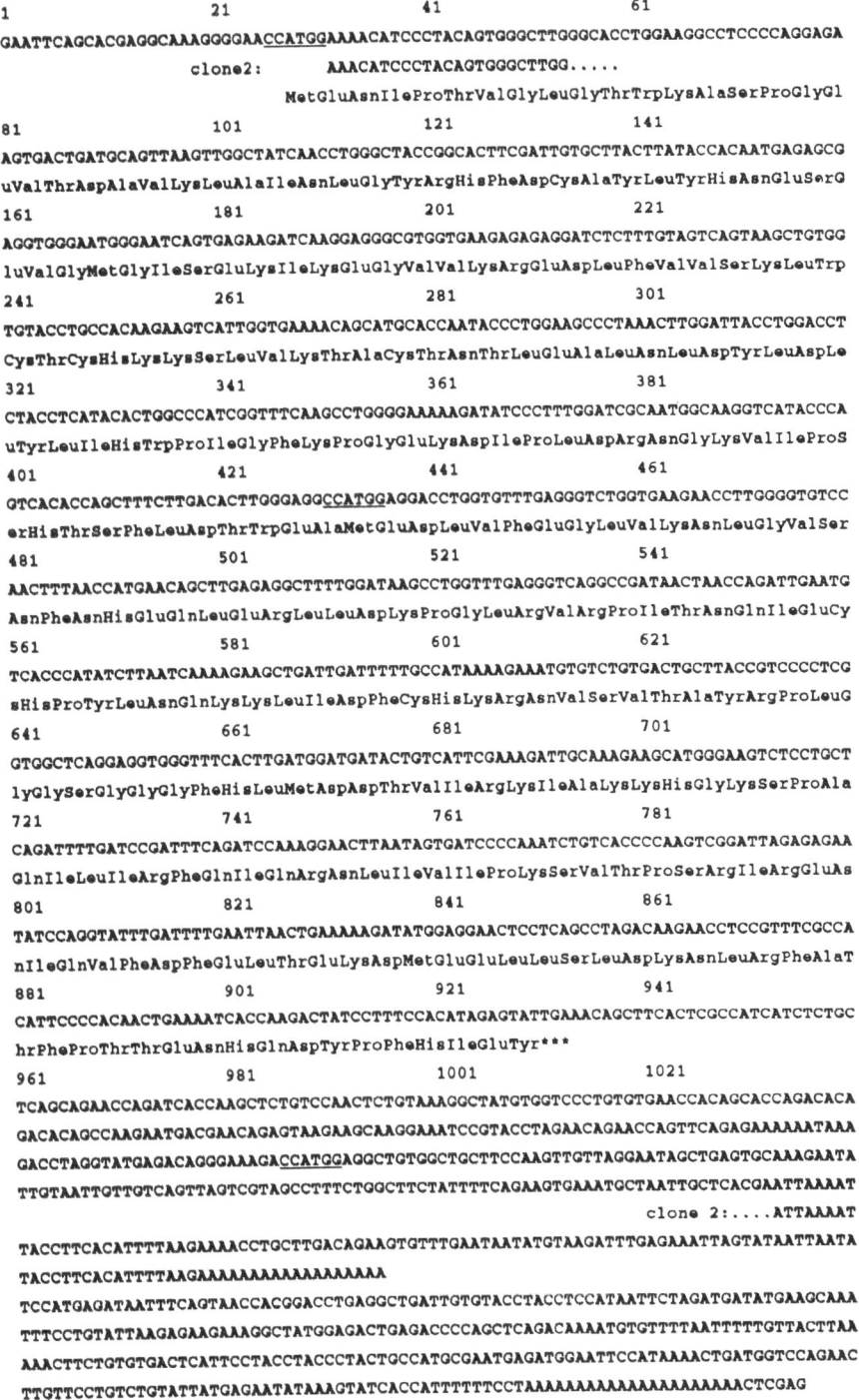

Figure 1. Nucleotide and deduced amino acid sequence of a novel murine aldo-keto reductase. The 5' and 3' end of a shorter but otherwise identical cDNA starting at position 33 and ending with a poly(A) tail at position 1294 is also shown (clone2). Putative adenylation signals, and NcoI sites are underlined.

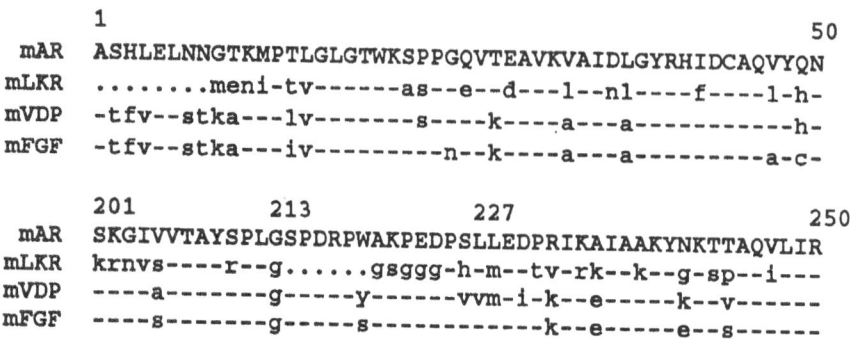

Figure 2. A computer assisted partial alignment with several aldo-keto reductases of the N-terminal domain and the six-residue area which is missing in the novel murine aldo-keto reductase. This "gap" is part of a loop that enfolds over the coenzyme in aldose reductase. Numbers follow the human aldose reductase sequence. Insertions (.) are introduced to maximize homology. Dashes indicate exact matches with the murine aldo-keto reductase. Note the Gly213 which is the "hinge" of the nucleotide enfolding loop (Wilson et al., 1992). Mouse aldose reductase, mAR; novel mouse liver keto reductase, mLKR; mouse vas deferens androgen regulated protein, mVDP; mouse fibroblast growth factor induced protein, mFGF.

ing six amino acid residues approximately in the middle of the sequence. In human aldose reductase these six residues initiate a loop that folds over the pyrophosphate of NADPH and is the structural basis for the slow release of the coenzyme in the catalytic reaction sequence of aldose reductase (Grimshaw et al, 1990;1995 Kubiseski et al 1992). The residues flanking this loop are conserved: Gly213 and Leu227 in aldose reductase from all species so far investigated correspond to Gly204 and Leu212 in the deduced mouse sequence (Fig 2). Interestingly, this shortened loop in the mouse protein might still be very flexible since five residues out of 10 are glycine residues. Extending the comparison to other members of the aldo-keto reductase superfamily shows that in this loop area other proteins have gaps of various sizes, ranging from 7 (*corynebacterium* 2,5-diketo-D-gluconic acid reductase) to 10 residues (*Leishmania* putative reductase). Unlike human chlordecone reductase, where 42 supposedly untranslated nucleotides 5' to the ATG could code for an amino acid sequence closely similar to the predicted actual amino acid sequence of prostaglandin F synthase (Bruce et al , 1994), translation of the nucleotides 5' of the ATG in the novel mouse sequence do not yield a sequence homologous to any N-terminus of a aldo-keto reductase. This indicates that the ATG initiation codon in the novel mouse cDNA is probably correctly identified and that clone 1 contains a complete coding sequence. Other pertinent residues that are extremely conserved and in the case of human aldose reductase (in parenthesis) form part of the active site or are involved in the mode of binding of the coenzyme (19) include Tyr39 (Tyr48), His101 (His110), Tyr200 (Tyr 209), and Lys253 (Lys262).

3.2. Expression and Physical Characterization of the Novel Aldo-Keto-Reductase

To characterize the novel mouse liver protein, we expressed its cDNA in the pET system, a system we have been using to produce large quantities of human aldose and aldehyde reductase (Bohren et al, 1991). Due to an internal NcoI site at position 431 of the cDNA sequence, partial digestion of the cDNA was used and ligation inserted a 1.1 kb

Figure 3. SDS-polyacrylamide gel electrophoresis of purified recombinant novel mouse keto reductase, mLKR (lane2) and human aldose reductase, hAR (lane 1). Molecular mass standards were also run (lane M).

mouse cDNA fragment into the pET expression vector pET11d. Maximal expression was obtained 6 h following induction with ß-D-thiogalactoside, with a yield of 20–25 mg per 3-liter culture. Protein purification was based on the procedure used successfully for human recombinant aldose and aldehyde reductase. The exceptionally tight binding of the mouse protein to the Matrex Orange A affinity chromatography material, however, made it necessary to elute the protein with a steep salt gradient instead of the typical pulse of NADPH. The expressed mouse protein eluted at approximately 2.5 M NaCl. The purified protein shows a single band on SDS-polyacrylamide gel electrophoresis (Fig.3). The gel also shows that the recombinant mouse protein has appropriately greater mobility than recombinant human aldose reductase (predicted M_r of 34.4 kD and M_r = 35.7 kD, respectively. Specific human aldose reductase antibodies did not recognize the mouse protein on Western blot analysis (not shown). Isolectric focusing gels show an isoelectric point of 6.7 (Fig.4a) which is slightly higher than the isoelectric point of human recombinant aldose reductase (pH = 6.2). Fig.4B shows an IEF gel stained for activity using 9,10-phenanthrenequinone as a substrate. In this method, the enzymatic reduction product, 9,10-phenanthrenehydroquinone reduces p-nitrotetrazolium blue to formazan that forms a blue precipitate.

Figure 4. Polyacrylamide gel isoelectric focusing of recombinant novel mouse keto reductase, mLKR (1) and human aldose reductase, hAR (2). Two different concentration of each protein were run. Marker proteins with characteristic isoelectric points are in lane M. Gel A is Coomassie-stained and gel B is activity-stained.

Table 1. Substrate specificity of novel mouse aldo-keto reductase (mLKR) compared to human aldose reductase (hAR)

	hAR			mLKR		
	k_{cat} s^{-1}	K_m mM	k_{cat}/K_m $M^{-1}s^{-1}$	k_{cat} s^{-1}	K_m mM	k_{cat}/K_m $M^{-1}s^{-}$
DL-Glyceraldehyde	0.53	0.03	18300	1.28	23	56
D-Glucuronate	0.48	5	96	0.1	138	0.7
D-Xylose	0.43	6.2	72	0.83	13000	0.1
p-Nitrobenzaldehyde	0.41	0.004	95000	0.63	2.6	245
Menadione	0	-	-	0.08	0.03	2700
9,10-phenanthrenequinone	0.5	0.02	25000	0.31	0.02	13500

3.3. Substrate Specificity

Table 1 shows that the recombinant mouse protein has reducing activity with a wide range of aldehydes and quinones. The catalytic effectiveness (k_{cat}/K_m) increases with increasing hydrophobicity and quinone character of the substrate. Thus 9,10-phenanthrenequinone is 20×10^3 times more efficiently reduced than D-glucuronate, while carbohydrates such as DL-glyceraldehyde and D-xylose are poor substrates. The recombinant mouse enzyme is specific for NADPH as coenzyme, since less than 5% of the NADPH-linked activity was detectable when NADH (0.2 mM) was used as the coenzyme. Overall, the substrate specificity is very different from that of human or murine aldose reductase.

3.4. Kinetic Implications

As already mentioned, the novel reductase has a gap approximately in the middle of the sequence when aligned to other aldo-keto reductases. Based on the atomic structure of human aldose reductase this gap results in a shortened nucleotide enfolding loop in the novel mouse keto reductase. The question arose as to what extent this nucleotide enfolding loop can influence the kinetics of aldo-keto reductases in general. Most of the diversity in terms of variation of amino acid residues among aldo-keto reductases is concentrated at the N and C-termini, and in this nucleotide enfolding loop. We may speculate what happens, if we were to delete those six residues in, e.g., human aldose reductase: Would such an engineered aldose reductase assume, at least in part, the kinetic properties of a loopless enzyme or be completely inactive? Assuming that there is no global disruption of the enzyme structure we would expect an increase in the nucleotide binding constants similar to the mouse novel keto reductase. Since release of NADP⁺, or more precisely, the isomerization of the enzyme (opening up of the nucleotide enfolding loop) is by far the most rate-limiting step in the overall reduction reaction of wild type human aldose reductase, an increase in k_{cat} can also be expected if the tight binding of the nucleotide is weakened. This is reminiscent of the characteristics of many mutants of aldose reductase. In the course of investigating human aldose reductase, several mutations were made by us and others (Bohren et al, 1991; Kubiseski et al, 1995) that affected coenzyme binding, i.e., C298A, K262M, K262R, D216A, and R268M. In each case the exceptionally tight binding of nucleotide in the wild type enzyme was weakened in the mutants with a concomitant increase in k_{cat} and $K_{mAldehyde}$ (Table 2). As described above, an increase in k_{cat}

Table 2. Kinetic parameters of mLKR and some aldose reductase mutants

Enzyme	Kd_{NADP+} (nM)	$K_m^{\dagger}$ (mM)	k_{cat} (s^{-1})
mLKR	124	23	1.28
hAR_{WT}[a]	6	0.03	0.53
hAR_{K262M}[b]	↑*	0.7	0.83
hAR_{K262R}[c]	↑*	2.2	3.8
hAR_{R268M}[d]	16000#	0.08	1.1
hAR_{C298A}[a,c]	60	0.6	4
hAR_{D216A}[c]	100	0.2	1.4

† DL-Glyceraldehyde
* K_{dNADP}+ not determined, but $K_{mNADPH} > 100\mu M$
# data adjusted to reflect same measuring conditions
a) from Ehrig et al (1994)
b) from Bohren et al (1991)
c) from Bohren et al (unpublished)
d) from Kubiseski et al (1995)

is easily explained, but what about an increase in $K_{mAldehyde}$? Does an increase in $K_{mAldehyde}$ necessarily mean that the mutation caused a change in the affinity for the substrate or could it also be due to a change in the chemistry of the reaction, or even due to an increase of the nucleotide release rate? Part of the answer might be found in the fact that in a complex mechanism such as the one followed by aldo-keto reductases, K_m does not denote binding affinity for a given substrate, but is a term defined by many rate constants of different steps of the specific mechanism.

To analyze this issue, we used the mechanism of Scheme 1 and the rate constant determined for the reduction and oxidation of D-xylose and xylitol, respectively, at pH 8.0, catalyzed by the wild-type human aldose reductase (Grimshaw et al 1995). We systematically varied certain rate constants while others were kept constant at the wild-type values shown in Scheme1 for the calculation of K_m and k_{cat} (Grimshaw et al, 1995). Figures 5A-5D illustrate some of the results in two and three-dimensional plots.

Figure 5 A shows that K_m and k_{cat} increase when the rate constant k_{11} was varied from 0.05 s^{-1} to 2.5 s^{-1} while all other constants were kept at wild-type values. Since k_{11}, the isomerization and release of NADP+ is the slowest step in the whole mechanism, it is

	forward		reverse
	k_1 1.2 x 10^8 $M^{-1}s^{-1}$	k_2	174 s^{-1}
	k_3 89 s^{-1}	k_4	0.83 s^{-1}
Mechanism of Human Aldose Reductase†	k_5 220 $M^{-1}s^{-1}$	k_6	25 s^{-1}
	k_7 130 s^{-1}	k_8	0.60 s^{-1}
	k_9 1.0 x 10^6 s^{-1}	k_{10}	5 x 10^6 $M^{-1}s^{-1}$
	k_{11} 0.23 s^{-1}	k_{12}	150 s^{-1}
	k_{13} 623 s^{-1}	k_{14}	2.0 x 10^8 $M^{-1}s^{-1}$
	k_{cat} 0.19 s^{-1}	K_m	1.0 mM

$$E \underset{k_2}{\overset{k_1}{\rightleftharpoons}} E \bullet NADPH \underset{k_4}{\overset{k_3}{\rightleftharpoons}} {}^*E \bullet NADPH \underset{k_6}{\overset{k_5}{\rightleftharpoons}} {}^*E \bullet NADPH \bullet RCHO$$

$$k_8 \uparrow\downarrow k_7$$

$$E \underset{k_{13}}{\overset{k_{14}}{\rightleftharpoons}} E \bullet NADP^+ \underset{k_{11}}{\overset{k_{12}}{\rightleftharpoons}} {}^*E \bullet NADP^+ \underset{k_9}{\overset{k_{10}}{\rightleftharpoons}} {}^*E \bullet NADP^+ \bullet RCH_2OH$$

†) from Grimshaw et al (1995)

Scheme 1

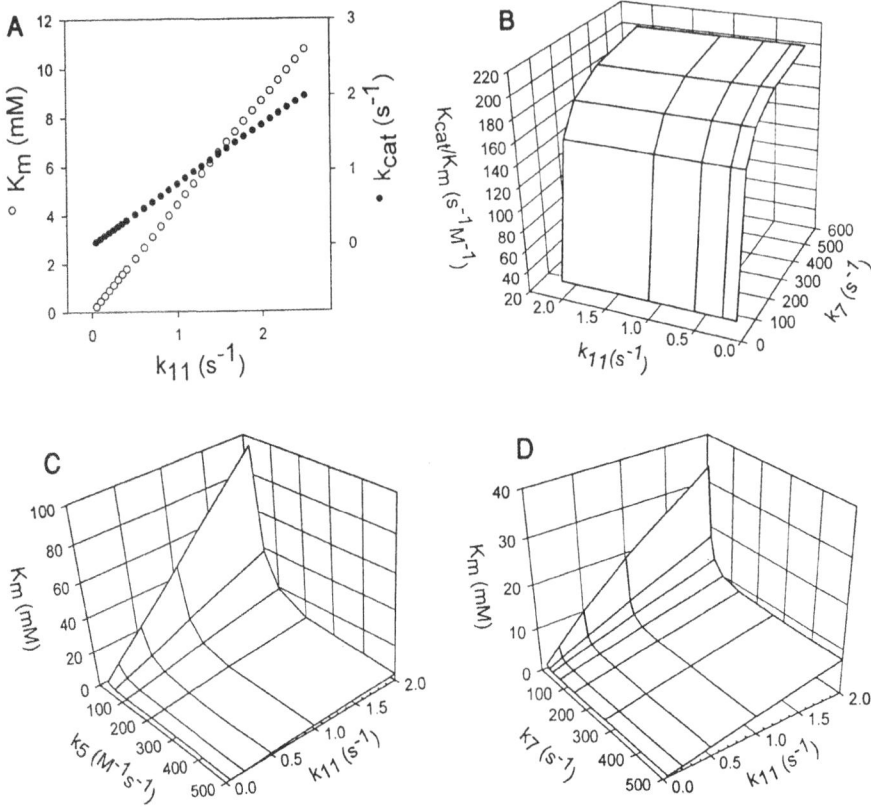

Figure 5. Plots from calculations based on the mechanism in Scheme 1 ($K_m = K_{mAldehyde}$).

clear that the turnover number or k_{cat}[†] increases with an increase of the isomerization rate. What is surprising is the increase of $K_{mAldehyde}$. This model partially reflects the situation observed with mLKR. As we have just demonstrated, higher k_{11} alone would explain to a certain degree higher values of $K_{dNucleotide}$, k_{cat} and K_m observed with this enzyme. However, the values of $k_{cat}/K_{mAldehyde}$ are much lower for mLKR than the ones for hAR (Table 2). This cannot be explained by an increase in k_{11} alone, as demonstrated in Figure 5B. This figure illustrates the well-known fact that k_{cat}/K_m does not include steps beyond addition of aldehyde or release of alcohol. Thus the catalytic effectiveness (k_{cat}/K_m) does not comprise the isomerization steps of the aldose reductase mechanism, for k_{cat}/K_m remains constant over all values of k_{11}, but varies over all values of k_7, the rate constant that together with k_8 determines the "chemistry" of the catalyzed reaction.

Thus, besides the shortening of the nucleotide enfolding loop and subsequent loosening of the binding of the nucleotide, other structural features of mLKR must be responsible for the different apparent kinetic parameters as compared to human aldose reductase.

---

† It seems actually misleading to call the turnover number (k_{cat}) "catalytic constant" in the case of aldose reductase, since the turnover number is not determined by catalytic steps in the mechanism, but by the isomerization, which happens to be the slowest step!

If we vary k_{11} and additionally k_5 (k_5 is the on-rate of substrate), assuming that those other structural features affect substrate binding, then the effect on K_m is cumulative and quite dramatic, and this is shown in plot 5C. This plot shows that in a hypothetical aldose reductase mutant an increase in k_{11} leads to an increase in K_m, unless the on-rate of aldehyde substrate is also increased.

A good example of such a case is the C298A mutant of human aldose reductase. Using transient kinetic data we have been able to identify precisely the origin of the 37-fold increase of $K_{mXylose}$ for the C298A mutant enzyme compared to the wild type enzyme. The increase in $K_{mXylose}$ is due largely to the 9-fold increase of the rate of $NADP^+$ release (k_{11} in scheme 1), the rate constant for substrate binding (k_5) accounts for only a factor of 5 (Grimshaw et al, 1995). Thus, the major effect on the $K_{mXylose}$ value for the C298A mutant arises from a change in the rate constant for a step which occurs at a point in the mechanism where neither aldehyde substrate nor alcohol product is involved!

A less tight binding of the nucleotide could also affect the alignment of the C4N carbon of the nicotinamide and thereby change the chemistry of the reaction, i.e. an increase or decrease of the of the rate of hydride transfer (k_7). This scenario is shown in figure 5C. Again, K_m increases considerably, with high values of k_{11} and low values of k_7. This is exactly the opposite of what we would expect from the simplest form of mechanism with only one central complex ($E+S \xleftarrow{k_{-1}} k_1 \rightarrow ES \; k_2 \rightarrow E + P$) where k_2 corresponds to k_7 in the more complex mechanism, and where K_m is defined as $K_m = (k_2+k_{-1})/k_1$. We would like to stress once more that the physical significance of K_m cannot be stated with any certainty in the absence of other experimental data concerning the relative magnitudes of the various microscopic rate constants. However, K_m is a useful constant that relates the overall rate of an enzyme-catalyzed reaction to the substrate concentration, and is therefore valuable for comparisons of different substrates or enzymes.

In conclusion, some unique properties of a novel murine keto reductase prompted us to think about and study again the mechanism of aldose reductase. While we were not yet able to pinpoint the details for the differences in kinetic properties between mLKR and hAR, it nevertheless became evident that variations in the nucleotide binding of aldo-keto reductases are sufficient to affect apparent kinetic properties like k_{cat} and K_m. This phenomenon was rationalized in a first approximation by simply analyzing possible changes in the basic mechanism of a wild-type enzyme. A prerequisite was the availability of this mechanism. The ultimate kinetic analysis of mLKR or any other homologue enzyme, however, can only be achieved experimentally by determination of its individual microscopic rate constants.

4. REFERENCES

Bohren KM, Bullock B, Wermuth B, Gabbay KH. (1989) Aldo-keto reductase superfamily: cDNAs and deduced amino acid sequences of human aldehyde and aldose reductases. J Biol Chem 264:9547–9551.

Bohren KM, Page JL, Shankar R, Henry SP, and Gabbay KH (1991) Expression of Human Aldose and Aldehyde Reductases: Site-directed Mutagenesis of a Critical Lysine-262. J. Biol. Chem. 266:24031–24037.

Bohren KM, Grimshaw CE, Lai C-J, Harrison DH, Ringe D, Petsko GA, Gabbay KH. (1994) Tyrosine-48 is the proton donor and histidine-110 directs substrate stereochemical selectivity in the reduction reaction of human aldose reductase: enzyme kinetics and the crystal structure of the Y48H mutant enzyme. Biochemistry 33:2021–2032.

Bruce NC, Willey DL, Coulson AFW, Jeffery J. (1994) Bacterial morphine dehydrogenase further defines a distinct superfamily of oxidoreductases with diverse functional activities. Biochemistry 299:805–811.

Cheng K-C, White PC, and Qin K-N (1991) Molecular Cloning and Expression of Rat Liver 3α-hydroxysteroid dehydrogenase. Mol. Endocrinol. 5, 823–828.

Dayhoff, MO, Schwartz, RM, and Orcutt, BC (1978) in Atlas of Protein Sequence and Structure (Dayhoff, MO ed) Vol 5 Suppl. 3 pp 345–352, National Biomedical Research Foundation, Washington, D.C.

Donohue, PJ, Alberts, GF, Hampton BS, and Winkles JA (1994) A delayed-early Gene activated by fibroblast growth factor-1 encodes a protein related to aldose reductase, J. Biol.Chem.269, 8604–8609.

Ehrig T, Bohren KM, Prendergast F G, and Gabbay KH (1994) Mechanism of aldose reductase inhibition: Binding of NADP⁺/NADPH and alrestatin-like inhibitors, Biochemistry 33, 7157–7165.

GCG, Genetics Computer Group, Program Manual for the Wisconsin Package, Version 8, September 1994, 575 Science Drive, Madison, Wisconsin, USA 53711.

Grimshaw CE, Shahbaz M, and Putney CG. (1990) Mechanistic basis fo nonlinear kinetics of aldehyde reduction catalyzed by aldose reductase. Biochemistry 29:9947–9955.

Grimshaw CE, Bohren KM, Lai C-J, and Gabbay KH (1995). Human aldose reductase: Rate constants for a mechanism including interconversion of ternary complexes by recombinant wild type enzyme. Biochemistry 34, 14356–14365.

Grimshaw CE, Bohren KM, Lai C-J, and Gabbay KH (1995). Human Aldose Reductase: Subtle Effects Revealed by Rapid Kinetic Studies of the C298A Mutant Enzyme. Biochemistry 34,14366–14373.

Gui T, Tanimoto T,kokai Y, and Nishimura C (1995) Presence of a closely related subgroup in the aldo-keto reductase family of the mouse. Eur. J. Biochem. 227., 448–453.

Kubiseski TJ, Hyndman DJ, Morjana NA, and Flynn TG. (1992) Studies on pig muscle aldose reductase. J Biol Chem 267:6510–6517.

Kubiseski TJ and Flynn TG (1995) Studies on human aldose reductase. Probing the role of arginine 268 by site-directed mutagenesis. J. Biol. Chem 270, 16911–16917.

Pailhoux EA, Martinez, A.,Veyssiere GM, Jean CG. (1990) cDNA cloning and protein homology with aldo-keto reductase superfamily. J. Biol. Chem. 256:19932–19936.

Varma, SD.,and Kinoshita, JH.(1974) The absence of cataracts in mice with congenital hyperglycemia, Exp.Eye. Res. 19, 577–582.

Varma SD, Mizuno F, Kinoshita JH (1977) Diabetic Cataracts and Flavonoids. Science 195, 205–206.

Watanabe, K, Fujii, Y, Nakayama, K Ohkubo, H, Kuramitsu, S Kagamiyama, H, Nakanishi, S and Hayaishi, O (1988) Structural similarity of bovine lung prostaglandin F synthase to lens ε-crystallin of the Proc.Natl. Acad.Sci. U.S.A. 85, 11–15.

Wilson DK, Bohren KM, Gabbay KH, Quiocho FA. (1992) An unlikely sugar substrate site in the 1.65 Å structure of human aldose reductase holoenzyme implicated in diabetic complications. Science 257:81–84.

53

A POTENTIAL ROLE FOR ALDOSE REDUCTASE IN STEROID METABOLISM

J. Mark Petrash,[1,2] Theresa M. Harter,[1] and Gary L. Murdock[3]

[1]Department of Ophthalmology and Visual Sciences
[2]Department of Genetics
[3]Department of Obstetrics and Gynecology
Washington University School of Medicine
St. Louis, Missouri 63110

1. INTRODUCTION

Aldose reductase (ALR2) is a monomeric NADPH-dependent reductase distinguished from other members of the aldo-keto reductase enzyme family in its ability to catalyze the reduction of a variety of hexoses and pentoses. The role of ALR2 in enhanced polyol synthesis in diabetic and galactosemic tissues is well documented (Kinoshita and Nishimura, 1988), as is the potential therapeutic use of ALR2 inhibitors to delay or prevent the onset and progression of metabolic complications leading to cataract and retinopathy (Sarges and Oates, 1993). New insights into the structure of ALR2, together with emerging data demonstrating that many individual aldo-keto reductases may participate in divergent metabolic pathways, lead us to question whether sugars represent the only physiologically-relevant substrate of this enzyme.

Several lines of evidence point to steroids as potential in vivo substrates of ALR2. Wermuth and Monder (1983) demonstrated almost 15 years ago that both aldehyde reductase and ALR2 efficiently catalyzed the reduction of synthetic isocorticosteroids containing a 20-hydroxy-21-aldehyde side chain. This steroid class was postulated to represent a metabolic intermediate in the conversion of corticosteroids to their corresponding acid metabolites via a so-called "long loop" pathway (Monder and Bradlow, 1977). While the existence of this metabolic pathway has not yet been definitively established, the tissue distribution pattern of ALR2 gene expression seems consistent with that of an enzyme involved in steroid metabolism. Since the original observation by Hers that ALR2 in seminal vesicles may participate in conversion of glucose to fructose (1960), many investigators have demonstrated by enzymatic and immunochemical methods that ALR2 is expressed in tissues involved in both synthesis of and sensitivity to steroid hormones (Ludvigson et al., 1982; Grimshaw and Mathur, 1989; Iwata et al., 1990; Vander Jagt et al., 1990) . Contrary to that expected for a receptor of monosaccharides, the active site of human ALR2 has been shown by high resolution X-ray crystallography studies to be lined with hydrophobic residues (Borhani et al., 1992; Wilson et al., 1992). Evidence linking ALR2 to steroid metabolism was further substantiated with the recent demonstration that bovine testicular 20α-hydroxysteroid dehydrogenase (20αHSD) and bovine lens ALR2 are encoded by the same gene (Warren et al., 1993).

To examine the potential role of ALR2 in steroid metabolism, we have cloned and overexpressed the genes encoding bovine and human ALR2 and have evaluated their ability to catalyze the reduction of various steroid substrates. Using a bromoacetoxysteroid as an affinity label, we present evidence that Cys-298 represents an important structural determinant of the steroid binding site. Since this reactive cysteine is located at the active site, these results provide further evidence that steroids could represent an important class of in vivo substrates for ALR2.

2. MATERIALS AND METHODS

2.1. Fractionation of Aldo-Keto Reductase Activities from Bovine Testicle Homogenates

Reductase activities were fractionated from a phosphate buffer extract of bovine testicular tissue using chromatofocusing (PBE 94, Pharmacia LKB). Reductase activities eluted from the chromatography matrix by a pH gradient (pH 4.0 - 7.0) were detected in column fractions by spectrophotometric assays utilizing either DL-glyceraldehyde (5 mM) or 17α-hydroxyprogesterone (0.18 mM) as carbonyl substrates. Enzyme assays were carried out at 23 °C in 50 mM potassium phosphate, pH 6.2, containing 0.4 M ammonium sulfate, 0.15 mM NADPH and substrate at the indicated concentration. Reductase activity in pooled fractions was further characterized by estimating the sensitivity to inhibition by zopolrestat and/or sorbinil (both kindly provided by Pfizer Central Research, Groton, CT) through measurement of the concentration of inhibitor required to bring about 50% inhibition (IC_{50}) using DL-glyceraldehyde as substrate.

2.2. Cloning and Overexpression of Bovine Aldose Reductase

A full length cDNA encoding bovine ALR2 was obtained using the polymerase chain reaction (PCR) on phage DNA extracted from a bovine retina pigment epithelium cDNA library (library generously provided by Dr. Debra Thompson, University of Michigan). ALR2 sequences were amplified from phagemid DNA using an iterative three step process. The first two amplification steps were carried out using nested sets of primers designed to anneal at vector sequences immediately upstream from the cDNA insertion site and downstream at ALR2 sequences identified in the bovine ALR2 cDNA clones described previously (Petrash and Favello, 1989). Products from the second amplification were subjected to direct PCR sequencing to yield the ALR2 mRNA structure corresponding to the 5' untranslated region extending 3' into the coding region. The resulting sequence was then used to design a third nested upstream primer to anneal at a site overlapping the presumed translation initiation codon. This upstream primer was combined with a nested downstream primer to amplify the region roughly corresponding to codons 1–56 of the ALR2 mRNA. The resulting product was then cleaved with NcoI and KpnI and ligated with a KpnI-digested cDNA fragment containing the remaining coding sequences (Petrash and Favello, 1989) to generate a full length bovine ALR2 cDNA fragment. Coding sequences were transferred into an expression vector similar to that used previously for expression of human ALR2 (Tarle et al., 1993) and the structure verified by nucleotide sequencing. Recombinant bovine ALR2 was overexpressed in Escherichia coli host cultures and purified using chromatofocusing and hydroxylapatite chromatography as described previously for human ALR2 (Tarle et al., 1993).

2.3. Overexpression of Murine FR-1

Quantities of FR-1 were purified from E. coli host cultures essentially as described previously (Wilson et al., 1995).

2.4. Preparation of Affinity Labeling Reagents

For inactivation studies, 16α-bromoacetoxyprogesterone (16αBAP) was synthesized as described previously (Sweet et al., 1972). For covalent modification and labeling, radioactivity was introduced into the inactivator with the use of ^{14}C-bromoacetic acid during synthesis (Figure 1). The purified [2'-^{14}C] 16α-BAP had a specific activity of 40.9 μCi/μmole.

ALR2 was purified from bovine testis on the basis of its activity as a 20αHSD using the procedure described previously (Pineda et al., 1989). For large scale labeling, the enzyme (3 μM) was dialyzed overnight against 20 mM potassium phosphate, pH 6.6 containing 20% glycerol and 4% ethanol to remove essentially all sulfhydryl reducing agents. The dialyzed material was then incubated in the presence of 30 μM [^{14}C]16α-BAP and NADPH for 12 hours at 23 °C. Reactions were terminated with the addition of excess β-mercaptoethanol, neutralized and unreacted [2'-^{14}C] 16α-BAP removed by exhaustive dialysis against ammonium bicarbonate buffer, pH 8.0. Lyophilized samples were treated with 50 mM NaOH followed by proteolytic digestion with trypsin as described previously (Murdock et al, 1988). The resulting peptides were separated by reverse phase HPLC and those containing radioactivity were pooled and further purified by a second cycle of RP-HPLC using different gradient elution conditions. Purified peptides were subjected to automated Edman degradation in an Applied Biosystems Model 470A gas phase sequenator with on-line PTH-amino acid analysis. PTH-amino acid derivatives, including carboxymethyl cysteine-PTH, were identified by co-elution with appropriate standards.

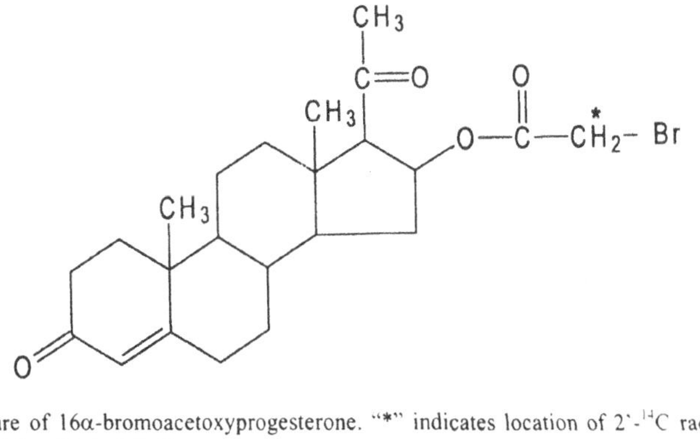

Figure 1. Structure of 16α-bromoacetoxyprogesterone. "*" indicates location of 2'-^{14}C radiolabel in the compound synthesized for affinity labeling.

2.5. Inactivation Studies

Enzymes were transferred by dialysis or column desalting procedures into potassium phosphate buffer, pH 6.6 - 7.0 immediately prior to affinity inactivation studies. All incubation mixtures contained 5% ethanol as cosolvent and 0.15 mM NADPH. Incubations were initiated by addition of 16α-BAP dissolved in ethanol and the reaction quenched by transferring aliquots at the indicated times to a buffer containing 10 mM dithiothreitol. Reductase activity in reaction mixtures was determined by a standard ALR2 assay (Tarle et al., 1993) using either DL-glyceraldehyde (1 mM) or 17α-hydroxyprogesterone (0.18 mM) as substrate.

3. RESULTS AND DISCUSSION

3.1. Fractionation of Aldo-Keto Reductases from Bovine Testicular Homogenates

To evaluate whether ALR2 and 20αHSD activities are expressed by the same gene product, we fractionated reductase and HSD activities present in bovine testis homogenates by isoelectric separation using chromatofocusing. Two peaks of ALR2-like activity (as defined by the ability to catalyze the NADPH-dependent reduction of DL-glyceraldehyde) were detected in column fractions. The first reductase peak (Peak I), which consti-

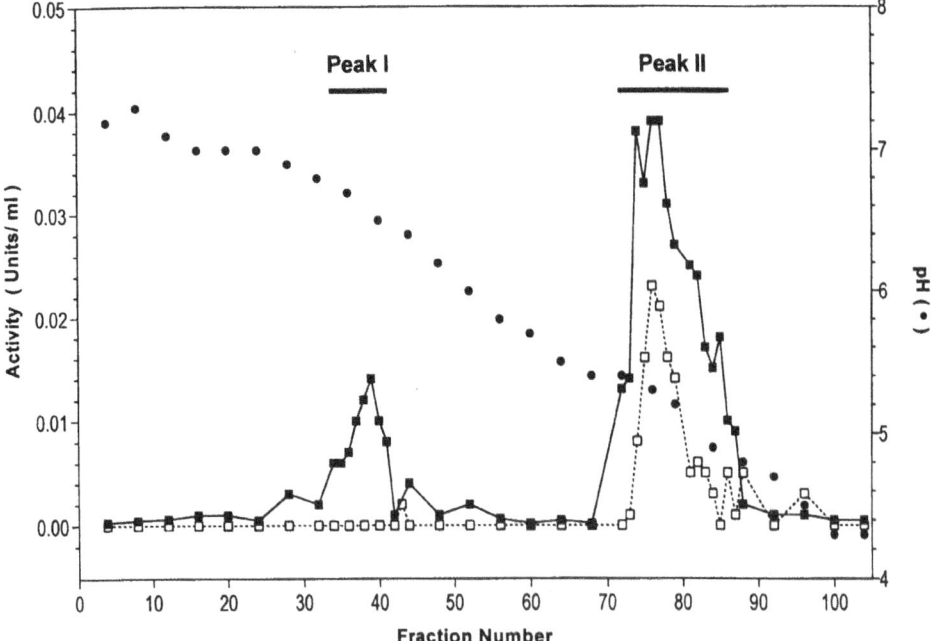

Figure 2. Fractionation of aldo-keto reductases by chromatofocusing. Elution of bovine testis aldo-keto reductases were monitored by enzyme assays using DL-glyceraldehyde (■) and 17α-hydroxyprogesterone (□). The pH gradient in column fractions is indicated by (●). Fractions pooled to constitute Peaks I and II are indicated by solid bars.

tuted approximately 11% of the total glyceraldehyde reductase activity applied to the column, eluted at a position corresponding to pH ~6.5. The second reductase peak (Peak II), which contained approximately 63% of starting glyceraldehyde reductase activity, eluted at a position corresponding to pH~ 5.4 (Figure 2). In contrast, a single peak of 20αHSD activity (as defined by the ability to catalyze the NADPH-dependent reduction of 17α-hydroxyprogesterone) was observed to elute at a position coincident with Peak II. We have tentatively determined that the reductase activity in Peak I corresponds to aldehyde reductase (glucuronate reductase). This tentative assignment was based on apparent K_m DL-glyceraldehyde and an IC_{50} sorbinil values that were similar to those published previously (Tanimoto et al., 1991). By similar criteria, reductase activity corresponding to Peak II was determined to represent ALR2. This pattern of elution in the pH gradient also is consistent with the pI values estimated from the amino acid compositions of aldehyde and aldose reductases. When probed with antibodies raised against bovine lens ALR2 (Petrash et al., 1991), a M_r~36,000 immunoreactive band was observed with material from peak II whereas no specific immunoreactivity was observed in lanes containing material from peak I (data not shown). Similar patterns in the distribution of reductase activities and segregation of 17α-hydroxyprogesterone reductase activity with ALR2 were observed with homogenates prepared from rat testes (data not shown). These results are consistent with the hypothesis that ALR2 and 20αHSD activities are expressed by the same gene product.

3.2. Cloning and Expression of Bovine Aldose Reductase

If ALR2 and 20αHSD activities are expressed by the same gene product, we hypothesized that purified recombinant bovine ALR2 should be capable of catalyzing the NADPH-dependent reduction of 17α-hydroxyprogesterone. To test this hypothesis, it was necessary to first clone the full length bovine ALR2 cDNA. It is somewhat surprising that the full length cDNA encoding bovine ALR2 had not been previously reported even though this enzyme has served for over 30 years as a prototype for kinetic and structural study of the aldo-keto reductase enzyme family (Sheaff and Doughty, 1976; Hayman and Kinoshita, 1965). Indeed, in our previous cloning studies we routinely failed to obtain full length cDNA clones despite extensive immunological and probe hybridization screening of expression libraries constructed from many bovine libraries including those from lens (Petrash and Favello, 1989) and testes (Warren et al., 1993). Efforts to establish the N-terminal sequence of ALR2 were similarly unsuccessful due to apparent blockage of the N-terminal residue (Petrash and Favello, 1989; Warren et al., 1993). With the aid of PCR amplification technology, we have now been successful in cloning the missing sequences and have established the complete sequence encoding bovine ALR2. The 315 residue primary sequence of bovine ALR2 is shown aligned with the sequence of human ALR2 in Figure 3. The deduced sequence of bovine ALR2 is identical with regard to the sequence of peptides produced by endoproteinase Lys-C treatment of bovine testicular enzyme isolated on the basis of its activity as a 20αHSD (Warren et al., 1993). Overall, the bovine and human ALR2 sequences are approximately 88% identical. As expected, the important catalytic residues identified in human ALR2, namely Tyr-48, His-110, Lys-77 and Asp-43, are conserved in the bovine sequence (Bohren et al., 1994; Tarle et al., 1993).

As with the enzymes purified from native bovine tissues, the kinetic parameters of recombinant bovine and human ALR2s were found to be highly similar. Both enzymes expressed activity with a broad range of carbonyl substrates. Of particular interest to this study was the relatively low Michaelis constant for 17α-hydroxyprogesterone, which was in the low micromolar range for both enzymes (Table 1). Indeed, the apparent catalytic ef-

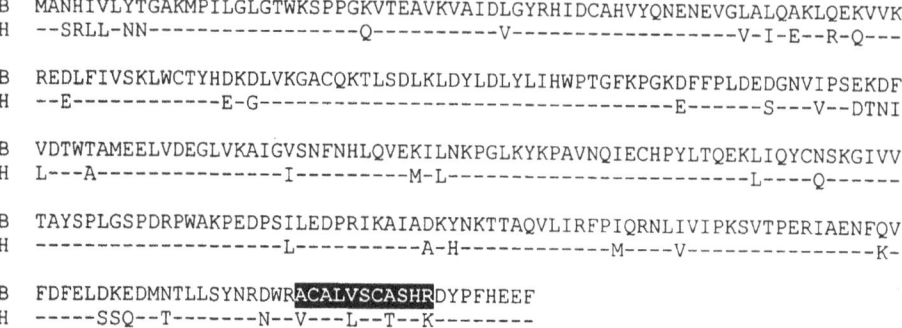

B MANHIVLYTGAKMPILGLGTWKSPPGKVTEAVKVAIDLGYRHIDCAHVYQNENEVGLALQAKLQEKVVK
H --SRLL-NN----------------Q----------V------------------V-I-E--R-Q---

B REDLFIVSKLWCTYHDKDLVKGACQKTLSDLKLDYLDLYLIHWPTGFKPGKDFFPLDEDGNVIPSEKDF
H --E------------E-G------------------------------------E------S---V--DTNI

B VDTWTAMEELVDEGLVKAIGVSNFNHLQVEKILNKPGLKYKPAVNQIECHPYLTQEKLIQYCNSKGIVV
H L---A---------------I---------M-L---------------------------L----Q------

B TAYSPLGSPDRPWAKPEDPSILEDPRIKAIADKYNKTTAQVLIRFPIQRNLIVIPKSVTPERIAENFQV
H --------------------L----------A-H------------M----V--------------K-

B FDFELDKEDMNTLLSYNRDWR__ACALVSCASHR__DYPFHEEF
H -----SSQ--T-------N--V---L--T--K--------

Figure 3. Alignment of bovine and human aldose reductases. The deduced primary structure of bovine retina ALR2 (B) is aligned with that deduced from cDNA clones encoding human placenta ALR2 (H). Sequence identities are indicated by "-". The highlighted region near the COO⁻ terminus of bovine ALR2 corresponds to a tryptic peptide identified by affinity labeling with [2'-^{14}C]-16α-bromoacetoxyprogesterone. The cDNA sequence of human ALR2 was determined from clones isolated from a placenta expression library (Petrash et al., 1992).

ficiency measured with 17α-hydroxyprogesterone was greater by 2–3 orders of magnitude than that measured with the presumed physiological substrate D-glucose (Table 1). The ALR2 inhibitor zopolrestat inhibited both enzymes with a K_i in the low nanomolar concentration range. These results provide further evidence that ALR2 and 20αHSD activities are expressed by the same gene product.

Utilization of 17α-hydroxyprogesterone as a substrate appears to be relatively specific to ALR2 in comparison with other catalytically active aldo-keto reductases. In our previous studies, we showed that the growth factor-inducible aldo-keto reductase designated FR-1, which shares approximately 69% sequence identity with murine ALR2 (Donohue et al., 1994), adopts an active site architecture strikingly similar to that determined for human ALR2 (Wilson et al., 1995). Using 17α-hydroxyprogesterone as a substrate, no detectable reductase activity could be measured with either purified recombinant FR-1 (Table 1) or with partially purified aldehyde reductase.

Table 1. Kinetic constants for aldo-keto reductases. "NDA" denotes no detectable activity observed under the given assay conditions

	Human ALR2	Bovine ALR2	Mouse FR-1
DL-glyceraldehyde			
K_m (μM)	67 ± 2	24 ± 2	920 ± 100
k_{cat} (min^{-1})	61	30	90
k_{cat}/K_m (mM^{-1}min^{-1})	912	800	98
17α-hydroxyprogesterone			
K_m (μM)	28 ± 5	19 ± 2	NDA
k_{cat} (min^{-1})	9	8	
k_{cat}/K_m (mM^{-1}min^{-1})	321	421	
D-glucose			
K_m (mM)	266 ± 30	22 ± 2	NDA
k_{cat} (min^{-1})	86	18	
k_{cat}/K_m (mM^{-1}min^{-1})	0.32	0.82	
K_i Zopolrestat (nM)	0.65 ± 0.1	1.8 ± 0.2	17 ± 10

3.3. Affinity Labeling Studies

Bovine and human ALR2 are rapidly inactivated by treatment with 16α-BAP, a bromoacetoxyprogesterone analog. To identify residues at the steroid binding site, we incubated native bovine ALR2 with [^{14}C]16α-BAP in the presence of NADPH. The inactivated enzyme was then subjected to proteolytic digestion using trypsin and the resulting peptides resolved by reverse phase HPLC. A major radioactive peak observed in the column eluate was purified to apparent homogeneity by repeated chromatography using high resolution gradient conditions. Automated Edman degradation of this radioactively-labeled peptide produced the following sequence: NH_2-Ala-Cys-Ala-Leu-Val-Ser-Cys-Ala-Ser-His-Arg-COO$^-$. Alignment of this sequence with the primary structure of bovine ALR2 revealed that it represents residues Ala-297 through Arg-307 (Figure 3). Substantially all of the radioactivity in the covalently modified protein originating from treatment with [^{14}C]16α-BAP eluted as a carboxymethyl-cysteine-PTH derivative in the second cycle of Edman degradation. This provides strong evidence that the major site of affinity labeling involved Cys-298. Crystallographic and kinetic studies have shown that this residue is located in the active site of human ALR2 (Petrash et al., 1992; Borhani et al., 1992; Wilson et al., 1992) and that it plays an important role in regulating the catalytic rate through its interactions with the nucleotide coenzyme. (Grimshaw et al., 1995). These results are consistent with the hypothesis that inactivation of ALR2 occurs through specific modification of Cys-298.

To test this hypothesis, we compared the susceptibility of wild type ALR2 with a mutated enzyme containing a serine substituted for Cys-298 (C298S). Previous studies from our lab and others demonstrated that the C298S mutant expresses robust catalytic activity with a broad range of substrates (Petrash et al., 1992; Bohren and Gabbay, 1993) and that the overall structural organization of the binary enzyme-nucleotide complex is substantially similar to wild type (Borhani et al., 1992). Treatment of wild type ALR2 with equimolar concentrations of 16α-BAP in the presence of NADPH resulted in rapid loss of approximately 75% of starting activity (Figure 4). In contrast, the C298S mutant was essentially insensitive to 16α-BAP even when treated with a 3-fold molar excess of the inactivator over a period of 2 hours or more. That no loss of activity was observed over this time period suggests that the covalent modification of ALR2 by 16α-BAP occured at a structurally specific site rather than by simply reacting with one or more solvent-exposed cysteine side chains.

4. CONCLUSIONS

Although these studies provide compelling evidence that ALR2 has the ability to catalyze the reduction of C-20 carbonyl of 17α-hydroxyprogesterone, further work will be required to establish whether this capability is physiologically relevant. Since ALR2 is a drug target of inhibitors designed to delay or prevent the onset of diabetic complications, studies to establish the potential role of this enzyme in steroid metabolism could have a profound impact on the indications for inhibitor therapy.

5. ACKNOWLEDGMENTS

Support for this work was provided by NIH grants EY05856, DK15708, P30EY02687, and a Diabetes Research and Training Center Grant (P60DK20579). Addi-

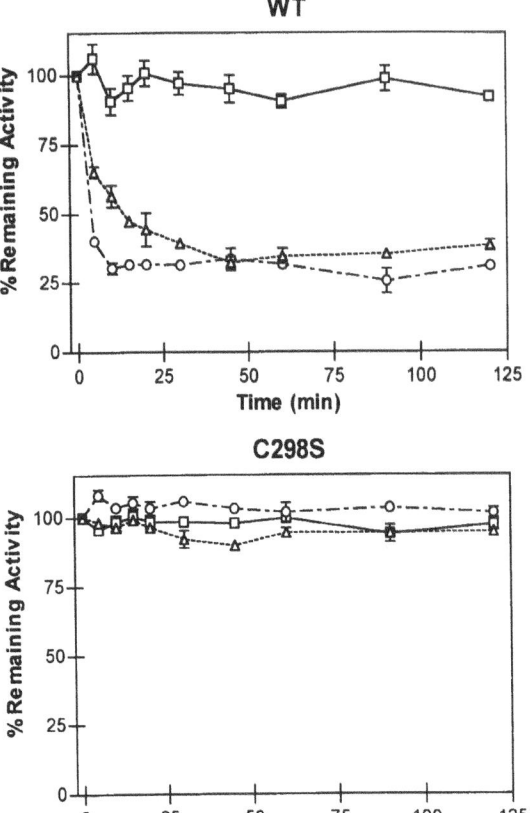

Figure 4. Inactivation of human ALR2 in the presence of 16αBAP. Enzymes (5 μM) were incubated in the presence of 150 μM NADPH and cosolvent alone (□), 5 μM 16αBAP (△) and 15 μM 16αBAP (O). Enzyme activity remaining was determined following incubation for the indicated times. Each data point represents the mean ± standard error obtained from triplicate measurements. Percent activity remaining over an incubation period of 120 min were measured for wild type (upper panel) and the C298S mutant (lower panel) taking the value at time zero as 100%.

tional support to the Department of Ophthalmology and Visual Sciences from Research to Prevent Blindness, Inc. is gratefully acknowledged.

6. REFERENCES

Bohren, K.M., Grimshaw, C.E., Lai, C.-J., Harrison, D.H., Ringe, D., Petsko, G.A., and Gabbay, K.H.: Tyrosine-48 is the proton donor and histidine-110 directs substrate stereochemical selectivity in the reduction reaction of human aldose reductase: Enzyme kinetics and crystal structure of the Y48H mutant enzyme. Biochemistry 33 (1994) 2021–2032.

Bohren, K.M. and Gabbay, K.H.: Cys298 is responsible for reversible thiol-induced variation in aldose reductase activity. Adv. Exp. Med. Biol. 328 (1993) 267–277.

Borhani, D.W., Harter, T.M., and Petrash, J.M.: The crystal structure of the aldose reductase·NADPH binary complex. J. Biol. Chem. 267 (1992) 24841–24847.

Donohue, P.J., Alberts, G.F., Hampton, B.S. and Winkles, J.A.: A delayed-early gene activated by fibroblast growth factor-1 encodes a protein related to aldose reductase. J. Biol. Chem. 269 (1994) 8604–8609.

Grimshaw, C.E., Bohren, K.M., Lai, C.J., and Gabbay, K.H.: Human aldose reductase - subtle effects revealed by rapid kinetic studies of the C298A mutant enzyme. Biochemistry 34 (1995) 14366–14373.

Grimshaw, C.E. and Mathur, E.J.: Immunoquantitation of aldose reductase in human tissues. Anal. Biochem. 176 (1989) 66–71.

Hayman, S. and Kinoshita, J.H.: Isolation and properties of lens aldose reductase. J. Biol. Chem. 240 (1965) 877–882.

Hers, H.G.: L'aldose-reductase. Biochim. et Biophys. Acta 37 (1960) 120–126.

Iwata, N., Inazu, N., and Satoh, T.: The purification and properties of aldose reductase from rat ovary. Arch. Biochem. Biophys. 282 (1990) 70–77.

Kinoshita, J.H. and Nishimura, C.: The involvement of aldose reductase in diabetic complications. Diabetes-Metabolism Rev. 4 (1988) 323–337.

Ludvigson, M.A., Waites, G.M.H., and Hamilton, D.W.: Immunocytochemical evidence for the specific localisation of aldose reductase in rat sertoli cells. Biol. Reprod. 26 (1982) 311–317.

Monder, C. and Bradlow, H.L.: Carboxylic acid metabolites of steroids. J. Steroid Biochem. 8 (1977) 897–908.

Murdock, G.L., Warren, J.C., and Sweet, F.: Human placental estradiol 17β-dehydrogenase: Evidence for inverted substrate orientation ("wrong-way" binding) at the active site. Biochemistry 27 (1988) 4452–4458.

Petrash, J.M., DeLucas, L.J., Bowling, E., and Egen, N.: Resolving isoforms of aldose reductase by preparative isoelectric focusing in the Rotofor. Electrophoresis 12 (1991) 84–90.

Petrash, J.M., Harter, T.M., Devine, C.S., Olins, P.O., Bhatnagar, A., Liu, S.Q., and Srivastava, S.K.: Involvement of cysteine residues in catalysis and inhibition of human aldose reductase -site-directed mutagenesis of cys-80, cys-298, and cys-303. J. Biol. Chem. 267 (1992) 24833–24840.

Petrash, J.M. and Favello, A.D.: Isolation and characterization of cDNA clones encoding aldose reductase. Curr. Eye Res. 8 (1989) 1021–1027.

Pineda, J.A., Murdock, G.L., Watson, R.J., and Warren, J.C.: Stereospecificity of hydrogen transfer by bovine testicular 20 alpha-hydroxysteroid dehydrogenase. J. Steroid Biochem. 33 (1989) 1223–1228.

Sarges, R. and Oates, P.J.: Aldose reductase inhibitors: Recent developments. Prog. in Drug. Res. 40 (1993) 99–161.

Sheaff, C.M. and Doughty, C.C.: Physical and kinetic properties of homogeneous bovine lens aldose reductase. J. Biol. Chem. 251 (1976) 2696–2702.

Sweet, F., Arias, F., and Warren, J.C.: Affinity labeling of steroid binding sites. Synthesis of 16 -bromoacetoxyprogesterone and its use for affinity labeling of 20 -hydroxysteroid dehydrogenase. J. Biol. Chem. 247 (1972) 3424–3433.

Tanimoto, T., Ohta, M., Tanaka, A., Ikemoto, I., and Machida, T.: Purification and characterization of human testis aldose and aldehyde reductase. Internat. J. Biochem. 23 (1991) 421–428.

Tarle, I., Borhani, D.W., Wilson, D.K., Quiocho, F.A., and Petrash, J.M.: Probing the active site of human aldose reductase. Site-directed mutagenesis of Asp-43, Tyr-48, Lys-77, and His-110. J. Biol. Chem. 268 (1993) 25687–25693.

Vander Jagt, D.L., Hunsaker, L.A., Robinson, B., Stangebye, L.A., and Deck, L.M.: Aldehyde and aldose reductases from human placenta. Heterogeneous expression of multiple enzyme forms. J. Biol. Chem. 265 (1990) 10912–10918.

Warren, J.C., Murdock, G.L., Ma, Y., Goodman, S.R., and Zimmer, W.E.: Molecular-cloning of testicular 20-alpha-hydroxysteroid dehydrogenase - Identity with aldose reductase. Biochemistry. 32 (1993) 1401–1406.

Wermuth, B. and Monder, C.: Aldose and aldehyde reductase exhibit isocorticosteroid reductase activity. Eur. J. Biochem. 131 (1983) 423–426.

Wilson, D.K., Bohren, K.M., Gabbay, K.H., and Quiocho, F.A.: An unlikely sugar substrate site in the 1.65 Å structure of the human aldose reductase holoenzyme implicated in diabetic complications. Science 257 (1992) 81–84.

Wilson, D.K., Nakano, T., Petrash, J.M. and Quiocho, F.A.: 1.7 Å structure of FR-1, a fibroblast growth factor-induced member of the aldo-keto reductase family, complexed with coenzyme and inhibitor. Biochemistry 34 (1995) 14323–14330.

54

HYDROXYSTEROID DEHYDROGENASES

New Drug Targets of the Aldo-Keto Reductase Superfamily

Trevor M. Penning

Department of Pharmacology
University of Pennsylvania School of Medicine
Philadelphia, Pennsylvania 19104–6084

1. INTRODUCTION

Mammalian hydroxysteroid dehydrogenases (HSDs) regulate the occupancy of steroid hormone receptors by interconverting active hormones with their cognate inactive metabolites. In this manner, they work as molecular switches that control steroid hormone action. Specificity is achieved by reducing carbonyl groups to hydroxyl groups in a positional and stereoselective manner on the steroid nucleus or steroid side-chain. cDNA cloning indicates that HSDs belong to two distinct protein phylogenies: the aldo-keto reductase (AKR) superfamily (Pawlowski et al., 1991; Lacy et al., 1993; Warren et al., 1993; Mao et al., 1994) and the short-chain dehydrogenase/reductase family (SDR, formerly known as the short-chain alcohol dehydrogenase family) (Krozowski, 1994; Jörnvall et al., 1995).

HSDs that belong to the AKR superfamily which act as molecular switches include 3α-HSD and 20α-HSD. For example, prostatic 3α-HSD regulates the amount of androgen that can bind to the androgen receptor. It catalyzes the conversion of the potent androgen 5α-dihydrotestosterone (5α-DHT) ($K_d = 10^{-11}$ M for the androgen receptor) to the weak androgen 3α-androstanediol ($K_d = 10^{-6}$ M for the androgen receptor) (Liao et al., 1973; Taurog et al., 1975; Jacobi et al., 1977). Evidence exists that prostatic 3α-HSD works in concert with 5α-reductase to maintain high levels of 5α-DHT required for normal and abnormal growth of the prostate (Jacobi et al., 1978; Isaacs, 1983). Inhibitors of this enzyme provide an opportunity for lowering 5α-DHT levels and may be useful in the treatment of benign prostatic hyperplasia and prostatic carcinoma. In another example, 20α-HSD converts progesterone into the hormonally inactive 20α-hydroxyprogesterone. Since 20α-HSD activity is elevated in luteolysis and the termination of pregnancy (Strauss and Stambaugh, 1974; Mao et al., 1994), this enzyme could play a fundamental role in regulating progesterone levels and controlling progesterone action. Therefore, inhibitors of the enzyme may help prevent luteolysis and would be progestational maintaining agents.

HSDs that belong to the SDR family which act as molecular switches include 11β-HSD and 17β-HSD. The type II 11β-HSD protects the mineralocorticoid receptor (MR) in the renal tubules from occupancy by the glucocorticoids (Funder et al., 1988; Mercer and Krozowski, 1992; Krozowski et al., 1994; Mune et al., 1995) by converting cortisol to the biologically inactive cortisone. Metabolism to cortisone prevents cortisol from exerting its mineralocorticoid activity. In this regard, cortisol and aldosterone have equal affinities for the MR receptor, and cortisol is at least 100 times more abundant; thus, without the action of 11β-HSD, the MR receptor would be flooded with cortisol. The specificity of mineralocorticoid response appears to be determined by 11β-HSD rather than by the receptor. Inhibitors of 11β-HSD lead to apparent mineralocorticoid excess (Stewart et al., 1987; Monder et al., 1989) which suggests that these compounds could be clinically useful in the treatment of hypotension.

In estrogen target tissues, type I 17β-HSD converts estrone (a weak estrogen) to 17β-estradiol (a potent estrogen) (Poutanen et al., 1993). Inhibitors of this enzyme would be anti-estrogenic and may be useful in the management of hormonally dependent breast cancer. In contrast, in androgen target tissues, type III 17β-HSD converts androstenedione (a weak androgen) to testosterone (a potent androgen) (Andersson et al., 1995). Inhibitors of this activity would be anti-androgenic and may be useful in the management of androgen dependent tumors.

Thus, HSDs are bona-fide drug targets. However, to employ structure-based inhibitor design in a rationale manner, it is important to understand the structural basis of steroid recognition and what governs the exquisite positional and stereochemical preference of these enzymes for their substrates. This chapter will discuss these issues as they relate to 3α-HSD and 20α-HSD and will draw on structure-function relationships established for rat liver 3α-HSD.

2. RESULTS AND DISCUSSION

2.1. Purification and Properties of Rat Liver 3α-HSD

Our group was the first to purify a mammalian 3α-HSD from any source. We obtained homogeneous enzyme from rat liver cytosol (Penning et al., 1984). The enzyme is a monomer with a molecular weight 37 kDa and is slightly acidic with a pI = 6.0. Its complete kinetic mechanism has been solved at physiological pH (Askonas et al., 1991). The enzyme catalyzes a sequential ordered bi-bi mechanism in which pyridine nucleotide binds first and leaves last. Thus, there is an obligatory requirement to bind co-factor before steroid. The individual rate constants indicate that the rate determining step is the formation of the E·NAD(H) binary complex. Isomerization of the central complex occurs readily and favors the oxidation direction three-fold. 3α-HSD is a non-metalloenzyme and the presence of a general acid is invoked to polarize the acceptor carbonyl so that hydride transfer can proceed. The stereochemistry of hydride transfer has been established. It is the 4-*pro(R)*-hydrogen that is transferred from the A-face of the co-factor to the β-face of the steroid, so that a 3α-axial alcohol is formed (Askonas et al., 1991).

2.2. Molecular Cloning and Assignment to the Aldo-Keto Reductase Superfamily

The availability of homogeneous 3α-HSD led to the development of a polyclonal antibody, immunoscreening of a female Sprague-Dawley rat liver cDNA library and the successful isolation of a full length cDNA clone (Pawlowski et al., 1991). The open-read-ing frame was 966 nucleotides in size and predicted a protein of 322 amino acids in length. This sequence shares high similarity with the full-length clone for human liver type I 3α-HSD (DD4 or chlordecone reductase) and human liver type II 3α-HSD, ob-tained by Khanna et al., 1995. Type I and type II 3α-HSDs differ in their K_m values for 5α-DHT, the type I enzyme has a $K_m = 1.2$ μM and the type II enzyme has a $K_m = 19$ μM. There is also high sequence identity with full length cDNA clones isolated for human pro-static 3α-HSD (Dufort et al., 1995; Lin et al., 1995). Based on this high sequence identity structural information to be presented on rat liver, 3α-HSD is likely to be highly relevant to these human isoforms.

When these sequences were used as a search query in GENBANK, the search re-vealed high sequence identity with members of the aldo-keto reductase (AKR) superfa-mily. This represented the first time that HSDs had been assigned to this superfamily. Cluster analysis indicates that mammalian HSDs form a cluster of the AKR superfamily. Other HSDs assigned to this cluster include rat and rabbit ovarian 20α-HSD (Lacy et al., 1993; Mao et al., 1994), murine liver 17β-HSD (Type A stereospecificity) (Deyashiki et al., 1995) and human type V 17β-HSD (Ferdinand Labrie and Stefan Andersson, personal communications). This analysis indicates that the HSDs, as well as prostaglandin $F_{2\alpha}$ syn-thase and rho-crystallin, belong to the same cluster, Figure 1. As more HSDs are cloned, more are anticipated to be assigned to the AKR superfamily.

Other drug targets that belong to the AKR superfamily include the aldose reductases and the aldehyde reductases. These enzymes form their own individual clusters. The al-dose reductases catalyze the formation of sorbitol from glucose. In diabetics, the accumu-lation of this hyperosmotic sugar leads to diabetic retinopathy and neuropathy. Thus, aldose reductase inhibitors are widely sought to attenuate these complications of diabetes. One report shows that bovine testicular 20α-HSD is identical to aldose reductase (Warren et al., 1993). A caveat associated with that work is that the isolated cDNA was not ex-pressed and shown to act as both a 20α-HSD and an aldose reductase. The aldehyde re-ductases work in concert with the monoamine oxidases to metabolize monoamines and inhibitors may have anti-depressant action. Therefore, it is not surprising that existing al-dose reductase inhibitors lack the desired specificity, since inhibition of multiple enzymes that belong to the AKR superfamily is possible . The issue of inhibitor specificity could be resolved via structure-based drug design. A pre-requisite would be a knowledge of the structures of enzymes of each cluster of the AKR superfamily.

2.3. X-Ray Crystal Structure of Rat Liver 3α-HSD

Rat Liver 3α-HSD is the most thoroughly characterized HSD of the AKR superfa-mily; details relating its structure to function serve as a good template for other HSDs that are ascribed to this superfamily. 3α-HSD is highly abundant and represents 1% of the sol-uble protein in rat liver, which led to its crystallization by vapor diffusion (Hoog et al., 1994). The crystals obtained diffracted x-rays beyond 3.0 A resolution and the phasing problem was solved by molecular replacement, using the X-ray coordinates for human placental aldose reductase complexed with NADPH provided by Dr. Florante Quiocho (Wilson et al., 1992).

The structure of the apoenzyme was solved with a nominal resolution of 3.1 Å (Hoog et al., 1994). The ribbon diagram shows that the enzyme adopts an $(\alpha/\beta)_8$-barrel or triose-phos-phate isomerase (TIM) barrel, Figure 2. This protein fold is characterized by an alternating ar-rangement of α-helix and β-strands with the β-strands forming the staves of the barrel in the

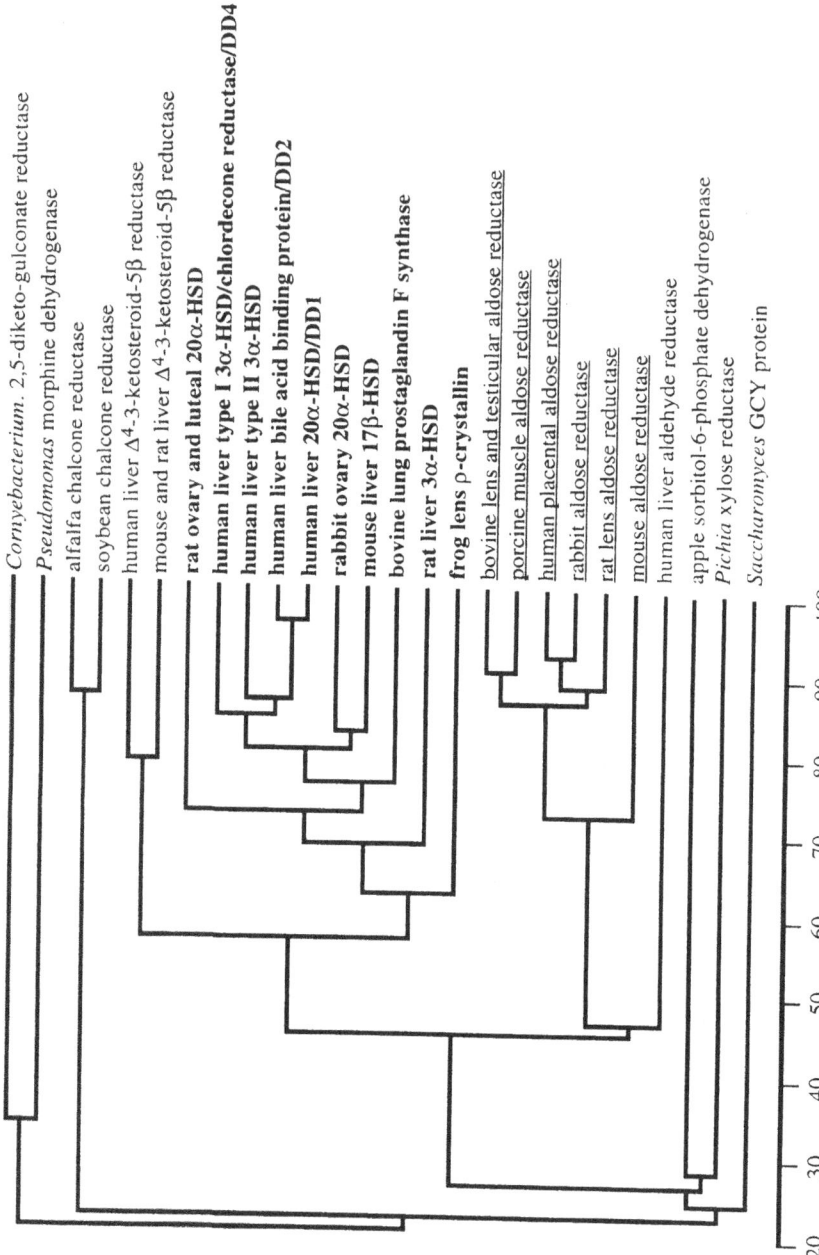

Figure 1. Cluster analysis of the aldo-keto reductase superfamily, analysis was performed using PILEUP in the GCG package.

center of the structure. When the structure is rotated 90° around the barrel axis, several large loops appear at the back of the structure which comprise a large cleft where apolar residues exist. The residues on one side of the cleft include: Leu54, Phe118, Phe129, Trp86. On an opposing loop (loop B) Tyr216 and Trp227 exist. The C-terminal loop, which starts after residue 308, is undefined in the electron density and may also make up a portion of this cleft. This cleft is considered to be the presumptive steroid hormone binding site; it is 11 Å deep and sufficient in size to accommodate a steroid ligand. NADPH was modeled into the structure using the coordinates from the ADR·NADPH binary complex. In close proximity to the C-4 position of the nicotinamide ring, four charged residues exist at the base of the pocket which comprise a putative catalytic tetrad: Asp50, Tyr55, Lys84, and His117. On the basis of this model, a catalytic mechanism was proposed in which Tyr55 would act as the general acid; by polarizing the carbonyl at the C3 position of a 3-ketosteroid, it could facilitate hydride transfer. The effective pKa of the tyrosine could be lowered by the adjacent Lys84, which in turn is salt linked to Asp50, Figure 3.

Recently, the structure of the binary complex of E·NADP$^+$ was solved to 2.7 Å by (Bennett et al., 1996). It was found that the cofactor resides in a very similar position as modeled into the apoenzyme structure. It lies perpendicular to the barrel axis with the nicotinamide ring facing the core of the barrel and the adenine ring lying at the periphery. The orientation of the nicotinamide ring is *anti-* with respect to the adjacent nicotinamide ribose moiety. The structure also reveals the atomic interactions that are involved in cofactor binding. The B-face of the nicotinamide ring is stacked against Tyr216, so that the A-face is presented to the active site cavity which allows the transfer of 4-*pro(R)* hydrogen to proceed. The stereochemistry of the nicotinamide ring is also maintained by hydrogen bonds between the carboxamide group and Ser166, Asn167 and Gln190. Furthermore, the 2'-phosphate of the adenine ring is accommodated by hydrogen bonds with Ser271 and Arg276, as well as by a counterion with Arg276. These interactions may in part explain the nucleotide preference of the enzyme: NADPH (K_d=190 μM), and NADH (K_d=165 μM). The binary-complex structure also reveals for the first time the presence of a highly ordered water molecule at the active site that is within hydrogen bond distance of His117 and Tyr55. The position of the water molecule may mimic the position of the substrate carbonyl. In other structures of the AKR superfamily, a similar water molecule is observed and is displaced by the oxygen atoms in several inhibitors, e.g., glucose-6-phos-

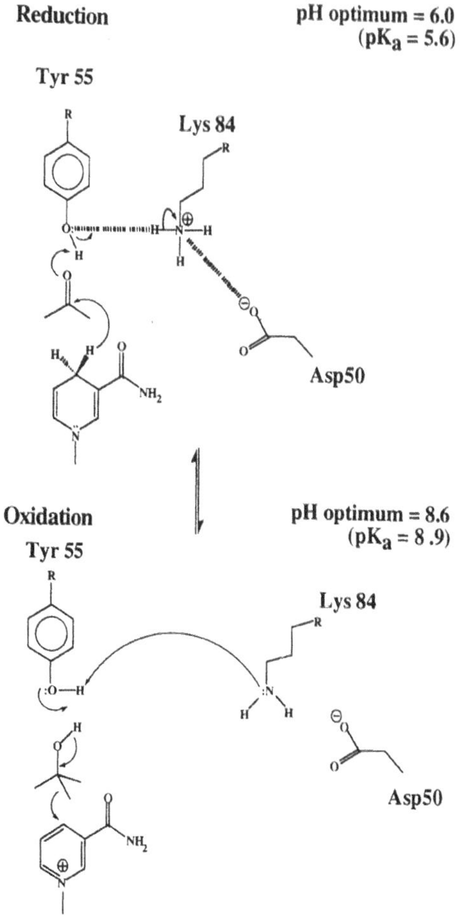
Reduction
pH optimum = 6.0
(pKa = 5.6)
Tyr 55
Lys 84
Asp50
Oxidation
pH optimum = 8.6
(pKa = 8 .9)
Tyr 55
Lys 84
Asp50

Figure 3. Proposed catalytic mechanism for rat liver 3α-HSD.

phate, cacodylate and zopolrestat (Wilson et al., 1993; Harrison et al., 1994). The residues that comprise the catalytic tetrad (Asp50, Tyr55, Lys84 and His117) are clear targets for site-directed mutagenesis.

2.4. Comparison of HSDs in the SDR and AKR Superfamilies

HSDs that belong to the SDR family differ in their protein folds from HSDs that belong to the AKR family. In the SDR family, the proteins function as multimers. Each monomer varies in molecular mass from 25–37 kDa and is characterized by a fold that consists of 7-stranded parallel β-sheets surrounded with 3-helices on either side of the β-sheets. This arrangement provides a scaffold for a Rossmann fold. The active site of these enzymes contains a conserved catalytic motif (Tyr-X-X-X-Lys), Figure 4.

Two HSDs that share this structure are the 3α,20β-HSD from *Streptomyces* (Ghosh et al., 1991; Ghosh et al., 1994a; Ghosh et al., 1994b) and the human type I 17β-HSD (Ghosh et al., 1995). Despite the low sequence identity in this family, type I and type II 11β-HSDs are anticipated to have the same overall structure (Jörnvall et al., 1995). When

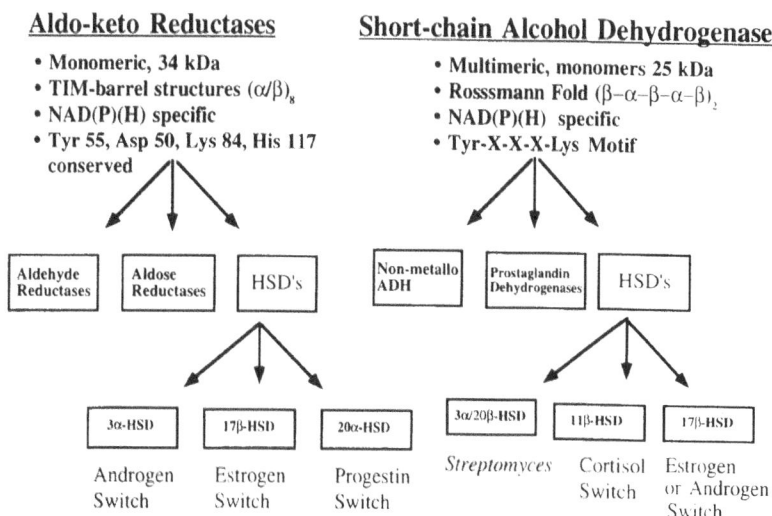

Figure 4. Comparison of two HSD protein families, aldo-keto reductases (AKRs) and the short-chain dehydrogenases/reductases (SDR) formerly short-chain alcohol dehydrogenase (SCAD). Reproduced from Penning (1996) Endocrin. Rev. In press, with permission from the Endocrine Society.

the Tyr and Lys in the catalytic motif have been mutated in SDR family members, it results in an inactive enzyme (Ensor and Tai, 1991; Obeid and White, 1992; Chen et al., 1993). It has been argued that in this family, Tyr may act as the general acid with its pKa being effectively lowered by the adjacent Lys (Chen et al., 1993). In contrast, HSDs that are AKRs are monomeric, with a molecular mass of 34–39 kDa, adopt an $(\alpha/\beta)_8$ barrel fold and, instead of binding cofactor across a Rossmann fold, are predicted to form a "salt-linked" safety belt for binding the pyrophosphate bridge of the dinucleotide. They also contain a conserved catalytic motif, Asp50, Tyr55, Lys84 and His117, which is invariant in HSD members of the AKR superfamily (numbering of residues is in accord with 3α-HSD). Even though 3α-HSD is not a member of the SDR family, it still maintains the conserved catalytic motif (Tyr-X-X-X-Lys) between positions 205 and 209. Other HSDs, e.g., the mammalian 3β-HSD/ketosteroid isomerase, also contain this motif. Thus, Tyr205 and Lys209 are also candidate residues for site-directed mutagenesis, Table 1.

2.5. Site-Directed Mutagenesis in 3α-HSD

The full length cDNA for rat liver 3α-HSD was subcloned into a prokaryotic expression vector to yield pKK3α-HSD (Pawlowski and Penning, 1994). Transformation of DH5α cells led to high levels of overexpression of 3α-HSD in *E. coli*. Wild-type recombinant 3α-HSD (wt r3α-HSD) was purified to homogeneity from *E. coli* sonicates using DEAE-cellulose and Sepharose-Blue chromatography and obtained in milligram quantities for crystallography. The kinetic parameters of the recombinant enzyme were almost indistinguishable from the enzyme that was purified from rat liver cytosol (Pawlowski and Penning, 1994). The following mutants were constructed using a PCR based approach: Y55F, Y55S, Y205F, K84M, K84R, D50E, D50N, and H117A (Pawlowski and Penning, 1994; Schlegel and Penning, 1995). The fidelity of the mutants was verified by re-sequencing the cDNA to ensure that no fortuitous mutations were introduced by the DNA

Table 1. Alignment of the Tyr-X-X-(Ser)-Lys consensus sequence of the short chain dehydrogenases/reductases (SDRs) in HSDs

	203														217
3α-HSD+	Leu 176	Asp	Tyr	Cys	Lys	**Ser**	**Lys**	Asp	Ile	Ile	Leu	Val	Ser	Tyr	Cys 190
11β-HSD	Ala 153	Ser	Tyr	Ser	Ala	**Ser**	**Lys**	Phe	Ala	Leu	Asp	Gly	Phe	Phe	Ser 167
17β-HSD	Asp 268	Val	Tyr	Cys	Ala	**Ser**	**Lys**	Phe	Ala	Leu	Glu	Gly	Leu	Cys	Glu 282
3β-HSD/KSI*	Leu 151	Asn	Tyr	Thr	Leu	**Ser**	**Lys**	Glu	Phe	Gly	Leu	Arg	Leu	Asp	Ser 165
3α-HSD‡	Leu 149	Ala	Tyr	Ala	Gly	**Ser**	**Lys**	Tyr	Ala	Val	Thr	Cys	Leu	Ala	Arg 163
3β-HSD‡	Ala 150	Gly	Tyr	Ser	Ala	**Ser**	**Lys**	Ala	Ala	Val	Ser	Ala	Leu	Thr	Arg 164
3α20β-HSD	Ser 157	Ser	Tyr	Gly	Ala	**Ser**	**Lys**	Trp	Gly	Val	Arg	Gly	Leu	Ser	Lys 171
7α-HSD	Ile 148	Ala	Tyr	Gly	Thr	**Ser**	**Lys**	Ala	Ala	Ile	Asn	Tyr	Leu	Thr	Lys 162
15-OH PGDH	Pro 150	Val	Tyr	Cys	Ala	**Ser**	**Lys**	His	Gly	Ile	Val	Gly	Phe	Thr	Arg 164
DADH	Pro 43	Val	Tyr	Ser	Gly	Thr	**Lys**	Ala	Ala	Val	Val	Asn	Phe	Thr	Ser 57
cis-DD	Val	Leu	Tyr	Thr	Ala	Gly	**Lys**	His	Ala	Val	Ile	Gly	Leu	Ile	Lys

3α-HSD+, rat liver 3α-hydroxysteroid dehydrogenase, which is not a SDR; 3β-HSD /KSI*, human placental 3β-hydroxysteroid dehydrogenase/ketosteroid isomerase; 11β-HSD, from rat liver; 17β-HSD, from human placenta; 3α-HSD ‡, from *Pseudomonas* sp., 3β-HSD ‡ from *Pseudomonas testosteroni*; 3α,20b-HSD, *Streptomyces hydrogenans*; 7α-HSD from *Eubacterium*; 15-OH PGDH 15-hydroxyprostaglandin dehydrogenase from human placenta; DADH, *Drosophila* alcohol dehydrogenase; *cis*-DD, *cis*-benzene dihydrodiol dehydrogenase from *Pseudomonas putida*. Reproduced from Penning et al., Steroids 61. In press (1996) Copyright Elsevier Science, Inc.

polymerase used in the PCR reactions. Each mutant enzyme was purified to homogeneity as judged by SDS-PAGE and immunoblot analysis. The immunoblots indicated that the correct protein had been purified even when it was inactive.

The mutant enzymes were kinetically characterized by their ability to perform steroid oxidation (androsterone plus NAD$^+$) and steroid reduction (androstanedione plus NADH). It was found that the Y205F mutant gave identical kinetic constants to wt r3α-HSD. This result differs from that observed for SDR family members and represents the first time that the Tyr-X-X-X-Lys sequence has been mutated with retention of enzyme activity. These results can be explained by our structural model, since Tyr205 is on the periphery of the structure buried within an α-helix.

The Y55F, Y55S, K84R, and K84M mutants were inactive with steroid substrates, implying that they were residues essential for enzyme activity. By contrast, D50N, D50E, and H117A, although severely impeded enzymes, still performed steroid oxidoreduction; catalytic efficiencies were 1/30 th, 1/30 th and 1/500 th of the wild type enzyme, respectively. The ability to retain activity in these mutants suggests that these residues are not essential for catalysis, Table 2.

To address whether mutants were inactive because they can no longer form productive binary and ternary complexes, the ability of each mutant to bind NADPH was tested by titrating the tryptophan fluorescence of the protein. In addition, the ability of each mutant to bind testosterone to the corresponding E·NADH complex was measured by equilibrium dialysis. It was found that each of the mutants were able to bind cofactor virtually unimpeded. The K_d values for NADPH varied by only two-fold versus wt r3α-HSD. Since the K_d for NADPH is in the nanomolar range, these data also imply that there was no global change in the protein structure. Therefore, loss of enzyme activity was not due to the inability to bind co-factor. The majority of the mutants were unable to bind testosterone with high affinity. Thus, the H117A, D50N, and D50E mutants were unable to bind steroid at the limit of detection of the assay >> 50 μM. Of the inactive mutants, only the

Table 2. Kinetic comparison of wild-type r3α-HSD and mutants of the catalytic tetrad

ENZYME	OXIDATION * ANDROSTERONE			REDUCTION * ANDROSTANEDIONE		
	k_{cat} (min^{-1})	K_m (μM)	k_{cat}/K_m (min^{-1} μM^{-1})	k_{cat} (min^{-1})	K_m (μM)	k_{cat}/K_m (min^{-1} μM^{-1})
WT	35 ± 3	3.5 ± 1.6	10	32 ± 2	<1.5	----
Y55F	----	----	----	----	----	----
Y55S	----	----	----	----	----	----
K84M	----	----	----	----	----	----
K84R	----	----	----	----	----	----
D50N	13.0 ± 0.8	43 ± 5	0.30	----	----	----
D50E	6.7 ± 0.4	20 ± 3	0.34	----	----	----
H117A	1.7 ± 0.2	95 ± 11	0.02	----	----	----

(---) means non-detectable

* Oxidation reactions were performed at pH 9.0 and reduction reactions were performed at pH 6.0 using a triple buffer system (50 μM sodium pyrophosphate, 50 μM sodium phosphate and 50 μM AMSO) to maintain ionic strength.

Y55F and Y55S mutants were capable of binding steroid but the K_d for testosterone was increased by 10–30 fold. This increase in K_d is insufficient to explain the complete loss of enzyme activity observed in these mutants. Since the Tyr55 mutants can still form binary and ternary complexes but are unable to perform steroid oxidoreduction, this residue becomes the strongest candidate for the general acid as originally proposed, Figure 3. These data support a mechanism in which Tyr55 is the general acid and has its pKa effectively lowered by Lys84. A similar mechanism in which there is reliance on a Tyr/Lys pair has been presented for aldose reductase (Wilson et al., 1992).

2.6. Convergent Evolution to a Common Mechanism for HSD Catalysis

It is a remarkable coincidence that a Tyr/Lys pair is essential for 3α-HSD catalysis and that the same residues are required for HSD catalysis by SDR family members. In this regard, the nicotinamide rings in the crystal structures of 3α-HSD (an AKR) and 3α,20β-HSD (a SDR) were superimposed taking into account that the stereochemistry of hydride transfer is 4-*pro(R)* and 4-*pro(S)*, respectively. It was found that the arrangement of the Tyr/Lys pair was positionally conserved across family. Thus the hydroxyl group of the tyrosines in both proteins came to within 0.5 Å of each other. These modeling data point to convergent evolution, where different protein folds have been utilized to generate the same constellation of active site residues so that a common mechanism for HSD catalysis can be achieved (Bennett et al., 1996), Figure 5.

2.7. Steroid Recognition in HSDs That Belong to the AKR Superfamily: Backwards Binding of Secosteroids

In order to design inhibitors that will be selective for one HSD of the AKR superfamily, it is important to understand the structural basis that determines whether a member of this family functions as a 3α-HSD, 17β-HSD or 20α-HSD. In considering issues of

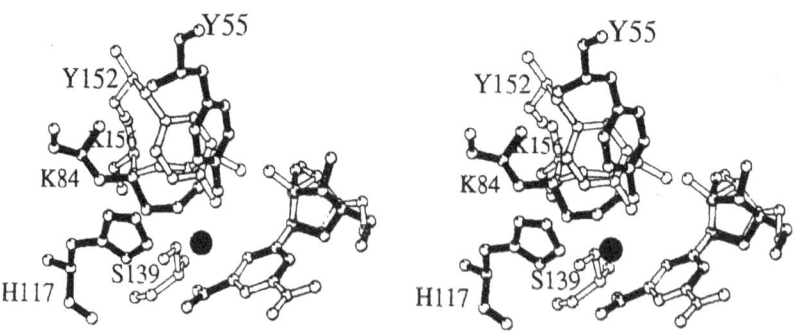

Figure 5. Superimposition of the AKR and SDR active sites. The binary complex 3α-HSD·NADP⁺ (an AKR) was superimposed with that of 3α,20β-HSD·NAD⁺ from *Streptomyces hydrogenans* (a SDR). The superimposition was based on the position of the nicotinamide ring (not including the carboxamide substituent at the C3 position). The reason for this exclusion is that 3α-HSD displays 4-*pro(R)* hydride transfer while 3α,20β-HSD displays 4-*pro(S)* hydride transfer, so that although the nicotinamide rings lie in the same plane they are flipped 180° relative to each other and the carboxamide groups do not superimpose. A close-up of the catalytic residues is shown in stereo, residues from 3α-HSD are black, and residues from 3α,20β-HSD are white.

steroid recognition, it was important to determine whether 3α-HSD would bind steroids backwards for two reasons. First, aldose reductase has been reported to reduce the C21 aldehyde of isocorticosteroids (Wermuth and Monder, 1983). Second, 17β-HSD and 20α-HSD are members of the AKR superfamily with high sequence identity to 3α-HSD and are predicted to have similar 3-dimensional structures. In these studies it is important to realize that rat liver 3α-HSD has no demonstrable 20α-HSD activity.

To determine whether 3α-HSD could bind steroids backwards, we used a series of A-ring and D-ring secosteroids in which acetylenic alcohols were introduced at the equivalent of the C3 and C17 positions of the steroid nucleus (e.g., 3*(R,S)*-1,10-seco-5α-estr-1-yne-3,17β-diol **1** and 17*(R,S)*-17-hydroxy-14,15-secoandrost-4-en-15-yn-3-one, **3**). These alcohols, if oxidized by 3α-HSD, would generate the corresponding acetylenic ketones, **2** and **4**, which would then inactivate the enzyme by forming Michael adducts with an enzyme thiol, Figure 6. It was found that when alcohols **1** and **3** were incubated with the enzyme in the presence of NAD$^+$ there was a time and concentration dependent inactivation of the enzyme. The obligatory requirement for NAD$^+$ for enzyme inactivation suggested that both acetylenic alcohols were turned over by the enzyme showing that backwards binding and catalysis occurred (Schlegel et al., 1994). Of the panel of site-directed mutants available, the C217A mutant was the only enzyme form that was resistant to enzyme inactivation, suggesting that C217 is the point of covalent attachment.

2.8. Steroid Recognition in HSDs That Belong to the AKR Superfamily: Role of Tryptophans and Loop Structures

Up until now, no functional studies have been performed on the apolar cleft in 3α-HSD that is postulated to be involved in binding steroid hormone. One side of the cleft is

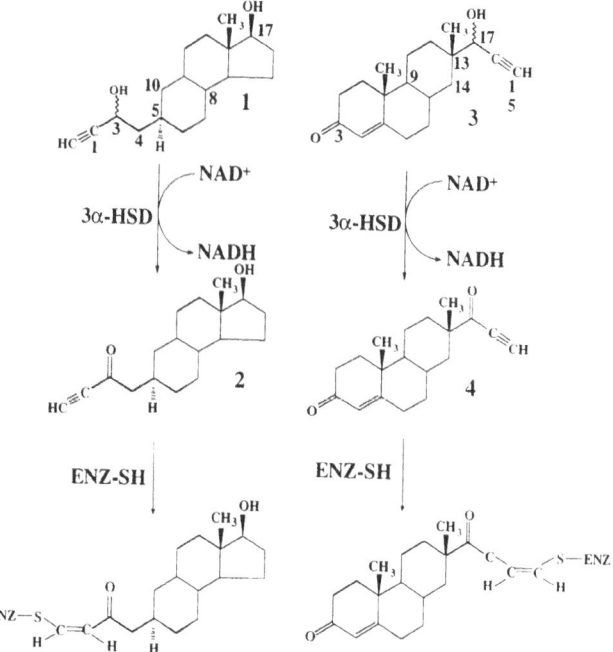

Figure 6. Mechanism-based inactivation of 3α-HSD by secosteroid acetylenic alcohols **1** and **3**.

lined by hydrophobic residues including W86; opposing this side is the flexible loop B which contains W227. Site-directed mutagenesis was performed on each of these trypto-phans, in addition the remaining tryptophan in the enzyme, W148, which lies on the pe-riphery of the structure on an α-helix was also mutated.

Each of the tryptophans was individually mutated to tyrosine, the mutants were puri-fied to homogeneity and their kinetic properties determined (Jez and Penning, 1995). It was found that the W148Y mutant was kinetically similar to wt r3α-HSD which is consis-tent with its position in the structural model. In contrast, the W86Y mutant affects both NADH and steroid binding. In the oxidation direction, the K_m for androsterone was in-creased 3-fold and the K_m for NAD$^+$ was increased 6-fold; similarly, in the reduction di-rection, the K_m for androstanedione was increased 6-fold and the K_m for NADH was increased 8-fold. The W227Y mutant affected steroid binding dramatically. The K_m for androsterone was increased 9-fold and the K_m for androstanedione was increased 43-fold. These alterations in steroid affinity were confirmed by equilibrium dialysis measurements where the binding of testosterone to the E·NADH complex was measured in the W86Y and W227Y mutants. It was found that the former mutant displayed an effective increase in the K_d for testosterone of 8-fold while the later mutant displayed an effective increase in the K_d for testosterone of 22-fold. These data provide the first evidence that residues which make up the apolar cleft and the opposing loop B are involved in steroid binding. The data can be interpreted with the aid of the structural model. W86 is on one side of the apolar cleft. It is close to the catalytic tetrad and is within 10 Å of the nicotinamide ring of the cofactor; therefore, mutation of this residue affects the binding of both steroid and co-substrate. W227 is on the flexible loop B and opposes W86, and although much of this loop is undefined, in the electron-density the ability of the W227 mutant to have such a dramatic effect on steroid binding indicates that W227 and its associated loop interact with the face or edge of the steroid. Similar movements in the loops of other AKR mem-bers are believed to be important in binding inhibitors.

Based on the structure of the binary-complex, the position of the water molecule at the active site, the secosteroid data, and the results of site-directed mutagenesis experi-ments, a model of how androstanedione, a C_{19} 3-ketosteroid, binds to the apolar pocket is proposed. In this model, the C3 ketone has been modeled where the water was found at the active site and is in close proximity to Tyr55. The α-face of the steroid points to W86 so that the stereochemistry of hydride transfer is maintained, and the β-face of the steroid points towards W227. Based on distance measurements, it is likely that the C- and D-rings of the steroid interact with this residue. The secosteroid data implies that the positions of the A- and B-rings are interchangeable with the positions of the C- and D-rings. Absent from the structural model is Cys217 which is alkylated by the acetylenic ketones. This residue is within the vicinity of this site, but steric hindrance by other residues restricts its accessibility to the cleft.

2.9. Determinants of Steroid Specificity

To examine the determinants of steroid specificity in the AKR superfamily, the resi-dues that reside in the apolar clefts of 3α-HSD, 17β-HSD and 20α-HSD have been com-pared using the three-dimensional structure of 3α-HSD as a template. It was found that the apolar pockets of these enzymes are very similar, Table 3. They contain the same catalytic tetrad as well as Trp86 and Trp227. These tryptophans are found in AKRs that do not bind steroids, therefore these residues, although important in substrate binding, do not deter-mine preference for steroids. An examination of the residues in the pockets of the HSDs

Table 3. Residues in the steroid binding pockets of HSDs that belong to the aldo-keto reductase superfamily.

Residue[a]	50	52	54	55	84	86	117	118	120	128	129	137	139	227	306	308	310
Protein																	
rat liver 3α-HSD	D	A	L	Y	K	W	H	F	M	F	F	L	F	W	N	A	Y
human type I 3α-HSD	D	A	L	Y	K	W	H	F	M	P	L	V	F	W	V	M	F
human type II 3α-HSD	D	A	L	Y	K	W	H	S	M	L	S	V	F	W	F	S	S
mouse liver 17β-HSD	D	A	M	Y	K	W	H	F	M	Y	L	L	Y	W	I	G	S
rat ovary 20α-HSD	D	S	L	Y	K	W	H	F	V	L	L	L	L	C	F	A	M
rabbit ovary 20α-HSD	D	A	F	Y	K	W	H	F	T	I	I	A	F	W	V	A	F

[a]residue numbering is relative to rat liver 3α-HSD

residues 306, 308 and 310 are at the C-terminus

indicate that the difference in positional and stereochemistry is either determined by alterations in three or four amino acids or by a steroid induced conformational change in the loop structures. These conformational changes may occur in the C-terminal loop of AKRs where the sequence identity is less striking. It is unlikely that any single point mutation in these apolar clefts will confer a different positional and stereo-specificity on an HSD. Rather, a combination of point mutations may be required to engineer alternate activity. This idea may be better conceived with a chimeric approach in which the loops that mark the boundaries of the apolar cleft in 3α-HSD could be mixed and matched with loops of HSDs with different positional and stereochemistry. Such an approach would compensate for changes in loop conformation that may occur on binding steroid.

3. CONCLUSIONS

Although there is much structural information on HSDs that belong to the AKR superfamily, we still do not understand the basis of ligand recognition and what determines whether an AKR will function as a 3α-HSD, a 17β-HSD or a 20α-HSD. From the crystal structure of the aldose reductase zopolrestat complex (Wilson et al., 1993), it is clear that over 110 contacts are made between the inhibitor and residues in the protein. This level of detail may be required for the HSDs and can be obtained only by determining the structure of a ternary complex E·NAD(P)H·Steroid. Such information will provide better templates for structure-based inhibitor design.

4. ACKNOWLEDGMENTS

I would like to acknowledge the contributions of my colleagues to this work. In particular, I thank Drs. Mitchell Lewis, Susan Smith-Hoog and Melanie Bennett for structural information on 3α-HSD; Dr. John Pawlowski, Brian Schlegel and Joseph Jez for site-directed mutagenesis experiments; and Dr. Douglas F. Covey at Washington University for providing the secosteroids.

5. REFERENCES

Andersson, S., Geissler, W.M., Patel, S. and Wu, L., 1995, The molecular biology of androgenic 17β-hydroxysteroid dehydrogenases, *J.Steroid.Biochem.Mol.Biol.* 53:37–39.

Askonas, L.J., Ricigliano, J.W. and Penning, T.M., 1991, The kinetic mechanism catalysed by homogeneous rat liver 3α-hydroxysteroid dehydrogenase. Evidence for binary and ternary dead-end complexes containing non-steroidal anti-inflammatory drugs, *Biochem.J.* 278:835–841.

Bennett, M.J., Schlegel, B.P., Jez, J.M., Penning, T.M. and Lewis, M., 1996, Structure of 3α-hydroxysteroid/dihydrodiol dehydrogenase complexed with NADP⁺, *Biochemistry, In Press*

Chen, Z., Jiang, J.C., Lin, Z.G., Lee, W.R., Baker, M.E. and Chang, S.H., 1993, Site-specific mutagenesis of *Drosophila* alcohol dehydrogenase: evidence for involvement of tyrosine-152 and lysine-156 in catalysis, *Biochemistry* 32:3342–3346.

Deyashiki, Y., Ohshima, K., Nakanishi, M., Sato, K., Matsuura, K. and Hara, A., 1995, Molecular cloning and characterization of mouse estradiol 17β-dehydrogenase (A-specific), a member of the aldoketoreductase family, *J Biol Chem* 270:10461–10467.

Dufort, I., Soucy, P., Zhang, Y. and Luu-The, V., 1995, Cloning and characterization of human type 2 3α-hydroxysteroid dehydrogenase from human prostatic cDNA library, Proceedings Fifth International Congress on Hormones and Cancer, Abstract, 190.

Ensor, C.M. and Tai, H.H., 1991, Site-directed mutagenesis of the conserved tyrosine 151 of human placental NAD(+)-dependent 15-hydroxyprostaglandin dehydrogenase yields a catalytically inactive enzyme, *Biochem.Biophys.Res.Commun.* 176:840–845.

Funder, J.W., Pearce, P.T., Smith, R. and Smith, A.I., 1988, Mineralocorticoid action: target tissue specificity is enzyme, not receptor, mediated, *Science* 242:583–585.

Ghosh, D., Erman, M., Wawrzak, Z., Duax, W.L. and Pangborn, W., 1994b, Mechanism of inhibition of 3α,20β-hydroxysteroid dehydrogenase by a licorice-derived steroidal inhibitor, *Structure* 2:973–980.

Ghosh, D., Pletnev, V.Z., Zhu, D.W., Wawrzak, Z., Duax, W.L., Pangborn, W., Labrie, F. and Lin, S.X., 1995, Structure of human estrogenic 17b-hydroxysteroid dehydrogenase at 2.20 A resolution, Structure 3:503–513.

Ghosh, D., Wawrzak, Z., Weeks, C.M., Duax, W.L. and Erman, M., 1994a, The refined three-dimensional structure of 3α,20β-hydroxysteroid dehydrogenase and possible roles of the residues conserved in short-chain dehydrogenases, *Structure* 2:629–640.

Ghosh, D., Weeks, C.M., Grochulski, P., Duax, W.L., Erman, M., Rimsay, R.L. and Orr, J.C., 1991, Three-dimensional structure of holo 3α,20β-hydroxysteroid dehydrogenase: a member of a short-chain dehydrogenase family, *Proc.Natl.Acad.Sci.U.S.A.* 88:10064–10068.

Harrison, D.H., Bohren, K.M., Ringe, D., Petsko, G.A. and Gabbay, K.H., 1994, An anion binding site in human aldose reductase: mechanistic implications for the binding of citrate, cacodylate, and glucose 6-phosphate, *Biochemistry* 33:2011–2020.

Hoog, S.S., Pawlowski, J.E., Alzari, P.M., Penning, T.M. and Lewis, M., 1994, Three-dimensional structure of rat liver 3α-hydroxysteroid/dihydrodiol dehydrogenase: a member of the aldo-keto reductase superfamily, *Proc.Natl.Acad.Sci.U.S.A.* 91:2517–2521.

Isaacs, J.T., 1983, Changes in dihydrotestosterone metabolism and the development of benign prostatic hyperplasia in the aging beagle, *J.Steroid.Biochem.* 18:749–757.

Jacobi, G.H., Moore, R.J. and Wilson, J.D., 1977, Characterization of the 3α-hydroxysteroid dehydrogenase of dog prostate, *J.Steroid.Biochem.* 8:719–723.

Jacobi, G.H., Moore, R.J. and Wilson, J.D., 1978, Studies on the mechanism of 3α-androstanediol-induced growth of the dog prostate, *Endocrinology* 102:1748–1758.

Jez, J.M. and Penning, T.M., 1995, The role of tryptophans in ligand recognition by rat liver 3α–hydroxysteroid dehydrogenase, Int. Symposium on DHEA Transformation into Androgens and Estrogens in Target Tissues: Intracrinology, Abstract, 11.

Jörnvall, H., Persson, B., Krook, M., Atrian, S., Gonzalez-Duarte, R., Jeffery, J. and Ghosh, D., 1995, Short-chain dehydrogenases/reductases (SDR), *Biochemistry* 34:6003–6013.

Khanna, M., Qin, K.N., Wang, R.W. and Cheng, K.C., 1995, Substrate specificity, gene structure, and tissue-specific distribution of multiple human 3α-hydroxysteroid dehydrogenases, *J Biol Chem* 270:20162–20168.

Krozowski, Z., 1994, The short-chain alcohol dehydrogenase superfamily: variations on a common theme, *J.Steroid.Biochem.Mol.Biol.* 51:125–130.

Krozowski, Z.S., Provencher, P.H., Smith, R.E., Obeyesekere, V.R., Mercer, W.R. and Albiston, A.L., 1994, Isozymes of 11β-hydroxysteroid dehydrogenase: which enzyme endows mineralocorticoid specificity?, *Steroids* 59:116–120.

Lacy, W.R., Washenick, K.J., Cook, R.G. and Dunbar, B.S., 1993, Molecular cloning and expression of an abundant rabbit ovarian protein with 20α-hydroxysteroid dehydrogenase activity [published erratum appears in Mol Endocrinol 1993 Sep;7(9):1239], *Mol.Endocrinol.* 7:58–66.

Liao, S., Liang, T., Fang, S., Castaneda, E. and Shao, T.C., 1973, Steroid structure and androgenic activity. Specificities involved in the receptor binding and nuclear retention of various androgens, *J.Biol.Chem.* 248:6154–6162.

Lin, H.-K., Jez, J.M. and Penning, T.M., 1995, Cloning of a human prostate cDNA with high sequence identity to rat liver 3α–hydroxysteroid/dihydrodiol dehydrogenase, Int. Symposium on DHEA Transformation into Androgens and Estrogens in Target Tissues: Intracrinology, Abstract, 37.

Mao, J., Duan, W.R., Albarracin, C.T., Parmer, T.G. and Gibori, G., 1994, Isolation and characterization of a rat luteal cDNA encoding 20α-hydroxysteroid dehydrogenase, *Biochem.Biophys.Res.Commun.* 201:1289–1295.

Mercer, W.R. and Krozowski, Z.S., 1992, Localization of an 11β-hydroxysteroid dehydrogenase activity to the distal nephron. Evidence for the existence of two species of dehydrogenase in the rat kidney, *Endocrinology* 130:540–543.

Monder, C., Stewart, P.M., Lakshmi, V., Valentino, R., Burt, D. and Edwards, C.R., 1989, Licorice inhibits corticosteroid 11β-dehydrogenase of rat kidney and liver: *in vivo* and *in vitro* studies, *Endocrinology* 125:1046–1053.

Mune, T., Rogerson, F.M., Nikkila, H., Agarwal, A.K. and White, P.C., 1995, Human hypertension caused by mutations in the kidney isozyme of 11β-hydroxysteroid dehydrogenase, *Nat.Genet.* 10:394–399.

Obeid, J. and White, P.C., 1992, Tyr-179 and Lys-183 are essential for enzymatic activity of 11β-hydroxysteroid dehydrogenase, *Biochem.Biophys.Res.Commun.* 188:222–227.

Pawlowski, J.E., Huizinga, M. and Penning, T.M., 1991, Cloning and sequencing of the cDNA for rat liver 3α-hydroxysteroid/dihydrodiol dehydrogenase, *J.Biol.Chem.* 266:8820–8825.

Pawlowski, J.E. and Penning, T.M., 1994, Overexpression and mutagenesis of the cDNA for rat liver 3α-hydroxysteroid/dihydrodiol dehydrogenase. Role of cysteines and tyrosines in catalysis, *J.Biol.Chem.* 269:13502–13510.

Penning, T.M., Mukharji, I., Barrows, S. and Talalay, P., 1984, Purification and properties of a 3α-hydroxysteroid dehydrogenase of rat liver cytosol and its inhibition by anti-inflammatory drugs, *Biochem.J.* 222:601–611.

Poutanen, M., Miettinen, M. and Vihko, R., 1993, Differential estrogen substrate specificities for transiently expressed human placental 17β-hydroxysteroid dehydrogenase and an endogenous enzyme expressed in cultured COS-m6 cells, *Endocrinology* 133:2639–2644.

Schlegel, B., Pawlowski, J.E., Hu, Y., Scolnick, D.M., Covey, D.F. and Penning, T.M., 1994, Secosteroid mechanism-based inactivators and site-directed mutagenesis as probes for steroid hormone recognition by 3α-hydroxysteroid dehydrogenase, *Biochemistry* 33:10367–10374.

Schlegel, B. and Penning, T., 1995, Site-directed mutagenesis of the catalytic triad of rat liver 3α-hydroxysteroid dehydrogenase, Int. Symposium on DHEA Transformation into Androgens and Estrogens in Target Tissues: Intracrinology, Abstract, 10.

Stewart, P.M., Wallace, A.M., Valentino, R., Burt, D., Shackleton, C.H. and Edwards, C.R., 1987, Mineralocorticoid activity of liquorice: 11β-hydroxysteroid dehydrogenase deficiency comes of age, *Lancet* 2:821–824.

Strauss, J.F., III Jr. and Stambaugh, R.L., 1974, Induction of 20α-hydroxysteroid dehydrogenase in rat corpora lutea of pregnancy by prostaglandin F-2α, *Prostaglandins* 5:73–85.

Taurog, J.D., Moore, R.J. and Wilson, J.D., 1975, Partial characterization of the cytosol 3α-hydroxysteroid: NAD(P)$^+$oxidoreductase of rat ventral prostate, *Biochemistry* 14:810–817.

Warren, J.C., Murdock, G.L., Ma, Y., Goodman, S.R. and Zimmer, W.E., 1993, Molecular cloning of testicular 20α-hydroxysteroid dehydrogenase: identity with aldose reductase, *Biochemistry* 32:1401–1406.

Wermuth, B. and Monder, C., 1983, Aldose and aldehyde reductase exhibit isocorticosteroid reductase activity, *Eur.J.Biochem.* 131:423–426.

Wilson, D.K., Bohren, K.M., Gabbay, K.H. and Quiocho, F.A., 1992, An unlikely sugar substrate site in the 1.65 A structure of the human aldose reductase holoenzyme implicated in diabetic complications, *Science* 257:81–84.

Wilson, D.K., Tarle, I., Petrash, J.M. and Quiocho, F.A., 1993, Refined 1.8 A structure of human aldose reductase complexed with the potent inhibitor zopolrestat, *Proc Natl Acad Sci U S A* 90:9847–9851.

55

PHYSIOLOGICAL SUBSTRATES OF HUMAN ALDOSE AND ALDEHYDE REDUCTASES

David L. Vander Jagt, Jose E. Torres, Lucy A. Hunsaker, Lorraine M. Deck, and Robert E. Royer

Department of Biochemistry
University of New Mexico
School of Medicine
Albuquerque, New Mexico 87131

1. INTRODUCTION

The aldo-keto reductase family of NADPH-dependent oxidoreductases is widely distributed in man and in animals (Wirth and Wermuth, 1985; Wermuth, 1985; Grimshaw and Mathur, 1989). Aldose reductase (EC 1.1.1.21; ALR2) and aldehyde reductase (EC 1.1.1.2; ALR1), monomeric members of the aldo-keto reductase family, exhibit broad, overlapping specificities, consistent with a role in detoxification of aldehydes (Grimshaw, 1992). ALR2 has received special attention due to its putative role in the etiology of diabetic complications (Gabbay, 1975; Kador and Kinoshita, 1985). Most crystallographic and mechanistic studies have focused on ALR2. ALR2 is comprised of an α/β-barrel without a classic Rossman fold at the dinucleotide binding site (Rondeau et al., 1992; Wilson et al., 1992); ALR2 utilizes an ordered mechanism with NADPH binding preceeding aldehyde binding, followed by hydride transfer, alcohol release and rate-determining isomerization of the ALR2-NADP binary complex prior to NADP release (Grimshaw et al., 1990; Kubisecki et al., 1992). ALR1 is similar structurally to ALR2 (El-Kabbani et al., 1995) and utilizes a similar ordered mechanism (Davidson and Flynn, 1979; Daly and Mantle, 1982). However, the rate determining reaction for ALR1 appears to involve earlier steps in the reaction sequence including hydride transfer from the reduced nicotinamide ring to the aldehyde as well as proton transfer (Barski et al., 1995). Thus, k_{cat} values for ALR1 are more sensitive to variations in substrate than is observed with ALR2.

The question of physiological substrates of ALR1 and ALR2 has not been extensively addressed. The emphasis that has been placed on ALR2 results from the ability of ALR2 to catalyze the reduction of glucose to sorbitol, which is an important reaction under conditions of hyperglycemia. ALR2 continues to be a major target for development of drugs that may be useful in preventing diabetic complications, an effort now guided by the availability of the crystal structure of an ALR2-cofactor-inhibitor ternary complex (Petrash et al., 1994; Wilson et al., 1993). It has been assumed that the only important physiological reaction catalyzed by ALR2 is the reduction of glucose to sorbitol. However, in a number of diabetic complications, especially those involving nerves and kidney, sorbitol does not reach high levels. Nevertheless, inhibitors of ALR2 appear effective in preventing diabetic complications that involve these sites, at least in experimental models of diabetes. In addition, inhibitors of ALR2 can prevent extracellular changes in diabetes, such as basement membrane thickening. These observations suggest that aldehydes in addition to glucose may be important in the development of diabetic complications.

In view of the overlapping specificities of ALR1 and ALR2 toward aldehyde substrates, we compared these two reductases in a study of physiological aldehydes. We also probed further into the question of the different rate determining steps of ALR1 and ALR2 by examining the effects of ring substituents in aromatic aldehydes on the substrate properties of ALR1 and ALR2.

2. MATERIALS AND METHODS

2.1. Chemicals

D,L-Glyceraldehyde, glucose and glucuronic acid were from Sigma. Glyoxal, acrolein and substituted benzaldehydes were from Aldrich and were purified by distillation. Methylglyoxal was generated from pyruvic aldehyde dimethylacetal (Aldrich) by treatment with sulfuric acid followed by azeotropic distillation with water. Methylglyoxal was standardized with the glyoxalase system (Vander Jagt, 1989). 3-Deoxy-D-glucosone, D-glucosone, 3-deoxy-L-xylosone and L-xylosone were synthesized by literature procedures (Hirsch et al., 1992). The diacetal of 4-hydroxynonenal was synthesized (Esterbauer and Weger, 1962); 4-hydroxynonenal was generated by treatment of the diacetal with citric acid (Mitchell and Petersen, 1991). Malondialdehyde was formed by hydrolysis of 1,1,3,3-tetraethoxypropane (Aldrich) and purified as the sodium salt (Marnett and Tuttle, 1980).

2.2. Enzyme Assays and Kinetics

Kinetic analysis of aldose reductase and aldehyde reductase was carried out at pH7, 25°C, in 0.1 M sodium phosphate buffer. K_m and k_{cat} values were determined by nonlinear regression analysis with the Enzfitter program. Human ALR2 was purified from skeletal muscle and ALR1 was purified from kidney (Vander Jagt et al., 1990a,b; 1992; Robinson et al., 1993). Reductase activity was extracted from the 100,000xg supernatant fraction with Red SepharoseCL-6B, followed by chromatofocusing on PBE 94 (Pharmacia LKB). Final purification was accomplished by chromatography on a Bio-Gel HPHT HPLC column (Bio-Rad).

3. RESULTS AND DISCUSSION

3.1. Evaluation of the Rate Determining Steps of ALR1 and ALR2

The question of differences in rate determining steps of ALR1 and ALR2 was probed by examining the effects of ring substituents on the rates of reduction of a series of

Table 1. Comparison of the kinetic parameters of Aldehyde Reductase (ALR1) and Aldose Reductase (ALR2) with para-substituted benzaldehydes as substrates

Para-substitutent	kcat (min⁻¹)		Km (µM)		Ratio ALR2/ALR1[b]
	ALR1	ALR2	ALR1	ALR2	
H	92	148[a]	850	7	195
CF₃	2,080	110	980	1	52
Cl	520	160[a]	1,400	7	61
NO₂	2,040	160[a]	160	3	4
CH₃	190	94	2,500	15	82
OCH₃	98	60	1,300	13	61

[a]Vander Jagt et al., 1995
[b]Ratio of kcat/Km values for ALR2 to kcat/Km values for ALR1

substituted benzaldehydes (Table 1). For ALR2, k_{cat} values show very little sensitivity to changes in substituents; k_{cat} values vary less than three-fold. K_m values ranged from 1 to 15 µM among the substituted benzaldehydes. A plot of log k_{cat}/K_m values for ALR2 against the Hammett σ-constants is shown in figure 1; the slope, 1.2, suggests a modest increase in catalytic efficiency with increased polarization of the aldehyde functional group. For ALR1, k_{cat} values exhibited a 23 fold range and generally were larger than for ALR2. K_m values were also larger, ranging from 850 to 2500 µM. A plot of log k_{cat}/K_m vs σ is shown in figure 2. The slope, 2.2, suggests that ALR1 exhibits a more pronounced dependence of catalytic efficiency on the polarization of the aldehyde functional group. A comparison of the ratio of k_{cat}/K_m for ALR2 and ALR1 (Table 1) reflects the generally higher catalytic efficiency of ALR2. However, for some substrates of ALR1, such as p-nitrobenzaldehyde, the increase in k_{cat} that is provided by having a polarized aldehyde functional group almost offsets the increase in K_m that ALR1 exhibits compared to ALR2 such that the catalytic efficiencies of the two enzymes with this substrate are comparable.

The results summarized in Table 1 and figures 1 and 2 are consistent with previous studies concerning the rate determining steps of ALR1 and ALR2. The lack of sensitivity of k_{cat} for ALR2 to changes in substrate is in agreement with kinetic and spectroscopic data that support NADP release, and associated protein conformational changes, as the main rate limiting step (Grimshaw et al., 1990; Kubisecki et al., 1992). However, the mod-

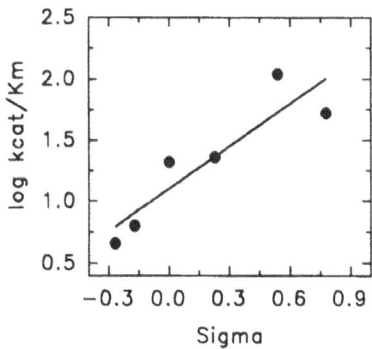

Figure 1. Hammett plot of log kcat/Km vs. sigma substituent constants for the reduction of substituted benzaldehydes catalyzed by ALR2. Slope = 1.2 (r = 0.90).

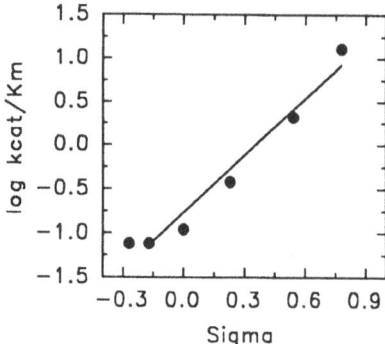

Figure 2. Hammett plot of log kcat/Km vs. sigma substituent constants for the reduction of substituted benzalde-hydes catalyzed by ALR1. Slope = 2.2 (r = 0.98).

est dependence of k_{cat}/K_m on aldehyde polarity suggests that the redox step itself may be partially rate determining. The stronger dependence of k_{cat}/K_m for ALR1 on aldehyde polarity is consistent with reports that for ALR1 the transfer of hydride to the aldehyde carbonyl is primarily rate limiting and that the subsequent transfer of a proton to the alcoholate product is also partially rate limiting, conclusions drawn from isotope effect data (Barski et al., 1995).

3.2. Comparison of ALR1 and ALR2 with Substituted Phenylglyoxals as Substrates

The effects of ring substituents on aldehyde reactivity can be predicted to be mark-edly different when one compares substituted benzaldehydes with substituted phenyl-glyoxals. Interposition of a ketone functional group between the aromatic ring and the aldehyde functional group essentially blocks the electronic substituent from having any ef-fect on the polarity of the aldehyde, as reflected in an essentially constant carbonyl stretching frequency of 1727 ± 2 cm^{-1} for substituted phenylglyoxals (Vander Jagt et al., 1972). ALR1 and ALR2 were compared using a series of substituted phenylglyoxals as substrates (Table 2). For both ALR1 and ALR2 there are no significant substituent effects either on k_{cat} or on K_m. In addition, k_{cat}/K_m ratios are very similar for both ALR1 and

Table 2. Comparison of the kinetic parameters of aldehyde reductase (ALR1) and aldose reductase (ALR2) with para-substituted phenylglyoxals as substrates

	kcat (min^{-1})		Km (μM)		Ratio
Para-substitutent	ALR1	ALR2	ALR1	ALR2	ALR2/ALR1[b]
H	2,110	60 [a]	190	4	1.4
Cl	1,860	45 [a]	120	1	2.9
Br	1,630	100 [a]	100	2	3.1
CH$_3$	1,200	119 [a]	190	2	9.4
OCH$_3$	1,300	45 [a]	230	2	4.0

[a]Vander Jagt et al., 1995
[b]Ratio of kcat/Km values for ALR2 to kcat/Km values for ALR1

Table 3. Physiological substrates of aldehyde reductase (ALR1) and aldose reductase (ALR2)

Aldehyde	kcat (min^{-1})		Km (µM)		Ratio ALR2/ALR1 [a]
	ALR1	ALR2	ALR1	ALR2	
glucose	246[b]	64[b]	2.5×10^6	7×10^4	9.3
glucuronate	1,250[b]	81[b]	3,600	2,800	0.08
glyceraldehyde	560[b]	120[b]	3,400	16	46
glycoaldehyde	150	106[b]	4,200	112	26
3-deoxyglucosone [c]	1,400	122	2,400	47	4.4
glucosone	340	92	4,000	167	6.5
3-deoxyxylosone	1,470	167	2,700	78	4.0
xylosone	280	123	2,600	234	4.9
glyoxal	760	154[b]	48,000	514	19
methylglyoxal	210	142[b]	1,200	8	101
4-hydroxynonenal	-- [d]	102[b]	-- [d]	22	--
acrolein	800	87[b]	41,000	80	56

[a]Ratio of kcat/Km values for ALR2 to kcat/Km values for ALR1
[b]Vander Jagt et al., 1990, 1992, 1995
[c]Preliminary data in Vander Jagt et al., 1994
[d]Low solubility precluded determination of kinetic parameters

ALR2. Thus the substrate selectivity observed with substituted benzaldehydes (Tables 1) disappears with substituted phenylglyoxals.

3.3. Comparison of ALR1 and ALR2 with Physiological Substrates

The kinetic parameters of ALR1 and ALR2 were compared with a series of physiological aldehydes that included 2-hydroxy-aldehydes, 2-oxo-aldehydes and α,β-unsaturated aldehydes. The results are summarized in Table 3. The 2-hydroxy-aldehydes include glucose, glucuronic acid, glyceraldehyde and glycoaldehyde. Glycoaldehyde has recently been identified as an intermediate in the complex Maillard chemistry involved in modification of proteins by sugars and may be important in the formation of Advanced Glycation Endproducts in the development of diabetic complications (Glomb and Monnier, 1995). The 2-oxo-aldehydes include 3-deoxyglucosone, glucosone, 3-deoxyxylosone, xylosone, glyoxal and methylglyoxal. The osones, especially 3-deoxyglucosone, have received extensive study recently as bifunctional crosslinking agents that may be formed as intermediates in nonenzymatic glycation reactions (Knecht et al., 1992; Bucala and Cerami, 1992). Glyoxal, like glycoaldehyde, may also be important in the formation of Advanced Glycation Endproducts (Glomb and Monnier, 1995). Methylglyoxal likewise has been implicated in the development of diabetic complications (Vander Jagt et al., 1992). The α,β-unsaturated aldehydes include 4-hydroxynonenal and acrolein, products of lipid degradation that are elevated during oxidative stress (Esterbauer et al., 1991). The pattern of substrate properties of these aldehydes as substrates for ALR1 and ALR2 is similar to that observed with the aromatic aldehydes. ALR2 exhibits a small range of k_{cat} values; ALR1 shows a somewhat larger range of k_{cat} values; and K_m values are smaller with ALR2. The selectivity generally favors ALR2 with the exception of glucuronic acid which favors ALR1 13-fold. ALR2 exhibits highest selectivity for methylglyoxal. The ability of ALR1 and ALR2 to catalyze the reduction of the osones in Table 3 has also recently been described by other investigators (Feather et al., 1995).

The results summarized in Table 3 extend the number of physiological aldehydes that are known substrates of ALR1 and ALR2. ALR1 and ALR2 catalyze the reduction of isocortico- steroids, which are the best physiological substrates identified to date (Wermuth and Monder, 1983). ALR1 and ALR2 also catalyze the reduction of aldehydes derived from metabolism of biogenic amines (Turner and Tipton, 1972; Tabakoff et al., 1973). Recently it was reported that isocaproaldehyde, derived from the sidechain cleavage of cholesterol, is an excellent substrate of ALR2 (Matsuura et al., 1996). Thus it appears that almost any endogenously generated aldehyde is a substrate of ALR1 and/or ALR2.

3.4. Malondialdehyde as a Substrate of ALR1 and ALR2

Malondialdehyde, a reactive bifunctional crosslinking agent produced by oxidative degradation of unsaturated lipids, is isoelectronic with methylglyoxal. However, malondialdehyde exists in physiological solution as an enolate anion. Malondialdehyde was examined as a substrate of ALR1 and ALR2. Neither of these enzymes catalyzed the reduction of malondialdehyde at concentrations up to 10 mM, suggesting that malondialdehyde is one of the exceptions to the rule that most endogenous aldehydes are substrates of ALR1 and ALR2.

4. ACKNOWLEDGMENTS

This work was supported by NIH grant DK43238 and by grant RR08139 from the Minority Biomedical Support Program, National Institutes of Health.

5. REFERENCES

Barski, O.A., Gabbay, K.H., Grimshaw, C.E. and Bohren, K.M.: Mechanism of human aldehyde reductase: Characterization of the active site pocket. Biochemistry 34 (1995) 11264–11275.

Bohren, K.M., Grimshaw, C.E., Lai, C-J., Harrison, D.H., Ringe, D., Petsko, G.A. and Gabbay, K.H.: Tyrosine-48 is the proton donor and histidine-110 directs substrate stereochemical selectivity in the reduction reaction of human aldose reductase: Enzyme kinetics and crystal structure of the Y48H mutant enzyme. Biochemistry 33 (1994) 2021–2032.

Bucala, R. and Cerami, A.: Advanced glycosylation: chemistry, biology and implications for diabetes and aging. Adv. Pharmacol. 23 (1992) 1–34.

Daly, A.K. and Mantle, T.J.: The kinetic mechanism of the major form of ox kidney aldehyde reductase with D-glucuronic acid. Biochem. J. 205 (1982) 381–388.

Davidson, W.S. and Flynn, T.G.: Kinetics and mechanism of aldehyde reductase from pig kidney. Biochem. J. 177 (1979) 595–601.

El-Kabbani, O., Judge, K., Ginell, S.L., Myles, D.A.A., DeLucas, L.J. and Flynn, T.G.: Structure of porcine aldehyde reductase holoenzyme. Nature Structural Biology 2 (1995) 687–692.

Esterbauer, H. and Weger, W.: Synthesis of homologous 4-hydroxy-2-alkenals II. Monatsh. Chem. 98 (1967) 1994–2000.

Esterbauer, H., Schaur, R.J. and Zollner, H.: Chemistry and biochemistry of 4-hydroxynonenal, malondialdehyde and related aldehydes. Free Radical Biol. Med. 11 (1991) 81–128.

Feather, M.S., Flynn, T.G., Munro, K.A., Kubisecki, T.J. and Walton, D.J.: Catalysis of reduction of carbohydrate 2-oxoaldehydes (osones) by mammalian aldose reductase and aldehyde reductase. Biochim. Biophys. Acta 1244 (1995) 10–16.

Glomb, M.A. and Monnier, V.M.: Mechanism of protein modification by glyoxal and glycoaldehyde, reactive intermediates of the Maillard reaction. J. Biol. Chem. 270 (1995) 10017–10026.

Grimshaw, C.E.: Aldose reductase: Model for a new paradigm of enzymic perfection in detoxification catalysis. Biochemistry 31 (1992) 10139–10145.

Grimshaw, C.E. and Mather, E.J.: Immunoquantitation of aldose reductase in human tissues. Anal. Biochem. 176 (1989) 66–71.

Grimshaw, C.E., Shahbaz, M. and Putney, C.G.: Mechanistic basis for non-linear kinetics of aldehyde reduction catalyzed by aldose reductase. Biochemistry 29 (1990) 9947–9955.

Hirsch, J., Petrokova, E. and Feather, M.S.: The reaction of some dicarbonyl sugars with aminoguanidine. Carbohydrate Res. 232 (1992) 125–130.

Kador, P.F. and Kinoshita, J.H.: Role of aldose reductase in the development of diabetes- associated complications. Am. J. Med. 79 (1985) 8–12.

Knecht, K.J., Feather, M.S. and Baynes, J.W.: Detection of 3-deoxyfructose and 3- deoxyglucosone in human urine and plasma: evidence for intermediate stages of the Maillard reaction in vivo. Arch. Biochem. Biophys. 294 (1992) 130–137.

Kubisecki, T.J., Hyndman, D.J., Morjana, N.A. and Flynn, T.G.: Studies of pig muscle aldose reductase . Kinetic mechanism and evidence for a slow conformational change upon coenzyme binding. J. Biol. Chem. 267 (1992) 6510–6517.

Marnett, L.J. and Tuttle, M.A.: Comparison of the mutagenicities of malondialdehyde and the side products formed during its chemical synthesis. Cancer Res. 40 (1980) 276–282.

Matsuura, K., Deyashiki, Y., Bunai, Y., Ohya, I. and Hara, A.: Aldose reductase is a major reductase for iso-caproaldehyde, a product of side-chain cleavage of cholesterol, in human and animal adrenal glands. Arch. Biochem. Biophys. 328 (1996) 265–271.

Mitchell, D.Y. and Petersen, D.R.: Inhibition of rat hepatic mitochondrial aldehyde dehydrogenase-mediated acet-aldehyde oxidation by trans-4-hydroxy-2-nonenal. Hepatology 13 (1991) 728–734.

Petrash, J.M., Tarle, I., Wilson, D.K. and Quiocho, F.A.: Aldose reductase catalysis and crystallography. Insights from recent advances in enzyme structure and function. Diabetes 43 (1994) 955–959.

Rondeau, J-M., Tete-Favier, F., Podjarny, A., Reymann, J-M., Barth, P., Biellmann, J-F. and Moras, D.: Novel NADPH-binding domain revealed by the crystal structure of aldose reductase. Nature 355 (1992) 469–472.

Robinson, B., Hunsaker, L.A., Stangebye, L.A. and Vander Jagt, D.L.: Aldose and aldehyde reductases from human kidney cortex and medulla. Biochim. Biophys. Acta 1203 (1993) 260–266.

Tabakoff, B., Anderson, R. and Alivisatos,, S.G.A.: Enzymatic reduction of "biogenic" aldehydes in brain. Mol. Pharmacol. 9 (1973) 428–437.

Turner, A.J. and Tipton, K.F.: The characterization of two reduced nicotinamide-adenine dinucleotide phosphate-linked aldehyde reductases from pig brain. Biochem. J. 130 (1972) 765–772.

Vander Jagt, D.L., Han, L-P and Lehman, C.H.: Effects of substituents on the rates of disproportionation of substituted phenylglyoxals in alkaline solution. J. Org. Chem. (1972) 4100–4104.

Vander Jagt, D.L.: The Glyoxalase System, in Glutathione: Chemical, Biochemical, and Medical Aspects-Part A, D. Dolphin, R. Poulson, and O. Avramovic, eds., 1989, J. Wiley, New York, pp. 598–641.

Vander Jagt, D.L., Hunsaker, L.A., Robinson, B., Stangebye, L.A. and Deck, L.M.: Aldehyde and aldose reductases from human placenta. J. Biol. Chem. 265 (1990a) 10912–10918.

Vander Jagt, D.L., Robinson, B., Taylor, K.K. and Hunsaker, L.A.:Aldose reductase from human skeletal and heart muscle. J. Biol. Chem. 265 (1990b) 20982–20987.

Vander Jagt, D.L., Robinson, B., Taylor, K.K. and Hunsaker, L.A.: Reduction of trioses by NADPH-dependent aldo-keto reductases. Aldose reductase, methylglyoxal, and diabetic complications. J. Biol. Chem. 267 (1992) 4364–4369.

Vander Jagt, D.L., Hunsaker, L.A., Deck, L.M., Chamblee, B.B. and Royer, R.E.: Ketoaldehyde detoxification enzymes and protection against the Maillard reaction, in Maillard Reactions in Chemistry, Food, and Health, T.P. Labuza, G.A. Reineccius, V.M. Monnier, J. O'Brien, and J.W. Baynes, eds., 1994, The Royal Society of Chemistry, Cambridge, pp. 314–318.

Vander Jagt, D.L., Kolb, N.S., Vander Jagt, T.J., Chino, J., Martinez, F.J., Hunsaker, L.A. and Royer, R.E.: Substrate specificity of human aldose reductase: identification of 4- hydroxynonenal as an endogenous substrate. Biochim. Biophys. Acts 1249 (1995) 117–126.

Wermuth, B., Aldo-keto reductases, in: Enzymology of Carbonyl Metabolism 2: Aldehyde Dehydrogenase, Aldo Keto Reductases, and Alcohol Dehydrogenase, T.G. Flynn and H. Weiner, eds., 1985, A.R. Liss, New York, pp. 209–230.

Wilson, D.K., Bohren, K.M., Gabbay, K.H. and Quiocho, F.A.: An unlikely sugar substrate site in the 1.65A structure of the human aldose reductase holoenzyme implicated in diabetic complications. Science 257 (1992) 81–84.

Wilson, D.K., Tarle, I., Petrash, J.M. and Quiocho, F.A.: Refined 1.8A structure of human aldose reductase complexed with the potent inhibitor zopolrestat. Proc. Natl. Acad. Sci. USA 90 (1993) 9847–9851.

Wirth, H-P. and Wermuth, B.: Immunochemical characterization of aldo-keto reductases from human tissues. FEBS Lett. 187 (1985) 280–282.

56

ALDOSE REDUCTASE AS DIHYDRODIOL DEHYDROGENASE

Naphthoquinone Formation by Rat Lens Aldose Reductase

Sanai Sato,[1] Katsumi Sugiyama,[1] and Deborah Carper[2]

[1]Laboratory of Ocular Therapeutics and
[2]Laboratory of Mechanisms of Ocular Diseases
National Eye Institute
National Institutes of Health
Bethesda, Maryland 20892–1850

1. INTRODUCTION

Animals fed with naphthalene quickly develop cataracts. The mechanism of this cataract has been explained by the formation of the biologically active compound 1,2-naphthoquinone. Naphthalene is first oxidized to naphthalene-1,2-oxide and then converted to naphthalene-1,2,-dihydrodiol in liver. Naphthalene-1,2-dihydrodiol is carried to the lens and further metabolized into 1,2-naphthoquinone through 1,2-dihydroxy-naphthalene. Because 1,2-naphthoquinone quickly reacts with various biologically important compounds, the formation of 1,2-naphthoquinone is considered to cause various biochemical changes such as depletion of glutathione and modification of lens proteins, which eventually leads to the formation of cataract (van Heyningen, 1979; Xu et al., 1992a,b). Recently, some aldose reductase inhibitors have been reported to prevent the formation of cataract in naphthalene-fed rats (Hockwin et al., 1984–85; Tao et al., 1991; Xu et al., 1992a). This has renewed the interest in naphthalene cataract and also raised the question of whether aldose reductase is involved in naphthalene metabolism in rat lens.

Aldose reductase (alditol:NADP$^+$ 1-oxidoreductase, EC 1.1.1.21) is a member of a broad family of aldo-keto reductases which utilizes NADPH to reduce a wide range of aromatic and aliphatic aldehydes and ketones to their corresponding alcohols (Turner and Flynn, 1982; Wermuth, 1985). Recently, Hara et al. (1991) have reported that aldose reductase possesses dihydrodiol dehydrogenase activity and utilizes naphthalene-1,2-dihydrodiol as substrate. Moreover, in rat lens, the dehydrogenase activity assayed with naphthalene-1,2-dihydrodiol as substrate is associated with aldose reductase (Sato, 1993). Based on this evidence, it has been suggested that, in rat lens, aldose reductase catalyzes the oxidation of naphthalene-1,2-dihydrodiol to 1,2-dihydroxynaphathalene which is eventually auto-oxidized to 1,2-naphthoquinone.

The role of aldose reductase in the formation of naphthalene cataract, however, is still controversial. The reason is that, in *in vivo* experiments, hydantoin containing inhibitors AL 1576 and sorbinil prevented cataract formation while non-hydantoin inhibitors failed to prevent cataract (Tao et al., 1991; Xu et al., 1992a). Recently, Lou et al. (1996) have also reported similar results with an *in vitro* dual cataract model induced by both naphthalene-1,2-dihydrodiol and galactose, in which hydantoin containing aldose reductase inhibitors AL 1576, AL 4114 and sorbinil successfully prevented cataract changes induced by both naphthalene-1,2-dihydrodiol and galactose, while tolrestat, a non-hydantoin inhibitor, prevented lens changes induced only by galactose but not by naphthalene-1,2-dihydrodiol.

Since it has been well established that the formation of 1,2-naphthoquinone is the key process in the formation of naphthalene cataract, we investigated the formation of naphthoquinone from naphthalene-1,2-dihydrodiol by rat lens *in vitro* and studied the effects of various aldose reductase inhibitors on this 1,2-naphthoquinone formation.

2. MATERIALS AND METHODS

2.1. Chemicals

All reagents utilized in this experiment were reagent grade. 1,2-naphthoquinone and NAD^+ were purchased from Sigma Chemical Co. (St. Louis, MO). Naphthalene-1,2-dihydrodiol was a gift from Alcon Laboratories (Ft. Worth, TX). Aldose reductase inhibitors Al 1576, sorbinil, tolrestat and Ponalrestat were gifts from Alcon Laboratories (Ft Worth, TX), Pfizer Central Research (Groton, CT), Wyeth-Ayest Research Inc. (Princeton, NJ) and ICI Americas (Wilmington, DE), respectively. Frozen Sprague-Dawley rat eyes were commercially obtained from Bristrol (Indiana, IN).

Rat lens aldose reductase utilized in this experiment was the recombinant rat lens enzyme expressed in *E. coli* and purified as previously described (Old et al., 1990).

2.2. *In Vitro* Incubation of Whole Rat Lens Soluble Proteins

Six eyes were thawed, the lenses carefully removed, and then homogenized with 1 ml of 20 mM phosphate buffer, pH 7.5 containing 7 mM 2-mercaptoethanol by glass homogenizer. After centrifugation at 10,000 xg for 10 min, the supernatant was applied to a PD-10 desalting column (Pharmacia-LKB, Uppsala, Sweden).

40 µl of the protein fraction from the PD-10 column (approximately 2 mg of protein) was incubated in 100 µl of the reaction mixture consisting of 0.1 M phosphate buffer, pH 7.5, 4 mM NAD^+, 7 mM 2-mercaptoethanol and either 1 mM naphthalene-1,2-dihydrodiol or 0.5 mM 1,2-naphthoquinone at 22°C for 12 hours. At the end of incubation, proteins were precipitated by adding 200 µl of ice cold methanol. After centrifugation at 10,000 xg for 5 min, the obtained precipitate was dissolved with 10% SDS and analyzed with a Shimadzu UV-2101PC spectrophotometer (Shimadzu Corp., Kyoto, Japan).

2.3. *In Vitro* Incubation of Rat Lens Aldose Reductase

Recombinant rat lens aldose reductase was incubated in 100 µl of incubation mixture containing 20 mM phosphate buffer, pH 7.5, 7 mM 2-mercaptoethanol, 2 mM NAD^+

and either 1 mM naphthalene-1,2-dihydrodiol or 1,2-naphthoquinone at 22°C for 30 min. After the incubation, the reaction was stopped by adding 200 μl of ice cold methanol. After centrifugation at 10,000 xg for 5 min, the supernatant was filtered with a molecular weight 3,000 cut-off membrane Centricon SR3 (Amicon, Beverly, MA) and the filtrate was analyzed by reverse-phase HPLC.

2.4. High Performance Liquid Chromatography (HPLC)

A Pharmacia Smart System (Pharmacia-LKB, Uppsala, Sweden) was equipped with a μRPC C2/C18 PC 3.2/3 column (Pharmacia-LKB, Uppsala, Sweden) and operated at 0.25 ml/min at room temperature. The column was equilibrated at 95:5 (water:methanol) and a linear solvent gradient developed to 75:25 over 10 min and then to 20:80 over 10 min. The column was returned to equilibrium conditions over 3 min. The eluent was monitored at 245 nm.

3. RESULTS

Although naphthalene-1,2-dihydrodiol is colorless in solution, 1,2-naphthoquinone displays a dark yellowish color and absorbance around 400 nm. Since 1,2-naphthoquinone-protein adducts also display a similar dark yellow color, the amount of protein adducts formed with 1,2-naphthoquinone can be spectrophotometrically quantified. Taking advantage of this, the formation of 1,2-naphthoquinone from naphthalene-1,2-dihdyrodiol in rat lenses was investigated with rat lens proteins incubated with naphthalene-1,2-dihydrodiol in the presence of NAD⁺. The spectra obtained from the rat lens proteins incubated with naphthalene-1,2-dihydrodiol displayed an absorbance around 450 nm, which was exactly the same as that observed with the proteins incubated with 1,2-naphthoquinone (Figure 1). This result strongly suggests that naphthalene-1,2-dihydrodiol is converted to 1,2-naphthoquinone and then forms 1,2-naphthoquinone-protein adducts in rat lens. This protein modification occurred only in the presence of either NAD⁺ or NADP⁺ (data not shown), indicating that the conversion of naphthalene-1,2-dihdyrodiol to 1,2-naphthoquinone is enzymatically catalyzed. This protein modification was almost completely eliminated by aldose reductase inhibitors Al 1576, sorbinil, tolrestat and Ponalrestat (Figure 2), while none of the inhibitors prevented the protein modification induced by 1,2-naphthoquinone (data not shown).

To confirm that aldose reductase generates 1,2-naphthoquinone from naphthalene-1,2-dihydrodiol, purified recombinant rat lens aldose reductase was incubated with 1 mM naphthalene-1,2-dihydrodiol in the presence of NAD⁺ and 2-mercaptoethanol. The 2-mercaptoethanol adduct with 1,2-naphthoquinone was analyzed by HPLC (Figure 3). HPLC analysis of the mixture of naphthalene-1,2-dihydrodiol and aldose reductase displayed only a single peak of unmetabolized naphthalene-1,2-dihydrodiol before starting the incubation (data not shown). After incubation for 30 min, aldose reductase clearly generated a new peak from naphthalene-1,2-dihydrodiol which exactly corresponded to the major peak generated from 1,2-naphthoquinone incubated under the same conditions (Figure 3). Aldose reductase inhibitors Al 1576, sorbinil, tolrestat and Ponalrestat significantly reduced the formation of this new peak of naphthalene-1,2-dihydrodiol metabolite (data not shown).

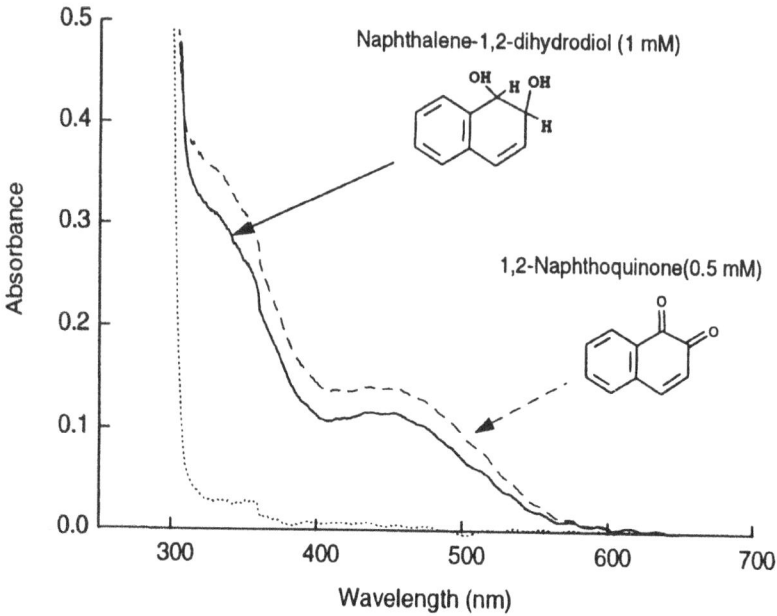

Figure 1. Spectra obtained from rat lens proteins incubated without (dotted line) or with either 1 mM naphthalene-1,2-dihydrodiol (solid line) or 0.5 mM 1,2-naphthoquinone (dashed line).

4. DISCUSSION

Ingested naphthalene is metabolized to naphthalene-1,2-dihydrodiol in either liver or ciliary body. The importance of naphthalene-1,2-dihydrodiol in the formation of cataract has been well documented by Xu et al. (1992a; 1992b). Naphthalene-1,2-dihydrodiol is the only naphthalene metabolite detected in aqueous humor and lens. Moreover, rat lenses incubated in medium containing naphthalene-1,2-dihydrodiol develop similar cataractous changes to those observed in naphthalene-fed rats. Based on this evidence, it was proposed that naphthalene-1,2-dihydrodiol is metabolized to 1,2-naphthoquinone through the intermediate metabolite 1,2-dihydroxynaphthalene in lens, which can lead to the formation of cataract (Figure 4). Xu et al. (1992b) have also demonstrated that the aldose reductase inhibitor Al 1576 prevents cataract formation induced by naphthalene-1,2-dihydrodiol, but not 1,2-naphthoquinone. This evidence suggests that aldose reductase inhibitors prevent naphthalene cataract in rats by inhibiting the conversion of naphthalene-1,2-dihydrodiol to 1,2-naphthoquinone.

It is well established that naphthalene-1,2-dihydrodiol is catalyzed to 1,2-naphthoquinone by the enzyme dihydrodiol dehydrogenase (*trans*-1,2-dihydrobenzene-1,2-diol dehydrogenase, EC 1.3.1.20) (Smithgall et al., 1988). It has also been demonstrated that aldose reductase possesses dihydrodiol dehydrogenase activity (Matsuura et al., 1987; Hara et al., 1989; Hara et al., 1991). This raises the question of which enzyme, dihydrodiol dehydrogenase or aldose reductase, is important in the metabolism of naphthalene-1,2-dihydrodiol to 1,2-naphthoquinone in rat lens. The rat lens contains a large amount of aldose reductase. However, dihydrodiol dehydrogenase is not present in rat lens (Hara et

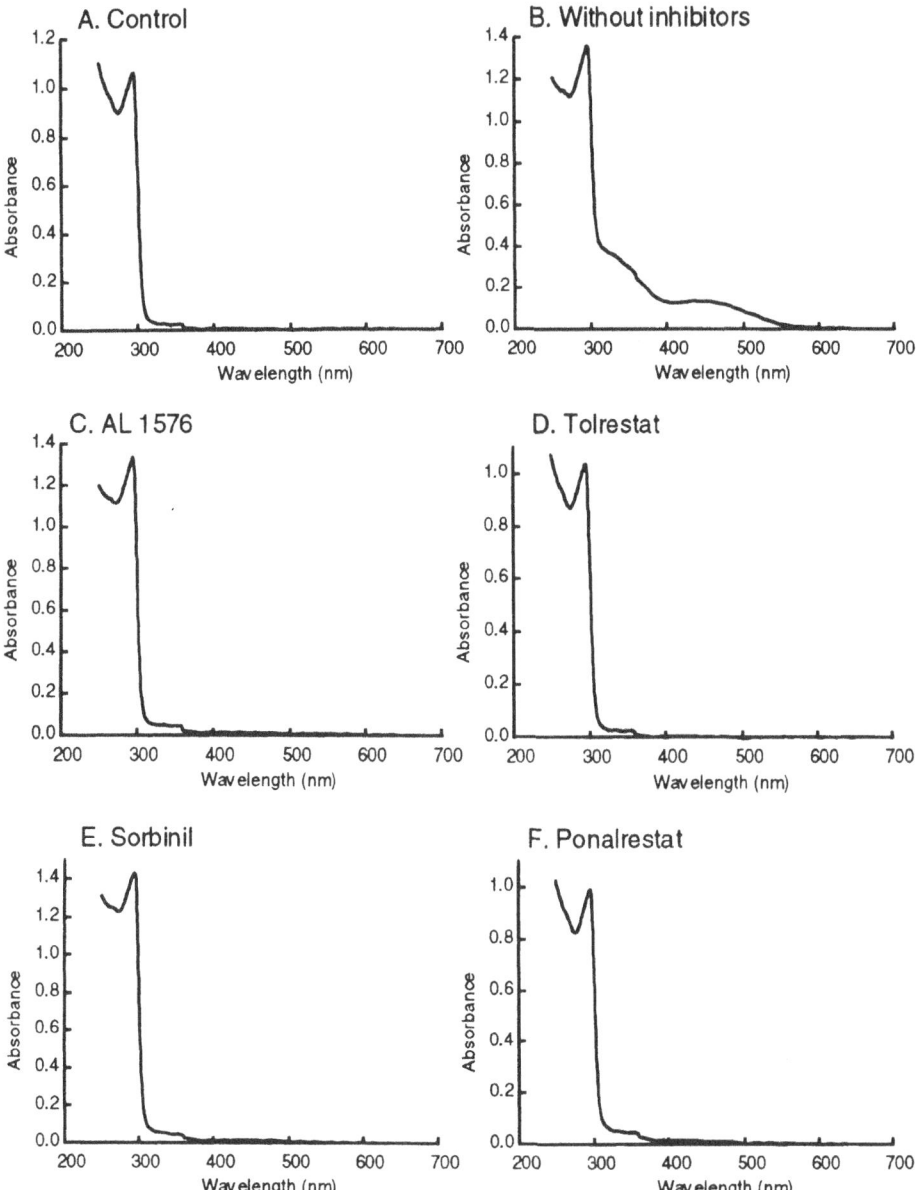

Figure 2. Spectra obtained from rat lens proteins incubated with 1 mM naphthalene-1,2-dihydrodiol in either absence (B) or presence of 10 µM aldose reductase inhibitors AL 1576 (C), tolrestat (D), sorbinil (E) and Ponalrestat (F). (A) illustrates the spectra obtained from rat lens proteins incubated without naphthalene-1,2,-dihydrodiol.

al. 1991; Sato, 1993). Moreover, the dehydrogenase activity assayed with naphthalene-1,2-dihydrodiol as substrate was found to originate from aldose reductase in rat lens (Sato, 1993). The present study also demonstrates that the metabolite of naphthalene-1,2-dihydrodiol by rat lens aldose reductase is 1,2-naphthoquinone. These data support the idea that aldose reductase plays a key role in conversion of naphthalene-1,2-dihydrodiol to 1,2-

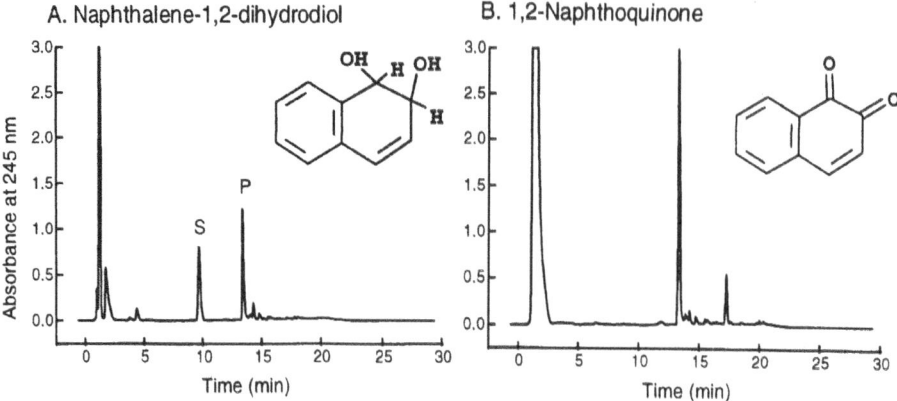

Figure 3. Elution profiles from HPLC of rat lens aldose reductase incubated with either 1 mM naphthalene-1,2-di-hydrodiol (A) or 1,2-naphthoquinone (B) for 30 min. Experimental details are described in *Material and Methods.* Peaks S and P indicate the substrate naphthalene-1,2-dihydrodiol and its metabolite by aldose reductase, respectively.

naphthoquinone in rat lens. This may also explain the fact that aldose reductase inhibitors prevent naphthalene cataract in rats.

The controversy on the role of aldose reductase in naphthalene cataract is mainly derived from the results that the non-hydantoin group of aldose reductase inhibitors such as tolrestat, Ponalrestat and FK 366 failed to prevent naphthalene cataract in both *in vivo* and *in vitro* lens culture studies despite the fact that the doses of inhibitors utilized were enough to prevent sugar cataract. However, the present study with the *in vitro* system using whole rat lens soluble proteins clearly demonstrates that the rat lens metabolizes naph-

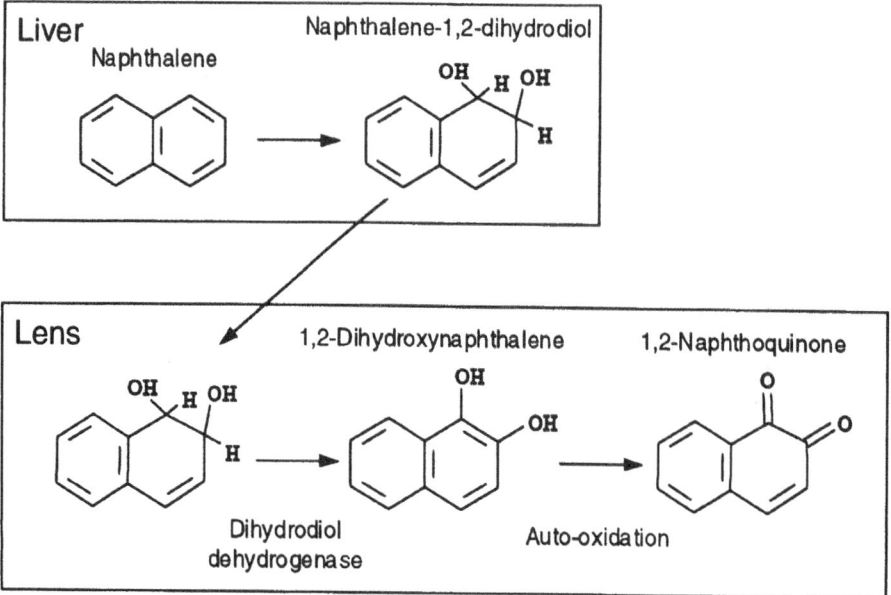

Figure 4. Metabolism of naphthalene (Xu et al. 1992a).

thalene-1,2-dihydrodiol to form 1,2-naphthoquinone-proteins adducts. Moreover, this protein modification was almost completely inhibited not only by the hydantoin containing aldose reductase inhibitors Al 1576 and sorbinil, but also by the non-hydantoin inhibitors tolrestat and Ponalrestat. Since the formation of 1,2-naphthoquinone is the key to formation of cataract, the discrepancy in efficacy of aldose reductase inhibitors on naphthalene cataract between *in vivo* and *in vitro* experiments may be explained by the differences in pharmacokinetics and pharmacodynamics of drugs.

5. REFERENCES

Hara, A., Harada, T., Nakagawa, M., Matuura, K., Nakayama, T., and Sawada, H.: Isolation from pig lens of two proteins with dihydrodiol dehydrogenase and aldehyde reductase activities. Biochem. J. 264 (1989) 403–407.

Hara, A., Nakayama, T., Harada, T., Kanazu, T., Shinoda, M., Deyashiki, Y., and Sawada, H.: Distribution and characterization of dihydrodiol dehydrogenases in mammalian ocular tissues. Biochem. J.275 (1991) 113–119.

van Heyningen, R.: Naphthalene cataract in rats and rabbits: a resume. Exp. Eye Res. 28 (1979) 435–439.

Hockwin, O., Wegener, A., Sisk, D.R., Dohrmann, B., and Kruse, M.: Efficacy of AL01576 in preventing naphthalene cataract in three rat strains. A slit-lamp and scheimpflug photographic study. Lens Research 2 (1984–85) 321–335.

Lou, M.F., Xu, G.T., Zigler, S. and York, B.: Inhibition of naphthalene cataract in rats by aldose reductase inhibitors. Curr. Eye Res. 15 (1996) 423–432.

Matsuura, K., Hara, A., Nakayama, T., Nakagawa, M., and Sawada, H.: Purification and properties of two multiple forms of dihydrodiol dehydrogenase from guinea-pig testis. Biochem. Biophys. Acta 912 (1987) 270–277.

Old, S.E., Sato, S., Kador, P.F., and Carper, D.A.: *In vitro* expression of rat lens aldose reductase in *Escherichia coli*. Proc. Natl. Acd. Sci. U.S.A. 87 (1990) 4942–4945.

Sato, S.: Aldose reductase the major protein associated with naphthalene dihydrodiol dehydrogenase activity in rat lens. Invest. Ophthalmol. Vis. Sci. 34 (1993) 3172–3178.

Smithgall, T.E., Harvey, R.G., and Penning, T.M.: Spectroscopic identification of *ortho*-quinones as the products of polycyclic aromatic *trans*-dihydrodiol oxidation catalyzed by dihydrodiol dehydrogenase. A potential route of proximate carcinogen metabolism. J. Biol. Chem. 263 (1988) 1814–1820.

Tao, R.V., Takahashi, Y., and Kador, P.F.: Effect of aldose reductase inhibitor on naphthalene cataract formation in the rat. Invest. Ophthalmol. Vis. Sci. 32 (1991) 1630–1637.

Turner, A. and Flynn, T.G.: Nomenclature of aldehyde reductases. Prog. Clin. Biol. Res. 114 (1982) 1295–1302.

Wermuth, B.: Aldo-keto reductases. Prog. Clin. Biol. Res. 174 (1985) 209–230.

Xu, G.T., Zigler, J.S. Jr, and Lou, M.F.: The possible mechanism of naphthalene cataract in rat and its prevention by an aldose reductase inhibitor (AL01576). Exp. Eye Res. 54 (1992a) 63–72.

Xu, G.T., Zigler, J.S. Jr, and Lou, M.F.: Establishment of a naphthalene cataract model in vitro. Exp. Eye Res. 54 (1992b) 73–81.

57

GENE REGULATION OF ALDOSE REDUCTASE UNDER OSMOTIC STRESS

Takeshi Iwata,[1] Saverio Minucci,[2] Michelle McGowan,[1] and Deborah Carper[1]

[1]Laboratory of Mechanisms of Ocular Diseases
National Eye Institute
[2]Laboratory of Molecular Growth Regulation
National Institute of Child Health and Human Development
National Institutes of Health
Bethesda, Maryland 20892

1. INTRODUCTION

Aldose reductase (AR) catalyzes the reduction of various sugars to their respective sugar alcohols, including glucose to sorbitol. The accumulation of sorbitol from excess glucose in diabetic patients is believed to be responsible for initiating diabetic complications, such as retinopathy and cataract (van Heyningen, 1959; Kinoshita, 1974). Inhibition of AR by AR inhibitors in diabetic animal models has been shown to reduce the accumulation of sorbitol and prevent or retard these diabetic complications.

Sorbitol, which is proposed to damage cell function in diabetic patients, is an important and beneficial osmolyte in normal cellular osmoregulation. Sorbitol is found in many organisms from bacteria and plants (Pritchard et al., 1986) to mammals. In mammals, the elevation of renal medullary extracellular hypertonicity during antidiuresis has been known to be adjusted by accumulation of osmolytes, such as sorbitol, myoinositol, betaine, and glycerophosphorylcholine. Expression of AR has been found in various tissues and is especially high in kidney inner medulla cells.

The elevation of AR synthesis in cultured rabbit renal inner medullary epithelial cells grown in hyperosmotic medium was reported by Moriyama et al., 1989. Kaneko et al., 1990 found that the osmotic induction of AR was a general response shared by other cell types and species, including rat renal mesangial cells and Chinese hamster ovary cells. The human AR promoter, up to -609 bp from the transcriptional start site using Hep G2 cells, was analyzed by Wang et al., 1993. DNA protection in a CGGAAA/G motif (-186 to -146), and a GC-rich region (-87 to -31) was observed by DNase I footprinting, and gel mobility assay. However, no increase in chloramphenicol acetyltransferase (CAT) activity was observed in cells grown under osmotic stress with any of the promoter constructs used, including the largest -4.2 Kb to +31 construct. Using the luciferase assay system, Ferraris et al., 1994 observed a 40-fold increase in promoter activity with a -3.6 Kb to +34 bp rabbit AR construct under osmotic stress.

Investigation of AR function is not only important in relation to finding the mechanism of cell volume regulation but also in understanding the regulation of AR in the diabetic patient.

2. EXPERIMENTAL PROCEDURES

2.1. Promoter Isolation and Luciferase Reporter Construction

The rat and human AR promoters were isolated from a phage genomic library and a cosmid genomic library, respectively, by cDNA probe hybridization screening (Graham et al., 1991). Varying sizes of the rat AR promoter were subcloned into the pGL3 luciferase reporter vector (Tropix, Bedford, MA).

2.2. Cell Culture and Plasmid Transfection

Cells from a normal rat liver cell line (Clone 9, ATCC, Rockville, MD) were cultured in six 35 mm diameter culture wells in 90 % Ham's F-12K medium supplemented with 10 % fetal bovine serum and 50 ug/ml gentamycin. When cells were 60 % confluent, 1.7 µg of luciferase construct plasmid and 0.7 µg of pSV-ß-galactosidase plasmid were transfected by the $CaCl_2$ precipitation method without osmotic shock. After 48 h, the medium was changed. Three plates received the same Ham's F-12K medium, while the other three plates received Ham's F-12K medium containing 150 mM NaCl (hypertonic medium; 600 mOsmol/Kg). Cells were harvested after 18 h and assayed for luciferase and ß-galactosidase activity (Tropix kit, Bedford, MA).

2.3. *In Vivo* Exonuclease III Mediated Footprinting

In vivo exonuclease III (Exo III) mediated footprinting was carried out as described by Archer et al., 1992, with slight modifications. Rat liver cells were cultured in 10 cm diameter dishes to 60 % confluency. The cells were then transfected with 10 µg of the luciferase construct (pRAR21–1) containing 1 Kb of upstream AR promoter sequence. After Asp718 or Hind III restriction enzymes and Exo III digestion on isolated nuclei, the DNA was extracted. Total DNA was digested with mung bean nuclease and 20 µg of the DNA was used for linear PCR. Primers used for linear PCR were 5'-AAG CAT GAC CCA GCA GAA GGA GA-3' (primer 1) for coding strand and 5'-AGT TGC CCC AAG AAC AAT GGC GGA A-3' (primer 2) for noncoding strand.

2.4. Ligation Mediated-PCR *in Vivo* Footprinting

Ligation-mediated (LM)-PCR was performed as described previously (Dey et al., 1992; Garrity et al 1992). Primers used for analysis were primer 2 (sequence shown above), 5'-TGA ACA GGC AGA ATC CCA TA-3' (primer 3), 5'-CTG AAA TAA TCG GAG TTG CCC CA-3' (primer 4), 5'-TAC AGC AAT GAT TCC AAG TTA GGG GA CAC-3' (primer 5), 5'-AAG GGG ACA CAA AAT AAA TGA ATC AGT TGG TAT TGG GTA G-3' (primer 6). After the LM-PCR, labeled products were resolved in a 6 % sequencing gel.

2.5. Whole Cell Extract Preparation and Electrophoretic Mobility Shift Assay (EMSA)

Whole cell extracts were prepared as previously described (Minucci et al., 1994). Rat liver cells were cultured in 10 cm diameter dishes to 90 % confluency. The cells were then exposed to normal medium or hypertonic medium for 30 min or 3 h. After treatment, cells were pelleted, washed with PBS, and resuspended in 2 packed cell volumes of 20 mM Hepes buffer, pH 7.9, containing 0.42 M NaCl, 1.5 mM $MgCl_2$, 0.2 mM EDTA, 0.5 mM DTT, 25 % v/v glycerol, and 2 mM proteinase inhibitor AEBSF. The cells were snap frozen in ethanol/dry ice. The frozen cells were quickly thawed for homogenization and centrifuged for 1 h at 34,000 g. The supernatant was used for EMSA without dialysis. Extracts (6 µg of protein) were incubated with 10 pmol of ^{32}P-labeled oligonucleotide (10^5 cpm per reaction mixture) in EMSA binding buffer containing 4 µg of poly (dI-dC) per ml, 20 mM EDTA, 5 mM dithiothreitol, and 5 % (vol/vol) glycerol for 30 min at 4°C. The oligomers used were wild type osmotic response element (ORE) 5'-GAG GGG TGT TGG AAG AGT GCC AAA TTT CCG CCA TT-3', a two guanosine mutated oligomer-osmotic response element mutation2 (OREM2) 5'-GAG GGG TGT TG(A) AAG AGT (A)CC AAA TTT CCG CCA TT-3', and an 8 bp mutated oligomer osmotic response element mutation8 (OREM8) 5'-GAG GGG TGT TG(A GGA GAC A)CC AAA TTT CCG CCA TT-3'.

3. RESULTS AND DISCUSSION

The rat and human AR gene promoters were cloned and their sequences were compared for similarities and differences up to 3 Kb from the transcriptional initiation start site. Distinct regions of homology were observed between +1 and -1,400 bp. Based on these clustered regions of homology, constructs were made using different sizes of the rat AR promoter linked to the luciferase report gene. Each construct was transfected into 6

Figure 1. Rat AR promoter constructs and luciferase reporter gene expression assays of transfected normal and osmotically-stressed rat liver cells. Left panel shows size of AR promoter constructs 1 to 11. Right panel shows the luciferase activity (relative luciferase unit) of constructs 1 to 11 which were transfected into rat liver cells for 48 h and subsequently grown in normal medium (filled bar) or hyper-osmotic medium (open bar) for 18 h.

Figure 2. Exo III mediated DNA footprinting of promoter region necessary for osmotic response. Lane 1; linear PCR with primer 1 on Hind III digested control template, 2; linear PCR with primer 1 on Hind III/Exo III digested transient template, 3; linear PCR with primer 2 on Asp718 digested control template, 4; linear PCR with primer 2 on Asp718/Exo III digested transient template. Solid lines indicate the stop position of Exo III. Sequence between the two solid lines are protected region.

wells of normal rat liver cells. Forty eight hours after transfection, one-half of the cells were grown in normal osmotic medium, while the other half were grown in hyper-osmotic medium (600 mOsmol/Kg) for an additional 18 h. The cells were then harvested and assayed for luciferase reporter gene expression. In Fig. 1, constructs 1 and 2 and constructs 5–6 had high luciferase activity when cells were osmotically stressed (open bars), indicating possible positive osmotic response elements for the rat AR promoter. Minimal osmotic response was observed with constructs shorter than construct 7. Luciferase activity with constructs 6 and 7, and deletion construct 11, indicated an approximate 200 bp core region for the AR osmotic response. Basal promoter activity also was affected by the various AR promoter constructs (Fig. 1, filled bars). Constructs 1 to 4 indicate a possible negative element within this region for normal promoter activity.

To see whether a transcription factor is bound to the 200 bp region involved in increased luciferase gene expression during osmotic stress, Exo III mediated DNA footprinting was performed (Fig. 2). The result shows that a 31 bp region of the transiently transfected plasmid, between -926 and -895, was protected from Exo III digestion in the rat liver cell nucleus.

DMS *in vivo* footprinting was performed to determine whether the same transcription factor is bound to the endogeneous AR promoter *in vivo* (Fig. 3). Intensity of bands produced by LM-PCR between *in vitro* methylated DNA and *in vivo* methylated DNA were compared. Within the 31 bp of the Exo III-protected region, two guanosines at positions -915 and -908 were protected from methylation under both normal and osmotically-stressed conditions. Differences in the length of osmotic stress, either 30 min or 3 h, did not effect the band intensity of these two guanosines.

Electrophoretic mobility shift assays (EMSA) on whole cell extracts from stressed (30 min or 3 h) or non-stressed cells was used to confirm transcription factor binding. In addition, the effect of nucleotide mutations on EMSA was examined. A 31 bp double-stranded Exo III-protected sequence was used as wild-type probe (ORE). Mutations of the wild-type probe were also used. One probe (OREM2) contained adenosine replacements of the two guanosines that were protected from DMS methylation, while the other probe (OREM8) contained an additional six nucleotide replacements (pyrimidine:pyrimidine or purine:purine change) between these two guanosines. As seen in Fig. 4, the wild-type probe ORE showed several intense complexes. OREM2 also showed complexes, but protein binding appeared to be less than with the ORE probe. The eight base pair-mutated OREM8 probe showed strong inhibition of protein binding.

Two *cis*-elements responsible for osmoregulation of AR have been identified. In addition, one of the elements has been identified within 200 bp and has been shown to be adjacent to a putative transcription binding factor present in both normal and hyper-osmotic conditions. Comparison of this *cis*-element with the human AR promoter sequence revealed two guanosines and five nucleotides between them that are shared between rat and human. This sequence of 8 bp is not registered in the transcription factor database. The fact that the putative transcription binding factor is stationed immediately adjacent to this 8 bp site may indicate a regulatory switching mechanism that includes this protein in a complex with additional specific factors involved in osmotic response.

4. ACKNOWLEDGMENTS

The authors like to thank Dr. Anup Dey for helpful technical suggestions on LM-PCR DMS *in vivo* footprinting.

Figure 3. DMS ligation mediated-PCR *in vivo* DNA footprinting of 200 bp promoter involved in increased osmotic response. Lane 1; LM-PCR of *in vitro* DNA, 2; LM-PCR of DNA from cells in normal osmolarity, 3; LM-PCR of DNA from cells in hyper-osmolarity for 30 min, 4; LM-PCR of DNA from cells in hyper-osmolarity for 3 h.

Figure 4. Gel mobility shift assay of the 31 bp Exo III protected region with whole cell extract. Lane 1; ORE free probe (no extract added), 2; ORE/normal osmolarity, 3; ORE/hyper-osmolarity for 30 min. 4; ORE/hyper-osmolarity for 3 h, 5; OREM2 free probe, 6; OREM2/normal osmolarity, 7; OREM2/hyper-osmolarity for 30 min, 8; OREM2/hyper-osmolarity for 3 h, 9; OREM8 free probe, 10; OREM8/normal osmolarity. 11; OREM8/hyper-osmolarity for 30 min, 12; OREM8/hyper-osmolarity for 3 h.

5. REFERENCES

Archer, T.K., Lefebvre, P., Wolford, R.G., and Hanger, G.L.: Transcription factor loading on the MMTV promoter. Science 255 (1992) 1573–1576.

Dey, A., Thornton, A.M., Lonergan, M., Weissman, S.M., Chamberlain, J.W., and Ozato, K.: Occupancy of upstream regulatory sites in vivo coincides with major histocompatibility complex class I gene expression in mouse tissues. Mol.Cell.Biol. 12 (1992) 3590–3599.

Ferraris J.D., Williams, C.K., Martin, B.M., Burg, M.B., and Garcia-Perez, A.: Cloning, genomic organization, and osmotic response of the aldose reductase gene. Proc.Natl.Acad.Sci.USA 91 (1994) 10742–10746.

Garrity, P.A., and Wold, B.J.: Effects of different DNA polymerases in ligation-mediated PCR: enhanced genomic sequencing and in vivo footprinting. Proc.Natl.Acad.Sci.USA 89 (1992) 1021–1025.

Graham, C., C.Szpirer, G.Levan, and D.Carper.: Characterization of the aldose reductase-encoding gene family in rat. Gene 107 (1991) 259–267.

Kaneko, M., Carper, D., Nishimura, C., Millen, J., Bock, M., and Hohman, T.: Induction of aldose reductase expression in rat kidney mesangial cells and chinese hamster ovary cells under hypertonic conditions. Exp.Eye.Res. 188 (1990) 135–140.

Kinoshita, J.H.: Mechanisms initiating cataract formation. Proctor Lecture. Invest.Ophthalmol. 13 (1974) 713–724.

Minucci, S., Zand, D.J., Dey, A., Marks, M.S., Nagata, T., Grippo, J.F., and Ozato, K.: Dominant negative retinoid X receptor b inhibits retinoic acid-responsive gene regulation in embryonal carcinoma cells. Mol.Cell.Biol. 14 (1994) 360–372.

Moriyama, T., Garcia-Prez, A., and Burg, M.B.: Osmotic regulation of aldose reductase protein synthesis in renal medullary cells. J.Biol.Chem. 264 (1989) 16810–16814.

Pritchard, H.W., Grout, B.W.W., and Short, K.C.: Osmotic stress as a pregrowth procedure for cryopreservation 2. Water relations and metabolic state of sycamore and soybean cell suspensions. Annals of Botany 57 (1986) 371–378.

Wang, K., Bohren, K.M., and Gabbay, H.: Characterization of the human aldose reductase gene promoter. J.Biol.Chem. 268 (1993) 16052–16058.

van Heyningen, R.: Formation of polyols by the lens of the rat with 'sugar' cataract. Nature 184 (1959) 194–195.

58

CHARACTERIZATION OF AN ELEMENT RESEMBLING AN ANDROGEN RESPONSE ELEMENT (ARE) IN THE HUMAN ALDOSE REDUCTASE PROMOTER

Barbara U. Ruepp,[1] Kurt M. Bohren,[1] and Kenneth H. Gabbay[1,2]

[1]Department of Pediatrics, Molecular Diabetes,
   and Metabolism Section
[2]Department of Cell Biology
Baylor College of Medicine
Houston, Texas 77030

1. INTRODUCTION

Aldose reductase has been implicated in a number of diabetic complications including background retinopathy (microaneurysms) and cataracts. The formation of microaneurysms is independently related to duration of disease and hyperglycemia, and is found in post-pubertal patients (Murphy et al., 1990). They are thought to be caused by the loss of pericytes resulting from the activity of aldose reductase. The mechanisms regulating the expression of the enzyme in specific tissues and in pathological conditions are unknown.

A protein highly homologous to the human aldose reductase, the mouse vas deferens protein (MVDP) was demonstrated to be regulated by androgens (Martinez et al., 1990). MVDP is expressed in the deferent duct and other tissues of the mouse (Lau et al., 1995). An androgen response element (ARE) has been identified in the 5'-flanking region of the MVDP gene (Jean et al., 1994).

Sequence analysis of the human aldose reductase promoter (Wang et al., 1993) revealed a sequence AGGACAAGGTATTCG present at position -396 to -382 which is homologous to the consesus ARE. To investigate possible hormonal regulation of the aldose reductase gene, chloramphenicol acetyltransferase- (CAT-) constructs containing fragments extending up to 600 bp upstream of the transcription start site were transfected into HepG2 cells and subjected to testosterone and progesterone stimulation.

2. METHODS

2.1. Plasmid Construction

A 640 bp and a 336 bp fragment were generated from a subclone of human aldose reductase genomic DNA by the polymerase chain reaction (PCR). The distal primers (-609 → +31) 5'-GGAAAGCTTGCTGAACCACACCT-3' and (-336 → +31) 5'-CAAAAGCTTTCTTCTGGGCTCTTAATG-3' were constructed with a HindIII site and a PstI site was introduced in the upstream primer (+31 → -336) and (+31→ -609): 5'-TGGCTGCAGCGCTCCCCAGACCC-3'.

The PCR products were digested with HindIII and PstI, purified on 1.5% agarose gels and inserted into the HindIII/PstI site of the pCATbasic vector (Promega).

The candidate region between -609 and -282 containing the putative ARE was cloned into the HindIII/BamHI site of the pCATpromoter plasmid consisting of the heterologous SV40 promoter upstream of the chloramphenicol acetyltransferase (CAT) reporter gene. The sequence of interest was obtained by PCR amplification using a similarly designed set of oligonucleotides. Clones were screened by restriction enzyme analysis for the appropriate insert size and the sequences of all constructs were confirmed by sequencing. A control plasmid consisting of the promoter region of the mouse mammary tumour virus (pMMTV-CATbasic) was obtained from Dr. Marco Marcelli (Baylor College of Medicine) and a 360 bp HinfI fragment containing repeats of the hormone regulatory element (HRE) was cloned into the SV40promoter driven CAT-vector yielding pMMTVHRE-CATprom construct.

2.2. Cell Culture and DNA Transfection

DNA constructs and control plasmids were grown and purified by standard procedures. HepG2 cells were maintained in DMEM supplemented with 10% fetal bovine serum, 5 mM L-glutamine, 50 U/ml penicillin and 50 µg/ml streptomycin, seeded at a density of 1.5×10^6 cells per dish and were transfected 12 hours later using the calcium phosphate precipitation method with 10 µg of the pARPn plasmid constructs and with 1 µg of the pRSV-Luciferase plasmid containing the luciferase gene under the control of the SV40-promoter to correct for transfection efficiency. HepG2 cells were provided with the androgen and the progesterone receptor by co-transfecting 2 µg plasmid vectors coding for the androgen and progesterone receptor pCMVAR and pRSVPR. Transfections of cells with CAT constructs of the promoter region of the long terminal repeat of the mouse mammary tumor virus, pMMTV-CATbasic, and of a 360 bp fragment containing several repeats of the hormone response element (HRE), pMMTV-HRECATprom, were included in each experiment as positive controls for androgen and progesterone induction. Following a 6 hour exposure to the DNA precipitate, the cells were shocked with 15% glycerol and subsequently incubated in growth media supplemented with 10% charcoal stripped fetal bovine serum in the presence or absence of 10^{-7} M dihydrotestosterone or 10^{-7} M progesterone as indicated in the figure legend for 48 hours.

2.3. Luciferase and CAT Assay

The transfected cell cultures were harvested by scraping and the cells were lysed by three freeze-thaw cycles. 10 µl of the cell extracts were assayed for luciferase activity and CAT activity was determined using the thin-layer chromatography method (Gorman et al.

1982). The acetylated chloramphenicol products were separated by TLC, visualized by autoradiography and quantitated by liquid scintillation counting.

3. RESULTS

The amino acid and the cDNA sequences as well as the gene structures of the human aldose reductase and of the MVDP have been described earlier (Bohren et al. 1989; Pailhoux et al. 1990; Graham et al. 1991, and Pailhoux et al. 1992).

Sequence comparison of the human aldose reductase and the MVDP gene revealed an overall sequence homology of 80% in the coding regions whereas the introns and the 5'-flanking region varied considerably. Despite these differences, a 15 bp sequence resembling an androgen response element was located at position -396 to -382 of the human aldose reductase promoter suggesting an androgen-regulated expression of the enzyme analogous to the androgen-dependent gene expression of the MVDP. Fig. 1 shows the sequence of a 15 bp putative element which shares a high degree of identity with the consesus sequence of the androgen response element (ARE). Five nucleotides in the first half site and four nucleotides in the second half site of the imperfect palindromic structure are conserved between the putative ARE of the aldose reductase promoter and the consesus ARE.

To determine the effect of androgens on the aldose reductase promoter the CAT-constructs shown in Fig. 2A were transfected into HepG2 cells and subjected to dihydro-testosterone and progesterone stimulation.

Control studies with homologous and heterologous MMTV-CAT constructs enhanced the transcription of the CAT-reportergene approximately 15- (MMTV-CATb) and 9-fold (MMTV-HRE-CATp) in response to androgen and 50- and 2-fold upon progesterone stimulation (Fig. 2B). Conversely, the CAT-gene expression under the control of a 600 and, as expected of a 300 bp portion of the aldose reductase promoter remained unchanged in the presence and absence of the two steroid hormones. Consistent with these findings androgen and progesterone stimulation did not affect the CAT-activity of the heterologous pARARE-CAT construct containing the putative ARE enhancer sequence. In summary, none of the aldose reductase promoter constructs was responsive to androgen or progesterone stimulation indicating that the ARE of the aldose reductase promoter is not functional.

4. CONCLUSIONS

The sequence data and the results from the transfection experiments indicate that the nucleotide differences between the human aldose reductase ARE, the MVDP-ARE and the

Figure 1. Sequence alignment of the aldose reductase ARE, the consensus ARE and the MVDP ARE. The conserved nucleotides between the AR-ARE and the consesus ARE are underlined.

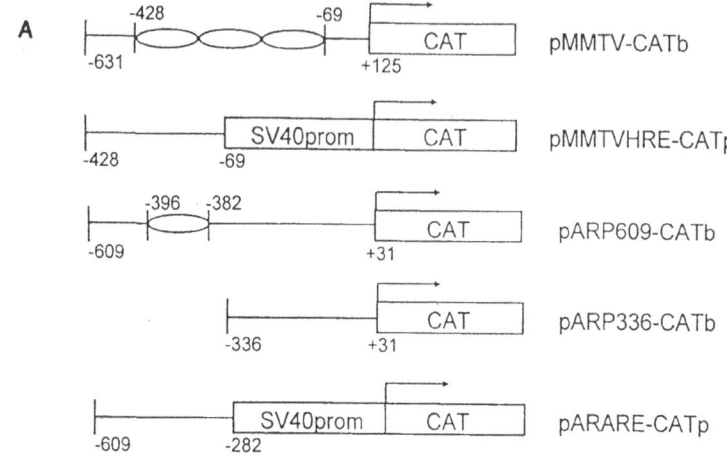

B	Dihydrotestosterone				Progesterone			
hormone	-	+		+	-	+		+
receptor co-transf.	-		+	+	-		+	+
pCATbasic	0.7	0.9	0.4	0.7	0.6	0.5	0.4	0.5
pCATpromoter	1.7	2.3	2.1	2.6	5.8	6.2	2.8	6.7
pMMTV-CATb	0.4	0.3	6.6	92.7	0.4	0.3	0.6	26.9
pMMTVHRE-CATp	8.2	9.0	9.5	84.3	8.2	9.3	7.3	12.8
pARP609-CATb	4.4	4.3	6.2	5.2	4.4	4.2	9.8	8.5
pARP336-CATb	4.5	4.9	4.1	3.8	4.5	5.0	4.9	3.7
pARARE-CATp	8.9	8.3	9.5	6.5	8.9	7.9	10.6	11.1

Figure 2. Androgen and progesterone induction of various aldose reductase promoter constructs. A): Schematic presentation of the aldose reductase promoter and control CAT-constructs used for transfection into HepG2 cells. B): CAT-activities of cells transiently transfected with the plasmids shown in Fig. 2A. The cell cultures were transfected as described (see methods) and exposed to media supplemented with 10^{-7} M dihydrotestosterone and progesterone for 48 hours.

consesus ARE seem to be essential for enhancer activity. Thus, this raises the possibility that a single nucleotide mutation in the aldose reductase ARE due to genetic variations could convert the "pseudo ARE" into a fully active transcriptional regulatory element.

Further in vitro experiments of appropriate mutations of the aldose reductase ARE and patient-oriented studies will be needed to determine if genetic variations in subjects in this gene region could alter the susceptibility of the aldose reductase gene to induction by androgens.

5. REFERENCES

Bohren, K.M., Bullock, B., Wermuth, B. and Gabbay K.H. (1989), The aldo-keto reductase superfamily: cDNAs and deduced amino acid sequences of human aldehyde and aldose reductases, J. Biol. Chem. 264, 9547 - 9551.

Fabre, S., Manin, M., Pailhoux, E., Veyssiere, G. and Jean, C. (1994), Identification of a functional androgen response element in the promoter of the gene for the androgen- regulated aldose reductase-like protein specific to the mouse vas deferens, J. Biol. Chem. 269, 5857 - 5864.

Gorman, C.M., Moffat, L.F., and Howard, B.H. (1982), Recombinant genomes which express chloramphenicol acetyltransferase in mammalian cells, Mol. Cell. Biol.2, 1044 -1051.

Graham, A., Brown, L., Hedge, P.J., Gammack, A.J. and Markham, A.F. (1991), Structure of the human aldose reductase gene, J. Biol. Chem. 266, 6872 -6877.

Lau, E. T., Cao, D., Lin, C., Chung, S. K. and Chung, S. S. (1995), Tissue-specific expression of two aldose reductase-like genes in mice: abundant expression of mouse vas deferens protein and fibroblast growth factor-regulated protein in the adrenal gland, Biochem. J. 312, 609 - 615.

Martinez, A., Pailhoux, E., Berger, M.and Jean, C. (1990), Androgen regulation of the mRNA encoding a major protein of the mouse vas deferens, Mol. Cell. Endocrinol. 72, 201 -211.

Murphy, R. P., Nanda, M., Plotnick, L., Enger, C., Vitale, S. and Pratz, A. (1990), The relationship of puberty to diabetic retinopathy, Arch. Ophthalmol. 108, 215 - 218.

Pailhoux, E.A., Martinez, A., Veyssiere, G.M. and Jean, C.G. (1990), Androgen- dependent protein from mouse vas deferens, J. Biol. Chem. 265, 19932 - 19936.

Pailhoux, E., Veyssiere, G., Fabre, S., Tournaire, C. and Jean, C. (1992), the genomic organization and DNA sequence of the mouse vas deferens androgen-regulated protein gene. J. Steroid. Biochem. Molec. Biol. 42, 561 - 568.

Wang, K., Bohren, K.M. and Gabbay K.H. (1993), Characterization of the human aldose reductase gene promoter, J. Biol. Chem. 268, 16052 - 16058.

59

CLONING, SEQUENCING, AND ENZYMATIC ACTIVITY OF AN INDUCIBLE ALDO-KETO REDUCTASE FROM CHINESE HAMSTER OVARY CELLS

T. Geoffrey Flynn,[1] David J. Hyndman,[1] Reiko Takenoshita,[1] Nathalie Vera,[1] and Stephen Pang[2]

[1]Department of Biochemistry and
[2]Department of Anatomy
Queen's University
Kingston, Ontario, Canada K7L 3N6

1. INTRODUCTION

To characterize the calpain inhibitor I-sensitive protease(s) involved in the degradation of HMG-CoA reductase, Simoni's group (Inoue et al., 1993) attempted to isolate Chinese hamster ovary (CHO) cells resistant to this peptide aldehyde [*N*-acetyl-leucyl-leucyl-norleucinal (ALLN)]. Instead of inducing a protease, a 35 kDa protein was overexpressed which gave tryptic fragments with a high degree of sequence identity to members of the aldo-ketoreductase superfamily. This superfamily is a rapidly growing group of monomeric oxidoreductases which include the aldehyde and aldose reductases as well as a number of hydroxysteroid dehydrogenases.

We have used the rapid amplification of cDNA ends (RACE) technique to isolate and sequence this new reductase from CHO cells. It shows highest sequence identity to two other inducible members of the family; FGF-1 regulated (FR-1) protein and mouse vas deferens protein (MVDP). The expressed protein exhibited a greater ability to reduce ketones than the aldehyde or aldose reductases and a strong preference for aromatic aldehydes.

2. MATERIALS AND METHODS

2.1. Materials

Calpain inhibitor I was purchased from Calbiochem (La Jolla, CA). Daunomycin was purchased from Rhône-Poulenc Rorer (Montreal, Que). NADPH and NADH were from Boehringer Mannheim Canada (Laval, Que). All steroids were from Steraloids Inc. (Wilton, NH). All other substrates were from Sigma (St. Louis, MI).

2.2. Cell Culture

CHO-K1 cells were grow essentially as described by Inoue et al. (1993). Briefly, cells were grown in minimal essential medium/Hams F_{12} (1:1) supplemented with 10% fetal calf serum, 8 mM HEPES, 2 x penicillin-streptomycin at 37°C in a 5% CO_2 atmosphere. Initial selection for ALLN used 10 μM ALLN and 20 μM verapamil for one month. Media was changed every 2–3 days with replating weekly. The concentration of ALLN was increased by 20 μM increments at monthly intervals up to a final concentration of 70 μM. Cells were collected and stored at -70°C until needed. Control CHO-K1 cells were obtained from American Type Culture Collection and grown in the absence of ALLN or verapamil.

2.3. Cloning of CHO Reductase

Total RNA was isolated from induced cells by the TRIzol method (Chomczynski and Sacchi, 1987). First strand synthesis was primed with the RACE adapter T_{17} primer using MMLV reverse transcriptase (Frohman, 1990). Initial fragments were generated using 3'-RACE PCR with primers to the pig (5'-CAACCTGAAGCTGGACT) or human (5'-CTCAACAACGGCGCCAAGATGCCCA) aldose reductase sequence. Subsequently gene specific primers were used with 5'-RACE PCR to obtain the 5'-end cDNA. The entire cDNA was sequenced in both directions using overlapping gene specific primers. Sequencing was performed by the dye terminator method using an Applied Biosystems model 377A sequencer. The entire CHO reductase cDNA was amplified using primers to the 5' and 3' untranslated region containing restriction sites for EcoRI and XhoI respectively. This fragment was digested with EcoRI and XhoI and ligated into pCRII (Invitrogen) similarly treated. The plasmid CHOpCRII was transformed into CJ236 cells. Introduction of an NdeI site at the N-terminal methionine was performed using the Kunkel method (Sambrook et al., 1989). Successful mutagenesis was screened by NdeI digestion and positive clones were resequenced to ensure introduction of no other mutations. The NdeI-XhoI fragment was subcloned into the pET16b vector (Novagen) to generate an inframe histidine leader sequence for HisTag purification.

2.4. Northern Blot Hybridization

Total RNA was isolated by the TRIzol method as described above. Total RNA was treated with glyoxal (Sambrook et al., 1989) and run on a 1% agarose gel in 10 mM sodium phosphate, pH 7.0 and transferred to a nylon membrane (Zeta-Probe, Bio-Rad). Prehybridization was done at 37°C for at least 1 hr in 6 x SSC, 5 x Denhardts, 0.5% SDS and 100 μg/ml salmon sperm DNA. Hybridization was performed overnight at 42°C in the same solution containing at least $1x10^7$ cpm/ml of the probe at $1x10^8$ cpm/μg. The probes were either an 800 bp hamster glyceraldehyde-3-phosphate dehydrogenase (GAPDH) cDNA or the full length CHO reductase cDNA prepared by nick-translation (Nick Translation System, Gibco BRL). Final membrane washes were at 0.1 SSC, 0.5% SDS, 60 min, 70°C with exposure for 16 hrs at 25°C with Dupont Reflection NEF film and one intensifying screen.

2.5. Expression and Purification of CHO Reductase

The CHOpET16b plasmid was transformed into BL21(DE3) cells containing pLysS. One half-litre cultures were grown in LB supplemented with 50 µg/ml of ampicillin and 34 µg/ml chloramphenicol at 30°C to an absorbance of 0.5 at 600 nm. IPTG was added to a final concentration of 1 mM and the cultures were grown for a further 3 hrs. The cells were harvested by centrifugation at 1000 x g for 15 min. The pellets were resuspended at 30 ml per litre of culture in bind buffer (5 mM imidazole, 0.5 M NaCl, 20 mM Tris pH 7.9) and stored at -20°C. Cells were thawed and PMSF added to 1 mM. Cells were disrupted by sonication and the insoluble fraction separated by centrifugation at 12000 x g for 20 min at 4°C. The supernatant was applied to a 5 ml chelating sepharose column (Pharmacia) previously charged with nickel sulfate and washed with bind buffer. The loaded column was washed with 60 mM imidazole, 0.5 M NaCl, 20 mM Tris pH 7.9 (wash buffer) and CHO reductase was eluted using a linear gradient from wash buffer to elute buffer (1 M imidazole 0.5 M NaCl, 20 mM Tris pH 7.9). The pooled protein was dialysed against 10 mM sodium phosphate buffer pH 8.0 containing 1 mM EDTA and stored at 4°C or -20°C.

2.6. SDS-Page analysis

Protein samples (approximately 10 µg) from the purification of CHO reductase were run on 12% SDS-polyacrylamide minigels and stained with Coomassie Brilliant Blue R. Gels were destained in 18% methanol, 9% acetic acid and preserved in BioDesignGel-Wrap (BioDesign Inc., New York) after soaking in 10% methanol, 5% glycerol.

2.7. Enzyme Kinetics

CHO Reductase was assayed at 25°C in 1 mL of 0.1 M sodium phosphate pH 7.0 using a Hewlett Packard 8452A diode array spectrophotometer. Apparent K_M values for substrate were assayed in the presence of 200 µM NADPH. Apparent K_M values for cofactor were assayed in the presence of 1 mM pyridine-3-aldehyde. NADH was assayed in a 5mm path-length cell at 370 nm with correction for the different $\Delta\epsilon$ at that wavelength. All steroids were assayed in the presence of 2.5% acetonitrile. Kinetic constants were determined using the Marquardt algorithm (Marquardt, 1963) supplied with the spectrophotometer.

3. RESULTS

3.1. Nucleotide Sequence of CHO Reductase

The initial 3'-RACE reactions were performed on induced CHO-K1 total RNA using either a primer for pig ALR2 at position 300 of the pig coding region (Kubiseski et al., 1993) or a primer to human ALR2 at position 30 of the human sequence (Nishimura et al., 1990). Both these reactions yielded readable sequence that was different from known ALR2 sequences. Specific primers were used to sequence the sense and antisense strands and to obtain the 5'-untranslated region using 5'-RACE. The full length cDNA sequence generated by RACE PCR consisted of 1268 nucleotides with a 5' non-coding region of 43 nucleotides, an open reading frame of 951 nucleotides and a 3' untranslated region of 274 nucleotides. The open reading frame encoded a protein of 316 amino acids (Fig. 1). The original peptide fragments sequenced by Inoue et al. (1993) match exactly with the sequence determined from cDNA sequencing.

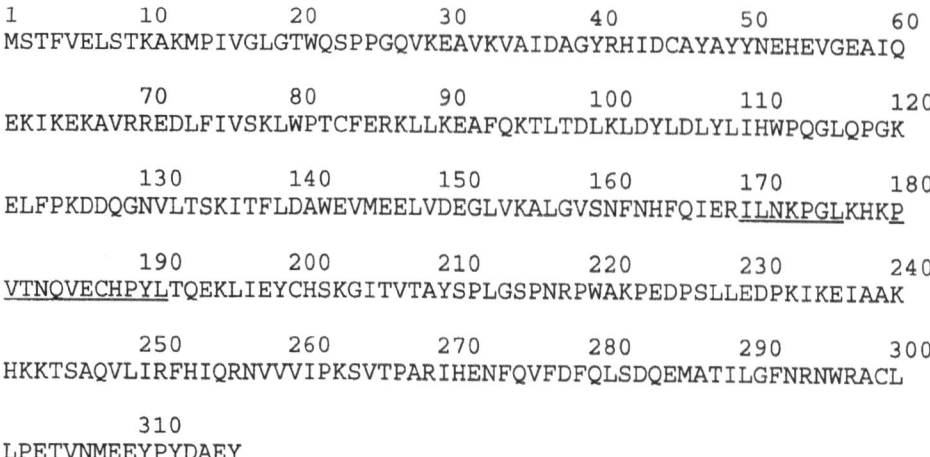

1		10		20		30		40		50		60

MSTFVELSTKAKMPIVGLGTWQSPPGQVKEAVKVAIDAGYRHIDCAYAYYNEHEVGEAIQ

	70		80		90		100		110		120

EKIKEKAVRREDLFIVSKLWPTCFERKLLKEAFQKTLTDLKLDYLDLYLIHWPQGLQPGK

	130		140		150		160		170		180

ELFPKDDQGNVLTSKITFLDAWEVMEELVDEGLVKALGVSNFNHFQIER<u>ILNKPGL</u>KHK<u>P</u>

	190		200		210		220		230		240

<u>VTNQVECHPYL</u>TQEKLIEYCHSKGITVTAYSPLGSPNRPWAKPEDPSLLEDPKIKEIAAK

	250		260		270		280		290		300

HKKTSAQVLIRFHIQRNVVVIPKSVTPARIHENFQVFDFQLSDQEMATILGFNRNWRACL

	310

LPETVNMEEYPYDAEY

Figure 1. Deduced amino acid sequence of CHO reductase. Underlined sequences indicate regions determined by protein sequence analysis by Inoue et al. (1993).

3.2. Sequence Characteristics

The protein sequence of CHO reductase showed high identity to members of the aldo-ketoreductase superfamily. The highest identity was to the mouse fibroblast growth factor-1 regulated (FR-1) protein at 92% (Donohue et al., 1994). CHO reductase was 80% identical to MVDP (Pailhoux et al., 1990). The aldose reductases as a group are approximately 70% identical while the aldehyde reductases are approximately 47% identical. CHO reductase contained the same residues, that have been implicated in the catalytic mechanism, as in other members of the aldo-ketoreductase family. These include Asp-44, Tyr-49, Lys-78 and His-111.

3.3. Induction of CHO Reductase mRNA in CHO-K1 Cells

RNA gel blot hybridization analysis from cells treated with ALLN at 70 μM concentration confirmed the induction of CHO reductase (Fig. 2). No change in GAPDH signal was seen with induced cells.

3.4. Expression of CHO Reductase in BL21(DE3) Cells

The CHO reductase cDNA was modified to include a HisTag leader sequence for rapid purification of the protein. Figure 3 shows a representative SDS-PAGE of a CHO reductase protein preparation. The nickel-sepharose column was very selective in removing only the CHO reductase protein from the cell supernatant and yielded essentially pure protein from a single column. Typical protein yields were 20–25 mg of CHO reductase from a 500 mL culture.

3.5. Kinetic Analysis

The expressed CHO reductase was active when tested with a number of common substrates for the aldo-ketoreductases. The K_M for the cofactors NADPH and NADH were

Figure 2. Northern blot analysis of CHO reductase induction.

Figure 3. Protein gel of a typical CHO reductase enzyme preparation. Two and ten μg of the eluted CHO reductase was applied in the two right lanes.

Table 1. Kinetic constants for CHO reductase. Numbers in parentheses after data values in the ALR1 and ALR2 columns indicate the sources for the data: 1; (Cromlish and Flynn, 1983a), 2; (Morjana and Flynn, 1989), 3; (Cromlish and Flynn, 1985), 4; (Cromlish and Flynn, 1983b), 5; (Rees-Milton and Flynn, unpublished results)

Substrate	CHO Reductase		ALR1 K_M (m)M	ALR2 K_M (mM)
	k_{cat} (s^{-1})	K_M (mM)		
NADPH		0.013	0.0028 (1)	0.0023 (2)
NADH		1.1	NR (1)	1.2 (1)
ALLN	1.09	8.4	NDF	NDF
DL-glyceraldehyde	1.02	28	4.5 (3)	0.072 (2)
pyridine-3-aldehyde	3.68	0.065	2.4 (4)	0.009 (2)
daunomycin	0.47 (pH 8.5)	0.043	NR (3)	NR (3)
17α-hydroxypro-gesterone 21-aldehyde	0.68	0.02	0.46 (5)	0.02 (3)
17α-hydroxyprogesterone	0.02	0.08	NR (5)	NDF

very similar to values obtained for aldehyde and aldose reductase (Table 1). ALLN which was used to induce the protein was a very poor substrate with a mM K_M. In general aliphatic substrates such as the commonly used DL-glyceraldehyde were poor substrates. This also included glucose which was only converted at a level of 2–3 times background up to a concentration of 1.2 M. Small aromatic aldehydes such as pyridine-3-aldehyde were better substrates with micromolar K_M values and k_{cat} values 2–10 fold higher than for the aliphatic substrates. The ketone daunomycin was a substrate for the CHO reductase. Though the k_{cat} value was low this aromatic substrate showed a micromolar K_M value. The pH optimum for daunomycin was 8.5 (fig. 4) while all other substrates, as exemplified by pyridine-3-aldehyde, showed a pH optimum around 6. CHO reductase was not significantly inhibited by sodium valproate or sodium barbitone at 1 mM. The diagnostic ketone substrate for the carbonyl reductases, menadione, was not a substrate for CHO reductase. CHO reductase was also able to catalyze the reduction of steroid aldehydes and ketones. The affinities were high (low μM) while the turnover rates were again less than 1 s^{-1}. The enzyme showed selectivity for aldehydes or ketones on ring D (eg. 17α-hydroxyprogesterone) since no reaction was seen with steroids, such as testosterone, containing a ketone in ring A.

4. DISCUSSION

We have isolated and sequenced a full-length cDNA from Chinese hamster ovary cells which encodes CHO reductase, a new member of the aldo-ketoreductase superfamily. Initial characterization of this protein was described by Inoue et al. (1993) and the sequence presented here confirms their assignment to the aldo-ketoreductase superfamily. The enzyme showed highest protein sequence identity to the FR-1 protein and to MVDP. FR-1, MVDP and now CHO reductase comprise a subgroup of the aldo-ketoreductases that are inducible. FR-1 is induced by cell exposure to acidic fibroblast growth factor (FGF-1) (Donohue et al., 1994) and thus may play a role in the mitogenic action of FGF-1. MVDP expression in the vas deferens is dependent on the presence of testosterone (Pailhoux et al., 1990) and androgen response elements have been isolated in the MVDP

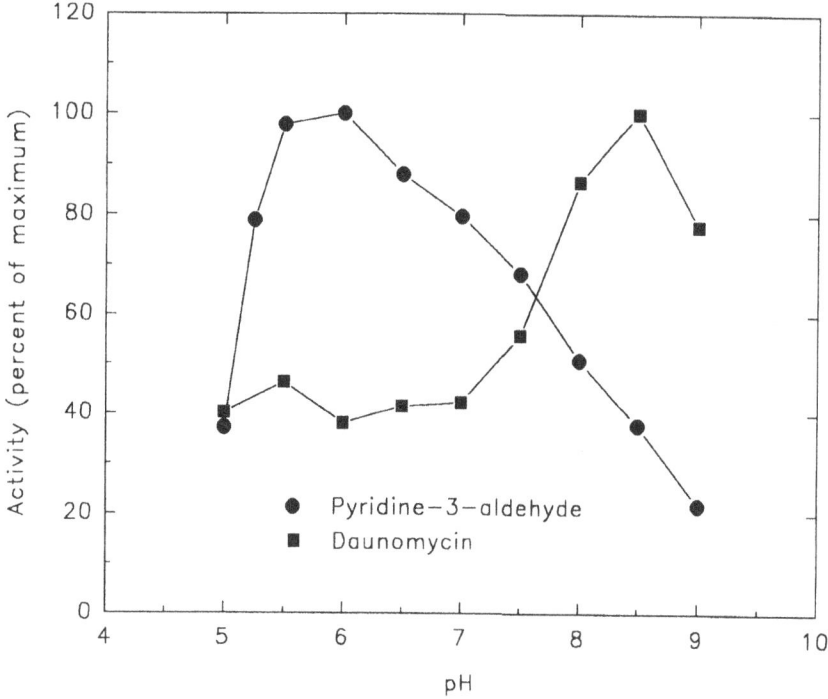

Figure 4. pH profile of CHO reductase for daunomycin and pyridine-3-aldehyde. Activities were assayed in a mixed buffer containing 50 mM each of MES, MOPS and EPPS adjusted to the indicated pH.

gene (Fabre et al., 1994). Both of these murine aldo-ketoreductases appear to be expressed in preference to the murine aldose reductase. In the present study CHO reductase was induced by the presence of an aldehyde containing protease inhibitor. This induction was also in preference to the Chinese hamster aldose reductase (data not shown). Though not the only members of the family that are regulated at the transcription level it is tempting to speculate that this is a feature of this subgroup related to a common origin.

Expressed CHO reductase was functional and displayed kinetics distinct from other members of the aldo-ketoreductase family. Despite its close sequence identity to the aldose reductases it showed an affinity for aliphatic substrates more closely comparable to the aldehyde reductases. Conversely, for aromatic substrates its affinity was comparable with aldose reductase. Its low sensitivity to inhibition by valproate or barbitone was also similar to that seen with the aldose reductases (Cromlish and Flynn, 1985). The ability of CHO reductase to catalyze the reduction of ketones such as daunomycin or 17α-hydroxyprogesterone makes it kinetically distinct from the aldehyde and aldose reductases. However, the low rate of turnover indicates that aldehydes are still the preferred substrates as seen with the aldehyde and aldose reductase members of the family. Despite its ability to reduce daunomycin with a pH optimum of 8.5 this enzyme does not appear to be the daunorubicin reductase described by Felsted et al. (1977) which was sensitive to barbitone.

The underlying structural reasons for the subtle kinetic differences between CHO reductase and the aldose reductases can partially be explained by analysis of the crystal structures of both these proteins. Comparison of the ternary complexes of human aldose reductase (Wilson et al., 1993) and FR-1 (Wilson et al., 1995) complexed with NADP and the inhibitor zopolrestat show that the FR-1 substrate site is wider due to the more removed position of the loop containing Cys-299. By analogy, since FR-1 and CHO reductase are identical in this region of the protein sequence (92% identical overall), it could be argued that CHO reductase and human aldose reductase also share this same difference at the substrate site. Since there are a large number of aromatic residues involved in the binding of zopolrestat to both FR-1 and aldose reductase, including Tyr-49, Trp-80, Trp-112, Phe-123 and Tyr-310, the primary binding energy for aromatic substrates such as pyridine-3-aldehyde or steroids could be π-bond stacking interactions. This would account for the similar K_M values using aromatic substrates for CHO reductase and aldose reductase. However, the larger substrate pocket in CHO reductase may not clamp small aliphatic substrates as firmly as does aldose reductase, resulting in a lower affinity. An extension of this proposal is that the larger substrate pocket in CHO reductase may allow for more movement of larger substrates such as daunomycin or the ketone steroids to allow for reduction of these ketones.

5. REFERENCES

Chomczynski, P., and Sacchi, N. Single-step method of RNA isolation by acid guanidinium thiocyanate-phenol-chloroform extraction. Anal. Biochem. 162 (1987) 156–159.

Cromlish, J.A., and Flynn, T.G. Pig muscle aldehyde reductase: identity of pig muscle aldehyde with pig lens aldose reductase and with the low-Km aldehyde reductase of pig brain and pig kidney. J. Biol. Chem. 258 (1983a) 3583–3586.

Cromlish, J.A., and Flynn, T.G. Purification and characterization of an enzymically active cleavage product of pig kidney aldehyde reductase. Biochem. J. 209 (1983b) 597–607.

Cromlish, J.A., and Flynn, T.G. Identification of pig brain aldehyde reductases with the high-Km aldehyde reductase, the low-Km aldehyde reductase and aldose reductase, carbonyl reductase, and succinic semialdehyde reductase. J. Neurochem. 44 (1985) 1485–1493.

Donohue, P.J., Alberts, G.F., Hampton, B.S., and Winkles, J.A. A delayed-early gene activated by fibroblast growth factor-1 encodes a protein related to aldose reductase. J. Biol. Chem. 269 (1994) 8604–8609.

Fabre, S., Manin, M., Pailhoux, E., Veyssiere, G., and Jean, C. Identification of a functional androgen response element in the promoter of the gene for the androgen-regulated aldose reductase-like protein specific to the mouse vas deferens. J. Biol. Chem. 269 (1994) 5857–5864.

Felsted, R.L., Richter, R., and Bachur, N.R. Rat liver aldehyde reductase. Biochem. Pharmacol. 26 (1977) 1117–1124.

Frohman, M.A. RACE: Rapid amplification of cDNA ends. PCR Protocols: A guide to methods and applications (Innis, M.A., Gelfand, D.H., Sninsky, J.J. & White, T.J. eds.), (1990) 28–38.

Inoue, S., Sharma, R.C., Schimke, R.T., and Simoni, R.D. Cellular detoxification of tripeptidyl aldehydes by an aldo-keto reductase. J. Biol. Chem. 268 (1993) 5894–5898.

Kubiseski, T.J., Green, N.C., and Flynn, T.G. Location of an essential arginine residue in the primary structure of pig aldose reductase. Adv. Exp. Med. Biol. 328 (1993) 259–265.

Marquardt, D.W. An algorithm for least-squares estimation of nonlinear parameters. J. Soc. Indust. Appl. Math. 11 (1963) 431–441.

Morjana, N.A., and Flynn, T.G. Aldose reductase from human psoas muscle. Purification, substrate specificity, immunological characterization, and effect of drugs and inhibitors. J. Biol. Chem. 264 (1989) 2906–2911.

Nishimura, C., Matsuura, Y., Kokai, Y., Akera, T., Carper, D., Morjana, N., Lyons, C., and Flynn, T.G. Cloning and expression of human aldose reductase. J. Biol. Chem. 265 (1990) 9788–9792.

Pailhoux, E.A., Martinez, A., Veyssiere, G.M., and Jean, C.G. Androgen-dependent protein from mouse vas deferens. cDNA cloning and protein homology with the aldo-keto reductase superfamily. J. Biol. Chem. 265 (1990) 19932–19936.

Sambrook, J., Fritsch, E.F., and Maniatis, T. Molecular Cloning: A Laboratory Manual.(1989) Cold Spring Harbor Laboratory Press, Cold Spring Harbor, NY

Wilson, D.K., Tarle, I., Petrash, J.M., and Quiocho, F.A. Refined 1.8 A structure of human aldose reductase complexed with the potent inhibitor zopolrestat. Proc. Natl. Acad. Sci. USA 90 (1993) 9847–9851.

Wilson, D.K., Nakano, T., Petrash, J.M., and Quiocho, F.A. 1.7 A structure of FR-1, a fibroblast growth factor-induced member of the aldo-keto reductase family, complexed with coenzyme and inhibitor. Biochemistry 34 (1995) 14323–14330.

60

D-FRUCTOSE-MEDIATED STIMULATION OF BOVINE LENS ALDOSE REDUCTASE ACTIVATION BY UV-IRRADIATION

Tadashi Mizoguchi,[1] Takeharu Ogura,[1] Kiyohito Yagi,[1] and Peter F. Kador[2]

[1]Faculty of Pharmaceutical Sciences
Osaka University
Suita 565, Japan
[2]National Eye Institute
National Institutes of Health
Bethesda Maryland 20892

1. INTRODUCTION

Reports from a number of laboratories have demonstrated that UV-irradiation can cause protein damage in the ocular lens and the inactivation of lens enzymes (Li et al., 1990; Zigman et al., 1991; Rafferty et al., 1993; Hightower, 1995). Glycation-induced damage to structural proteins—hemoglobin and γ-crystalline (Pennington and Harding, 1994), and enzymes—glucose 6-phosphate dehydrogenase (Ganea and Harding, 1995), Cu-Zn-superoxide dismutase (Arai et al., 1987), and glutathione reductase (Blakytny and Harding, 1992), glucokinase (Zahner et al., 1990) has been widely studied. Although, glycation of proteins is thought to play a major role in aging, diabetes, and cataract formation (van Boekel, 1991), the reaction between protein and glucose is considered a slow process. We are interested in the UV-irradiation of lens enzymes in the presence of a high concentration of sugars as a means of elucidating the morphological and physiological sequellae in cataract formation. Here, we report on the enhancement of bovine lens aldose reductase activation induced by UV-irradiation under relatively high D-fructose conditions. This phenomenon can be induced by the generation of active oxygen species, and may result from intramolecular modifications of the enzyme at thiol groups.

2. MATERIALS AND METHODS

NADPH was obtained from Sigma Chemical Co. D,L-Glyceraldehyde, D-fructose, and 5,5'-dithiobis (2-nitrobenzoic acid) were purchased from Wako Pure Chemical Co. Dithiothreitol was from Nacalai Tesque, Pharmalyte from Pharmacia Fine Chemicals, and

Enzymology and Molecular Biology of Carbonyl Metabolism 6
edited by Weiner *et al.* Plenum Press, New York, 1996

Dyematrex gel Red A from Amicon, Inc. All other chemicals were of reagent grade. Fresh bovine eyes were obtained from a local slaughterhouse and the lenses were removed and kept frozen until needed. Aldose reductase from bovine lens was purified routinely by the series of chromatographic procedures described previously (Tanimoto et al., 1990), except that Dyematrex gel Red A and Pharmalyte were used instead of Matrex gel Orange A and Mono P, respectively. The isoelectric point of the bovine lens aldose reductase obtained with Pharmalyte was 4.59, which was close to the pI of 4.85 previously reported (Sheaff and Doughty, 1976).

Enzyme activity was measured as previously described with D,L-glyceraldehyde as a substrate (Tanimoto et al., 1990). When irradiated in the presence of sugars, enzyme-sugar mixtures were employed for enzyme assay without sugar removal. The sugars used did not affect the enzyme activity in the dark. Protein was determined by the Coomassie blue binding assay (Bradford, 1976), using bovine serum albumin as a standard. Thiol groups in the enzyme were estimated using the DTNB reagent (Ellman, 1959). UV irradiation was done with a UV lamp (National GL-15, Japan). The UV output incident upon the sample cuvettes with 1.0-cm path quartz was 0.07 mW/cm^2 at an ambient temperature of 24–27°C. The photodestruction of tryptophan residue in aldose reductase was monitored by determining the intrinsic fluorescence of tryptophan (Pajot, 1976). The fluorescent scopoletin-peroxidase assay (Corbett, 1989) was used to determine the hydrogen peroxide generated during UV-irradiation in the presence of D-fructose.

3. RESULTS AND DISCUSSION

The concentrations of glucose, sorbitol, and fructose in the normal lens range between 0.3 -2.0 mM, dependent on the sorbitol pathway, but in diabetes their concentrations are 4.5, 9.5, and 12 mM, respectively (Jedziniak et al., 1981). UV-Irradiation in the presence of a high concentration of D-fructose (30 mM) resulted in an initial increase followed by a decrease in the activity of bovine lens aldose reductase (Fig.1). Activation of the enzyme was also induced by UV-irradiation in the absence of D-fructose, but to a lesser extent than in its presence. The same result was obtained with pig lens aldose reductase (data not shown). The net maximal increase in enzyme activity occurred within 10 min, showing that D-fructose causes rapid but transient stimulation of the enzyme activation induced by UV-irradiation. Neither glucose nor sorbitol stimulated activation of the enzyme. Since the UV-activation of aldose reductase was accelerated dose-dependently by D-fructose (Fig. 2), the transient stimulation of the enzyme activation induced by UV-irradiation was specific for D-fructose.

Figure 3 shows that D-fructose did not stimulate aldose reductase activation mediated by the photosensitized reaction in the presence of rose bengal, which is well known to generate singlet oxygen (Goosey et al., 1981; Glaser et al., 1988). Singlet molecular oxygen thus seems not to participate in the UV-irradiation stimulation process in the presence of D-fructose. There are a number of reports in the literature concerning the generation of hydrogen peroxide accompanied by other active oxygen species in UV-photooxidation (see Andley and Clark, 1989). In the present study, the scopoletin-fluorescence assay during UV-irradiation revealed increased in generation of hydrogen peroxide by D-fructose (Fig. 4). Hydrogen peroxide is therefore considered to be a candidate for the active oxygen species generated during UV-irradiation in the presence of D-fructose. According to glycation theory (Wolff et al., 1989), glucose has the ability to generate diverse free radicals, including hydroxy radicals, in the presence of trace

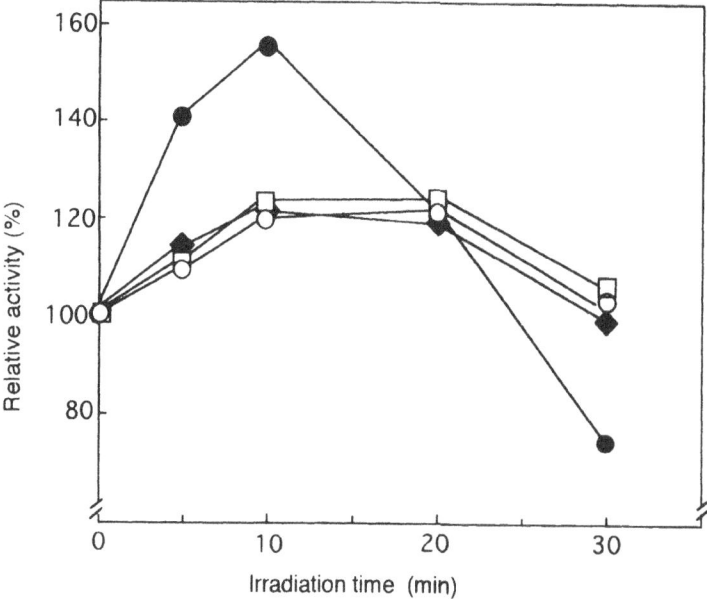

Figure 1. Activation of bovine lens aldose reductase by UV-irradiation in the presence of 30 mM sugars: no addition (O), D-fructose (●), D-glucose (◆), and sorbitol (□). The activity with no irradiation in the absence of sugars is taken as 100%.

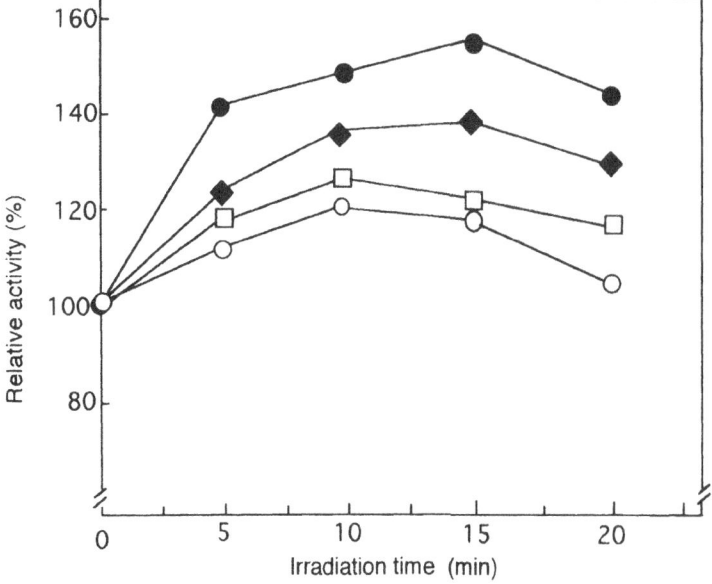

Figure 2. Activation of bovine lens aldose reductase by UV-irradiation with varying amounts of D-fructose added: 0 (O), 10 (□), 30 (◆), and 50 mM (●). The activity with no irradiation in the absence of D-fructose is taken as 100%.

Figure 3. Rose bengal-sensitized photoreaction of bovine lens aldose reductase in the absence (open square) and the presence (shaded square) of 30 mM D-fructose. The activity with no irradiation in the absence of D-fructose is taken as 100%.

amounts of transition metals. The photodynamic reaction induced by UV-irradiation in the presence D-fructose is unique, and differs from glucose autooxidation.

Bovine lens aldose reductase involves two cysteine residues (Liu et al., 1992). Estimation of the thiol groups in the enzyme in the presence of 6.4 M guanidine, pH 8.0, revealed that the decrease in the enzyme thiols mediated by UV-irradiation was significantly high in the presence of D-fructose (Fig. 5). Without UV-irradiation, D-fructose was inef-

Figure 4. Hydrogen peroxide formation during the UV-irradiation of bovine lens aldose reductase in the presence of 30 mM D-fructose: reductase alone (●) reductase and D-fructose (◆), and D-fructose alone (○).

Figure 5. Decrease in SH groups of bovine lens aldose reductase by UV-irradiation in the presence of 30 mM D-fructose for 10 min: no irradiation without (dark shaded) or with (lightly shaded) D-fructose and irradiation without (squares) or with (dots) D-fructose. The number of SH groups without UV-irradiation is taken as 100%.

fective on the enzyme thiols. Judging from the decrease in the enzyme thiols under UV-irradiation and in the presence of D-fructose within a short time, oxidation of the thiol groups seems to differ from enzyme glycation by D-fructose during UV-irradiation. Accordingly, the transient acceleration of aldose reductase activation induced by UV-irradiation in the presence of D-fructose may be primarily attributed to the photodynamic oxidation of the enzyme thiol groups. The disulfide bond formation of bovine serum albumin under long-term exposure to natural sun-light (including UV-light) has been shown by SDS-PAGE analysis (Watanabe et al., 1991). Under short-term UV-irradiation in the presence of D-fructose, the SDS-PAGE pattern of aldose reductase (Fig. 6) was no different from that of the native enzyme, indicating there was no crosslinking of the enzyme among inter-molecules through disulfide bonds. Microenvironmental changes at the active site of the enzyme molecules due to the photodynamic oxidation of thiol groups is suggested, but, details of the accelerated UV-activation of the enzyme mediated by D-fructose remain to be elucidated. Thiol-dependent oxidative activation of aldose reductase by the Fe(II)/Fe(III) redox system (Giannessi et al., 1993), and activation of the enzyme me-

Figure 6. SDS-PAGE profiles of bovine lens aldose reductase by UV-irradiation in the presence of 30 mM D-fructose. The numbers of lanes correspond to UV-irradiation time from left to right: 0 (lane 1), 5 (lane 2), 10 (lane 3), and 20 min (lane 4). Molecular weight standard is lane 5.

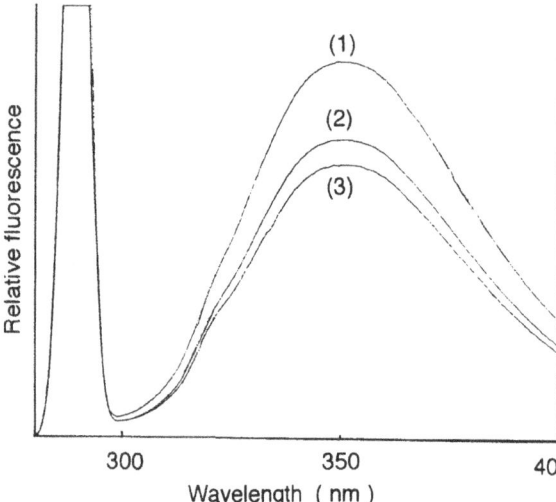

Figure 7. Decrease in tryptophan intrinsic fluorescence of bovine lens aldose reductase during UV-irradiation in the presence of 30 mM D-fructose. The numbers on the fluorescence spectra correspond to: no UV-irradiation (1), and UV-irradiation in the absence (2), and in the presence of D-fructose (3) for 10 min.

diated by active oxygen species (Del et al., 1987) have previously been shown, but these occurred in the order of hours rather than minutes. Many details of the action of UV-irradiation on proteins, peptides, and amino acids have been revealed, and have sufficiently elucidated the photochemical reaction of aromatic amino acid residues including tryptophan, tyrosine, phenylalanine, and histidine (see Vladimirov et al., 1970). Exposing aldose reductase to UV-light resulted in a slight decrease in the intrinsic tryptophan fluorescence intensity, while the coexistence of D-fructose increased the reduction of the fluorescence intensity (Fig. 7). The photodestruction of tryptophan residues in the enzyme molecules may be involved in a decrease in the activity of bovine aldose reductase induced by prolonged UV-irradiation rather than the rapid and transient activation of the enzyme under UV-irradiation in the presence of D-fructose. Oxidative damage to lens constituents is a major factor in cataract. Lens aldose reductase activation mediated by UV-irradiation may cause high lens D-fructose levels via the sorbitol pathway. The findings described above suggests that high lens D-fructose levels may be a risk factor for cataract formation in patients exposed to high levels of light.

4. REFERENCES

Andley, U.P. and Clark, B.A. (1989). The effects of near-UV radiation on human lens beta-crystallins: protein structural changes and the production of O^{2-} and H_2O_2. *Photochem. Photobiol.* **50**: 97–105.

Arai, K., Maguchi, S., Fujii, S., Ishibashi, H., Oikawa, K., and Taniguchi, N. (1987). Glycation and inactivation of human Cu-Zn-superoxide dismutase. Identification of the in vitro glycated sites. *J. Biol. Chem.* **262**: 16969–72.

Blakytny, R. and Harding, J.J. (1992). Glycation (non-enzymic glycosylation) inactivates glutathione reductase. *Biochem. J.* **228**: 303–307.

Bradford, M.M. (1976). A rapid and sensitive method for the quantitation of microgram quantities of protein utilizing the principle of protein-dye binding. *Anal Biochem* **72**: 248–54.

Corbett, J.T. (1989). The scopoletin assay for hydrogen peroxide. A review and a better method. *J Biochem. Biophys. Methods* **18**: 297–307.

Del, C.A., Camici, M., and Mura, U. (1987). In vitro modification of bovine lens aldose reductase activity. *Biochem Biophys Res Commun* 148:369–75.

Ellman, G.L. 1959. Tissue sulfhydryl groups. *Arch. Biochem. Biophys* **82**: 70–77.

Ganea, E. and Harding, J.J. 1995. Molecular chaperones protect against glycation-induced inactivation of glucose-6-phosphate dehydrogenase. *Eur. J. Biochem.* **231**: 181–5.

Giannessi, M., Del, C.A., Cappiello, M., Voltarelli, M., Marini, I., Barsacchi, D., Garland, D., Camici, M., and Mura, U. (1993). Thiol-dependent metal-catalyzed oxidation of bovine lens aldose reductase. I. Studies on the modification process. *Arch. Biochem. Biophys.* **300**: 423–9.

Glaser, E., Cadenas, E., Andell, S., and Ernster, L. (1988). Inhibition of the mitochondrial F1-ATPase by rose bengal mediated photooxidation Interaction of the Fe2+ chelate of bathophenanthroline with the sensitizer. *Acta. Chem. Scand.* **42**: 175–82.

Goosey, J.D., Zigler, J.J., Matheson, I.B., and Kinoshita, J.H. (1981). Effects of singlet oxygen on human lens crystallins in vitro. *Invest. Ophthalmol. Vis. Sci.* **20**: 679–83.

Hightower, K.R. 1995. The role of the lens epithelium in development of UV cataract. *Curr. Eye. Res.* **14**: 71–8.

Jedziniak, J.A., Chylack, L.J., Cheng, H.M., Gillis, M.K., Kalustian, A.A., and Tung, W.H. (1981). The sorbitol pathway in the human lens: aldose reductase and polyol dehydrogenase. *Invest. Ophthalmol. Vis. Sci.* **20**: 314–26.

Li, D.Y., Borkman, R.F., Wang, R.H., and Dillon, J. (1990). Mechanisms of photochemically produced turbidity in lens protein solutions. *Exp. Eye Res.* **51**: 663–9.

Liu, S.Q., Bhatnagar, A. and Srivastava, S.K. (1992). Carboxymethylation-induced activation of bovine lens aldose reductase. *Biochim. Biophys. Acta* **1120**: 329–36.

Pajot, P. 1976. Fluorescence of proteins in 6-M guanidine hydrochloride. A method for the quantitative determination of tryptophan. *Eur. J. Biochem.* **63**: 263–9.

Pennington, J. and Harding, J.J. (1994). Identification of the site of glycation of gamma-II-crystallin by (^{14}C)-fructose. *Biochim. Biophys. Acta.* **1226**: 163–7.

Rafferty, N.S., Zigman, S., McDaniel, T., and Scholz, D.L. (1993). Near-UV radiation disrupts filamentous actin in lens epithelial cells. *Cell Motil. Cytoskeleton* **26**: 40–8.

Sheaff, C.M. and Doughty, C.C. (1976). Physical and kinetic properties of homogenous bovine lens aldose reductase. *J. Biol. Chem.* **251**: 2696–702.

Tanimoto, T., Sato, S., and Kador, P.F. (1990). Purification and properties of aldose reductase and aldehyde reductase from EHS tumor cells. *Biochem. Pharmacol.* **39**: 445–53.

van Boekel, M.A.M. (1991). The role of glycation in aging and diabetes mellitus. *Mol. Biol. Rep.* **15**:57–64.

Vladimirov, Y.A., Roshchupkin, D.I., and Fesenko, E.E. (1970). Photochemical reactions in amino acid residues and inactivation of enzymes during U.V.-irradiation. A Review. *Photochem. Photobiol.* **11**: 227–46.

Watanabe, Y., Horii, I. Nakayama, Y. and Osawa, T. (1991). Effect of cysteine on bovine serum albumin (BSA) denaturation induced by solar ultraviolet (UVA, UVB) irradiation. *Chem. Pharm. Bull. Tokyo* **39**: 1796–801.

Wolff, S.P., Bascal, Z.A., and Hunt, J.V. (1989). "Autoxidative glycosylation": free radicals and glycation theory. *Prog. Clin. Biol. Res.* **304**: 259–75.

Zahner, D., Ramirez, R., and Malaisse, W.J. (1990). Kinetic behaviour of liver glucokinase in diabetes. II. Possible role of non-enzymatic protein glycation. *Diabetes Res.* **14**: 109–15.

Zigman, S., Paxhia, T., McDaniel, T., Lou, M.F., and Yu, N.T. (1991). Effect of chronic near-ultraviolet radiation on the gray squirrel lens in vivo. *Invest. Ophthalmol. Vis. Sci.* **32**: 1723–32.

61

METABOLIC INACTIVATION AND EFFLUX OF DAUNORUBICIN AS COMPLEMENTARY MECHANISMS IN TUMOR CELL RESISTANCE

Michael Soldan and Edmund Maser

Department of Pharmacology and Toxicology
Philipps-University of Marburg
Karl-von-Frisch Str. 1
35033 Marburg, Germany

1. INTRODUCTION

Multidrug-resistance is a major obstacle in cancer therapy with antineoplastic chemotherapeutics. Drug-resistant tumor cells frequently display overexpression of a plasma membrane phosphoglycoprotein termed P-170. P-170 is thought to act as an energy depend-end efflux pump (Juliano and Ling, 1976) that lowers the intracellular accumulation of structurally and functionally unrelated antineoplastic agents such as anthracyclines, vinca alkaloides and taxol, and, thus, confers multidrug resistance. Anthracyclines, like daunorubicin (DRC) are the most valuable cytostatic agents in clinical chemotherapy, but their usefulness is limited by intrinsic or aquired resistance towards these drugs. However, it is known that P-170 cannot fully account for the chemoresistance phenomenon in multidrug resistant tumor cells (Booser and Hortobagyi, 1994). In addition to an increasing efflux of chemotherapeutic agents, enzymatic drug inactivation can also contribute to a higher drug resistance (Deffie et al., 1988). In the case of DRC, carbonyl reduction leads to 13-hydroxy-daunorubicinol (DRCOL), the major metabolite of DRC with a significant lower antineoplastic potency compared to the parent drug (Schott and Robert, 1989).

Increased P-170 expression causes a lower intracellular DRC concentration, whereas a higher carbonyl reduction of DRC leads to the 13-hydroxy metabolite with a significant lower toxicity. We propose that enzymatic inactivation by carbonyl reduction serves as a complementary means in DRC detoxification. The combined action of both mechanisms in tumor cells may result in an additive resistance towards higher DRC. This situation becomes especially important, when high DRC concentrations exhaust the capacity of the P-170 efflux pump and intracellular drug concentrations increase dramatically.

In our studies, a direct comparison between P-170 mRNA expression and DRC carbonyl reduction after a 72 hour incubation with increased concentrations of DRC confirmed that cancer cells are able to use both detoxification mechanisms

Enzymology and Molecular Biology of Carbonyl Metabolism 6
edited by Weiner *et al.* Plenum Press, New York, 1996

2. EXPERIMENTAL

Two human pancreas carcinoma cell lines were used in this study: the parent DRC sensitive cell line (PaCa sen) and its DRC resistant descendant (PaCa res). Both cell lines were grown over 72 hours in the presence of increasing concentrations of DRC (10^{-4}, 10^{-3}, 10^{-2}, 10^{-1}, and 1µg/ml) in the culture media. DRC carbonyl reduction was determined in cytosol as described previously (Soldan et al. 1996). Inhibition characteristics of DRC carbonyl reduction with diagnostic inhibitors were studied with equimolar concentrations of DRC and inhibitor. RT-PCR of P-170 mRNA was performed as discribed previously (Murphy et al., 1990).

3. RESULTS

3.1. DRC Metabolism

DRC carbonyl reducing activity is mainly located in the cytosol. Both cell lines, PaCa sen and PaCa res, have the same activity of DRCOL formation when they were grown in the absence of DRC. Supplementation of DRC to the culture media led to a weak decrease of DRC reducing activity at low DRC concentrations and to an increase of activity at high DRC concentrations. Highest activity is reached at 10^{-1} µg/ml DRC supplementation in PaCa sen and at 1 µg/ml in PaCa res. Combined, these results demonstrate that DRC carbonyl reduction in cytosol is inducible by itself in a concentration dependend manner. The decrease of reductive activity in PaCa sen at 1 µg/ml DRC in culture media is in accordance with a strong impairment of viability at this high DRC concentration (Fig. 1).

3.2. Inhibition of DRC Carbonyl Reduction

DRC carbonyl reduction has been shown to be mediated by carbonyl reductase (EC 1.1.1.184), aldehyde reductase (EC 1.1.1.2) and one isoform of dihydrodiol dehydrogenase (EC 1.3.1.20) (Ohara et al., 1995). In order to obtain information about a possible DRC linked change in the normal pattern of enzymes involved in DRC reduction, we performed an inhibitor characterization at equimolar concentrations of the substrate and diagnostic inhibitors in the homogenate of both cell lines. Untreated and pretreated PaCa sen and PaCa res, grown with 10^{-1} µg/ml DRC in culture media, were compared with respect to altered inhibition. Phenobarbital, the diagnostic inhibitor of aldehyde reductase (EC 1.1.1.2), inhibited DRC carbonyl reduction in sensitive and resistant cells for about 30%. When the cells were grown in the presence of DRC, phenobarbital inhibition disappeared. NAD(P)H (quinone acceptor) oxidoreductase (EC 1.6.99.2), previously known as DT-diaphorase, catalyzes the reduction of quinones to hydroquinones and is inhibited by dicoumarol. In our studies, dicoumarol inhibition could only be observed in the resistant descendant, and DRC pretreatment abolished the inhibitory effect. Glycyrrhetinic acid, the inhibitor of several hydroxysteroid dehydrogenases, and indomethacin, the inhibitor of dihydrodiol dehydrogenase, aldose reductase and 3α-hydroxysteroid dehydrogenase, could only inhibit DRC carbonyl reduction in the range of 20% up to 30% and did not change their inhibitory effect significantly after DRC pretreatment.

Only rutin and quercitrin, both potent inhibitors of carbonyl reductase (EC 1.1.1.184), led to a strong decrease of DRC reducing activity. This inhibition is more effective in PaCa res than in PaCa sen, and is significantly diminished by DRC pretreatment. These results suggest that DRC reduction in the choosen tumor cells is mediated

Figure 1. Effect of DRC pretreatment on DRC carbonyl reduction in the cytosolic fraction of pancreas carcinoma cells. Sensitive and resistant cells were cultured in the absence or presence of increasing concentrations of DRC. Specific activities are expressed as nmol/mg protein DRCOL formed in 60 minutes. Each bar represents the mean of 4–12 determinations.

mainly by carbonyl reductase. The enzymatic pattern of DRC carbonyl reduction, which occurs in untreated PaCa, seems to change after the pretreatment with DRC. The number of inhibitors with an effect on DRC carbonyl reduction becomes smaller, hence DRC pre-treated cells probably reduce the expression of certain enzymes involved (Fig. 2).

3.3. Glycoprotein P-170 mRNA Expression

Expression of P-170, a plasma membrane glycoprotein involved in multidrug resistance and encoded by mdr genes, was investigated in both PaCa cell lines. Mdr mRNA

Figure 2. Effect of specific inhibitors on DRC carbonyl reduction in pancreas carcinoma cells. Sensitive and resistant cells were cultured in the absence or presence of 0.1 µg/ml DRC for 72 hours. Inhibitor linked characterisations are performed at equimolar concentrations of DRC and inhibitor. Activities are expressed as per cent of control. Abbreviations used are: phen = phenobarbital, glyc = glycyrrhetinic acid, quer= quercitrin, indo = indomethacin, and dic = dicoumarol.

levels were the same in sensitive and resistant PaCa, when the cells were untreated. The PaCa sen had a strong, concentration dependent induction of P-170 expression after DRC pretreatment, whereas the resistant descendent had only a weak induction. The concentration dependend induction of P-170 showed nearly the same pattern as DRC carbonyl reductase induction. Both mechanisms had their maximum induction at 10^{-1} µg/ml DRC in culture media. The reason for the weak P-170 mRNA induction could be that other resistance mechanisms become evident. For example, the expression of "multidrug resistance associated protein" (MRP), which is also an energy dependend transmembrane pump, may get more importance compared to P-170. The PaCa sen which had never been in contact with DRC before, showed a significant stronger induction of P-170. This may be because P-170 is more important for PaCa sen than for the resistant cells (Fig. 3).

Figure 3. Effect of DRC pretreatment on P-170 expression of pancreas carcinoma cells. Sensitive and resistant cells were cultured in the absence or presence of increasing concentrations of DRC. Activities are expressed as per cent of control. The internal standard used was β_2-microglobulin.

3.4. Alterations Caused by P-170 Blocking

P-170 mRNA expression and DRC carbonyl reduction are both inducible by DRC supplementation to the culture media in a concentration dependend manner. This simultaneous induction of both mechanisms could occur accidentally or as a consequence of the increasing intracellular DRC concentrations. P-170 lowers the amount of intracellular DRC, but cannot care for a DRC free cell. Increasing extracellular DRC does also raise intracellular DRC, especially when P-170 efflux capacity is exhausted. If there is really an association between P-170 linked DRC efflux and detoxification of remaining intracellular DRC via carbonyl reduction, blocking of P-170 must also cause alterations of DRC reducing activity.

Verapamil is a P-170 modifying agent which is able to block the efflux of DRC (Tsuruo et al., 1984). We used 20 µM verapamil, a subtoxic concentration for the PaCa cells to block P-170, and pretreated both cell lines again with the same concentrations of DRC (10^{-4} up to 1 µg/ml).

DRC and verapamil supplementation to the culture media cause two main reactions. Activities of both mechanisms, DRC carbonyl reduction and P-170 expression, are higher at low DRC concentrations but significant weaker at high DRC supplementation concentrations.

The possible reason for increasing reductive activity and P-170 expression at low dose DRC supplementation could be a higher intracellular DRC concentration, compared to the unblocked experiment, because the efflux of DRC is blocked by verapamil. The maxima of induction are significant weaker.

These two main alterations can be observed in PaCa sen and PaCa res. Hence, intervention in the P-170 efflux mechanisms causes variations in the DRC reductase system, probably due to changes in intracellular DRC concentrations (Fig. 4).

Figure 4. Effect of verapamil and DRC pretreatment on DRC carbonyl reduction and P-170 expression in pancreas carcinoma cells. Sensitive and resistant cells were cultured in the absence or presence of 20 μM verapamil and in the absence or presence of increasing concentrations of DRC. Activities of DRC carbonyl reduction are expressed as nmol/mg protein DRCOL formed in 60 minutes and activities of P-170 expression are expressed as per cent of control.

4. DISCUSSION

DRC, a member of the anthracycline family, is a valuable cytostatic agent in cancer therapy. Resistance against this drug often occurs together with an overexpression of the efflux pump P-170 (Juliano and Ling, 1976) . It is known that P-170 mediates multidrug resistance by lowering intracellular drug concentrations (Safa, 1992). When we added low DRC concentrations to the culture media, P-170 expression increased markedly in the PaCa cell lines. DRC carbonyl reduction, a proposed additional DRC detoxifying mechanism, showed no increase of activity at this concentrations. DRC carbonyl reduction to the less toxic metabolite DRCOL becomes apparent, when the DRC concentration added to the culture media reaches a concentration of 10^{-2} μg/ml. This seemes to be the minimal concentration for an induction of DRC carbonyl reducing enzymes, mainly carbonyl reductase. At this DRC concentration, diffusion through the cell membrane seems to be so strong that P-170 efflux is not sufficient to protect the cancer cells from increasing intracellular drug accumulations. In this situation, a second protection mechanism, which is the phase 1 metabolism of DRC to the 13-hydroxyalcohol DRCOL becomes important. Determination of DRC- and DRCOL-toxicity by the MTT-test indicated that DRCOL is about 10-fold less toxic in PaCa sen. Moreover no measurable toxicity at the highest concentration choosen (20 µg DRC/ml culture media) was detectable in PaCa res. Additionally, the alcohol metabolites formed are subject to an increased glucuronidation as well as renal elimination (Gessner et al., 1990). This additional intracellular detoxification of penetrating DRC via enzymatic activity seems to enable cancer cells to reach a higher level of anthracycline resistance. The PaCa sen, which had never been treated with DRC before, showed the strongest increase of P-170 expression and DRC reducing activity at the same maxima of DRC concentrations added to the culture media. PaCa res, in contrast, showed

Figure 5. Proposed mechanisms of complementary metabolic inactivation and efflux of DRC.

only a weak induction of P-170 and a weaker induction of DRC reducing activity as compared to PaCa sen. A possible explanation is that cancer cells, often treated with DRC, express already other resistant mechanisms, which become more important than P-170 for the PaCa res. The alterations caused by verapamil linked blocking of P-170 showed that both mechanisms are associated, presumably not directly, but the intracellular DRC concentration seems to be the decisive fact.

In conclusion, our studies provide evidence that enzymatic inactivation of DRC by carbonyl reduction represents an intracellular detoxification mechanism, which, in addition to P-170 mediated drug efflux, serves as a complementary means in the chemoresistance of cancer cells towards anthracyclines (Fig. 5).

5. REFERENCES

Booser, D.J. and Hortobagyi, G.N.: Anthracycline antibiotics in cancer therapy. Focus on drug resistance. Drugs 47 (1994) 223–258.

Deffie, A.M., Alam, T., Seneviratne, C., Beenken, S. W., Batra, J. L., Shea, T. C., Henner, W. D. and Goldenberg, G. J.: Multifactorial resistance to adriamycin: relationship of DNA repair, glutathione transferase activity, and P-glycoprotein in cloned cell lines of adriamycin sensitive and resistant P388 leukemia. Cancer Res. 48 (1988) 3595–3602.

Gessner, T., Vaughan, L. A., Beehler, B. C., Bartels, C. J. and Baker, R. M.: Elevated pentose cycle and glucuronyltransferase in daunorubicin-resistant P388 cells. Cancer Res. 50 (1990) 3921–3927.

Juliano, R. L. and Ling, V.: A surface glycoprotein modulating drug permeability in Chinese hamster ovary cell mutants. Biochem. Biophys. Acta 455 (1976) 152–162.

Murphy, L. D., Herzog, C. E., Rudick, J. B., Fojo, A. T. and Bates, S. E.: Use of polymerase chain reaction in the quantitation of mdr-1 gene expression. Biochemistry 29 (1990) 10351–10356.

Ohara, H., Miyabe, Y., Deyashiki, Y., Matsuura, K. and Hara, A.: Reduction of drug ketones by dihydrodiol dehydrogenases, carbonyl reductase and aldehyde reductase of human liver. Biochem. Pharmacol. 50 (1995) 221–227.

Safa, A. R.: Photoaffinity labelling of P-glycoprotein in multidrug-resistant cells. Cancer Invest. 10 (1992) 295–305.

Schott, B. and Robert, J.: Comparative activity of anthracycline 13-hydroxymetabolites against glioblastoma cells in culture. Biochem. Pharmacol. 38 (1989) 4069–4074.

Soldan, M., Netter, K. J. and Maser, E.: Induction of daunorubicin carbonyl reducing enzymes by daunorubicin in sensitive and resistant pancreas carcinoma cells. Biochem. Pharmacol. 51 (1996) 117–123.

Tsuruo, T., Lida, H., Tsukagoshi, S. and Sakurai, Y.: Overcoming of vincristine resistance in P388 leukemia in vivo and in vitro through enhanced cytotoxicity of vincristine and vinblastine by verapamil. Cancer Res. 41 (1981) 1967–1972.

62

CLONING AND EXPRESSION OF cDNA ENCODING BOVINE LIVER DIHYDRODIOL DEHYDROGENASE 3, DD3

Tomoyuki Terada,[1] Hideki Adachi,[1] Hirofumi Nanjo,[1] Naomi Fujita,[1] Tatsuya Takagi,[2] Jun-ichi Nishikawa,[1] Masayoshi Imagawa,[1] Tsutomu Nishihara,[1] and Masatomo Maeda[1]

[1]Faculty of Pharmaceutical Sciences
Osaka University
1–6 Yamada-oka
[2]Genome Information Research Center
Osaka University
3–1 Yamada-oka, Suita
Osaka 565, Japan

1. INTRODUCTION

Dihydrodiol dehydrogenase (DDH) [EC 1.3.1.20] has been suggested to play an important physiological role in the metabolic detoxication of polycyclic aromatic hydrocarbons (PAHs) (Glatt et al., 1979). The pathways for metabolic activation and detoxication of a typical PAH, benzo[a]pyrene, are shown in Scheme 1. Environmental PAH is mainly activated through the action of liver P-450. The resulting 7,8-epoxide moiety is hydrolyzed by epoxide hydrolase in microsome. Then, the 7,8-dihydrodiol of PAH is further oxygenated to the 7,8-dihydrodiol-9,10-epoxide form by P-450 (Penning, 1993). This compound is assumed to be an ultimate carcinogen which binds covalently to DNA. Alternatively, DDH is reported as an important converting enzyme of the 7,8-dihydrodiol moiety to less active 7,8-dione (dicarbonyl). The dicarbonyl compound is further detoxicated with nonenzymatic ghlutathione-conjugation reaction (Penning, 1993).

In the recent two decades, the unique properties of DDHs from many mammalian species and tissues have been reported with respect to substrate specificities, inhibitor sensitivities, immunological cross reactions, relative molecular masses and their structures (Vogel et al., 1980; Bolcsak and Nerland, 1983; Hara et al., 1986; Matsuura et al., 1987; Nakagawa et al., 1989; Penning and Sharp, 1990; Hara et al., 1991; Klein et al., 1992; Mizoguchi et al., 1992; Nanjo et al., 1992; Nanjo et al., 1995). Since the primary structure of rat liver 3a-hydroxysteroid/dihydrodiol dehydrogenase [EC 1.1.1.50] was determined as a member of DDHs (Pawlowski et al., 1991), the sequences of many DDHs have been re-

Scheme 1. Possible involvement of dihydrodiol dehydrogenase in the metabolic detoxication of polycyclic hydrocarbon.

ported (Cheng et al., 1991 ; Stolz et al., 1993 ; Deyashiki et al., 1994). The results indicated that they can be actually classified into an aldo-keto reductase superfamily including aldehyde reductase [EC 1.1.1.2], aldose reductase [EC 1.1.1.21] and prostaglandin F synthase [EC 1.1.1.188] (Watanabe et al., 1988; Bohren et al., 1989 ; Winters et al., 1990).

In our studies on the carbonyl metabolism in bovine liver, we have purified 5 aldo-keto reductases, namely aldehyde reductase/dihydrodiol dehydrogenase 2 (DD2), 3α-hydroxysteroid/dihydrodiol dehydrogenase (DD1), 3α/17β-hydroxysteroid dehydrogenase (carbonyl reductase), cytochrome c reductase and dihydrodiol dehydrogenase 3 (DD3) (Terada et al., 1985; Mizoguchi et al., 1992; Nanjo et al., 1992; Terada et al., 1993). Among the three DDHs, enzymatic properties (substrate specificity and inhibitor sensitivity) and relative molecular mass (Mr) of DD3 are distinct from the other two enzymes (DD1 and DD2) (Table 1). We have already reported the peptide sequences and the important cysteine residue (Cys-193) for the activity of DD3 (Nanjo et al., 1995).

To further clarify the structure and function of DD3 we cloned and expressed DD3 cDNA in the present study. The similarity of DD3 in the amino acid sequence with those

Table 1. Properties of multiple forms of bovine liver dihydrodiol dehydrogenase

Property	DD1	DD2	DD3
Mr	35.0 kDa	36.5 kDa	35.5 kDa
Typical substrate	Androsterone	D-Glucuronic acid	BDD
Km-value for BDD (mM)	2.3	20	0.18
$Kcat/Km$ for BDD	3.8	1.0	260

Mr, Relative molecular mass ; BDD, *trans*-benzene dihydrodiol.

of the other aldo-keto reductases such as bovine liver and lung prostaglandin F synthases and rat liver 3α-hydroxysteroid dehydrogenase clearly indicated that DD3 belongs to a member of aldo-keto reductase superfamily.

2. MATERIALS AND METHODS

Bovine liver λgt11 5'-stretch cDNA library (adult male) was purchased from Clonetech. Geneclean II kit, Sequenase Version 2.0 Sequencing kit, ΔTth DNA Sequencing kit and restriction endonucleases were the products of Funakoshi, Amersham-USB, TOYOBO and NEB, respectively. pKK 223–3 expression vector and Q-Sepharose were obtained from Pharmacia. DyeMatrex Gel Red A, Ultrogel AcA 44 and Immobilon P transfer membrane were supplied from Amicon, IBF and MILIPORE, respectively. Pyridine nucleotides, (S)-(+)-1-indanol and cholic acid derivatives were from WAKO Pure Chemical Industries and Nacalai Tesque. The dehydrogenase activity was determined by the increase of A_{340} nm due to the reduction of $NADP^+$ at 25°C. The assay mixture contained 100 mM glycine/NaOH, pH 9.5, 0.2 mM $NADP^+$, 2.0 mM $trans$-naphthalene dihydrodiol and an appropriate amount of the enzyme.

3. RESULTS AND DISCUSSION

3.1. Cloning of DD3

The cDNA was synthesized from bovine liver poly(A)$^+$ RNA using TaKaRa cDNA Synthesis Kit according to the instruction manual. Targeted cDNA was amplified by forward primer, (5'-GCG GAA TTC CTG GAC TAT GTC GAT CTC-3') and reverse primer (3'-GTG GGA CTC ATG GGT AAA CTT AAG GCG-5') based on the peptide sequences of DD3 homologous with those of bovine liver prostaglandin F synthase (vPGFS). The PCR product was digested with Eco RI and cloned into pBluescript KS (+) (Starategene) using TaKaRa Ligation Kit Ver. 2.0. The resulting plasmid was introduced into JM109. Sequence determination of two different fragments (about 700 bp) were demonstrated that one is identical and the other (DDp) is similar to that of vPGFS (Kuchinke et al., 1992). Deduced amino acid sequence of DDp was highly homologous to those of vPGFS and DD3 (Fig. 1). We screened a cDNA library using an amplified DDp sequence as a probe by plaque hybridization technique. From 3 x 10^5 plaques, 14 positive clones (DD$_A$-DDM) carrying 0.8 to 1.5 kbp Eco RI inserts were obtained. Two of the clones (DD$_B$, 1,158 bp; DD$_L$, 887 bp) had an overlapping 706 bp sequence (Fig. 2). Combined 1,334 bp sequence including poly A tail had an 969 bp open reading frame encoding DD3. This nucleotide size is close to that of the 1.4 kb DD3 mRNA detected by Northern blot hybridization. Translated sequences of the rest of the clones do not show any significant similarity to DD3.

The amino acid sequence of DD3 (323 amino acid) was compared with those of other aldo-keto reductases. As shown in Table 2, DD3 shows significantly high homologies with many aldo-keto reductases such as bovine liver prostaglandin F synthase (vPGFS) (Watanabe et al., 1988), bovine lung prostaglandin F synthase (gPGFS) (Kuchinke et al., 1992), human liver bile acid binding protein (hBABP) (Stolz et al., 1993), human liver dihydrodiol dehydrogenase 4 (hDD4) (Deyashiki et al., 1994), rabbit ovary 20α-hydroxysteroid dehydrogenase (rbHSD) (Lacy et al., 1993) and rat liver 3α-hy-

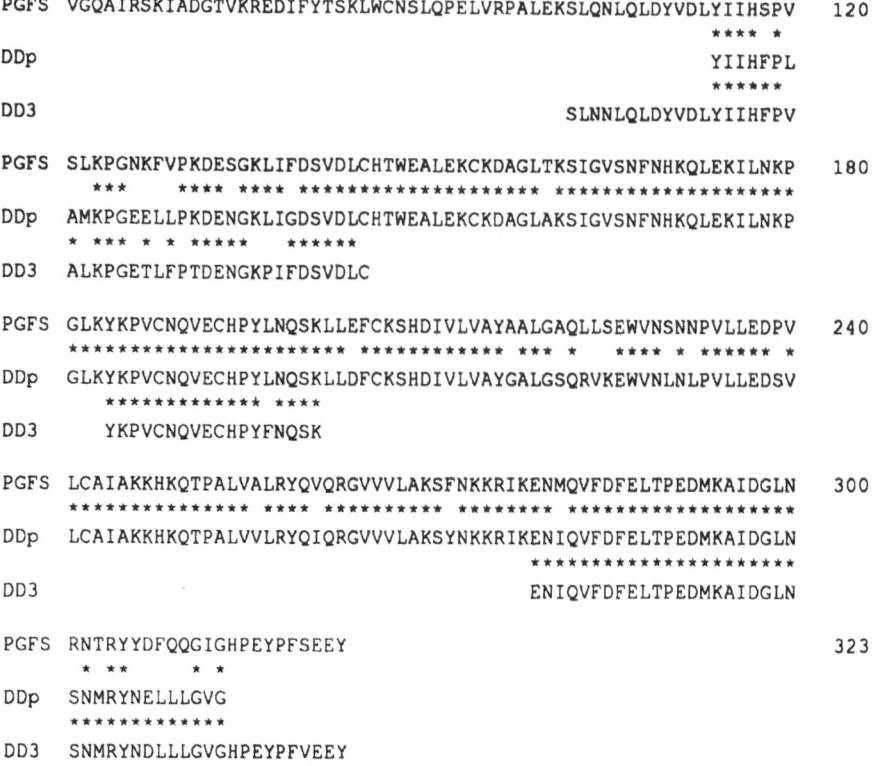

Figure 1. Amino acid sequence homology between DDp and PGFS (Kuchinke et al., 1994). Amino acid sequence deduced from DDp cDNA was aligned with that of PGFS. The peptide sequences of DD3 were also aligned.

droxysteroid dehydrogenase (rHSD) (Pawlowski et al., 1991). Amino acid residues of these proteins are conserved in the entire sequences (Fig. 3). Furthermore, the hydropathy profile of DD3 closely resembled to that of rHSD (data not shown).

3.2. Expression and Purification of Recombinant DD3 (rDD3)

To insert dd3 fragment into pKK223–3 expression vector, *Eco* RI site was shifted in front of the initiation codon of dd3 DNA using two primers 5'-TTC GAA TTC ATG GAT CCC AAA GGC-3' and 3'-GTG GGA CTC ATG GGT AAA CTT AAG GCG-5' by PCR

Figure 2. Restriction maps of DDB and DDL.

Table 2. Homology score in amino acid sequence between aldo-keto reductases

	DD3	vPGFS	gPGFS	hBABP	hDD4	hDD2	rbHSD	rHSD
DD3	—	86.4	86.7	78.3	76.0	78.3	75.1	69.6
vPGFS		—	99.1	77.4	74.1	77.4	75.1	69.3
gPGFS			—	77.7	74.5	77.7	75.4	69.6
hBABP				—	82.9	100.0	80.1	69.3
hDD4					—	82.9	77.7	69.9
hDD2						—	80.1	69.3
rbHSD							—	72.6
rHSD								—

DD3, bovine liver dihydrodiol dehydrogenase 3 ; vPGFS, bovine liver prostaglandin F synthase ; gPGFS, bovine lung prostaglandin F synthase ; hBABP, human liver bile acid binding protein ; hDD4, human liver dihydrodiol dehydrogenase 4 ; hDD2, human liver dihydrodiol dehydrogenase 2 ; rbHSD, rabbit ovary 20α-hydroxysteroid dehydrogenase ; rHSD, rat liver 3α-hydroxysteroid dehydrogenase. The score numbers were expressed as percentages of the conserved amino acid.

with the linearized dd3-pBluescript KS (+) plasmid (pBS-dd3) as a template. The *Eco* RI-I I fragment of PCR product was inserted into pBluescript KS (+). The *Eco* RI-I I fragment of the construct was substituted with corresponding region of pBS-dd3. The *Eco* RI fragment of the resulting plasmid (pBS-dd3') was inserted into the *Eco* RI site of pKK223–3. This pKK-dd3' expression plasmid was introduced into various *E.coli* strains as listed in the Table 3. Cells were precultured at 37°C in L-Broth containing 2 % glucose with 100 µg/ml of ampicilin. After 50 fold dilution, the recombinant DD3 protein (rDD3) was induced for 16 h in the same medium containing 0.1 mM IPTG instead of glucose. The homogenate was prepared and its enzyme activity was determined with *trans*-naphthalene dihydrodiol, 1-acenaphthenol or taurochenodeoxycholic acid as a typical substrate. Since JM101 cells showed the highest activities and expression, we used this strain as a suitable host to express rDD3. Cells from 4 l culture were suspended in buffer A (10 mM Tris/HCl, pH 7.4, with 0.1 mM DFP, 1 mM EDTA and 5 mM DTT), and then sonicated

Table 3. Expression of rDD3 in various *E.coli* strains. The rDD3 was expressed in various *E.coli* strains and the enzyme activities were determined according to the standard assay system. The amounts of rDD3 was analyzed by Western-blotting using rabbit anti-DD3 anti serum (below)

Substrate	Specific activity (units/mg) $\times 10^3$								
	JM83	JM101	JM105	JM109	DH1	DH5α	TG1	TB1	LE392
NDD	4.7	5.4	2.7	2.6	2.4	3.5	4.4	3.3	3.5
1-Acenaphthenol	4.5	9.0	3.9	7.4	5.6	7.0	8.3	4.3	5.7
TCDCA	5.6	7.5	1.7	6.7	3.2	3.8	5.8	7.6	6.0

NDD, *trans*-naphthalenedihydrodiol; TCDCA, taurochenodeoxycholic acid.

```
DD3      1:MDPKGQRVKLNDGHFIPVLGFGTFAPREVPKSEALEVTKFAIEAGFRHIDSAHLYQNEEQ
vPGFS    1:MDPKSQRVKFNDGHFIPVLGFGTYAPEEVPKSEALEATKFAIEVGFRHVDSAHLYQNEEQ
gPGFS    1:MDPKSQRVKLNDGHFIPVLGFGTYAPEEVPKSEALEATKFAIEVGFRHVDSAHLYQNEEQ
hBABP    1:MDSKYQCVKLNDGHFMPVLGFGTYAPAEVPKSKALEATKLAIEAGFRHIDSAHLYNNEEQ
hDD4     1:--PKYQRVELNDGHFMPVLGFGTYAPPEVPRNRAVEVTKLAIEAGFRHIDSAYLYNNEEQ
rbHSD    1:MDPKFQRVALSDGHFIPVLGFGTYAPEEVPKSKAMEATKIAIDAGFRHIDSAYFYKNEKE
            *  *    **** ******* ** ***    *  * **  **   **** ***   *  **

DD3     61:VGQAIRSKIADGTVKREDIFYTSKLWSTSLRPELVRPALEKSLNNLQLDYVDLYIIHFPV
vPGFS   61:VGQAIRSKIADGTVKREDIFYTSKLWCNSLQPELVRPALEKSLQNLQLDYVDLYIIHSPV
gPGFS   61:VGQAIRSKIADGTVKREDIFYTSKLWCNSLQPELVRPALEKSLQNLQLDYVDLYIIHSPV
hBABP   61:VGLAIRSKIADGSVKREDIFYTSKLWCNSHRPELVRPALERSLKNLQLDYVDLYLIHFPV
hDD4    61:VGLAIRSKIADGSVKREDIFYTSKLWCTFFQPQMVQPALESSLKKLQLDYVDLYLLHFPM
rbHSD   61:VGLAIRSKIADGTVKREDIFYTSKLWCTFHRPELVRPSLEDSLKNLQLDYVDLYIIHFPT
            ** ********* ***********   *   * * **  **  ********   *  *

DD3    121:ALKPGETLFPTDENGKPIFDSVDLCRTWEALEKCKDAGLTKSIGVSNFNHKQLEKILNKP
vPGFS  121:SLKPGNKFVPKDESGKLIFDSVDLCHTWEALEKCKDAGLTKSIGVSNFNHKQLEKILNKP
gPGFS  121:SLKPGNKFVPKDESGKLIFDSVDLCHTWEALEKCKDAGLTKSIGVSNFNHKQLEKILNKP
hBABP  121:SVKPGEEVIPKDENGKILFDTVDLCATWEAVEKCKDAGLAKSIGVSNFNRRQLEMILNKP
hDD4   121:ALKPGETPLPKDENGKVIFDTVDLSATWEVMEKCKDAGLAKSIGVSNFNCRQLEMILNKP
rbHSD  121:ALKPGVEIIPTDEHGKAIFDTVDICATWEAMEKCKDAGLAKSIGVSNFNRRQLEMILNKP
            ***    *  **  **   ** **    ***    ******** *********   *** *****

DD3    181:GLKYKPVCNQVECHPYFNQSKLLDFCKSHDIVLVAYGALGSQRLKEWVNPNLPFLLEDPV
vPGFS  181:GLKYKPVCNQVECHPYLNQSKLLEFCKSHDIVLVAYAALGAQLLSEWVNSNNPVLLEDPV
gPGFS  181:GLKYKPVCNQVECHPYLNQSKLLEFCKSHDIVLVAYAALGAQLLSEWVNSNNPVLLEDPV
hBABP  181:GLKYKPVCNQVECHPYFNQRKLLDFCKSKDIVLVAYSALGSHREEPWVDPNSPVLLEDPV
hDD4   181:GLKYKPVCNQVECHPYLNQSKLLDFCKSKDIVLVAHSALGTQRHKLWVDPNSPVLLEDPV
rbHSD  181:GLKYKPVCNQVECHPYLNQGKLLEFCKSKGIVLVAYSALGSHREPEWVDQSAPVLLEDPL
            ***************   ** *** ***   *****  ***       **    * *****

DD3    241:?LSAIAKKHRQTPALVALRYQIQRGVVVLAKSYNKKRIKENIQVFDFELTPEDMKAIDGLN
vPGFS  241:LCAIAKKHKQTPALVALRYQVQRGVVVLAKSFNKKRIKENMQVFDFELTPEDMKAIDGLN
gPGFS  241:LCAIAKKHKQTPALVALRYQVQRGVVVLAKSFNKKRIKENMQVFDFELTPEDMKAIDGLN
hBABP  241:LCALAKKHKRTPALIALRYQLQRGVVVLAKSYNEQRIRQNVQVFEFQLTSEEMKAIDGLN
hDD4   241:LCALAKKHKRTPALIALRYQLQRGVVVLAKSYNEQRIRENIQVFEFQLTSEDMKVLDGLN
rbHSD  241:IGALAKKHQQTPALIALRYQLQRGIVVLAKSFTEKRIKENIQVFEFQLPSEDMKVIDSLN
            * ****  **** ****  ***.  *** ******       **  * *** * *   * **   * **

DD3    301:SNMRYNELLLGVGHPEYPFVEEY
vPGFS  301:RNTRYYDFQQGIGHPEYPFSEEY
gPGFS  301:RNIRYYDFQKGIGHPEYPFSEEY
hBABP  301:RNVRYLTLDIFAGPPNYPFSDEY
hDD4   301:RNYRYVVMDFLMDHPDYPFSDEY
rbHSD  301:RNFRYVTADFAIGHPNYPFSD--
              * **        *  ***
```

Figure 3. Multi-sequence alignment of members of aldo-keto reductases. Underlines and double underlines indicate high-homologous regions and aldo-keto reductase motifs, respectively. Conserved residues are asterisked. See legend to Table 2 for abbreviations of the enzyme names.

Table 4. Substrate specificities of rDD3 and DD3. Enzyme activities of rDD3 and DD3 were determined at the indicated concentration of substrate. The relative activities for *trans*-benzene dihydrodiol were shown

Substrate	Concentration (mM)	Relative activity (%)	
		rDD3	DD3
[Dehydrogenase]			
trans-Benzene dihydrodiol	2.0	100	100
trans-Naphthalene dihydrodiol	2.0	61	68
1-Acenaphthenol	0.5	230	200
(S)-(+)-1-Indanol	0.5	74	75
[Reductase]			
1S-(+)-Camphorquinone	1.0	10	8
1R-(+)-Camphorquinone	1.0	10	7
4-Benzoylpyridine	1.0	5	3
p-Nitrobenzaldehyde	1.0	3	5

(Model W-220F HEAT SYSTEM ULTRASONICS, Inc.). The supernatant of 20,000 x g for 30 min centrifugation was subjected to an ammonium sulfate fractionation. The 40–70 % precipitate was collected and dissolved in buffer B (5 mM Tris/HCl, pH 7.4, with 1 mM EDTA and 5 mM DTT). rDD3 was purified by successive column chromatographies on Ultrogel AcA 44 gel filtration, Q-Sepharose anion exchange and DyeMatrex Red A affinity. Three M NaCl eluate, being dialyzed against buffer B, was further purified on HA-Ultrogel column chromatography. The active peak fraction contained a homogeneous protein as determined by 12.5 % SDS-PAGE. The molecular size of purified rDD3 is the same as that of tissue enzyme (DD3).

The relative activities of rDD3 and DD3 for various substrates are essentially the same (Table 4). Furthermore, both enzymes show similar Km-values for the substrates (*trans*-benzene dihydrodiol, *trans*-naphthalene dihydrodiol, 1-acenaphthenol, (S)-(+)-1-indanol and taurochenodeoxycholic acid) and the coenzymes (NADP+ and NADPH).

These results suggest that the catalytic activities of rDD3 and DD3 are mutually indistinguishable. Not only in the substrate specificity but also in the inhibitor sentivity, rDD3 shows similarities to DD3: the steroids (androsterone, 5β-androstan-3α-ol-17-one, 5α-androstane-3,17-dione and 5b-androstane-3,17-dione) are the competitive inhibitors and gave similar Ki-values for rDD3 and DD3, they are known as typical substrates of 3α-hydroxysteroid dehydrogenase. These kinetic results strongly suggest that the structure of rDD3 purified from bacterial cells is essentially the same as that of DD3.

4. CONCLUSION

We succeeded in the cloning, expression and purification of bovine liver cytosolic dihydrodiol dehydrogenase 3. Sequence of dd3 reveals a high homology with those of aldo-keto reductases such as bovine liver and lung prostaglandin F synthases, human dihydrodiol dehydrogenases, human bile acid binding protein and rat 3α-hydroxysteroid dehydrogenase in the nucleotide and amino acid levels. Recombinant enzyme shows similar properties to the liver enzyme in the relative molecular mass, immunological cross reaction, substrate specificity and inhibitor sensitivity. Furthermore, the computer simulation on the basis of primary structure and crystallographic analyses of

Figure 4. Cartoon of the three-dimensional structures of rat liver 3α-hydroxysteroid dehydrogenase (rat HSD/DDH) (Hoog et al., 1994) and bovine liver dihydrodiol dehydrogenase 3 (bovine DD3). The structure was simulated with *Insight II* computer program.

human aldose reductase (Rondeau et al., 1992; Wilson et al., 1992) and rat 3α-hydroxysteroid dehydrogenase (Hoog et al., 1994) predicted that DD3 has the typical α/β barrel structure as shown in Fig. 4. Much more studies on the reaction mechanism, essential residues for the activity, higher ordered structure and gene expression mechanism are needed to clarify the physiological role of DD3 in detail.

5. ACKNOWLEDGMENTS

This work was supported in part by a grant-in-aid for Scientific Research from the Ministry of Education, Science, Sports and Culture of Japan (08672509) and a Grant from Institute of the Cataract Foundation in Japan.

6. REFERENCES

Bohren, K.M., Bullock, B., Wermuth, B. and Gabbay, K.H. (1989) The aldo-keto reductase superfamily. cDNA and deduced amino acid sequences of human aldehyde and aldose reductases. *J.Biol.Chem.* 264: 9547–9551.

Bolcsak, L.E. and Nerland, D.E. (1983) Purification of mouse liver benzene dihydrodiol dehydrogenases. *J.Biol.Chem.* 258: 7252–7255.

Cheng, K.C., White, P.C. and Qin,K.N. (1991) Molecular cloning and expression of rat liver 3α-hydroxysteroid dehydrogenase. *Mol.Endcrinol.* 5: 823–828.

Ciaccio, P.J. and Tew, K.D. (1994) cDNA and deduced amino acid sequences of a human colon dihydrodiol dehydrogenase. *Biochim.Biophys.Acta.* 1186: 129–132.

Deyashiki, Y., Ogasawara, A., Nakayama, T., Nakanishi, M., Miyabe, Y., Sato, K., and Hara, A. (1994) Molecular cloning of two human liver 3α-hydroxysteroid/dihydrodiol dehydrogenase isozymes that are identical with chlordecone reductase and bile-acid binder. *Biochem.J.* 299: 545–552.

Glatt, H.R., Vogel, K., Bentley, P. and Oesch, F. (1979) Reduction of benzo(a)pyrene mutagenicity by dihydrodiol dehydrogenase. *Nature* 277: 319–320.

Hara, A., Hasebe, K., Hayashibara, M., Matsuura, K., Nakayama, T. and Sawada, H. (1986) Dihydrodiol dehydrogenases in guinea pig liver. *Biochem.Pharmacol.* 35: 4005–4012.

Hara, A., Nakayama, T., Harada, T., Kanazu, T., Shinoda, M., Deyashiki, Y., and Sawada, H. (1991) Distribution and characterization of dihydrodiol dehydrogenases in mammalian ocular tissues. *Biochem.J.* 275: 113–119.

Hoog, S.S., Pawlowski, J.E., Alzari, P.M., Penning, T.M. and Lewis, M. (1994) Three-dimensional structure of rat liver 3 alpha-hydroxysteroid/dihydrodiol dehydrogenase: a member of the aldo-keto reductase superfamily. *Proc.Natl.Acad.Sci.U.S.A.* 91: 2517–2521.

Hou, Y.T., Xia, W., Pawlowski, J.E. and Penning, T.M. (1994) Rat dihydrodiol dehydrogenase: complexity of gene structure and tissue-specific and sexually dimorphic gene expression. *Cancer.Res.* 54: 247–255.

Klein, J., Thomas, H., Post, K., Wšrner, W. and Oesch, F. (1992) Dihydrodiol dehydrogenase activities of rabbit liver are associated with hydroxysteroid dehydrogenases and aldo-keto reductases. *Eur.J.Biochem.* 205: 1155–162.

Kuchinke, K., Barski, O., Watanabe, K. and Hayaishi, O. (1992) A lung type prostagladin F synthase is expressed in bovine liver : cDNA sequence and expression in *E.coli*. *Biochem.Biophys.Res.Commun.* 183: 1238–1246.

Lacy, W.R., Washenick, K.J., Cook, R.G., & Dunbar, B.S. (1993) Molecular cloning and expression of abundant rabbit ovarian protein with 20α-hydroxysteroid dehydrogenase activity. *Mol.Endocr.* 7: 58–66.

Matsuura, K., Hara, A., Nakayama, T., Nakagawa, M., Sawada, H. (1987) Purification and properties of two multiple forms of dihydrodiol dehydrogenase from guinea- pig testis. *Biochim.Biophys.Acta.* 912: 270–277.

Mizoguchi, T., Nanjo, H., Umemura, T., Nishinaka, T., Iwata, C., Imanishi, T., Tanaka, T., Terada, T. and Nishihara, T. (1992) A novel dihydrodiol dehydrogenase in bovine liver cytosol: purification and characterization of multiple forms of dihydrodiol dehydrogenase. *J.Biochem.(Tokyo)* 112: 523–529.

Nakagawa, M., Harada, T., Hara, A., Nakayama, T. and Sawada, H. (1989) Purification and properties of multiple forms of dihydrodiol dehydrogenase from monkey liver. *Chem.Pharm.Bull.* 37: 2852–2854.

Nanjo, H. Adachi, H., Aketa, M., Mizoguchi, T., Nishihara T. and Terada, T. (1995) Role of cysteine in the alteration of bovine liver dihydrodiol dehydrogenase 3 activity. *Biochem.J.* 310: 101–107.

Nanjo, H., Terada, T. Umemura, T. Nishinaka, T. Mizoguchi, T. and Nihishihara, T. (1992) Characterization of bovine liver cytosolic 3α-hydroxysteroid dehydrogenase and its aldo-keto reductase activity. *Int.J.Biochem.* 24: 815–820.

Pawlowski, J., Huizinga, M. and Penning, T.M. (1991) Cloning and sequncing of the cDNA for rat liver 3α-hydroxysteroid/dihydrodiol dehydrogenase. *J.Biol.Chem.* 266: 8820–8825.

Penning, T.M. (1993) Dihydrodiol dehydrogenase and its role in polycyclic aromatic hydrocarbon metabolism. *Chem.Biol.Interact.* 89: 1–34.

Penning, T.M. and Sharp, R.B. (1990) Characterization of dihydrodiol dehydrogenase in human liver and lung. *Carcinogenesis* 11: 1203–1208.

Qin, K.N., Khanna, M. and Cheng, K.C. (1994) Structure of a gene coding for human dihydrodiol dehydrogenase/bile acid-binding protein. *Gene* 149: 357–361.

Rondeau, J.M., Téte-Favier, F., Podjarny, A., Reymann, J.M., Barth, P., Biellmann, J.F. and Moras, D. (1992) Novel NADPH-binding domain revealed by the crystal structure of aldose reductase. *Nature* 355: 469–72.

Sato, S. (1993) Aldose reductase the major protein associated with naphthalene dihydrodiol dehydrogenase activity in rat lens. *Invest.Ophthalmol.Vis.Sci.* 34: 3172–3178.

Stolz, A., Rahimi-Kiani, M., Ameis, D., Chan, E., Ronk, M., and Shively, J.E. (1991) Molecular structure of rat hepatic 3α-hydroxysteroid dehydrogenase. A member of the oxidereductase gene family. *J.Biol.Chem.* 266: 15253–15257.

Terada, T., Kohno, T., Samejima, T., Hosomi, S., Mizoguchi, T. and Uehara, K. (1985) Purification and properties of beef liver aldehyde reductase catalyzing the reduction of D-erythrose 4-phosphate. *J.Biochem.* (Tokyo) 97: 79–87.

Terada, T., Niwase, N., Koyama, I., Imamura, M., Shinagawa, K., Toya, H. and Mizoguchi, T. (1993) Purification and characterization of carbonyl reductase from bovine liver cytosol and microsome. The cytosolic enzyme has a novel 3α/17β- hydroxysteroid dehydrogenase activity. *Int.J.Biochem.* 25: 12331239.

Vogel, K., Bentley, P., Platt, K.L. and Oesch, F. (1980) Rat liver cytoplasmic dihydrodiol dehydrogenase. Purification to apparent homogeneity and properties. *J.Biol.Chem.* 255: 9621–9625.

Watanabe, K., Fujii, Y., Nakayama, K., Ohkubo, H., Kuramitsu, S., Kagamiyama, H., Nakanishi, S. and Hayaishi, O. (1988) Structural similarity of bovine lung prostaglandin F synthase to lens e-crystalin of European common frog. *Proc.Natl.Acad.Sci.U.S.A.* 85: 11–15.

Wilson, D.K., Bohren, K.M., Gabbay, K.H. and Quiocho, F.A. (1992) An unlikely sugar substrate site in the 1.65 Å structure of the human aldose reductase holoenzyme implicated in diabetic complications. *Science* 257: 81–84.

63

SITE-DIRECTED MUTAGENESIS OF RESIDUES IN COENZYME-BINDING DOMAIN AND ACTIVE SITE OF MOUSE LUNG CARBONYL REDUCTASE

Masayuki Nakanishi,[1] Hiroyuki Kaibe,[2] Kazuya Matsuura,[2] Mikio Kakumoto,[2] Nobutada Tanaka,[3] Takamasa Nonaka,[3] Yukio Mitsui,[3] and Akira Hara[2]

[1]Department of Applied Chemistry, Faculty of Engineering
Gifu University, Gifu 501–11, Japan
[2]Biochemistry Laboratory
Gifu Pharmaceutical University
Gifu 502, Japan
[3]Department of BioEngineering
Nagaoka University of Technology
Niigata 940–21, Japan

1. INTRODUCTION

Tetrameric carbonyl reductase (CR) [EC 1.1.1.184] in mouse, guinea-pig and pig lungs catalyzes the reduction of various carbonyl compounds and the oxidation of secondary alcohols with NADP(H) as preferable coenzymes to NAD(H) (Nakayama, 1986; Matsuura, 1988; Oritani, 1992). The cDNAs for pulmonary CRs of pig and mouse have been cloned (Nakanishi, 1993; 1995), and the cDNA for mouse lung CR is identical with that for an mRNA which is increased in differentiation of murine adipocytes (Navre and Ringold, 1988). Their deduced amino acid sequences, composed of 244 residues (85% identity between them), are structurally related to members of the short-chain dehydrogenase/reductase (SDR) family, which includes a large number of prokaryotic and eukaryotic enzymes with different specificities for coenzymes and substrates (Jörnvall, 1995). The pulmonary CRs conserve two sequences, Gly-X-X-X-Gly-X-Gly and Tyr-X-X-X-Lys, which have been demonstrated to be in the coenzyme-binding domain and the active site, respectively, by site-directed mutagenesis (Jörnvall, 1995; and references cited therein) and X-ray crystallographic studies (Varughese, 1994; Ghosh, 1994; 1995; Tanaka, 1996b) of several NAD(H)-dependent SDR family proteins. The region around the former sequence forms a βαβ-fold which is characteristic of the coenzyme binding

Table 1. Comparison of partial sequences around the coenzyme binding fold of the
SDR family proteins, grouped according to their coenzyme specificity

Enzyme	Sequence from the numbered residue
NADP(H)-dependent	↓ ↓↓
Mouse lung CR	9 RALVTGAGKGIGRDTVKALHAS.GAKVVAVTRT
Pig lung CR	9 RALVTGAGKGIGRDTVKALHVS.GARVVAVTRT
Human CR	7 VALVTGGNKGIGLAIVRDLCRLFSGDVVLTARD
Rat CR	7 VALVTGANKGIGFAIVRDLCRQFAGDVVLTARD
Pig 20β-HSD	7 VALVTGANKGIGFAIVRDLCRKFLGDVVLTARD
Mouse 11β-HSD	36 KVIVTGASKGIGREMAYHLSKM.GAHVVLTARS
Tropinone R-I	23 TALVTGGSKGIGYAIVEELAGL.GARVYTCSRN
Tropinone R-II	11 TALVTGGSRGIGYGIVEELASL.GASVYTCSRN
Eubacterium 7α-HSD	7 VILVTASTRGIGLAIAQACAKE.GAKVYMGARN
Tetrahydroxynaphthalene R	30 VALVTGAGRGIGREMAMELGRRGCKVIVNYANS
Protochlorophyllide R	90 NVVVTGASSGLGLATAKALAETGKWNVIMACRD
NAD(H)-dependent	ββββ αααααααααααα ββββ
3α,20β-HSD	8 TVIITGGARGLGAEAARQAVAA.GARVVLADVL
E. coli 7α-HSD	13 CAIITGAGAGIGKEIAITFATA.GASVVVSDIN
Dihydropteridine R	9 RVLVYGGRGALGSRCVQAFRAR.NWWVASIDVV
Human 15-HPGD	7 VALVTGAAQGIGRAFAEALLLK.GAKVALVDWN
Drosophila ADH	9 VIFVAG.LGGIGLDTSKQLLKRDLKNLVILDRI

The consensus glycines are illustrated in bold face, and Lys-17, Thr-38 and Arg-39 of mouse
lung CR by arrows. The secondary structure of the βαβ-fold is indicated above the *Streptomyces*
3α,20β-hydroxysteroid dehydrogenase (3α,20β-HSD) sequence on the basis of its crystal
structure.   Abbreviations: HSD, hydroxysteroid dehydrogenase; R, reductase; 15-HPGD, 15-
hydroxyprostaglandin dehydrogenase; ADH, alcohol dehydrogenase.

fold in dehydrogenases of other families (Wierenga, 1985;1986). The Tyr residue in the
latter sequence plays the most essential role as the catalyst in the reaction mechanism, and
the side-chain of the Lys residue stabilizes the ionized form of the Tyr.

We have recently solved the three dimensional structure of the ternary complex of
mouse lung CR with NADPH and 2-propanol (Tanaka, 1996a). The comparison of crystal
structures between NADP(H)-preferring CR and NAD(H)-preferring SDR enzymes, together wiih sequence alignment (Table 1), has proposed a new aspect on coenzyme preference of the SDR family enzymes. The positively charged environment made by a pair of
basic residues (Lys-17 and Arg-39 in CR) is the determinant for the NADP(H)-preference,
whereas an Asp is conserved at the C-terminus of the second β-strand of the βαβ-fold of
the NAD(H)-dependent enzymes and is responsible for the coenzyme specificity. In
mouse lung CR, the Asp is substituted with Thr-38 which also interacts with 2'-phosphate
of NADPH. For the reaction mechanism of the enzyme, we also proposed a subsidiary
role of Ser-136 as a stabilizer of the reaction species in addition to the major roles of the
conserved Tyr and Lys residues (Tyr-149 and Lys-153, respectively, in mouse lung CR).

We undertook a series of studies employing site-directed mutagenesis to confirm the roles of Lys-17, Thr-38, Arg-39, Ser-136, Tyr-149 and Lys-153 in the coenzyme-binding and catalysis of CR.

2. MATERIALS AND METHODS

The wild-type mouse lung CR (WT) was expressed in *E. coli* (JM 105) from a plasmid pKK223–3 containing the cDNA encoding the enzyme as described (Nakanishi, 1995). Mutagenesis was performed according to the overlap-extension method (Horton, 1993), using *Pfu* polymerase and primer pairs which contained the codons of altered amino acids. Fifteen mutated CRs were expressed: K17H (Lys-17Å→His), K17R (Lys-17Å→Arg), K17S (Lys-17Å→Ser), T38D (Thr-38Å→Asp), R39A (Arg-39Å→Ala), R39D (Arg-39Å→Asp), KSRA (double mutation of Lys-17Å→Ser and Arg-39Å→Ala), KSTD (double mutation of Lys-17Å →Ser and Thr-38Å→Asp), S136A (Ser-136Å→Ala), S136C (Ser-136Å→Cys), Y149F (Tyr-149Å→Phe), Y149S (Tyr-149Å→Ser), Y149W (Tyr-149Å→Trp), K153H (Lys-153Å→His) and K153R (Lys-153Å→Arg). The production of the recombinant proteins in the *E. coli* cells was examined by Western blot analysis of the cell lysates using an antibody against mouse lung CR, and the recombinant proteins, except S136C, Y149S and Y149W, were purified essentially by the method for purification of WT (Nakanishi, 1995). Protein concentration of the purified enzymes was determined by Bradford's method (1976) using bovine serum albumin as the standard, and protein purity was examined by SDS/polyacrylamide gel electrophoresis. The fluorescence study of NADP(H) binding was carried out as described (Matsuura, 1988). The pyridine-3-aldehyde (P3A) reductase and cyclohex-2-en-1-ol (CHX) dehydrogenase activities of the enzymes were determined spectrophotometrically with saturating concentrations (3–200 × K_m) of NADP(H) or NAD(H) and the substrates in 0.1 M potassium phosphate, pH 7.0 (Nakanishi, 1995). One unit of enzyme activity was defined as the amount of enzyme catalyzing the formation and oxidation of 1 μmol of NAD(P)H per min at 25 °C.

3. RESULTS AND DISCUSSION

3.1. Residues in Coenzyme-Binding Domain

In addition to single mutations of Lys-17, Thr-38 or Arg-39, the double mutations of both Lys-17 and Thr-38 or Arg-39 were performed to ensure their roles in the coenzyme binding. The effects of the mutations on the kinetic constants for the NADP(H)-linked and NAD(H)-linked reactions are summarized in Table 2.

The replacement of Lys-17 with a basic residue, Arg, did not considerably affect the kinetic constants, but that with His or Ser resulted mainly in increases in the K_i and K_m for NADP(H). The mutagenesis of Arg-39 to Ala or Asp similarly affected the K_i and K_m values for NADP(H). The net effect of the combined mutations, KSRA, also appeared in the constants for NADP(H) which increased by 2 to 3 orders of magnitude. However, all the mutations produced no or little changes in the kinetic constants for the NAD(H)-linked activity. Therefore, it is evident that both Lys-17 and Arg-39 interact with the 2'-phosphate of NADP(H).

Arg-39 is well conserved in NADP(H)-dependent enzymes of the SDR family, but Lys-17 *is substituted* with Arg in several enzymes, such as *Datura* tropinone reductase (Nakajima, 1993), *Eubacterium* 7α-hydroxysteroid dehydrogenase (Baron, 1991) and

Table 2. Alteration of kinetic constants by mutations of Lys-17, Thr-38 and Arg-39

Constant	Value for WT	Ration of the constant for mutant enzyme to WT							
		K17H	K17R	K17S	R39A	R39D	KSRA	T38D	KSTD
K_i NADPH	1.2 µM	5.1	1.3	7.4	4.5	48	92	180	1000
K_m NADPH	1.0 µM	4.5	1.2	7.2	3.5	52	1000	380	710
K_m P3A	16 µM	10	2.1	14	1.5	8	ND	ND	ND
k_{cat} NADPH	1.2 s^{-1}	3.2	1.4	4.9	1.2	0.4	1.6	1.9	0.2
K_m NADH	60 µM	11	0.7	2.2	2.3	1.2	3.3	0.2	1.0
K_m P3A	550 µM	5.0	1.0	1.0	1.9	1.0	4.5	0.4	1.1
k_{cat} NADH	1.6 s^{-1}	3.7	1.5	2.1	0.4	0.3	0.3	1.9	2.2
K_i NADP$^+$	1.0 µM	30	6.8	92	18	500	ND	1900	ND
K_m NADP$^+$	3.0 µM	11	2.7	8.3	7.7	400	930	960	1200
K_m CHX	280 µM	1.4	1.2	2.0	1.1	2.5	3.2	3.4	10
k_{cat} NADP$^+$	2.4 s^{-1}	2.1	1.7	2.0	1.5	0.5	0.4	0.7	0.04
K_i NAD$^+$	3500 µM	1.2	1.6	0.6	1.1	1.0	ND	0.09	ND
K_m NAD$^+$	1800 µM	2.2	0.6	0.6	2.0	1.4	1.5	0.1	0.2
K_m CHX	700 µM	1.2	0.8	1.1	0.9	1.6	1.4	0.4	0.6
k_{cat} NAD$^+$	1.4 s^{-1}	1.1	1.5	1.3	1.5	0.9	0.5	2.5	2.2

NAD(P)$^+$ and NADPH were competitive inhibitors with respect to NADPH and NADP$^+$, respectively. ND, the value could not be determined because of high K_m value for NADPH.

Magnaporthe tetrahydroxynaphthalene reductase (Vidal-Cros, 1994) in Table 1. The almost silent effect of the K17R mutation on the kinetic constants for NADP(H) suggests that the presence of a positively charged residue, either Lys or Arg, at this position is important for the tight binding of NADP(H). Of the NADP(H)-dependent SDR family proteins, some enzymes, such as protochlorophyllide reductase (Baker, 1994) and tetrahydroxynaphthalene reductase in Table 1, lack one of the two basic residues at positions 17 and 39. The significant effect of KSRA, compared to that of the single mutation of Lys-17 or Arg-39, also suggests that the remaining one positively charged residue contributes to the coenzyme specificity of the two SDR family enzymes, although other residues of the enzymes would be involved in the interaction with the 2'-phosphate group.

Some NAD(H)-dependent enzymes such as 3α, 20β-hydroxysteroid dehydrogenase (Marekov, 1990) and *Drosophila* alcohol dehydrogenase (Thatcher and Sawyer, 1980) conserve an Arg residue at positions corresponding to Lys-17 and Arg-39, respectively, of mouse lung CR sequence (Table 1). For the enzymes, site-directed mutagenesis and crystallographic studies have shown that Asp (corresponding to the residue at position 38 of mouse lung CR sequence) confers the coenzyme specificity upon the two NAD(H)-dependent enzymes by forming hydrogen bonds to the hydroxyl groups of the adenine ribose of NAD(H) (Chen, 1991; Ghosh, 1994). Since the Asp of the NAD(H)-dependent SDR enzymes also occupies the spatially similar position to Thr-38 of mouse lung CR which hydrogen-bonds to the 2'-phosphate of NADPH through a water molecule (Tanaka, 1996a), Thr-38 was mutated to Asp. This single mutation resulted in switch of coenzyme specificity of CR from NADP(H) to NAD(H). The K_i and K_m for NADP(H) increased more than 180-fold and those for NAD(H) decreased to less than one fifth, without large change in the k_{cat}. The ratios of the K_m values for the NADP(H)-linked to the NAD(H)-linked activities increased by more than 4 orders of magnitude: The values of K_m (NADPH/NADH) were 0.017 and 42 for WT and T38D respectively, and the respective values of K_m (NADP$^+$/NAD$^+$) were 0.002 and 29. In addition, the combined mutation of Lys-17 and Thr-38 to Ser and Asp, respectively, converted NADP(H)-preferring CR to almost NAD(H)-dependent CR that retained the high NAD(H)-linked activities but showed much lower NADP(H)-linked activity.

Table 3. Alteration of kinetic constants by mutation of Ser-136, Tyr-149 or Lys-153

Constant	Value for WT	Ratio of the constant for mutant enzyme to WT			
		S136A	Y149F	K153R	K153H
K_m NADPH	1.0 μM	2	ND	5	75
K_m P3A	16 μM	200	ND	580	24
k_{cat} NADPH	1.2 s^{-1}	0.01	0.0003	0.2	0.03

ND, the value could not be determined because of the low activity, and the k_{cat} value of Y149F was calculated from the specific activity with 0.2 mM NADPH and 20 mM P3A as the coenzyme and substrate, respectively.

Thus, the present results not only confirmed the roles of Lys-17, Thr-38 and Arg-39 in NADP(H)-binding of mouse lung CR, but also provides clear evidence for the key role of Asp at the C-terminus of the second β-strand of the βαβ-fold in the NAD(H) specificity of the SDR enzymes. Of the SDR enzymes with the same coenzyme-binding fingerprint of Gly-X-X-X-Gly-X-Gly, the Asp is the most critical determinant for coenzyme preference for NAD(H) over NADP(H), and both the substitution of the Asp with other residue and presence of the basic residues are important for the NADP(H)-specificity, although other residue(s), like Thr-38 in mouse lung CR, may also be involved in the tight binding of NADP(H) depending on the respective enzymes.

3.2. Residues in Active Site

The three dimensional structure of the ternary complex of mouse lung CR has proposed a catalytic mechanism composed of the Ser-Tyr-Lys triad, in which Ser-136 and Tyr-149 act as a stabilizer of reaction species and base catalyst, respectively, and Lys-153 plays dual roles in the binding to the nicotinamide ribose moiety of NADPH and in the interaction with Tyr-149 (Tanaka, 1996a). Ser136 was replaced with Ala or Cys, Tyr-149 with Phe, Try or Ser, and Lys-153 with Arg or His. While the production of all the recombinant enzymes in the lysates of *E. coli* cells was confirmed with Western blot analysis, the enzyme activity was detected only for K153R and K153H. In addition to the two enzymatically active enzymes, S136A and Y149F were purified by detecting the proteins on SDS/polyacrylamide gel electrophoresis.

When the P3A reductase activity was assayed with high concentrations of the substrate (20 mM) and the purified enzymes, it was detected for S136A and Y149F, and the respective k_{cat} values were 1% and 0.03%, respectively, of the that for WT (Table 3). The S136A mutation increased the K_m for P3A 200-fold with little change in the K_m for NADPH, which supports the critical role of Ser-136. Although the kinetic constants of Y149F could not be determined because of its low activity, the binding of NADP(H) was evidenced by the fluorescence study: The protein fluorescence at 336 nm (excitation at 280 nm) of the enzyme was quenched by the addition of NADP(H), and that of NADPH was shifted from 460 nm (excitation at 340 nm) to 453 nm with enhancement of the fluorescence intensity by the addition of the enzyme, as reported for the native enzyme from guinea pig lung (Matsuura, 1988). The residual activity of Y149F seems to be inconsistent with the essential role of Tyr-149 shown by the crystallographic study. However, such a low activity has also been detected by the mutation of the conserved Tyr to Ala for human 17β-hydroxysteroid dehydrogenase (Puranen, 1994) and by that to Phe for rat dihydropteridine reductase (Kiefer, 1996), in which the role of the conserved Tyr as the catalytic residue has been shown by crystallographic studies (Ghosh, 1995; Varughese, 1994). The Y149F mutation may alter the orientation of the functional residues in the active site, and another residue or a water molecule might act as the proton donor.

The mutations of K153H and K153R also resulted in a large decrease in the k_{cat}, but only K153H increased the K_m for NADPH significantly. When the CHX dehydrogenase activity was assayed at pH range from 6.0 to 11.0, K153R, similarly to WT, showed a broad pH optimum at pH 10, but the activity of K153H was detected above pH 9.5 and increased up to pH 11.0. The increase in the K_m for the coenzyme and the shift of the pH optimum by the K153H mutation support the dual roles of Lys-153 in the binding of NADPH and in lowering the pK_a value of Tyr-149. The replacement of Lys-153 with the positively charged Arg may not considerably affect the binding of the coenzyme, but must influence the environment in the active site as evidenced by the large increase in the K_m for the substrate.

4. ACKNOWLEDGMENT

This research was supported by a Grant-in-Aid for Encouragement of Young Scientists from the Ministry of Education, Science and Culture of Japan.

5. REFERENCES

Baker, M.: Protochlorophyllide reductase is homologous to human carbonyl reductase and pig 20β-hydroxysteroid dehydrogenase. Biochem. J. 300 (1994) 605–607.

Bradford, M.M.: A rapid and sensitive method for the quantitation of microgram quantities of protein utilizing the principle of protein-dye binding. Anal. Biochem. 72 (1976) 248–254.

Baron, S.F., Franklund, C.V. and Hylemon, P.B.: Cloning, sequencing, and expression of the gene coding for bile acid 7α-hydroxysteroid dehydrogenase from *Eubacterium sp.* strain VPI 12708. J. Bacteriol. 173 (1991) 4558–4569.

Chen, Z., Lee, W.R. and Chang, S.H.: Role of aspartic acid 38 in the cofactor specificity of *Drosophila* alcohol dehydrogenase. Eur. J. Biochem. 202 (1991) 263–267.

Ghosh, D., Pletnev, V.Z., Zhu, D.-W., Wawrzak, Z., Duax, W.L., Pangborn, W., Labrie, F. and Lin, S.-X.: Structure of human estrogenic 17β-hydroxysteroid dehydrogenase at 2.20 Å resolution. Structure 3 (1995) 503–513.

Ghosh, D., Wawrzak, Z., Weeks, C.M., Duax, W.L. and Erman, M.: The refined three-dimensional structure of 3α, 20β-hydroxysteroid dehydrogenase and possible roles of the residues conserved in short-chain dehydrogenases. Structure 2 (1994) 629–640.

Horton, R.M.: In vitro recombination and mutagenesis of DNA. In White, B.A. (Ed.), PCR Protocols, Humana Press, New Jersey, 1993, pp. 251–261.

Jörnvall, H., Persson, B., Krook, M. and Atrian, S., Gonzalez-Duarte, R., Jeffery, J. and Ghosh, D.: Short-chain dehydrogenases/reductases (SDR). Biochemistry 34 (1995) 6003–6013.

Kiefer, P.M., Varughese, K.I., Su, Y., Xuong, N.-H., Chang, C.F., Gupta, P., Bray, T. and Whiteley, J.M.: Altered structural and mechanistic properties of mutant dihydropteridine reductases. J. Biol. Chem. 271 (1996) 3437–3444.

Marekov, L., Krook, M. and , Jörnvall, H.: Prokaryotic 20β-hydroxysteroid dehydrogenase is an enzyme of the short-chain, non-metalloenzyme alcohol dehydrogenase type. FEBS Lett. 266 (1990) 51–54.

Matsuura, K., Nakayama, T., Nakagawa, M., Hara, A. and Sawada, H.: Kinetic mechanism of pulmonary carbonyl reductase. Biochem. J. 252 (1988) 17–22.

Nakajima, K., Hashimoto, T. and Yamada, Y.: Two tropinone reductases with different stereospecificities are short-chain dehydrogenases evolved from a common ancestor. Proc. Natl. Acad. Sci. USA 90 (1993) 9591–9595.

Nakanishi, M., Deyashiki, Y., Nakayama, T., Sato, K. and Hara, A.: Cloning and sequence analysis of a cDNA encoding tetrameric carbonyl reductase of pig lung. Biochem. Biophys. Res. Commun. 194 (1993) 1311–1316.

Nakanishi, M., Deyashiki, Y., Ohshima, K. and Hara, A.: Cloning, expression and tissue distribution of mouse tetrameric carbonyl reductase. Identity with an adipocyte 27-kDa protein. Eur. J. Biochem. 228 (1995) 381–387.

Nakayama, T., Yashiro, K., Inoue, Y., Matsuura, K., Ichikawa, H., Hara, A. and Sawada, H.: Characterization of pulmonary carbonyl reductase of mouse and guinea pig. Biochim. Biophys. Acta 882 (1986) 220–227.

Navre, M. and Ringold, G.M.: A growth factor-repressible gene associated with protein kinase C-mediated inhibition of adipocyte differentiation. J. Cell Biol. 107 (1988) 279–286.

Oritani, H., Deyashiki, Y., Nakayama, T., Hara, A., Sawada, H., Matsuura, K., Bunai, Y. and Ohya, I.: Purification and characterization of pig lung carbonyl reductase. Arch. Biochem. Biophys. 292 (1992) 539–547.

Puranine, T.J., Poutanine, M.H., Peltoketo, H.E., Vihko, P.T. and Vihko, R.K.: Site-directed mutagenesis of the putative site of human 17β-hydroxysteroid dehydrogenase type 1. Biochem. J. 304 (1994) 289–293.

Tanaka, N., Nonaka, T., Nakanishi, M., Deyashiki, Y., Hara, A. and Mitsui, Y.: Crystal structure of the ternary complex of mouse lung carbonyl reductase at 1.8 Å resolution: the structural origin of coenzyme specificity in the short-chain dehydrogenase/reductase family. Structure 4 (1996a) 33–45.

Tanaka, N., Nonaka, T., Tanabe, T., Yoshimoto, T., Tsuru, D. and Mitsui, Y.: Crystal structure of the binary and ternary complexes of 7α-hydroxysteroid dehydrogenase from *Escherichia coli*. Biochemistry 35 (1996b) 7715–7730.

Thatcher, D.R. and Sawyer, L.S.: Secondary-structure prediction from the sequence of *Drosophila* melanogaster (fruitfly) alcohol dehydrogenase. Biochem. J. 187 (1980) 884–886.

Varughese, K.I., Xuong, N.H., Kiefer, P.M., Matthews, D.A. and Whiteley, J.M.: Structural and mechanistic characterization of dihydropteridine reductase: A member of the Tyr-(Xaa)$_3$-Lys-containing family of reductases and dehydrogenases. Proc. Natl. Acad. Sci. USA 91 (1994) 5582–5586.

Vidal-Cros, A., Viviani, F., Boccara, M. and Gaudry, M.: Polyhydroxynaphthalene reductase involved in melanin biosynthesis in *Magnaporthe grisea*. Eur. J. Biochem. 219 (1994) 985–992.

Wierenga, P.K., De Maeyer, M.C.H. and Hol, W.G.J.: Interaction of pyrophoshate moieties with a helixes in dinucleotide binding proteins. Biochemistry 24 (1985) 1346–1357.

Wierenga, P.K., Terpstra, P. and Hol, W.G.J.: Prediction of the occurrence of the ADP-binding βαβ-fold in proteins, using an amino acid sequence fingerprint. J. Mol. Biol. 187 (1986) 101–107.

64

AN ESSENTIAL CYSTEINE (CYS-227) IN HUMAN CARBONYL REDUCTASE IS INVOLVED IN GLUTATHIONE BINDING

Josiane Tinguely, Elsbeth Ernst, and Bendicht Wermuth

Department of Clinical Chemistry
University of Berne
Inselspital, 3010 Berne, Switzerland

1. INTRODUCTION

Cytosolic carbonyl reductase (secondary-alcohol:NADP$^+$ oxidoreductase, EC 1.1.1.184) is a monomeric member of the short-chain dehydrogenase/reductase superfamily (Wermuth, 1992). It catalyzes the NADPH-dependent reduction of a variety of endogenous and xenobiotic carbonyl compounds, including prostaglandins, steroids and pterins as well as many quinones derived from polycyclic aromatic hydrocarbons and the anthracycline antibiotic daunorubicin (Ahmed et al., 1978; Wermuth, 1981; Jarabak et al., 1983; Nakayama et al., 1985; Park et al., 1991). Its natural substrate(s) and physiological role, however, are not known. Evidence from various studies suggests that glutathione (GSH) may act as a cofactor or modulator of enzyme activity and specificity. Cagen and Pisano (1979) first showed that prostaglandin A$_1$, which by itself is not a substrate, is efficiently metabolized in the form of its GSH adduct by prostaglandin 9-keto reductase (identical to carbonyl reductase) from chicken heart and suggested that the natural substrates of the enzyme may be the GSH conjugates of carbonyl compounds. A similar preference for the GSH adduct of prostaglandin A$_1$ was subsequently demonstrated for carbonyl reductase from various mammalian sources (Toft and Hansen, 1979; Chang and Tai, 1981) including man (Wermuth, 1981; Feldman et al., 1981). More recently, the GSH adducts of several quinones of polycyclic aromatic hydrocarbons were found to be both substrates and inhibitors of human placental carbonyl reductase (Chung et al., 1987a). A specific GSH binding site close to, or overlapping with, the active site was postulated by Jarabak and coworkers based on the observation that oxidized glutathione (GSSG) and other oxidation products of GSH are potent inhibitors of carbonyl reductase from human placenta (Feldman et al., 1981; Chung et al., 1987b).

Little is known about the structural properties of the GSH binding site. Using glutathione thiosulfonate which covalently reacts with sulfhydryl groups as an affinity ligand, Jarabak and coworkers identified a cysteine residue at the GSH binding site of human placental carbonyl reductase which was essential for enzyme activity (Chung et al., 1987c). The presence of an essential cysteine residue in human carbonyl reductase (designated as aldehyde reductase 1 by the authors) was first suggested by Ris and von Wartburg (1973) based on the observation that incubation of the partially purified enzyme from brain with 4-OH-mercuribenzoate (pMB) abolished the enzyme activity. We subsequently showed that the coenzyme and substrate do not protect the enzyme from inactivation, locating the residue outside of the active site (Wermuth, 1981). The relationship between this residue and the residue identified by Jarabak's group has not been established.

Carbonyl reductase from all species whose sequence has so far been determined, i.e. man (Wermuth et al. 1988), pig (Tanaka et al., 1992), rat (Wermuth et al., 1995) and rabbit (Gonzales et al. 1995), contains five conserved cysteine residues at positions 26, 122, 150, 226 and 227. In order to investigate which of the five residues is (are) essential for enzyme activity, we replaced each one of them by alanine. For this purpose, the cDNA for human placental carbonyl reductase was mutated by site-directed mutagenesis and the mutant enzyme expressed in E. coli. In addition, the stoichiometry of carbonyl reductase inactivation by pMB was investigated. The results indicate that a single residue, Cys-227, is essential for enzyme activity and is also needed for GSH binding.

2. EXPERIMENTAL PROCEDURES

2.1. Site-Directed Mutagenesis and Expression of the Mutant Enzymes

Recombinant human carbonyl reductase was obtained by overexpression in E.coli BL21(De3) containing the cDNA in the expression vector pET-11a (Novagen), followed by chromatography on DEAE-cellulose and 2'-ADPR-Sepharose (Bohren et al., 1994). Mutations replacing Cys by Ala were created by the PCR-mediated, primer-directed method of Higuchi et al. (1988). For the first PCR the sense and antisense primers containing the desired mutation were paired with primers distal to unique restriction sites: XbaI (pET11a-397) and KpnI (cDNA-353) for C26, KpnI and NcoI (cDNA-792) for the

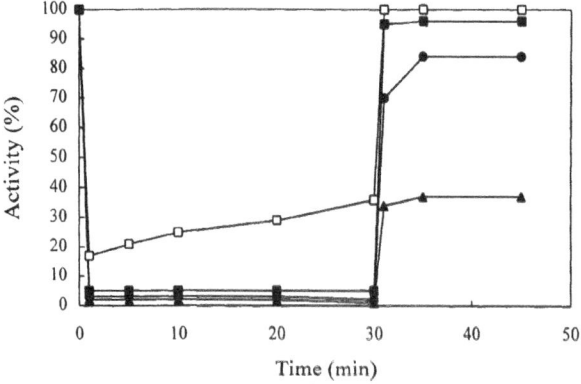

Figure 1. Inactivation of recombinant human carbonyl reductase by 4-OH-mercuribenzoate. Carbonyl reductase (1.5 μM) was incubated with 1 (□), 2 (■), 4 (●) and 16 (▲) equivalents, respectively, of 4-OH-mercuribenzoate. After 30 min, 1 mM dithiothreitol was added. At the times indicated by the symbols aliquots of < 1% of the assay mixture were withdrawn and analyzed for enzyme activty.

other cysteine residues. Amplimers obtained in the second PCR were digested with KpnI and XbaI or NcoI and ligated into correspondingly digested wild-type plasmid, and the mutant enzymes were overexpressed in E. coli and purified as described for the wild-type enzyme.

2.2. Enzyme Assay and Inactivation

Carbonyl reductase activity was determined spectrophotometrically by recording the decrease of NADPH absorbance at 340 nm and 30°C. The assay mixture consisted of 80 mM sodium phosphate buffer (pH 7.0), 0.05 mM NADPH and 0.25 mM menadione. Reactions were initiated by the addition of enzyme.

Modification of cysteine residues by pMB was carried out in 50 mM Tris phosphate (pH 7.5) at 25°C and reagent and enzyme concentrations as indicated in the text. In reactivation experiments, 1 mM dithiothreitol (DTT) was added to the reaction mixture after the time indicated in the text. At intervals, aliquots were withdrawn from the incubation mixture and immediately assayed for enzyme activity. Controls containing all components except pMB were routinely included and set as 100 % activity at each time point.

3. RESULTS AND DISCUSSION

3.1. Inactivation by pMB

Recombinant human carbonyl reductase was incubated with 1, 2, 4 and 16 equivalents of pMB, respectively, and the activity was determined at specified times as indicated in Fig. 1. One equivalent of pMB abolished the enzyme activity by about 85 %. This number further increased to 95 % and more at twice equimolar and higher concentrations of pMB. With all concentrations of pMB, the loss of enzyme activity was essentially complete within 1 min. Longer incubation periods resulted in partial recovery of the enzyme treated with 1 equivalent of pMB, whereas no recovery or even a further decrease of activity was observed when 2 or more equivalents of pMB had been added. Addition of 1 mM dithiothreitol (DTT) to the inactivated enzyme restored the enzyme activity. Essentially complete reactivation was observed at all pMB concentrations if DTT was added 5 minutes after the addition of pMB. Similarly, enzyme treated with 1 or 2 equivalents of pMB showed almost complete recovery if DTT was added after 30 minutes. In contrast, the recovery after 30 minutes was only 80 and 35 % for the enzyme incubated with 4 and 16 equivalents of pMB, respectively.

The results suggest the presence of a single essential cysteine residue which is highly susceptible to modification by pMB. Secondary events, e.g. conformational changes, and/or the modification of additional cysteine residues probably lead to irreversible inactivation at higher pMB concentrations and extended incubation times.

3.2. Effect of Cysteine Replacement

In order to investigate which of the five cysteine residues is essential, each one of them was replaced by alanine by site-directed mutagenesis. The mutant enzymes were expressed in E. coli and purified to homogeneity by DEAE-cellulose and 2'-ADPR-Sepharose affinity chromatography. Four of the the five mutant enzymes, C26A, C122A, C150A and C226A, showed normal enzyme activity. In contrast, the specific activity of

Figure 2. Isoelectric focusing of C227A (lanes 1–3) and wild-type (lanes 4–6) human recombinant carbonyl reductase. Isoelectric focusing on thin-layer polyacrylamide gels (Pharmacia) in the pH range 3.5–9 was carried out as described by the manufacturer. Protein bands were made visible by silver staining. Lanes 1 and 4 controls; lanes 2 and 5, + 0.1 mM NADPH; lanes 3 and 6, + 0.1 mM NADPH, 10 mM GSH.

the fifth mutant, C227A, was only 2 % of that of the wild-type enzyme under standard assay conditions. The Michaelis constants for NADPH and menadione were increased 4- and 12-fold, respectively, relative to the wild-type enzyme.

To investigate the effect of cysteine replacement on the GSH binding site, we took advantage of an earlier observation that binding of GSH to the binary enzyme-coenzyme complex shifts the isoelectric point from pH 8.5 to about 6 (Bohren et al., 1994). Fig. 2 shows the migration pattern of C227A and wild-type carbonyl reductase after isoelectric focusing on a thin-layer polyacrylamide gel in the absence and presence of GSH and coenzyme. In agreement with our earlier results, the isoelectric point of the wild-type enzyme was shifted to about pH 6 in the presence of NADPH and GSH. Similarly, the mutant enzymes C26A, C122A, C150A and C226A focused in the acidic pH range in the presence of GSH and NADPH (not shown). In contrast, no pH shift was detectable with C227A, suggesting that Cys-227 is needed for GSH binding.

Further evidence of GSH binding to carbonyl reductase and the involvement of a cysteine residue was recently obtained in a collaborative study with Joe DePierre and coworkers in Stockholm who had shown that an enzyme from rat ovaries with both GSH S-transferase and carbonyl reductase activities and high sequence similarity to human carbonyl reductase tightly binds to GSH-Sepharose (Toft et al., 1994). Together with DePierre's laboratory we showed that the GSH S-transferase activity was due to contamination by classical GSH S-transferases and confirmed the identity of the enzyme with carbonyl reductase (unpublished results). In the course of these studies we found that recombinant human carbonyl reductase tightly binds to GSH-Sepharose. No elution of the enzyme was detectable by the addition of 2 mM DTT, 10 μM NADP$^+$, 1 M NaCl and 5 mM GSH, respectively to the standard elution buffer consisting of 100 mM Tris/Cl (pH 9.1). Elution was finally achieved by a mixture of 5 mM each of GSH and S-hexylglutathione. Unlike the native enzyme, the C227A mutant was not retained on a GSH-Sepharose column confirming the involvement of Cys-227 in GSH binding. The significance of this reaction, however is unclear, as is the role of Cys-227 for enzyme function. In the primary structure, the cysteine is adjacent to a proline residue (Pro-228) which is highly conserved in more than 70 members of the short-chain dehydrogenase/reductase superfamily (Jörnvall et al., 1995) and is thought to be involved in substrate orientation and binding. A similar function of the cysteine preceding the proline residue in the primary structure would explain the large increase in the Km for menadione of the C227A mutant. Further work, however, will be necessary to elucidate its effect on enzyme activity and the role of GSH for enzyme function.

4. ACKNOWLEDGMENTS

This work was supported by grants from the Swiss National Science Foundation and the Foundation for the Advancement of Scientific Research at the University of Berne.

5. REFERENCES

Ahmed, N.K., Felsted, R.L. and Bachur, N.R.: Heterogeneity of anthracycline antibiotic carbonyl reductases in mammalian livers. Biochem. Pharmacol. 27 (1978), 2713–2719.

Bohren, K.M., Wermuth, B., Harrison, D., Ringe, D., Petsko, G.A. and Gabbay, K.H.: Expression, crystallization and preliminary crystallographic analysis of human carbonyl reductase. J. Mol. Biol. 244 (1994) 659–664.

Cagen, L.M. and Pisano, J.J.: The glutathione conjugate of prostaglandin A_1 is a better substrate than prostaglandin E for partially purified avian prostaglandin E 9-ketoreductase. Biochim. Biophys. Acta 573 (1979) 547–551.

Chang, D.G.-B. and Tai, H.-H.: Prostaglandin 9-ketoreductase/type II 15-hydroxyprostaglandin dehydrogenase is not a prostaglandin specific enzyme. Biochem. Biophys. Res. Commun. 101 (1981) 898–904.

Chung, H., Harvey, R.G., Armstrong, R.N. and Jarabak, J.: Polycyclic aromatic hydrocarbon quinones and glutathione thioethers as substrates and inhibitors of the human placental NADP-linked 15-hydroxyprostaglandin dehydrogenase. J. Biol. Chem. 262 (1987a) 12448–12451.

Chung, H.. Fried, J., Williams-Ashman, E. and Jarabak, J.: Glutathione mixed disulfide inhibitors of the human placental NADP-linked 15-hydroxyprostaglandin dehydrogenase. Prostaglandins 33 (1987b) 383–390.

Chung, H., Fried, J. and Jarabak, J.: Irreversible inhibition of the human placental NADP-linked 15-hydroxyprostaglandin dehydrogenase/9-ketoprostaglandin reductase by glutathione thiosulfonate. Prostaglandins 33 (1987c) 391–402.

Feldman, R., Luncsford, A., Heinrikson, R.L., Westley, J. and Jarabak, J.: Glutathione-related inhibition of prostaglandin metabolism. Arch. Biochem. Biophys. 211 (1981) 375–381.

Gonzalez, B., Sapra, A., Rivera, H., Kaplan, W.D., Yam, B. and Forrest, G.L.: Cloning and expression of the cDNA encoding rabbit liver carbonyl reductase. Gene 154 (1995) 297–298.

Higuchi, R., Krummel, B. and Saiki, R.K.: A general method of in vitro preparation and specific mutagenesis of DNA fragments: Study of protein and DNA interaction. Nucl. Acids Res. 16 (1988) 7351–7367.

Jarabak, J., Luncsford, A. and Berkowitz, D.: Substrate specificity of three prostaglandin dehydrogenases. Prostaglandins 26 (1983) 849–868.

Jörnvall, H., Persson, B., Krook, M., Atrian, S., Gonzàlez-Duarte, R., Jeffery, J. and Ghosh, D.: Short-chain dehydrogenases/reductases (SDR). Biochemistry 34 (1995) 6004–6013.

Nakayama, T., Hara, A., Yashiro, K. and Sawada H.: Reductases for carbonyl compounds in human liver. Biochem. Pharmacol. 34 (1985) 107–117.

Park, Y.S., Heizmann, C.W., Wermuth, B., Levine, R.A., Steinerstauch, P., Guzman, J. and Blau, N.: Human carbonyl and aldose reductases: New catalytic functions in tetrahydrobiopterin biosynthesis. Biochem. Biophys. Res. Commun. 175 (1991) 738–744.

Ris, M.M. and von Wartburg, J.-P.: Heterogeneity of NADPH-dependent aldehyde reductase from human and rat brain. Eur. J. Biochem. 37 (1973) 69–77.

Tanaka, M., Ohno, S., Adachi, S., Nakajin, S., Shinoda, M. and Nagahama, Y.: Pig testicular 20ß-hydroxysteroid dehydrogenase exhibits carbonyl reductase-like structure and activity. J. Biol. Chem. 267 (1992) 13451–13455.

Toft, B.S. and Hansen, H.S.: Metabolism of prostaglandin E_1 and of glutathione conjugate of prostaglandin A_1 (GSH-prostaglandin A_1) by prostaglandin 9-ketoreductase from rabbit kidney. Biochim. Biophys. Acta 574 (1979) 33–38.

Toft, E., Söderström, M., Bengtsson, M. and DePierre, J.W.: A novel 34 kDa glutathione-binding protein in mature rat ovary. Biochem. Biophys. Res. Commun. 201 (1994) 149–154.

Wermuth, B.: Purification and properties of an NADPH-dependent carbonyl reductase from human brain: relationship to prostaglandin 9-ketoreductase and xenobiotic ketone reductase. J. Biol. Chem. 256 (1981) 1206–1213.

Wermuth, B.: NADP-dependent 15-hydroxyprostaglandin dehydrogenase is homologous to NAD-dependent 15-hydroxyprostaglandin dehydrogenase and other short-chain alcohol dehydrogenases. Prostaglandins 44 (1992) 5–9.

Wermuth, B., Bohren, K.M., Heinemann, G., von Wartburg, J.P. and Gabbay, K.H.: Human carbonyl reductase: Nucleotide sequence analysis of a cDNA and amino acid sequence of the encoded protein. J. Biol. Chem. 263 (1988) 16185–16188.

Wermuth, B., Mäder Heinemann, G. and Ernst, E.: Cloning and expression of carbonyl reductase from rat testis. Eur. J. Biochem. 228 (1995) 473–479.

65

CARBONYL REDUCING 17β-HYDROXYSTEROID DEHYDROGENASE FROM THE FILAMENTOUS FUNGUS *Cochliobolus lunatus*

Tea Lanišnik Rižner,[1] Matjaž, Zorko,[1] Jasna Peter-Katalinić,[2] Kerstin Strupat,[3] and Marija Žakelj-Mavrič[1]

[1]Institute of Biochemistry, Medical Faculty
University of Ljubljana
Vrazov trg 2, 1000 Ljubljana, Slovenia
[2]Physiologisch-Chemisches Institut
Rheinische Friedrich-Wilhelms-Universitaet
Nussallee 11, D-53115 Bonn, Germany
[3]Institut fuer medizinische Physik und Biophysik
Universitaet Muenster
Robert-Koch-Strasse 31
D-48149 Muenster, Germany

1. INTRODUCTION

Hydroxysteroid dehydrogenases (HSDs) which catalyse reversible oxidoreduction of ketosteroids and their respective hydroxysteroids have often been found to have additional substrate specificities towards nonsteroidal ketones, aldehydes and quinones (Maser, 1995). In mammals, such pluripotent enzymes were found in peripheral (Boutin, 1986; Hara et al., 1986; Hara et al. 1986a; Sawada et al., 1988; Klein et al., 1992; Maser and Bannenberg, 1994; Deyashiki et al., 1994; Ohara et al., 1994; Maser, 1995; Oppermann et al., 1995) as well as steroidogenic tissues (Tanaka et al., 1992; Nakajin et al., 1994; Jarabak et al., 1996). They were supposed to be involved in the detoxification of xenobiotic carbonyl compounds in addition to their role in the metabolism of endogenous steroids, prostaglandins, and quinones (Maser and Bannenberg, 1994; Oppermann et al., 1995). In microorganisms it has been shown for some bacterial HSDs that they have the versatility to accept steroidal as well as non-steroidal substrates (Oppermann et al., 1993; Maser, 1995). On the other hand, pluripotency of fungal HSD has not been discovered. Recently, 17β-hydroxysteroid dehydrogenase (17β-HSD) from the fungus *Cochliobolus lunatus* was purified and some of its molecular and kinetic properties were characterized (Lanišnik Rižner et al., 1996). In the present paper we provide further characterization of this enzyme and also some data indicating that it is a fungal enzyme with carbonyl reductase activity.

Enzymology and Molecular Biology of Carbonyl Metabolism 6
edited by Weiner *et al.* Plenum Press, New York, 1996

2. MATERIALS AND METHODS

2.1. Fungal Species

Cochliobolus lunatus m 118 was obtained from the strain collection of the Friedrich Schiller University of Jena, Germany. It was grown as already described (Plemenitaš et al., 1988).

2.2. 17β-HSD Purification

After 42 hours of growth the mycelium of *Cochliobolus lunatus* was filtered, frozen in liquid nitrogen, homogenized, and resuspended in 50 mM Tris/HCl buffer, pH 8.0; 0.1mM EDTA, 0.1mM dithiothreitol (DTT), 20% glycerol (100g of mycelium/ 200ml of buffer). The cytosol was prepared by successive centrifugation as previously described (Lanišnik Rižner et al., 1996). Ammonium sulphate precipitation yielded a concentrated enzyme preparation which was used either for further purification or for chemical modification of the enzyme and kinetic studies. The purification procedure also included gel chromatography and affinity chromatography. After gel chromatography on Sephacryl S 300, the dialyzed enzyme preparation was applied to a 2'5'ADP Sepharose column, and the purified enzyme was eluted with 0.2 M NaCl in 10 mM Tris/HCl buffer, pH 8.0; 0.1 mM EDTA, 0.1 mM DTT, 20% glycerol (Lanišnik Rižner et al., 1996).

The purification procedure was performed at 4° C. Proteins were determined by measuring absorbance at 280 nm or by the bicinchoninic acid (BCA) assay (Brown et al., 1989).

2.3. Enzyme Assays

The enzyme activity was determined using either the chromatographic method during the purification procedure and chemical modification of the enzyme or spectrophotometric method during the kinetic studies (Lanišnik Rižner et al., 1996).

2.4. Molecular Mass Determination

UV-MALDI mass spectra were obtained on a home-built single-stage reflectron time-of flight instrument. Ions were extracted with 12kV acceleration potential under soft-extraction conditions (400V/mm). The mass spectrometer was equipped with a postacceleration detector at -15kV postacceleration followed by a secondary electron multiplier (EMI 9643, Electron Tube Ltd.) as described previously (Hillenkamp et al., 1991). A frequency-tripled Nd-YAG laser (J:K: Laser Ltd., Rugby, Warwickshire, England) was used at a wavelength of 355 nm and the pulse duration was 15 ns. Detector signals were preamplified by a factor of ten and digitized by a LeCroy 9400 (Chesnut Ridge, N.Y.USA) at 10–80 ns time intervals depending on the mass range covered. Four single spectra were accumulated to increase the signal to noise ratio. Positively charged ions were recorded.

The DHB matrix used was obtained by mixing the solutions of 2,5-dihydroxybenzoic acid (2,5-DHB) and 2-hydroxy-5-methoxybenzoic acid in a ratio of 9:1 in acetonitrile/water (2:1, v/v) containing 0.1% trifluoroacetic acid (Karas et al., 1993). The 17β-HSD analyte (0.5 μl) and DHB matrix (2.5 μl) solutions were mixed on the target and dried in a stream of air at ambient temperature. Spectra were taken from the crystalline rim (Strupat et al., 1991). 17β-HSD spectra were calibrated with the standard proteins cytochrome c (horse heart, 12 360 Da) and myoglobin (whale sperm, 17 199 Da) in a DHB matrix.

2.5. Chemical Modifications

Chemical modifications of Lys, Tyr, Cys, and His residues of 17β-HSD were performed. The chemical modification of Lys was done using 2,4,6-trinitrobenzenesulfonic acid (TNBS), 3.5 mM final conc., and 5 mM KH_2PO_4 buffer, pH 7.4; 0.15 M KCl (Nakayama et al., 1992). For modification of Tyr tetranitromethane, 8 mM final conc., and 50 mM Tris buffer, pH 8.0 were used (Krook et al., 1992). The modification of Cys was performed with 5,5'dithio-bis (2-nitrobenzoic acid) (DTNB), 454μM final conc., and 36 mM KH_2PO_4 buffer, pH 7.0; 0.15 M KCl (Nakayama et al., 1992). For His diethylpyrocarbonate (DEP), 70μM final conc., and 36 mM KH_2PO_4, pH 6.0; 0.15 M KCl were used (Nakayama et al., 1992). To 1 ml of the enzyme preparation the same volume of buffer with the chemical reagent was added. Incubations were performed for 30 min at room temperature.

Chemical modifications were also performed in the presence of the coenzyme NADPH, 1mM final conc., or NADPH and the substrate androstenedione, each 1mM final conc., in order to investigate the ability of these molecules to protect 17β-HSD against the modification. The determination of the enzyme activity after chemical modifications was performed using a chromatographic method (Lanišnik Rižner et al., 1996).

2.6. Carbonyl Reductase Activity Measurement and Competition between Steroidal and Non-Steroidal Substrates

Different carbonyl-group containing substances were tested as possible substrates for *Cochliobolus lunatus* 17β-HSD. Relative enzyme activity in the presence of 100 μM NADPH and the potential substrates: androstenedione, estrone, menadione, 9,10-phenathrenequinone, acetoacetate, and p-nitrobenzaldehyde (100 μM each) was determined spectrophotometrically as described in section 3. Competition between steroid and quinone substrates of 17β-HSD was studied by following the activity of 17β-HSD in the presence of two substrates, androstenedione and menadione. The concentration of one substrate was kept constant, while the concentration of the other was varied from zero to 100 μM and *vice versa*. The constant concentration of menadione was 75 μM and that of androstenedione 50 μM. The concentration of coenzyme NADPH was always 100 μM.

2.7. Antibodies

Polyclonal antibodies against 3α-HSD from the bacterium *Comamonas testosteroni* were kindly given by Dr.Udo C.T. Oppermann. Polyclonal antibodies against 17β-HSD from *Cochliobolus lunatus* were raised in a rabbit. The animal twice received 50 μg of antigen at a monthly interval first with complete Freund's adjuvant, later with incomplete Freund's adjuvant, by subcutaneous injection.

2.8. Immunoblot Analysis

17β-HSD samples were denatured in Laemmli (Laemmli, 1970) sample buffer for 5 min at 90°C and applied to 12% acrylamide slab gel. Electrophoresis was performed using

Bio-Rad Mini-Protean II Cell for 45 min at 200 V. Samples were transfered from the gel to the nitrocellulose membrane for 60 min at 100 V using a Bio-Rad Trans Blot Cell. Polyclonal antibodies against 3α-hydroxysteroid dehydrogenase (3α-HSD) from the bacterium *Comamonas testosteroni* diluted 1:100 and incubated overnight were used as primary antibodies.The secondary was a conjugate with alkaline phosphatase diluted 1:3000 and incubated for 2 hours. Colour was developed using 5-bromo-4-chloro-3-indolyl phosphate and nitroblue tetrazolium.

3. RESULTS AND DISCUSSION

17β-HSD activity was detected in the filamentous fungus *Cochliobolus lunatus* (Plemenitaš et al., 1988) and later in all representatives of different fungal classes (Lanišnik et al., 1992). 17β-HSD from *Cochliobolus lunatus* was purified and its approximate subunit molecular mass was determined by SDS PAGE to be 28 kDa (Lanišnik Rižner et al., 1996). Precise determination of molecular mass was done by mass spectromety. In figure 1 the UV-MALDI mass spectrum of the 17β-HSD preparation is shown.

The singly charged molecular ion of 17β-HSD was determined to be at m/z = 28 297 Da $(M+H)^+$ containing a shoulder at m/z = 28 703 $(M+H)^+$, perhaps originating from the same protein due to the partial post-translational modification. The doubly charged 17β-HSD molecular ion at m/z = 14 160 $(M+2H)^{2+}$ was also accompanied by the shoulder, both roughly half of the respective monocharged species. Additional ions in the mixture at

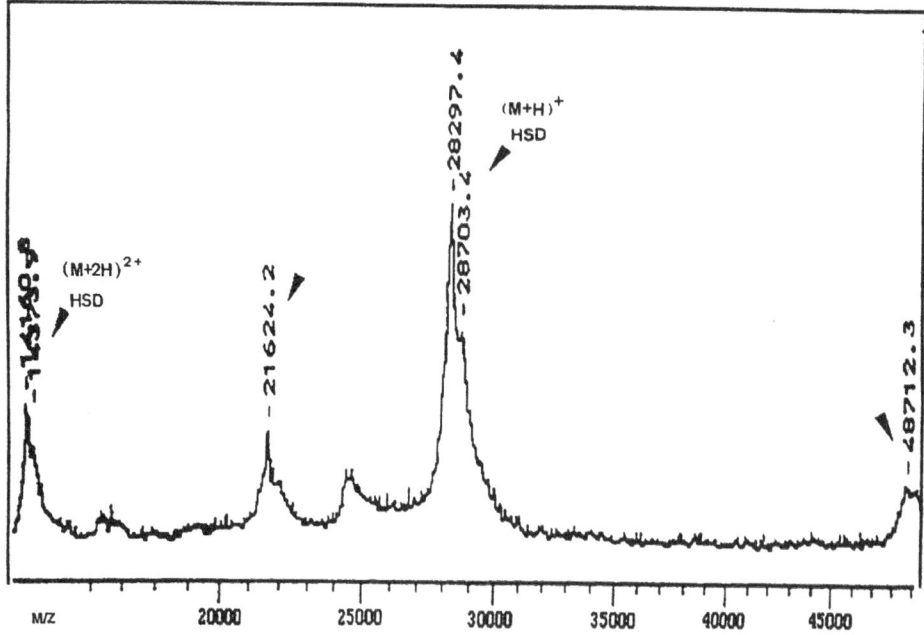

Figure 1. Positive ion UV-MALDI spectrum of a 17β-HSD preparation. Matrix: DHB (see text); laser wavelength, 355 nm; 4 single spectra were summed up. 17β-HSD appears at m/z = 28 297 $(M+H)^+$ containing a shoulder at m/z = 28 703 $(M+H)^+$. The doubly charged HSD molecular ion at m/z=14 160 $(M+2H)^{2+}$ is accompanied by the shoulder. Additional ions at m/z=21 624 and 48 712 cannot be attributed to the HSD proteins.

Table 1. Chemical modifications of Lys, Tyr, His, and Cys from 17β-HSD from the fungus *Cochliobolus lunatus*. In the table, residual enzyme activity after chemical modification of these amino acids is presented. Sample 1: enzyme preparation + reagent; sample 2: enzyme preparation + NADPH + reagent; sample 3: enzyme preparation + NADPH + androstenedione + reagent

amino acid	Lys			Tyr			His			Cys		
reagent	TNBS			tetranitromethane			DEP			DTNB		
concentration	3.5 mM			8 mM			70μM			454μM		
sample	1	2	3	1	2	3	1	2	3	1	2	3
residual activity (%)	10	49	94	0	0	100	0	20	92	65	150	160

TNBS is 2,4,6-trinitrobenzenesulfonic acid
DEP is diethylpyrocarbonate
DTNB is 5,5'dithio-bis(2-nitrobenzoic acid)

m/z = 21 624 and 48 712 Da together with its doubly charged species of low abundance at around 24 400 Da cannot be attributed to the 17β-HSD protein.

Partial amino acid sequence of the 17β-HSD had previously been determined and the enzyme was found to belong to the short-chain alcohol dehydrogenase (SCAD) super-family (Lanišnik Rižner et al., unpublished results). In order to analyze the role of amino acids known to be important for the activity of the enzymes from the SCAD superfamily (Inano, 1988; Murdock et al., 1986; Nakayama et al., 1992; Baker, 1994), chemical modifications of Lys, Tyr, His, and Cys were performed. The results are presented in Table 1.

After chemical modifications of Tyr and His, 17β-HSD activity could not be detected, while after chemical modification of Lys only 10% of the initial enzyme activity was found. When chemical modifications of these amino acids were performed in the presence of the co-enzyme NADPH, enzyme activity could not be detected for Tyr and only 20% of the initial enzyme activity for His and 49% for Lys, showing that the coenzyme could not protect these amino acids from modifications. However, when the experiment was performed in the presence of the coenzyme NADPH and the substrate androstenedione almost 100% of the initial activity was found. These results show that the presence of coenzyme and substrate protects Lys, Tyr, and His from the modifications and, therefore, Lys, Tyr, and His may be considered to be present in the substrate binding site of 17β-HSD. This is in accordance with the results obtained for 3α,20β-HSD from *Streptomyces hydrogenans*, the model enzyme of SCAD su-perfamily (de Pouplana and Fothergill-Gilmore, 1994). The chemical modification of Cys represents a special case. As shown in Table 1, after the modification of Cys only 65% of in-itial enzyme activity was determined. In the presence of coenzyme or coenzyme and substrate chemical modification resulted in increased 17β-HSD activity 150 and 160%,respectively, possibly due to conformational changes.

The relative activity of 17β-HSD in the presence of different substrates is shown in Table 2. As can be seen this enzyme can use at least four different types of organic mole-cules as substrates: keto-steroids, quinones, aldehydes and non-steroidal ketones. In the presence of 100 μM concentration of NADPH and 100 μM concentration of each substrate the highest relative activity was observed with quinones. These results demonstrate the pluripotency of our 17β-HSD, also showing that a steroidal structure is not a necessary re-quirement of the molecules which are substrates of this enzyme. It seems that a suitably

Table 2. Relative activity of 17β-HSD in the presence of different substrates in 100μM concentration. The coenzyme is NADPH, 100μM. The activities are normalized to the activity in the presence 100μM androstenedione and 100μM NADPH which was taken as 100%

substrate	structure	relative activity
estrone		184%
androstenedione		100%
menadione		355%
9,10-phenanthrene-quinone		2144%
p-nitrobenzaldehyde		182%
acetoacetate	CH_3COCH_2COOH	27%

positioned carbonyl group is an essential feature of the substrates, while the rest of the molecule to which the group is bound can have very different three-dimensional structures.

The competition between two different substrates of 17β-HSD, one a steroid, androstenedione, and one a quinone, menadione, is shown in Fig.2. It is obvious from this figure that with increasing concentrations of menadione androstendione is displaced from the active site of the enzyme. The displacement of menadione with increasing concentrations of androstenedione was also observed (results not shown). These findings demonstrate that keto-steroids and quinones are reduced by the same enzyme. Additionally, they also strongly indicate the binding of keto-steroid and quinone substrates to the same region in the active centre of our 17β-HSD.

As 17β-HSD from *Cochliobolus lunatus* was found to exhibit carbonyl reductase activity, immunocharacterization was performed using polyclonal antibodies against 3α-HSD from the bacterium *Comamonas testosteroni*. The carbonyl reductase activity of this bacterial enzyme, which shares structural and functional properties with vertebrate liver microsomal carbonyl reductases, has also been detected (Opperman et al., 1992). The results presented in Figure 3 show that polyclonal antibodies against 3α-HSD from *Comamonas testosteroni* did not cross-react with our 17β-HSD, while on the contrary

Figure 2. Competition between androstenedione and menadione for the active centre of 17β-HSD from *Cochliobolus lunatus.*Curve A was obtained in the absence of androstenedione and curve B in the presence of androstenedione in a constant concentration of 75 μM. The concentration of coenzyme NADPH was always 100 μM.

polyclonal antibodies against 17β-HSD from *Cochliobolus lunatus* clearly recognized the enzyme. Thus, 17β-HSD from *Cochliobolus lunatus* and 3α-HSD from *Comamonas testosteroni*, both HSDs with carbonyl reductase activity, may be considered immunochemically distinct enzymes.

   17β-HSD from *Cochliobolus lunatus* is thus the first fungal HSD found to exhibit hydroxysteroid dehydrogenase as well as carbonyl reductase activity. Amino acids Lys,

Figure 3. Western analysis. Polyclonal antibodies against 3α-HSD from *Comamonas testosteroni* (A) and against 17β-HSD from *Cochliobolus lunatus* (B) were used for immunocharacterization of 17β-HSD from *Cochliobolus lunatus.*

His and Tyr are suggested to be present in the substrate binding site of our 17β-HSD/carbonyl reductase. The enzyme differs immunochemically from the carbonyl reducing 3α-HSD from *Comamonas testosteroni*.

4. ACKNOWLEDGMENTS

The authors wish to thank Dr.U.T.Opperman for his kind gift of antibodies and Prof. F. Hillenkamp for his cooperation.Technical support of M. Marušić and K. Makovec is gratefully acknowledged. The work was supported by the Ministry of Science and Technology of Slovenia.

5. REFERENCES

Baker, M.E., 1994, Sequence analysis of steroid- and prostaglandin-metabolising enzymes: Application to understanding catalysis, Steroids, 59, 248–258.

Boutin, J.A., 1986, Camhoroquinone reduction: another reaction catalyzed by rat liver cytosol 3α-hydroxysteroid dehydrogenase, Biochim.Biophys.Acta, 870, 463–472.

Brown, R.E., Jarvis, K.L. & Hyland, K.J., 1989, Protein measurement by bicinchoninic acid: elimination of interfering substances, Anal.Biochem., 180, 136–139.

Deyashiki, Y., Ogasawara, A., Nakayama, T., Nakanishi, M., Miyabe, Y., Sato, K. & Hara A., 1994, Molecular cloning of two human liver 3α-hydroxysteroid/dihydrodiol dehydrogenase isoenzymes that are identical with chlordecone reductase and bile-acid binder, Biochem.J., 299, 545–552.

Hara, A., Hasebe, K., Hayashibara, M., Matsuura, K., Nakayama, T. & Sawada, H., 1986, Dihydrodiol dehydrogenases in guinea pig liver, Biochem.Pharmacol., 35, 4005–4012.

Hara, A., Kariya, K., Nakamura, M., Nakayama, T. & Sawada, H., 1986 a, Isolation of multiple forms od indanol dehydrogenase associated with 17β-hydroxysteroid dehydrogenase activity from male rabbit liver, Arch. Biochem. Biophys., 249, 225–236.

Hillenkamp, F., Karas, M., Beavis, R.C. & Chait,B.T., 1991, Matrix-assisted laser desorption/ionization mass spectrometry of biopolymers, Anal.Chem., 63, 1193A-1203A.

Inano, H., 1988, Chemical modification of lysine residues at active site of human placental estradiol 17β-dehydrogenase, Biochem.Biophys.Res. Commun., 152, 2, 789–793.

Jarabak, R., Harvey, R.G. & Jarabak, J., 1996, Polycyclic aromatic hydrocarbon quinone-mediated oxidation reduction cycling catalyzed by a human placental 17β-hydroxysteroid dehydrogenase, Arch. Biochem.Biophys., 327, 174–180.

Karas,M., Ehring, H., Nordhoff, E., Stahl, B., Hillenkamp, F., Grhel, M. & Krebs, B., 1993, Matrix-assisted laser desorption/ionization mass spectrometry with additives to 2,5-dihydroxybenzoic acid, Org.Mass Spectrom., 28, 1476.

Klein, J., Post, K., Seidel, A., Frank, H., Oesch, F. and Platt, K.L., 1992, Quinone reduction and redox cycling catalysed by purified rat liver dihydrodiol/3α-hydroxysteroid dehydrogenase, Biochem.Pharmacol., 44, 341–349.

Krook, M., Prozorovski, V., Atrian, S., Gonzalez-Duarte, R. & Joernvall, H., 1992, Short-chain dehydrogenases, Proteolysis and chemical modification of prokaryotic 3α/20β-hydroxysteroid, insect alcohol and human 15-hydroxyprostaglandin dehydrogenases, Eur. J. Biochem., 209, 233–239.

Laemmli, U.K., 1970, Cleavage of structural proteins during the assembly of the head of bacteriophage T4, Nature, 227, 680–685.

Lanišnik, T., Žakelj-Mavrič M. & Beliè, I., 1992, Fungal 17β-hydroxysteroid dehydrogenase, FEMS Microbiol.Lett., 99, 49–52.

Lanišnik Rižner, T., Žakelj-Mavrič, M., Plemenitaš, A. & Zorko, M., 1996, Purification and characterization of 17β-hydroxysteroid dehydrogenase from the filamentous fungus *Cochliobolus lunatus*, J.Steroid Biochem. Molec.Biol., 59, in press.

Maser, E. & Bannenberg, G., 1994, 11β-Hydroxysteroid dehydrogenase mediates reductive metabolism of xenobiotic carbonyl compounds, Biochem. Pharmacol., 47, 1805–1812.

Maser, E., 1995, Xenobiotic carbonyl reduction and physiological steroid oxidoreduction, Biochem.Pharmacol., 49, 421–440.

Murdock, G.L., Chin, C.C. & Warren, J.C., 1986, Human placental estradiol 17β-dehydrogenase: Sequence of a histidine-bearing peptide in catalytical region, Biochemistry, 25, 641–646.

Nakajin, S., Fujita, Y., Ohno, S., Uchida, M., Aoki, M. and Shinoda, M., 1994, Purification and characterization of 3α/β-hydroxysteroid dehydrogenase from mature porcine testicular cytosol, J.Steroid.Biochem.Molec.Biol., 48, 249–256.

Nakayama, T., Tanabe, H., Deyashiki, Y., Shinoda, M., Hara, A. & Sawada, H., 1992, Chemical modification of cysteinyl, lysyl and histidyl residues of mouse liver 17β-hydroxysteroid dehydrogenase, Biochim.Biophys.Acta, 1120, 144–150.

Ohara, H., Nakayama, T., Deyashiki, Y., Hara, A., Miyabe, Y. & Tsukada, F., 1994, Reduction of prostaglandin D_2 to 9α,11β-prostaglandin F_2 by a human liver 3α-hydroxysteroid/dihydrodiol dehydrogenase isozyme, Biochim. Biophys. Acta, 1215, 59–65.

Oppermann, U.C.T., Maser, E., Hermans, J.J.R., Koolman, J. & Netter, K.J., 1992, Homologies between enzymes involved in steroid and xenobiotic carbonyl reduction in vertebrates, invertebrates and procaryonts, J.Steroid. Biochem.Molec.Biol., 43, 665–675.

Oppermann, U.C.T., Netter, K.J. & Maser, E., 1993, Carbonyl reduction by 3α-hydroxysteroid dehydrogenase from Comamonas testosteroni - New properties and its relationship to the SCAD family. In Enzymology and Molecular Biology of Carbonyl Metabolism 4, edited by H.Weiner, Plenum Press, New York pp. 379–390.

Oppermann, U.C.T., Netter, K.J. & Maser, E., 1995, Cloning and primary structure of murine 11β-hydroxysteroid dehydrogenase/microsomal carbonyl reductase, Eur.J.Biochem., 227, 202–208.

Plemenitaš, A., Žakelj-Mavrič, M. & Komel, R., 1988, Hydroxysteroid dehydrogenase of Cochliobolus lunatus, J.Steroid Biochem, 29, 371–372.

de Pouplana, R.L. & Fothergill-Gilmore, L.A., 1994, The active site architecture of a short chain dehydrogenase defined by site-directed mutagenesis and structure modeling, Biochemistry, 33, 7047–7055.

Sawada, H., Hara, A., Nakayama, T., Nakagawa, M., Inoue, Y., Hasebe, K. and Zhang, Y.P., 1988, Mouse liver dihydrodiol dehydrogenases: Identity of the predominant and a minor form with 17β-hydroxysteroid dehydrogenase and aldehyde reductase, Biochem. Pharmacol., 37, 453–458.

Strupat, K., Karas, M. & Hillenkamp, F., 1991, 2,5-Dihydroxybenzoic acid: A new matrix for laser desorption/ionization mass spectrometry, J.Mass Spectrom.Ion.Proc., 111, 89.

Tanaka, M., Ohno, S., Adachi, S., Nakajin, S., Shinoda, M. & Nagahama, Y., 1992, Pig testicular 20β-hydroxysteroid dehydrogenase exhibits carbonyl reductase-like structure and activity, J.Biol.Chem., 267, 13451–13455.

66

A NOMENCLATURE SYSTEM FOR THE ALDO-KETO REDUCTASE SUPERFAMILY

Joseph M. Jez,[1] T. Geoffrey Flynn,[2] and Trevor M. Penning[3]

[1]Department of Biochemistry and Biophysics
[3]Department of Pharmacology
University of Pennsylvania Medical School
Philadelphia, Pennsylvania 19104
[2]Department of Biochemistry
Queen's University
Ontario, Canada, K7L 3N6

1. INTRODUCTION

The aldo-keto reductases (AKRs) represent a growing superfamily of oxidoreductases (Bohren et al., 1989; Bruce et al., 1994). Proteins of the AKR superfamily are monomeric $(\alpha/\beta)_8$-barrel proteins, about 320 amino acids in length, which bind NAD(P)(H) without a Rossmann-fold motif (Rondeau et al., 1992; Wilson et al., 1992 & 1995; Hoog et al., 1994, El-Kabbani et al., 1995). Found in mammals, amphibians, plants, yeast, protozoa, and bacteria, the AKRs metabolize a range of substrates including aliphatic aldehydes, monosaccharides, steroids, prostaglandins, polycyclic aromatic hydrocarbons, and isoflavinoid phytoalexins. To date, at least thirty-nine proteins have been cloned and characterized as members of the superfamily, and additional genes have been identified that potentially code for AKR proteins. The rapid progress in identifying new AKRs has lead to some problem in the naming of these proteins.

Substrate specificity has formed the basis of previous naming schemes for the AKRs. For example, enzymes capable of carbonyl metabolism where named as follows: aldehyde reductase (ALR1; EC 1.1.1.2), aldose reductase (ALR2; EC 1.1.1.21), and carbonyl reductase (ALR3; EC 1.1.1.184) (Turner and Flynn, 1982). However, cDNA cloning indicates that ALR3 is not an AKR but is a member of the short-chain alcohol dehydrogenase/ reductase superfamily (Jornvall et al., 1995). Similarly, the broad substrate specificity of some AKR proteins has also lead to single members of the superfamily having multiple names. For example, 3α-hydroxysteroid dehydrogenase (3α-HSD; EC 1.1.1.213) is also known as dihydrodiol dehydrogenase (EC 1.3.1.20), dehydroascorbate reductase (EC 1.6.5.4), and chlordecone reductase (EC 1.1.1.225) (Pawlowski et al., 1991; Del Bello et al., 1994; Winters et al., 1990).

Enzymology and Molecular Biology of Carbonyl Metabolism 6
edited by Weiner *et al.* Plenum Press, New York, 1996

Table 1. Members of the aldo-keto reductase superfamily

Abbreviation	Enzyme	Species	Accession	Reference
Cb_25dkg	2,5-diketo-D-gulonic acid reductase	*Corynebacterium* sp.	GB: M12799	Anderson et al., 1985
Cb2_25dkg	2,5-diketo-D-gulonic acid reductase	*Corynebacterium* sp.	SP: P15339	Grindley et al., 1988
Lei_putRed	putative reductase	*Leishmania major*	SP: P22045	Samaras et al., 1989
Ps_Mordh	morphine dehydrogenase	*Pseudomonas putida*	GB: M94775	Willey et al., 1993
Alf_ChalRed	chalcone reductase	*Medicago sativa*	PIR: S48851	Sallaud et al., 1995
Soy_ChalRed	chalcone reductase	*Glycine max*	SP: P26690	Welle et al., 1991
Hum_3o5bred	Δ^4-3-ketosteroid-5β-reductase	*Homo sapiens*—liver	PIR: S41120	Kondo et al., 1994
Rat_3o5bred	Δ^4-3-ketosteroid-5β-reductase	*Rattus norvegicus*—liver	SP: P31210	Onishi et al., 1994
Rat_3aHSD	3α-hydroxysteroid dehydrogenase	*Rattus norvegicus*—liver	SP: P23457	Pawlowski et al., 1991
Hum_3aHSDII	3α-hydroxysteroid dehydrogenase type II	*Homo sapiens*—liver		Khanna et al., 1995
Hum_BABP	bile acid binding protein (also dihydrodiol dehydrogenase 2 and 3α-hydroxysteroid dehydrogenase)	*Homo sapiens*—liver	PIR: S43843	Hara et al., 1996 & Stolz et al., 1993
Hum_ChlorRed	chlordecone reductase (also dihydrodiol dehydrogenase 4 and 3α-hydroxysteroid dehydrogenase type I)	*Homo sapiens*—liver	PIR: S43844	Winters et al., 1990
Mou_17bHSD	17β-hydroxysteroid dehydrogenase	*Mus musculus*—liver	PIR: A56424	Deyashiki et al., 1995
Hum_20aHSD	20α-hydroxysteroid dehydrogenase (also dihydrodiol dehydrogenase 1)	*Homo sapiens*—liver	PIR: A53436	Hara et al., 1996 & Stolz et al., 1993
Rat_20aHSD	20α-hydroxysteroid dehydrogenase	*Rattus norvegicus*—ovary	PIR: S43842	Miura et al., 1994
Rab_20aHSD	20α-hydroxysteroid dehydrogenase	*Oryctolagus cuniculus*—ovary	PIR: A45366	Lacy et al., 1993
Bov_Pgs	prostaglandin F synthase	*Bos taurus*—lung	GB: J03570	Watanabe et al., 1988
Frog_Rho	rho-crystallin	*Rana catesbeiana*	GB: X87724	Tomarev et al., 1984
Bov_ADR	aldose reductase	*Bos taurus*—lens/testis	SP: P16116	Petrash & Favello, 1989
Por_ADR	aldose reductase	*Sus scrofa*—lens	SP: P80276	Jaquinod et al., 1993
Hum_ADR	aldose reductase	*Homo sapiens*—placenta	GB: J04795	Bohren et al., 1989
Rav_ADR	aldose reductase	*Oryctolagus cuniculus*—kidney	SP: P15122	Garcia-Perez et al., 1989
Rat_ADR	aldose reductase	*Rattus norvegicus*—lens	PIR: A60603	Carper et al., 1987
Mou_ADR	aldose reductase	*Mus musculus*—mouse	SP: 45376	Gui et al., 1995
Mou_VDP	androgen-dependent vas deferens protein	*Mus musculus*	GB: 81448	Pailhoux et al., 1990
Mou_FR1	fibroblast growth factor induced protein	*Mus musculus*	SP: P45377	Donohue et al., 1994
Hum_ALR	aldehyde reductase	*Homo sapiens*—liver	SP: 14550	Wermuth et al., 1987
Por_ALR	aldehyde reductase	*Sus scrofa*	GB: U46064	Flynn et al., 1995
Rat_ALR	aldehyde reductase	*Rattus norvergicus*—liver	GB: D10853	Takahasi et al., 1993

Table 1. Members of the aldo-keto reductase superfamily (*continued*)

Bar_ALR	aldehyde reductase	*Hordeum vulgare*	GB: Z48360	Bartels et al., 1991
Brgr_ALR	aldehyde reductase	*Bromus inermis*	PIR: JQ2253	Lee & Chen, 1993
App_S6Pdh	sorbitol phosphate dehydrogenase	*Malus domestica*	SP: P28475	Kanayama et al., 1992
Pic_XylRed	xylose reductase	*Pichia stipitis*	SP: P31867	Amore et al., 1991
Klu_XylRed	xylose reductase	*Kluyveromyces lactic*	SP: P49378	Billard et al., 1994
Spor_ALR	aldehyde reductase	*Sporidiobolus salmonicolor*	GB: U26463	Kita, 1995
Sac_GCY	GCY protein	*Saccharomyces cerevisiae*	SP: P14065	Oechsner et al., 1988
Bov_Shaker	β2 subunit of *Shaker* K⁺ channel	*Bos taurus*	GB: X70661	Scott et al., 1994
Rat_Shaker	β1 subunit of *Shaker* K⁺ channel	*Rattus norvergicus*	GB: X70662	Rettig et al., 1994
Rat_AFBred	aflatoxin inducible aldehyde reductase	*Rattus norvergicus*—liver	GB: X74673	Ellis et al., 1993

Ascession numbers are from the GenBank (GB), the Protein Identification Resource (PIR), and the SwissProt (SP) databases.

To address these issues, we propose a new nomenclature system for the AKR superfamily. We present how the known AKRs can be divided into seven distinct families and how these divisions can form the basis for a naming system based on amino acid sequence identity. The proposed system is systematic and expandable, so that as new members of the superfamily are identified they can be assigned a unique name.

2. METHODS

2.1. Identification of AKR Proteins

The list of AKRs was compiled by screening the GenBank (GB), the Protein Identification Resource (PIR), and SwissProt (SP) databases with the BLAST program (Altschul et al., 1990; Gish & States, 1993) using the amino acid sequence of rat liver 3α-HSD as a search query (Pawlowski et al., 1991). Additional sequences were also included from the available literature. However, only sequences which encode for proteins that have been identified and characterized were included in the final list of 39 AKR members.

2.2. Multiple Sequence Alignments and Cluster Analysis

Alignments used PILEUP from the GCG system (Devereux et al., 1984). PILEUP clusters sequences through a pairwise comparison to determine which proteins have the highest similarity, and generates the alignment using the Needleman-Wunsch algorithm (Needleman & Wunsch, 1970). Preference was given to minimize the number of gaps introduced in the alignments by setting the gap and gap length penalties to 5.0 and 0.25, respectively. Alignments were conducted without weighting the comparisons based on predicted or known secondary structure. Cluster analysis using PILEUP constructed a dendrogram based on the amino acid multiple sequence alignment and indicates the relative percent identity among proteins of the AKR superfamily.

3. THE ALDO-KETO REDUCTASES

Using the amino acid sequence of rat liver 3α-HSD as a search query, 39 proteins were identified as members of the AKR superfamily (Table 1). The mammalian aldehyde reductases, aldose reductases, and hydroxysteroid dehydrogenases represent the majority of the superfamily. However, aldehyde and xylose reductases from plants, yeast, and bacteria have also been identified as AKRs. In addition, the bovine and rat *Shaker* K$^+$ channel β-subunit are included in the superfamily. The widespread distribution of AKR proteins in both prokaryotes and eukaryotes suggests that this superfamily is an ancient one.

Multiple sequence alignment of the 39 amino acid sequences (Figure 1) indicates that strict conservation of amino acid side chains occurs at 10 positions in the primary structure of all the AKRs -- G45, D50, E58, G62, K84, P119, G164, P186, Q190, S271 (numbering is that of rat liver 3α-HSD). Two of these residues (D50 and K84) form part of the active site and two are involved in binding of nicotinamide cofactor (Q190 and S271). The remaining other conserved residues may play structural roles in forming the barrel core, since they are found within the β-strands, α-helices, and short loops of the barrel. Overall, the sequences forming the eight β-strands and the eight α-helices are well conserved within separate clusters of AKR proteins. The greatest variation in amino acid sequence occurs in the three large loops found on the C-terminal side of the barrel structure (the A- and B-loops and the C-terminal loop). Among (α/β)$_8$-barrel proteins, the barrel core is generally conserved with changes occurring in the size and sequence of the loops required for catalysis and ligand binding (Farber and Petsko, 1990). The AKRs seemingly follow this architectural pattern of maintaining the barrel scaffold while tailoring substrate specificity through modification of the loops near the active site.

4. SEVEN DISTINCT FAMILIES OF ALDO-KETO REDUCTASES

Cluster analysis demonstrates that the AKRs fall into seven families within the superfamily (Figure 2). The largest and most thoroughly studied family, AKR1, primarily consists of the mammalian aldehyde reductases, aldose reductases, and the hydroxysteroid dehydrogenases and the plant aldehyde reductases. The members of the AKR2 and AKR3 families are found in apple and a number of yeast species and utilize substrates similar to the aldehyde and aldose reductases of AKR1. The chalcone reductases of soybean and alfalfa that belong to the AKR4 family synthesize isoflavinoid phytoalexins in conjunction with chalcone synthase as an induced response to pathogen attack. The AKR5 family consists of the prokaryotic and protozoan AKRs. In the AKR6 family, the β-subunit of the *Shaker* K$^+$ channel may be derived from an ancestral AKR protein and may function in the inactivation of the channel by altering sensitivity to pyridine nucleotide levels (McCormack and McCormack, 1994). Finally, the sole member of the AKR7 family is an ethoxyquin-inducible aldehyde reductase that metabolizes aflatoxin B1. These seven distinct families within the AKR superfamily provide a straightforward way of approaching how to devise a new nomenclature system for the AKRs.

5. THE PROPOSED NOMENCLATURE SYSTEM

The divisions within the cluster analysis serve as a natural starting point for a new nomenclature system for the AKR superfamily. Unlike the P450 superfamily nomencla-

Figure 1. Multiple sequence alignment of the AKR superfamily. The multiple sequence alignment was generated using PILEUP from the GCG program suite. Above the alignment, the secondary structure of rat liver 3α-HSD is noted (Hoog et al., 1994). B1, B2, H1, and H2 are β-sheets and α-helices not involved in forming the core (α/β)$_8$-barrel structure. Also, the residues corresponding to the three loops on the C-terminal side of the barrel are noted. Invariant residues are indicated in bold-face. Amino acids involved in catalysis (c), NAD(P)(H) binding (n), and substrate binding (s) are as indicated. Abbreviations for the proteins are as in Table 1.

(continued)

```
                        LOOP A                              α4          β5          α5                                      β6          α6         β7        LOOP B
Hum_ALR      RGD.......NPFFPKNA  DGTICYDSTH  YKETWKALEA  LVAKGLVQAL  QLSNFWSRQI  DDI.LSVAS   V..RPAVLQV  ECHPYLAQNE  LIAHCQ.AR  GLEVTAYSP  LGS.SDR.  AWRDP
Pci_ALR      RGD.......NPFFPKNA  DGTIRYDATH  YKDTWKALEA  LVAKGLVRAL  QLSNFSSRQI  DDV.LSVAS   V..RPAVLQV  ECHPYLAQNE  LIAHCQ.AR  GLEVTAYSP  LGS.SDR.  AWRDP
Pac_ALR      RGD.......NPFFPKNA  DGTVAKYDSTH YKETWKALEA  LVAKGLVKAL  QLSNFSSRQI  DDV.LSVAS   V..RPAVLQV  ECHPYLTQEK  LIQYCQ.AR  GIVVTAYSP  LGS.PDR.  PWAKP
Hum_ADR      PGK.......EFFPLDA   SGNVVPSDTH  FLDTWAAMEQ  LVDEGLVKAI  GVSNFNPLQI  ERI.LNKPG   LKYKPAVNQI  ECHPYLTQEK  LIEYCH.SK  GIVVTAYSP  LGS.PDR.  PWAKP
Rab_ADR      HGS.......DFFPLDA   SGNVIPSDTD  FVDTWTAMEQ  LVDEGLVKTI  GVSNFNPLQI  ERI.LNKPG   LKYKPAVNQI  ECHPYLTQEK  LIEYCH.CK  GIVVTAYSP  LGS.PDR.  PWAKP
Mou_ADR      PGP.......DFFPLDA   SGNVIPSDTD  FVDTWTAMEQ  LVDEGLVKAI  GVSNFNPLQI  ERI.LNKPG   LKYKPAVNQI  ECHPYLTQEK  LIEYCN.SK  GIVVTAYSP  LGS.PDR.  PWAKP
Rat_ADR      PGP.......DFFPLDA   SGNVIPSDTD  FVDTWTAMEQ  LVDEGLVKAI  GVSNFNPLQI  EKI.LNKPG   LKYKPAVNQI  ECHPYLTQEK  LIQYCN.SK  GIAVTAYSP  LGS.PDR.  PWAKP
Bov_ADR      PGK.......DFPPLDE   DGNVIPSEKD  FVDTWTAMEE  LVDEGLVKAI  GVSNFNHLQV  ERL.LNKPG   LKHHKPVTNQI ECHPYLTQEK  LIQYCN.SK  GIAVTAYSP  LGS.PDR.  PWAKP
Por_ADR      PGK.......DFPPLDG   DGNVVPDESD  FVETWEAMEE  LVDQGLVKAL  GVSNFNHLQV  ERI.LNKPG   LKHHKPVTNQI ESHPYLTQEK  LIQTCH.SK  GIAVTAYSP  LGS.PDR.  PWAKP
Mou_VDP      AGN.......ALLPKDN   KGKVLLSKST  FLDAWEAMEE  LVDQGLVKAL  QISNFNHFQI  ERL.LNKPG   LKHHKPVTNQI ECHPYFNQRK  LLDFCK.SK  GIVSVTAYSP LGS.PDR.  PYAKP
Mou_FR1      PGK.......ELPPKDD   QGRILTSKTT  FLEAWEGMEE  LVDQGLVKAL  QVSNFNHRLL  ERL.LNKPG   LKYKPVCNQV  ECHPYFNQRK  LLDFCK.SK  GISVTAYSP  LGS.PDR.  PSAKP
Hum_20aHSD   PGE.......EVIPKDE   NGKILFDTVD  LCATWEAVEK  CKDAGLAKSI  GVSNFNHRLL  EMI.LNKPG   LKYKPVCNQV  ECHPYFNQRK  LLDFCK.SK  DIVLVAYSA  LGSHREE.  PWVDP
Hum_BABP     PGE.......EVIPKDE   NGKILFDTVD  LCATWEAMEK  CKDAGLAKSI  GVSNFNHRQL  EMI.LNKPG   LKYKPVCNQV  ECHPYFNRSK  LLDFCK.SK  DIVLVAYSA  LGSHREE.  PWVDP
Hum_3aHSD2   PGE.......ELSPTDE   NGKVIFDIVD  LCTTWEAMEK  CKDAGLAKSI  GVSNFNRRQL  EII.LNKPG   LKYKPVCNQV  ECHPYFNRSK  LLDFCK.SK  GIVLVAHSA  LGTQRHK.  LAVDP
Hum_ChlorRed PGE.......TPLPKDE   NGKLIYDAVD  LSATWEVMEK  CKDAGLAKSI  GVSNFNCRQL  EMI.LNKPG   LKYKPVCNQV  ECHPYLANQSK LLEFCK.SK  GIVLVAYSA  LGSHREP.  QWVDQ
Rab_20aHSD   PGV.......EIIPTDE   HGKALFDTVD  ICATWEAMEK  CKDAGLAKSI  GVSNFNHRQL  EKI.LNKPG   LKYKPVCNQV  ECHPYLANQGK LLDFCR.SK  DIVLVAYSA  LGSHREK.  QWVDQ
Mou_17bHSD   PGE.......NTLPKDE   NGKLIYDAVD  ICDTWEAMEK  CKDAGLAKSI  GVSNFNRRQL  EKI.LKKPG   LKYKPVCNQV  ECHPYLANQSK LLEFCK.SK  DIVLVAYSA  LGSHREK.  QWVDQ
Bov_Pgs      PGN.......KFVPKDE   SGKLIFDSVD  LCHTWEALEK  CKDAGLTKSI  GVSNFNHKQL  EKI.LNKPG   LKYKPVCNQV  ECHLYLNQSK  LLAYCK.SH  DIVLVAYCA  LGAQLLS.  YC1NE
Rat_20aHSD   PGD.......ELLPQDE   HGNLLLDTVD  LCDTWEANEK  CKDAGLAKSI  GVSNFNRRQL  EKI.LNKPG   LKHRPVCNQV  ECHLYLNQSK  MLAYCK.MN  DIVLVAYGA  LGTQRTYK. TWVDQ
Rat_3aHSD    PGD.......IFFPRDE   HGKLLFETVD  ICDTWEAMEK  CKDAGLAKSI  GVSNFNCRQL  EKI.LNKPG   LKYKPVCNQV  ECHLVLANQSK MLDCK.SH  DIVLVAVSA  LGSHRDR.  NWVDL
Frog_Rho     PSG.......ASDPSDK   DKFFIYDNVD  LCATWEALEA  LVDQGLVKSL  GVSNFNRRQL  EKT.LNKPG   LKYKPVCNQV  ECHPYTQPR   LLKFCQ.QR  DIVTAYSP   LGTSRNP.  IWVNV
Rat_JoSbRed  PGD.......EYYPKDE   NGKVIYHKSN  LCATWEAMEA  LVDEGLVKSL  GVSNFNRRQL  EVI.LNKPG   LKYKPVTNQV  ECHPYTQTK   LLEVSA.SS  MTSFIVAYSP LGTCRNP.  LAVNV
Hum_JoSbRed  PGE.......P         PEAGEVLEPD  MEGVWKEMEN  LVKDQLVKDI  QVCNTYVTKL  NRL.LOSA.   KIPPAVCQM   EMHPGWKNDK  ILEACK.KH  GIHATAYSP  L....CS.  .....
Bar_ALR      DGAHK.....P         PEAGEVLEPD  MEGVWKEMEN  LVKDQLVKDI  QLSNYELFLT  RDC.L..AY   SKIKPAVSQF  ETHPYFQRDS  LVKFCM.KH  GVLPTAHTP  LGGAA.    AMK
Brg_ALR      DGAHK.....ASLLGED   KVLDIDVTIS  LQQTWEDMEK  TVSLGLVRSI  GVSNFPGALL  LDG.L..RG   ATIKPSVLQV  EHPYLQQPR   LIEFAQ.SR  GIAVTAYSS  FGPQSFV.  ELOGG
App_S6fdh    FVPLEEKYPP GFYVCGKGD..NPDYEDVP   ILETWKALEK  LVDQGKIKSL  QISNFSGALI  QDL.L..RG   ARIKPVALQI  EHRPYLTQEK  LIKVK.NA   GIQVVAYSS  FGPVSFL.  ELDNK
Pic_XylRed   PVPFDEKYPP GFYTGKEDEA KGHIEEEQVP   LLDTWRALEK  LVDQGKIKSL  GISNFSGALI  QDL.L..RG   LASQG       NKLTPAANQV EIHPLLPODE  LINFCK.SK  GIVVEAYSP  LGN.      ....
Klu_XylRed   PA..YIKNED ILSVFFKKDG SRAVDITWNN   FIKTWELMQE  LPKTGKTKAV  GVSNFSINNL  KDL.LASQG   .VTPSVNQI   ERHPLLQPE   LIAHHK.AK  NIHITAYSP  LGN.      ....
Sac_GCY      PEGDITQMLF .PKAND    KEVKLDLEVS  LVDTWKAMVK  LLDTGKVKAV  GVSNFDAKMV  DAI.IEATG   .VTPSVNQI   ERHPLLLQPE  LIAHHK.AK  NIHITAYSP  LGN.      ....
Sporo_ALR    PGKFSF....P         IDVADLLPFD  VKGVWESMEE  SLKLGLTKAI  GVSNFSVKKL  ENL.LSVA.   TVLPAVNQV   EDRLAWQQKK  LREFCN.AH  GIVLTAFSP  LRKGASR.  ....
Alf_Chalked  PGKFSF....P         IEVEDLLPFD  VKGVWESMEE  CQKLGLTKAI  GVSNFSVKKL  QNL.LSVA.   TIRPVVDDV   EDRLAWQQKK  LREFCK.EN  GIIVTAFSP  LRKGASR.  ....
Soy_Chalked  PGKFSF....P         IQVEDLLPFD  VKGVWESMEE  CLKLGLTKAI  GVSNFSVKKL  QNL.LSVA.   TIRPAVNQV   EDRLAWQQKK  LREFCT.AN  GIVLATFSP  LRKGASR.  ....
Gly_PKR      .......RG  KDILSKEGKK  YLDSWRAFEQ  LYKEKKVRAI  GVSNFNIHHL  EDV.LAMCT   .VTPVVNQV   ELHPLNNQAD  LRAFCD.AK  QIKVEAMSP  LG.       ....
Lei_DutRed   .........TKDWNA     TIQSWKAAEK  ILGEGRARAI  GVCNFLEDQL  DEL.IAASD   .VVPAVNQI   ELHPYFAQKP  LLAKNR.AL  GIVTEAWSP  IGCHQRW.  ....
Ps_Mordh     .........TPAADN     YVHAWEKMIE  LRAAGLTRSI  GVSNHLVPHL  KTL.IDETG   .VVPAVNQI   ELHPAYQRE   ITDWAA.AH  DWRIESWGP  LG.       ....
Cb_25dkg     .........NPSVGR     WLDTWRGMID  AREAGLVRSI  GVSNFTEPML  KTL.IDETG   .VTPAVNQV   ELHPYFPQAA  LRAFHD.EH  GIRTESWSP  LAR.      QDGDN
Cb_25dkg2    .........NTP        MEETVRAMTH  VINQGMAMYW  GTSRWSSMEI  MEAYSVARQF  NLIPPICEDA  .VTPVMNQV   EYHMFQREXV  EVQLPELFHK IGVGAMTWSP LACGIVSGKY DSGIPPYSRA SLKGY...... QMLED
Bov_Shaker   .........NTP        MEETVRAMTH  VINQGMAMYW  GTSRWSAMEI  MEAYSVARQF  NNIPPVCEDA  .VTPVMNQV   ETHLPQREKV  EVQLPELYHK IGVGAMTWSP LACGIISGKY GNGVPESSRA SLKGY...... QMLKE
Rat_Shaker   .........GTP        IEETLQACHH  VNQEGKFVEL  GLSNYSWEV   AEICTLCKKN  GWIMPTVYGG  MTRNAITRQ.V ETELFPCLRH FGLRFYAFNP LAGGLLTGRY KYQKDGKNP  ESRFGNFPS  QL1MORTWKE

                  s    s    s          s     s     s                                 nn          n                                        n                     s
```

Figure 1. Multiple sequence alignment of the AKR superfamily. The multiple sequence alignment was generated using PILEUP from the GCG program suite. Above the alignment, the secondary structure of rat liver 3α-HSD is noted (Hoog et al., 1994). B1, B2, H1, and H2 are β-sheets and α-helices not involved in forming the core $(\alpha/\beta)_8$-barrel structure. Also, the residues corresponding to the three loops on the C-terminal side of the barrel are noted. Invariant residues are indicated in bold-face. Amino acids involved in catalysis (c), NAD(P)(H) binding (n), and substrate binding (s) are as indicated. Abbreviations for the proteins are as in Table 1.

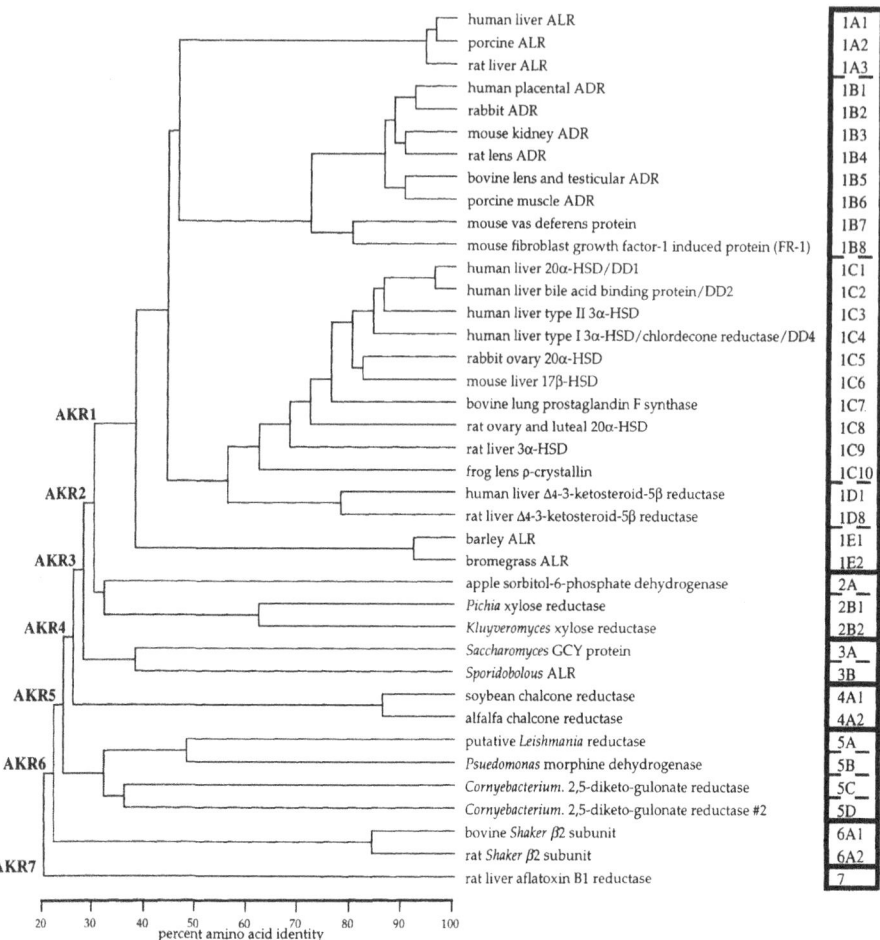

Figure 2. Cluster analysis of the AKR superfamily. The dendrogram is based on the pairwise-sequence alignments performed by the GCG program PILEUP and indicates the percent amino acid identity among proteins of
the AKR superfamily.

ture system, which uses genes as the defining feature, we suggest that amino acid sequences represent the best method for delineating this system, since the gene structures of
many AKR proteins are unknown. However, the naming of AKRs will use a method similar to the P450 superfamily.

We recommend the root symbol "AKR" for *A*ldo-*K*eto *R*eductase; an Arabic number
designating the family; a letter indicating the subfamily when multiple subfamilies exist;
and an Arabic numeral representing the unique protein sequence. For example, AKR1A1
would be the first AKR in family 1, subfamily A. For historical reasons, the AKR1A subfamily represents the aldehyde reductases and the AKR1B subfamily represents the aldose
reductases.

An AKR protein from a particular family will be defined as having ≤ 40% amino
acid identity with an AKR from any other family. The seven AKR families described
above are consistent with this cut-off. Within a particular family, subfamilies will be de-

fined by ≥ 60% identity in amino acid sequence among subfamily members. All the families except for AKR4, AKR6, and AKR7 have multiple subfamilies. For example, family
AKR1 can be divided into the following subfamilies: A) mammalian aldehyde reductases;
B) mammalian aldose reductases; C) the hydroxysteroid dehydrogenases; D) the Δ^4–3-ke-
tosteroid-5β-reductases; and E) plant aldehyde reductases. Finally, numbering of subfamily members will be in an arbitrary fashion for the sequences known to date with new
additions being numbered chronologically.

Allelic variation may occur between superfamily members. We propose that proteins with ≥ 97% amino acid sequence identity are alleles of the same gene unless different catalytic activities are demonstrated or nontranslated regions are clearly divergent and
indicate distinct genes. For example, although human dihydrodiol dehydrogenase 2 (bile
acid binding protein) and human dihydrodiol dehydrogenase 1 are 98% identical in amino
acid sequence, the substrate specificity of each protein is different.

Finally, in order to distinguish between a gene and a protein, the designation for a
gene of the AKR superfamily should be noted in italics. For example, the gene *AKR1A1*
encodes the protein AKR1A1.

The new designations for the 39 known AKR proteins are shown in Figure 2.

6. SUMMARY

As new members of the AKR superfamily are identified the need for a systematic
and expandable nomenclature has risen, especially since some members of the superfamily have multiple names based on substrate specificity. We have proposed a nomenclature system for the AKR superfamily that is similar to the P450 system but based on
amino acid sequence comparisons instead of nucleotide sequence comparisons. Our system uses percent amino acid identities to delineate families and subfamilies within the
larger superfamily. Although there are not as many AKRs as P450s, having a flexible nomenclature system will allow for easy incorporation of new proteins into the superfamily.

7. ENDNOTE

The adoption of this nomenclature system was agreed to by the participants of the
8th International Workshop on the Enzymology and Molecular Biology of Carbonyl Metabolism held in Deadwood, SD between June 29 and July 3, 1996.

A web site has been set up so investigators can receive or add sequence data. It can
be reached at http/pharme26.upenn.edu.

8. REFERENCES

Altschul, S.F., Gish, W., Miller, W., Myers, E.W., & Lipman, D.J. (1990) Basic local alignment search tool. *J. Mol. Biol.* **215**: 403–410

Amore, R., Kotter, P., Kuster, C., Ciriacy, M., & Hollenberg, C.P. (1991) Cloning and expression in Saccharomyces cerevisiae of the NAD(P)H-dependent xylose reductase encoding gene (XYL1) from the xylose-assimi-
lating yeast *Pichia stipitis*. *Gene* **109**: 89–97

Anderson, S., Berman-Marks, C., Lazarus, R., Miller, J., Stafford, K., Seymour, J., Light, D., Rastetter, W., &
Estell, D. (1985) Production of 2-keto-L-gulonate, an intermediate in L-ascorbate synthesis, by a genetically modified *Erwinia herbicola* . *Science* **230**: 144–149

Bartels, D., Engelhardt, K., Roncarati, R., Schneider, K., Rutts, M., & Salamini, F. (1991) An ABA and GA modulated gene expressed in the barley embryo encodes an aldose reductase related protein. EMBO J. 10: 1037–1043

Billard, P.P., Menart, S.S., Fleer, R.R., & Bolotin-Fukuhara, M.M. (1994) Isolation and characterization of the gene encoding xylose reductase from *Kluyveromyces lactis*. *Gene* 162: 93–97

Bohren, K.M., Bullock, B., Wermuth, B., & Gabbay, K.H. (1989) The aldo-keto reductase superfamily *J. Biol. Chem.* 264: 9547–9551

Bruce, N.C., Willey, D.L., Coulson, A.F.W., & Jeffrey, J. (1994) Bacterial morphine dehydrogenase further defines a distinct superfamily of oxidoreductases with diverse functional activities. *Biochem. J.* 299: 805–811

Carper, D., Nishimura, C., Shinohara, T., Dietzchold, B., Wistow, G., Craft, C., Kador, P., & Kinoshita, J.H. (1987) Aldose reductase and ρ-crystallin belong to the same protein superfamily as aldehyde reductse. *FEBS Letters* 220: 209–213

Del Bello, B., Maellaro, E., Sugherini L., Santucci, A., Comporti, M., & Casini, A.F. (1994) Purification of NADPH-dependent dehydroascorbate reductase from rat liver and its identification with 3α-hydroxysteroid dehydrogenase. *Biochem. J.* 304: 385–390

Devereux, J., Haeberil, P., & Smithies, O. (1984) A comprehensive set of sequence analysis programs for the VAX. *Nucleic. Acids Res.* 12: 387–395

Deyashiki, Y., Ohshima, K., Nakanishi, M., Sato, K., Matsuura, K., & Hara, A. (1995) Molecular cloning and characterization of mouse estradiol 17β-dehydrogenase (A- specific), a member of the aldo-keto reductase family. *J. Biol. Chem.* 270: 10461–10467

Donohue, P.J., Alberts, G.F., Hampton, B.S. & Winkles, J.A. (1994) A delayed-early gene activated by fibroblast growth factor-1 encodes a protein related to aldose reductase. *J. Biol. Chem.* 269: 8604–8609

El-Kabbani, I., Judge, K., Ginell, S.L., Myles, D.A.A., DeLucas, L.J., & Flynn, T.G. (1995) Structure of porcine aldehyde reductase holoenzyme. *Nature Structural Biology* 2: 687–692

Ellis, E.M., Judah, D.J., Neal, G.E., & Hayes, J.D. (1993) An ethoxyquin-inducible aldehyde reductase from rat liver that metabolizes aflatoxin B1 defines a subfamily of the aldo-keto reductases. *Proc. Natl. Acad. Sci. USA* 90: 10350–10354

Farber, G.K. & Petsko, G.A. (1990) The evolution of α/β-barrel enzymes. *Trends in Biochem. Sciences* 15: 228–234

Flynn, T.G., Green, N.C., Bhatia, M.B., & El-Kabbani, O. (1995) Structure and mechanism of aldehyde reductase *Adv. Exp. Med. & Bio.* 372: 193–201

Garcia-Perez, A., Martin, B., Murphy, H.R., Uchida, S., Murer, H., Cowley, B.D., Handler, J.S., & Burg, M.B. (1989) Molecular cloning of cDNA coding for kidney aldose reductase. *J. Biol. Chem.* 264: 16815–16821

Gish, W. & States, D.J. (1993) Identification of protein coding regions by database similarity search. *Nature Genetics* 3: 266–272

Grindley, J.G., Payton, M.A., De Pol, H., & Hardy, K.G. (1988) Conversion of glucose to 2-keto-L-gulonate, an intermediate in L-ascorbate synthesis, by a recombinant strain of *Erwinia citreus*. *Appl. Environ. Microbiol.* 54: 1770–1775

Gui, T., Tanimoto, T., Kokai, Y., & Nishimura, C. (1995) Presence of a closely related subgroup in the aldo-keto reductase superfamily of the mouse. *Eur. J. Biochem.* 227: 448–453

Hara, A., Matsuura, K., Tamada, Y., Sato, K., Miyabe, Y., Deyashiki, Y., & Ishida, N. (1996) Relationship of human liver dihydrodiol dehydrogenases to hepatic bile- acid binding protein and an oxidoreductase of human colon cells. *Biochem. J.* 313: 373–376.

Hoog, S.S., Pawlowski, J.E., Alzari, P.M., Penning, T.M., & Lewis, M. (1994) Three- dimensional structure of rat liver 3α-hydroxysteroid/dihydrodiol dehydrogenase: a member of the aldo-keto reductase superfamily. *Proc. Natl. Acad. Sci. USA* 91: 2517–2521

Jaquinod, M., Potier, N., Klarskov, K., Reymann, J.M., Sorokine, O., Kieffer, S., Barth, P., Andriantomanga, V., Biellman, J.F., & Van Dorsselaes, A. (1993) Sequence of pig lens aldose reductase and electrospray mass spectroscopy of non- covalent and covalent complexes. *Eur. J. Biochem.* 218: 893–903

Jornvall, H., Persson, M., Krook, M., Atrian, S., Gonzalez-Duarte, R., Jeffrey, J., & Ghosh, D. (1995) Short-chain dehydrogenases/reductases (SDR). *Biochemistry* 34: 6003–6013

Kanayama, Y., Moni, H., Imaseki, H., & Yamaki, S. (1992) Nucleotide sequence of a cDNA encoding NADP-sorbitol-6-phosphate dehydrogenase from apple. *Plant Physio.* 100: 1607–1608

Khanna, M., Qin, K.N., Wang, R.W., & Cheng, K.C. (1995) Substrate specificity, gene structure, and tissue-specific distribution of multiple human 3a- hydroxysteroid dehydrogenases. J. Biol. Chem. 270: 20162–20168

Kita, K. (1995) Cloning, sequence analysis, and expression of aldehyde reductase from red yeast *Sporobolomyces salmonicolor*. GenBank submission

Kondo, K., Kai, M., Setoguchi, Y., Eggertsen, G., Sjoblom, P., Setoguchi, T., Okuda, K., & Bjorkhem, I. (1994) Cloning and expression of cDNA of human Δ⁴-3- oxosteroid-5β-reductase and substrate specificity of the expresed enzyme. *Eur. J. Biochem.* 219: 357–363

Lacy, W.R., Washenick, K.J., Cook, R.G., & Dunbar, B.S. (1993) Molecular cloning and expression of an abundant rabbit ovarian protein with 20α-hydroxysteroid dehydrogenase activity. *Mol. Endo.* **7**: 58–66

Lee, S.P. & Chen, T.H.H. (1993) Molecular cloning of abscisic acid-responsive mRNAs expressed during induction of freezing tolerance in bromegrass suspension culture. *Plant Physio.* **101**: 1089–1092

McCormack, T. & McCormack, K. (1994) Shaker K⁺ channel β-subunits belong to an NAD(P)H-dependent oxidoreductase superfamily. *Cell* **79**: 1133–1135

Miura, R., Shiota, K., Noda, K., Yagi, S., Ogawa, T., & Takahashi, M. (1994) Molecular cloning of cDNA for rat ovarian 20α-hydroxysteroid dehydrogenase (HSD1). *Biochem. J.* **299**: 561–567

Needleman, S.B. & Wunsch, C.D. (1970) A general method applicable to the search for similarities in the amino acid sequence of two proteins. *J. Mol. Biol.* **48**: 443–453

Oechsner, U., Magodolen, V., & Bandlow, W. (1988) A nuclear yeast gene (GCY) encodes a polypeptide with high homology to a vertebrate lens protein. *FEBS Letters* **238**: 123–128

Onishi, Y., Noshiro, M., Shimosato, T., & Okuda, K. (1994) Molecular cloning and sequence analysis of cDNA encoding Δ⁴-3-ketosteroid-5β-reductase of rat liver. *FEBS Letters* **283**: 215–218

Pailhoux, E.A., Martinez, A., Veyssiere, G.M., Jean, C.G. (1990) Androgen-dependent protein from mouse vas deferens. *J. Biol. Chem.* **265**: 19932–19936

Pawlowski, J.E., Huizinga, M., & Penning, T.M. (1991) Cloning and sequencing of the cDNA for rat liver 3α-hydroxysteroid/dihydrodiol dehydrogense. *J. Biol. Chem.* **266**: 8820–8825

Petrash, J.M. & Favello, A.D. (1989) Isolation and characterization of cDNA clones encoding aldose reductase. *Exp. Eye Res.* **8**: 1021–1027

Rettig, J., Heinemann, S.H., Wunder, F., Lorra, C., Parcej, C.N., Dolly, J.O., & Pongs, O. (1994) Inactivation properties of voltage-gated K⁺ channels altered by presence of beta-subunit. *Nature* **369**: 289–294

Rondeau, J.M., Tete-Favier, F., Podjarny, A., Reymann, J.M., Barth, P., Biellman, J.F., & Moras, D. (1992) Novel NADPH-binding domain revealed by the crystal structure of aldose reductase. *Nature* **355**: 469–472

Sallaud, C., El Turk, C., Bigerre, L., Sevin, H., Welle, R., & Esnault, R. (1995) Nucleotide sequence of three chalcone reductase genes from alfalfa. *Plant Physiology* **108**: 869–870

Samaras, N. & Spithill, T.W. (1989) The developmentally regulated P100/11E gene of *Leishmania major* shows homology to a superfamily of reductase genes. *J. Biol. Chem.* **264**: 4251–4254

Scott, V.E., Rettig, J., Parcej, D.N., Keen, J.N., Findlay, J.B., Pongs, O., & Dolly, J.O. (1994) Primary structure of a beta-subunit of alpha-dendrotoxin-sensitive K⁺ channels from bovine brain. *Proc. Natl. Acad. Sci. USA* **91**: 1637–1641

Stolz, A., Hammond, L., Lou, H., Takikawa, H., Ronk, M., & Shively, J.E. (1993) cDNA cloning and expression of the human bile-acid binding protein. *J. Biol. Chem.* **268**: 10448–10457

Takahasi, M., Fujii, J., Teshima, T., Suzuki, K., Shiba, T., & Taniguchi, N (1993) Identity of a major 3-deoxyglucosone reducing enzyme with aldehyde reductase in rat liver established by amino acid sequencing and cDNA expression. *Gene* **127**: 249–253

Tomarev, S.I., Zinovieva, R.D., Dolgilevich, S.M., Luchin, S.V., Krayev, A.S., Skyryabin, K.G., & Gause, G.G. (1984) A novel type of crystallin in the frog eye lens. *FEBS Letters* **171**: 297–301

Turner, A.J. and Flynn, T.G. (1982) The nomenclature of aldehyde reductases in *Enzymology of Carbonyl Metabolism: Aldehyde Dehydrogenase and Aldo/Keto Reductase* (Weiner, H., and Wermuth, B., eds.) pp. 401–402, Alan Liss, New York

Warren, J.C., Murdock, G.L., Ma, Y., Goodman, S.R., & Zimmer, W.E. (1993) Molecular cloning of testicular 20α-hydroxysteroid dehydrogenase: identity with aldose reductase. *Biochemistry* **32**: 1401–1406

Watanabe, K., Fujii, Y., Nakayama, K., Ohkubo, H., Kuramitsu, S., Kagamiyama, H., Nakanishi, S., & Hayaishi, O. (1988) Structural similarity of bovine lung prostaglandin F synthase to lens ε-crystallin of the European common frog. *Proc. Natl. Acad. Sci. USA* **85**: 11–15

Welle, R., Schroder, G., Schiltz, E., Grisebach, H., & Schroder, J. (1991) Induced plant responses to pathogen attack *Eur. J. Biochem.* **196**: 423–430

Wermuth, B., Omar, A., Forster, A., di Francesco, G., Wolf, M., von Wartbuth, J.P., Bullock, B., & Gabbay, K.H. (1987) Primary structure of aldehyde reductase from human liver. *Prog. Clin. Biol. Res.* **232**: 297–307

Willey, D.L., Caswell, D.A., Lowe, C.R., & Bruce, N.C. (1993) Nucleotide sequence and over-expression of morphine dehydrogenase, a plasmid-encoded gene from *Pseudomonas putida* M10. *Biochem. J.* **290**: 539–544

Wilson, D.K., Bohren, K.M., Gabbay, K.H., & Quiocho, F.A. (1992) An unlikely sugar substrate site in the 1.65 Å structure of the human aldose reductase holoenzyme implicated in diabetic complications. Science 257: 81–84

Wilson, D.K., Nakano, T., Petrash, J.M., & Quiocho, F.A. (1995) 1.7 Å Structure of FR-1, a fibroblast growth factor-induced member of the aldo-keto reductase family, complexed with coenzyme and inhibitor. *Biochemistry* **34**: 14323–14330

Winters, C.J., Molowa, D.T., & Guzelian, P.S. (1990) Isolation and characterization of cloned cDNAs encoding human liver chlordecone reductase. *Biochemistry* **29**: 1080–1087

APPENDIX A
ALCOHOL DEHYDEGENASES

Compliled by Bengt Persson

Department of Medical Biochemistry and Biophysics
Karolinska Institutet
S-171 77 Stockholm, Sweden

The following are alcohol dehydrogenases whose sequence or gene has been deposited in a database. The search was completed in September, 1996.

Species	Swissprot code /Genbank accession number
CLASS I	
Mammals	
human alpha	ADHA_HUMAN
human beta	ADHB_HUMAN
human gamma	ADHG_HUMAN
Hamadryas baboon	ADH_PAPHA
rhesus macaque	ADH_MACMU
horse E	ADHE_HORSE
horse S	ADHS_HORSE
rabbit	ADH1_RABIT
rat	ADHA_RAT
mouse	ADHA_MOUSE
deer mouse	ADH1_PERMA
Geomys bursarius	L15463
Gossypium hirsutum	U49061
Birds	
brown kiwi	ADH1_APTAU
ostrich	ADH1_STRCA
chicken	ADH1_CHICK
quail	ADH3_COTJA
Reptiles	
American alligator	ADH1_ALLMI

Indian cobra (Naja naja)	ADH1_NAJNA
Lizard Uromastix hardwickii A	ADHA_UROHA
Lizard Uromastix hardwickii B	ADHB_UROHA
Amphibians	
Mediterranian green frog (Rana perezi)	ADH1_RANPE
Fish	
Baltic cod	ADH_GADCA

CLASS II
human	ADHP_HUMAN
rat	(1)
ostrich	ADH2_STRCA

CLASS III
human	ADHX_HUMAN
horse	ADHX_HORSE
rat	ADHX_RAT
mouse	ADHX_MOUSE
lizard Uromastix hardwickii	ADHX_UROHA
cod	(2)
Atlantic hagfish	ADHX_MYXGL
octopus	(3)
Drosophila melanogaster	ADHX_DROME
Entamoeba histolytica	ADH1_ENTHI
Caenorhabditis elegans	U18781
pea	(4)
Arabidopsis thaliana	X82647
Candida maltosa	FADH_CANMA
Saccharomyces cerevisiae	FADH_YEAST
Pseudomonas putida	FADH_PSEPU
Paracoccus denitrificans	FADH_PARDE
Rhodobacterium. sphaeroides	L47326
Pasteurella piscicida	ADH3_PASPI
Escherichia coli	X73835
Haemophilus influenzae	ADH3_HAEIN

CLASS IV
human	ADH7_HUMAN
rat	ADH7_RAT
mouse	U20257

CLASS V
human	ADH6_HUMAN

CLASS VI
deer mouse	ADH2_PERMA

CLASS P (plants)
pea	ADH1_PEA
Phaseolus acutifolius 1F	Z23170
Phaseolus acutifolius 1CN	Z23171

petunia	ADH1_PETHY
Arabidopsis thaliana	ADH_ARATH
barley 1	ADH1_HORVU
barley 2	ADH2_HORVU
barley 3	ADH3_HORVU
maize 1	ADH1_MAIZE
maize 2	ADH2_MAIZE
pearl millet	ADH1_PENAM
rice 1	ADH1_ORYSA
rice 2	ADH2_ORYSA
potato 1	ADH1_SOLTU
potato 2	ADH2_SOLTU
potato 3	ADH3_SOLTU
tomato 1	S75487
tomato 2	ADH2_LYCES
strawberry	ADH_FRAAN
apple	ADH_MALDO
clover	ADH1_TRIRP
petunia	U25536
Pinus banksiana	U48369,U48370,U48374,U48375,U48376

CLASS Y (yeast)

Saccharomyces cerevisiae 1	ADH1_YEAST
Saccharomyces cerevisiae 2	ADH2_YEAST
Saccharomyces cerevisiae 3	ADH3_YEAST
Saccharomyces cerevisiae 5	ADH5_YEAST
Schizosaccharomyces pombe	ADH_SCHPO
Candida albicans	ADH1_CANAL
Kluyveromyces lactis 1	ADH1_KLULA
Kluyveromyces lactis 2	ADH2_KLULA
Kluyveromyces lactis 3	ADH3_KLULA
Kluyveromyces lactis 4	ADH4_KLULA
Kluyveromyces marxianus	ADH1_KLUMA
Zymomonas mobilis	ADH1_ZYMMO
Emericella nidulans 1	ADH1_EMENI
Emericella nidulans 2	Z48000
Emericella nidulans 3	ADH3_EMENI
Aspergillus flavus	ADH1_ASPFL

BACTERIA

Bacillus stearothermophilus 1	ADH1_BACST
Bacillus stearothermophilus 2	ADH2_BACST
Bacillus stearothermophilus 3	ADH3_BACST
Alcaligenes eutrophus	ADH_ALCEU
Mycobacterium bovis	ADH_MYCBO
Sulfolobus solfataricus	ADH_SULSO
Methylobacterium extorquens	L48340

REFERENCES (when not in a database)

1. Höög, J.-O. (1995) FEBS Lett. **368**, 445–448.
2. Danielsson, O., Shafqat, J., Estonius, M., El-Ahmad, M. & Jörnvall, H. (1996) Biochemistry, **35**, 14561–14568.
3. Kaiser, R., Fernandez, M. R., Parés, X. & Jörnvall, H. (1993) Proc. Natl. Acad. Sci. USA **90**, 11222–11226.
4. Shafqat, J., El-Ahmad, M., Danielsson, O., Martinez, M. C., Persson, B., Parés, X. & Jörnvall, H. (1996) Proc. Natl. Acad. Sci. USA **93**, 5595–5599.

APPENDIX B
ALDEHYDE DEHYDROGENASE GENES

Vasilis Vasiliou

University of Colorado Health Sciences Center
Department of Pharmaceutical Sciences
4200 East Ninth Avenue
Denver, Colorado 80262

This chapter serves as a reference list of accession numbers for aldehyde dehydrogenase (*ALDH*) genes, as December 24, 1996 (GenBank, version 97.0, 10/21/96; GenBank Updates version 97.0⁺, 12/24/96).

ALDHs ON INTERNET

The data in this chapter, as well as extensive sequence alignments, classification of *ALDH* genes including phylogenetic trees, monthly updated list of accession numbers, and chromosomal location of *ALDH* genes are available on the Internet at the "Vasiliou Laboratory Home Page" (http://www.uchsc.edu/sp/sp/alcdbase/alcdbase.html).

Aldehyde Dehydrogenase Gene Superfamily

GenBank name / Definition	Acession Number	Species	Date of Submission	Length (bp)
Gb_ba:Abcaldh Membrane-bound aldehyde dehydrogenase gene, complete cds and flanks.	D00521	A.polyoxogenes	7/91	2,683 bp
Gb_ba:Afaacod Acetaldehyde dehydrogenase II (acoD) gene, complete cds.	M74003	*Alcaligenes eutrophus*	3/92	2,822 bp
Gb_ba:Bacaldht aldhT gene for aldehyde dehydrogenase, complete cds.	D13846	*B. stearothermophilus*	2/94	1,975 bp
Gb_ba:Bsadhaldh Aldehyde dehydrogenase (partial).	Z25544	*B.stearothermophilus*	4/94	1,471 bp

Gb_ba:Bacaldht   D13846   *B.stearothermophilus*   2/94   1,975 bp
B. stearothermophilus *aldhT* gene for aldehyde dehydrogenase, complete cds.

Gb_ba:Cloaad   L14817   *Clostridium acetobutylicum*   4/94   4,800 bp
Clostridium acetobutylicum aldehyde-alcohol dehydrogenase (*aaD*) gene, 2 complete
cds.

Gb_ba:Mc137   Z33096   *Mycoplasma capricolum*   4/94   875 bp
M.capricolum DNA for CONTIG MC137 encoding for aldehyde dehydrogenase.

Gb_ba:Bsu47861   U47861   *B.subtilis*   9/96   4,106 bp
Bacillus subtilis gbsAB operon, glycine betaine aldehyde dehydrogenase *GbsA*,
alcohol dehydrogenase *GbsB* genes, complete cds.

Gb_ba:Clocatla   L21902   *Clostridium kluyveri*   9/93   6,575 bp

Succinate semialdehyde dehydrogenase (*sucD*)complete

Gb_ba:Ecoaldhq3   M38433   *E.coli*   5/91   2,883 bp
Aldehyde dehydrogenase (ALDH) gene, complete cds.

Gb_ba:Ecobetb   M77739   *E.coli*   9/91   1,854 bp
Betaine aldehyde dehydrogenase (*betB*) gene, complete cds.

Gb_ba:Ecosusedeh M88334   *E.coli*   12/93   5,378 bp
Succinic semialdehyde dehydrogenase (*gabD*) gene, complete cds.

Em_ba:Eckl2a   S56953   *E.coli*, K-12   5/ 92   1,764 bp
 [Genomic 1764 nt] *alD*=aldehyde dehydrogenase gene.

Gb_ba:Ecbet   X52905   *E.coli*   12/93   7,412 bp
E.coli betaine aldehyde dehydrogenase *betB* gene.

Gb_ba:Ecoaldb   L40742   *E.coli*   6/95   1,933 bp
E.coli aldehyde dehydrogenase (*aldB*) gene, complete cds.

Gb_ba:Ecoaldhq3   M38433   *E.coli*   8/94   2,883 bp
E.coli (clones λ-ht5.3b and pEco4.1) ALDH gene, complete cds.

Gb_ba:Ecoaldb   L40742   *E.coli*   6/95   1,933 bp
E.coli aldehyde dehydrogenase (*aldB*) gene, complete cds.

Gb_ba:Ppdmpcd   X52805   *Pseudomonas putida*   12/90   2,493 bp
dmpC and *dmpD* genes for 2-hydroxymuconic semialdehyde dehydrogenase

Gb_ba:Ppu09250   U09250   *Pseudomonas putida* F1   5/94   3,271 bp
Acylating aldehyde dehydrogenase (*todI*) complete cds.

Gb_ba:Pseaksd   M69158   *Pseudomonas putida*   6/92   3,430 bp
α-ketoglutarate semialdehyde dehydrogenase gene, complete cds.

Gb_ba:Pseksda   M69159   *Pseudomonas putida*   6/92   2,289 bp
α-ketoglutarate semialdehyde dehydrogenase gene, complete cds.

Gb_ba:Psemmsrab   M84911   *Pseudomonas aeruginosa*   12/92   5,417 bp
Methylmalonate semialdehyde dehydrogenase (*mmsA*) gene, complete cds.

Gb_ba:Psepada   L23214   *Pseudomonas putida*   8/93   1,501 bp
ALDH gene--high specificity for p-hydroxybenzaldehyde, complete cds.

Gb_ba:Pseterp   M91440   *Pseudomonas sp.*   7/92   6,620 bp
Aldehyde dehydrogenase (*terpE*) gene, complete cds.

Gb_ba:Psu01825   U01825   *Pseudomonas sp.*   5/94   1,241 bp
2-hydroxymuconic semialdehyde dehydrogenase (*bphG*) gene, partial cds.

Gb_ba:Pooct   X65936   *Pseudomonas oleovorans*   6/93   9,092 bp
P.oleovorans aldehyde dehydrogenase TF4-1L (+ OCT) plasmid alkBFGHJKL genes.

```
Gb_ba:Rmu39940    U39940       Rhizobium meliloti       1/96        3,599 bp
Rhizobium meliloti betaine aldehyde dehydrogenase (betB) gene, complete cds.
```

```
Gb_ba:Hiu32731    U32731       Haemophilus influenzae  9/96         10,066 bp
Haemophilus influenzae aldehyde dehydrogenase (aldH; from bases 8459..9754 of the
complete genome).
```

```
Gb_ba:Vhu39638    U39638       Vibrio harveyi           2/96        1,649 bp
Vibrio harveyi fatty aldehyde dehydrogenase (aldH) gene, complete cds.
```

```
Gb_ba:Sycslrf     D64004       Synechocystis sp         4/96        134,199 bp
Synechocystis sp. slr0072 gene encoding for an ALDH (from bases 29944..31314 )
homologous to microsomal ALDH.
```

```
Gb_ba:Vibtagalda  M60658       Vibrio cholerae          10/93       2,068 bp
Aldehyde dehydrogenase gene, complete Cds,
```

```
Gb_ba:Mtcy4c12    Z81360       M.tuberculosis           10/96       38,396 bp
Mycobacterium tuberculosis cosmid SCY04C12 containing an aldehyde dehydrogenase
gene encoding for a protein similar to GABD_ECOLI P25526 succinate-semialdehyde.
```

```
Gb_ba:Sycslrf     D64004       Synechocystis sp.        9/96        134,199 bp
Synechocystis sp. PCC6803 complete genome, containing aldehyde dehydrogenase
"hisB" gene (from bases 29944..31314)
```

```
Gb_pl:Aaalta2     X78227       A.alternata              3/94        1,647 bp
AltA2 mRNA for aldehyde dehydrogenase.
```

```
Gb_pl:Ahbadh      X69770       A.hortensis              7/94        1,513 bp
mRNA for betaine-aldehyde dehydrogenase.
```

```
Gb_pl:Ahbadhg     X69772       A.hortensis              7/94        276 bp
BadH gene for betaine-aldehyde dehydrogenase.
```

```
Gb_pl:Asnaldaa    M32351       A.niger                  9/90        3,497 bp
Aldehyde dehydrogenase (aldA) gene, complete cds.
```

```
Gb_pl:Chclah3     X78228       Cladosporium herbarum   7/96         1,660 bp
C.herbarum ClaH3 mRNA for aldehyde dehydrogenase.
```

```
Gb_pl:Emealda     M16197       A.nidulans               8/87        2,804 bp
aldA gene encoding aldehyde dehydrogenase, complete cds.
```

```
Gb_pl:Sobadhg     X69771       S.oleracea               7/94        649 bp
BADH gene for betaine-aldehyde dehydrogenase.
```

```
Gb_pl:Spibadh     M31480       Spinach                  9/90        1,797 bp
Betaine-aldehyde dehydrogenase (BADH) mRNA, complete CDs.
```

```
Gb_pl:Psrnagapn   X75327       Pisum sativum            6/94        1,631 bp
P.sativum    (Rosakrone)   mRNA    for    NADP'-specific    nonphosphorylating,
glyceraldehyde-3-phosphate dehydrogenase (GAPDH), a member of the ALDH gene
superfamily with no sequence homology to phosphorylating GAPDH.
```

```
Gb_pl:Zmrnagpn    X75326       Zea mays                 5/94        1,746 bp
Z.mais    (maize)   mRNA    for    nonphosphorylating   glyceraldehyde-3-phosphate
dehydrogenase, a member of the ALDH gene superfamily with no sequence homology
to phosphorylating GAPDH.
```

```
Gb_pl:Schom2a     X15649       S.cerevisiae             11 /89      1,885 bp
HOM2 gene for aspartic beta-semialdehyde dehydrogenase (ASADH)
```

```
Gb_pl:Yscaldhaa   M57887       S.cerevisiae             6/91        1,953 bp
Aldehyde dehydrogenase (ALDH) gene, complete CDs.
```

```
Gb_pl:Scu56604    U56604       S.cerevisiae             5/96        2,744 bp
Saccharomyces cerevisiae cytosolic aldehyde dehydrogenase gene, complete cds.
```

Gb_pl:Scu56605   U56605   *S.cerevisiae*   5/96   2,661 bp
S.cerevisiae precursor aldehyde dehydrogenase gene, nuclear gene encoding
mitochondrial protein, complete cds.

Gb_pl:Sce3612   U18814   *S.cerevisiae*   12/94   8,443 bp
S.cerevisiae chromosome V lambda clone 3612 and cosmid 9747 containing a putative
aldehyde dehydrogenase (from bases 953..2515)similar to aldehyde dehydrogenase
(NAD+) from *Cladosporium herberum* [GenBank Accession Number X78228].

Gb_pl:Scu39205   U39205   *S.cerevisiae*   3/96   37,947 bp
S.cerevisiae chromosome XVI, left arm, cosmid 8460, containing an ALDH homolog
(from bases 9754..11256) similar to *S. cerevisiae* Yer073p: Swiss-Prot Accession
Number P40047, and to *Aspergillus niger* aldehyde dehydrogenase aldA: Swiss-Prot
Accession Number P41751"

Gb_pl:Sc8520X   Z49705   *S.cerevisiae*   6/95   41,200 bp
S.cerevisiae chromosome XIII cosmid 8520, containing an aldehyde dehydrogenase
gene YM8520.18c (from bases 35673..37193), with a 91.5% identical to downstream
gene, YM8520.19c (*ALD2*).

Gb_pl:Sc9718   Z49702   *S.cerevisiae*   5/95   35,811 bp
S.cerevisiae chromosome XIII cosmid 9718 containing an aldehyde dehydrogenase
gene YM9718.09c (from bases19579..21177).

Gb_pl:Scald2   X85987   *S.cerevisiae*   11/96   2,079 bp
S.cerevisiae NAD(P)'-dependent aldehyde dehydrogenase gene (*ALD2*).

Gb_in:Celf42g9   U00051   *C.elegans*   3/96   34,008 bp
C. elegans cosmid F42G9 containing an ALDH gene "F42G9.5"" similar to betaine
aldehyde dehydrogenase; complement [join (18579..18709,18760..18836,18886..19123,
19172..19837,20148..20408,21047..21191,21775..22320)].

Gb_in:CelF54D8   U12966   *C.elegans*   8/94   37,017 bp
C.elegans cosmid F54D8, containing an ALDH gene "F54D8.3" highly similar to
ALDH-E2; [join(1896..1951, 2036..2129, 2842..2970, 3018..3214, 3435..3480,
3556..3832, 3879..4625, 4687..4775)].

Gb_in:Ebaldh   X95396   *Enchytraeus buchholzi* 9/96   2,048 bp
E.buchholzi mRNA for a cadmium inducible aldehyde dehydrogenase (aldH) gene.

Gb_in:Enhaldh1x   L05667   *Entamoeba histolytica* 7/94   2,009 bp
Aldehyde dehydrogenase 1 (*ALDH1*) gene, complete cds.

Gb_in:Ltaldde   Z31698   *L.tarentolae*   4/94   1,934 bp
(UC) gene for aldehyde dehydrogenase.

Gb_pr:Hsaldh   X75425   *Human*   12/93   1,635 bp
mRNA for aldehyde dehydrogenase (using γ-aminobutyraldehyde).

Gb_pr:Humaldh2   K03001   *Human*   6/89   1,603 bp
Mitochondrial Class 2 ALDH (ALDH2) mRNA.

Gb_pr:Hsaldhi1   X05409   *Human*   2/91   1,989 bp
Mitochondrial Class 2 ALDH (ALDH2) mRNA.

Gb_pr:Hsaldhir   Y00109   *Human*   2/91   1,980 bp
mRNA for mitochondrial aldehyde dehydrogenase (ALDH 2).

Gb_pr:Humaldde   M77477   *Human*   7/92   1,625 bp
Stomach Class 3 ALDH (ALDH3) mRNA, complete cds.

Gb_pr:Humaldhiii M74542   *Human*   8/91   1,636 bp
Class 3 ALDH (ALDH3) mRNA, complete cds.

Gb_pr:Humaldhal   K03000   *Human*   5/91   1,560 bp
Cytosolic ALDH1 mRNA.

Gb_pr:S61235   S61235   *Human*   7/93   971 bp
ALDH1=aldehyde dehydrogenase 1 {5' region}, Genomic, 971 nt.

Gb_pr:Hummsadha   M93405   *Human*   10/92   1,423 bp
Methylmalonate semialdehyde dehydrogenase gene, complete cds.

Gb_pr:Hummtald   M63967   *Human*   1/93   6,074 bp
Mitochondrial aldehyde dehydrogenase x gene(ALDH5), complete cds.

Gb_pr:Hsu07919   U07919   *Human*   10/95   3,442 bp
Aldehyde dehydrogenase 6 (ALDH6) mRNA, complete cds.

Gb_pr: Hsu10868   U10868   *Human*   12/95   2,790 bp
Aldehyde dehydrogenase (ALDH7) mRNA, complete cds.

Gb_pr:Hsu37519   U37519   Human   1/96   2,827   bp
Aldehyde dehydrogenase (ALDH8) mRNA, complete cds.

Gb_pr:Humssadh   L34820   *Human*   5/95   1,071 bp
NAD$^+$-dependent succinate-semialdehyde dehydrogenase (SSADH) mRNA, 3'end.

Gb_pr:Humfald   L47162   *Human*   5/96   1,791 bp
Human fatty aldehyde dehydrogenase (FALDH) mRNA, complete cds.

Gb_ro:Mmu07235   U07235   *Mouse*   3/94   1,954 bp
Mitochondrial aldehyde dehydrogenase (AHD5) mRNA, complete cds.

Gb_ro:Musahd2f   M74570   *Mouse*   10/92   2,038 bp
Cytosolic aldehyde dehydrogenase II (AHD2) mRNA, complete cds.

Gb_ro:Musahd2u   M74571   *Mouse*   2/94   714 bp
Aldehyde dehydrogenase II gene, exon 1.

Gb_ro:Mmu12785   U12785   *Mouse*   8/94   1,722 bp
Mouse dioxin-inducible cytosolic ALDH3(AHD4) mRNA, complete cds.

Gb_ro:Mmu14390   U14390   *Mouse*   8/96   2,997 bp
Mouse Microsomal aldehyde deydrogenase (AHD3) mRNA, complete cds.

Gb_ro:Rata1d   J03637   *Rat*   9/88   1,725 bp
Tumor-associated ALDH3 mRNA, complete cds.

Gb_ro:Ratalddeha L11282   *Rat*   8/93   1,363 bp
ALDH3 gene(clone ALDH-NL2), complete exon 1.

Gb_ro:Ratalddehb L11283   *Rat*   8/93   654 bp
ALDH3 gene(clone ALDH-NL2), complete exon 2.

Gb_ro:Rataldh   M23995   *Rat*   12/89   2,024 bp
Phenobarbital inducible ALDH (PB-ALDH) mRNA, complete cds.

Gb_ro:Rataldh5pb L24074   *Rat*   12/93   1,375 bp
Phenobarbital inducible ALDH (PB-ALDH) gene, 5' flank.

Gb_ro:Ratmad   M73714   *Rat*   12/91   2,977 bp
Microsomal aldehyde dehydrogenase mRNA, complete cds.

Gb_ro:Ratmsadha   M93401   *Rat*   10/92   2,059 bp
Methylmalonate semialdehyde dehydrogenase mRNA, complete cds.

Gb_ro:Rnadr   X14977   *Rat*   2/91   1,889 bp
Mitochondrial aldehyde dehydrogenase (ALDH2) mRNA.

Gb_ro:Rnu60063   U60063   *Rat*   6/96   2,240 bp
Retinal dehydrogenase mRNA, complete cds.

Gb_ro:Rat10HCO   M59861   *Rat*   6/95   3,109 bp
10-formyltetrahydrofolate dehydrogenase mRNA,complete cds.

Gb_ro:Rataldha   L42009   *Rat*   4/96   1,806 bp
ALDH mRNA with high specificity for retinal, complete cds.

Gb_ro:Ratssadh   L34821   *Rat*   4/95   1,731 bp
Succinate-semialdehyde dehydrogenase (SSADH) mRNA, 3' end.

Gb_om:Bovaldhyda   L36128   *Bos taurus*   8/95   2,071 bp
Bos taurus aldehyde dehydrogenase mRNA present in amacrine cells of bovine
retina, complete cds.

Gb_om:Bovmmsa   L08643   *Bos taurus*   6/93   1,687bp
methylmalonate semialdehyde dehydrogenase mRNA, complete cds.

Gb_vr:Ggadhr   X58869   *Gallus gallus*   6/96   1,876 bp
Chicken mRNA for aldehyde dehydrogenase highly expressed in the undifferentiated
chick retina.

Gb_om:S61045   S61045   *Cattle*   7/ 93   379 bp
Cornea aldehyde dehydrogenase (ALDH5) [Partial, 379 nt].

Gb_:Oau12761   U12761   *Ovis aries*   1/96   2,080 bp
Ovis aries cytosolic ALDH1 mRNA, complete cds.

Gb_om:Octomegcry   L06902   *Octopus dofleini*   8/93   1,799 bp
Octopus dofleini ω-crystallin, complete cds.

Gb_om:Ommomegcry   L06903   *Ommastrephes sloanei*   8/93   3,831 bp
Ommastrephes sloanei ω-crystallin gene, complete cds.

Gb_om:U02483   U02483   *Elephantulus edwardi*   6/96   2,065 bp
Elephantulus edwardi ALDH1/η-crystallin mRNA, complete cds.

Gb_om:Mpu03906   U03906   *Macroscelides proboscideus*   6/96 2,053 bp
Macroscelides proboscideus ALDH I/η-crystallin mRNA, complete cds.

Gb_om:Mpu40486   U40486   *Macroscelides proboscideus*   6/96 1,119 bp
Macroscelides proboscideus cytoplasmic aldehyde dehydrogenase ALDH1-nl mRNA,
partial cds.

INDEX

The manufacturer's authorised representative in the EU is Springer
Nature Customer Service Centre GmbH, Europaplatz 3, 69115 Heidelberg,
Germany. If you have any concerns regarding our products, please
contact ProductSafety@springernature.com

Printed and bound by CPI Group (UK) Ltd, Croydon, CR0 4YY

24/04/2026

02096348-0018